CELLULAR BIOENERGETICS: Role of Coupled Creatine Kinases

Developments in Molecular and Cellular Biochemistry

Series Editor: Naranjan S. Dhalla, Ph.D., FACC

1. V.A. Najjar (ed.): *Biological Effects of Glutamic Acid and Its Derivatives.* 1981   ISBN 90-6193-841-4

2. V.A. Najjar (ed.): *Immunologically Active Peptides.* 1981   ISBN 90-6193-842-2

3. V.A. Najjar (ed.): *Enzyme Induction and Modulation.* 1983   ISBN 0-89838-583-0

4. V.A. Najjar and L. Lorand (eds.): *Transglutaminase.* 1984   ISBN 0-89838-593-8

5. G.J. van der Vusse (ed.): *Lipid Metabolism in Normoxic and Ischemic Heart.* 1989   ISBN 0-7923-0479-9

6. J.F.C. Glatz and G.J. van der Vusse (eds.): *Cellular Fatty Acid-Binding Proteins.* 1990
    ISBN 0-7923-0896-4

7. H.E. Morgan (ed.): *Molecular Mechanisms of Cellular Growth.* 1991   ISBN 0-7923-1183-3

8. G.J. van der Vusse and H. Stam (eds.): *Lipid Metabolism in the Healthy and Diseased Heart.* 1992
    ISBN 0-7923-1850-1

9. Y. Yazaki and S. Mochizuki (eds.): *Cellular Function and Metabolism.* 1993   ISBN 0-7923-2158-8

10. J.F.C. Glatz and G.J. van der Vusse (eds.): *Cellular Fatty-Acid-Binding Proteins, II.* 1993
    ISBN 0-7923-2395-5

11. R.L. Khandelwal and J.H. Wang (eds.): *Reversible Protein Phosphorylation in Cell Regulation.* 1993
    ISBN 0-7923-2637-7

12. J. Moss and P. Zahradka (eds.): *ADP-Ribosylation: Metabolic Effects and Regulatory Functions.* 1994
    ISBN 0-7923-2951-1

13. V.A. Saks and R. Ventura-Clapier (eds.): *Cellular Bioenergetics: Role of Coupled Creatine Kinases.* 1994
    ISBN 0-7923-2952-X

SPRINGER SCIENCE+BUSINESS MEDIA, B.V.

CELLULAR BIOENERGETICS:
Role of Coupled Creatine Kinases

edited by

V. A. SAKS

Laboratory of Bioenergetics
Institute of Bioenergetics
Institute of Chemical and Biological Physics
Ravala 10, Tallinn EE 0001, Estonia

Centre de Physiologie et Physiopathologie Cellulaires
Université Joseph Fourier
2280, rue de la Piscine
F-38041 Grenoble Cedex, France

and

RENÉE VENTURA–CLAPIER

Laboratoire de Cardiologie Cellulaire et Moléculaire
INSERM CJF 92-11
Université de Paris-Sud, Faculté de Pharmacie
5, rue Jean-Baptiste Clement
F-92296 Chatenay-Malabry Cedex, France

Reprinted from *Molecular and Cellular Biochemistry*, Volumes 133/134 (1994)

SPRINGER SCIENCE+BUSINESS MEDIA, B.V.

Library of Congress Cataloging-in-Publication Data

Cellular bioenergetics: role of coupled creatine kinases / edited by
   V.A. Saks and Renée Ventura-Clapier.
      p. cm. -- (Developments in molecular and cellular biochemistry: 13)
   "Reprinted from Molecular and cellular biochemistry."
   ISBN 978-1-4613-6119-0   ISBN 978-1-4615-2612-4 (eBook)
   DOI 10.1007/978-1-4615-2612-4
   1. Creatine kinase.   2. Energy metabolism.   I. Saks, V.A.
II. Ventura-Clapier, Renée.   III. Molecular and cellular biochemistry.   IV. Series:
Developments in molecular and cellular biochemistry: v. 13.
QP606.073045 1994
571.87'6 -- dc20

ISBN 978-1-4613-6119-0

Printed on acid-free paper

CONTENTS

Molecular and Cellular Biochemistry **133/134**: 1, 1994.
© 1994 *Kluwer Academic Publishers.*

Preface

The first book in the English language all about one single enzyme – creatine kinase – was published, to our knowledge, in 1980, and that was *Heart Creatine Kinase*, edited by William E. Jacobus and Joanne S. Ingwall. Since that time, the progress of studies of creatine kinase systems in different cells has been very impressive. There have been several publications covering some part of the topic, such as *Myocardial and Skeletal Muscle Bioenergetics*, edited by N. Brautbar (Plenum Press, 1986), a monograph by S. Lyslova and V. Stephanov *Phospagen Kinases* (CRC Press, 1991) and several review articles. However, the main part of information on the functional role, structure and molecular biology of creatine kinase is scattered in several hundreds of publications. The intention of the editors of this volume was to invite all authors who are most active at the present time in the experimental research of creatine kinases to summarize their work and make the information of the 'state of the art' easily available to a wider scientific audience, especially to young investigators who are just now entering this field of research. Since the functional role of coupled creatine kinases is directly related to the phenomenon of compartmentation and structural organization of metabolic networks, we also invited experts in these related areas to contribute to this volume. Almost all invited authors responded enthusiastically and the results of this collective work are presented here for the attention of the interested reader. We hope that the volume will be of interest for many biochemists, biophysicists, physiologists as well as medical research workers.

The work of the editors in the preparation of this volume was possible due to the generous help from 'Institut de la Santé et de la Recherche Médicale', France. Permanent support from Prof. N. Dhalla, President of the International Society for Heart Research, is gratefully recognized.

Valdur A. Saks and Renée Ventura-Clapier
Laboratoire de Cardiologie Cellulaire et Moleculaire
INSERM CJF 92-11
Université de Paris-Sud, Faculté de Pharmacie
5, rue Jean-Baptiste Clement
F-92296 Chatenay-Malabry Cedex, France

PART I

MUSCLE ENERGY METABOLISM

Molecular and Cellular Biochemistry **133/134**: 5–8, 1994.
© 1994 *Kluwer Academic Publishers*.

'My work is not a piece of
writing designed to meet
the taste of an immediate public,
but was done to last forever'

Thucydides, 460–400 BC, Athens
'The Peloponnesian War'

I-1 Introduction: history of the problem

Rather controversial and interesting history of phosphocreatine (PCr) research is already well described in several books and reviews [1–5]. In considering the development of ideas of physiological role of PCr in muscle, we may see movement of scientific opinions from one extreme position to another. Sometimes it looks like movement of a pendulum. After works of Lundsgaard which ended the 'lactate theory' of contraction, there was a 'phosphagen' era, when phosphocreatine was considered as a direct source of energy for contraction [1, 2, 6]. Subsequent discoveries of creatine kinase reaction by Lohmann [7], ATPase reaction of myosin by Engelhardt and Lyubimova [8] and direct demonstration of utilization of ATP in muscular contraction by Davies group [9, 10], however, reduced PCr to the role of simple energy buffer, a source for extra ATP supplied, if necessary, via equilibrium creatine kinase reaction in cytoplasm at increased workloads. This is a point of view which still occupies the minds of many learned men and can be found in very recent publications. However, already at the time of important works by Davies [1, 9, 10], there were investigators who began considering alternative points of views. One of those pioneers who started pushing the PCr pendulum back again was Samuel Bessman. At that time, his ideas were mostly based on good intuition and logical consideration of the mechanisms of acceptor control of respiration and they led him to formulation of the concept of phosphocreatine shuttle [11, 12, see 4, 5 for reviews). Very slowly, the experimental material also started to accumulate in favour of such more active role of phosphocreatine, and the pendulum of opinions started slowly moving again from the shadow, energy reserve area, in direction of the central role of PCr in muscle cells. A decisive event was the discovery of mitochondrial form of creatine kinase by Jacobs, Heldt and Klingenberg in 1964 [13]. In this issue, you can find many papers solely devoted to this isoenzyme of creatine kinase which is encoded by two separate genes different from those for M or B type creatine kinases (chapter IV-1). The central role of this creatine kinase isoenzyme is emphasized by Fig. 3 in chapter IV-2 by Mühlebach et al.

who have traced back in the evolutionary time-scale the moment when mi-CK appeared – a very long time ago, and surely not for nothing. Why it happened became clear, however, only after 1964. Again, S. Bessman in cooperation with Fonyo [14] was first experimentally to demonstrate its functional role by showing that creatine is able to exert acceptor control of respiration. New independent line of support for the central role of phosphocreatine and creatine kinase system was found in studies of metabolically inhibited or ischemic hearts by Gercken and Schlette and Gudbjarnason and coworkers [15, 16], when contraction was shown to stop at high cellular levels of ATP, but in parallel to the tissue level of PCr. These two works attracted attention to the phenomenon of ATP compartmentation in the cells as a basis for the important role of creatine kinase system. This concept – compartmentation of adenine nucleotides and other metabolites – is now well verified (see chapters I-3 and III-1) and most obviously will serve as a basis for our understanding of cellular bioenergetics and integrated cellular metabolic systems.

After discovery of mi-CK and compartmentation of adenine nucleotides, it became clear that for real understanding of the physiological role of PCr in the cells, detailed biochemical information of the behavior of CK isoenzymes in cellular structures was necessary.

From the very beginning, and up to our days, an important role in these studies was played by the Switzerland group of Eppenberger, Wallimann, Perriard and their coworkers. Their work is reviewed comprehensively and they are active authors of this volume (chapters I-3, I-4, II-1, II-4, III-2, IV-2) with impressive presentation of new data including presentation of the primary structures of 26 guanidino kinases (chapter IV-2).

Important contribution into these studies has been made by William E. Jacobus who, in cooperation with Albert Lehninger, published in 1973 a paper on mi-CK which is still very well read as a fine phenomenological description of control of mitochondrial respiration by mitochondrial creatine kinase [17]. Equally important was the contribution by another American group – Ingwall and Bittl, who have performed important series of energy flux measurements by NMR magnetization saturation transfer technique. Some contribution into the field of creatine kinase research has been made in our laboratories by applying physiological methods to show the importance of myofibrillar creatine kinase for control of muscular contraction (chapter II-5) and in studies of the mechanisms of functional coupling of creatine kinases with other cellular systems (reviewed in chapter III-1). It is not possible to list here the names of all investigators who have worked for finding the solution of the problems, but whose contribution is recognized in the lists of references. The results of all these works are summarized by the Scheme 1, which shows connection of creatine kinase to the systems of energy production (mitochondria) and energy utilization (myofibrils and cellular membranes) in cardiac cells. Because of their specific localization and interaction with other cellular systems by a mechanism of direct substrate channelling in multienzyme complexes, these creatine kinase isoenzymes integrate energy metabolism by forming efficient phosphocreatine pathway (shuttle) [4, 5], or circuit (see chapter III-2) for intracellular energy transport. Several versions of this Scheme will appear in other chapters of this volume (chapters III-1 and III-2) emphasizing in details its different aspects.

This scheme is only one of the good examples of the importance of compartmentation, both of enzymes and substrates in the cells. In literature, we can see rapidly increasing number of publications on compartmentation of different substances including even cations, in the cells (see chapter III-1 for review). There is no doubt that this development will strengthen in the future and reflects the complexity of intracellular reality. In this volume, classical approaches and methods for studies of muscle energetics are summarized by M. Osbakken (chapter I-2), and as emphasized by the author, no clear answer was found for intracellular regulatory mechanisms of energy metabolism. What seems now important to do is not only to use new techniques of research but first of all to account for the complex nature of the intracellular medium with characteristic compartmentation phenomena, multienzyme organized structures and direct interaction between their components – substrate – product channeling. In other words, it is probably time now to reject the idea of intracellular medium being a simple aqueous solution as an historical necessity but outlived simplification. Instead, it is time to incorporate into the metabolic research the very serious achievements of physical chemistry of organized mul-

tienzyme systems. This volume contains an important work by Kholodenko, Cascante and Westerhoff (chapter VI-1) which leads the reader into the field of these ideas. This new approach contains a novel and potentially powerful theory of metabolic control in organized systems which may be much more efficient than that in diluted solutions and may lead us to a better understanding of muscle biochemistry and biochemical regulation. Another example of theoretical treatment of substrate channelling is given in chapter VI-2.

In the experimental part of this volume there is a group of papers from three different laboratories consistently describing the rather new and potentially very important phenomenon for metabolic regulation – the low permeability of the mitochondrial outer membrane for adenine nucleotides and its possible role in regulation of cellular respiration in cells in vivo (chapters II-1, II-2 and III-1). This is a very new and unexpected group of results, but a suspicion that the data collected in the experiments with mitochondria *in vitro* are not very helpful for explaining the *in vivo* phenomena because of the loss of some important factors during isolation has been clearly expressed in recent publications and the papers collected in this volume show that such a suspicion was highly justified. If confirmed in further studies, these results may open new perspectives for studies of cellular respiration *in vivo*.

Chapter IV-3 of this volume contains description of a new technology of creatine kinase gene manipulation. Even if we take into account very significant adaptive mechanisms which may compensate for the switching off of the creatine kinase gene and change the initial metabolic situation (see chapters I-4 and III-2), this is a very promising new area of research. Gene structure and expression for creatine kinase isoenzymes have been studied in details and are described in this volume by Payne and Strauss (chapter IV-1) and Mühlebach et al. (chapter IV-2). These studies are important for understanding of the developmental changes and pathological alteration of the creatine kinase systems (chapters V-1, V-2, V-3).

Very interestingly, detailed studies of smooth muscle showed high degree of compartmentation of metabolic systems in these cells (chapter I-3) in spite of rather low rates of energy turnover, and the presence of creatine kinases which seem to be actively involved in the energy transport (chapters I-3 and III-3). Thus, it is clear that, in many tissues, the creatine kinases play more important role than that of a simple energy buffer. In muscle and brain mitochondria, the coupled creatine kinase reac-

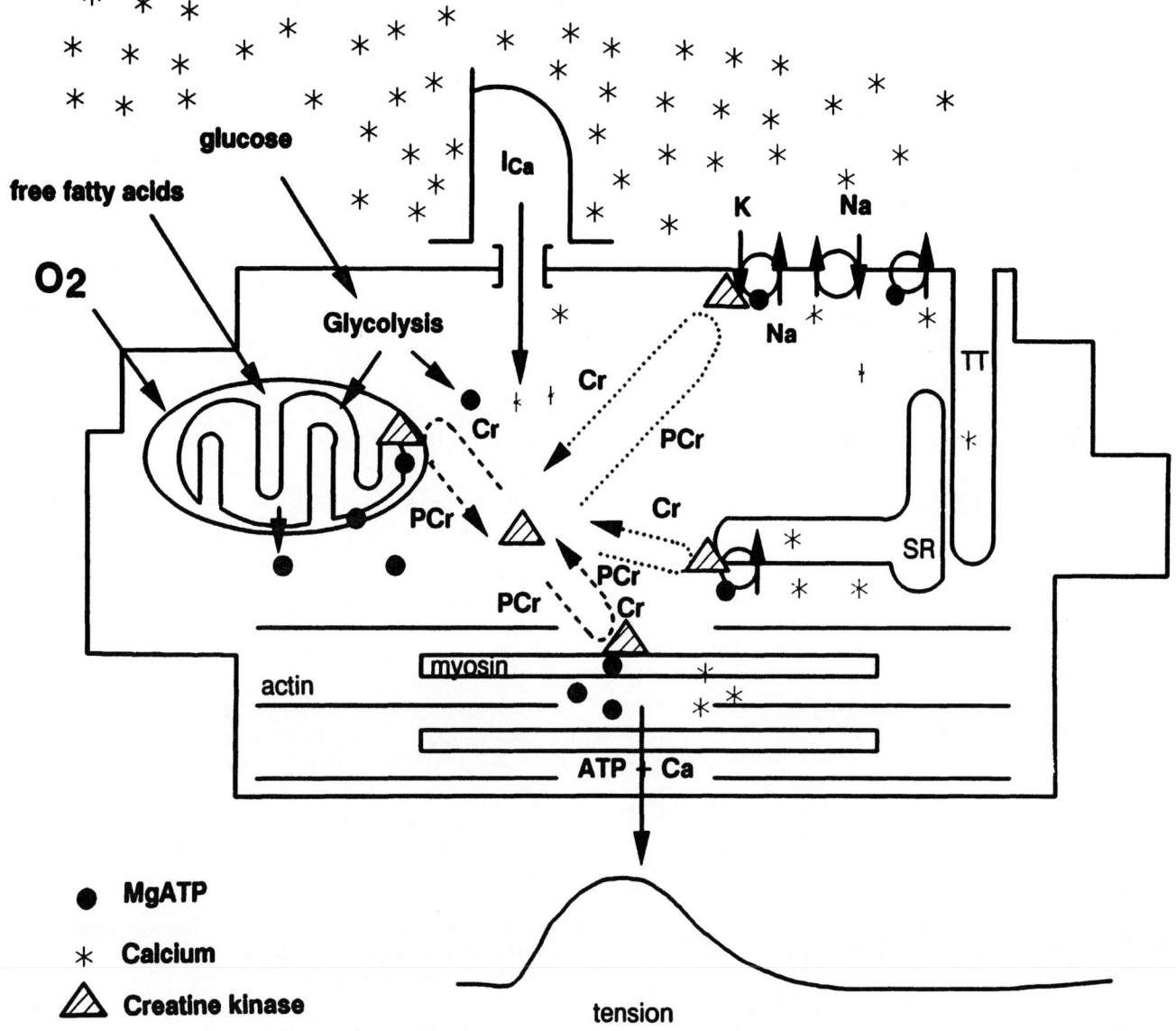

Fig. 1. Excitation-contraction coupling in adult mammalian cardiac cells and creatine kinase pathway. Calcium entry into the cell is triggered by action potential and activates contractile proteins. Creatine kinase isoenzymes are bound to sites of energy production (inner mitochondrial membrane and glycolytic clusters in cytosol) and to sites of energy utilization (myofibrils, sarcoplasmic reticulum, plasma membrane) in close association with ATPases. SR, sarcoplasmic reticulum; TT, T tubules; I_{Ca}, calcium current.

tion seems to function as an effective feedback signal amplifier (chapter III-1). However, its functional interaction with the adenylate kinase system as well as with hexokinase should be taken into account in explaining the cellular mechanisms of metabolic regulation by substrate channelling via different specialized pathways (chapters II-1, II-2 and II-6). In this way we may approach more realistic understanding of the role of cre-

atine kinases in the cells, and find a steady central position for the pendulum of opinions.

Taken together, all these works give not only comprehensive account for the results of the research of creatine kinase systems, but also formulate new concepts concerning the structural and functional organization of integrated metabolic systems and their regulation in the cell. We hope that many readers will find the experimen-

8

tal material and the ideas described in this book interesting and useful for their own research.

Renée Ventura-Clapier
Laboratoire de Cardiologie Cellulaire
   et Moleculaire
INSERM CJF 92-11
Universite de Paris-Sud
Faculte de Pharmacie
5, rue Jean-Baptiste Clement
F-92296 Chatenay-Malabry Cedex
France

Valdur Saks
Laboratory of Bioenergetics
Institute of Chemical and
   Biological Physics
Ravala 10
Tallinn EE 0001
Estonia

References

1. Mommaerts WFHM: Energetics of muscular contraction. Physiol Rev 49: 427–508, 1969
2. Ivanov II, Korovkin BF, Pinaev GP: Biochemistry of Muscles. Editsina, Moscow, 1977, p. 3–29 (in Russian)
3. Jacobus WE, Ingwall JS (Eds): Heart creatine kinase. The integration of isozymes for energy distribution. Williams and Wilkins, Baltimore/London, 1980, p. 1–8
4. Bessman SP, Geiger PJ: Transport of energy in muscle. The phosphorylcreatine shuttle. Science 211: 448–452, 1981
5. Bessmann SP, Carpenter CL: The creatine-creatine phosphate energy shuttle. Ann Rev Biochem 54: 831–862, 1985
6. Lundsgaard E: Untersuchungen uber Muskelkontraktion ohne Milchsaurebildung. Biochem Z 217: 162–177, 1930
7. Lohmann K: Uber die enzymatische aufspaltung der kreatinphosphorsaure; Zugleich ein beitrag zum chemismus der muskelkontraktion. Biochem Z 271: 264–277, 1934
8. Engelhardt W, Lyubimova M: Myosin and adenosine triphosphatase. Nature 144: 668–669, 1939
9. Cain DF, Davies RE: Breakdown of adenosine triphosphate durding a single contraction of working muscle. Biochem Biophys Res Comm 8: 361–366, 1962
10. Infante AA, Daves RE: The effect of 2,4-dinitrofluorobenzene on the activity of striated muscle. J Biol Chem 240: 3996–4001, 1965
11. Bessman SP: A contribution to the mechanism of diabetes mellitus. In: V.A. Najjar (ed). Fat metabolism. John Hopkins Press Baltimore, 1954, p. 133–137
12. Bessman SP: A molecular basis for the mechanism of insulin action. Am J Med 40: 740–749, 1966
13. Jacobs H, Heldt HV, Klingenberg M: High activity of creatine kinase in mitochondria from muscle and brain and evidence for a separate mitochondrial isoenzyme of creatine kinase. Biochem Biophys Res Comm 16: 516–521, 1964
14. Bessman SP, Fonyo A: The possible role of the mitochondrial bound creatine kinase in regulation of mitochondrial respiration. Biochem Biophys Res Comm 22: 597–602, 1966
15. Gercken G, Schlette U: Metabolite status of the heart in acute insufficiency due to 1-fluoro-2,4-dinitrobenzene. Experientia 24: 17–19, 1968
16. Gudbjarnason S, Mathes P, Ravens KG: Functional compartmentation of ATP and creatine phosphate in heart muscle. J Mol Cell Cardiol 1: 325–339, 1970
17. Jacobus WE, Lehninger A: Creatine kinase of rat heart mitochondria. J Biol Chem 248: 4803–4810, 1973

Molecular and Cellular Biochemistry **133/134**: 9–11, 1994.
© 1994 *Kluwer Academic Publishers.*

I-1a A brief summary of the history of the detection of creatine kinase isoenzymes

Hans M. Eppenberger
Institute for Cell Biology, ETH-Hönggerberg, CH-8093 Zürich, Switzerland

In the late nineteen fifties and early nineteen sixties multiple forms of more and more enzymes were described. They were also called 'isoenzymes, isozymes, isoforms of enzymes, etc.'. Markert and Moller [1] introduced for the first time these terms for multiple molecular forms of a given enzyme existing in a single organism or tissue and having identical or at least nearly identical catalytic activities. At the time, it was an operational definition which did not in itself indicate anything about the genetical or structural basis of multiple forms of enzymes. Nevertheless, the term 'isoenzyme' has turned out to be a useful and fortunate choice since isoenzymes have subsequently been shown to be generated by a variety of mechanisms.

One of the enzymes appearing as isoforms was creatine kinase (= CK), officially called ATP: Creatine Phosphotransferase (2.7.3.2), catalysing the reaction: $MgADP^- + PCr^{2-} + H^+ \leftrightarrows MgATP^{2-} + Cr$.

CK had been isolated, purified and partially crystallized from rabbit skeletal muscle by the group of Kuby in 1954 [2]; but only in 1962 Dance and Watts [3] suggested that active CK could be a dimeric molecule. At the same time CK also started to become an important marker for the detection of carriers for Duchenne muscular dystrophy. This was, at the time, the main reason for the group of Richterich and Aebi at the University Hospital of Berne to have a closer look at this enzyme. One member of this group, Monika Eppenberger, homogenized different organs from a rabbit and, by analogy to what was known from lactate dehydrogenase isoenzymes, placed the homogenates on an agar gel plate for separating proteins by electrophoresis. Isoenzymes had usually been separated by zone electrophoresis on various carriers and identified subsequently by specific staining. Starch, agar, agarose, and polyacrylamide gels, but also cellulose polyacetate foils and dextrans like Sephadex were the most commonly used carrier media. Figure 1 is a re-production of, to our knowledge, the first separation of CK isoenzymes [4] from different tissues of rabbit (top fig.). In parallel, also a distinct change of the CK-isoenzyme pattern during rat development could be demonstrated (bottom fig.). An isoenzyme pattern, like the one shown, could most easily be explained by the combination of two subunits either to form a homologous or a heterologous dimer; we coined the nomenclature commonly used since then, namely BB-CK for the isoenzyme dimer mainly found in brain (B = brain) and MM-CK for the dimer specifically found in skeletal muscle (M = muscle) [4–7].

The heterodimer composed of one M- and one B-subunit frequently appeared in a transitory fashion during fetal and neonatal development of skeletal muscle, but also persisted, e.g. in rat and human heart, throughout adult life [5–8]. Meanwhile, B-subunits have also been found in organs and tissues different from brain or neural tissues, but M-subunits are still expressed specifically in cross-striated muscle tissues (see review 9). The number of CK-isoenzymes increased further; a specific CK-subunit type in mitochondria (Mi-CK) was first revealed by Jacobs et al. [10] and recently in this laboratory two isoforms of Mi-CK, namely Mi_a-CK, which is specific for cross-striated muscle, and an ubiquitous Mi_b-CK have been detected [11]. In addition, both isoforms of Mi-CK were shown – in strong contrast to B- and M-CK – to form octamers [12].

M. Eppenberger et al. [13] detected particularly in birds a doublette of the BB-CK, which much later in our laboratory was shown in chickens to consist of two distinct subspecies, Ba- and Bb-CK, both with different isoelectric points [14], which arise by differential splicing of a unique B-CK gene [15].

From the very beginning, much research was dedicated to studying the changes in CK-isoenzyme patterns during development [5]. Such isoenzyme transitions

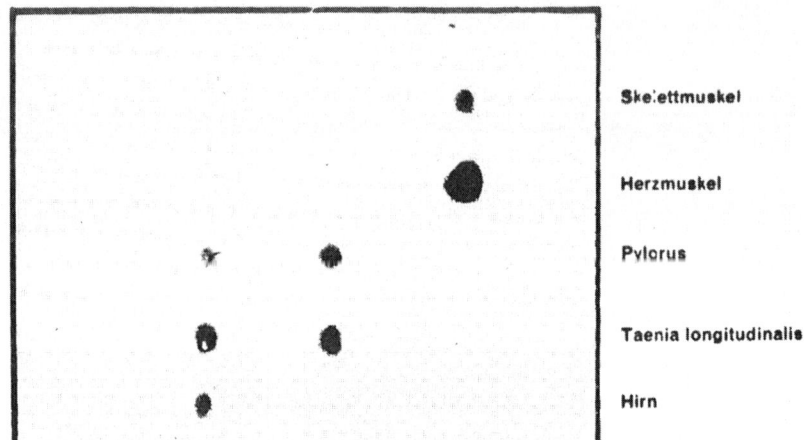

Abb. 1. Elektrophorese der Kreatinkinasen einiger Organe des Kaninchens.

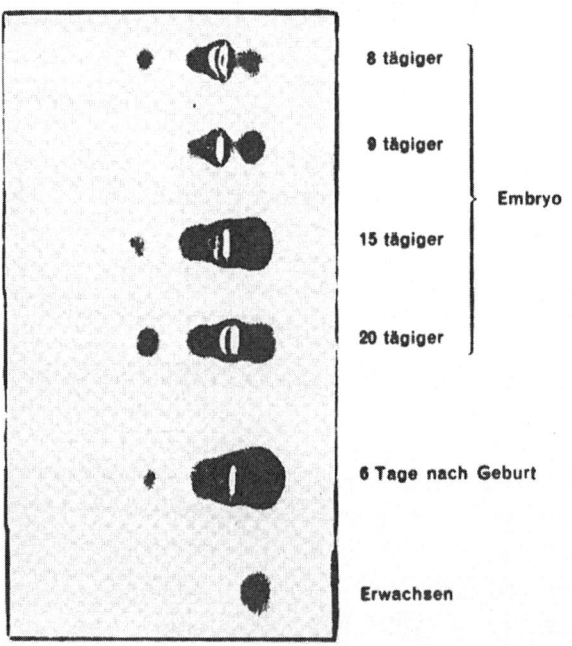

Abb. 2. Elektrophoretisch aufgetrennte Kreatinkinase der Skelettmuskulatur der Ratte (Laufzeit 20 Minuten).

Fig. 1. Reproduction of the original agar gel electrophoresis by A. Burger, M. Eppenberger, U. Wiesmann and R. Richterich [4]. Separation of tissue homogenates from rabbit (top) and of homogenates of rat skeletal muscle from different developmental stages (bottom). Detection of CK-activity by the 'gel-overlay technique' as described in [5].

point directly to the molecular events involved in cell differentiation, whereby features of specialized cells during differentiation may be correlated directly with the stage-specific isoenzyme pattern. By using cell cultures in the early seventies, we could easily follow CK-isoenzyme changes during differentiation of skeletal muscle cells and we tried at that time to relate the switch from BB-CK to MM-CK during development to other events of skeletal muscle differentiation, like the fusion of the mononucleated presumptive myoblasts to post-mitotic multinucleated myotubes or to the appearance of myofibrils and of their muscle specific protein building blocks [16, 17]. Clearly the onset of M-CK expression and the heavy increase of MM-CK seemed to be correlated with the development of functional muscle. Today it is well known that the M-CK promoter is under the control of myogenic transcription factors, like MyoD-I, and thus parallels the upregulation of other muscle specific protein isoforms [18, 19].

It was, of course, very interesting and promising when we could demonstrate that only MM-CK was able to interact with the M-band region of a myofibillar sarcomer, thus suggesting a functional importance for this specific CK-isoenzyme [20, 21]. Later on, it could be shown that the C-terminal half of the M-CK molecule is responsible for specific binding of M-CK to the sarcomeric M-line [22]. As it has been summarized in our reviews [9, 23], many reasons speak for a key function of the different CK-isoenzymes in energy metabolism and energy homoeostasis.

In conclusion, it certainly can be said that the study of CK-isoenzymes consistently provided new and exciting aspects about this enzyme family for ongoing research. The many chapters in this book strongly support the accuracy of this statement.

References

1. Markert CL, Moller F: Multiple forms of enzymes: Tissue, ontogenetic and species specific patterns. Proc Nat Acad Sci U.S. 45: 753–763, 1959
2. Kuby SA, Noda L, Lardy HA: Adenosine-triphosphate – creatine transphosphorylase. I. Isolation of the crystalline enzyme from rabbit muscle. J Biol Chem 209: 191–201, 1954
3. Dance N, Watts DC: Comparison of creatine phosphertransferase from rabbit and brown-hare muscle. Biochem J 84: 114P–115P, 1965
4. Burger A, Eppenberger M, Wiesmann U, Richterich R: Isoenzyme der Kreatinkinase. Helv Physiol Acta 21: c&–C10, 1963
5. Eppenberger HM, Eppenberger M, Richterich R, Aebi H: The ontogeny of creatine kinase isoenzymes. Develop Biol 10: 1–16, 1964
6. Dawson DM, Eppenberger HM, Kaplan NO: Creatine kinase: Evidence for a dimeric structure. Biochem Biophys Res Comm 21: 346–353, 1965
7. Eppenberger HM, Dawson DM, Kaplan NO: The comparative enzymology of creatine kinases. I. Isolation and characterization from chicken and rabbit tissues. J Biol Chem 242: 204–209, 1967
8. Thomson AR, Eveleigh JW, Miles BJ: Amino acid sequence around the reactive Thiol groups of adenosine triphosphate – creatine phosphotransferase. Nature 203: 267–269, 1964
9. Wallimann T, Wyss M, Brdiczka D, Nicolay K, Eppenberger HM: Intracellular compartmentation, structure and function of creatine kinase isoenzymes in tissues with high & fluctuating energy demands: the 'phosphocreatine circuit' for cellular energy homeostasis. Biochem J 281: 21–40, 1992
10. Jacobs H, Heldt WH, Klingenberg M: High activity of CK in mitochondria from muscle and brain. Evidence for a separate mitochondrial isoenzyme of CK. Biochem Biophys Res Comm 16: 516–521, 1964
11. Schlegel J, Wyss M, Schürch U, Schnyder T, Quest A, Wegmann G, Eppenberger HM, Wallimann T: Mitochondrial Creatine Kinase from cardiac muscle and brain are two distinct isoenzymes but both from octameric molecules. J Biol Chem 263: 16963–16969, 1988
12. Schlegel J, Wyss M, Eppenberger HM, Wallimann T: Functional studies with the octameric and dimeric form of mitochondrial creatine kinase. J Biol Chem 265: 9221–9227, 1990
13. Eppenberger ME, Eppenberger HM, Kaplan NO: Evolution of creatine kinase. Nature 214: 239–241, 1967
14. Quest A, Eppenberger HM, Wallimann: Purification of brain-type creatine kinase (B-CK) from several tissues of the chicken: B-CK Sunspecies. Enzyme: 33–42, 1989
15. Wirz T, Brändle U, Soldati T, Hossle JP, Perriard JC: A unique chicken B-creatine kinase gene gives rise to two B-creatine kinase isoproteins with distinct M termini by alternative splicing. J Biol Chem 265: 11656–11666, 1990
16. Turner DC, Maier V, Eppenberger HM: Creatine kinase and aldolase isoenzyme transitions in cultures of chick skeletal muscle cells. Develop Biol 37: 63–89, 1974
17. Perriard JC, Perriard ER, Eppenberger HM: Detection and relative quantitation of mRNA for creatine kinase isoenzymes in RNA from myogenic cell cultures and embryonic chicken tissues. J Biol Chem 253: 6529–6535, 1978
18. Jaynes JB, Chamberlain JS, Buskin JN, Johnson JE, Hauschka SD: Transcriptional regulation of the muscle creatine kinase gene and regulated expression in transfected mouse myoblasts. Mol Cell Biol 6: 2855–2864, 1986
19. Lassar AB, Buskin JN, Lockshon D, Davis RL, Apone S, Hauschka SD, Weintraub H: MyoD is a sequence-specific DNA binding protein requiring a region of myc homology to bind to the muscle creatine kinase enhancer. Cell 58: 823–831, 1989
20. Turner DC, Wallimann T, Eppenberger HM: A protein that binds specifically to the M-line of skeletal muscle is identified as the muscle form of creatine kinase. Proc Natl Acad Sci USA 70: 702–705, 1973
21. Wallimann T, Schlösser T, Eppenberger HM: Function of M-line bound creatine kinase as intramyofibrillar ATP regenerator at the receiving end of the phosphorylcreatine shuttle in muscle. J Biol Chem 259: 5238–5246, 1985
22. Schäfer BW, Perriard JC: Intracellular targeting of isoproteins in muscle cytoarchitecture. J Cell Biol 106: 1161–1170, 1988
23. Wyss M, Smeitink J, Wevers R, Wallimann T: Mitochondrial creatine kinase: a key enzyme of energy metabolism. Biochim Biophys Acta 1102: 119–166, 1992

Molecular and Cellular Biochemistry **133/134**: 13–37, 1994.
© 1994 *Kluwer Academic Publishers.*

I-2 Metabolic regulation of *in vivo* myocardial contractile function: multiparameter analysis

Mary D. Osbakken
The Departments of Medicine and Biochemistry/Biophysics, University of Pennsylvania, Philadelphia, PA 19104

Abstract

To gain insite into the mechanisms of myocardial regulation as it relates to the interaction of mechanical and metabolic function and perfusion, intact animal models were instrumented for routine physiological measurements of mechanical function and for measurements of metabolism (^{31}P NMR, NADH fluorescence (redox state)) and perfusion (^2H NMR and Laser doppler techniques). These techniques were applied to canine and cat models of volume and/or pressure loading, hypoxia, ischemia and cardiomyopathic states. Data generated using these techniques indicate that myocardial bioenergetic function is quite stable under most loading conditions as long as the heart is not ischemic. In addition, these data indicate that there is no universal regulator and that different biochemical regulators appear to mediate stable function under different physiological and pathophysiological conditions: for example; during hypoxia, NADH redox state appears to play a regulatory role; and in pressure loading, ADP, phosphorylation potential and free energy of ATP hydrolysis as well as NADH redox state appear to be regulatory. (Mol Cell Biochem **133/134**: 13–37, 1994)

Key words: ^{31}P, NADH, Ions, Perfusion, Creatine Kinase, Animal Models, Cardiac Disease

Introduction

The heart is an amazing organ; an organ which works continuously while other organs rest. The heart can be considered to be a sack which collects blood from the systemic and pulmonary vasculature; or a pump which functions to maintain the necessary supply of nutients (oxygen and other metabolic substrate such as carbohydrate, amino acids and lipids) to the rest of the body. But the heart is much more. It is not only versitile with respect to the substrate which it can use; it can adapt to use different substrate depending on what is available and on the conditions under which it functions. It responds to increased and decreased demand depending on external and internal signals. It is a neuroendocrine organ which synthesizes, secretes and responds to neurohumors. While the heart is all these things, the interaction of and regulation of myocardial metabolism and mechanical function are still incompletely understood and open to debate by many experts in the field.

Our laboratory has spent the last 10 years studying this topic; i.e., that of the metabolic regulation of myocardial mechanical function (and vice versa) using a variety of animal models and loading conditions, and using a variety of routine (heart rate (HR), systolic and diastolic blood pressure (SBP, DBP), cardiac output (CO), oxygen consumption (MVO_2), stroke volume (SV)) and recently developed (NADH fluorometry, ^{31}P NMR, piezoelectric crystal determination of wall thickness and shortening, 2H NMR to measure myocardial perfusion, laser doppler measurement of tissue perfusion, and K^+,

Address for offprints: M. Osbakken, Director, Clinical Research, Bristol-Myers Squibb, P.O. Box 4000, Princeton, NJ 08543-4000 USA

Ca^{2+}, and Na^+ electrode (in a multiprobe assembly) measurement of ion flux) physiological monitoring tools.

Cardiac physiology

The role of metabolism of any organ is to supply adequate ATP to support 'work load' [1]. In the case of the myocardium, workload consists of maintenance of membrane ionic equilibrium and generation of contractile activity. Because the energy requirements of ionic pumps are relatively constant from beat to beat, the observed change in ATP demand is a function of muscle contraction-relaxation, i.e., the mechanical work of the heart [2]. In cardiac muscle, mechanical work is a good measure of the state of oxidative phosphorylation; i.e., V/V_{max}. Metabolic control of mechanical work can be described by relating P_i/PCr or PCr/ATP (which can be measured by ^{31}P NMR spectroscopy and represents the concentration of free ADP [3, 4]) to work (in this case, heart rate x mean arterial blood pressure product) using an algorithm which is based on classical Michaelis-Menten kinetics. The following equations, developed by Chance et al. [3] for skeletal muscle, formally describe the relationship between mechanical work and metabolism: $V/V_{max} = 1/(1 + (0.6)/(P_i/PCr))$ (Eq 1); $V = (P_i/PCr)/(1 + (0.6)/(P_i/PCr)$ (Eq 2).

Equation (1) describes a hyperbolic relationship, while Eq (2) describes a sigmoid relationship between phosphorylation potential and V/V_{max}. In Eq. (2), we assume that P_i/PCr is a measure of Vmax. Both equations describe 'transfer functions' [2] of the kinetics of oxidative phosphorylation or work/cost relationship. Equation (1) represents a smooth transition between tight and no feedback control of phosphorylation by work. In contrast, (Eq 2) represents a step function between tight and no control. A similar relationship probably exists for cardiac muscle, although the range of experimental data is obviously smaller, since the heart is never close to basal state (V_0) $(V_0$ = basal rate of phosphorylation).

Until recently, there has been no good method to simultaneously compare both mechanical and metabolic function of the heart in the intact animal (which is necessary in order to understand basic cardiac function as it exists under in vivo conditions). With the introduction of ^{31}P NMR spectroscopy techniques, this type of investigation has become possible [5–23].

Potential regulators of myocardial function

General model

Myocardial perfusion, metabolism and mechanical function are closely interrelated and regulated [24–29]. Generally, increases in mechanical function are accompanied by increases in perfusion and metabolism and vice versa. These closely regulated processes are designed to maintain stable function during a variety of external stimuli and/or stresses, thus allowing the heart a wide range of metabolic and mechanical recruitment and/or reserve.

Maintenance of appropriate cardiac function in the face of changing work load is necessary for normal organ function [30]. The presence of the many humoral and autonomic feedback systems which regulate cardiac function in the intact organism allow fine tuning of metabolic responses. However, these complex interactions in intact organisms make it difficult to determine the causal basis of regulation.

An effective myocardial biochemical regulator should fulfill certain conditions. It should change with physiological intervention in the direction which can be interpreted by the feedback mechanism as a signal which can correctly change output or input. Specific models should have the appropriate functional dependence on defined variables. This has been amply described by Chance and co-workers [31]. For example, if [ADP] regulates mitochondrial oxygen utilization, $\dot{V}O_2$, will be expected to exhibit the well known functional dependence;

$$\dot{V}O_2 = \frac{\dot{V}O_{2max}[ADP]}{K_m + [ADP]} \tag{3}$$

where K_m and $\dot{V}O_{2max}$ are constants. This quantitative model of [ADP] action [32] can be compared to the actual cardiac behavior. The form of dependence of $\dot{V}O_2$ on ADP given in Eq. (1) is such that the regulation of $\dot{V}O_2$ is not very tight; for example, for a one percent change in $\dot{V}O_2$ at [ADP] = K_m, Eq (3) requires a two percent change in [ADP].

Extensive studies of isolated cardiac mitochondria have suggested that most of the intermediate products of metabolism may be important regulators; included in this list are ATP, ADP, P_i, $ATP/(ADP \cdot P_i)$, and NAD (H). Other proposed regulators are PCr (in the regulation of the phosphocreatine shuttle), Ca^{2+}, various substrates, numerous neurohumoral agents, and thermodynamic availability of energy represented by ΔG (all important in the conversion of chemical to mechanical energy). Thus, there are many potential steps in the regulatory process of myocardial mechanical function and metabolism. Recent work on the intact myocardium [26] suggests that no single regulator, whether dynamic or thermodynamic, is prominent under all physiological conditions.

Enzyme biochemistry

Myocardial enzyme biochemistry has been extensively studied in order to clarify the mechanisms of regulation, which allow a wide range in work performed while maintaining the heart in a stable condition [33]. The PCr content in muscle, brain, and heart is thought to act as a source of short-term energy through the creatine kinase reaction and as an integral component of the system for transfer of energy from mitochondria to the cytoplasm [34]. This reaction is regulated by temperature [35] and through its substrates [36, 37], [Mg-ATP^{2-}], [Mg-ADP$^-$], [H$^+$], [PCr^{2-}], [Cr]. Whether the flux of PCr to ATP through this reaction is actually an important part of the regulatory system for the myocardium remains speculative. Furthermore, it remains to be shown how changes in the creatine kinase reaction rates are utilized in the overall regulation of myocardial metabolism and function.

Techniques used to study in vivo myocardial function

^{31}P NMR to evaluate bioenergetics

The use of ^{31}P NMR to examine bioenergetic phenomena in the living organism has recently been documented by us and others [3, 5–24, 38–41]. Ratios of PCr/ATP and P$_i$/PCr (used to estimate ADP [3]) observed by ^{31}P NMR were used to determine metabolic stability during the changes in mechanical function. These parameters, reflect changes in high energy phosphate synthesis and use and provide information concerning creatine kinase kinetics. In addition, changes in the PCr/ATP and P$_i$/PCr ratios can be used to determine the heart's ability to

maintain it's high energy phosphate content during increased mechanical work (as long as ATP does not break down to adenosine and leave the cell). Relevant details concerning the use of ^{31}P NMR data to delineate in vivo bioenergetic function in animal models will be presented throughout this paper.

^{31}P NMR saturation transfer to measure enzyme function [33, 41]

Saturation transfer experiments measure the rate constant, K$_f$, of the creatine kinase (CK) reaction: PCr^{2-} + Mg-ADP$^-$ + H$^+$ → Mg-ATP^{2-} + Cr, or more simply PCr → ATP. Additional sources of ATP arise from the resynthesis of ATP by oxidative phosphorylation and glycolysis via the general reaction: ADP + P$_i$ → ATP. The fluxes of P$_i$ → ATP and ATP → P$_i$ have been experimentally shown to be closely matched in isolated perfused rat hearts under varying work loads [26, 42]. Saturation transfer methods are described as follows.

Acquisition of ^{31}P NMR saturation transfer spectra

After the resonant frequencies of γP-adenosine triphosphate (γP-ATP) and phosphocreatine (PCr) are determined from control ^{31}P spectra, saturation transfer measurements can be done by progressively increasing saturation times (for example, 0.1 sec to 5 sec) of γP-ATP, while monitoring the change in the PCr area. The saturation pulse widths vary (for example, using a 2.7 Tesla magnet at 46.9 MHz, pulse widths varied from 70–150 μsec) depending on the anatomy of the 'coil interface' with the heart. A sufficiently long relaxation delay is chosen (10 sec) to provide fully relaxed spectra. Cardiac and respiratory gating are used to triggor both saturation and acquisition radiofrequencies.

Determination of K$_f$ and flux from saturation transfer data [26, 33, 42]

Change in the areas of the PCr peaks obtained during progressive saturation was used to determine the forward rate constant, K$_f$, of creatine kinase and the unidirectional flux from PCr to ATP. The following standard equations are used to calculate K$_f$ and flux.

$$\frac{d\,M_{PCr}}{dt} = \frac{M_{oPCr} - M_{PCr}}{T_1} - K_f\,M_{PCr}, \qquad (1)$$

16

whose solution is:

$$M_{PCr} = M_{oPCr} [1 - K_f t_p(1-e^{-t/t}p)] \qquad (2)$$

$$\text{where } \frac{1}{tp} = \frac{1}{T_1} + K_f \qquad (3)$$

M_{PCr} = magnetization of PCr

M_{oPCr} = magnetization of PCr before saturation has been applied to the γP-ATP

T_1 longitudinal relaxation

K_f =   forward rate constant of creatine kinase

τ_p =   time constant

t =   saturation time

A least square fit is performed using equations 2 and 3 to solve for τ_p, T_1, and K_f. Flux is the product of M_{oPCr} and K_f.

NADH fluorometry to evaluate redox state [38]

The principle of NADH monitoring from the surface of the heart is based on excitation light (366 nm) which is passed from the light source in a fluorometer to the heart via a bundle of quartz optical fibers [33, 38, 43–47]. The fluorescent emitted light (450 nm), together with the reflected light at the excitation wavelength, is transferred to the fluorometer via a second optical fiber bundle. The changes in the reflected light are used to correct for hemodynamic artifacts appearing in the NADH measurement. Calibration of the system is done as follows: the reflectance and fluorescence signals obtained from the photomultipliers (RCA-931B) are calibrated to a standard signal (0.5 V) by variation of photomultiplier dynode voltage obtained from the high voltage power supply. The standard 0.5 V signal is set to give a half-scale deflection on the recorder or the computer monitor (2.5 cm) with pen resting in mid scale. The gain is increased as required by a factor of two or four to give 50% or 25% of the full scale respectively. The changes in fluorescence and reflectance signals are calculated relative to the calibrated signals under normoxic conditions. This type of calibration is not an absolute one, but provides reliable and reproducible results from different animals and between different laboratories [44, 48–50]. NADH fluorescence measurements (corrected fluorescence, CF) are recorded as percent changes in fluorescence (F) and reflectance (R) and are calculated relative to calibrated signals obtained under normoxic conditions [38, 43, 44, 47]: i.e., CF = F – R.

NADH fluorometry has been widely used to estimate redox state and oxidation and oxygenation state in dif-

ferent organs since Chance, et al. introduced the technique several decades ago [51, 52]. Although this technique can provide very interesting *in vivo* data, it is not without problems. The NADH fluorescence signal can be influenced by motion, blood flow and volume changes, and by hemoglobin, cytochrome, and myoglobin absorption of light. To control for motion artifact, the light guide holder can be sutured to the myocardium to minimize relative motion of the light guide with respect to the heart. To determine the contribution of blood flow and volume and hemoglobin concentration to NADH fluorescence signal, NADH measurements can be made before and after the heart is perfused with an oxygen carrying fluorocarbon. If the responses to an intervention are the same before and after perfusion with the fluorocarbon, then the fluorescence changes can be attributed to changes in myocardial redox state and not due to change in blood flow, tissue blood volume or hemoglobin absorption. Procedures such as these have been done in this laboratory and in other laboratories (unpublished data and [53]) and indicate that valid measurements of relative changes in myocardial redox state can be made with NADH fluorescence techniques. Artifacts which might be due to cytochrome absorption are likely to be minimal because cytochromes absorb at different wavelengths than NADH.

In addition, as discussed by Chance et al. [51], the effects of the oxy-deoxy transition of hemoglobin on the NADH fluorescence signal is minimal due to equal and opposite effects of 366 nm and 450 nm wavelengths (response at 366 nm is subtracted from the response at 450 nm). Thus any decrease in absorption of excitation light is corrected for by an increase of transmission at the emission wavelength. Since the absorption spectra of myoglobin are very similar to those of hemoglobin at the two wavelengths mentioned, myoglobin absorption effects on NADH fluorescence measurements would also, by analogy, be very small. In addition, since evaluation of relative and not absolute changes are made during an intervention, *in vivo*, NADH fluorescence measurements give a reasonable estimate of changes in redox state, as also concluded by others [43, 54].

Dueterium (^{2}H) NMR to evaluate myocardial perfusion [38, 55]

The use of ^{2}H NMR to measure organ perfusion has been developed only recently [55–58]. Despite its great potential, it has not yet been widely applied to the heart.

The technique is based on monitoring the washout of a perdeuterated saline solution (0.9% $^w/_v$ NaCl in D$_2$O injected into the left ventricle) from the left ventricular myocardium with ^{2}H NMR [55–58]. This technique has the advantage that the tracer is quickly redistributed, so that intravascular deuterium contributes only minimally to the myocardial signal obtained during washout from the myocardium. Therefore, repeated sequential injections and washout measurements can be made so that real time sequential perfusion measurements can be reliably made using this relatively non-toxic tracer. The data are analyzed using the Kety-Schmidt algorithm [59].

The major problem with this type of washout technique is that it takes several minutes to obtain the data used for calculation of perfusion, because tracer washout must be followed until a stable decreased level of tracer (in this case, perdeuterated saline) is achieved. Therefore, the organism must be in a reasonably stable physiological state throughout the measurements (washout measurements can be obtained in between 2–5 min depending on physiological state of the heart). However, the technique does have the advantage of yielding a perfusion value within a few minutes after the conclusion of each data acquisition.

Laser doppler measurement of relative tissue perfusion [45, 46]
Another method which can provide real time tissue perfusion measurements is based on the laser doppler technique [60, 61]. This technique has been calibrated against an H$_2$ clearance probe [62] (LD2) and against ^{14}C iodoantipyrine autoradiography techniques [63], two well established methods to quantitatively monitor flow. LDF provides relative changes in flow which correlate well with relative changes in flow measured with the two other approaches.

Two types of LDF systems (TSI, Inc. St. Paul, MN and Perimed, Inc. NJ and Sweden) were used interchangeably, and gave similar results. Briefly, the laser doppler principle depends on delivering laser generated light (780 nm) to the heart via a single optical fiber. Light is then returned to the monitoring instrument via two collection fibers. Moving red blood cells scatter the instant photons and shift their frequency according to the doppler principle. Stationary tissue scatters light, but does not impart doppler shifts to the photons. The photodetector senses both the scattered and the shifted light. The resultant electrical signal, containing power and frequency information, is generated by a heterodyned mixing process. Blood flow is directly proportional to a linearized product of blood volume and velocity. This meth-

od allows determination of percent change in flow from baseline during an intervention. The laser doppler flow meter was used in some experiments to obtain relative changes in perfusion or flow units (0–100%) from baseline during each intervention and are presented as percent change from baseline [60–65].

Multiprobe assembly to measure combined NADH fluorescence, laser doppler flow and extracellular ion flux [46]
The technology of the multiprobe apparatus (MPA) was developed by Mayevsky et al. [66, 67] and until recently has only been used in rat and gerbil brains. Recently we adapted this technique (and assembly) for use in the *in vivo* dog heart. The MPA used in these studies contains two sets of fiberoptic light guides (one for NADH fluorometry and one for laser doppler tissue flow measurements) and a series of electrodes (Na$^+$, K$^+$ and Ca^{2+}) to measure extracellular ion fluxes [66, 67, 69].

Light guides. Light guides used to measure NADH redox state [49, 50, 68] and blood flow via laser doppler methods [60–65] were placed in the center hole of the MPA [46, 48], were covered by a rigid brass tube, glued in place in the MPA with 5 minute epoxy, and served as an axis to hold the cannula and the connector holder at a fixed convenient distance [48, 66]. The theory of the function of these two techniques is presented above and will not be duplicated here.

Surface electrodes to monitor K^+_e, Na^+_e and Ca$^{2+}_e$. Electrodes to measure extracellular K$^+$, Na$^+$, and Ca^{2+} (K^+_e, Na^+_e, and Ca$^{2+}_e$) were made by World Precision Instruments (WPI, Sarasota, FL, USA) using 1.2 mm diameter polyvinyl chloride (PVC) tubing (Cole Flex, W. Babylon, N.Y., USA). Ion sensitive and DC electrodes were connected and sealed to an electrode holder containing Ag/AgCl pellets (WPI, Inc.) using 5 minute epoxy as a sealant. The electrodes were assembled in a cannula and glued separately to the cannula so that if needed, replacement of electrodes would not disturb the other electrodes or light guides connected to the holder [48, 66].

The electrode pairs are placed on the surface of the heart. A DC current for each ion is generated from the monitoring system. The signals for each ion are obtained by subtracting the DC current from the current obtained from the measuring electrodes. Sampling from the electrodes is done continuously using a custom built electrometer and a computer-based data acquisition and analysis system [48]. Millivolt readings are converted to mEq/L or millimoles/L using the Nernst equation. Calculation of extracellular ion concentrations is based on calibration of each electrode before and after each measurement, using standard solutions [66, 67, 69].

Piezoelectric crystal measurement of myocardial fractional shortening [70]

To obtain measurements of contractile activity, two pairs of piezoelectric crystals are placed on the heart [70–74]: 1) the short axis pair, which consists of one crystal sutured to the posterior wall and one crystal sutured to the anterior wall. 2) The wall thickness pair, which consists of one crystal plunged into the anterior surface of the myocardium to the level of the endocardium, and a second crystal aligned to the anterior surface of the heart directly over the endocardial crystal. Because the radiofrequency of the piezoelectric crystals interfer with the NMR signal, two types of gating are used: 1) ECG and respiratory gating to triggor the NMR radiofrequency on: 2) gating on of the piezoelectric crystals, 200 msec after the NMR radiofrequency is gated on. The crystals are then gated off when the NMR radiofrequency is on. Piezoelectric crystal data are used to determine fractional shortening (FS) via the following formula:

FS = EDID – ESID/EDID, where EDID = end diastolic inner dimension and ESID = end systolic inner dimension. EDID = EDD-2 (WT$_1$) and ESID = ESD-2 (WT$_2$) where EDD = end diastolic dimension, ESD = end systolic dimension, WT$_1$ = wall thickness at end diastole, and WT$_2$ = wall thickness at end systole.

Animal models

General

Open chest models [39]

After animals (dogs and cats) were anesthetized (Innovar-Vet and/or Nembutal), tracheal intubation was performed to maintain the airway and provide a route for positive pressure ventilation. Cannulae were placed in the femoral artery and vein (systemic blood pressure monitoring and administration of fluids and drugs, respectively) and via the carotid artery into the left ventricle (to monitor left ventricular pressure (LVP)). Thermodilution catheters were placed in the pulmonary ar-

teries via the external jugular veins to measure cardiac output. Left lateral thoracotomies or median sternotomies were performed to expose the heart. Pericardial cradles were constructed to support each heart. Double-tuned NMR surface coils (^{1}H – 116 MHz and ^{31}P – 49.6 MHz) were secured to the left ventricle with cyanoacrylate. If NADH fluorometry, laser doppler tissue flow, or extracellular ion flux (alone or combined) were measured alone or simultaneously with NMR, specific transducers were secured to the heart using specially designed delrin holders which were sutured and/or glued to the heart. For experiments where NMR measurements were made, animals were placed in plexiglass holders for introduction into the magnet. Pulse widths and relaxation delays were chosen appropriate to the animal preparation and specific experiment. Cardiac and respiratory gating were used to minimize motion artifacts from the NMR signal.

Closed chest models [75]

Placement of cardiac windows. Under general anesthesia, two ribs (approximately 8–10 cm portion of rib from sternum to vertebral origin) and associated skeletal muscle were surgically removed under sterile conditions. A Silastic tube filled with methylene disphosphonic acid was sewn to the pericardium for later use as a cardiac marker and to calibrate NMR signal. A piece of Marlex mesh (Bard, Inc. Billerrica, MA) was sutured between the two exposed ribs. Fascia and skin were resutured. Vascular access ports (Norfolk Medical Products, Skokie, IL) were placed in the femoral arteries and veins to monitor systemic blood pressure, arterial blood gases and for fluid and drug administration, respectively. Dogs were allowed to recover 3–6 weeks before NMR experiments were performed. Dogs with cardiac windows were physiologically stable for 6 months to 4 years, depending on the particular chronic physiological insult they were exposed to.

NMR experimental setup in chronic closed chest models. Dogs were anesthetized and intubated as for the open chest experiments. Blood pressure and arterial blood gas measurements and fluid and drug administration were made by introducing stainless steel needles (connected to saline filled canulae) into the vascular access ports. Dogs were placed in the left lateral decubitus position on a plexiglass platform which held an NMR coil in a slight recess. The animal was positioned so that the heart (point of maximal impulse) lay directly over the surface coil. The platform was introduced into the magnet (2.7 Tesla) for NMR measurements. This arrangement allowed sequential acquisition of NMR spectroscopy in the intact dog exposed to chronic pathophysiological interventions or during progression of inherited disease processes. Pulse sequences were designed and adapted for each specific experiment.

Models of disease

Volume overload [39, 76]

There are well recognized changes in cardiac structure and mechanical function resulting from acute and chronic volume overload [77–91]. Ventricular chamber volumes and muscle mass increase [78, 79]. Contractile function as measured by ejection fraction, ventricular emptying rate, cardiac output, shortening velocity, ventricular stress, end systolic pressure/volume relationship, stroke work index, and oxygen consumption can be increased or decreased depending on the magnitude and/or duration of volume overload [77–91]. Capillary density and myocardial blood flow distribution may also be altered, with endocardial flow less than epicardial flow. This may contribute to altered cardiac function [80, 84, 86]. Myocyte size, number of mitochondria per myocyte and capillary density also increase [85]. Mechanical function changes induce metabolic changes, such as increased protein synthesis [87–89], increased rate of oxidative phosphorylation [90–92] and changes in actin-activated myosin ATPase [88]. Changes in bioenergetic function during volume overload are not as well defined as mechanical function changes.

Acute: in vivo open chest cat (*^{31}P nuclear magnetic resonance (NMR)*) *[39]*. To study the effects of acute volume loading on myocardial metabolic and mechanical function, seven cats were volume loaded via anastomosis of the abdominal aorta to the vena cava (AVS). Mechanical function was evaluated with heart rate × systolic blood pressure product (HR × SBP). In these studies, ^{31}P nuclear magnetic resonance (NMR) was used to estimate phosphorylation potential (PP) (P$_i$/PCr) and adenosine diphosphate (ADP) concentration (PCr/ATP), both potential regulators of oxidative phosphorylation [3]. The following equation was used to define this function; PP = (ATP)/(ADP)(P$_i$) = (PCr)/(P$_i$) × K$_{CK}$ (H$^+$)/(Cr), where K$_{CK}$ is the equilibrium constant of creatine kinase, H$^+$ is the hydrogen ion concentration, and Cr is creatine [3, 93]. Shunts were opened for 1–2 h during which time PCr, ATP, P$_i$, and HR × SBP were moni-

tored. P$_i$/PCr and PCr/ATP changes were correlated with HR × SBP.

Acute volume loading produced by means of the AVS [86, 90, 94] was associated with variable changes in HR × SBP, some cats with an increase and some with a decrease in HR × SBP. These responses were probably closely related to the animal's baseline homeostatic condition and/or ability of reflex recruitment secondary to volume loading. Metabolic responses to volume overload as reflected by changes in P$_i$/PCr and PCr/ATP generally paralleled mechanical responses; i.e., an increase in HR × SBP was associated with an increase in P$_i$/PCr and decrease in PCr/ATP and a decrease in HR × SBP was associated with a slight increase, decrease, or no change in P$_i$/PCr and a slight decrease, increase or no change in PCr/ATP. When group data were analyzed, the P$_i$/PCr ratio was generally increased during volume loading and appeared to be linearly related to HR × SBP (Fig. 1), even though this relationship was not linear in each individual animal. This indicates that the P$_i$/PCr ratio may be a measure of metabolic induction and/or stability during myocardial loading conditions.

Thus, changes in P$_i$/PCr and/or PCr/ATP ratios may regulate the increase in oxidative phosphorylation which results from increasing the work of the heart; i.e., phosphorylation potential as estimated by P$_i$/PCr, and ADP concentration as estimated by PCr/ATP appear to be regulators of myocardial bioenergetics during acute volume loading. These metabolic changes may act as messengers for enzyme induction to ultimately increase oxidative phosphorylation (state 4 to state 3: inactive to active state; although because the heart is always working, it is always in some sense in state 3) and protein synthesis, both which are compensatory mechanisms associated with more chronic loading conditions. Small changes in P$_i$/PCr and/or PCr/ATP ratios may indicate that there is considerable metabolic reserve and thus myocardial function is stable. Larger changes in these ratios may indicate that myocardial metabolism is severely taxed and there is minimal metabolic reserve, thus making myocardial function relatively less stable.

Chronic: in vivo closed chest dog (*^{31}P NMR*) *[76]*. Several previous studies report that there are few, if any, measurable metabolic changes during chronic volume overload [91, 95], while other studies report that there are decreases in Na,K ATPase activity and high energy phosphate metabolites such as ATP and PCr [88, 90]. None of these studies were done sequentially in intact organisms. In most studies, acquisition of metabolic da-

20

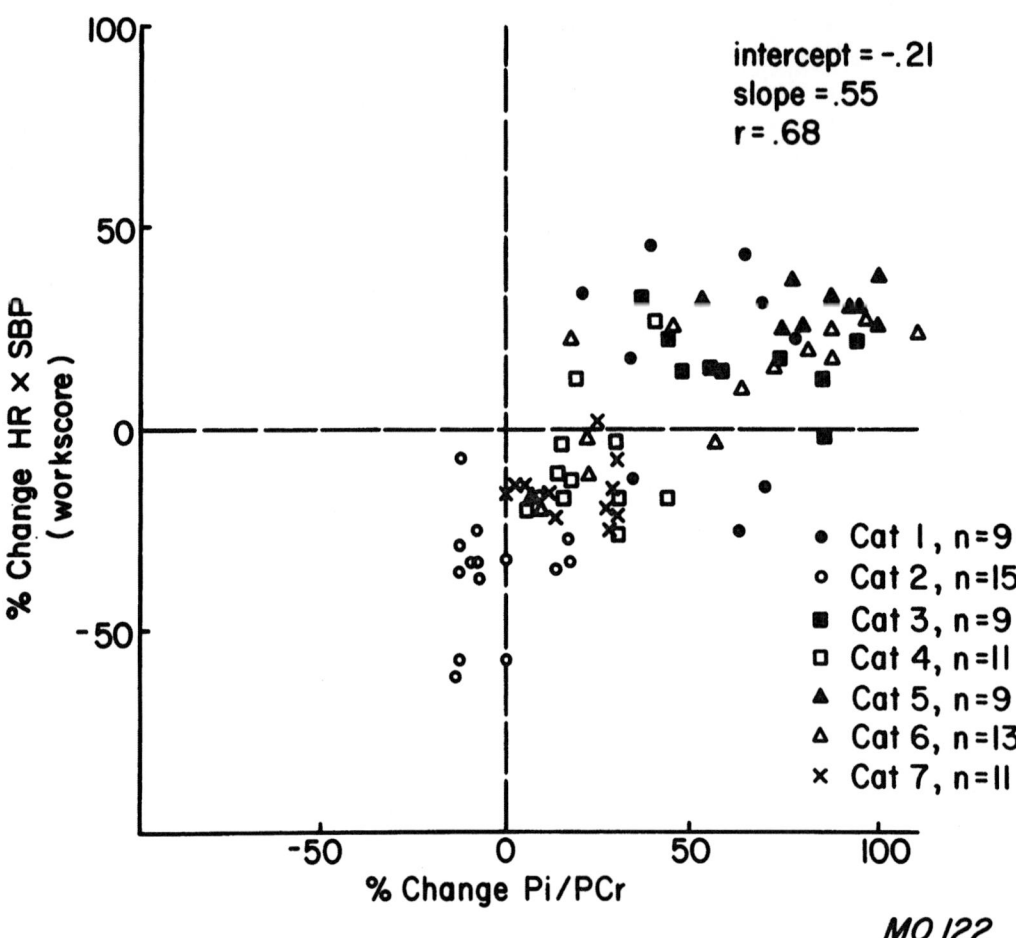

Fig. 1. Composit relationship of workload to P$_i$/PCr ratio during acute volume overload (arterialvenous shunt) in a group of 7 cats. These combined data show a general response of *in vivo* bioenergetic mechanisms to increased myocardial work.

ta for each time point in the volume load process was obtained in different sets of animals. In addition, results obtained using *in vitro* biochemical analysis may not accurately describe the *in vivo* conditions. Thus, there is precident for definition of metabolic phenomena associated with chronic volume overload in a non-destructive manner in the *in vivo* organism.

The purpose of the present study was to investigate the interaction between myocardial bioenergetic and mechanical function during progressive volume overload [1–17 months] in a canine model, using *in vivo* ^{31}P NMR spectroscopy. To this end, surgical anastomosis of the abdominal aorta to the vena cava was done in six dogs previously prepared with cardiac windows [75]. These dogs were studied sequentially with non-invasive ^{31}P MRS spectroscopy to determine whether demonstrable changes in high energy phosphate bioenergetics accompany the documented structural and mechanical function changes associated with volume overload.

Initial application of volume loads to the heart caused a decrease in PCr/ATP (over the first 3 months) with gradual return to baseline levels by month 15 [76] (Fig. 2). Similar small decreases in PCr/ATP were found in dogs subjected to acute volume overload [96]. This indicates that the response to the initial insult was bioenergetic decompensation with later compensation as the heart adapted to the volume load. This response may be a reflex response to an increased ATP demand; i.e., an increased demand for ATP may induce an increased use of PCr to buffer ATP. Thus, initially PCr may decrease. Later enzymatic induction and/or more efficient use of ATP may return the bioenergetic state to baseline, and PCr buffering may become less important.

Superimposition of acute pressure loads on the chronic volume loaded hearts were not associated with significant changes in PCr/ATP, similar to the response of acutely volume loaded dogs [96]. This demonstrates that the chronic volume loaded hearts possess a large metabolic reserve, which allowed for stable myocardial function during combined loading conditions and that bioenergetic regulation is efficient. At least in this model,

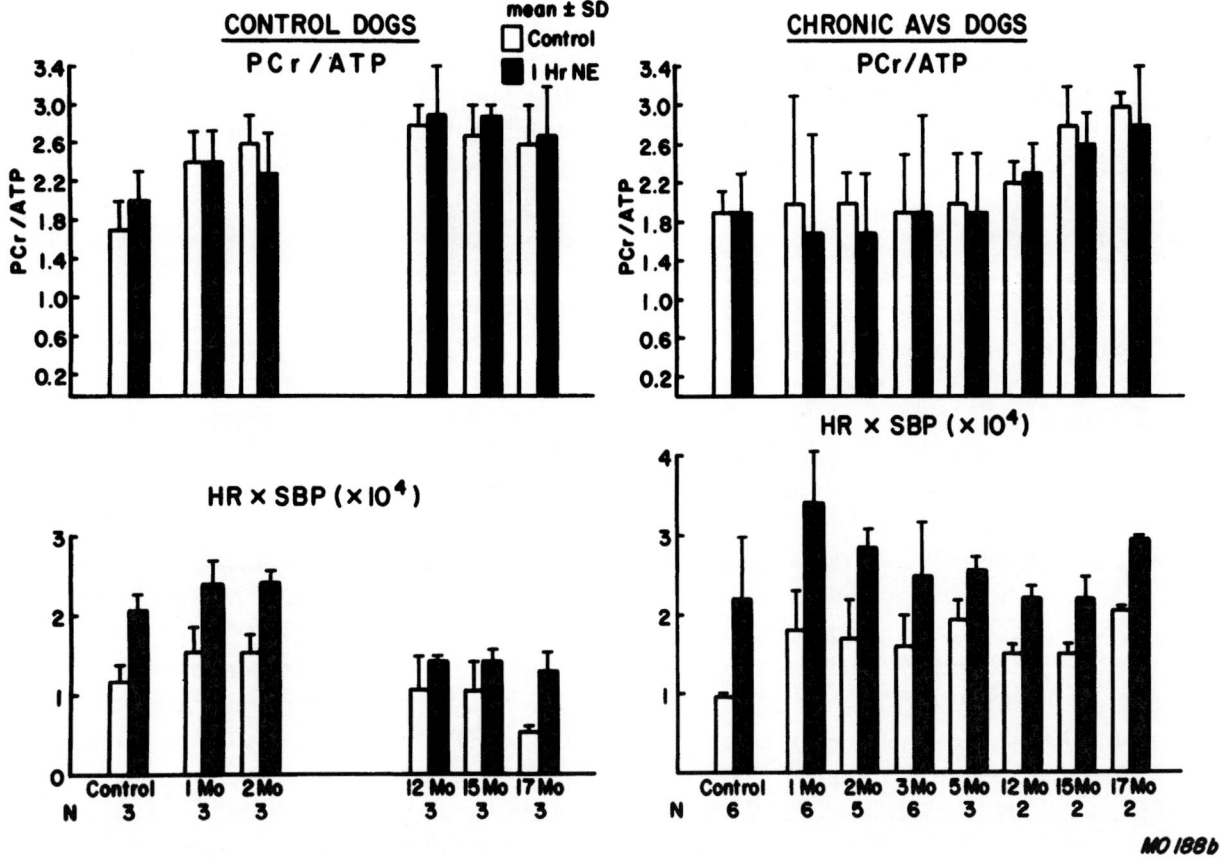

Fig. 2. Composit bioenergetic (PCr/ATP) and mechanical function (HR × SBP) data obtained in chronic volume loaded dogs. To determine the dog's metabolic reserve, after baseline measurements were obtained, NE (1 μg/kg/min) was infused for 60 min.

any decrement in mechanical function was not due to bioenergetic failure.

Pressure overload [41]

That myocardial hypertrophy secondary to acute and chronic pressure loading is associated with metabolic changes has been demonstrated in a number of animal models, including renovascular hypertension [97–101], aortic stenosis [102] pulmonary stenosis [103], and spontaneous hypertension [104]. There can be an increase in V_3 isomyosin [97, 98, 102, 103] and decrease in mitochondrial creatine kinase [104]. Other enzyme changes that also can occur include decreases in b-hydroxy acyl CoA dehydrogenase [100], citrate synthase [100, 101], 3-oxoacid-CoA transferase [101], acetoacetyl-CoA synthase [101], and ATPase activities [98, 105] and increases in hexokinase [100], lactate dehydrogenase [100, 101], phosphorylase [101], and phosphofructokinase [101] activities. There also are increased rates of glucose use [101] and decreased rates of ketone body use [101]. The decreased ATPase activity was associated with improvement of thermodynamic efficiency and required less oxygen for efficient oxidative phosphorylation [98]. In ad-

dition, the rate of Ca^{+2} uptake and binding by the sarcoplasmic reticulum can be decreased [99]. Most of these previous studies evaluated enzyme activity on biopsy samples and/or heart extracts and therefore it is difficult to determine how these changes relate to *in vivo* myocardial mechanical function.

Acute: in vivo open chest dog *(^{31}P NMR, ^{31}P NMR saturation transfer, NADH fluorometry, and 2H NMR perfusion measurements) [33, 38, 44].* Combined *in vivo* ^{31}P NMR, ^2H NMR and NADH fluorometry were used to evaluate left ventricular oxidative phosphorylation and perfusion during application of increasing workloads with norepinephrine, (NE, 1 mg/kg/min). Simple PCr/ATP ratios were measured in 24 dogs [44], while creatine enzyme kinetics were measured in an additional 9 dogs [33]. NADH fluorometry measurements were made in all dogs [33, 38, 44] and ^2H perfusion measurements were made in 9 dogs [38, 55]. NADH data collected during pressure loading were normalized to the maximal NADH response obtained by changing the anesthesia gasses from O_2/N_2 (21/79) to 100% N_2. Myocardial mechanical work was evaluated with HR × SBP product, cardiac output (CO), systolic pressure × stroke volume (P × V) and oxygen consumption (MVO$_2$). Baseline NADH fluorometry, ^{31}P-NMR (PCr/ATP ratio and/or saturation transfer measurements), ^{2}H NMR perfusion and cardiac function measurements were performed simultaneously in each dog, after which blood pressure was increased with NE.

Pressure loading with NE was associated with increased HR × SBP product (Con = $2.02 \times 10^4 \pm 0.6$, NE = $3.6 \times 10^4 \pm 0.86$); P × V work (Con = $1.9 \times 10^3 \pm 1.3$, NE = $3.2 \times 10^3 \pm 2$); CO (l/min) (Con = 1.6 ± 0.5, NE = 2.4 ± 0.2); and MVO$_2$ (ml/min/100 gm) (Con = 9.7 ± 2.5, NE = 15.8 ± 3.3), P < 0.05. In all cases, myocardial perfusion (ml/min/100 gm) increased appropriately to support the increased workload (Con = 87 ± 10, NE = 131 ± 20) P < 0.05. Pressure loading was associated with an increase in oxidative phosphorylation as evidenced by a decrease in NADH ($-25\% \pm 6.0$) and small, but significant decrease in PCr (Con = 3.56 ± 0.76, NE = 2.5 ± 1.3; P < 0.05), with minimal changes in PCr/ATP (Con = 2.4 ± 0.2, NE = 2.37 ± 0.44). In addition, the forward rate constant (K_f) of creatine kinase and the flux of PCr to ATP (flux) both increased (K_f: 0.5 ± 0.17 to 0.92 ± 0.5; flux: 1.7 ± 0.9 to 2.2 ± 1.3, P < 0.05).

Our data show that under normoxic conditions, NE induced increased workloads produced a decrease in PCr (to a new equilibrium value) associated with increases in both the rate constant of creatine kinase and the flux of PCr to ATP. The decrease in PCr (through the creatine kinase reaction) suggests an increase in ADP, which is not directly measurable. Although, as noted by Kupriyanov et al. [106], this change in ADP in the fully coupled system is not expected to alter the flux through the creatine kinase reaction (i.e., is not felt to regulate the reaction in the physiological range), it nevertheless is consistent with such a regulatory role. At the same time, NADH became oxidized, as measured by a decrease in NADH fluorescence, which indicates that oxidative phosphorylation and/or glycolysis were increased and suggests that NADH redox state may also be regulatory.

The NE induced decrease in PCr reported above [33], is in contrast to earlier work [24, 107], where minimal changes in PCr occurred. Some of the prior data (considered to be preliminary, in that the number of experiments and the physiological controls were not optimal) was obtained in pigs [107] and could be due to species differences. On the other hand, the work of Katz et al. [24] on dogs is carefully done; however, the protocols were somewhat different from ours. Different drugs (epinephrine and phenylephrine [24] vs NE in our work) were used to induce increased cardiac work and MVO$_2$. Drug infusion was maintained for only 10 min in the earlier work (not the 40 min used in our studies), and therefore altered workloads were maintained for shorter times. The baseline MVO$_2$ and HR × SBP were also different (both were higher) in the animals used in our studies compared to the earlier work [24]. Despite the slightly different results obtained during physiological manipulation derived in the present study compared to the Katz work [24], the final conclusions (vida infra) are similar for both studies; i.e., changes in PCr/ATP can be regulatory.

Chronic: in vivo open chest dog (*^{31}P NMR saturation transfer*) [41]. A few recent studies have evaluated real time myocardial bioenergetics in an *in vivo* model of hypertensive hypertrophy [97–99, 104]. For example, in a rat model of spontaneous hypertension [104], the baseline forward rate constant of creatine kinase and flux of PCr to ATP were found to be similar to control for 12 and 18 month old rats. However, when hearts from 12 and 18 month hypertensive rats were exposed to increased workloads, the 18 month hearts decompensated (no increase in flux of PCr to ATP) while the 12 month hearts responded appropriately (increase in flux of PCr to ATP). These data suggest that in the older rats, the creatine kinase system may be critical for maintenance of mechanical work and that compromise of this system may be a premonitory event of heart failure.

Because of these previous data and because the heart is dependent on oxidative phosphorylation to support continuous contractile activity, we decided to evaluate bioenergetic function (evaluation of creatine kinase kinetics with ^{31}P NMR saturation transfer techniques) in chronically hypertrophied dog hearts [41]. In our model of canine renovascular hypertension, we found creatine kinase kinetics similar to those reported for the 18 month old spontaneous hypertensive rat model [104]. The baseline forward rate constant of creatine kinase and the flux of PCr to ATP were similar to those found in control hearts (K_f: Con = 0.5 ± 0.2; Hypertensive = 0.4 ± 0.2; Flux: Con = 1.9 ± 0.9; Hypertensive = $1.6 + 1.4$); however, during similar norepinephrine induced increases in mechanical work, the hearts from hypertensive dogs did not significantly increase the forward rate constant of creatine kinase or flux of PCr to ATP, while these parameters were increased significantly in hearts from control dogs K_f: Con 1.0 ± 0.5; Hypertensive = 0.5 ± 0.2; Flux: Con = 2.5 ± 1.3; Hypertensive = 1.3 ± 0.6). Since only one hypertensive dog had evidence of scattered myocardial fibrosis, and all hypertensive dogs had similar changes in creatine kinase kinetics (i.e. smaller change in the forward rate constant and flux of PCr to ATP) during norepinephrine infusion, it is unlikely that the apparent metabolic changes were due merely to an anatomical lesion (i.e. loss of cardiomyocytes due to necrosis and ultimate fibrosis). These data suggest that hypertensive hearts work less per gram heart muscle (similar increases in heart rate × systolic blood pressure from baseline in hearts which weigh more, produced during lower synthesis of ATP via the creatine kinase shuttle) and that they work more efficiently than control hearts because they can maintain lower ADP levels (as indicated by the lower forward rate constant of creatine kinase) than controls.

Our present studies may have clinical relevance. Previous echocardiography studies [108, 109] demonstrated that the chronic renovascular hypertension model had diastolic dysfunction (characterized by a decreased early to atrial (E/A) inflow velocity determined by Doppler, increased atrial filling fraction, decreased peak rates of wall thinning and filling determined by sonomicrometer, and prolonged time constant of isovolumetric relaxation). It is possible that the subtle metabolic changes found in the moderate myocardial hypertrophy of the present study are the basis for the aforementioned diastolic dysfunction; that is, the decrease in forward rate constant and flux (or velocity) of PCr to ATP might translate into the inability to increase ATP during increased demand. A subtle decrease in the rate of synthesis of ATP (which is necessary for both muscle contraction and relaxation) may delay myocardial relaxation during diastole and thus contribute to increased wall stiffness and compliance changes found in the earlier studies [108, 109], even though global mechanical function is within normal limits. If these metabolic abnormalities progress, they may be a basis for heart failure.

Thus, an understanding of metabolic changes which occur during hypertensive processes in an animal model may provide insite into the pathogenesis of heart failure secondary to similar processes in a clinical patient population. These studies also point out the fact that even in early stages of hypertensive heart disease, when there is no evidence of compromised hemodynamic or mechanical function, there are metabolic abnormalities. One possible interpretation of our data and that of Bittl et al. [104] is that the creatine kinase system has a large safety factor, such that even the inability to increase the for-

ward rate constant of creatine kinase and the flux of PCr to ATP during stress does not significantly affect global myocardial mechanical function until these values are decreased to less than 50% of baseline.

Combined acute volume and pressure loading: in vivo open chest dogs (^{31}P NMR and piezoelectric crystal measurement of contractile function) [70, 96]

A number of studies were done to evaluate the effects of combined volume and pressure loads on left ventricular metabolic and mechanical function. In all cases, bioenergetic parameters (PCr/ATP) were evaluated with ^{31}P NMR. In 13 dogs, mechanical function was evaluated with routine HR × SBP, CO, P × V and MVO_2 [96], while fractional shortening (FS) was obtained with 2 sets of piezoelectric crystals in 7 additional dogs (3 dogs were subjected to pressure loading, 4 dogs to volume loading, and 1 dog to both) [70]. Volume loads were applied by shunting the abdominal aorta to the vena cava using polyethylene tubing (5 mm inner diameter). A plastic regulator allowed shunts to be opened and closed. Dogs were heparinized (100 units/kg) to prevent shunts from clotting. To study the effects of pressure loading, a norepinephrine infusion (1 µg/kg/min) was administered.

Data were analyzed using 2-way analysis of varience of bioenergetic and mechanical function data with respect to time after loading, graphic analysis of percent change from control of mechanical and metabolic parameters, and using a graphic correlation approach relating metabolic parameters to mechanical function parameters. This third approach may be described using a Michaelis-Menten type of algorithm as mentioned previously: $V/V_{max} = 1/(1 + (Km/S))$, where V = velocity of reaction, here estimated by mechanical work; V_{max} = maximum velocity, here estimated by maximal mechanical work; K_m = affinity constant determined empirically from *in vitro* experiments; S = substrate (ADP concentration) estimated by PCr/ATP. This Michaelis-Menten algorithm was translated into a 'transfer function', which relates mechanical function (estimate of velocity of reaction) with metabolic function (estimate of substrate concentration), as defined for ^{31}P NMR studies by Chance et al. [3, 93]. Mechanical function (velocity of reaction) was estimated by HR × SBP, CO, P × V, MVO_2 and (FS); metabolic function (substrate concentration) was estimated by PCr/ATP (used as an estimate of ADP concentration).

Mechanical function measured with piezoelectric crystals point out the variability of contractile response to similar loading conditions [70]. Pressure loading (50–

Table 1. Fractional shortening (FS) and bioenergetic data (PCr/ATP) during pressure and/or volume loading

Pressure load					Volume load				
% Change from control					% Change from control				
Dog #	Intervention	FS	PCr/ATP	HR × SBP	Dog #	Intervention	FS	PCr/ATP	HR × SBP
1	NE: 10 min	− 24	− 38	+ 39	4	AVS 10 min	− 11	− 61	− 44
	50 min	− 33	− 25	+ 131	5	AVS 10 min	+ 97	0	+ 6
2	NE: 10 min	+ 47	− 31	− 9		50 min	+ 52	− 32	0
	50 min	+ 64	− 21	− 5	6	AVS 10 min	+ 45	− 44	+ 10
3	NE: 10 min	+ 24	− 15	+ 67		50 min	+ 32	0	+ 18
	50 min	+ 33	− 30	+ 217	7	AVS 10 min	− 37	0	+ 24
						50 min	+ 4	− 11	− 34
					7	AVS + NE			
						10 min	+ 84	− 42	− 15
						50 min	+ 34	+ 34	+ 16

NE: Norepinephrine (1 µg/kg/min).
AVS: Abdominal aorta vena cava shunt.
HR × SBP: Heart rate × systolic blood pressure.
PCr/ATP: Phosphocreatine/adenosine triphosphate.

80 mm Hg above control) was associated with a decrease in FS in one dog and an increase in FS in 2 dogs (Table 1). In all 3 cases, the PCr/ATP ratio decreased. Two volume loaded dogs had an increase in FS, while 2 others had a decrease. In volume loading, the changes in PCr/ATP were variable. In the dog with both pressure and volume loading, FS increased, while PCr/ATP decreased. The different responses to similar loading conditions are probably related to the baseline neurohumoral, metabolic and contractile homeostatic equilibrium of each dog.

These data show that even large changes in mechanical function are generally associated with only small changes in bioenergetics as measured by the PCr/ATP ratio [70, 96]. This indicates that there is indeed good metabolic regulation of high energy phosphates during wide changes in mechanical function and this metabolic regulation maintains stable myocardial function, and that change in ADP, as estimated by PCr/ATP, is not the only regulator of myocardial metabolic function. Thus, nonischemic, nonhypoxic myocardium is metabolically stable (i.e., minimal changes in PCr/ATP for up to 190% increase in $P \times V$, up to 118% increase in HR × SBP, up to 168% increase in CO, and up to 97% increase in MVO$_2$).

Our data are consistent with those presented by From et al. [110] and Ugurbil et al. [26] during glucose or glucose and insulin infusion in perfused rat hearts and by Katz et al. [24] during *in vivo* ^{31}P data acquisition from dogs; that is, a 2 to 3-fold change in work load (MVO$_2$)

was associated with minimal change in (ADP), ATP/ADP and phosphorylation portential.

In summary, these data demonstrate that it is difficult to define heart contractile work with any one parameter and thus, it is important to measure a variety of parameters when evaluating mechanical function of the heart during different types of loading conditions. Another important observation was made; that is, when data such as those collected in our investigation are grouped and analyzed with routine statistical methods (in this case a 2-way analysis of variance), the relationship between metabolic and mechanical function changes can be obscured. This results because changes in PCr/ATP occur at different times after mechanical loading in the different animals. This phenomenon points out the physiological variability which can at times invalidate global statistical analysis, and demonstrates the importance of evaluation of individual data from each animal separately to demonstrate the metabolic responses to mechanical loading.

Muscular dystrophy cardiomyopathy: in vivo closed chest dog *(^{31}P NMR saturation transfer) [111]*
While Duchenne muscular dystrophy is recognized to cause skeletal muscle myopathy, the cardiac involvement is less well recognized. Cardiac lesions include intermittant foci of mineralization, myocyte hypercontraction, and linear and anastomosing fibrosis [112]. On ultrastructural analysis a number of abnormalities have

Table 2. 3 Year data point for bioenergetic and contractile function in each chronic muscular dystrophy (MD) dog

	Control dogs: N = 6			MD Dogs: N = 5		
	Base	NE	% Change	Base	NE	% Change
K_f	0.5 ± 0.1	1.0 ± 0.5	1.26 ± 3	0.33 ± 0.09	0.38 ± 0.11*	0.15 ± 0.03*
PCr	3.5 ± 0.4	2.7 ± 0.4	− 0.21 ± 0.04	2.91 ± 0.17*	2.84 ± 0.23	− 0.03 ± 0.007*
Flux	1.9 ± 0.5	2.5 ± 0.3	0.31 ± 0.04	0.97 ± 0.26*	1.07 ± 0.35*	0.07 ± 0.03*
HR × SBP	1.8 ± 0.8	3.1 ± 0.6	0.9 ± 0.1	1.62 ± 0.5	3.05 ± 0.88	0.96 ± 0.5
MVO_2	8.8 ± 3.2	14.1 ± 6.4	0.7 ± 0.25	8.04 ± 2.0	18.88 ± 3.6	0.76 ± 0.3

* $P < 0.05$, CON vs MD. CON = Control dogs; MD = Muscular Dystrophy Dogs; NE = Norepinephrine (1 µg/kg/min); % Change = NE induced change: NE-Con/NE); K_f = Forward rate constant of creatine kinase; PCr = Phosphocreatine; Flux = PCr → ATP; HR × SBP = Heart rate × systolic blood pressure; MVO_2 = Oxygen consumption.

been identified and include endomysial fibrosis, decreased myofibrillar density, large mitochondria, intracytoplasmic myelin figures, lipid droplets and lipofuscin deposits. These morphological abnormalities translate into clinical conditions such as arrythmias and heart failure, and ultimately cardiac related death [113].

Because Duchenne cardiomyopathy in the canine model is similar to that found in humans [112, 113], we decided to evaluate the interaction of bioenergetic and mechanical function in this model using ^{31}P NMR saturation transfer techniques to study creatine kinase kinetics and relate this to echocardiographic contractile function and to HR × SBP. To this end, the forward rate constant (K_f) and flux of PCr to ATP were determined in 5 MD dogs (supplied from an XMD breeding colony maintained at the College of Veterinary Medicine, Cornell University), and compared with 6 control dogs.

Each MD dog was a descendent of a single affected male and had hyperechoic regions on echocardiography, consistent with a diagnosis of the cardiomyopathy associated with muscular dystrophy [113]. Dogs in this study were followed chronically in a closed chest condition. Therefore, cardiac windows were created (2–3 ribs and accompanying skeletal muscle were removed, marlex mesh was sutured between exposed ribs, the fascia and skin were sutured closed and dogs were allowed to recover 1 month before NMR studies were begun) so that dogs could be studied sequentially (over 1–3 years) in the closed chest mode during progression of their disease [75].

At each time point, after echocardiographic determination of ventricular function and after baseline measurements of creatine kinase kinetics, physiological loading with norepinephrine (1 µg/Kg/Min) was applied (infusion over 30 min). During this intervention, mechanical function was monitored using HR × SBP and MVO_2. At each time point (1–3 years) in the progression

of the MD cardiomyopathic process, there was no significant degradation in mechanical function documented either by echocardiography (regional wall motion abnormalities or ejection fraction) or HR × SBP. However, baseline PCr was less in MD dogs compared to CON at each point in the disease process. This could be due to replacement of muscle mass with fibrotic tissue. In addition, baseline K_f of creatine kinase and flux of PCr to ATP were greater in CON than in MD. NE loading induced an increase in K_f and flux in CON but not in MD. (Table 2 shows data from the final time point in each dog).

These data indicate that the cardiomyopathy associated with Duchenne's Muscular dystrophy is associated with decreased PCr and with altered creatine kinase kinetics. This bioenergetic alteration did not cause apparent deficits in mechanical function as evidenced at rest by echocardiographic determined regional wall motion and ejection fraction and by the cardiomyopathic heart's ability to mechanically recruit HR × SBP during NE increased workloads. These data suggest that either creatine kinase has a large safety factor (such that a large decrease in K_f or flux of PCr to ATP must exist before mechanical dysfunction is observed) or that the CK system is not important in overall maintenance of bioenergetic processes which support mechanical function.

Interventions on control animals

Species differences: in vivo *open chest dogs versus cats* (^{31}P *NMR) [114]*
One of our early studies was designed to explore the possibility of a bioenergetic difference between dog (adapted for running) and cat (adapted for sprinting and stalking) heart [114]. Cardiac metabolism was studied with ^{31}P NMR in 7 dogs and 4 cats. Cardiac work loads

26

were changed by pacing the heart at 4, 4.5, and 5 Hz. We found that the cat heart was more sensitive to increased work load than the dog heart, as indicated by a significantly greater increase in P_i/PCr in cat heart for a similar increase in heart rate $\times$ blood pressure 'work' (Cat P_i/PCr: Con = 0.29 $\pm$ 0.07; Max work = 0.68 $\pm$ 0.3; Dog P_i/PCR: Con = 0.29 $\pm$ 0.15; Max work = 0.3 $\pm$ 0.14). According to our observations, unloaded dog hearts were much closer to V_0 than cat hearts. In dogs, increasing work load corresponded to the initial portions of the rectangular hyperbola characteristic of the Michaelis-Menten kinetics, suggesting that the resting point of the dog heart is close to V_0 (baseline). In cats, on the other hand, the resting metabolic rate is closer to V_{max}; thus, cats have less recruitment ability. These data suggest that it is possible that animals adapted to different life styles have cardiovascular systems which are metabolically and/or mechanically adapted for different quality and/or quantity of body work. This phenomenon may be related to differences in skeletal muscle function, cardiac microcirculation [2, 3] and/or to intrinsic differences in myocardial enzyme kinetics.

Hypoxia: in vivo *open chest dogs (^{31}P NMR, ^{31}P NMR saturation transfer, NADH fluorometry, and ^2H NMR to measure tissue perfusion) [33, 38, 44]*
Combined in vivo NADH fluorometry, ^{31}P NMR (PCr/ATP, K_f of creatine kinase and flux of PCr to ATP) and ^2H NMR were used to evaluate oxidative phosphorylation and tissue perfusion, respectively, during hypoxia (inspired gasses, N_2/O_2 = 15/1) [33, 38, 44]. Hypoxia was associated with increased HR $\times$ SBP product (2.02 $\times$ $10^4 \pm 0.6$, to $3.6 \times 10^4 \pm 1.3$), P $\times$ V work ($1.9 \times 10^3 \pm 1.3$ to $3.6 \times 10^3 \pm 2.6$) and MVO$_2$ (ml/min/100 gm) (9.7 $\pm$ 2.5 to 16.9 $\pm$ 4.5), P < 0.05. Myocardial perfusion (ml/min/100 gm) increased appropriately to support the increased workload (60 $\pm$ 12 to 182 $\pm$ 14) [38]. During hypoxia, oxidative phosphorylation was depressed as evidenced by an increase in NADH (+ 160% $\pm$ 39) from baseline, with an associated decrease in PCr/ATP (2.65 $\pm$ 0.35 to 1.97 $\pm$ 0.66, P < 0.05). During the hypoxic state we also found slight decreases in K_f and flux of PCr to ATP (K_f: 0.5 $\pm$ 0.17 to 0.36 $\pm$ 0.15; flux: 1.7 $\pm$ 0.9 to 1.3 $\pm$ 0.77; P < 0.05) similar to findings of Bittl et al. [115].

These data suggest that both ADP (as estimated by PCr/ATP) and NADH may be metabolic regulators during hypoxia. If we consider the work of Cleland [36, 37, 116] and Morrison and James [117] on the kinetics of the creatine kinase reaction *in vitro*, the forward rate constant, K_f, will increase as ADP increases. This relationship has been demonstrated experimentally *in vivo* by Bittl et al. [115] and Blum and Johnson [118]. In our study, hypoxia induced a decrease in K_f, suggesting a decrease in ADP. It is possible, as suggested by Bittl et al. [115], that the kinetics of the creatine kinase reaction may change variably during hypoxia. Assuming no change in intracellular pH, they found an increase in ADP during hypoxia in rat hearts. In their experiments, HR, SBP, and thus HR $\times$ SBP all fell significantly during hypoxia (in contrast to the HR $\times$ SBP responses reported above). Since Bittl et al. [115] also found a decrease in K_f under these conditions, they postulate a hypoxia induced change in creatine kinase kinetics. This, however, remains speculative.

Ischemia: in vivo *open chest dogs (multiprobe apparatus; K^+, Ca^{2+}, and Na^+ surface electrodes, NADH fluorometry and laser doppler tissue flow) [45, 46]*
Myocardial ischemia and reperfusion injury are related to a number of functional abnormalities associated with failure of bioenergetic processes and their control mechanisms [8, 10, 119-128]. Since the heart depends primarily on oxidative metabolism, the initial abnormalities result from the oxygen deficit [8, 10, 128]. Oxidative phosphorylation is quickly depressed and glycolysis and/or the phosphocreatine shuttle function are recruited to maintain high energy phosphates [121]. Even though the quantity of ATP produced via these processes is not usually sufficient to maintain normal global physiological function, it is sufficient to maintain normal ion pump function. Eventually, pH changes (lactic acidosis and other acids) [122, 123, 125, 126] combined with energy failure compromise enzyme function [120, 128] (phosphorylases, sodium-potassium ATPases, sodium-calcium ATPases) causing ion disequilibrium [129-141], which allows sodium and possibly calcium to leak into the cells and potassium to leak out of the cells. On reperfusion, return of oxygen to the tissue may provide the substrate for generation of toxic oxygen metabolites (oxygen free radicals, singlet oxygen, H_2O_2 and OH^-) which can further damage membrane and enzyme function [142-145]. This can also further damage bioenergetic processes [146]. As a result, further ion disequilibrium can result which may provide the basis for abnormal mechanical and metabolic function in the post-ischemic period [121, 126, 127, 135-137].

Previous investigators have used surface electrodes to measure changes in K^+_e, and H^+_e during ischemia/reperfusion and other interventions [129-135]. There is some evidence that changes in K^+_e may be involved in the reg-

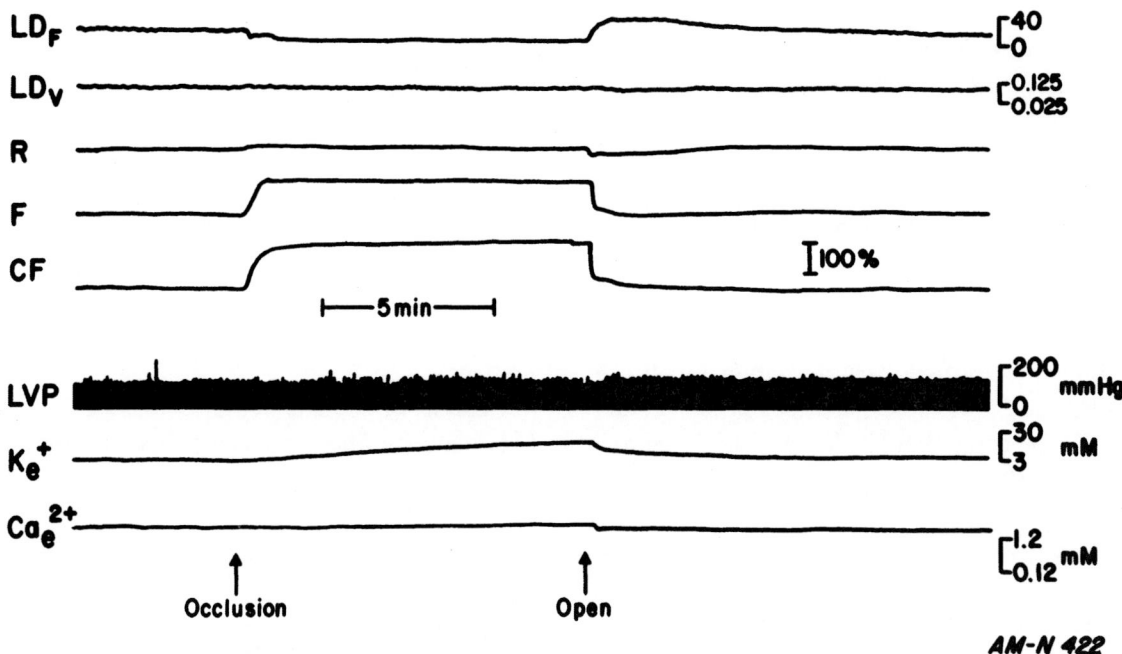

Fig. 3. Response of the heart to ischemia/reperfusion. Note that during ischemia, LD_f decreases, LD_v changes minimally, CF increases, and K^+_e and Ca^{2+}_e increase. These changes are reversed upon reperfusion. Also note the initial hyperemic response upon reperfusion, as monitored by $LD_f \cdot LD_f =$ % change from baseline in tissue flow measured by the laser doppler flow (LDF) system; $LD_v =$ % change from baseline in tissue volume measured by the laser doppler flow (LDF) system; R = reflectance measured by the NADH fluorometer; F = fluorescence (uncorrected) measured by the NADH fluorometer; CF = corrected fluorescence (R–F) reported as % change from baseline; LVP = left ventricular pressure in mm Hg; $K^+_e =$ response of potassium electrode in mM; $Ca^{2+}_e =$ response of the calcium electrode in mM.

ulation of myocardial metabolism (including redox state), perfusion, and contractile function [135]. However, very little information has been obtained under *in vivo* conditions to support the possible regulatory role of K^+_e in any of these functions. In addition, no one has previously simultaneously monitored two or more ions.

We [141], as well as other investigators [136–140], have used nuclear magnetic resonance (NMR) techniques to monitor changes in Ca^{2+}_i and Na^+_i, in conjunction with changes in high-energy phosphates, during ischemia/reperfusion and other pathophysiological processes. While these studies are of interest, the shift reagents used to monitor changes in Ca^{2+}_i and Na^+_i change the physiological state (perfusion and contractile function) of the heart. Therefore, it is difficult to use data obtained in this manner to determine the effects of either Ca^{2+}_i or Na^+_i in regulation of myocardial perfusion or contractile function.

We [33, 38, 44, 46, 47] and others [43, 49, 50, 52, 147] have also used NADH fluorescence to determine changes in the redox state of the heart under different physiological and pathophysiological conditions. While a number of these studies have been done in conjunction with ^{31}P NMR and ^2H NMR techniques to evaluate the interaction between oxidative phosphorylation and perfu-

sion, respectively, with O_2 consumption, redox state and contractile function [38, 47], no studies have been done to investigate the interaction of ion transport with this composit of parameters. Because changes in ion content and ion transport are thought to signal change in perfusion, and metabolic and mechanical function, it is important to simultaneously and accurately measure ion content and fluxes in conjunction with the other physiological parameters.

Thus, in order to more fully understand the 'complete equation', i.e., the balance between O_2 supply and demand, in conjunction with ions which may act as signalers/effectors of change in mechanical function and perfusion and vice versa, a multiprobe assembly (MPA) [48, 66, 67, 69, 148] has been developed which simultaneously monitors ion flux (K^+, Ca^{2+} and Na^+) [46, 66, 129–135], tissue perfusion [60–65], and redox state [33, 38, 43, 44, 47, 49, 50, 53, 68, 147]. Using this apparatus, we have demonstrated that we can obtain simultaneous information concerning tissue volume (LDv) and perfusion (LD_f), NADH redox state, and ion flux (K^+_e, Ca^{2+}_e and Na^+_e) during physiological interventions such as ischemia/reperfusion (2 min ischemia) and hypoxia (5–15 min hypoxia). Because of the nature of the probe, these data are obtained only from the surface of the heart and may not reflect the heterogeneity that may exist throughout the myocardium.

Using these techniques, generally the epicardial tissue flow – NADH redox relationship is clear. That is, as tissue flow (or perfusion) decreases, NADH increases; and vice versa, when tissue flow increases, NADH decreases (Fig. 3). However, the responses of extracellular ions during ischemia/reperfusion and hypoxia are not always as clear and deserve a more detailed explanation.

The K^+_e generally increased during either ischemic or hypoxic insult and returned to baseline upon discontinuation of the insult. However, the magnitude of the changes in K^+_e are somewhat smaller than those reported by other investigators [129–135]. This may be related to the fact that our measurements were made with surface electrodes placed on the epicardium, whereas measurements by other investigators were made with electrodes introduced into the myocardium. The electrodes used for intramyocardial placement may have damaged some myocytes during their introduction. Damaged cells may have an exaggerated response to further insults, such as those of ischemia/reperfusion or hypoxia. It is also possible that tissue heterogeneity may contribute to the different K^+_e response.

While there are no other data in the literature concerning surface electrode measurement of Ca^{2+}_e or Na^+_e in the intact perfused and beating heart, in vivo measurements of these ions have been made in the brain during ischemia/reperfusion and hypoxia [48, 66, 67, 69, 148]. In the brain, either insult caused a marked increase in Ca^{2+}_e and decrease in Na^+_e. In the heart, there were minimal increases in Ca^{2+}_e during ischemia, but no change during hypoxia. With respect to the myocardial Na^+_e response to either ischemia/reperfusion or hypoxia, no obvious fluctuation was seen during either insult. There are several possible explanations for these responses. It is possible that there are differences in the kinetic response of different organs to similar insults. The heart (myocytes) may have more stable membrane function (supported for longer periods of ischemia and/or hypoxia by glycolysis), thus, preventing rapid and dramatic fluxes of ions in the short time periods used in the present studies. We have demonstrated this to be the case in isolated cardiomyocytes [149–151]; i.e., that it takes at least 15 minutes of ischemia and often longer (especially during hypoxia) to effect changes in Ca^{+2}_i and Na^+_i. Therefore, it may take longer than 15 minutes of ischemia or hypoxia to produce a large change in ion millieu in vivo, as well. Alternatively, there may be a diffusion problem; ions may not penetrate to where the electrodes are measuring, or the electrodes may not be sufficiently sensitive to sense small and/or delayed flux of ions into and out of the surrounding tissue. In addition, with respect to Ca^{+2}, it is possible that the changes are primarily intracellular (into and out of sarcoplasmic reticulum, and contractile apparatus), and thus may not be sensed by surface electrodes. With respect to the Na^+_e, the external millieu of the heart is replete in Na; therefore, small changes in Na^+_e due to insult may be averaged out by the baseline (or normally large) Na^+_e content.

With the above data in mind, we must think of tissue heterogeneity. There is evidence that the response of the myocardium to ischemia is heterogeneous in terms of perfusion (hyperemia; perfusion through collateral vessels), metabolism (aerobic vs glycolytic) and mechanical function (stunning, hyberbnation) changes. Even though there is indirect evidence of this heterogeneous physiology, it is often difficult to measure these differences in vivo. Therefore, we decided to address several aspects of the heterogeneity question using intra-myocardial measurements of NADH fluorescence (NADH/NAD redox state) in conjunction with laser doppler techniques to estimate tissue perfusion. The goal was to evaluate regional gradients or patterns of response of the heart during ischemia [45]. To do this, needle fiberoptic light guides (0.45 mm in diameter) were introduced various distances into the myocardium and used to measure regional changes in NADH redox state and perfusion compared to surface changes (obtained with 2 mm surface fiberoptic light guides), during 2 min periods of ischemia produced with a balloon occluder placed around the proximal left anterior descending coronary artery (LAD) (Fig. 4).

Gradients of redox state and tissue perfusion were documented during ischemia produced by a balloon occluder placed around the LAD in 7 dogs. Complete occlusion was documented by a transonic flow probe which was placed around the LAD. Reperfusion occured upon deflation of the balloon occluder. Measurements were made on the epicardium and at 4 intramyocardial levels.

The range of responses to ischemia-reperfusion of intramyocardial perfusion and NADH redox state was wide: These responses might parallel changes on the surface, or could be significantly different (depending on the position) (Fig. 5a & b). This is due to perfusion through collateral vessels, and heterogeneous tissue response due to variability of low flow and/or hyperemic responses to ischemia. NADH and perfusion responses

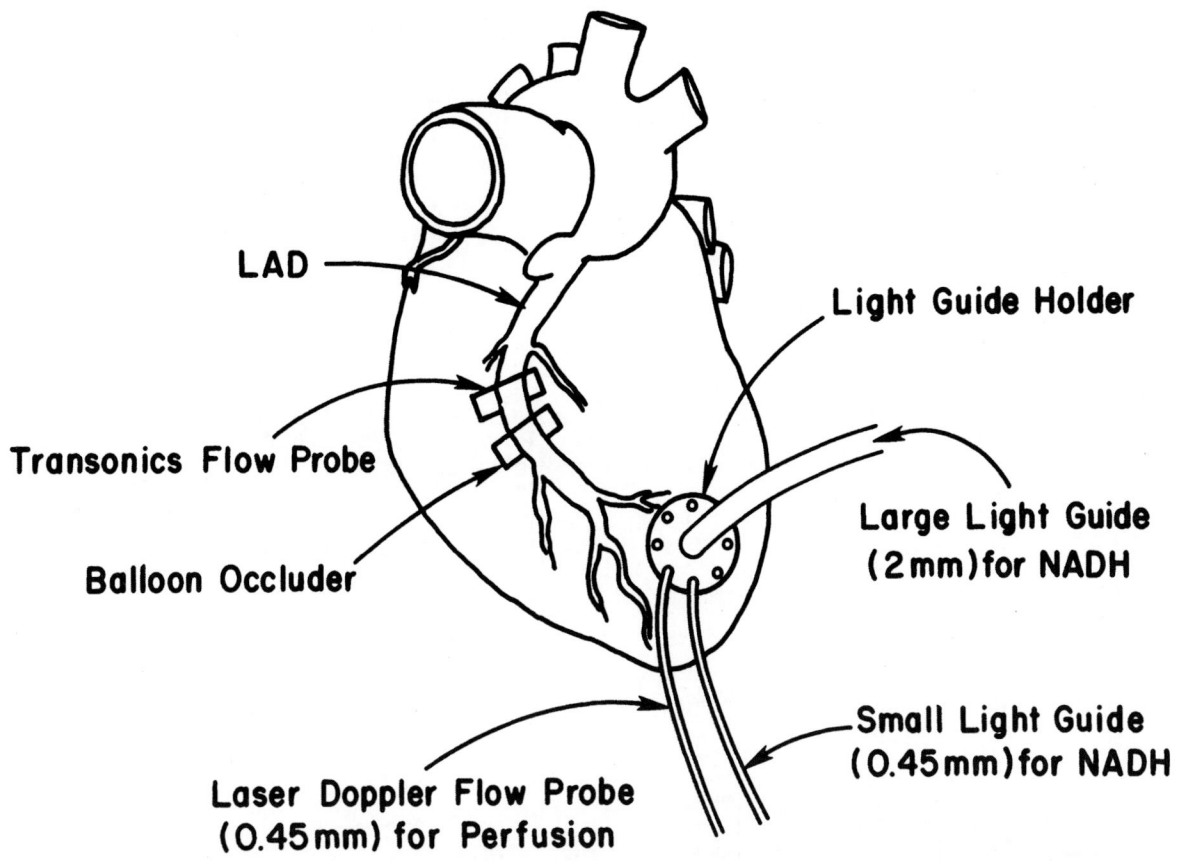

Fig. 4. Diagram of heart instrumented with 2 sets of light guides for measurement of NADH redox state (2 mm light guide and 0.45 mm light guide) and laser doppler tissue perfusion (0.45 mm light guide). Note that the light guides are held in place by a delrin holder which is sutured to the heart. The large light guide measures surface fluorescence. The small light guides can be used to measure intramyocardial, as well as, surface fluorescence and perfusion, respectively. Also indicated are the balloon occluder used to produce ischemia and the Transonics flow probe used to measure coronary flow.

could be in phase or out of phase, which also may be related to tissue heterogeneity, both in collateral vessel perfusion responses and metabolic responses to changes in perfusion.

These studies show that myocardial ischemia and reperfusion induce heterogeneous responses in perfusion and metabolism. The gradients of change are not necessarily linear across the myocardium. This metabolic heterogeneity is most likely due to a perfusion heterogeneity related to collateral vessels, but may also be a metabolic phenomenon related to change in substrate presentation and/or use. Thus, global regulation of myocardial function must be a composit of independent regulatory factors. Depending on the region and parameters monitored, different forms of regulation can be interpreted as dominant.

Potential neurohumoral regulators: acetylcholine (ACH): in vivo open chest dog (^{31}P NMR saturation transfer, NADH fluorometry and ^2H NMR to measure tissue perfusion [47]

In liver mitochondria, acetylcholine (Ach) administration can stimulate a-ketoglutarate (KG) oxidation and substrate level phosphorylation [152]. This effect is related to a decrease in succinate oxidation, and is opposite to the effect of epinephrine, which stimulates succinate oxidation. These findings indicate that there is reciprocal regulation of metabolism by endogenous neurohumors.

The hemodynamic (dose related vasodilation and/or vasoconstriction) [153–157] and inotropic and chronotropic (dose related positive and/or negative) effects [153, 154, 156–160] of ACH on the heart are well documented. Because of the intimate involvement of ACH in these hemodynamic processes, we decided to investigate the possibility that ACH could also induce either primary (unrelated to mechanical function and perfusion) or secondary (related to changes in mechanical

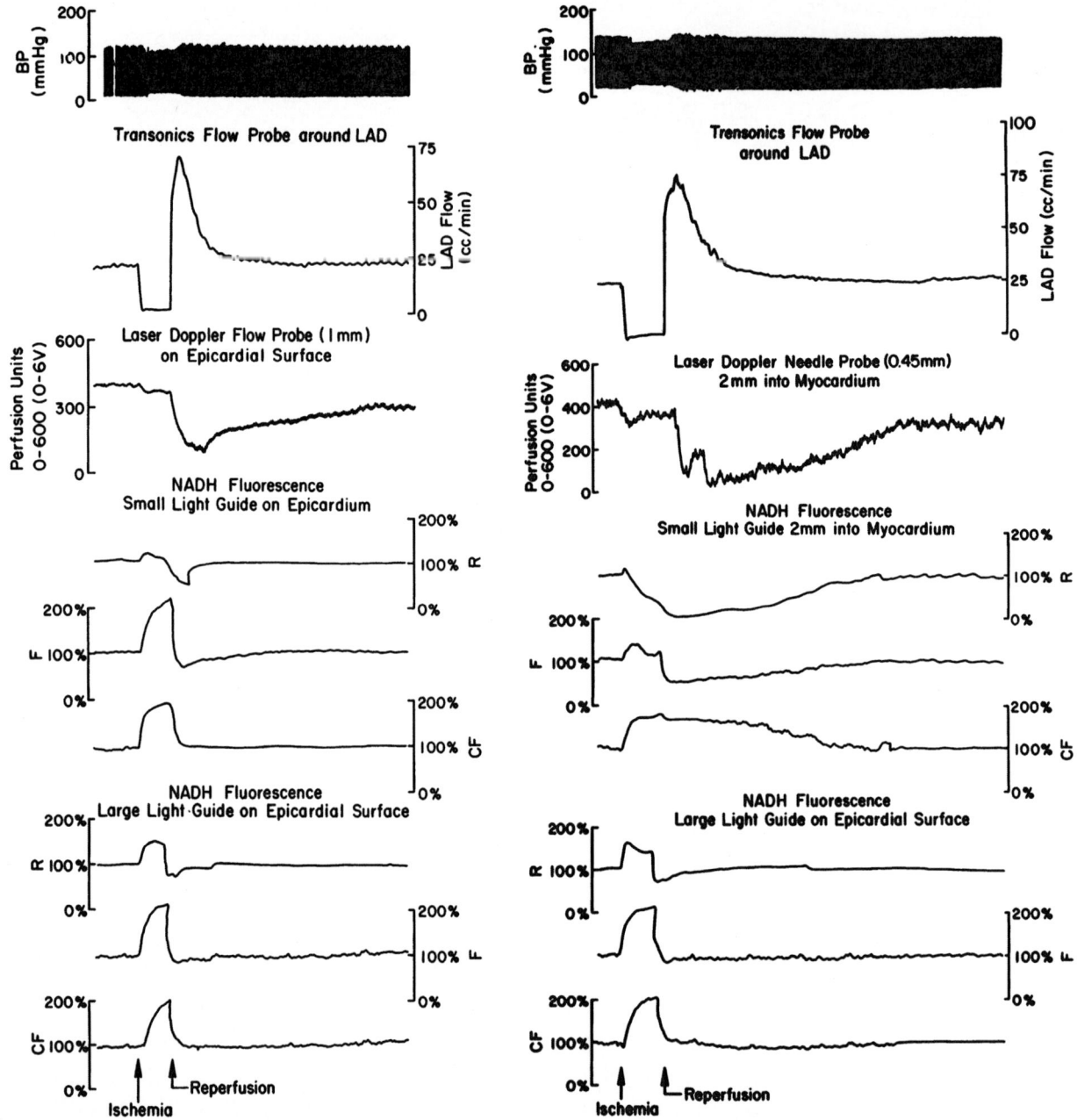

Fig. 5. Data which demonstrate the heterogeneous response of the heart to ischemia. a) All light guides are on the surface of the heart. b) The large light guide (2 mm diameter, NADH fluorescence) was placed on the surface and the 2 small light guides (0.45 mm diameter, NADH fluorescence and laser doppler perfusion) were introduced 2 mm into the myocardium. Note the differences in the intramyocardial NADH redox state and perfusion responses to ischemia.

function or perfusion) metabolic effects in the myocardium.

To this end, the effects of ACH on myocardial metabolism and perfusion were evaluated in 13 open chest dogs. Specifically, the forward rate constant (K_f) of creatine kinase and the flux of PCr to ATP were determined using ^{31}P NMR saturation transfer. NADH/NAD redox state was determined with NADH fluorescence techniques. Myocardial perfusion was determined using washout rates of perdeuterated saline (using a Kety-

Schmidt model [38, 55, 59]) in conjunction ^{2}H NMR. Mechanical function was estimated using HR × SBP, CO, and MVO$_2$. Acetylcholine infusion generally produced decreases in K_f and flux of PCr to ATP, associated with variable but slight increases in HR × SBP, CO and MVO$_2$ (Fig. 6). NADH fluorescence increased and myocardial perfusion decreased. Immediately after ACH infusion was complete (1–30 min), K_f and flux continued to decrease, while HR × SBP and MVO$_2$ were variable but still with an upward trend, and CO tended to de-

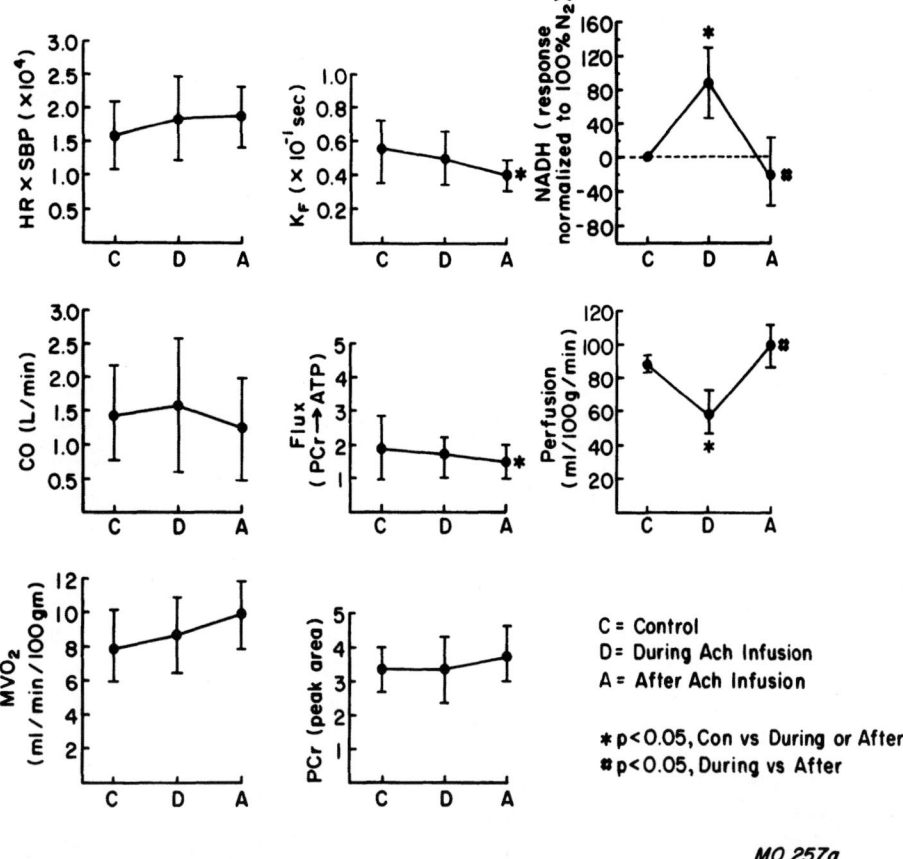

MO 257a

Fig. 6. Myocardial bioenergetic, mechanical, and perfusion responses to acetylcholine (ACH) administration. Note the biphasic responses of NADH redox state and perfusion, in contrast to the steady decline in the forward rate constant of creatine kinase (K_f) and flux of PCr to ATP during and after ACH infusion. HR × SBP = heart rate × systolic blood pressure; CO = cardiac output (l/min); MVO_2 = oxygen consumption (ml/min/100 gm heart); PCr = phosphocreatine peak area; NADH = nicotinamide adenine dinucleotide (normalized to 100% N_2); Perfusion = ml/min/100 gm (^{2}H NMR).

crease. NADH fluorescence decreased and myocardial perfusion increased. These data suggest that ACH infusion is associated with metabolic effects, i.e. induction of changes in creatine kinase enzyme kinetics with resulting changes in the flux of PCr to ATP.

These data do not permit definition of the exact mechanism of ACH's metabolic effect. However, these metabolic effects do not appear to be related to catecholamine mediated responses [158], which might occur secondary to ACH induced catecholamine release. In similar parallel experiments using norepinephrine (NE) infusion we found that NE produced increases in HR × SBP and MVO_2 are associated with concomitant increases in K_f of creatine kinase and flux of PCr to ATP, which are just the opposite of the Ach induced metabolic effects [33].

In summary endogenous, ACH release may be directly involved in regulation of myocardial metabolic function. Because the metabolic changes found in this study were not directly correlated with mechanical function and/or perfusion (and in fact were often negatively cor-

related with these parameters) it is possible that Ach acts as a primary as well as secondary regulator of myocardial metabolism. Our data indicate that this regulatory effect may be related to change in the forward rate constant, K_f of creatine kinase and flux of PCr to ATP, as well as secondary, related to Ach induced changes in hemodynamics and mechanical function, with resulting changes in NADH redox state.

Summary and conclusions

Metabolic correlations of myocardial mechanical function have been topics of interest to cardiovascular physiologists for decades. Early studies which investigated substrate utilization and/or availability in disease states, indicated that there was little or no correlation between disease state and oxidizable substrates such as fatty acids, glucose, or lactate [161]. Other studies indicated that failing heart had abnormal high energy metabolism, i.e., decreased mitochondrial oxidative phosphorylation and NADH-linked respiration (indicating that there was mitochondrial damage) [92, 162, 163], which resulted in decreased adenosine triphosphate (ATP) and phosphocreatine (PCr) concentrations [164–166]. Several investigators suggested that this deficit was due to uncoupling of oxidative phosphorylation [167]. Some investigations, which involved *in vitro* evaluation of enzyme activity from failing hearts [161], demonstrated that isocitrate dehydrogenase activity was impaired in heart failure (again indicating depressed oxidative phosphorylation mechanisms). Similar studies which evaluated glyceraldehyde phosphate dehydrogenase demonstrated an increase in heart failure states, which indicated that an augmentation of anaerobic metabolism was necessary to supplement decreased oxidative metabolism.

More recent work using mitochondria [27, 168] and perfused heart [26, 110, 169–172] preparations indicated that there may be a number of different regulators of myocardial metabolism, each effective under different physiological conditions. Adenosine diphosphate (ADP) concentration, phosphorylation ratio (ATP/ADP × Pi), intramitochondrial reducing power (NAD/NADH), cellular respiration rate, H^+ concentrations, inorganic phosphate (P_i), ATP/ADP, creatine (Cr), and citric acid cycle (TCA) substrate availability have all been implicated as potential regulators of oxidative phosphorylation. These studies indicate that there is no one universal regulator which functions under all conditions and that regulation can be controlled and/or influenced by more than one regulator at a time [38, 41, 44]. It has been proposed that changes or abnormalities in myocardial response to different regulators may be associated with extreme loading conditions and disease states [170–172].

The understanding of biological control mechanisms under *in vivo* conditions is quite important and a natural extention of delineation of these mechanisms under *in vitro* conditions. Until recently it has not been possible to study the control of bioenergetics in the living organism. The development of NMR spectroscopy and NADH fluorometry techniques have provided the means to evaluate these important biochemical phenomena *in vivo*.

In summary

To gain insite into the mechanisms of myocardial regulation as it relates to the interaction of mechanical and metabolic function and perfusion, intact animal models were instrumented for routine physiological measurements of mechanical function (arterial, venous, and ventricular canulae to measure heart rate × systolic blood pressure, cardiac output, and oxygen consumption) and for measurements of metabolism (NMR surface coils for measurement of bioenergetics (^{31}P) and fiberoptic light guides for measurement of NADH fluorescence (redox state)) and perfusion (^2H NMR and Laser doppler techniques). Studies using these techniques in canine and cat models of volume and/or pressure loading, hypoxia, ischemia and cardiomyopathic states indicate that myocardial bioenergetic function is quite stable under most loading conditions as long as the heart is not ischemic. Results from these *in vivo* studies indicate that there is no universal regulator and that different biochemical regulators appear to mediate this stable function under different physiological and pathophysiological conditions. Adenosine diphosphate (ADP), adenosine monophosphate (AMP), phosphorylation potential PP = (ATP/ADP · Pi), free energy of hydrolysis (dG/

$$de = \Delta G_{obs} + RT \ln \frac{(ADP)\,(Pi)}{(ATP)},$$ NADH/NAD redox

state, P_i concentration, pH (H^+), oxygen content and substrate availability and use are all potential myocardial regulators. In addition, change in enzyme kinetics (studied with ^{31}P NMR saturation transfer techniques) are induced by certain substrates and/or physiological conditions, which in turn may also act as regulators. Similar work loads under different conditions appear to induce different myocardial regulators to come into play to maintain stable function: for example; during hypoxia, NADH redox state appears to play a regulatory role; in pressure loading, ADP, PP and free energy of ATP hydrolysis as well as NADH redox state, appear to be regulatory. Because there is significant overlap in the function of these regulators, it is difficult to control for all the variables and thus evaluate the effect of each individual variable in *in vivo* models.

References

1. Bessman SP, Gieger PJ: Transport of energy in muscle: the phosphocreatine shuttle. Science 211: 448–452, 1981
2. Chance B, Leigh JS, Clark BJ, Maris J, Kent J, Nioka S, Smith D: Control of oxidative metabolism and oxygen delivery in human skeletal muscle: a steady-state analysis of the work/energy cost transfer function. Proc Natl Acad Sci, USA 82: 8384–8388, 1985
3. Bassingthwaigthe JB, Yipintso T, Harvey RB: Microvasculature

of the dog left ventricular myocardium. Microvas Res 7: 229–249, 1974

4. Rapaport S, Guest GM: Distribution of acid-soluble phosphorus in the blood cells of various vertebrates. J Biol Chem 138: 269–282, 1941

5. Lange R, Ingwall J, Hale SL, Alker KJ, Braunwald E, Kloner RA: Preservation of high-energy phosphates by verapamil in reperfused myocardium. Circulation 70: 743–741, 1984

6. Flarerty JT, Weisfeldt ML, Bulkley BH, Gardner TJ, Gott VL, Jacobus WE: Mechanisms of ischemic myocardial cell damage assessed by phosphorus-31 nuclear magnetic resonance. Circulation 65: 561–570, 1982

7. Fossel ET, Morgan HE, Ingwall JS: Measurement of changes in high-energy phosphates in the cardiac cycle using gated ^{31}P nuclear magnetic resonance. Proc Natl Acad Sci. USA 77: 3654–3658, 1980

8. Gadian DG, Hoult DI, Radda GK, Seeley PJ, Chance B, Barlow C: Phosphorus nuclear magnetic resonance studies on normoxic and ischemic cardiac tissue. Proc Natl Acad Sci. USA 73: 4446–4448, 1976

9. Garlick PB, Radda GK, Seeley PJ: Phosphorus NMR studies on perfused heart. Biochem Biophys Res Commun 74: 1256–1262, 1977

10. Hollis DP, Nunnally RL, Jacobus WE, Taylor GJ: Detection of regional ischemia in perfused beating hearts by phosphorus nuclear magnetic resonance. Biochem Biophys Res Commun 75: 1086–1091, 1977

11. Salany JM, Pieper GM, Wu S, Tood GL, Clayton FC, Eliot RS: ^{31}P nuclear magnetic resonance measurement of cardiac pH in perfused guinea-pig hearts. J Mol Cell Cardiol 11: 601–610, 1979

12. Nakazawa M, Katano Y, Imai S, Matsushita K, Ohuchi M: Effects of l- and d-propranolol on the ischemic myocardial metabolism of the isolated guinea pig heart, as studied by ^{31}P-NMR. J Cardiovas Pharmacol 4: 700–704, 1984

13. Bailey IA, Radda GK, Seymour AML, Williams SR: The effects of insulin on myocardial metabolism and acidosis in normoxia and ischemia. A ^{31}P-NMR study. Biochim Biophys Acta 720: 17–27, 1982

14. Peiper GM, Murray WJ, Salhany JM, Nu ST, Eliot RS: Salient effects of l-carnitine on adenine-nucleotide and coenzyme A acylation in the diabetic heart perfused with excess palmitic acid. A phosphorus-31 NMR and chemical extract study. Biochim Biophys Acta 803: 241–249, 1984

15. Pernot AC, Ingwall JS, Menasche P, Grousset G, Bercot M, Piurnica A, Fossel ET: Evaluation of high-energy phosphate metabolism during cardioplegic arrest and reperfusion: a phosphorus-31 nuclear magnetic resonance study. Circulation 67: 1296–1303, 1983

16. Ligeti L, Barlow C, Chance B, Kovach AGB, O'Connor M: ^{31}P NMR spectroscopy of brain and heart. Adv Exp Med Biol 159: 281–292, 1983

17. Matthews PM, Williams SR, Seymour AM, Schwartz A, Dube G, Gadian DG, Radda GK: A ^{31}P-NMR study of some metabolic and functional effects of the inotropic agents epinephrine and ouabain, and the ionophore RO2-2985 (X537A) in the isolated, perfused rat heart. Biochim Biophys Acta 720: 163–171, 1982

18. Neurohr KL, Gollin G, Barrett EJ, Shulman RG: In vivo ^{31}P-NMR studies of myocardial high energy phosphate metabolism during anoxia and recovery. FEBS Let 159: 207–210, 1983

19. Grove TH, Ackerman JJH, Radda GK, Bore PJ: Analysis of rat heart in vivo by phosphorus nuclear magnetic resonance. Proc Natl Acad Sci. USA 77: 299–302, 1980

20. Whitman GJR, Chance B, Bode H, Maris J, Haselgrove J, Kelley R, Clark BJ, Harken AH: Diagnosis and therapeutic evaluation of a pediatric case of cardiomyopathy using phosphorus-31 nuclear magnetic resonance spectroscopy. JACC 5: 745–749, 1985

21. Kantor HL, Briggs RW, Balaban RS: In vivo ^{31}P nuclear magnetic resonance measurements in canine heart using a catheter coil. Circ Res 55: 261–266, 1984

22. Koretsky AP, Wang S, Murphy-Boesch J, Klein MP, James TL, Weiner MW: ^{31}P NMR spectroscopy of rat organs, in situ, using chronically implanted radiofrequency coils. Proc Natl Acad Sci. USA 80: 7491–7495, 1983

23. Bottomley PA: Noninvasive study of high-energy phosphate metabolism in human heart by depth-resolved ^{31}P NMR spectroscopy. Science 229: 769–772, 1985

24. Katz LA, Swain JA, Portman MA, Balaban RS: Relation between phosphate metabolites and oxygen consumption of heart in vivo. Am J Physiol 256 (Heart Circ Physiol 25): H256–274, 1989

25. Camacho SA, Parmley WW, James TL et al.: Substrate regulation of the nucleotide pool during regional ischemia and reperfusion in an isolate rat heart preparation: a phosphorus-31 magnetic resonance spectroscopy analysis. Cardiovas Res 22: 193–203, 1988

26. Ugurbil K, Kingsley-Hickman PB, Sako EY, Zimmer S, Mohana-Krisman P, Robitaille PML, Thomo WJ, Johnson A, Foker JE, From AHL: ^{31}P NMR studies of the kinetics and regulation of oxidative phosphorylation in the intact myocardium. Ann NY Acad Sci 508: 265–286, 1987

27. Gyulai L, Roth Z, Leigh JS, Chance B: Bioenergetic studies of mitochondrial oxidative phosphorylation using ^{31}P NMR. J Biol Chem 260: 3947–3954, 1985

28. Katz LA, Koretsky AP, Balaban RS: Activation of dehydrogenase activity and cardiac respiration: a ^{31}P NMR study. Am J Physiol 255 (Heart Circ Physiol 24): H185–H188, 1988

29. Rooke A, Fiegl ED: Work as a correlate of canine left ventricular oxygen consumption and the problems of catecholamine oxygen wasting. Circ Res 50: 273–286, 1982

30. Blum H, Ivanics T, Zhang D, Wrobwelski K, Osbakken M: Effect of temperature and coronary flow on the metabolic and mechanical function of the isolated rat heart. Cardiology 82: 238–248, 1993

31. Chance B, Lanoue K: Metabolic control in exercising skeletal muscle. Am J Physiol 258: R288–R289, 1990

32. Chance B, Williams CM: The respiratory chain and oxidative phosphorylation. Adv Enzymol 17: 65–134, 1956

33. Osbakken M, Blum H, Wang DJ, Doliba N, Ivanics T, Zhang D, Mayevsky A: In vivo mechanisms of myocardial functional stability during physiological interventions. Cardiology 79(1): 1–14, 1991

34. Meyer RA, Sweeney HC, Kushmerick MJ: A simple analysis of the 'phosphocreatine shuttle'. Am J Physiol 246: C365–C377, 1984

35. Eldar H, Degani H: ^{31}P NMR studies of the thermodynamics and kinetics of the creatine kinase reaction. Mag Res Med 11: 121–126, 1989

36. Morrison JF, Cleland WW: Isotope exchange studies of the mechanism of the reaction catalyzed by adenosine triphosphate. Creatine phosphokinase. J Biol Chem 241: 673–683, 1966

37. Schimerlik MI, Cleland WW: Inhibition of creatine kinase by chromium nucleotide. J Biol Chem 248: 8418–8423, 1973

38. Osbakken M, Mitchell MD, Zhang D, Mayevsky A, Chance B: *In vivo* correlation of myocardial metabolism, perfusion and mechanical function during increased cardiac work. Cardiovasc Res 25: 749–756, 1991

39. Osbakken M, Young M, Huddle J, Closter J, Prammer M, Chance B: Acute volume loading studied in cat myocardium with ^{31}P nuclear magnetic resonance. Mag Res Med 7: 143–155, 1989

40. Bittl JA, Delayre J, Ingwall: Rate equation for creatine kinase predicts the *in vivo* reaction velocity: ^{31}P NMR surface coil studies in brain, heart, and skeletal muscle of the living rat. Biochemistry 26: 6083–6090, 1987

41. Osbakken M, Douglas PS, Ivanics T, Zhang D, Van Winkle T: Creatine kinase kinetics in a canine model of chronic hypertension induced cardiac hypertrophy. J Am Coll Cardiol 19(1): 223–228, 1992

42. Spenser RGS, Balshi JA, Leigh JS, Ingwall JS: ATP synthesis and degradation rates in perfused rat heart: ^{31}P NMR double saturation transfer measurements. Biophys J 54: 921–929, 1988

43. Chance B, Salkowitz IA, Kovach GB: Kinetics of mitochondrial flavoprotein and pyridine nucleotide in perfused heart. Am J Physiol 223: 207–218, 1972

44. Osbakken M, Mayevsky A, Ponomarenko I, Zhang D, Duska C, Chance B: Combined *in vivo* NADH fluorescence and ^{31}P NMR to evaluate oxidative phosphorylation. J Appl Cardiol 4: 305–313, 1989

45. Osbakken M, Ivanics T, Zhang D, Alter C: Gradients of redox state and perfusion across the myocardium during ischemia. FASEB, Journal Abstracts 5(5): Part II; 3934, March 15, 1991

46. Osbakken MD, Mayevsky A, Zhang D, Ivanics T: Multiparameter analysis of the ischemic and hypoxic myocardium. Clin Res 39(2): 158A, 1991

47. Osbakken M, Doliba N, Mitchell MD, Ivanics T, Zhang D, Mayevsky A: Acetylcholine: Is it a myocardial metabolic regulator? J Appl Cardiol 5: 357–366, 1990

48. Mayevsky A, Flamm ES, Pennie W, Chance B: A fiber optic based multiprobe system for intraoperative monitoring of brain functions. SPIE Proc 1431: 330–313, 1991

49. Mayevsky A: Brain NADH redox state monitored *in vivo* by fiber optic surface fluorometry. Brain Res Rev 7: 49–68, 1984

50. Mayevsky A, Chance B: Intracellular oxidation reduction state measured in situ by a multichannel fiber-optic-surface fluorometer. Science 217: 537–540, 1982

51. Chance B, Oshima N, Sugano T, Mayevsky A: Basic principles of tissue oxygen determination from mitochondrial signals. In: Bicker H, Bruley DF eds. Oxygen transport to tissue. New York: Plenum Press, 1973; 277–292

52. Chance B, Barlow C: Ischemic areas in perfused hearts: measurement by NADH fluorescence photography. Science 193: 909–910, 1972

53. Rosenthal M, Sick TJ: Measurement of metabolic activity associated with ion shifts. In: Boulton AA, Baker GB, Wolz W Eds. Neuro-Methods: the neuronal microenvironment. Clifton, NJ: Humana Press, 1988, 187–245

54. Steenbergen C, Deleeuw G, Narlow C, Chance B, Williamson JR: Heterogeneity of the hypoxic state in perfused rat heart. Circ Res 41: 606–615, 1977

55. Mitchell MD, Osbakken M: Estimation of myocardial perfusion using deuterium nuclear magnetic resonance. Mag Res Imaging 9(4): 545–552, 1991

56. Ackerman JJH, Ewy CS, Kim SG, Shalwitz RA: Dueterium magnetic resonance *in vivo:* the measurement of blood flow and tissue perfusion. Ann NY Acad Sci 508: 89–98, 1986

57. Kim SJ, Ackerman JJH: Multicompartment analysis of blood flow and tissue perfusion employing D2O as freely diffusable tracer: a novel deuterium NMR technique demonstrated via application with murine RIF-1 tumor. Mag Res Med 8: 410–426, 1988

58. Mitchell MD, Clark BJ, Leigh JS: Simultaneous *in vivo* phosphorus metabolic spectroscopy and deuterium flow measurement (abstract). Proc Mag Res Med (New York) 6: 427, 1987

59. Kety SS, Schmidt CF: The nitrous oxide method for quantitative determination of cerebral blood flow in man: theory, procedure and normal values. J Clin Invest 27: 476–483, 1948

60. Arbit E, DiResta GR, Bedford RF, Shah NK, Galicih JH: Intraoperative measurement of cerebral and tumor blood flow with laser-Doppler fluorometry. Neurosurg 24: 166–170, 1989

61. Haberl RL, Heizer ML, Ellis EF: Laser-Doppler assessment of brain microcirculation: effect of local alterations. Amer J Physiol 256 (heart Circ Physiol 25): H1255–H1260, 1989

62. Bonner R, Nossal R: Model for laser-Doppler measurements of blood flow in tissue. Appl Optics 20: 2097–2107, 1981

63. Dirnagl U, Kaplan B, Jacewic M, Pukinelli W: Continuous measurement of cerebral cortical blood flow by laser-Doppler fluorometry in a rat stroke model. J Cereb Blood Flow and Meta 9: 589–596, 1989

64. Haberl RL, Heizer ML, Marmarow A, Ellis EF: Laser-Doppler assessment of brain microcirculation: effect of systemic alterations. Am J Physiol 256 (Heart Circ Physiol 25): H1247–H1254, 1989

65. Stern MD, Lappe DL, Bowen PD, Chimosky JE, Holoway GA, Keiser HR, Bowen PD: Continuous measurement of tissue blood flow by Laser-Doppler spectroscopy. Am J Physiol 232: H441–H448, 1977

66. Friedli CM, Sclarsky DS, Mayevsky A: Multiprobe monitoring of ionic, metabolic and electrical activities in the awake brain. Am J Physiol 243 (Regulatory Integrative Comp Physiol 12): R462–R469, 1982

67. Mayevsky A: Multiparameter monitoring of the awake brain under hyperbaric oxygenation. J Appl Physiol 54: 740–748, 1983

68. Barlow CH, Harden WR, Harken AH, Simson MB, Haselgrove JC, Chance B, O'Conner M, Austin G: Fluorescence mapping of mitochondrial redox changes in heart and brain. Crit Care Med 7: 402–406, 1979

69. Mayevsky A: Level of ischemia and brain functions in the mongolian gerbil *in vivo*. Brain Res 524: 1–9, 1990

70. Osbakken MD, Closter J, Huddell J, Chance B: Relationship between fractional shortening and high energy phosphate metabolism under different loading conditions. Proc Mag Res Med 5(2): 1019, 1987

71. Badke FR, Covell JW: Early changes in left ventricular regional dimensions and function during chronic volume overloading in the conscious dog. Circ Res 45: 420–428, 1979

72. Edwards Ch, Rankin JS, McHale PA, Ling D, Anderson RW: Effects of ischemia on left ventricular function in the conscious dog. Am J Physiol 240 (Heart Circ Physiol 9): H423–H420, 1981

73. Gallager KP, Kumada T, Kozoil JA, McKnown MD, Kemper WS, Ross J Jr: Significance of regional wall thickening abnormalities

relative to transmural myocardial perfusion in anesthetized dogs. Circulation 62: 1266–1274, 1980

74. Hurlbut TA, Scoles F, Buja LM, Willerson JT: Use of chronically-implanted ultrasonic crystals in the intact heart of the cat. Cardiovas Res 12: 387–390, 1978

75. Osbakken M, Ligeti L, Clark BJ, Bolinger L, Subramanian H, Schnall M, Leigh J, Chance B: Myocardial high energy phosphate metabolism in closed chest dog: creation of an animal model. Mag Res Med 3: 801–807, 1986

76. Osbakken MD, Pigott J, Ligeti L, Duska C, Ponomarenko I, Chance B: Myocardial bioenergetics of chronic volume overload studied with ^{31}P MRS. J Appl Cardiol 5(1): 39–51, 1990

77. Mehmel HC, Mazzoni S, Krayenbull HP: Contractility of the hypertrophied human left ventricle in chronic pressure and volume overload. Am Heart J 90: 236–240, 1975

78. Miller CAH, Kirklin JW, Swan HJC: Myocardial function and left ventricular volumes on acquired valvular insufficiency. Circulation 31: 374–384, 1965

79. Osbakken M, Bove AA, Spann JF: Left ventricular function in chronic aortic regurgitation with reference to end-systolic pressure, volumes, and stress relations. Am J Cardiol 47: 193–198, 1981

80. Malik AB, Geha AS: Cardiac function, coronary flow, and MVO2 in hypertrophy induced by pressure and volume-overloading. Cardiovas Res 11: 310–316, 1977

81. Wysten F, Flaming W, Schaper W: Cardiac function in the chronically volume overloaded canine heart. Basic Res Cardiol 72: 172–177, 1977

82. Newman HW: Contractile state of hypertrophied left ventricle in long standing volume overload. Am J Physiol 234: H88–H93, 1978

83. Suga H, Hisano R, Hirato S, Hayshi T, Ninonuja I: Mechanisms of higher oxygen consumption rate: Pressure-loaded vs volume-loaded heart. Am J Physiol 242 (Heart Circ Physiol 11): H942–H948, 1983

84. Hultgren PB, Bove AA: Myocardial blood flow and mechanics in volume overload-induced left ventricular hypertrophy in dog. Cardiovas Res 15: 522–528, 1981

85. Thomas DP, Phillips STJ, Bove AA: Myocardial morphology and blood flow distribution in chronic volume overload hypertrophy in dogs. Basic Res Cardiol 79: 379–388, 1984

86. Badke FR, White FC, Letiuler M, Covell J, Andreas J, Bloor C: Effects of experimental volume-overload hypertrophy on myocardial blood flow and cardiac function. Am J Physiol 241: H564–H570, 1981

87. Molic JM, Bercovici J, Swynghedauw B: Protein synthesis during systolic and diastolic overloading in rats: a comparative study. Cardiovas Res 15: 515–521, 1981

88. Scheuer J, Bhan AK: Cardiac contractile proteins: ATPase activity and physiological function. Circ Res 45(1): 1–12, 1979

89. Meerson FZ, Berger AM: The common mechanism of the heart's adaptation and deadaptation: hypertrophy and atrophy of the heart muscle. Basic Res Cardiol 72: 228–234, 1977

90. Moravec J, Moravec M, Hatt PY: Rate of pyridine nucleotide oxidation and cytochrome oxidase interaction and intracellular oxygen in hearts from rats with compensated volume overload. Pflugers Arch 392: 106–114, 1981

91. Cooper G, Paga KJ, Zujko KJ, Harrison CE, Coleman HM: Normal myocardial function and energetics in volume overload hypertrophy in the cat. Circ Res 32: 140–148, 1973

92. Raczniak TJ, Chesney CF, Allen JR: Oxidative phosphorylation and respiration of mitochondria from normal, hypertrophied and failing rat hearts. J Mol Cell Cardiol 92: 215–223, 1977

93. Chance B, Leigh JS, Kent J, McCully K, Nioka S, Clark BJ, Maris J, Graham T: Multiple controls of oxidative metabolism in living tissues as studied by phosphorus magnetic resonance. Proc Natl Acad Sci USA 83: 9458–9462, 1986

94. Hatt PY, Rakusan K, Gastineau P, LaPlace M, Cluzeaud F: Aorta Caval fistula in the rat. An experimental model of heart volume overload. Basic Res Cardiol 75: 105–108, 1980

95. Carey RA, Natarajan G, Bove AA, Coulson RL, Spann JF: Myosin adenosine triphosphatase activity in the volume-overload hypertrophied feline right ventricle. Circ Res 45 (1): 81–87, 1979

96. Osbakken M, Ligeti L, Huddell J, Duska C, Ponomarenko I, Chance B: In vivo myocardial bioenergetics during acute volume and/or pressure loading in a canine model: A ^{31}P NMR study. Cardiology 76(6): 405–417, 1989

97. Buser PT, Wagner S, Wu ST et al.: Verapamil preserves myocardial performance and energy metabolism in left ventricular hypertrophy following ischemia and reperfusion. Circulation 80: 1837–45, 1989

98. Buser PT, Wikman-Coffelt J, Wu ST, Derugin N, Parmley WW, Higgins CB: Postischemic recovery of mechanical performance and energy metabolism in the presence of left ventricular hypertrophy: a ^{31}P MRS study. Circ Res 66: 735–746, 1990

99. Wexler LF, Lorell BH, Momomura S, Weinberg EO, Ingwall JS, Apstein CS: Enhanced sensitivity to hypoxia-induced diastolic dysfunction in pressure-overload left ventricular hypertrophy in the rat: role of high energy phosphate depletion. Circ Res 62: 766–775, 1988

100. Koehler U, Medugorac I: Left ventricular enzyme activities of the energy-supplying metabolism in Goldblatt-II rats. Res Exp Med 185: 299–307, 1985

101. Taegtmeyer H, Overturf ML: Effects of moderate hypertension on cardiac function and metabolism in the rabbit. Hypertension 11: 416–426, 1988

102. LeCarpentier Y, Bugausky LB, Chemla D et al.: Coordinated changes in contractility, energetics, and isomyosins after aortic stenosis. Am J Physiol 252: H275–H282, 1987

103. Gibbs CL, Wendt IR, Kotsanas G, Young IR, Wooley G: Mechanical, energetic, and biochemical changes in long-term pressure overload of rabbit heart. Am J Physiol 259: H849–H859, 1990

104. Bittl JA, Ingwall JS: Intracellular high-energy phosphate transfer in normal and hypertrophied myocardium. Circulation 75 (suppl 1): I96–I101, 1987

105. Stephans MR, Leger JJ, Preteseille M, Swynghedauw B: The relationship of a decline in myofibrillar ATPase activity to the development of severe left ventricular hypertrophy in rat. Path Biol 27: 41–44, 1979

106. Kupriyanov VV, Steinschenider AY, Ruuge EK, Kapel'Ko VI, Zuevo MY, LaKomkin VL, Smiranov VN, Saks VA: Regulation of energy reaction velocity through the creatine kinase reaction in vitro and in perfused rat heart. ^{31}P NMR studies. Biochim Biophys Acta 805: 319–331, 1984

107. Martin JF, Guth BD, Greffey RH, Hoekenga DE: Myocardial creatine kinase exchange rates and ^{31}P NMR relaxation rates in intact pigs. Mag Res Med 11: 64–72, 1989

108. Douglas P, Berko B, Lesh M, Reicheck N: Alterations in diastolic

function in response to progressive left ventricular hypertrophy. J Am Coll Cardiol 13: 461–467, 1989

109. Douglas PS, Tallant B: Hypertrophy, fibrosis, and diastolic dysfunction in early canine experimental hypertension. J Am Coll Cardiol 17: 530–536, 1991

110. From AHL, Petein MA, Michurski SP, Zimmer SD, Ugurbil K: ^{31}P-NMR studies of respiratory regulation in the intact myocardium. FEBS Letter 206: 257–261, 1986

111. Osbakken MD, Zhang D, Valentine B, Cooper B, Li JY: Altered enzyme kinetics in muscular dystrophy cardiomyopathy. Circulation 82(4): 264, 1990

112. Valentine BA, Cummings JF, Cooper BJ: Development of Duchenne-type cardiomyopathy. Morphologic studies in a canine model. Am J Path 135(4): 671–678, 1989

113. Moise NS, Valentine BA, Brown CA, Erb HN, Beck KS, Cooper BJ, Gilmour RF: Duchenne's cardiomyopathy in a canine model: Electrocardiographic and echocardiographic studies. J Am Coll Cardiol 17: 812–20, 1991

114. Ligeti L, Osbakken MD, Clark BJ, Schnall M, Bolinger L, Subramanian H, Leigh J, Chance B: Cardiac transfer function relating energy metabolism to workload in different species as studied with ^{31}P NMR. Mag Res Med 4: 112–119, 1987

115. Bittle JA, Balschi JA, Ingwall JS: Contractile failure and high energy phosphate turnover during hypoxia; ^{31}P NMR surface caoil studies in living rat. Circ Res 60: 871–878, 1987

116. Cleland WW: The kinetics of enzyme-catalyzed reactions with two or more substrates or products. I, Nomenclature and rate equation. Biochem Biophys Acta 67: 104–137, 1963

117. Morrison JF, James E: The mechanism of the reaction catalyzed by adenosine triphosphate creatine phosphotransferase. Biochem J 97: 37–52, 1965

118. Blum H, Johnson RG Jr: *In vivo* creatine phosphokinase activity in rat resting skeletal muscle (abstract). Soc Mag Res Med, 9th An Meet, 1990, p897

119. Osbakken M, Ito K, Zhang D, Ponomarenko I, Ivanics T, Jahngan EGE, Cohn M: Creatine and cyclocreatine effects on ischemic myocardium: ^{31}P NMR evaluation of intact heart. Cardiology 80 (3): 184–195, 1992

120. Bailey IA, Williams SR, Radda GK, Gadian DG: Activity of phosphorylase in total global ischemia in the rat heart. Biochemistry 196: 171–178, 1981

121. Brooks WM, Haseler LJ, Clark K, Willis RJ: Relation between the phosphocreatine to ATP ratio determined by ^{31}P nuclear magnetic resonance spectroscopy and left ventricular function in underperfused guinea-pig heart. J Mol Cell Cardiol 18: 149–155, 1986

122. Rossi A, Martin J, DeLeires J: Phosphorus nuclear magnetic resonance studies of the energetic state and the intracellular pH of the isolated rat heart in the course of ischemia. J Physiol (Paris) 76: 902–905, 1980

123. Stein PD, Goldstein S, Sabbath HN, Liu ZQ, Helpern JA, Ewing JR, Lakier JB, Chopp M, LaPenna WF, Welch KMA: *In vivo* evaluation of intracellular pH and high-energy phosphate metabolites during regional myocardial ischemia in cat using ^{31}P nuclear magnetic resonance. Mag Res Med 3: 262–269, 1986

124. Bailey IA, Seymour AML, Radda GK: A ^{31}P-NMR study of the effects of reflow on the ischemic rat heart. Biochem Biophys Acta 637: 1–7, 1981

125. Brooks WM, Willis RJ: ^{31}P nuclear magnetic resonance study of the recovery characteristics of high-energy phosphate compounds and intracellular pH after global ischemia in the perfused guinea pig heart. J Mol Cell Cardiol 15: 495–502, 1983

126. Jacobus WE, Pores IH, Lucas SK, Weisfeldt ML, Flaherty JT: Intracellular acidosis and contractility in the normal and ischemic heart as examined by ^{31}P NMR. J Mol Cell Cardiol 14: 13–20, 1982

127. Whitman GJR, Keival RS, Seeholzer S, McDonald G, Simpson MD, Marker AH: Recovery of left ventricular function after graded cardiac ischemia as predicted by myocardial ^{31}P nuclear magnetic resonance. Surgery 97: 428–435, 1985

128. Chance B, Williams GR: Respiratory enzymes in oxidative phosphorylation. I. Kinetics of oxygen utilization. J Biol Chem 17: 383–393, 1955

129. Hill JL, Gettes LS, Lynch MR, Hebert NC: Determination of intravascular and myocardial K$^+$. Am J Physiol 235(4): H455–H459, 1978

130. Hill JL, Gettes LS: Effect of acute coronary artery occlusion on local myocardial extracellular K$^+$ activity in swine. Circulation 61: 768–778, 1980

131. Fleet WF, Johnson TA, Graebner CA, Gettes LS: Effect of serial brief ischemic episodes on extracellular K$^+$, pH, and activation in the pig. Circulation 72: 922–932, 1985

132. Fleet WF, Johnson TA, Graebner CA, Engle CL, Gettes LS: Effects of verapamil on ischemia-induced changes in extracellular K$^+$, pH, and local activation in the pig. Circulation 73: 837–846, 1986

133. Jenkins MG, Johnson TA, Engle C, Gettes JS: Metabolic protection by verapamil during graded coronary flow reduction independent of effect on baseline systolic function. Circulation 80: 1870–1877, 1989

134. Takano M, Qin Dayi, Noma A: ATP-dependent decay and recovery of K channels in guinea pig cardiac myocytes. Am J Physiol 258 (Heart Circ Physiol 27): H45–H50, 1990

135. Sagisaka K, Tamura M, Yamazaki I: The effect of K$^+$ concentration on the energy metabolism in perfused rat heart. J Biochem 95: 1091–1103, 1984

136. Koretsune Y, Marban E: Mechanism of ischemic contracture in ferret hearts: relative roles of [Ca^{2+}], elevation and ATP depletion. Am J Physiol 258 (Heart Circ Physiol 27): H9–H16, 1990

137. Steenbergen C, Murphy E, Watts JA, London RE: Correlation between cytosolic free calcium, contracture, ATP, and irreversible ischemic injury in perfused rat heart. Circ Res 66: 135–146, 1990

138. Marban E, Koretsune Y, Corretti M, Chacko VP, Kusuoka H: Calcium and its role in myocardial cell injury during ischemia and reperfusion. Circulation 80(suppl IV): IV17–IV22, 1989

139. Pierce GN, Maddaford TG, Kroeger EA, Cragoe EJ: Protection by benzamil against dysfunction and damage in rat myocardium after calcium depletion and repletion. Am J Physiol 258 (Heart Circ Physiol 27): H17–H23, 1990

140. Ponce-Hornos JE, Musi EA, Bonazzola P: Role of extracellular calcium on heart muscle energetics: effects of verapamil. Am J Physiol 258 (Heart Circ Physiol 27): H64–H72, 1990

141. Osbakken M, Ivanics T, Zhang D, Mitra R, Blum H: Isolated cardiomyocytes in conjunction with NMR spectroscopy techniques to study metabolism and ion flux. J Biol Chem 267 (22): 15340–15347, 1992

142. Tamura M, Oshino N, Chance B: Some characteristics of hydrogen- and alkylperoxide metabolizing systems in cardiac tissue. J Biochem 92: 1019–1031, 1982

143. Zweier JL, Flaherty JT, Weisfeldt ML: Direct measurement of free radical generation following reperfusion of ischemic myocardium. Proc Natl Acad Sci, USA 84: 1404–1407, 1987

144. Lucchesi BR: The role of oxygen-derived radicals in myocardial reoxygenation injury. The age of reperfusion 1: 1–4, 1989

145. van der Kraaij AMM, van Ejik HG, Koster JF: Prevention of postischemic cardiac injury by the orally active iron chelator 1,2-dimethyl-3-hydroxy-4-pyridone (L1) and the antioxidant (+)-cyanidanol-3. Circulation 80: 158–164, 1989

146. Osbakken M, Ivanics T, Zhang D, Kumar C: Singlet O_2 and superoxide anion effects on myocardial metabolism. J Mag Res Imaging, 1(2): P426 (Works in Progress): Page 36, 1991

147. Chance B, Oshino N, Sugano T, Mayevsky A: Basic principles of tissue oxygen determination from mitochondrial signals. In, Oxygen Transport to Tissue, ed. H. Beiker, DF Bouley. Plenum Press, New York, pp. 277–292, 1973

148. Mayevsky A, Zarchin N, Yoles E, Tannenbaum B: Oxygen supply to the brain under hypoxic and hyperoxic conditions. In: 'Oxygen Transport in Red Blood Cells', C. Niclau ed. Pergamon Press, pp. 119–132, 1986

149. Ivanics T, Blum H, Wroblewski K, Wang DJ, Osbakken M: Intracellular sodium in cardiomyocytes using ^{23}Na nuclear magnetic resonance. Biochim Biophysi Acta, in press 1994

150. Osbakken M, Zhang D, Ivanics T, Rawson N, Friedman M: Bioenergetics of sodium transport in models of disease. Proc Mag Res Med Abstract, P364, vol. 2, 1993

151. Osbakken M, Ivanics T, Zhang D, Blum H: Glycolysis is necessary for sodium transport in cardiomyocytes. Proc Mag Res Med, Abstract P2209, vol. 2, 1992

152. Kondrashova MN, Doliba NM: Polarographic observation of substrate-level phosphorylation and its stimulation of acetylcholine. FEB Letter 243(2): 153–155, 1989

153. Blumenthal MR, Wang H, Markee S, Wang SC: Effects of acetylcholine on the heart. Am J Physiol 214(6): 1280–1287, 1968

154. Daggett WM, Nugent GC, Carr PW, Powers PC, Harada Y: Influence of vagal stimulation on ventricular contractility, O2 consumption and coronary flow. Am J Physiol 212(1): 8–18, 1967

155. Hashimoto K, Shigei T, Imai S, Saito Y, Yago N, Uei I, Clark RE: Oxygen consumption and coronary vascular tone in the isolated fibrillating dog heart. Am J Physiol 198(5): 965–970, 1960

156. Feigl EO: Parasympathetic control of coronary blood flow in dogs. Circ Res 14: 509–519, 1969

157. Van Winkle G, Feigl E: Acetylcholine causes coronary vasodilation in dogs and baboons. Circ Res 65: 1580–1593, 1989

158. Higgins CB, Vatner SF, Braunwald E: Parasympathetic control of the heart. Pharmacol Rev 25(1): 119–155, 1973

159. Nuutinen EM, Nishiki K, Erechinska M, Wilson DF: Role of mitochondrial oxidative phosphorylation in regulation of coronary blood flow. Am J Physiol 242 (Heart Circ Physiol 12): H159–H169, 1982

160. Nuutinen EM, Wilson DF, Erechinska M: The effect of cholinergic agonists on coronary flow rate and oxygen consumption in isolated perfused rat heart. J Mol Cell Cardiol 17: 34–42, 1983

161. Bing RJ: Biochemical basis of myocardial failure. Hosp Pract 9: 93–112, 1983

162. Moravec J, Laplace M, Renault G, Corsin A, Hatt PY: Mitochondrial respiratory activity during early stages of pressure induced hypertrophy. Pathol Biol 27: 51–59, 1979

163. Schwartz A, Sordayl LA, Entman ML, Allen JC, Reddy YS, Goldstein MA, Luchi RJ, Wyboray LE: Abnormal biochemistry in myocardial failure. Am J Cardiol 32: 407–422, 1973

164. Katz AM, Brady AJ, Mechanical and Biochemical correlates of cardiac contraction. Part 1. Mod Concepts of Cardiovasc Dis 40: 39–44, 1971

165. Katz AM, Brady AJ, Mechanical and Biochemical correlates of cardiac contraction. Part 2. Mod Concepts of Cardiovasc Dis 40: 45–48, 1971

166. Meerson FZ, Pomoinistsky VD: The roles of high-energy phosphate compounds in development of cardiac hypertrophy. J Mol Cell Cardiol 4: 571–597, 1972

167. Meerson FZ: The failing heart: adaptation and deadaptation. New York, Raven Press, 1983.

168. Jacobus WE, Moreadith RW, Vandegaer KM: Mitochondrial Respiratory Control. Evidence against the regulation of respiration by extramitochondrial phosphorylation potentials or by [ATP]/[ADP] ratios. J Biol Chem 257: 2397–2402, 1982

169. Connett RJ: Analysis of metabolic control: new insights using scaled creatine kinase model. Am J Physiol 254: R949–R959, 1988

170. Gibbs C: The cytoplasmic phosphorylation potential: Its possible role in the control of myocardial respiration and cardiac contractility. J Mol Cell Cardiol 17: 727–731, 1985

171. Griesen J, Kammermeier H: Relationship of phosphorylation potential and oxygen consumption in isolated perfused rat heart. J Mol Cell Cardiol 12: 891–907, 1980

172. Nishiki K, Erechinska M, Wilson D: Energy relationships between cytosolic metabolism and mitochondrial respiration in rat heart. Am J Physiol 234: C73–C81, 1978

Molecular and Cellular Biochemistry **133/134**: 39–50, 1994.
© 1994 *Kluwer Academic Publishers.*

I-3 Compartmentation of ATP synthesis and utilization in smooth muscle: roles of aerobic glycolysis and creatine kinase

Y. Ishida[1], I. Riesinger[2], T. Wallimann[2] and R.J. Paul[3]

[1] *Mitsubishi Kasei Institute of Life Sciences, 11 Minamiooya, Machida-Shi, Tokyo 194, Japan;* [2] *Institute for Cell Biology, ETH-Hönggerberg, CH 8093 Zürich, Switzerland;* [3] *Department of Physiology & Biophysics, University of Cincinnati, College of Medicine, Cincinnati, OH 45267-0576, USA*

Abstract

The phosphocreatine content of smooth muscle is of similar magnitude to ATP. Thus the function of the creatine kinase system in this tissue cannot simply be regarded as an energy buffer. Thus an understanding of its role in smooth muscle behavior can point to CK function in other systems. From our perspective CK function in smooth muscle is one example of a more general phenomenon, that of the co-localization of ATP synthesis and utilization. In an interesting and analogous fashion distinct glycolytic cascades are also localized in regions of the cell with specialized energy requirements. Similar to CK, glycolytic enzymes are known to be localized on thin filaments, sarcoplasmic reticulum and plasma membrane. In this chapter we will describe the relations between glycolysis and smooth muscle function and compare and contrast to that of the CK system. Our goal is to more fully understand the significance of the compartmentation of distinct pathways for ATP synthesis with specific functions in smooth muscle. This organization of metabolism and function seen most clearly in smooth muscle is likely representative of many other cell types. (Mol Cell Biochem **133/134**: 39–50, 1994)

Key words: smooth muscle, creatine kinase, glycolysis, oxidative metabolism, compartmentation

Introduction

As the phosphocreatine content of smooth muscle is of similar magnitude to ATP, the function of the creatine kinase system in this tissue cannot simply be regarded as an energy buffer. Thus understanding its role in smooth muscle behavior can point to CK function in other systems. From our perspective CK function in smooth muscle is related to the co-localization of ATP synthesis and utilization.

In an interesting and analogous fashion distinct glycolytic cascades also appear to be localized in the regions of the cells with specialized energy requirements. Simi-lar to CK, glycolytic enzymes are known to be localized on the thin filaments, sarcoplasmic reticulum and plasma membrane. In this chapter we will describe the relations between glycolysis and smooth muscle function and compare and contrast to that of the CK system. Our goal is to more fully understand the significance of the compartmentation of ATP synthesis and utilization in smooth muscle.

Aerobic glycolysis and smooth muscle function

Vascular smooth muscle is primarily an oxidative tissue, that is, most of its ATP requirements (> 70%) are calculated to be met by oxidative phosphorylation [1]. However even under fully aerobic conditions, a substantial production of lactate is characteristic of smooth muscle metabolism. The rate of this aerobic lactate production in smooth muscle on a molar basis is similar in magnitude to that of oxygen consumption, and thus it would generate about 20% of the ATP. The production of lactate under aerobic conditions may be similar in other cell types, for example in cardiac or skeletal muscle [2]. However the rate of oxidative metabolism in these cell types is considerably higher than in smooth muscle and their aerobic glycolysis is thus a much smaller component of total metabolism and often overlooked.

The role of this anomalous aerobic glycolysis has long been subject of speculation [1]. Obvious explanations, such as lack of mitochondrial capacity or incomplete oxygenation do not appear applicable. Our work in this area began over a decade ago, when we reported a correlation between the rate of aerobic glycolysis and the activity of the Na-K pump in coronary artery [3]. Oxygen consumption, on the other hand, was found to be strongly correlated with the level of active isometric force, presumably reflecting the energy requirements of the actin-myosin ATPase. For example, ouabain, an inhibitor of the Na-K pump, abolished lactate production, while oxidative metabolism and also force were increased [4]. More recently, quantitative studies of this correlation were undertaken [5]. Na-K pump activity was quantitated by measuring K^+-uptake after readmission of K^+ to K^+-depleted carotid arteries. In the physiological range of pump rates, aerobic glycolysis is its primary energy source and lactate production is stoichiometrically linked to the Na-K pump, i.e., 2 K^+ ions per ATP per lactate. Strikingly, the readmission of K^+ was associated first with a stimulation of lactate production, followed by activation of oxidative metabolism only after the aerobic lactate production was saturated. This 'reverse' Pasteur effect is opposite to that expected from standard biochemistry in which glycolysis is activated only after oxidative capacity is exceeded.

The mechanisms underlying this functional compartmentation of aerobic glycolysis and membrane energy requirements would not appear compatible with the usual notion that glycolytic enzymes are uniformly distributed throughout the 'cytosol'. We investigated this coupling by studying the nature of the substrate underlying aerobic glycolysis. Smooth muscle has a substantial glucose uptake [6] as well as a fully competent mechanism for glycogen utilization [7]. Using radiolabelled glucose, we found that over 90% of the label appeared in the lactate pool. Under stimulated conditions, glucose uptake and glycogen catabolism occur at similar rates, suggesting that lactate could arise from either or both substrates. Surprisingly, the specific activity of lactate was identical to that of glucose, indicating that all of the lactate production originated from glucose, despite the entry of glucosyl moieties from glycogen into the glycolytic pathway [8]. Thus even in the absence of identifiable barriers and the small size of vascular smooth muscle cells relative to diffusion, glycolytic intermediates arising from glucose did not mix with the pool of intermediates originating from glycogen. Measurement of the specific activity of glucose-6-phosphate lent further support to this hypothesis [9]. Under anaerobic conditions, a detectable decrease in the specific activity of lactate was observed, indicating that unlabelled lactate arising from glycogen was produced. Recent studies of Hardin et al. [10] provide further support for the separation of glycogenolytic and glycolytic metabolism. In this study, carbon 13 NMR techniques were used to identify the fate of glycogen and glucose carbon. The NMR label in lactate again originated solely from glucose whereas the decrease in labelled glycogen was attributed to its oxidation to CO_2. Thus the functional compartmentation of glycolysis and membrane energetics and oxidative metabolism with force appears to reflect this compartmentation of glycolytic and glycogenolytic metabolism.

Perhaps at an initial glance, a most puzzling aspect of this compartmentation is that it is not absolute. It has long been known that ionic gradients in smooth muscle can be maintained in the absence of glucose [cf. 5]. It would appear that in the absence of glycolytic production of ATP, oxidative sources can be utilized. We would suggest then that rather than an all or none type phenomenon, compartmentation leads to a more efficient coupling between energy supply and transduction. This is more difficult to demonstrate and less well documented than the compartmentation phenomenon itself, but there are a number of studies to this end. Lynch and Balaban [11] showed that the V_{max} of the Na-K pump in MDCK cells was about twice as large in the presence of both oxidative and glycolytic substrates than with substrate for oxidative metabolism alone. We showed that the restoration of K^+ gradient in K^+-depleted carotid ar-

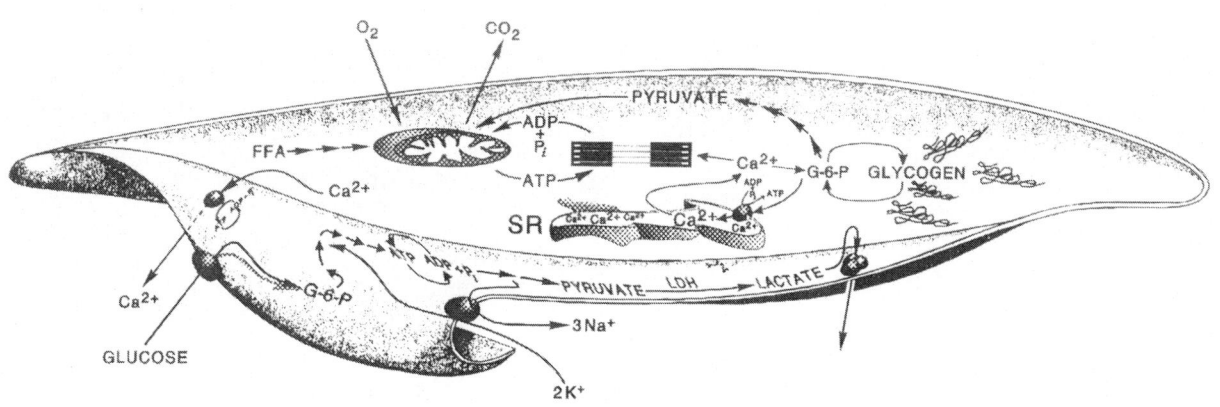

Fig. 1. Model of metabolic compartmentation. Localization of glycolytic enzyme cascades at the plasmalemma and sarcoplasmic reticulum. This schematic of enzyme localization is proposed to underlie the coupling of aerobic glycolysis and membrane energy requirements, whereas oxidative metabolism is more strongly correlated with the level of isometric force.

teries was faster in the presence of glucose, presumably reflecting a similar increase in efficacy of the Na-K pump [5]. In cardiac muscle, ATP from glycolytic sources has been reported to be more effective than that from oxidative sources in closing ATP-dependent K+-channels [12]. Thus coupling of glycolysis to membrane energetics may be a more general phenomenon not limited solely to smooth muscle.

Our working hypothesis is that these observations are attributable to co-localization of ATP synthesis and utilization systems. We tested this hypothesis using a model system, that of a purified plasma membrane preparation from smooth muscle [13]. Despite purification steps involving 0.6 M KCl, a complete glycolytic enzyme cascade was retained in this membrane preparation. Moreover, with fructose-1,6 diphosphate as substrate, this cascade was functional in terms of production of lactate. Of particular interest was whether this endogenous glycolytic cascade could support membrane energetics, which in this case was measured in terms of Ca^{2+}-pump activity. ATP, as expected, supported Ca^{2+}-uptake. Fructose-1,6 diphosphate also supported uptake, despite the fact that measured bath concentration of ATP produced by this substrate were low (10–30 μM). Moreover, the rate of Ca^{2+}-uptake supported by glycolysis was greater than that supported by ATP infused from an external pump at the same rate as produced by the glycolytic cascade [14]. This suggested that the local ATP concentration produced by the glycolytic cascade in the vicinity of the Ca^{2+}-ATPase might be of greater importance to pump function. We tested the importance of localized ATP production using an immobilized, hexokinase-based ATP trap. Ca^{2+}-uptake was not supported by ATP in the presence of the trap, but significant Ca^{2+}-uptake could be supported by fructose-1,6 diphosphate. Minimally

this indicates that ATP produced by the endogenous glycolytic cascade is more accessible to the pump, than to the ATP-trap in the solution. Thus in this model system, co-localization of both the glycolytic enzyme cascade and the Ca^{2+}-pump would appear to optimize function and may underlie the functional compartmentation observed in our studies of intact vascular tissue. A model summarizing our views of smooth muscle metabolic compartmentation based on these studies is shown in Fig. 1.

In addition to the preferential support of the Ca^{2+}-pump by fructose-1,6 diphosphate, we also found that phosphoenolpyruvate and importantly phosphocreatine could also support Ca^{2+}-pump activity in the presence of an ATP trap. There was substantial creatine kinase activity in these preparations [14] and its ability to support Ca^{2+}-pump function was similar to that reported for the sarcoplasmic reticulum in striated muscle [15–17]. It is possible that these different ATP producing systems might communicate to the Ca^{2+}-pump via a common, localized ATP pool as suggested by the work of Hoffman and colleagues [18]. Further discussion of the connection or disconnections between the creatine kinase system and glycolytic and oxidative metabolism will be pursued after the ensuing discussion of our studies of smooth muscle creatine kinase.

Compartmentation of creatine kinases in smooth muscle

Smooth muscle is characterized by a remarkably low tension cost, that is, (its rate of ATP utilization per unit isometric force maintained may be up to 1000-fold less than skeletal muscle. Though economical, smooth mus-

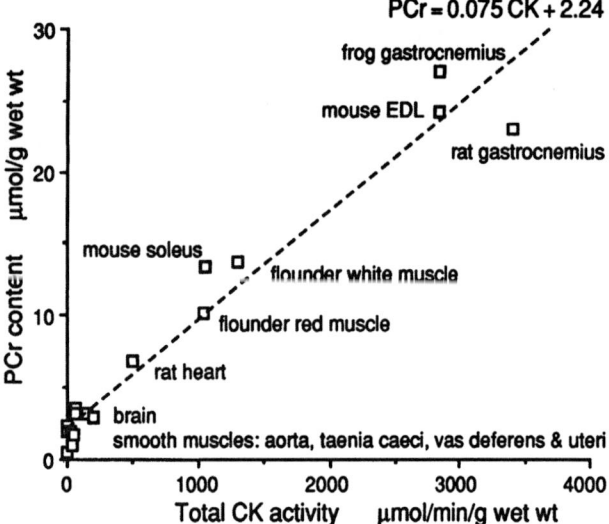

Fig. 2. Relationship between PCr content and total CK activity in various tissues. Values are taken from the review of Iyengar (1984). Our values for guinea pig aorta, taenia caeci and vas deferens are added: the CK activities (Ishida et al., 1991b), PCr for taenia (Ishida and Paul, 1990) and PCr values for aorta and vas deferens (0.5 and 1.9 μmol/g, respectively; Ishida, unpublished). A dashed line represents the linear regression line (r = 0.976).

cle is even more immediately dependent on a continuous supply of metabolic energy for normal function than skeletal muscle. This is primarily due to fact that the preformed pool of high-energy phosphate compounds, phosphocreatine (PCr) and ATP, is an order of magnitude lower in smooth muscle than in striated muscle. In particular, the reported PCr content of various smooth muscle is small, 0.1–4 μmol/g wet weight, and is about 0.5–2 times the content of ATP [1].

Iyengar [19] pointed out that the amount of PCr in a tissue is linearly related to its creatine kinase (CK) activity. Fig. 2 shows this linear relationship between PCr content and CK activity among various tissues including guinea pig smooth muscle. The intestinal smooth muscle of guinea pig taenia caeci contains similar amounts of PCr and CK as that found in rat brain. On the other hand, the vascular smooth muscle of the guinea pig aorta possesses much less. Presumably, the smaller amounts of PCr and CK in the aorta are related to the much smaller ATP utilization upon contraction of tonic, conduit arteries like hog coronary artery [20].

Lohmann equilibrium and high-energy phosphate compounds in smooth muscle

Muscle contraction requires ATP. Direct evidence for ATP utilization in the intact muscle was first shown by

experiments using 1-fluoro-2,4-dinitrobenzene (FDNB), a CK inhibitor [21]. Unless muscle metabolism is adequately suppressed, changes in ATP content directly associated with muscle contraction are not observed. Similarly, ATP and PCr levels of many smooth muscles are scarcely altered by the contractile stimulus in aerobic medium with substrate at 37° C [22, 23]. For example, ATP and PCr levels of guinea pig taenia caeci and porcine vascular smooth muscles (coronary and carotid arteries) stimulated for 30 min in high potassium were not significantly different from control [24–26]. Measurements of pyridine nucleotide, flavoprotein and fura-2-Ca^{2+} fluorescence suggest that the stimulation-induced elevation of the cytoplasmic Ca^{2+} concentration could independently activate glycogenolysis, oxidative phosphorylation and actomyosin ATPase; activation of the metabolic pathways preceded activation of the contractile proteins [27]. Therefore, a physiological temperature in tonic smooth muscle, the rate of aerobic ATP synthesis matches the increased ATP breakdown upon stimulation, resulting in virtually no change in ATP + PCr content of the muscle.

Rat portal vein, which is characterized by rapid phasic contractions and a relatively high ATPase activity, appears to be an exception in that PCr breakdown during contraction can be measured at 37° C [28]. At 23° C, at which resynthesis may be impeded, a loss of PCr, but not ATP, was reported using NMR techniques in the rabbit bladder as well as taenia coli [29]. These reported preferential losses of PCr in smooth muscle are well predicted by the Lohmann equilibrium. In contrast, at 18° C, after electrical stimulation, an initial ATP reduction followed by a decline in PCr was reported in the rabbit taenia coli and the initial phase of changes in ATP and PCr apparently deviated from the Lohmann equilibrium [30]. Whether this deviation directly correlates to the compartmentation of metabolites and enzymes is ambiguous, for at low temperatures the ATP utilization by the contractile proteins may start earlier than the onset of CK reaction.

Changes in ATP and PCr levels of smooth muscle were much more clearly observed when the muscle was exposed to poisons or metabolic inhibiting conditions such as hypoxia and glucose-depletion. Administration of 0.8 mM DNFB elicited a gradual decline in ATP, but not PCr, in electrically stimulated bovine carotid artery [31, 32]. Also, the presence of 0.4 mM DNFB for 60 min elicited a marked loss of ATP of 95% of control but did not alter the PCr level in non-stimulated guinea pig taenia caeci (Ishida and Paul, unpublished data). These results indicate that CK is capable of catalyzing the transfer of high energy phosphate between creatine and adenylate in non-DNFB treated smooth muscle.

Exposure of various smooth muscles to conditions inhibiting oxidative phosphorylation, such as hypoxia (N_2 bubbling instead of O_2), cyanide, etc., has been reported to elicit a loss of PCr [25, 29, 33–38]. In guinea pig taenia caeci with or without contractile stimulation, imposition of hypoxia elicited a significant ATP loss only after PCr was reduced to approximately 25% of control [34, 25]. This early loss in PCr followed by an ATP loss with a lag time is consistent with conversion of PCr to ATP and the equilibrium constant of Lohmann reaction.

In guinea pig taenia caeci exposed to hypoxia, raising the glucose concentration of the medium from 5.5 to 55 mM dose-dependently elevated the ATP content from $\sim$ 50 to $\sim$ 80% of the level under normoxia, and increased lactate production [39]. Whereas, under hypoxia, the PCr content of the taenia was not significantly elevated by high glucose concentrations [25, 37]. This preferential ATP production observed after increasing the glucose concentration under hypoxia seemed to follow the Lohmann equilibrium during the recovery process.

On the other hand, when oxygen was readmitted after hypoxia in the guinea pig taenia caeci, PCr increased rapidly and overshot the original level in 5 min after reoxygenation [37]. The ATP content also gradually increased and attained 75% of the original level. The relation between PCr and ATP in the taenia during reoxygenation, however, differed substantially from those after imposition of hypoxia and after increasing glucose under hypoxia. This rapid increase in PCr could not be predicted by the Lohmann equilibrium. Although the incomplete recovery ATP and the overshoot of PCr could be explained by the hypoxia-induced reduction of the total adenylate content of the muscle to 70% of control, the observed difference in the relations between PCr and ATP after imposition of hypoxia, or increasing glucose under hypoxia, and that after reoxygenation could not completely be defined by the effects of pH, loss of creatine and insufficient creatine kinase, as discussed previously [37].

When substrate, glucose in most experiments *in vitro*, was removed from the medium bubbled with oxygen, the PCr and ATP content of guinea pig taenia caeci smooth muscle also decreased. Under glucose-depleted conditions in the presence of oxygen, when the PCr and ATP of the muscle were partially reduced, a subsequent imposition of hypoxia was reported to elicit a further

rapid decrease in PCr and reoxygenation elicited a rapid recovery of the PCr content of guinea pig taenia caeci [40]. In our recent experiments, 60 min after exposure of taenia caeci to aerobic, glucose-depleted conditions in the presence of high potassium, the PCr content was still about 2.5 times greater, but the ATP content was smaller, than that in the muscle exposed to hypoxia for 15 min (Ishida, unpublished data). Similar to hypoxia, a gradual decrease in PCr of the taenia caeci was reported after glucose-depletion using NMR techniques [35]. These results also suggest that the relations between PCr and ATP after exposure of the muscle to substrate depletion are apparently different from that seen after imposition of hypoxia. Since an appreciable amount of oxygen consumption still persisted in the taenia in glucose-free medium [41], the rapid loss of PCr after hypoxia and its gradual loss after glucose depletion further suggest that the PCr production of the taenia is highly dependent on oxidative metabolism.

Similarly in vascular smooth muscle, the change in PCr content measured using NMR techniques was reported to be more resistant to substrate depletion compared to hypoxic conditions in hog carotid artery [36, 42] and in the sheep aorta [10]. Biochemical determination of PCr and ATP in the guinea pig aorta also showed similar results as seen in NMR studies (Ishida, unpublished data). A marked loss of PCr after hypoxia was also observed in the rabbit aorta [38]. In the sheep aorta, oxygen consumption was reported to increase after glucose removal [10]. In the hog coronary artery, a substantial amount of oxygen consumption was maintained even in the glucose-free medium plus 2-deoxyglucose, an inhibitor of glycolysis [24]. These results obtained in vascular and intestinal smooth muscles suggest that, under glucose-depleted conditions, the remaining oxidative metabolism effectively supplies PCr. These observations demonstrating a strong association between PCr production and oxidative phosphorylation led us to investigate the presence and localization of CK isoenzymes in smooth muscle.

Identification and localization of CK isoenzymes in smooth muscle

Since the early demonstration of CK in chicken gizzard smooth muscle [43], the isoforms of BB-, MB- and MM-CK have also been detected in smooth muscles [19, 44–49]. BB-CK is the predominant isoform in all smooth muscles investigated. In smooth muscle, some MB- or

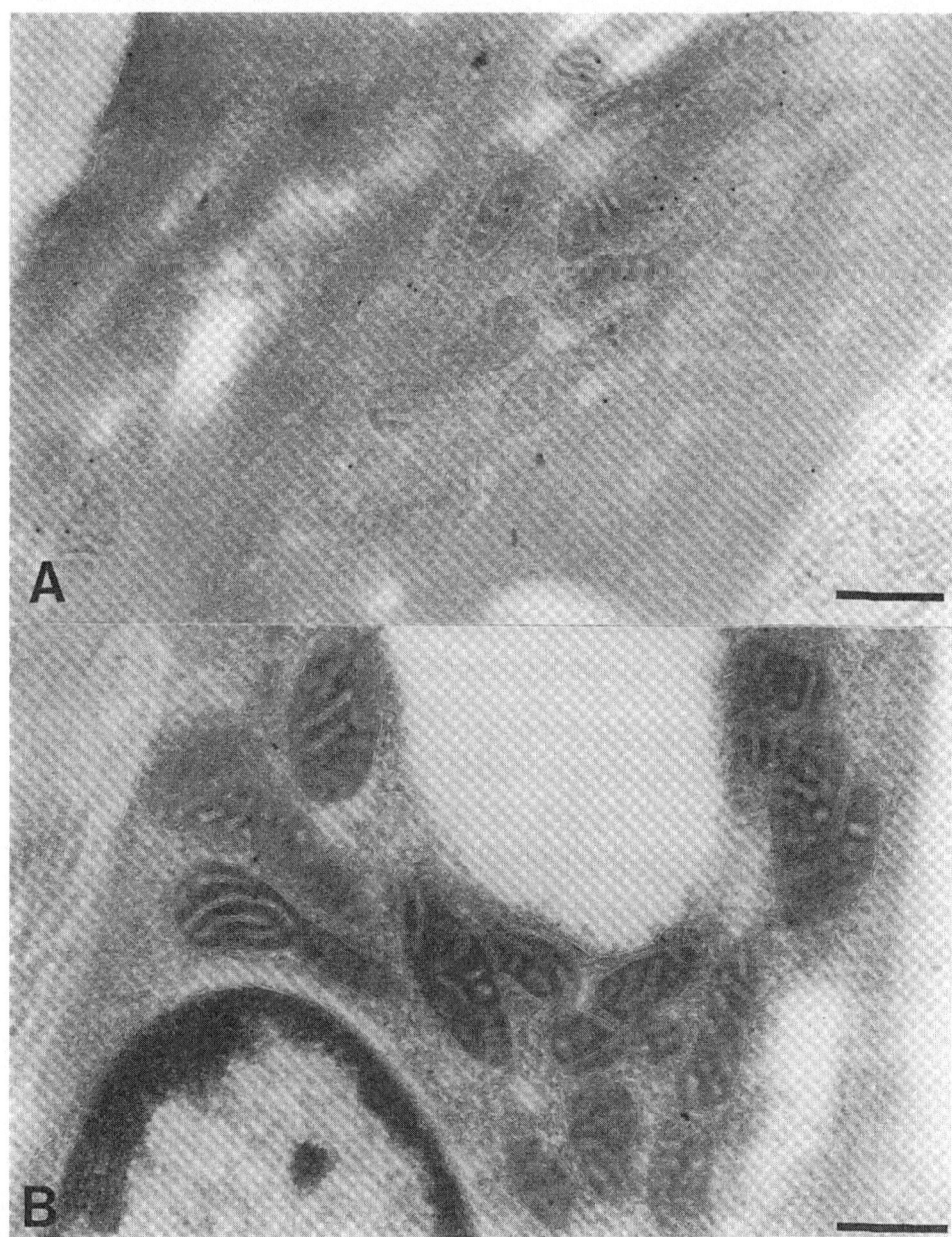

Fig. 3. Immunolocalization of mitochondrial creatine kinase (Mi-CK) in guinea pig aorta. Ultrathin cryosections were incubated with anti-chicken cardiac Mi-CK antibodies followed by goat anti-rabbit IgG coupled with 10 nm gold particles. Gold particles are specifically associated only with mitochondria. Bars indicate 0.5 µm in A and 0.3 µm in B.

even MM-CK could be found but when identified, they are of low abundance.

Another important CK isoform in terms of energetics is mitochondrial CK (Mi-CK). An appreciable amount of CK activity was detected in the enriched mitochondrial fractions of some smooth muscles [37, 50, 51]. Recently, using cellulose polyacetate electrophoresis, Mi-CK predominantly in the octameric form was identified in mitochondrial fractions prepared from guinea pig aorta, taenia caeci and vas deferens [48]. In the guinea

pig, smooth muscle Mi-CK migrated more slowly in cathodic direction than cardiac muscle Mi-CK, and it was relatively resistant to dimerizing conditions, suggesting that smooth muscle Mi-CK possesses properties more similar to the brain isoform than to cardiac Mi-CK.

Smooth muscle tissue usually contains a substantial number of nerve cells. Thus, we tried to directly show the localization of Mi-CK in smooth muscle mitochondria using immunoelectron microscopy [52]. Fig. 3 shows that immunogold labeling of cryosectioned guinea pig aorta, using an anti-chicken cardiac Mi-CK antibody. The immunogold specifically labeled the aortic mitochondria. Similarly, specific immunogold labeling of mitochondria was also observed in guinea pig taenia caeci and vas deferens. Fig. 3 further shows that a substantial number of gold particles seemed to be located in areas where the inner and outer mitochondrial membranes are in close proximity. These results are consistent with recent reports indicating that octameric Mi-CK may accumulate within the contact sites [53–56].

On the other hand, when cryosections of guinea pig taenia caeci were stained with anti-chicken BB-CK antibodies, immunogold labeling was observed in the cytosol. Although substantial amounts of BB-CK in the taenia caeci were rather soluble, thus cytosolic, the myofibrils of this smooth muscle were also clearly labeled, with a substantial amount of immunogold particles closely associated with the thick filaments of the taenia caeci. This is corroborated by a recent report which showed that after Triton X-100 skinning, only BB-CK, but not MB- or MM-CK, was left in guinea pig taenia caeci [49]. Thus, it will be interesting to test whether some of the BB-CK present in smooth muscle is indeed an integral protein of the contractile apparatus. This may be analogous to the MM-CK in skeletal muscle which is a genuine structural element of the M-line, i.e., representing the M4 m-bridges [57–59]. In addition, on these cryosections significant immuno-reactivity was also observed on mitochondria, presumably due to cross-reactivity of the anti-chicken BB-CK antibody with Mi-CK, a phenomenon observed previously (Riesinger, unpublished data).

The data show an isoenzyme-specific compartmentation of CK isoenzymes in smooth muscle and thus provide the cytological evidence for the phosphocreatine circuit model [60] in smooth muscle, a scheme for phosphagen transfer from mitochondria to the cytoplasm via Mi-CK at mitochondrial sites and cytosolic CK (predominantly BB-CK) at myofibrillar sites.

Physiological significance of CK compartmentation in smooth muscle

Compared with skeletal muscle, smooth muscle contractions are much longer in duration. Tonic smooth muscle contractions can last for hours and are accompanied by sustained increases in lactate production and oxygen consumption, suggesting a relatively tight coupling between contractile response and metabolism in smooth

muscle. In these tonic muscles, PCr may not be simply metabolized as a dead-end product in reactions of phosphorylation and dephosphorylation via ATP, although its concentration in smooth muscle cell is small. Interestingly, chicken gizzard smooth muscle is characterized with short lasting (< 1 min) contractile activity [61, 62]. It does not possess Mi-CK, it cannot be detected even in the mitochondrial fraction, but only BB-CK [52]. Presumably, the presence of Mi-CK in mitochondria and BB-CK in the cytoplasm of smooth muscle presumably provides evidence for an active role of PCr as a shuttle compound between the mitochondrial ATP producing site and the cytoplasmic energy transducing sites. PCr produced by mitochondria moves to the cytoplasm and recycles back to mitochondria as creatine after its phosphate is transferred to adenylate in the cytoplasm. This may facilitate the maintenance of tension in smooth muscle over long periods without fatigue.

Guinea pig taenia caeci exposed to high potassium contract and can maintain maximum levels of force for more than 60 min. With maximal stimulation, oxygen consumption of this muscle increased from 0.25 to 0.45 µmol/min/g wet wt at 37° C [25], and the maximum calculated ATP turnover rate was approximately 2.7 µmol/min/g wet wt. The CK activity of a mitochondrial fraction isolated from taenia caeci was approximately 1 µmol/min/mg mitochondrial protein at 37° C [37, 48]. Assuming a mitochondrial content of 5 mg/gwet wt, as estimated in the canine tracheal smooth muscle [63], the Mi-CK activity seems to well match the oxygen consumption rate in the taenia caeci.

In contrast, the total CK activity, ~ 150 µmol/min/g wet wt at 25° C in guinea pig taenia caeci, which is predominately cytosolic BB-CK activity, is much higher than the maximally activated ATP turnover rate (~ 5 µmol/min/g wet wt at 37° C) calculated from rates of oxygen consumption and lactate production [25, 48]. Recently, Clark et al. [49] reported that after Triton X-100 skinning, only BB-CK was left in the taenia caeci. Although the relevance of this high activity of BB-CK is not known at present, it will be interesting to investigate whether in smooth muscle a fraction of the cytosolic BB-CK is also compartmentalized and bound to subcellular locations with high energy turnover, taking a dual role in enzymatic reactions and structural formation of myofilaments and organelles [60, 64].

Clarifying a more direct role of CK isoenzymes for contractile function of smooth muscle is an ultimate goal of our studies. Earlier, Born [34] reported that the hypoxia-induced reduction in PCr content, but not ATP, was linearly correlated with spontaneously developed tension, in guinea pig taenia caeci. We also [25, 37] reported that after imposition of hypoxia, PCr loss was associated with reduction of active tension induced by high potassium. Based on several lines of evidence we concluded that the reduction in force with hypoxia was a straightforward consequence of energy limitation. Clark et al. [49] reported that substantial BB-CK remained after Triton X-100 skinning of the taenia caeci. This remaining BB-CK was sufficient to support a near maximal contraction when the skinned taenia caeci was bathed in a medium containing PCr and ADP. Moreover, BB-CK was reported to have a low Km for PCr compared with the MM-CK isoform [65, 66]. These results suggest that BB-CK could take a more direct role for contraction in taenia caeci, producing ATP though rephosphorylation of ADP by PCr in a localized contractile protein compartment. Since the contractile response of taenia caeci requires relatively more energy than vascular smooth muscles [20, 67], the production of PCr via Mi-CK may limit the BB-CK reaction at the contractile protein site, thus eliciting a relaxation of the taenia caeci.

Interestingly, the effects of hypoxia are dependent on the type of stimulation. For example, vasoconstrictions induced by high potassium, in contrast to agonists such as norepinephrine, show a different sensitivity to oxygen. In the rabbit aorta, hypoxia and cyanide preferentially inhibited norepinephrine-induced contraction but elicited only a small inhibition of a potassium-induced contraction [68]. This suggests that an energy limitation at the contractile protein level is not directly associated with the vasorelaxation induced by the inhibition of oxidative phosphorylation.

Covalent modification via phosphorylation [66] may also play a role in defining the function of BB-CK in smooth muscle. Recently, BB-CK has been shown to be phosphorylated by protein kinase C and this phosphorylation lowered the Km for PCr [69]. During a norepinephrine contraction in rabbit aorta, hypoxia inhibited the formation of inositol phosphate and myosin light chain phosphorylation in addition to reducing its PCr content [38, 70]. Therefore, in vascular smooth muscle inhibition of oxidative phosphorylation may cooperatively suppress inositol phosphate formation shortly after receptor stimulation and reduce the affinity of BB-CK for PCr in the vicinity of the contractile proteins. This could result in a more effective vasorelaxation in the presence of agonist than in the presence of high potassium. How-

ever, the exact role of CK isoenzymes in vascular smooth muscle remains speculative.

In contrast to the sharp loss of force and PCr induced by hypoxia, the level of ATP did not decline drastically and would appear to remain above that required for many cellular process, including in particular, contractile protein function in taenia caeci [25, 34, 37] and rabbit aorta [38]. ATP has been reported to be present in a bound form in the cell [71, 72]. Therefore, some ATP may exist in a form or functional compartment not readily accessible to the contractile proteins.

On the roles of aerobic glycolysis and creatine kinase in smooth muscle energetics

ATP originating from oxidative or glycolytic metabolism does not appear to be simply interchangeable in smooth muscle. Oxidative metabolism is more strongly associated with the ATP requirements of the contractile apparatus, where aerobic glycolysis is identified with membrane-related energy-dependent processes. One of the more striking examples of the differences between these two ATP sources can be found by comparing the effects of oxygen to those of high exogenous glucose in restoring phosphagen content after hypoxia. In either case, ATP levels are restored to about 70–80% of control. However, while added glucose under hypoxia restores ATP, PCr is little affected and remains depressed at about 20% of control. In contrast oxygen elicits a strong recovery of PCr to 120% of control and a similar recovery of ATP as glucose. What then is the significance of this observation? Is mitochondrial ATP preferred or essential for PCr formation? Perhaps, the constraints posed by hypoxia alter the reaction-diffusion dynamics in the cell such that the ATP produced by glycolysis is not as effectively converted to PCr. We have argued that a membrane-associated glycolytic cascade is located near ATP utilizing enzymes. In the absence of mitochondrial PCr, perhaps competition for glycolytic ATP in such a membrane compartment would favor the ion transport ATPases, rather than conversion to PCr by creatine kinase. This would be somewhat different than the PCr-circuit model [60] in which glycolysis is considered to be a second producing site of the PCr-circuit.

There is substantial evidence for coupling of glycolysis to the PCr-circuit. During anoxia, the patterns of phosphocreatine utilization and tissue acidification are similar in fish skeletal muscle [73]. In skeletal muscle, creatine kinase is co-localized with glycolytic enzymes in the I-band [74]. Interestingly, there is evidence that creatine kinase can couple to pyruvate kinase to form a diazyme complex [75]. Characteristic of these systems, however is a large pool of phosphocreatine and large changes in energy demand that far exceed the metabolic capacity for ATP synthesis. Another example is the coupling of the Na-K pump to the PC-circuit in the electric fish, Torpedo [76, 77]. In this case again, limited aerobic and anaerobic metabolic capacity in the face of large step changes in energy demand appear to have evolved a strategy whereby phosphocreatine pools are utilized to meet the transient demands followed by slower rates of recovery.

The role of the creatine kinase system in smooth muscle is not yet known with certainty. The differences in energy utilization and replenishment strategy are quite different from skeletal muscle. When generating maximal isometric force, the total ATP utilization of smooth muscle is often less than twice the basal rate [1]. The energy utilization during contraction in skeletal muscle may exceed basal rates by orders of magnitude. As noted above, in the steady state of contraction in smooth muscle, there often is no change in either ATP or PCr from prestimulus levels. Thus the term 'functional coupling' used to describe systems in which PCr provides the immediate source of ATP for skeletal muscle contraction or the Torpedo Na-K pump has a somewhat different meaning when used to describe the coupling between aerobic glycolysis and membrane function in smooth muscle. We have used the term 'preferential coupling' to describe this phenomenon [14]. Under normal physiological conditions, both oxidative and glycolytic metabolism function in parallel to provide for smooth muscle energy needs. However, glycolytic ATP appears to be the preferred source for membrane energy requirements. It is not known whether PCr plays an intermediate role in these energy transductions in smooth muscle.

Why then is there substantial creatine kinase in smooth muscle? As in skeletal muscle, CK and glycolytic enzymes in smooth muscle tend to show a similar localization pattern [14, 78]. The relatively low ATP requirements for contractility coupled with the fact that the PCr content is of similar magnitude to ATP would suggest that a role as an ATP buffer is limited. CK has also been ascribed a 'spatial' buffering function, whereby smaller gradients of ATP and ADP can maintain a given diffusive flux of high-energy phosphate [79]. However, in small diameter cells, as most smooth muscles, this also would not be anticipated to be a major factor. Perhaps this type of facilitated diffusion might be important in axial distribution of energy in smooth muscle.

Experimental tests of CK function in smooth muscle are limited. Ekmehag and Hellstrand [80] used β-guanidino propionic acid treatment to reduce PCr to about 1/10 control values. In rat portal vein, no changes in function or metabolism were detected. It can be argued, however, that even at these reduced PCr levels, a transport function for the CK system would be largely preserved. Likewise for glycolysis, smooth muscle cells can maintain normal function in the absence of glucose, provided other substrate is available. If as we have argued that a particular energy-dependent function might not be eliminated by inhibition of CK or glycolysis, but simply not optimized, one might only clearly define the importance of co-localization of energy transduction and utilization under conditions of stress.

A role for these systems in local regulation might perhaps be of more significance to smooth muscle [81]. Many energy-dependent cellular functions are differentially regulated. For example, relaxation of smooth muscle would involve a decrease in the actin-myosin ATPase but by an increase in Ca^{2+}-pump ATPase activity. Often these pathways involve common second messengers. For example, ADP is a common activator or inhibitor of many processes including metabolism, and localization of CK or glycolysis would restrict the effects of ADP to the region where a specific function was co-localized. The other side of this coin is that high-energy phosphagen produced in the local region would be at a higher concentration and perhaps more effective in a kinetic sense, if located in proximity to the utilization site. However both this regulatory and efficacy arguments are not unique to smooth muscle and remain speculative until more definitive experiments are devised.

Acknowledgments

This work was supported in part by NIH HL 23240 and 22619 (R.J.P.) and by the Swiss National Science Foundation, SNF Grant No. 31-26384.89 and 31-33907-92, and the Swiss Society for Muscle Diseases (T.W.) and by the fund for advanced research established by Dr. K. Imahori at Mitsubishi Kasei Institute of Life Sciences (Y.I.).

48

References

1. Paul RJ: Chemical energetics of vascular smooth muscle. In: DF Bohr, AP Somlyo, HV Sparks (eds.), Handbook of Physiology, 2, Vol. II. pp. 201–236. American Physiological Society, Bethesda, MD (1980)
2. Kushmerick MJ, Paul RJ: Aerobic recovery metabolism following single isometric tetani in frog sartorius at 0° C. J Physiol 254: 693–709, 1976
3. Paul RJ, Bauer M, Pease W: Vascular smooth muscle: Aerobic glycolysis linked to Na-K transport processes. Science 206: 1414–1416, 1979
4. Paul RJ: Aerobic glycolysis and ion transport in vascular smooth muscle. Am J Physiol 244: C399–C409, 1983
5. Campbell JD, Paul RJ: The nature of fuel provision for the Na$^+$, K$^+$-ATPase in porcine vascular smooth muscle. J Physiol 447: 67–82, 1992
6. Lynch RM, Paul RJ; Glucose uptake by porcine carotid artery: Relation to alterations in Na,K-transport. Am J Physiol 247 (Cell Physiol 16): C433–C440, 1984
7. Lynch RM, Kuettner CP, Paul RJ: Regulation of glycogen metabolism during tension generation and maintenance in vascular smooth muscle. Am J Physiol 257: C736–C742, 1989
8. Lynch RM, Paul RJ: Compartmentation of glycolysis and glycogenolysis in vascular smooth muscle. Science 222: 1344–1346, 1983
9. Lynch RM, Paul RJ: Compartmentation of carbohydrate metabolism in vascular smooth muscle: Evidence for at least two functionally independent pools of glucose-6-phosphate. Biochim Biophys Acta 887: 315–318, 1986
10. Hardin CD, Wiseman RW, Kushmerick MJ: Vascular oxidative metabolism under different metabolic conditions. Biochim Biophys Acta 1133: 133–141, 1992
11. Lynch RM, Balaban RS: Coupling of aerobic glycolysis and Na$^+$-K$^+$-ATPase in the renal cell line MDCK. Am J Physiol 253: C269–C276, 1987
12. Weiss JN, Lamp ST: Glycolysis preferentially inhibits ATP-sensitive K$^+$ channels in isolated guinea pig cardiac myocytes. Science 238: 67–69, 1987
13. Paul RJ, Hardin C, Wuytack F, Raeymaekers L, Casteels R: An endogenous glycolytic cascade can preferentially support Ca uptake in smooth muscle plasma membrane vesicles. FASEB J 3: 2298–3201, 1989
14. Hardin C, Raeymaekers L, Paul RJ: Comparison of endogenous and exogenous sources of ATP in fueling Ca uptake in smooth muscle plasma membrane vesicles. J Gen Physiol 99: 21–40, 1992
15. Rossi AM, Eppenberger HM, Volpe P, Cotrufo R, Wallimann T: Muscle-type MM creatine kinase is specifically bound to sarcoplasmic reticulum and can support Ca^{2+}-uptake and regulate local ATP/ADP ratios. J Biol Chem 265: 5258–5266, 1990
16. Läuger P: Ca^{2+}-pumps from sarcoplasmic reticulum. In: P Läuger (ed) Electrogenic Ion Pumps, Sinderland, MA, USA, pp 227–251, 1991
17. Korge P, Byre SK, Campbell KB: Functional coupling between SR-bound creatine kinase and Ca^{2+}-ATPase. Eur J Biochem 213: 923–980, 1993
18. Proverbio F, Shoemaker DG, Hoffmann JF: Functional consequences of the membrane pool of ATP associated with the human red blood cell Na/K pump. Prog Clin Biol Res 268A: 561–567, 1988
19. Iyengar MR: Creatine kinase as an intracellular regulator. J Muscle Res Cell Motil 5: 527–534, 1984
20. Paul RJ, Ishida Y, Rubanyi G: Vascular smooth muscle metabolism and mechanisms of oxygen sensing. In: JA Will, CA Dawson, EK Weir, CK Buckner (eds), The Pulmonary Circulation in Health and Disease. pp. 97–108. Academic Press, Orland, 1987
21. Cain DF, Davies RE: Breakdown of adenosine triphosphate during a single contraction of working muscle. Biochem Biophys Res Commun 8: 361–366, 1962
22. Butler TM, Davies RE: High energy phosphates in smooth muscle. In: DF Bohr, AP Somlyo, HV Sparks (eds), Handbook of Physiology, section 2, Vol. II. pp. 237–252. American Physiological Society, Bethesda, MD, 1980
23. Paul RJ: Smooth muscle energetics. Ann Rev Physiol 51: 331–349, 1989
24. Ishida Y, Hashimoto M, Paul RJ: Effects of substrate depletion and hypoxia on the energy metabolism and isometric force of the hog coronary artery. Fed Proc 46: 1096, 1987
25. Ishida Y, Paul RJ: Effects of hypoxia on high-energy phosphagen content, energy metabolism and isometric force in guinea-pig taenia caeci. J Physiol (Lond.) 424: 41–56, 1990
26. Krisanda JM, Paul RJ: High energy phosphate and metabolite content during isometric contraction in porcine carotid artery. Am J Physiol 244: C385–C390, 1983
27. Ozaki H, Satoh T, Karaki H, Ishida Y: Regulation of metabolism and contraction by cytoplasmic calcium in the intestinal smooth muscle. J Biol Chem 263: 14074–14079, 1988
28. Hellstrand P, Paul RJ: Phosphagen content, breakdown during contraction and oxygen consumption in rat portal vein. Am J Physiol 244: C250–C258, 1983
29. Hellstrand P, Vogel HJ: Phosphagens and intracellular pH in intact rabbit smooth muscle by ^{31}P-NMR. Am J Physiol 248: C320–C329, 1985
30. Butler TM, Siegman MJ, Mooers SV, Davies RE: Chemical energetics of single isometric tetani in mammalian smooth muscle. Am J Physiol 235: C1–C7, 1977
31. Daemers-Lambert C: Action du fluorodinitrobenzene sur le metabolisme phosphore du muscle lisse arteriel pendant la stimulation electrique (carotide de bovide). Angiologica 6: 1–12, 1969
32. Daemers-Lambert C: Mechanochemical coupling in smooth muscle. In: NL Stephens (ed), The Biochemistry of Smooth Muscle. pp. 51–82. University Park Press, Baltimore, MD, 1977
33. Furchgott RF, Shorr E: The effect of succinate on respiration and certain metabolic processes of mammalian tissues at low oxygen tensions in vitro. J Biol Chem 175: 201–215, 1948
34. Born GVR: The relation between the tension and the high-energy phosphate content of smooth muscle. J Physiol (Lond.) 131: 704–711, 1956
35. Nakayama S, Seo Y, Takai A, Tomita T, Watari H: Phosphorus compounds studied by ^{31}P nuclear magnetic resonance spectroscopy in the taenia of guinea-pig caecum. J Physiol (Lond.) 402: 565–578, 1988
36. Dillon PF, Fisher MJ, Adams GR, Clark JF: The use of ^{31}P-NMR to study smooth muscle. Prog Clin Biol Res 315: 439–449, 1989
37. Ishida Y, Paul RJ: Evidence for compartmentation of high energy phosphagens in smooth muscle. Prog Clin Biol Res 315: 417–428, 1989
38. Coburn RF, Moreland S, Moreland RS, Baron C: Rate-limiting energy-dependent steps controlling oxidative metabolism-con-

traction coupling in rabbit aorta. J Physiol (Lond.) 448: 473–492, 1992

39. Ishida Y, Takagi K, Urakawa N: Tension maintenance, calcium content and energy production of the taenia of the guinea-pig caecum under hypoxia. J Physiol (Lond.) 347: 149–159, 1984

40. Bueding E, Bulbring E, Gercken G, Hawkins JT, Kuriyarna H: The effect of adrenaline on the adenosinetriphosphate and creatine phosphate content of intestinal smooth muscle. J Physiol (Lond) 193: 187–212, 1967

41. Urakawa N, Ikeda M, Saito Y, Sakai Y: Effects of factors inhibiting tension development on oxygen consumption of guinea pig taenia coli. Jpn J Pharmacol 19: 578–576, 1969

42. Adams GR, Dillon PF: Glucose dependence of sequential norepinephrine contractions of vascular smooth muscle. Blood Vessels 26: 77–83, 1989

43. Geiger WB: The creatine phosphokinase of a visceral muscle. J Biol Chem 220: 871–877, 1956

44. Dawson DM, Fine IH: Creatine kinase in human tissues. Arch Neurol 16: 175–180, 1967

45. Allard D, Cabrol D: Etude electrophoretique des isozymes de la creatine phosphokinase dans les tissus de l'Homme et du Lapin. Pathol Biol 18: 847–950, 1970

46. Focant B: Isolemente et proprietes de la creatine-kinase de muscle lisse de boeuf. FEBS Lett 10: 57–61, 1970

47. Traugott C, Massaro EJ: An electrophoretic analysis of rodent creatine phosphokinase isoenzymes. Comp Biochem Physiol 42B: 255–262, 1972

48. Ishida Y, Wyss M, Hemmer W, Wallimann T: Identification of creatine kinase isoenzymes in the guinea-pig: Presence of mitochondrial creatine kinase in smooth muscle. FEBS Lett 283: 37–43, 1991b

49. Clark JF, Khuchua Z, Ventura-Clapier R: Creatine kinase binding and possible role in chemically skinned guinea-pig taenia coli. Biochim Biophys Acta 1100: 137–145, 1992

50. Jacobus WE, Lehninger AL: Creatine kinase of rat heart mitochondria: Coupling of creatine phosphorylation to electron transport. J Biol Chem 248: 4803–4810, 1973

51. Haas RC, Strauss AW: Separate nuclear genes encode sarcomere-specific and ubiquitous human mitochondrial creatine kinase isoenzymes. J Biol Chem 265: 6921–6927, 1990

52. Ishida Y, Honda H, Riesinger I, Wallimann T: Localization of creatine kinase isoenzymes in smooth muscle cells. Jpn J Physiol 41: S301, 1991a

53. Adams V, Bosch W, Schlegel J, Wallimann T, Brdiczka D: Further characterization of contact sites from mitochondria of different tissues: Topology of peripheral kinases. Biochim Biophys Acta 981: 213–225, 1989

54. Biermans W, Bernaert I, De Bie M, Nijs B, Jacob W: Ultrastructural localization of creatine kinase activity in the contact sites between inner and outer mitochondrial membranes of rat myometrium. Biochim Biophys Acta 974: 74–80, 1989

55. Lipskaya TY, Trofimova ME: Study on heart mitochondrial creatine kinase using a cross-linking bifunctional reagent. I. The binding involves the octameric form of the enzyme. Biochem Int 18: 1029–1039, 1989

56. Wyss M, Smeitink J, Wevers RA, Wallimann T: Mitochondrial creatine kinase: a key enzyme of aerobic energy metabolism. Biochim Biophys Acta 1102: 119–166, 1992

57. Strehler EE, Carlsson E, Eppenberger HM, Thornell LE: Ultrastructural localization of M-band proteins in chicken breast muscle as revealed by combined immunocytochemistry and ultracryotomy. J Mol Biol 166: 141–158, 1983

58. Wallimann T, Moser H, Eppenberger H: Isoenzymes specific localization of M-line-bound creatine kinase in myogenic cells. J Muscle Res Cell Motil 4: 429–441, 1983

59. Wallimann T, Eppenberger HM: Localization and function of M-line bound creatine kinase: M-band model and creatine phosphate shuttle. In: JW Shay (ed), Cell and Muscle Motility, pp 239–285, Plenum, NY, 1985

60. Wallimann T, Schnyder T, Schlegel J, Wyss M, Wegmann G, Rossi AM, Hemmer W, Eppenberger HM, Quest AFG: Subcellular compartmentation of creatine kinase isoenzymes of CK and octameric structure of mitochondrial CK: Important aspects of the phosphoryl-creatine circuit. Prog Clin Biol Res 315: 159–176, 1989

61. Fischer W, Pfitzer G: Rapid myosin phosphorylation transients in phasic contractions in chicken gizzard smooth muscle. FEBS Lett 258: 59–62, 1989

62. Ozaki H, Kasai H, Hori M, Sato K, Ishihara H, Karaki H: Direct inhibition of chicken gizzard smooth muscle contractile apparatus by caffeine. Naunyn-Schrniecleberg's Arch Pharmacol 341: 262–267, 1990

63. Stephens NL, Kroeger EA, Wrogeman K: Energy metabolism: Methods in isolated smooth muscle and methods at cellular and subcellular levels. In: Daniel EE, Paton DM (eds), Methods in Pharmacology, Vol. 3, Smooth Muscle. pp. 555–591. Plenum Press, New York, 1975

64. Wallimann T, Wyss M, Brdiczka D, Nicolay K, Eppenberger HM: Intracellular comparmentation, structure and function of creatine kinase isoenzymes in tissues with high and fluctuating energy demands: the 'phosphocreatine circuit' for cellular energy homeostasis. Biochem J 281: 21–40, 1992

65. Iyengar MR, Fluellen CE, Iyengar C: Creatine kinase from the bovine myometrium; purification and characterization. J Muscle Res Cell Motilit 3: 312–246, 1982

66. Quest AFG, Soldati T, Hemmer W, Perriard JC, Eppenberger HM, Wallimann T: Phosphorylation of chicken brain-type creatine kinase affects a physiologically important kinetic parameter and gives rise to protein microheterogeneity in vivo. FEBS Lett 269: 457–464, 1990

67. Ishida Y, Hashimoto M, Paul RJ: Does a limitation of energy supply to the contractile apparatus underlie the relaxation induced by hypoxia in smooth muscle? Prog Clin Biol Res 245: 463–464, 1989

68. Coburn RF, Grubb B, Aronson RD: Effect of cyanide on oxygen tension dependent mechanical tension in rabbit aorta. Circ Res 44: 368–378, 1979

69. Chida K, Tsunenaga M, Kasahara K, Kohno Y, Kuroki T: Regulation of creatine kinase-B activity by protein kinase-C. Biochem Biophys Res Commun 173: 346–350, 1990

70. Coburn RF, Baron C, Papadopoulos MT: Phosphoinositide metabolism and metabolism-contraction coupling in rabbit aorta. Am J Physiol 255: H1476–H1483, 1988

71. Liou RS, Anderson SR: Binding of ATP and of 1,N6-ethenoadenosine triphosphate to rabbit muscle phosphofructokinase. Biochemistry 17: 999–1004, 1978

72. Neidl C, Engel J: Exchange of ADP, ATP and 1:N6-ethenoadenosine 5′-triphosphate at G-actin. Equilibrium and kinetics. Eur J Biochem 101: 163–169, 1979

73. Van Waade A, Van Den Thillart G, Erkelens C, Addink A, Lugtenburg J: Functional coupling of glycolysis and phosphocreatine utilization in anoxic fish muscle. J Biol Chem 265: 914–923, 1990

50

74. Wegmann G, Zanolla E, Eppenberger HM, Wallimann T: In situ compartmentation of creatine kinase in intact sarcomeric muscle: the acto-myosin overlap zone as a molecular sieve. J Muscle Res Cell Motil 13: 420–435, 1992

75. Dillon PF, Clark JF: The theory of diazymes and functional coupling of pyruvate kinase and creatine kinase. J Theo Biol 143: 275–284, 1990

76. Walliman T, Wlazthony D, Wegmann G, Moser H, Eppenberger HM, Barrantes FJ: Subcellular localization of creatine kinase in torpedo electrocytes: association with acetylcholine receptor-rich membranes. J Cell Biol 100: 1063–1072, 1985

77. Blum H, Balshi JA, Johnson RG: Coupled *in vivo* activity of creatine phosphokinase and membrane bound (Na$^+$, K$^+$)-ATPase in the resting and stimulated electric organ of the electric fish narcine brasiliensis. J Biol Chem 266: 10254–10259, 1991

78. Hardin CD, Paul RJ: Localization of two glycolytic enzymes in guinea pig *taenia coli*. Biochim Biophys Acta 1134(3): 256–259, 1992

79. Meyer RA, Sweeney HL, Kushmerick MJ: A simple analysis of the 'phosphocreatine shuttle'. Am J Physiol 246: C365–C377, 1984

80. Ekmehag BL, Hellstrand P: Contractile and metabolic characteristics of creatine-depleted vascular smooth muscle of the rat portal vein. Acta Physiol Scand 133: 525–533, 1988

81. Lynch RM, Paul RJ: Functional compartmentation of carbohydrate metabolism. In: D Jones (ed), CRC Reviews Microcompartmentation, CRC Press, pp. 17–35, 1988

Molecular and Cellular Biochemistry **133/134**: 51–66, 1994.
© 1994 *Kluwer Academic Publishers.*

I-4 Creatine metabolism and the consequences of creatine depletion in muscle

Markus Wyss[1] and Theo Wallimann

Swiss Federal Institute of Technology, Institute for Cell Biology, ETH Hönggerberg, CH-8093 Zürich, Switzerland
[1] *Present address: Department of Transplant Surgery, Research Division, University Hospital, Anichstr. 35, A-6020 Innsbruck, Austria*

Abstract

Currently, considerable research activities are focussing on biochemical, physiological and pathological aspects of the creatine kinase (CK) – phosphorylcreatine (PCr) – creatine (Cr) system (for reviews see [1, 2]), but only little effort is directed towards a thorough investigation of Cr metabolism as a whole. However, a detailed knowledge of Cr metabolism is essential for a deeper understanding of bioenergetics in general and, for example, of the effects of muscular dystrophies, atrophies, CK deficiencies (e.g. in transgenic animals) or Cr analogues on the energy metabolism of the tissues involved. Therefore, the present article provides a short overview on the reactions and enzymes involved in Cr biosynthesis and degradation, on the organization and regulation of Cr metabolism within the body, as well as on the metabolic consequences of 3-guanidinopropionate (GPA) feeding which is known to induce a Cr deficiency in muscle. In addition, the phenotype of muscles depleted of Cr and PCr by GPA feeding is put into context with recent investigations on the muscle phenotype of 'gene knockout' mice deficient in the cytosolic muscle-type M-CK. (Mol Cell Biochem **133/134**: 51–66, 1994).

Key words: creatine kinase, mitochondrial myopathies, ragged red fibers, intramitochondrial inclusions, creatine analogues, phosphagen

Abbreviations: Cr – creatine; Crn – creatinine; PCr – phosphorylcreatine; CK – creatine kinase; M-CK – cytosolic muscle type CK isoenzyme; Mi-CK – mitochondrial CK isoenzyme; AGAT – L-arginine: glycine amidinotransferase; GAMT – S-adenosylmethionine: guanidinoacetate methyltransferase; Arg – arginine; Met – methionine; GPA – 3-guanidinopropionate = β-guanidinopropionate; PGPA – phosphorylated GPA; GBA – 3-guanidinobutyrate = β-guanidinobutyrate; CPEO – chronic progressive external ophthalmoplegia

Creatine metabolism

Although a diagram of the reactions involved in Cr metabolism (Fig. 1) seems simple, Cr metabolism is complicated by the fact that most tissues lack one or several of the reactions shown, necessitating a transport of intermediates between the tissues (through the blood) in order to allow the whole cascade to proceed. L-Arginine: glycine amidinotransferase (AGAT), the first enzyme in the two-step biosynthesis of Cr, catalyzes the transfer of the amidino group of Arg to glycine to yield L-ornithine and guanidinoacetate. The latter compound, by the ac-

Address for offprints: T. Wallimann, Swiss Federal Institute of Technology, Institute for Cell Biology, ETH Hönggerberg, CH-8093 Zürich, Switzerland

52

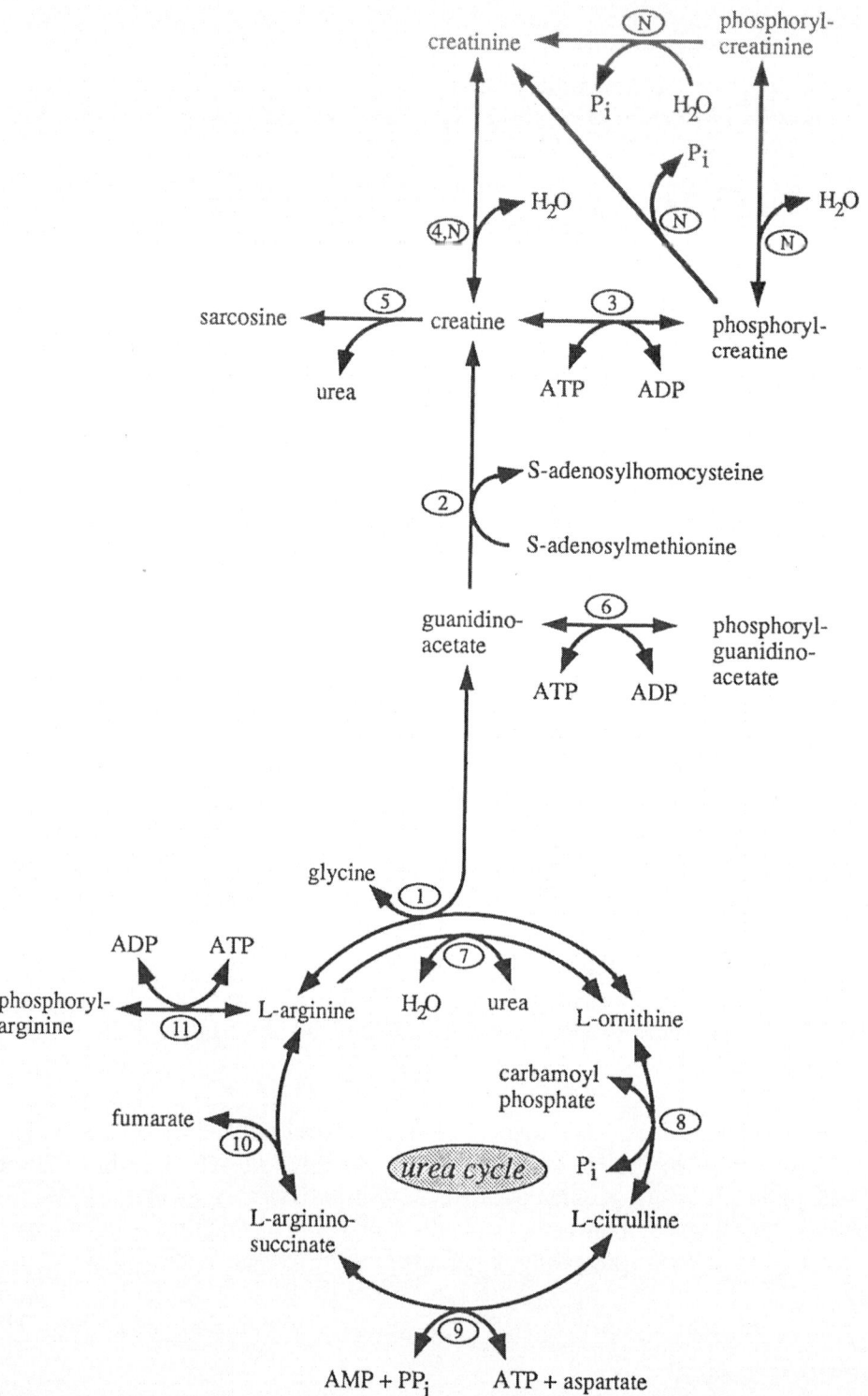

Fig. 1. Schematical representation of the reactions and enzymes involved in Cr metabolism. The respective enzymes are denoted by numbers: (1) L-arginine: glycine amidinotransferase (AGAT; EC 2.1.4.1); (2) S-adenosylmethionine: guanidinoacetate N-methyltransferase (GAMT; EC 2.1.1.2); (3) creatine kinase (CK, EC 2.7.3.2); (4) creatininase = creatinine amidohydrolase (EC 3.5.2.10); (5) creatinase = creatine amidinohydrolase (EC 3.5.3.3); (6) guanidinoacetate kinase = glycocyamine kinase (EC 2.7.3.1); (7) arginase (EC 3.5.3.1); (8) ornithıne carbamoyltransferase (EC 2.1.3.3); (9) arginınosuccinate synthetase (EC 6.3.4.5); (10) arginınosuccınate lyase (EC 4.3.2.1); (11) argınıne kınase (EC 2.7.3.3); (N) nonenzymatic reaction.

tion of S-adenosylmethionine: guanidinoacetate methyltransferase (GAMT), is then methylated at the amidino group to give Cr [3, 4]. In the course of evolution, both AGAT and GAMT seem to have evolved with the appearance of the lampreys [5]. While these enzyme activities were not detected in invertebrates, they were found in most, but not all vertebrates examined. Nevertheless, some invertebrate species (e.g. some annelids, echinoderms, hemichordates and urochordates, etc.) contain significant amounts of Cr, PCr and CK in their tissues, especially in spermatozoa [5–9], indicating that they either accumulate Cr from their environment or from the diet, or that the enzymes for Cr biosynthesis in these animals escaped detection so far.

Many of the lower vertebrates (fish, frogs and birds) express both AGAT and GAMT in their livers and often kidneys. In mammals, pancreas contains high levels of both enzymes, while kidneys have fairly high amounts of AGAT, but relatively lower levels of GAMT. On the contrary, all mammalian livers tested contain high levels of GAMT, but display only low levels of Cr, lack CK activity and consequently also PCr. While livers of cow, pig, monkey and man also have high amounts of AGAT, livers of common laboratory mammals such as the rat, mouse, dog, cat and rabbit were reported to lack AGAT activity (for reviews see [5, 10]). Based mostly on these latter findings and on the fact that nephrectomized animals display a drastically reduced rate of Cr synthesis [11, 12], it was postulated and is still largely accepted that the main route of Cr biosynthesis in mammals involves the formation of guanidinoacetate in the kidney, its transport through the blood and its methylation to Cr in the liver. Cr exported from the liver and transported again through the blood may then be taken up by the Cr-requiring tissues.

There is, however, experimental evidence not fully agreeing with this view. Immunofluorescence microscopy with antibodies against AGAT revealed significant amounts of this enzyme not only in rat kidney and pancreas, but also in liver [13]. The underestimation of rat liver AGAT in previous biochemical studies was most likely due to the high levels of liver arginase interfering with the AGAT assay. Furthermore, AGAT activity was detected in heart, lung, spleen, muscle, brain, testis and thymus, and the total amount of AGAT in these tissues was estimated to even approach the AGAT activity in kidney and pancreas [14]. In the rat, the highest specific AGAT activity was found in the decidua of pregnant females [10], but AGAT is absent from human placenta. GAMT activity, on the other hand, was also detected in rat spleen, heart and skeletal muscle as well as in human fetal lung fibroblasts and mouse neuroblastoma cells [15, 16], but the specific activities in these tissues are rather low. The GAMT activity in skeletal muscle, however, was calculated to have the potential to synthesize all Cr needed in this tissue [16]. Finally, feeding of rats and mice with GPA, a competitive inhibitor of Cr entry into cells, progressively decreases the concentrations of Cr and PCr in heart and skeletal muscle, but has only little influence on the Cr and PCr contents of brain [17; Riesinger, Haas & Wallimann, unpublished results]. One possible explanation is that brain contains its own Cr-synthesizing machinery [18]. To conclude, the detailed contribution of various tissues to total Cr synthesis on one hand as well as the relevance of guanidinoacetate and Cr transport through the blood and of Cr uptake by the cells on the other hand are still not fully understood, this being due to a lack of thorough investigations and to pronounced species differences.

A specific saturable, Na^+- and Cl^--dependent Cr transporter has been described for skeletal muscle, heart, fibroblasts, smooth muscle, neuroblastoma and astroglia cells, as well as for red blood cells and macrophages [16, 19–27]. Recent cloning of the Cr transporter cDNA, followed by Northern blot analysis, revealed the largest amounts of Cr transporter mRNA in kidney, heart and skeletal muscle, somewhat lower levels in brain, lung, epididymis and testis, but no Cr transporter mRNA at all in uterus, liver, small intestine and spleen [27].

As far as the tissue concentrations of Cr and PCr are concerned, the highest levels were observed in skeletal muscle, heart, spermatozoa and photoreceptor cells of the retina, somewhat lower amounts in brain, brown adipose tissue, intestine, seminal vesicles and seminal vesicle fluid, and only low levels in lung, spleen, kidney, liver, white adipose tissue, erythrocytes and serum [15, 28–36]. In skeletal muscles, [Cr] and [PCr] correlate with the glycolytic capacity, with resting type 2a and 2b fibers containing approx. 32 mM PCr and 7 mM Cr, and type 1 fibers containing approx. 16 mM PCr and 7 mM Cr [36, 37]. In serum and erythrocytes, on the other hand, [Cr] amounts to only 25–50 μM and 270–400 μM, respectively [35], implying that Cr has to be accumulated by most Cr-containing tissues against a large concentration gradient from the blood. Very likely, Cr uptake via the Cr transporter is driven by the electrochemical potential difference of extracellular versus intracellular $[Na^+]$.

As indicated in the upper part of Fig. 1, several path-

54

ways have to be considered for the biodegradation of Cr and PCr. *In vitro*, reversible and nonenzymatic cyclization of Cr to creatinine (Cr ↔ Crn) takes place, with the equilibrium of this spontaneous reaction being both pH- and temperature-dependent. Cr is favoured at low pH and low temperature, while Crn is favoured at higher temperatures and in alkaline solutions [38, 39]. In both directions, the reaction is monomolecular. Starting with pure Cr solutions, 1.0–1.3% of the Cr per day converts to Crn at pH 7.0–7.2 and 38 °C. *In vitro* studies on the stability of PCr revealed that this 'high-energy phosphate compound' is acid-labile, yielding P_i and either Cr or Crn upon hydrolysis. Both the rate of PCr hydrolysis and the ratio of Cr to Crn formed depend on temperature and pH and can additionally be influenced in a concentration-dependent manner by molybdate (for reviews see [7, 40]).

In contrast to the *in vitro* situation, studies with ^{15}N-labelled Cr and Crn clearly demonstrated that the conversion of Cr to Crn *in vivo* is an *irreversible* process [41]. An almost constant fraction of the body Cr (1.1%/day) and PCr (2.6%/day) nonenzymatically converts to Crn *in vivo*, giving an overall conversion rate for the total Cr pool (Cr + PCr) of approx. 1.7%/day [42; for a review see ref. 10]. Consequently, in a 70 kg-man containing approx. 120 g of total Cr, roughly 2g/day are converted into Crn and have to be replaced by Cr from the diet or by de-novo biosynthesis [10, 42, 43]. Since in contrast to Cr, no specific saturable uptake mechanism exists for Crn [24], and since Crn, most likely due to its nonionic nature, is membrane-permeable, Crn constantly diffuses out of the tissues into the blood and is excreted by the kidneys into the urine [29].

20–25% of the *in vivo* conversion of PCr into Crn may proceed via phosphorylcreatinine (PCrn) as an intermediate [44]. Accordingly, [PCrn] in rabbit white skeletal muscle was found to be 0.4% of [PCr]. In addition, commercial preparations of PCr contain 0.3–0.7% of PCrn. Although PCrn was proposed to be an obligatory intermediate of the CK reaction [45], this idea has to be dismissed due to the lack of any experimental evidence.

In contrast to the nonenzymatic conversion of Cr and PCr to Crn in vertebrates, various bacteria (Alcaligenes, Arthrobacter, Clostridium, Flavobacterium, Micrococcus, and Pseudomonas strains) were shown to (inducibly) express specific enzymes for the biodegradation of Cr and Crn ([46–51]; for further references see [52]). In some of these bacteria, for example, creatininase (Crn amidohydrolase) first converts Crn into Cr which then is further metabolized by creatinase (Cr amidinohydro-

lase) into urea and sarcosine. Even though creatinase was also detected in human skeletal muscle [53], this finding awaits confirmation and demonstration of its physiological relevance. Very interesting, however, is the indication that in Duchenne muscular dystrophy, the kinetic properties of human muscle creatinase are affected [54].

An important aspect of Cr metabolism to add is that in man, the daily utilization of methyl groups for Cr synthesis (in the GAMT reaction) and, consequently, also the daily loss of methyl groups due to Crn (and Cr) excretion approximately equal the daily intake of 'labile' methyl groups (Met + choline) on a normal, equilibrated diet containing Met and choline [55]. Even when de novo Met biosynthesis is taken into consideration, Cr biosynthesis still accounts for approx. 70% of the total utilization of 'labile' methyl groups. It might therefore be assumed that methyl group availability becomes limiting for Cr biosynthesis, at least under some physiological or pathological conditions. This is, actually, not the case, since a deficit in 'labile' methyl groups in man will normally be compensated by increased de novo Met biosynthesis.

Regulation of creatine metabolism

Despite the relatively simple scheme of Cr metabolism (Fig. 1), a variety of potential regulatory mechanisms have to be considered (for an extensive review see [10]), for instance allosteric regulation, covalent modification or alterations of expression of enzymes involved in Cr metabolism, or changes in the diffusion and transport properties (membrane barriers, transport proteins, blood transport) of intermediary metabolites.

Formation of guanidinoacetate is rate-limiting for Cr biosynthesis (see [10]). Consequently, the AGAT reaction is the most likely control step in the pathway, a hypothesis that has been proven by a great deal of experimental work. Most important in this respect is the feedback repression of AGAT by Cr, the endproduct of the pathway. An increase in endogenous or exogenously supplied Cr causes a parallel decrease in the mRNA content, the enzyme level, as well as the specific activity of AGAT, suggesting regulation of AGAT expression at a pretranslational level ([56, 57]; for a review see [10]). Feedback repression of AGAT by Cr is most pronounced in kidney and pancreas, the main tissues of guanidinoacetate formation, but is also observed in the decidua of the pregnant rat (see [10]). Cyclocreatine, Nacetimidoylsarcosine and N-ethylguanidinoacetate also display repressor activity like Cr, while Crn, PCr, N-methyl-3-guanidinopropionate and a variety of other compounds are ineffective [10, 58, 59]. L-Arg and guanidinoacetate have only 'apparent' repressor activity, since they have no effect on AGAT expression by themselves, but are readily converted into Cr which then acts as the true repressor. Since the half-life of AGAT in rat kidney is two to three days [56], the changes in the AGAT levels described here are rather slow processes, thus only allowing for long-term adaptations. Furthermore, immunological studies suggested the presence of multiple forms (isoenzymes?) of AGAT in rat kidney, of which only some are repressible by Cr [60].

The expression of AGAT was suggested to be modulated not only by Cr, but also by dietary and hormonal factors (for reviews see [10, 61]). Dietary deficiencies (fasting, protein-free diets) and diseases (vitamin E deficiency, spreptozotocin-induced diabetes; [62, 63]) decrease AGAT levels in liver, pancreas and kidney. During fasting and in vitamin E deficiency, however, at least part of the observed effect may be explained by the increased blood levels of Cr ([12]; see also [10]). Furthermore, kidney AGAT activity is reduced upon thyroidectomy or hypophysectomy of rats [64]. The original AGAT activities can be restored by injection of thyroxine or growth hormone, respectively. In contrast, injections of growth hormone into thyroidectomized rats and of thyroxine into hypophysectomized rats are without effect, implying that both hormones are necessary for maintaining proper levels of AGAT in rat kidney. Since enzyme activity, protein and mRNA contents are always affected to the same extent, regulation of AGAT expression by thyroid hormones and growth hormone also occurs at a pretranslational level, very similar to the feedback repression by Cr [57, 65]. Growth hormone and Cr have an antagonistic action on AGAT expression, as evidenced by identical mRNA levels and enzymatic activities of kidney AGAT in rats fed Cr and injected with growth hormone as compared to rats receiving neither of these compounds [57]. Finally, AGAT levels in rat testes and decidua are presumably under the control of sex hormones, with estradiol and diethylstilbesterol decreasing and testosterone increasing the AGAT levels (see [10]).

In contrast to the repression of AGAT, Cr does not interfere with the expression of GAMT or arginase in liver. Cr, Crn and PCr also do not act as allosteric regulators of the enzymatic activities of AGAT or GAMT in vitro [10], suggesting that feedback regulation is

achieved exclusively by the action of Cr on the rate of AGAT biosynthesis. AGAT is, however, very efficiently inhibited by ornithine, a fact that seems to be pathologically relevant [10, 66]. In gyrate atrophy of the choroid and retina, for instance, plasma ornithine concentrations are increased 10-20-fold due to deficiency of L-ornithine: 2-oxo-acid aminotransferase [67]. The increased ornithine concentrations, in turn, inhibit AGAT and thus decrease the rate of guanidinoacetate formation from Arg, resulting in drastically decreased serum concentrations of Cr and Crn [68].

As far as regulation of transport processes involving intermediates of Cr metabolism is concerned, many potential points of attack have to be considered, for instance uptake of Arg into mitochondria, release of guanidinoacetate from pancreas and kidney, uptake of guanidinoacetate into and release of Cr from the liver, uptake of Cr into the tissues, and penetration of ATP, ADP and PCr through the mitochondrial membranes. In chicken kidney and liver where AGAT is localized in the mitochondrial matrix, penetration of L-Arg through the inner membrane was found to occur only in respiring mitochondria and only in the presence of anions such as acetate or phosphate [69]. Consequently, the rate of Cr synthesis in the chicken may be influenced by the rate of penetration of Arg into the matrix space. The uptake of blood Cr by muscle was shown to be stimulated by insulin (see [10]). In contrast, the Cr transporter activity in rat and human myoblasts and myotubes is down-regulated by extracellular Cr [70].

The permeability itself as well as changes in permeability of the outer mitochondrial membrane may be critical for the stimulation of mitochondrial 'high-energy phosphate' synthesis and for the transport of these 'high-energy phosphates' between sites of ATP production and ATP utilization within the cell [71, 72]. Changes in the permeability of the outer mitochondrial membrane pore protein (VDAC) may be achieved by 'capacitive coupling' to the membrane potential of the inner mitochondrial membrane, leading to a voltage-dependent 'closure' of the pore (for a review see [73]), or by the recently discovered VDAC modulator protein which increases the rate of voltage-dependent channel closure by approximately 10-fold [74]. Since upon stimulation of mitochondrial respiration from state 4 to state 3, the number of contact sites between mitochondrial inner and outer membranes increases (see [73]), capacitive coupling between the two membranes may be favoured, and the pore protein may be shifted from its open, anion-selective to its closed, cation-selective state.

Finally, increased Cr levels in the blood cause increased guanidinoacetate excretion in the urine, presumably by inhibiting reabsorption of guanidinoacetate by kidney tubules (see [10]). To conclude, the theoretical possibilities of regulating Cr metabolism are manifold. Practically, the highest rates of Cr biosynthesis are observed in young, healthy, fast-growing vertebrates in optimal hormonal balance and under anabolic conditions with a high-quality, Cr-free food supply [10]. The best established regulatory mechanism in Cr metabolism so far is feedback repression of AGAT by Cr. An attractive further possibility, which is testible more easily now after the recent identification and cloning of the Cr transporter, represents a reversible up- and down-regulation of this transporter as a function of dietary Cr intake.

Interference with creatine uptake into muscle

Muscle phenotype after chronic creatine depletion

Animal models for cardiac hypertrophy [75, 76] and diseased human myocardium [77] are characterized by lowered overall intracellular [Cr] and [PCr] as well as by a higher vulnerability to hypoxia. Consequently, lowering of the energy reserves for ATP synthesis may render muscle more susceptible to failure. The same holds true also for a number of skeletal muscle myopathies [78] in which a disturbance in transport and handling of Cr is indicated.

Therefore, feeding of experimental animals with Cr analogues would seem a promising tool to test whether lowered levels of total Cr are related to pathological muscle function, and to investigate the physiological role of PCr, Cr and CK in intact muscle. In order to provide clear-cut answers, an 'ideal' Cr analogue should i) either completely inhibit Cr biosynthesis, ii) completely prevent Cr uptake by muscle and nerve *in vivo*, iii) completely and specifically inhibit CK activity *in vivo*, or iv) completely replace Cr and PCr, with the phosphorylated synthetic analogue possessing markedly different thermodynamic and kinetic properties relative to PCr [10]. Unfortunately, none of the Cr analogues studied so far fulfils any of these criteria.

The frequently used Cr analogues 3-guanidinopropionic acid (β-guanidinopropionic acid; GPA) and 3-guanidinobutyric acid (β-guanidinobutyric acid; GBA) both competitively inhibit the Cr transporter activity and thus reduce Cr import through the sarcolemma,

with GBA being somewhat less effective than GPA [10, 79–85]. Compared to GPA, however, GBA has the advantage not to be phosphorylated by CK *in vivo* [85]. Long-term feeding (6–10 weeks) of rats with GPA results in a marked decrease in PCr, Cr and ATP levels in skeletal muscle to approximately 10, 20 and 50% of normal, respectively [83, 86]. At the same time, GPA and its phosphorylated counterpart, PGPA, are accumulated at high concentrations, especially in white, fast-twitch skeletal muscles. In spite of the severely reduced PCr levels, these muscles continue to function reasonably well [79, 81], i.e. neither the initial peak tension nor the long-term steady-state force developed at low workloads are significantly reduced [83]. This, however, does not surprise the informed reader: i) Although the levels of PCr are decreased drastically, the calculated unidirectional flux from PCr to ATP via the CK reaction in resting analogue-loaded muscle ($0.5\ \mu\text{mole} \cdot \text{g}^{-1} \cdot \text{s}^{-1}$) is still several-fold greater than the steady-state ATP turnover rate at rest ($0.07\ \mu\text{mole} \cdot \text{g}^{-1} \cdot \text{s}^{-1}$) [83]. And ii), the rate of PGPA break-down by CK ($0.18\ \text{mM} \cdot \text{s}^{-1}$) is still in excess of the ATPase rate during a transition from low to high work load ($0.1\ \text{mM} \cdot \text{s}^{-1}$) [87]. This is due to GPA and PGPA both serving as substrates for CK to some extent [88]. The V_{max} values of CK for GPA and PGPA are approximately 0.3% and 0.01% of those for Cr and PCr, respectively [89].

Considering these facts, interpretations of results obtained with GPA-treated animals that have been put forward, like 'PCr is not essential for steady-state energy production' [83] or 'neither PCr nor the activity of CK is critical for aerobic metabolism' [81], must be considered with well founded dubiety. As a matter of fact, a closer look at Cr-depleted muscle reveals considerable deviations in contractile properties. In Cr-depleted rat diaphragm muscle during a burst of intense muscle activity (0.2-s tetanic stimulation every 0.5 s), the maximum isometric tension, rate of tension development and rate of relaxation decrease rapidly to reach a minimum about 3 s after the onset of activation [90, 91]. In contrast, normal muscles show a small decrease in tension and relaxation rate but an increase in the rate of tension development under these conditions. Similar findings, although interpreted differently, were made with rat skeletal muscle [83]. During 3-Hz stimulation, hypoxic tibialis anterior muscle of GPA-treated rats is characterized by a rapid decline in peak tension and by the absence of the so-called staircase phenomenon [88]. Post-tetanic stimulation, a common phenomenon seen in normal muscle after a 1-s tetanus, is largely reduced in GPA-loaded

EDL muscle [92]. Furthermore, GPA-loaded soleus muscle displays altered isometric twitch characteristics, in particular a decrease in the maximum velocity of shortening as well as a prolonged half-relaxation time [93]. A major effect of Cr depletion on excitation-contraction coupling, specifically on the relaxation rate, is observed in hearts of GPA- and GBA-treated animals, especially at higher work loads [94, 95]. Long-term feeding of rats with GPA and GBA also causes a decrease in the cytosolic phosphorylation potential [83, 85] and a decrease in the thermodynamic efficiency of cardiac energy metabolism [85]. Finally, the effects of Cr depletion can be markedly exacerbated by superimposing a thyrotoxicosis, e.g. by simultaneously feeding GPA and thyroid powder. Much more severe muscle degeneration is observed under these conditions than with GPA feeding or thyroid powder supplementation alone, indicating that high concentrations of Cr and PCr are essential for the maintenance of muscle integrity during periods of metabolic stress [96].

Many of the findings listed above indicate that muscle relaxation is affected in analogue-treated animals. This is most likely due to an impairment of proper Ca^{2+} handling, since i) muscle relaxation afforded by Ca^{2+}-sequestration into the sarcoplasmic reticulum (SR) depends critically on a highly negative ΔG for ATP hydrolysis in the cytosol [97, 98], and since ii) the ΔG for ATP hydrolysis is less negative in analogue-loaded than in control muscle, due mainly to an increase in [free ADP] [83, 85]. Fully in line with this interpretation, a fraction of the cytosolic CK is bound to the SR and is functionally coupled to the SR-Ca^{2+}-ATPase [99, 100]. In this location, CK was proposed to locally regenerate ATP and thus to maintain a high phosphorylation potential in the intimate vicinity of the Ca^{2+} pump of the SR [1].

Metabolic adaptation of muscle chronically depleted of creatine

The altered contractile properties of muscle, chronically depleted of Cr, are reflecting the effects of the defect itself (Cr, PCr and thus 'high-energy phosphate' deficiency) plus a superimposition of effects caused by compensatory, qualitative and quantitative adaptational changes induced by the Cr-depletion. The astonishing plasticity of muscle to adapt to specific requirements under physiological conditions is well documented and can be nicely demonstrated by chronic stimulation of fast glycolytic muscle fibers which are readily converted into slow oxidative fibers [101, 102]. Obviously, this muscle plasticity can also be provoked by artificial stimuli. In the case of Cr-depletion due to GPA or GBA feeding, fast-twitch skeletal muscle fibers adapt to a 'high-energy phosphate' deficit i) by reducing fiber diameter, resulting in reduced diffusion distances for 'energy' metabolites [103], ii) by increasing aerobic capacity [104, 105], iii) by decreasing their glycolytic potential, with glycogen content increasing at the same time, iv) by increasing the proportion of slow-twitch fibers in skeletal muscle [103], and v) by shifting the myosin isoform pattern from fast to slow isomyosins [106]. The finding of a similar shift in cardiac muscle from the faster ventricular myosin V_1 to the slower V_2 isoform [84] contrasts with an earlier study where no changes in ventricular isoforms were reported [106].

Consistent with the notion that the oxidative capacity of Cr analogue-treated muscle is increased, long-term GPA feeding causes a 60–67% increase in cytochrome c mRNA in rat soleus and white quadriceps muscle [104] as well as a 40–50% increase in cytochrome c, citrate synthase and hexokinase activity in rat plantaris muscle [105]. Furthermore, a 50% increase in the major glucose transporter isoform in skeletal muscle, GLUT-4, was seen in Cr-depleted muscle [105]. Since GLUT-4 is the major determinant of a muscle's maximal insulin-stimulated glucose transport capacity, all these findings point to a metabolic adaptation to increase the availability of energy sources (ATP) for proper cell function. Finally, AMP deaminase activity is significantly reduced, specifically in fast-twitch muscle fibers, to a level normally found in slow-twitch muscles [107].

To conclude, chronic depletion of Cr in muscle clearly results in a multiplicity of stratified metabolic adaptations. However, the very interesting questions of how and by which signalling cascades a low-energy stress situation is transmitted to bring about the induction of compensatory measures still await an answer. On the other hand, much care should be taken in the interpretation of results obtained with animal models of long-term Cr depletion.

In order to circumvent the problem of metabolic and structural adaptations, an acute *ex vivo* model has recently been established using perfusion of isolated hearts with 150 mM GPA [108]. In this short-term model, a linear accumulation of PGPA is accompanied by a 30% decrease of PCr over a 2 hr period. The increase in P_i and the decrease in ATP which occur concomitantly with PGPA accumulation indicate that ATP synthesis is not keeping up with ATP demand. Short-term GPA perfusion reduces the cardiac frequency and developed tension by approx. 40% and 10%, respectively. Perfusion with 150 mM mannitol instead of GPA results in a 15% decrease in cardiac frequency, with a similar decrease in ATP, increase in intracellular pH and a smaller rise in P_i being observed with mannitol compared to GPA. These results suggest that some of the effects observed with the acute GPA model are due to the hyperosmolarity caused by 150 mM GPA. Therefore, the scientific merit of this method of Cr depletion has still to be evaluated thoroughly.

Changes of mitochondrial structure seen in creatine-depleted muscle

As *in vivo*, muscle cells in culture seem to depend on Cr for proper differentiation and cell function [109]. Cultivation of adult rat cardiomyocytes *in vitro* in a Cr-deficient medium or in the presence of GPA results in marked morphological changes, mainly affecting mitochondria [110]. After 3–4 days in culture, a population of enlarged, rod-shaped mitochondria with characteristic crystalline intramitochondrial inclusions appears in these cells. This phenomenon is fully reversible if the cell culture medium is supplemented with Cr. The appearance of highly ordered intramitochondrial inclusions correlates with a low intracellular total Cr content of the cardiomyocytes [110]. Most importantly, the large, rod-shaped mitochondria react very strongly with specific anti-mitochondrial creatine kinase (Mi-CK) antibodies in immunofluorescence experiments. Higher-magnification immuno-electron microscopy showed that, in fact, the highly ordered intramitochondrial inclusions are heavily enriched for Mi-CK [110].

Very similar intramitochondrial inclusions have been observed in several animal models: in skeletal muscle of rats fed with a diet containing 1–2% GPA [111–113]; in adult ventricular cardiomyocytes, cultured for six days in serum-supplemented medium, followed by serum-free medium containing the α_1-adrenoceptor agonist, phenylephrine (stimulating protein synthesis) [114]; in ischemic rat skeletal muscle *in vivo* [115]; in organ cultures of rat diaphragm, most likely suffering from anoxic conditions [116]; as well as after acute *in vivo* administration of uncouplers of oxidative phosphorylation [117]. It is tempting to speculate that all of these intramitochondrial inclusions contain Mi-CK as main or even sole component [110].

GPA administration to rats for 6–10 weeks induces two types of intramitochondrial inclusions, mainly seen in enlarged subsarcolemmal mitochondria of red skeletal muscle and diaphragm [111–113, 118]; long ribbon-like peripheral inclusions (Fig. 2A), and regular staples of intracristae inclusions often arranged in packages consisting of two or four distinct tracks (Fig. 2B). Each track contains periodically arranged 'material' that is fitted between or connects two adjacent cristae membrane folds (Fig. 2B). The ensheathment of the periodic material by the cristae membrane can most clearly be seen at the ends of individual tracks (Fig. 2B, arrows). Often, if the tracks are very long, they tend to circularize or break (Fig. 2B, arrowhead). Detailed inspection of high-magnification electron micrographs suggests that the inclusions are generated by the close apposition of two folds of the mitochondrial inner membrane, with the crystalline material accumulating in between, first as one single layer [119] and then building up to more compact structures [111–113, 115, 118] (See also Fig. 2). The findings that i) the highly symmetrical octameric Mi-CK molecules are able to link two membranes and to stabilize such membrane contacts *in vitro* [120] and that ii) octameric Mi-CK under certain experimental conditions 'polymerizes' to form ribbon-like linear filaments [121] suggest that the intramitochondrial inclusions seen in GPA-treated experimental animals, like those observed in Cr-depleted cardiomyocytes in culture, also consist mainly of Mi-CK. This hypothesis has recently been corroborated by preliminary immunogold labelling experiments [118, Gorman et al., unpublished].

Mitochondrial creatine kinase is a major constituent of pathological inclusions seen in biopsies of human patients with mitochondrial myopathies

Most interestingly, mitochondrial inclusions are frequently seen in patients with so-called mitochondrial encephalomyopathies. These mitochondrial myopathy diseases [122] are characterized by the presence of mutations in the mitochondrial DNA which affect in some way or another ATP production by oxidative phosphorylation [123–125]. These defects are particularly revealed in tissues such as skeletal muscle, heart and brain which rely the most on oxidative phosphorylation. The structural hall-mark of these syndromes is the presence of 'ragged red muscle fibers' in muscle biopsies [126, 127]. Characteristic aspects of 'ragged red muscle fibers' are an accumulation of enlarged and abnormal mitochondria and the occurrence in these mitochondria of

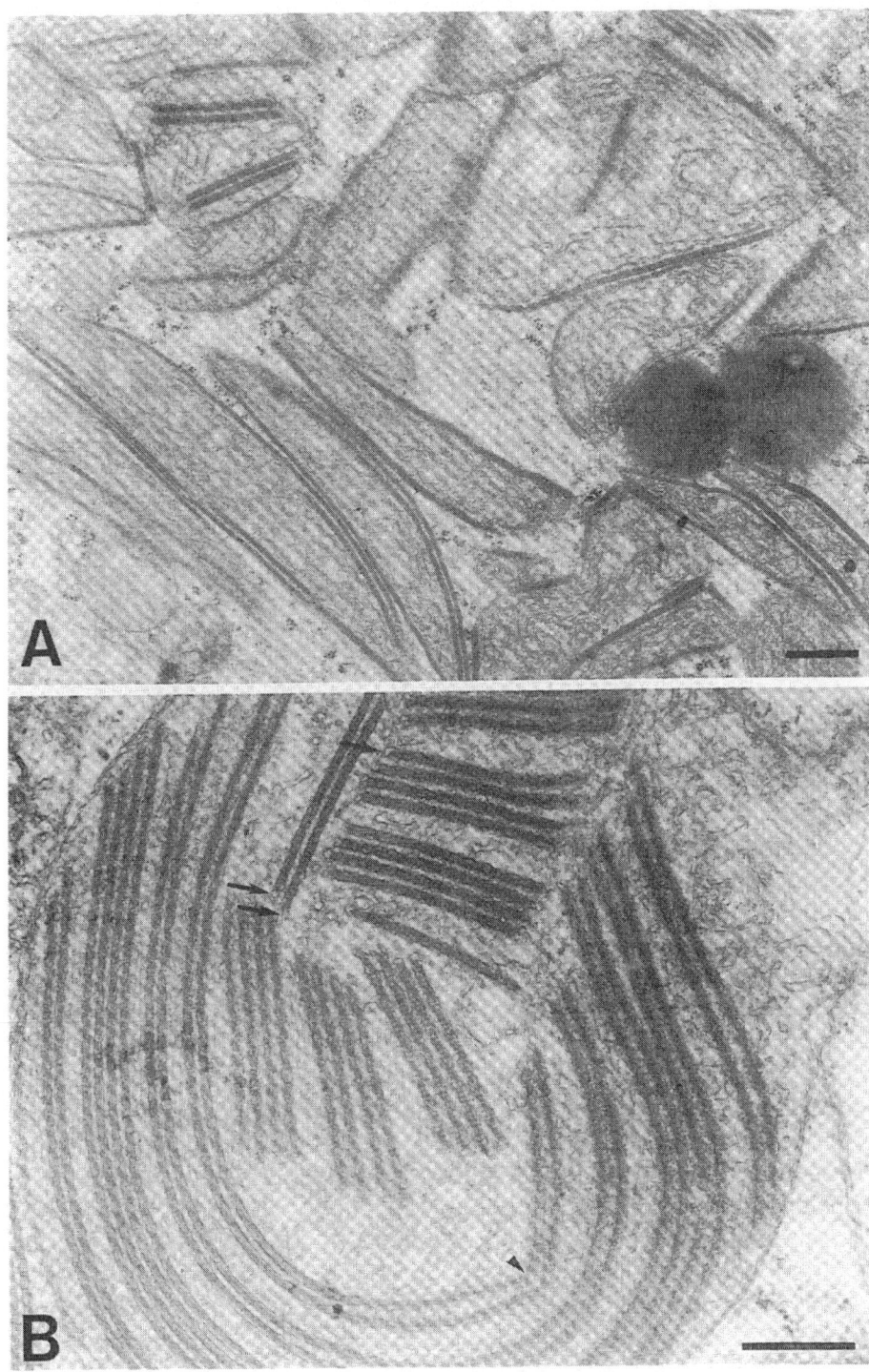

Fig. 2. GPA-induced morphological changes in muscle mitochondria. Intramitochondrial crystalline inclusions were induced in rat muscle by 10 weeks of supplementation of the normal rat food with 2% (w/w) of 3-guanidinopropionic acid (GPA), a creatine uptake inhibitor. A) Rat diaphragm muscle with enlarged mitochondria and inclusions located beneath the mitochondrial outer membrane (peripheral inclusions). B) Rat diaphragm muscle with enlarged roundish mitochondria and inclusions located between cristae membranes (intracristae inclusions). Note pairs of two or four 'ribbons' appearing either as short stacks or alternatively as long structures that are often curved or kinked (arrowhead). At the end of some of the ribbons, the cristae membrane enveloping the crystalline material can be identified (arrows). Bars = 0.4 μm (courtesy of C. Haas and Dr. I. Riesinger, see [118]).

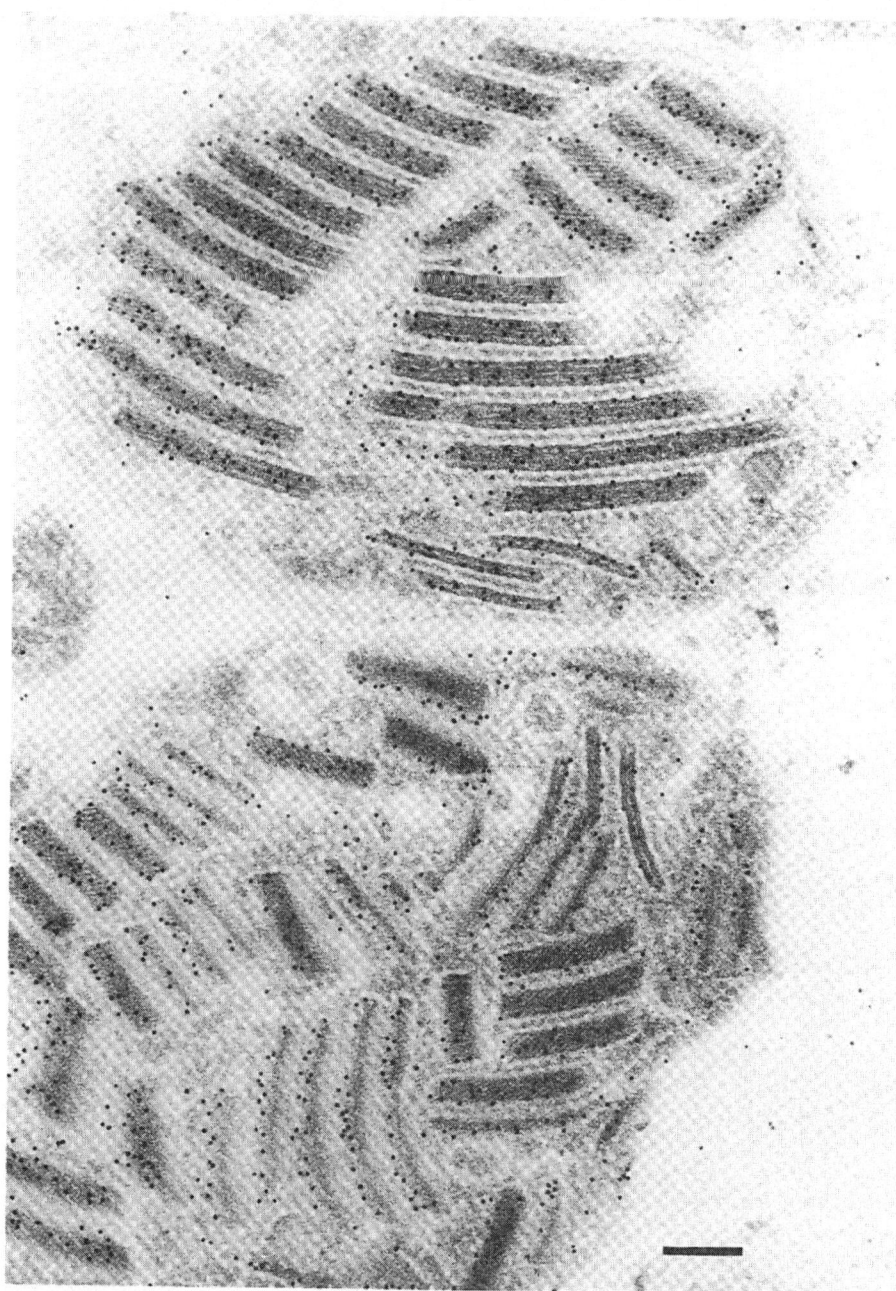

Fig. 3. Intramitochondrial inclusions in biopsies of patients with mitochondrial myopathies consist mainly of mitochondrial creatine kinase. A section of a skeletal muscle biopsy (from m. quadriceps) from a patient with chronic progressive external ophthalmoplegia (CPEO) was stained with specific anti-mitochondrial creatine kinase (Mi-CK) antibodies, followed by gold-conjugated second antibody. Note the strong and specific immunostaining of the intramitochondrial inclusions even after standard EM embedding techniques. Bar = 0.1 μm (courtesy of Prof. Ad M. Stadhouders, see [132]).

highly ordered crystal-like inclusions [128–130]. These crystals are very reminiscent of those seen in the experimental GPA-animal model described above.

The structural features of the crystalline intramitochondrial inclusions, observed in biopsy material mostly from patients with chronic progressive external ophthalmoplegia (CPEO), have been studied in great detail by electron microscopy of oriented thin sections [131]. Analysis of the electron micrographs by image processing revealed regularly packed, square-shaped building blocks of 10 nm side width and a central channel [132], features matching the structural characteristics of isolated Mi-CK octamers [121]. Recent immunogold labelling experiments showed that the mostly proteinaceous inclusions in CPEO patients react very strongly with specific anti-Mi-CK antibodies [132, 133] (see Fig. 3).

Functional aspects of intramitochondrial
Mi-CK-containing inclusions

An interesting aspect of a metabolic adaptation has recently been elucidated. In skeletal muscles of patients with defects in and low expression of mitochondrial genes involved in oxidative phosphorylation, nuclearly encoded mitochondrial proteins involved in ATP production are overexpressed to compensate for the respiratory deficiency [134]. Thus, one would like to propose that the Mi-CK-rich inclusions seen in patients with mitochondrial myopathies are generally formed as a response to a chronic cytosolic 'low energy' stress situation. It will be extremely interesting to find the sensors as well as signalling factors responsible for the induction of these compensatory mechanisms. Whether the Mi-CK inclusions in fact serve a compensatory ameliorating function, i.e. by increasing the export of 'high-energy phosphates' from mitochondria to the cytosol (increased V_{max}/K_m ratio), remains to be elucidated. However, this idea would be fully compatible with the proposed function of Mi-CK as an energy channelling molecule [2, 135]. The metabolic compensation hypothesis is also supported by enzyme histochemical methods, showing that Mi-CK is enzymatically active even within fully grown intramitochondrial crystals [132]. At an advanced stage, on the other hand, the crystallization of massive amounts of Mi-CK within cristae folds or between inner and outer mitochondrial membranes may have to be considered pathological itself, for it is difficult to envisage how such crystals could improve functional coupling of Mi-CK with the adenine nucleotide translocator of the inner and with porin of the outer membrane [2, 135, 136].

Transgenic null mutants for cytosolic M-CK reveal
similar phenotype and metabolic adaptation as seen
in creatine-depleted muscle

Very recently, a null mutation for the cytosolic muscle CK (M-CK) gene was created in transgenic mice [137]. These M-CK knockout mutant mice, despite the complete absence of M-CK mRNA as well as active M-CK enzyme, still express more or less normal levels of Mi-CK and show normal concentrations of free ATP, PCr

and P_i in their resting muscles. Most strikingly, the well-known symmetrical changes of [PCr] and [P_i] (a decrease in one is paralleled by an increase in the other) during exercise and recovery are similar in the mutants and controls. Nevertheless, upon closer examination of muscle performance, a clear physiological phenotype becomes apparent. The mutant mice lack the ability to perform burst muscle activity, that is, although their muscles reach normal initial peak tension, they are unable to maintain it for any length of time. Transgenic mice lacking M-CK compensate for the deficiency by structural and metabolic adaptations in their muscles, e.g. by increasing mitochondrial capacity and glycogen content in fast muscle fibres. The expansion of the mitochondrial network in mutant fast-twitch fibres, e.g. in the subsarcolemmal and intermyofibrillar space, results in a considerable reduction of diffusion distances between mitochondria and myofibrils. These findings point to an adaptation towards increased energy transport mediated directly via ATP and ADP [137].

Thus, M-CK-deficient transgenic mice (with no cytosolic CK, but normal substrate concentrations; [137]) and GPA- or GBA-treated animals (with normal CK activity, but decreased substrate concentrations [82, 83, 90–92]) display a notably similar phenotype as far as physiological muscle performance is concerned. In both animal models, i) the initial peak muscle tension reached is normal, ii) the peak force, however, cannot be maintained and declines rapidly after the onset of muscle stimulation, i.e. muscle burst activity is affected, and iii), marked compensatory adaptive changes take place, leading to an improvement of endurance performance of the muscles [137]. This muscle phenotype is exactly what one would expect if the high-energy phosphate buffering function, facilitated by PCr and cytosolic CK, were hampered. Therefore, both animal models provide additional confirmation for the postulated role of cytosolic M-CK as an immediate 'high-energy phosphate' buffer for short-time muscle activity (see [1, 136]).

However, the enhancement in aerobic capacity of M-CK-deficient mice is afforded mainly by an increase in the number of relatively large intermyofibrillar mitochondria, and only rarely are intramitochondrial inclusions seen [137]. By contrast, in GPA-treated animals, grossly enlarged mitochondria accumulate mostly in a clustered fashion in the subsarcolemmal space and frequently display prominent intramitochondrial crystals [111–113, 115, 118] (see Fig. 3). Although in the M-CK 'knock-out' animals, the fast-twitch type 2 muscle fibers have an increased intra-myofibrillar mitochondrial volume and an increased glycogenolytic/glycolytic potential, and in spite of the adaptation of the muscles to endurance performance, no signs of muscle fiber-type conversion, as seen in GPA-treated animals, have been observed in M-CK-deficient mice [137].

Concluding remarks

The accumulation of the unique type of intramitochondrial inclusions described above is a common denominator seen in a variety of diseases linked to defects in mitochondrial energy metabolism as well as in animal models with a depleted energy status. Since these inclusions contain Mi-CK as their major component or may even be 'pure' Mi-CK protein crystals, their occurrence in pathology points to the physiological importance of the CK/PCr system for cellular energetics. These findings further impose the intriguing question of how a defect in cellular energy metabolism (substrate *or* enzyme deficiency) can regulate and influence muscle plasticity towards long-term structural and metabolic adaptation.

Seen in the broad context of system physiology, organ and cell function, the pathways and regulation of Cr biosynthesis and degradation are very fascinating, but unexpectedly complex, with many basic questions still open. Cr and CK, however, have managed to make a come-back and obtain at least a glimpse at the lime-light of modern biological science. The surprising findings made with M-CK null-mutant transgenic mice and the results to be expected in the near future with 'knock-out' mutants of the other CK isoenzymes, which most likely also will display distinct phenotypes, have already provided a deeper insight in CK function *in vivo* and are likely to shed new light on Cr, PCr and CK function in the intact animal, respectively. Above all, these types of experiments will keep the next generation of cell, organ and system physiologists (if there are any left?) quite busy in the future.

Acknowledgments

We would like to thank Prof. Ad M. Stadhouders, Institute for Cell Biology at the University of Nijmegen, Holland for providing figure 3 and to C. Haas and Dr. I. Riesinger, Institute for Cell Biology, ETH Zürich, for the negatives for figure 2. Thanks are also due to Eddie O'Gorman for carefully reading the manuscript and to Dr. E. Gnaiger, Innsbruck, Austria, for discussion. This work was supported by the Swiss National Science Foundation (SNF grant 31-33907.92 to T.W.), a postdoctoral SNF training grant (823A-037106 to M.W.), the Swiss Society for Muscle Diseases and the Helmut Horten Foundation.

References

1. Wallimann T, Wyss M, Brdiczka D, Nicolay K, Eppenberger HM: Significance of intracellular compartmentation, structure and function of creatine kinase isoenzymes for cellular energy homeostasis: 'The Phospho-Creatine Circuit'. Biochem J 281: 21–40, 1992
2. Wyss M, Smeitink J, Wevers RA, Wallimann T: Mitochondrial creatine kinase: a key enzyme of aerobic energy metabolism. Biochim Biophys Acta 1102: 119–166, 1992
3. Bloch K, Schoenheimer R: The biological formation of creatine. J Biol Chem 133: 633–634, 1940
4. Bloch K, Schoenheimer R: The biological origin of the amidine group in creatine. J Biol Chem 134: 785–786, 1940
5. van Pilsum JF, Stephens GC, Taylor D: Distribution of creatine, guanidinoacetate and the enzymes for their biosynthesis in the animal kingdom. Biochem J 126: 325–345, 1972
6. Needham DM, Needham J, Baldwin E, Yudkin J: A comparative study of the phosphagens, with some remarks on the origin of vertebrates. Nature 110: 260–294, 1932
7. Ennor AH, Morrison JF: Biochemistry of the phosphagens and related guanidines. Physiol Rev 38: 631–674, 1958
8. Robin Y: Biological distribution of guanidines and phosphagens in marine annelida and related phyla from california, with a note on pluriphosphagens. Comp Biochem Physiol 12: 347–367, 1964
9. Watts DC: Evolution of phosphagen kinases. In: E. Schoffeniels (ed). Biochemical Evolution and the Origin of Life, North-Holland Publishing Company, 1971, pp 150–173
10. Walker JB: Creatine: biosynthesis, regulation, and function. Adv Enzymol 50: 177–242, 1979
11. Horner WH: Transamidination in the nephrectomized rat. J Biol Chem 234: 2386–2387, 1959
12. Fitch CD, Hsu C, Dinning JS: The mechanism of kidney transamidinase reduction in vitamin E-deficient rabbits. J Biol Chem 236: 490–492, 1961
13. McGuire DM, Gross MD, Elde RP, van Pilsum JF: Localization of L-arginine-glycine amidinotransferase protein in rat tissues by immunofluorescence microscopy. J Histochem Cytochem 34: 429–435, 1986
14. van Pilsum JF, Olsen B, Taylor D, Rozycki T, Pierce JC: Transamidinase activities, *in vitro*, of tissues from various mammals and from rats fed protein-free, creatine-supplemented and normal diets. Arch Biochem Biophys 100: 520–524, 1963
15. Yanokura M, Tsukada K: Decreased activities of glycine and guanidinoacetate methyltransferases and increased levels of creatine in tumor cells. Biochem Biophys Res Commun 104: 1464–1469, 1982
16. Daly MM: Guanidinoacetate methyltransferase activity in tissues and cultured cells. Arch Biochem Biophys 236: 576–584, 1985
17. Holtzman D, McFarland E, Moerland T, Koutcher J, Kushmer-

ick MJ, Neuringer LJ: Brain creatine phosphate and creatine kinase in mice fed an analogue of creatine. Brain Res 483: 68–77, 1989

18. Defalco AJ, Davies RK: The synthesis of creatine by the brain of the intact rat. J Neurochem 7: 308–312, 1961

19. Fitch CD, Shields RP: Creatine metabolism in skeletal muscle. I. Creatine movement across muscle membranes. J Biol Chem 241: 3611–3614, 1966

20. Seraydarian MW, Artaza L, Abbott BC: Creatine and the control of energy metabolism in cardiac and skeletal muscle cells in culture. J Mol Cell Cardiol 6: 405–413, 1974

21. Syllm-Rapoport I, Daniel A, Rapoport S: Creatine transport into red blood cells. Acta Biol Med Germ 39: 771–779, 1980

22. Syllm-Rapoport I, Daniel A, Starck H, Götze W, Hartwig A, Gross J, Rapoport S: Creatine in red cells: transport and erythropoietic dynamics. Acta Biol Med Germ 40: 653–659, 1981

23. Daly MM, Seifter S: Uptake of creatine by cultured cells. Arch Biochem Biophys 203: 317–324, 1980

24. Ku C-P, Passow H: Creatine and creatinine transport in old and young human red blood cells. Biochim Biophys Acta 600: 212–227, 1980

25. Loike JD, Somes M, Silverstein SC: Creatine uptake, metabolism, and efflux in human monocytes and macrophages. Am J Physiol 251: C128–C135, 1986

26. Möller A, Hamprecht B: Creatine transport in cultured cells of rat and mouse brain. J Neurochem 52: 544–550, 1989

27. Guimbal C, Kilimann MW: A Na$^+$-dependent creatine transporter in rabbit brain, muscle, heart and kidney. cDNA cloning and functional expression. J Biol Chem 268: 8418–8421, 1993

28. Baker Z, Miller BF: Studies on the metabolism of creatine and creatinine. II. The distribution of creatine and creatinine in the tissues of the rat, dog, and monkey. J Biol Chem 130: 393–397, 1939

29. Peters JP, van Slyke DD: Quantitative Clinical Chemistry, Interpretations, Vol I, 2nd ed, Williams & Wilkins Co, Baltimore, 1946

30. Berlet HH, Bonsmann I, Birringer H: Occurrence of free creatine, phosphocreatine and creatine phosphokinase in adipose tissue. Biochim Biophys Acta 437: 166–174, 1976

31. Wallimann T, Eppenberger HM: Localization and function of M-line-bound creatine kinase. M-band model and creatine phosphate shuttle. In: J.W. Shay (ed). Cell and Muscle Motility, Vol 6, Plenum Publishing Corp, 1985, pp 239–285

32. Wallimann T, Moser H, Zurbriggen B, Wegmann G, Eppenberger HM: Creatine kinase isoenzymes in spermatozoa. J Muscle Res Cell Motil 7: 25–34, 1986

33. Wallimann T, Wegmann G, Moser H, Huber R, Eppenberger HM: High content of creatine kinase in chicken retina: compartmentalized localization of creatine kinase isoenzymes in photoreceptor cells. Proc Natl Acad Sci USA 83: 3816–3819, 1986

34. Lee HJ, Fillers WS, Iyengar MR: Phosphocreatine, an intracellular high-energy compound, is found in the extracellular fluid of the seminal vesicles in mice and rats. Proc Natl Acad Sci USA 85: 7265–7269, 1988

35. Delanghe J, De Slypere J-P, De Buyzere M, Robbrecht J, Wieme R, Vermeulen A: Normal reference values for creatine, creatinine, and carnitine are lower in vegetarians. Clin Chem 35: 1802–1803, 1989

36. Kushmerick MJ, Moerland TS, Wiseman RW: Mammalian skeletal muscle fibers distinguished by contents of phosphocreatine, ATP and P$_i$. Proc Natl Acad Sci USA 89: 7521–7525, 1992

37. Harris RC, Hultman E: Muscle phosphagen status studied by needle biopsy. In: J.M. Kinney and H.N. Tucker (eds). Energy Metabolism: Tissue Determinants and Cellular Corollaries. Raven Press, NY, 1992, pp 367–379

38. Edgar G, Shiver HE: The equilibrium between creatine and creatinine, in aqueous solution. The effect of hydrogen ion. J Am Chem Soc 47: 1179–1188, 1925

39. Cannan RK, Shore A: The creatine-creatinine equilibrium. The apparent dissociation constants of creatine and creatinine. Biochem J 22: 920–929, 1928

40. Morrison JF, Ennor AH: N-Phosphorylated guanidines. In: P.D. Boyer, H. Lardy, and K. Myrbäck (eds). The Enzymes, 2nd ed, Vol 2, Academic Press, NY, 1960, pp 89–109

41. Bloch K, Schoenheimer R: Studies in protein metabolism. XI. The metabolic relation of creatine and creatinine studied with isotopic nitrogen. J Biol Chem 131: 111–119, 1939

42. Crim MC, Calloway DH, Margen S: Creatine metabolism in men: creatine pool size and turnover in relation to creatine intake. J Nutr 106: 371–381, 1976

43. van Hoogenhuyze CJC, Verploegh H: Beobachtungen über die Kreatininausscheidung beim Menschen. Zschr physiol Chem 46: 415–471, 1905

44. Iyengar MR, Coleman DW, Butler TM: Phosphocreatinine, a high-energy phosphate in muscle, spontaneously forms phosphocreatine and creatinine under physiological conditions. J Biol Chem 260: 7562–7567, 1985

45. Clark VM, Warren SG: Why do phosphagens function as phosphoryl-transfer reagents? Nature 199: 657–659, 1963

46. Akamatsu S, Kanai Y: Bacterial decomposition of creatinine. I. Creatinomutase. Enzymologia 15: 122–125, 1951

47. Akamatsu S, Miyashita R: Bacterial decomposition of creatine. III. The pathway of creatine decomposition. Enzymologia 15: 173–176, 1951

48. Szulmajster J: Bacterial fermentation of creatinine. I. Isolation of N-methyl-hydantoin. J Bacteriol 75: 633–639, 1958

49. Szulmajster J: Bacterial degradation of creatinine. II. Creatinine desimidase. Biochim Biophys Acta 30: 154–163, 1958

50. van Eyk HG, Vermaat RJ, Leijnse-Ybema HJ, Leijnse B: The conversion of creatinine by creatininase of bacterial origin. Enzymologia 34: 198–202, 1968

51. Forde A, Johnson DB: Preliminary studies on enzymes of creatinine degradation. Biochem Soc Trans 2: 1342–1344, 1974

52. Chang MC, Chang CC, Chang JC: Cloning of a creatinase gene from *Pseudomonas putida* in *Escherichia coli* by using an indicator plate. Appl Environm Microbiol 58: 3437–3440, 1992

53. Miyoshi K, Taira A, Yoshida K, Tamura K, Uga S: Presence of creatinase and sarcosine dehydrogenase in human skeletal muscle. Proposal for creatine-urea pathway. Proc Japan Acad 56B: 95–98, 1980

54. Miyoshi K, Taira A, Yoshida K, Tamura K, Uga S: Abnormalities of creatinase in skeletal muscle of patients with Duchenne muscular dystrophy. Proc Japan Acad 56B: 99–101, 1980

55. Mudd SH, Ebert MH, Scriver CR: Labile methyl group balances in the human: the role of sarcosine. Metabolism 29: 707–720, 1980

56. McGuire DM, Gross MD, van Pilsum JF, Towle HC: Repression of rat kidney L-arginine: glycine amidinotransferase synthesis by creatine at a pretranslational level. J Biol Chem 259: 12034–12038, 1984

57. van Pilsum JF, McGuire DM, Miller CA: The antagonistic action of creatine and growth hormone on the expression of the gene for rat kidney L-arginine: glycine amidinotransferase. In: P.P. De Deyn, B. Marescau, V. Stalon and I.A. Qureshi (eds). Guanidino Compounds in Biology and Medicine, John Libbey & Company Ltd, 1992, pp 147–151

58. Walker JB, Wang S-H: Tissue repressor concentration and target enzyme level. Biochim Biophys Acta 81: 435–441, 1964

59. Roberts JJ, Walker JB: Higher homolog and N-ethyl analog of creatine as synthetic phosphagen precursors in brain, heart, and muscle, repressors of liver amidinotransferase, and substrates for creatine catabolic enzymes. J Biol Chem 260: 13502–13508, 1985

60. Gross MD, Simon AM, Jenny RJ, Gray ED, McGuire DM, van Pilsum JF: Multiple forms of rat kidney L-arginine: glycine amidinotransferase. J Nutr 118: 1403–1409, 1988

61. Methfessel J: Transamidinase. Zschr inn Med 25: 80–84, 1970

62. van Pilsum JF, Wahman RE: Creatine and creatinine in the carcass and urine of normal and vitamin E-deficient rabbits. J Biol Chem 235: 2092–2094, 1960

63. Funahashi M, Kato H, Shiosaka S, Nakagawa H: Formation of arginine and guanidinoacetic acid in the kidney *in vivo*. Their relations with the liver and their regulation. J Biochem 89: 1347–1356, 1981

64. van Pilsum JF, Carlson M, Boen JR, Taylor D, Zakis B: A bioassay for thyroxine based on rat kidney transamidinase activities. Endocrinology 87: 1237–1244, 1970

65. McGuire DM, Tormanen CD, Segal IS, van Pilsum JF: The effect of growth hormone and thyroxine on the amount of L-arginine: glycine amidinotransferase in kidneys of hypophysectomized rats. Purification and some properties of rat kidney transamidinase. J Biol Chem 255: 1152–1159, 1980

66. Sipilä I: Inhibition of arginine-glycine amidinotransferase by ornithine. A possible mechanism for the muscular and chorioretinal atrophies in gyrate atrophy of the choroid and retina with hyperornithinemia. Biochim Biophys Acta 613: 79–84, 1980

67. Valle D, Kaiser-Kupfer MI, Del Valle LA: Gyrate atrophy of the choroid and retina: deficiency of ornithine aminotransferase in transformed lymphocytes. Proc Natl Acad Sci USA 74: 5159–5161, 1977

69. Grazi E, Magri E, Balboni G: On the control of arginine metabolism in chicken kidney and liver. Eur J Biochem 60: 431–436, 1975

70. Loike JD, Zalutsky DL, Kaback E, Miranda AF, Silverstein SC: Extracellular creatine regulates creatine transport in rat and human muscle cells. Proc Natl Acad Sci USA 85: 807–811, 1988

71. Gellerich FN, Schlame M, Bohnensack R, Kunz W: Dynamic compartmentation of adenine nucleotides in the mitochondrial intermembrane space of rat-heart mitochondria. Biochim Biophys Acta 890: 117–126, 1987

72. Gellerich FN, Khuchua ZA, Kuznetsov AV: Influence of the mitochondrial outer membrane and the binding of creatine kinase to the mitochondrial inner membrane on the compartmentation of adenine nucleotides in the intermembrane space of rat heart mitochondria. Biochim Biophys Acta 1140: 327–334, 1993

73. Brdiczka D: Contact sites between mitochondrial envelope membranes. Structure and function in energy- and protein-transfer. Biochim Biophys Acta 1071: 291–312, 1991

74. Liu M, Colombini M: Voltage gating of the mitochondrial outer membrane channel VDAC is regulated by a very conserved protein. Am J Physiol 260: C371–C374, 1991

75. Seppet EK, Adoyaan AJ, Kallikorm AP, Chernousova GB, Lyulina NV, Sharov VG, Severin VV, Popovich MI, Saks VA: Hormone regulation of cardiac energy metabolism. I. Creatine transport across cell membranes of euthyroid and hyperthyroid rat heart. Biochem Med 34: 267–279, 1985

76. Ingwall JS, Atkinson DE, Clarke K, Fetters JK: Energetic correlates of cardiac failure: changes in the creatine kinase system in the failing myocard. Eur Heart J 11 (Suppl B), 108–115, 1990

77. Ingwall JS, Kramer MF, Fifer MA, Lovell BH, Shemin R, Grossman W, Allen PD: The creatine kinase system in normal and diseased human myocardium. N Engl J Med 131: 1050–1054, 1985

78. Fitch CD: Significance of abnormalities of creatine metabolism. In: P. Rowland (ed), Pathogenesis of human muscular dystrophies. Excerpta Medica, Amsterdam, 1977, pp 328–336

79. Fitch CD, Jellinek M, Mueller E: Experimental depletion of creatine and phosphocreatine from skeletal muscle. J Biol Chem 249: 1060–1063, 1974

80. Fitch CD, Chevli R: Inhibition of creatine and phosphocreatine accumulation in skeletal muscle and heart. Metabolism 29: 686–690, 1980

81. Shoubridge EA, Radda GK: A ^{31}P nuclear magnetic resonance study of skeletal muscle metabolism in rats depleted of creatine with the analogue β-guanidino-propionic acid. Biochim Biophys Acta 805: 79–88, 1984

82. Shoubridge EA, Radda GK: A gated ^{31}P NMR study of tetanic contraction in rat muscle depleted of phosphocreatine. Am J Physiol 252: C532–C542, 1987

83. Meyer RA, Brown TR, Krilowicz BL, Kushmerick MJ: Phosphagen and intracellular pH changes during contraction of creatine-depleted muscle. Am J Physiol 250: C264–C274, 1986

84. Mekhfi H, Hoerter J, Lauer C, Wisnewsky C, Schwartz K, Ventura-Clapier R: Myocardial adaptation to creatine deficiency in rats fed with β-guanidinopropionic acid, a creatine analogue. Am J Physiol 258: H1151–H1158, 1990

85. Zweier JL, Jacobus WE, Korecky B, Brandejs-Barry Y: Bioenergetic consequences of cardiac phosphocreatine depletion induced by creatine analogue feeding. J Biol Chem 266: 20296–20304, 1991

86. Shoubridge EA, Jeffry FMH, Keogh JM, Radda GK, Seymour A-ML: Creatine kinase kinetics, ATP turnover, and cardiac performance in hearts depleted of creatine with the substrate analogue β-guanidinopropionic acid. Biochim Biophys Acta 847: 25–32, 1985

87. Conley KE, Kushmerick MJ: Buffering work transitions in myocardium: role of a poorly metabolized creatine analogue. Magn Res Med 2: 902, 1990

88. Fitch CD, Jellinek M, Fitts RH, Baldwin KM, Holloszy JO: Phosphorylated β-guanidinopropionate as a substitute for phosphocreatine in rat muscle. Am J Physiol 228: 1123–1125, 1975

89. Chevli R, Fitch CD: beta-Guanidinopropionate and phosphorylated beta-guanidinopropionate as substrates for creatine kinase. Biochem Med 21: 162–167, 1979

90. Mainwood GW, Alward M, Eiselt B: Contractile characteristics of creatine-depleted rat diaphragm. Can J Physiol Pharmacol 60: 120–127, 1982

91. Mainwood GW, Alward M, Eiselt B: The effects of metabolic inhibition on the contraction of creatine-depleted muscle. Can J Physiol Pharmacol 60: 114–119, 1982

92. Mainwood GW, de Zepetnek JT: Post-tetanic responses in creatine-depleted rat EDL. Muscle Nerve 8: 774–782, 1985

93. Petrofsky JS, Fitch CD: Contractile characteristics of skeletal muscle depleted of phosphocreatine. Pflügers Arch 384: 123–129, 1980

94. Korecky B, Brandejs-Barry Y: Effect of creatine depletion on myocardial mechanics. Basic Res Cardiol 82 (Suppl 2): 103–110, 1987

95. Kapelko VI, Kupriyanov VV, Novikova NA, Lakomkin VL, Steinschneider AYa, Severina MYu, Veksler VI, Saks VA: The cardiac contractility failure induced by chronic creatine and phosphocreatine deficiency. J Mol Cell Cardiol 20: 465–479, 1988

96. Otten JV, Fitch CD, Wheatley JB, Fischer VW: Thyrotoxic myopathy in mice: accentuation by a creatine transport inhibitor. Metabolism 35: 481–484, 1986

97. Hasselbach W, Oetliker H: Energetics and electrogenicity of the sarcoplasmic reticulum pump. Annu Rev Physiol 45: 325–339, 1983

98. Kammermeier H: Why do cells need phospho-creatine and a phospho-creatine shuttle. J Mol Cell Cardiol 19: 115–118, 1987

99. Rossi AM, Eppenberger HM, Volpe P, Cotrufo R, Wallimann T: Muscle-type MM-creatine kinase is specifically bound to sarcoplasmic reticulum and can support Ca^{2+}-uptake and regulate local ATP/ADP ratios. J Biol Chem 265: 5258–5266, 1990

100. Korge P, Byrd SK, Campbell KB: Functional coupling between sarcoplasmic-reticulum-bound creatine kinase and Ca^{2+}-ATPase. Eur J Biochem 213: 973–980, 1993

101. Pette D: Plasticity of Muscle. Walter de Gruyter, Berlin and New York, 1980

102. Pette D: The Dynamic State of Muscle Fibers. Walter de Gruyter, Berlin and New York, 1990

103. Shoubridge EA, Challiss JRA, Hayes DJ, Radda GK: Biochemical adaptation in the skeletal muscle of rats depleted of creatine with the substrate analogue β-guanidinopropionic acid. Biochem J 232: 125–131, 1985

104. Lai MM, Booth FW: Cytochrome c mRNA and α-actin mRNA in muscles of rats fed β-GPA. J Appl Physiol 69: 843–848, 1990

105. Ren JM, Semenkovich CF, Holloszy JO: Adaptation of muscle to creatine depletion: effect on GLUT-4 glucose transporter expression. Am J Physiol 264: C146–C150, 1993

106. Moerland TS, Wolf NG, Kushmerick MJ: Administration of a creatine analogue induces isomyosin transitions in muscle. Am J Physiol 257: C810–C816, 1989

107. Ren JM, Holloszy JO: Adaptation of rat skeletal muscle to creatine depletion: AMP deaminase and AMP deamination. J Appl Physiol 73: 2713–2716, 1992

108. Unitt JF, Radda GK, Seymour AM: The acute effects of the creatine analogue, β-guanidinopropionic acid, on cardiac energy metabolism and function. Biochim Biophys Acta 1143: 91–96, 1993

109. Young RB, Denome RM: Effect of creatine on contents of myosin heavy chain and myosin-heavy-chain mRNA in steady-state chicken muscle-cell cultures. Biochem J 218: 871–876, 1984

110. Eppenberger-Eberhardt M, Riesinger I, Messerli M, Schwarb P, Müller M, Eppenberger HM, Wallimann T: Adult rat cardiomyocytes cultured in creatine-deficient medium display large mitochondria with paracrystalline inclusions enriched for creatine kinase. J Cell Biol 113: 289–302, 1991

111. Ohira Y, Kanzaki M, Chen CS: Intramitochondrial inclusions caused by depletion of creatine in rat skeletal muscles. Jap J Physiol 38: 159–166, 1988

112. Gori Z, De Tata V, Pollera M, Bergamini E: Mitochondrial myopathy in rats fed with a diet containing beta-guanidine propionic acid, an inhibitor of creatine entry in muscle cells. Br J exp Pathol 69: 639–650, 1988

113. De Tata V, Cavallini G, Pollera M, Gori Z, Bergamini E: The induction of mitochondrial myopathy in the rat by feeding β-guanidinopropionic acid and the reversibility of the induced mitochondrial lesions: a biochemical and ultrastructural investigation. Int J Exp Pathol 74: 501–509, 1993

114. Pinson A, Schlüter KD, Zhou XJ, Schwartz P, Kessler-Icekson G, Piper HM: Alpha- and beta-adrenergic stimulation of protein synthesis in cultured adult ventricular cardiomyocytes. J Mol Cell Cardiol 25: 477–490, 1993

115. Hanzlikova V, Schiaffino S: Mitochondrial changes in ischemic skeletal muscle. J Ultrastruct Res 60: 121–133, 1977

116. Heine H, Schaeg G: Origin and function of 'rod-like structures' in mitochondria. Acta anat 103: 1–10, 1979

117. Melmed C, Karpati G, Carpenter S: Experimental mitochondrial myopathy produced by in vivo uncoupling of oxidative phosphorylation. J Neurol Sci 26: 305–318, 1975

118. Riesinger I, Haas C, Wallimann T: Mitochondrial inclusions induced by feeding a creatine analogue exhibit a high density of mitochondrial creatine kinase. EBEC Short Reports (Biochim Biophys Acta) 7: 140, 1992

119. Hall JD, Crane FL: A new structure in beef heart mitochondria. J Cell Biol 48: 420–425, 1971

120. Rojo M, Hovius R, Nicolay K, Wallimann T: Mitochondrial creatine kinase mediates contact formation between mitochondrial membranes. J Biol Chem 266: 20290–20295, 1991

121. Schnyder T, Winkler H, Gross H, Eppenberger HM, Wallimann T: Structure of the mitochondrial creatine kinase octamer: High resolution shadowing and image averaging of single molecules and formation of linear filaments under specific staining conditions. J Cell Biol 112: 95–101, 1991

122. Zeviani M, Bonilla E, DeVivo DC, DiMauro S: Mitochondrial Diseases. Neurol Clin 7: 123–156, 1989

123. Harding AE: Neurological disease and mitochondrial genes. TINS 14: 132–138, 1991

124. Wallace, DC: Diseases of the mitochondrial DNA. Annu Rev Biochem 61: 1175–1212, 1992

125. Wallace DC: Mitochondrial diseases: genotype versus phenotype. TIG 9: 128–133, 1993

126. DiMauro S, Bonilla E, Zeviani M, Nakagawa M, DeVivo DC: Mitochondrial myopathies. Ann Neurol 17: 521–538, 1985

127. Morgan-Hughes JA, Ayes DJ, Cooper M, Clark JB: Mitochondrial myopathies: Deficiencies localized to complex I and II of the respiratory chain. Biochem Soc Trans 13: 648–650, 1985

128. Stadhouders AM: Mitochondrial ultrastructural changes in muscular diseases. In: H.F.M. Busch, F.G.I. Jennekens and H.R. Scholte (eds). Mitochondria and Muscular Diseases. Mefar b.v. Beetsterzwaag, The Netherlands, 1981, pp 113–132

129. Sarnat HB: Muscle pathology and histochemistry. Am Soc Clin Pathol Press, 1983, Chicago, USA

130. Schmalbruch H: The fine structure of mitochondrial abnormalities in muscle diseases. In: G. Scarlato and C. Cerri (eds). Mitochondrial Pathology in Muscle Diseases. Piccin Medical Books, Padua, Italy, 1983, pp 40–56

131. Farrants GW, Hovmöller S, Stadhouders AM: Two types of mitochondrial crystals in diseased human skeletal muscle fibers. Muscle Nerve 11: 45–55, 1988

132. Stadhouders AM, Jap PHK, Winkler HP, Eppenberger HM,

66

Wallimann T: Mitochondrial creatine kinase: a major constituent of pathological inclusions seen in mitochondrial myopathies. Proc Natl Acad Sci USA (in press), 1994

133. Smeitink J, Stadhouders A, Sengers R, Ruitenbeek W, Wevers R, ter Laak H, Trijbels F: Mitochondrial creatine kinase containing crystals, creatine content and mitochondrial creatine kinase activity in chronic progressive external ophthalmoplegia. Neuromusc Disord 2: 35–40, 1992

134. Heddi A, Lestienne P, Wallace DC, Stepien G: Mitochondrial DNA expression in mitochondrial myopathies and coordinated expression of nuclear genes involved in ATP production. J Biol Chem 268: 12156–12163, 1993

135. Wyss M, Wallimann T: Metabolite channelling in aerobic energy metabolism. J Theor Biol 158: 129–132, 1992

136. Wallimann T: Dissection of the role of creatine kinase. The phenotype of gene 'knock out' mice deficient in a creatine kinase isoform sheds new light on the physiological function of the phosphocreatine circuit. Current Biol 4, 42–46: 1994

137. van Deursen J, Heerschap A, Oerlemans F, Ruitenbeek W, Jap P, ter Laak H, Wieringa B: Skeletal muscle of mice deficient in muscle creatine kinase lack burst activity. Cell 74: 621–631, 1993

PART II

SUBSTRATE AND CREATINE KINASE ISOENZYME COMPARTMENTATION

Molecular and Cellular Biochemistry **133/134**: 69–83, 1994.
© 1994 *Kluwer Academic Publishers.*

II-1 The importance of the outer mitochondrial compartment in regulation of energy metabolism

Dieter Brdiczka and Theo Wallimann[1]

Faculty of Biology University of Konstanz, D-78434 Konstanz, Germany and [1] Department of Cell Biology ETH Zürich, CH-8093 Zürich, Switzerland

Abstract

Substitution of physiologically present macromolecules during isolation of mitochondria and investigation of their functions led to a significant change in regulation of oxidative phosphorylation. The differences compared to conventionally isolated mitochondria were that stimulation of oxidative phosphorylation appeared to rather depend on the activity of peripheral kinases than on the addition of free ADP. The localisation of peripheral kinases such as hexokinase and mitochondrial creatine kinase are described as well as the effects of macromolecules on the regulation of bound hexokinase and of oxidative phosphorylation via this enzyme. (Mol Cell Biochem **133/134:** 69–83, 1994)

Key words: mitochondria, brain, liver, macromolecules, localisation, hexokinase, mitochondrial creatine kinase, oxidative phosphorylation

Introduction

Different levels of muscle function and mitochondrial regulation

Muscle cells are excitable as nerve cells, and have a highly specialised contractile apparatus which relates to the cytoskeleton of other cells. In addition skeletal muscle cells resemble hepatocytes as they take up glucose to form glycogen- and amino acids to build proteins stores. All these processes depend directly or indirectly on ATP, provided mainly by oxidative phosphorylation. Membrane depolarisation and repolarisation during excitation, muscle contraction and relaxation as well as Ca^{2+} release and sequestration are processes with a time scale of milliseconds while the synthesis of glycogen and proteins proceeds in seconds. To avoid large ATP/ADP fluctuations during the former fast and energy consum-

ing processes creatine/phosphocreatine is used as an energy buffer and energy transferring system. By this way, high ATP turnover during excitation increases ADP levels only locally but the creatine level (instead of ADP) globally. The latter would then serve as a signal to activate mitochondrial metabolism. We hypothesise that during rest, metabolites other than free ADP also regulate the mitochondrial activity. When oxidative phosphorylation generates the energy needed for building up glycogen and protein stores, the mitochondrial metabolism may be activated by metabolites which are utilised in these pathways, such as glucose, glycerol, UDP and GDP. The activation is performed via specific kinases which are organised at the mitochondrial surface and directly communicate with the inner mitochondrial compartment. This mitochondrial coupling of kinases, such

Address for offprints: Dr D. Brdiczka, Faculty of Biology University of Konstanz, D-78434 Konstanz, Germany

70

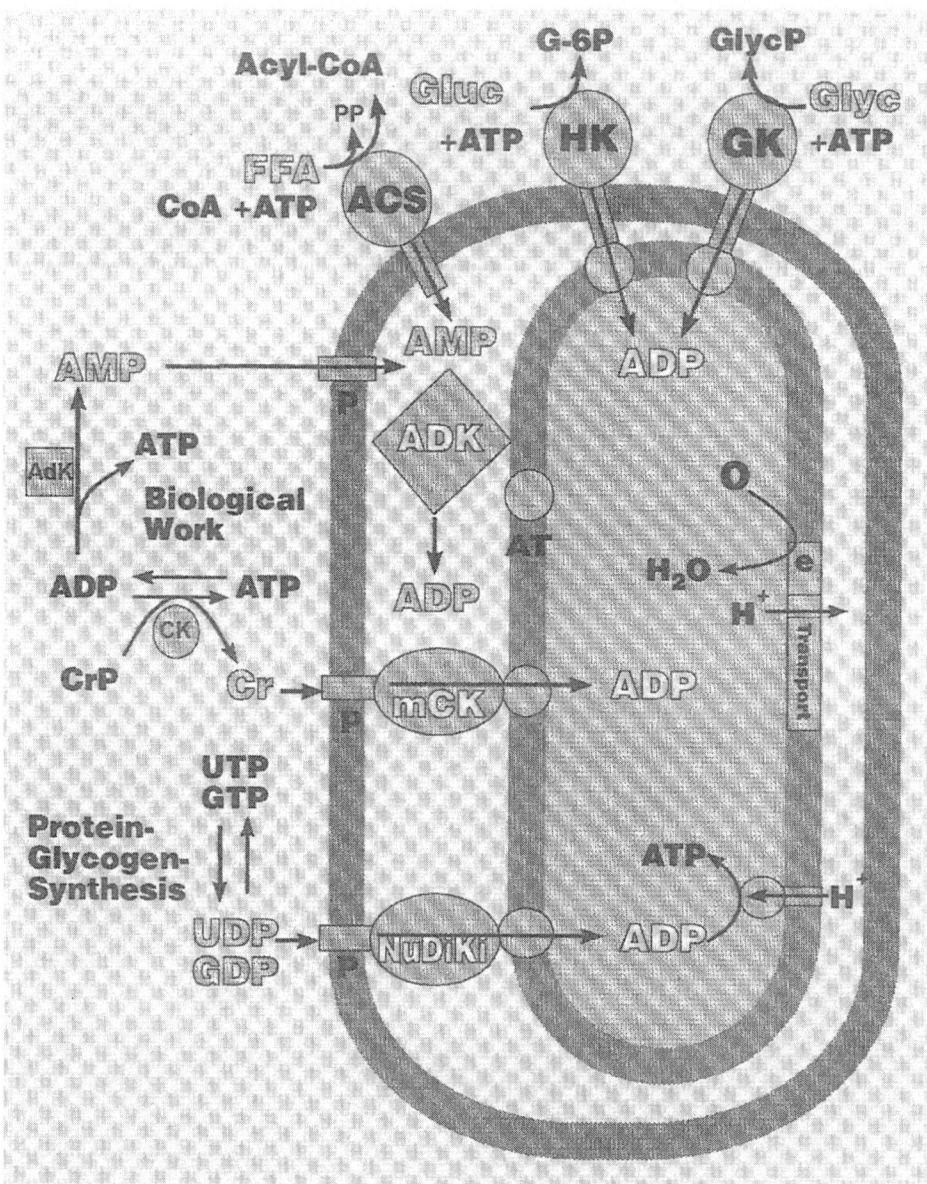

Fig. 1. Scheme showing the organisation and function of several kinases at the mitochondrial surface. On group of kinases named 'energy consuming kinases' is bound to the outer membrane pore protein: acyl-CoA synthase (ACS), hexokinase (HK), and glycerol kinase (GK). A second group named 'energy transferring' kinases is located between the two envelope membranes: adenylate kinase (ADK), mitochondrial creatine kinase (mCK), nucleoside-diphosphate kinase (NuDiKi). Some kinases (HK, GK, mCK, NuDiKi), which were enriched in the contact site fraction, are thought to form complexes with the outer membrane pore (P) and the adenine nucleotide translocator (AT). To keep the scheme clear, the reactions of several kinases are shown incompletely in abbreviated form.

as known for hexokinase, may accomplish an additional function in permanently working muscles like heart or in nerve cells which depend more or less on blood born substrates, it may regulate the cellular substrate (glucose) uptake according to the mitochondrial activity. In summary, the coupling of kinases to the inner compartment serves to transmit information in both directions either about increased ATP turnover and/or extramitochondrial substrate level directed towards the mito-

chondria or about mitochondrial activity directed towards the substrate supplying metabolic pathways.

Specific organisation of kinases at the mitochondrial surface

Two groups of kinases are found at the mitochondrial surface (Fig. 1). **Energy consuming kinases** such as acyl-

CoA-synthase (ACS), hexokinase (HK) and glycerolkinase (GK) of which the latter two specifically bind to the outer membrane pore protein [1, 2]. A second group, for example adenylate kinase (ADK) creatine kinase (CK) and nucleoside diphosphate kinase (NuDiKi), is located between the two envelope membranes. These kinases are called **energy transmitting kinases** because their products are energy rich and able to form ATP.

During kinetic analyses these kinases appeared to be functionally coupled to the inner mitochondrial ATP [1, 3–9], meaning that they interact directly with the adenine nucleotide translocator in the inner membrane. The structural basis of this coupling may be the contact sites between the two mitochondrial boundary membranes in which the kinases may form complexes with the adenylate translocator and the outer membrane pore [10]. During preparation and characterisation of contact sites it was learned that the structure of these complexes was effected by conventional isolation in sucrose media.

Conventional isolation of the mitochondria destroys the structure of the outer compartment

The function of the outer mitochondrial compartment in regulation of mitochondrial metabolism has so far not been regarded. The reason for this blind spot in our knowledge was that this compartment is severely disintegrated during the standard isolation of mitochondria. The commonly used media adjust the physiological osmotic pressure by sugars such as mannitol or sucrose. Because exclusively the inner membrane is semipermeable for the sugar molecules, these media preserve the structure and function of the inner membrane matrix compartment. The structure of the outer mitochondrial compartment, however, is strongly altered during isolation as the approximately 30% solution of cytosolic proteins which do not permeate the outer membrane is completely removed. The structural changes caused by isolation compared to the mitochondrial morphology *in situ* are known since a long time [11]. They were visible as swelling of the intercristae space and the space between the two envelope membranes. However, specific functions of the intact structure serving to estimate the role of the outer compartment were missing completely. We supposed that contact sites between the two boundary membranes represent reminiscent structures of the physiological organisation of the outer mitochondrial compartment in intact mitochondria [10]. A function of

contact sites in regulation of transport and exchange processes either of precursor proteins [12] or metabolites [13] has been shown, which emphasises the importance of the outer compartment in regulation of mitochondrial activity.

These considerations led us try to restore an intact outer compartment by substituting the physiologically present proteins with dextrans [14, 15]. Based on the experience with contact sites, we were able to define several structural and functional criteria to prove the intactness of the outer compartment: i) narrow space between cristae and the two boundary membranes, ii) increase in contact site frequency in freeze fractured mitochondria, iii) changes in binding affinity of proteins to the mitochondrial surface, iv) changes in kinetic properties of enzymes located in the outer compartment, v) reduction of exchange rate of adenine nucleotides.

The experiments were designed in two ways: the structure of the outer compartment was either restored in conventionally isolated mitochondria by subsequent addition of macromolecules or the structure was preserved by the presence of macromolecules during isolation. By the first method, it was possible to study the effect of gradual restoration of the outer compartment, while the other method served to analyse physiological binding of proteins to the surface of mitochondria.

Results and discussion

Effect of macromolecules on the structure of the outer compartment

Conventionally isolated liver mitochondria were subjected to rapid freezing and freeze fracture (Fig. 2). The presence of 10% dextran Mr 70 kDa led to a significant reduction of the space between the two envelope- and the crista membranes. The cristae visible in cross fractures of control mitochondria were swollen and exhibited the same, pears like, shape as in thin sections, whereas they had a long, thin structure in dextran treated mitochondria. The close attachment of the envelope membranes under dextran suggested changes in protein organisation in the outer compartment and rose the question whether this would also affect contact formation.

We previously developed a method in freeze fractured mitochondria to estimate the frequency of contact sites [16]. In mitochondrial freeze fractures the fracture plane often jumps between the two boundary mem-

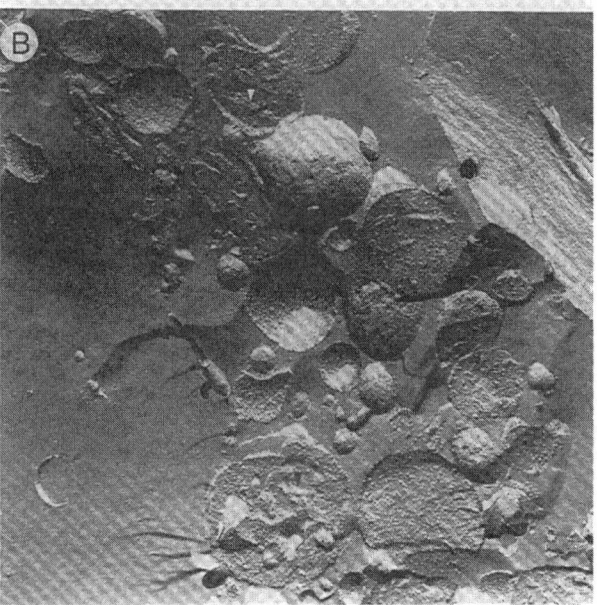

Fig. 2. Surveys of the isolated mitochondrial fraction from rat liver. (A) Mitochondria in the presence of 10% Dextran-70, (B) control mitochondria. The convex fractures faces show fracture plane deflections. Large, white arrowheads point to smooth patches of the exoplasmic face of the outer membrane laying on top of the particle rich protoplasmic face of the inner membrane. In contrast to the control, cross fractured mitochondria in the presence of Dextran-70 exhibit narrow crista like structures (small arrowheads) and no space between the two envelope membranes. Magnification 25,000. *Methods:* The mitochondrial samples were incubated in 113 mM sucrose, 10 mM K_2HPO_4 (pH 7.4), 5 mM $MgCl_2$, 12.5 mM succinate. The oxidation rate was stimulated by addition of 10 mM ADP (state 3). Mitochondria were directly subjected to rapid freezing by the sandwich technique as described by Knoll *et al.* [50]. The time to cryofixation was approximately 15 s. The samples were broken in a Balzers BAF 301 freeze etch device at – 100° C and (2–7) 10^{-7} Torr, followed by Pt/C and C shadowing. For electron microscopy, a Hitachi H7000 instrument at 75 KV was used. The nomenclature of the exposed membranes follows that of Branton *et al.* [51].

Table 1. Regulation of contact sites and bound hexokinase activity in hepatocytes and isolated liver mitochondria

Effector	Relative contact site frequency	Relative bound hexokinase activity	Ref.
State 4			
(O2, Pi, substrate present)	1	1	16, 17, 41
State 3			
(O2, Pi, substrate,			
ADP present)	3.85	4–5	16, 17, 41
ADP (Pi lacking)	3.85	4–5	41
Atractyloside	3.85	nd	41
10% Dextran 70 state 4	3.70	nd	14
10% Dextran 70 state 3	6.15	6	14
DNP (uncoupler)	0.15	0.2	16, 23
20% Glycerol	0.12	nd	16
Fatty acids	1.35	0.24	17
Glucagon (β-receptor)	2.23	0.64	18
Epinephrine (α1-receptor)	6.54	nd	18

As a means of quantifying the difference in fracture-plane deflections, the length (L) of the edge where the fracture plane deflects was measured as it related to the corresponding mitochondrial area. In convex fractures, the length of the edge of the exoplasmic face of the outer membrane was measured, whereas, in concave fractures, the deflection line on the exoplasmic face of the inner membrane was determined. These measurements of L were expressed as length (μm) per unit of mitochondrial fractured membrane area. In every population of mitochondria, there are some which are completely void of fracture plane deflections. The number being dependent upon the metabolic state of the whole sample. To compensate for these differences we adjusted our calculation of fracture plane deflections by first determining on survey pictures the total area of mitochondria with no deflections. M_s and those with deflections M_p, and then normalising these values by the expression $M_p(M_p + M_s)$. The final value for quantification of freeze fracture deflections L_p was then calculated from the equation: L_p $(\mu m/\mu m^2) = L\, M_p(M_s + M_p)$, and the statistical differences obtained by applying the U-Test. The measurements were made in the areas where the curvature was low to avoid large distortions of the measured edge lines. The data in the Table are contact site frquency and bound hexokinase activity relative to the values observed in mitochondria in state 4 (set to 1). The data were taken from the quoted publications.

branes. As a result one observes for example in convex fractures shown in Fig. 2A (large arrowheads) a frequent change between the smooth inner leaflet (exoplasmic face) of the outer membrane and the particle rich inner leaflet (protoplasmic face) of the inner membrane. The contact site analysis was based on the assumption that the contact sites were responsible for the jumping of the fracture plane and that the frequency of fracture plane deflections between the two boundary membranes would therefore correlate with that of the contact sites. By determination of fracture plane deflections in different mitochondrial states it was observed that phosphorylating mitochondria in state 3 exhibited a 3 to 4 times higher frequency of contact sites than mitochondria in state 4 (Table 1). In the presence of 10% dextran Mr 70 kDa the frequency of fracture plane deflections was already higher in state 4 and comparable to that of control mitochondria in state 3, while full activation of mitochondrial respiration in state 3 led to a further increase of contacts (Table 1). On the whole, electron microscopy showed that dextran reduced the space between boundary and crista membranes and induced contact site formation.

Composition of the contact sites

The observation that the frequency of contact sites was dependent on the functional state of mitochondria pointed to a dynamic nature of these structures. Indeed the contacts appeared to be subjected to hormonal and metabolite control (Table 1). While glucagon and free fatty acids decreased, epinephrine and insulin increased the frequency of contacts in hepatocytes [17, 18]. In all cases of different contact frequency we met a positive correlation between contacts and the amount of surface bound hexokinase (see column 3 Table 1 and ref. 17–19).

Contact site analysis by electron microscopy
The distribution of hexokinase was therefore analysed by means of specific antibodies at the electron microscopic level, it was observed in intact liver and brain mitochondria that hexokinase was preferentially located in attachment points between the two boundary membranes [20, 21] (schematically shown in Fig. 3). Because of this location hexokinase is not removed by disruption of the outer membrane using digitonin which leaves the contacts intact. When liver mitochondria were treated with hexokinase antibodies and subsequently incubated with 100 μg digitonin, the gold grains, representing hexokinase, were found inside the outer membrane vesicles which remained attached to the inner boundary membrane (Fig. 4A–C and schematically Fig. 3) while the surface of the inner envelope membrane was not immuno-reactive.

Contact site analysis by isolation
Based on the electron microscopic observation we judged hexokinase as a specific marker enzyme of contacts, while attempting to isolate the contact sites from osmotically disrupted mitochondria. By sucrose gra-

dient centrifugation of the membrane fragments we isolated from brain, kidney and liver mitochondria a fraction of intermediate density besides outer or inner membrane fractions. In this intermediate fraction hexokinase activity was concentrated, which led us to use hexokinase antibodies and precipitate the bound enzyme attempting to identify presumably associated components [22]. The precipitate derived from the contact site fraction of brain and kidney mitochondria contained activity of marker enzymes of outer- (mono-amine oxidase) and inner membrane (succinate dehydrogenase) and in addition to hexokinase, activity of creatine kinase and nucleoside-diphosphate kinase while adenylate kinase was absent [21, 22].

Electron microscopic localisation of mitochondrial creatine kinase
These results suggested that contacts may reflect the functional interaction of components in the outer mitochondrial compartment (schematically shown in Fig. 3). Based on kinetic analyses of surface bound kinases such as glycerol kinase [5–7] and hexokinase [3, 4, 6], surface proteolysis analysis [22, 23], electron microscopy [20, 21], and interaction with specific antibodies [22], a structural organisation of the functional complexes was postulated as depicted in Fig. 3. It was assumed that the peripheral kinases might interact with the adenylate translocator via the pore protein, while kinases in the intermembrane space such as creatine kinase might functionally couple porin and outer membrane with the adenylate translocator and inner membrane [13, 24]. The capacity of the octamer of creatine kinase to connect two artificial membranes supported this idea [25].

In order to localise mitochondrial creatine kinase by specific antibodies, it was necessary to overcome the outer membrane barrier. This was performed either by reaction of the antibodies with sections of Lowicryl embedded samples of chicken retina (post-embedding) or by pre-treatment of isolated rat brain mitochondria with digitonin (pre-embedding). It was known from previous investigations that about 60% of the mitochondrial creatine kinase as well as hexokinase remained fixed to the inner membrane matrix fraction after incubation with 100–300 μg digitonin per mg of mitochondrial protein [13].

The application of the two different techniques resulted in the observation of two fractions of mitochondrial creatine kinase. In sections of retina, decorated post embedding (Fig. 5), the enzyme was located at the periphery between the two envelope membranes and along the

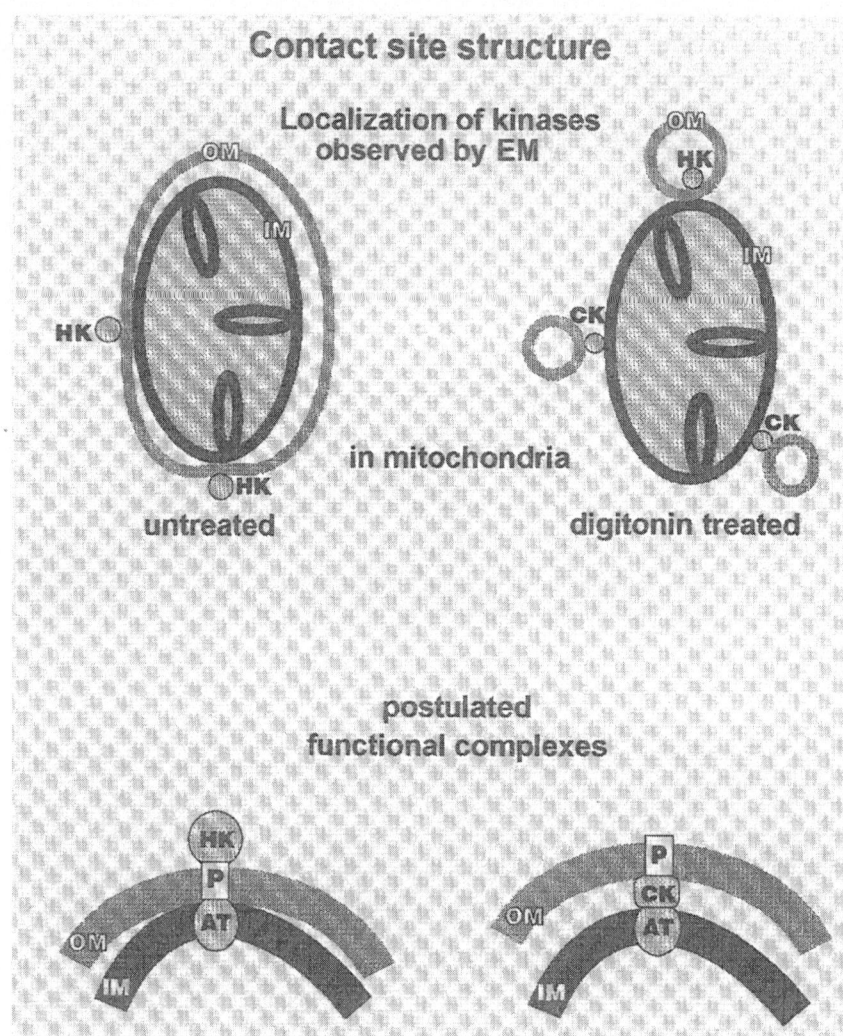

Fig. 3. The upper part of the figure schematically depicts the localisation of hexokinase in intact liver and brain mitochondria as observed by electron microscopy in Ref. 20 and 21 and the localisation of hexokinase in liver- and creatine kinase in brain mitochondria after digitonin treatment (shown in Fig. 4A–C and D). The lower part of the figure presents the postulated functional complexes derived from the electron microscopic and kinetic analyses. HK = hexokinase, CK = creatine kinase, AT = adenylate translocator, P = porin, OM = outer membrane, IM = inner membrane.

mitochondrial cristae, while pre-embedding staining of digitonin treated mitochondria led to an exclusive peripheral immuno reactivity (Fig. 4D). Closer analysis [58] of the distribution of the peripheral creatine kinase resulted in the observation that 43% was found in the contact points where after digitonin treatment remnants of the outer membrane remained attached (Fig. 4D large arrowheads). Since small fragments of the outer membrane, not capable of forming vesicles (Fig. 4D small arrowheads), may not have been identified as contacts, this relatively high amount of gold grains in identified attachment points means that presumably most of the peripheral mitochondrial creatine was organised in the contact sites.

Despite the exclusive peripheral localisation of mito-chondrial creatine kinase found by Biermans et al. [26] and by contact site isolation [13, 21, 22], we concluded on the basis of our results with post embedding staining (Fig. 5), and in agreement with Wegman et al. [27] and Hemmer et al. [28], that the enzyme has a second intra-mitochondrial location along the surface of the crista membranes in brain mitochondria. The mainly peripheral localisation of mitochondrial creatine kinase in the case of pre-embedding labelling can easily be explained by the fact that the enzyme in the intracristae space was not accessible to antibodies although the mitochondria were treated with digitonin (Fig. 4D). As described earlier, we were able to separate from disrupted brain mitochondria a contact site fraction (containing peripheral creatine kinase) and an inner memrane-matrix fraction

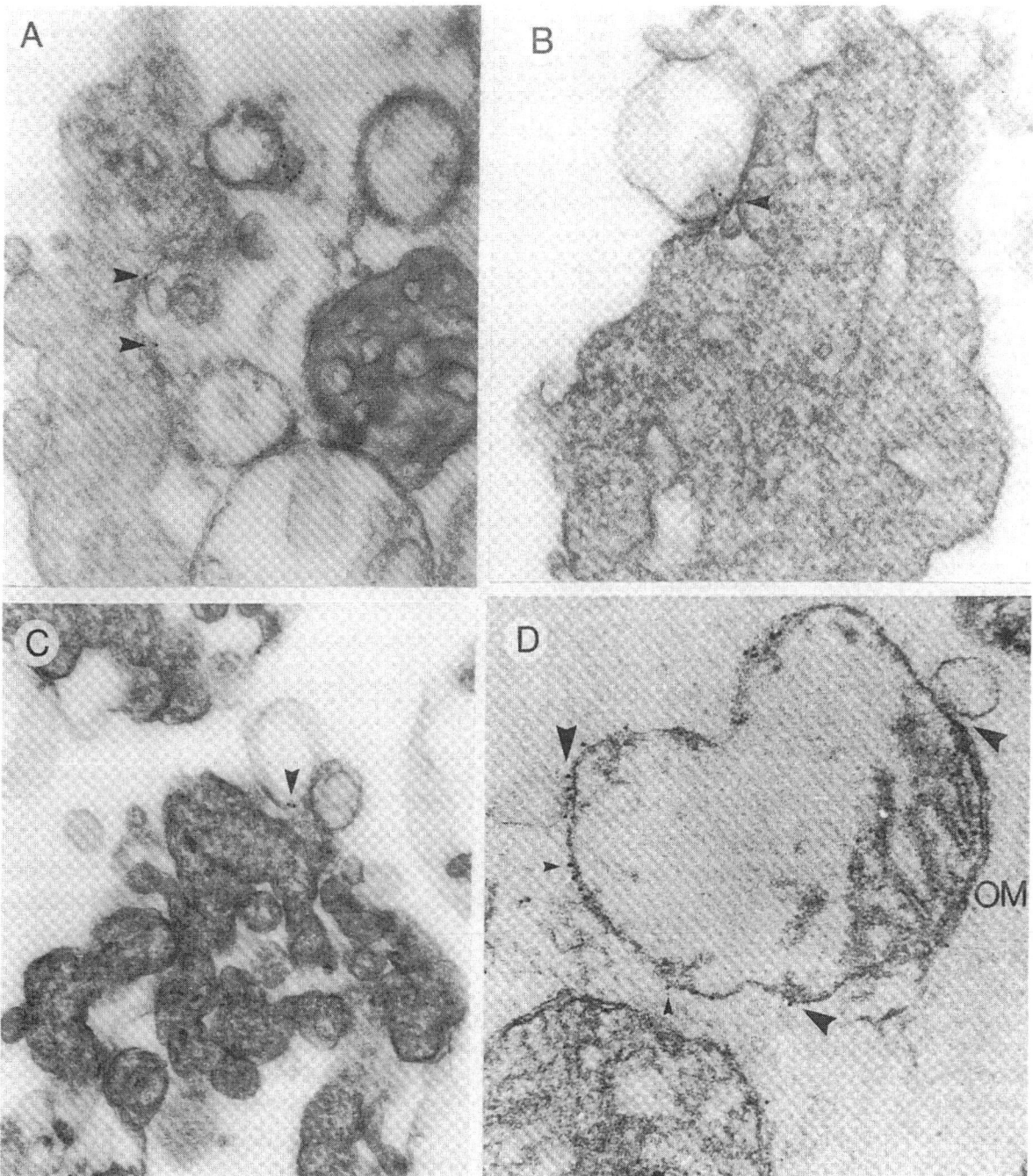

Fig 4 Preembedding labelling of hexokinase and mitochondrial creatine kinase in isolated digitonin treated liver and brain mitochondria A B C) Intact liver mitochondria labelled with hexokinase antibodies and protein A gold and subsequently treated with 100 μg digitonin/mg of mitochondrial protein The gold grains left at the surface of the inner membrane were exclusively found in vesicles of the outer membrane which remained bound to the inner envelope at the contact sites (arrowheads) No label was attached to the free surface of the inner membrane Magnification 70 000 D) Brain mitochondria treated with 200 μg/mg digitonin and subsequently incubated with antibodies against mitochondrial creatine kinase and protein A gold Parts of the outer membrane remained attached to the inner membrane either forming vesicles (large arrowheads) or small patches (small arrowheads) It is striking that creatine kinase is often located (large arrowheads) between outer membrane vesicles and the inner membrane In general only parts of the free accessible inner membrane are labelled whereas the outer membrane (OM) is free of gold grains Magnification 60 000 *methods* Mitochondria from rat liver were incubated intact for 10 min at room temperature with polyclonal antiserum against brain hexokinase whereas brain mitochondria were treated with 200 μg digitonin per mitochondrial protein and than incubated with antibodies against brain mitochondrial creatine kinase The antibodies were diluted in blocking solution (0 2% gelatine and 0 5% bovine serum albumin in sucrose Hepes medium) Samples were sedimented by 45 sec centrifugation in a tabletop centrifuge followed by a washing step in sucrose Hepes medium Mitochondria were then incubated for 10 min at room temperature with protein A gold diluted in blocking solution and were again centrifuged and washed in sucrose Hepes medium Liver mitochondria were subsequently treated with 100 μg digitonin per mg The mitochondrial sediments of liver and brain were fixed for 45 min on ice with 2 5% glutaraldehyde in 0 1 M cacodylate buffer pH 7 4 washed several times in 0 1 M cacodylate 0 15 M sucrose and post fixed for 2 hours with 1% OsO_4 in 0 1 M cacodylate 0 11 M sucrose After washing in 0 1 M cacodylate with 0 15 M sucrose the samples were subjected to stepwise dehydration with acetone and embedded in Spurr s low viscosity resin

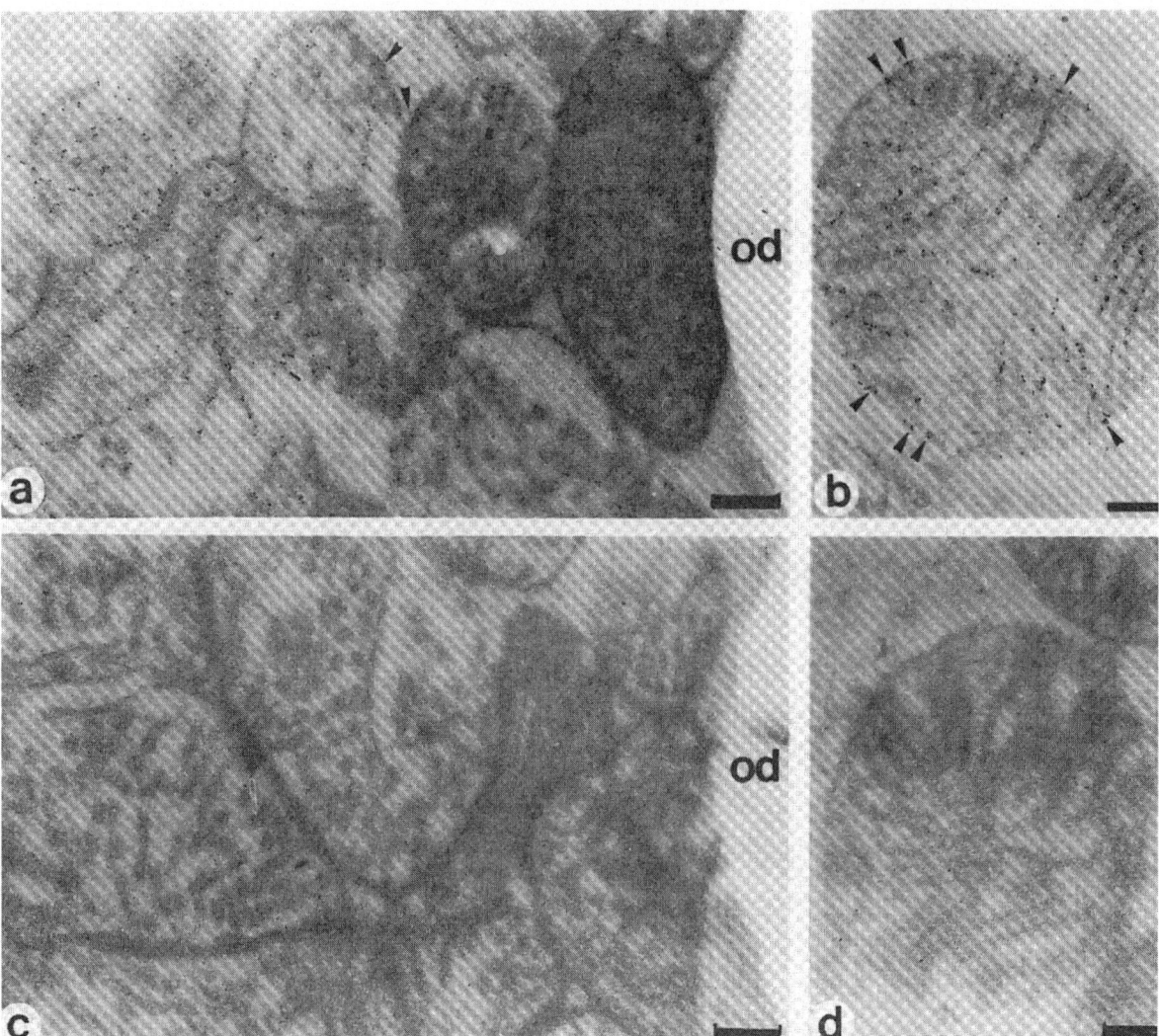

Fig. 5. Post-embedding immuno-gold labelling of mitochondrial creatine kinase in chicken retina photoreceptor cells. a) shows at lower magnification parts of the inner segment at the level of the ellipsoid portion, where mitochondria are clustered. b) single mitochondrion in the same region at higher magnification. Note the dual localisation of mitochondrial creatine kinase along the entire cristae membranes as well as at the periphery, where inner and outer membrane are in close proximity (see arrow-heads in a) and b)). c) and d) depict control sections of the same tissue region as above stained with pre-immune serum and the same second antibody as used in a) and b). There is no significant immuno-gold labelling. od = oil droplets of chicken photo receptor cells. Note the dual localisation of mitochondrial creatine kinase at the periphery and along the cristae. Bar = 250 nm. *Methods:* Chicken eye cups were prefixed with 3% paraformaldehyde/0.1% glutaraldehyde in PBS for 30 min at 0° C. Retinas were removed and fixed for an additional 2 h in the same fixative. Dehydration and embedding in Lowicryl K4M were carried out according to [52] in a low-temperature embedding apparatus at − 35° C. Infiltration with liquid Lowicryl K4M resin was performed at − 35° C as described in [53]. 70 nm sections were cut and immuno labelled according to [27]. An affinity-purified rabbit anti-chicken antibody against mitochondrial creatine kinase was used followed by goat anti-rabbit IgG coupled to 5 nm colloidal gold particles. Sections were stained with 2% aqueous uranyl acetate for 5 min followed by lead citrate for 1 min and examined in a JEOL 100 electron microscope.

(containing central creatine kinase). However we determined creatine kinase activity almost exclusively in the contact site fraction, suggesting that, like the antibodies, also access of substrates to the enzyme in the inter-cristae space was restricted [21, 22].

In view of these differences in accessibility between the peripheral and central creatine kinase activity one might assign a specific structural and regulatory func-

tion to the peripheral mitochondrial creatine kinase in situations of short and rapid energy requirements, while the activity of mitochondrial creatine kinase located along the cristae membranes may be recruited in the case of prolonged and large energy requirement. In addition, it is tempting to speculate that the relative proportion of cristae versus contact site creatine kinase may change in a dynamic way depending on the actual requirements in energy flux [26, 29].

Effect of macromolecules on the binding of hexokinase

It was observed that hexokinase had a higher affinity to isolated contact sites than to outer membrane [23, 30] which agreed with the electron microscopic localisation of hexokinase in zones of boundary membrane contacts, and the correlation of bound hexokinase activity with the frequency of contacts summarised in Table 1. In general these findings indicated on one hand a specific regulatory role of the contact sites in complex formation with hexokinase and on the other hand suggested that hexokinase binding could be used to look into changes in contact site frequency. We, therefore, used hexokinase binding to follow the effect of macromolecules on the reconstitution of contact sites and the structure of the outer mitochondrial compartment, respectively [14].

It was found that the half maximal saturation by hexokinase of the surface of liver mitochondria was significantly lower in the presence of 10% dextran with a Mr of 70 kDa than compared to the control. In general the binding of hexokinase to contact sites appeared to be a co-operative process and led to a 5 to 10 fold activation [14, 20, 31] of the enzyme. When, however, contacts were suppressed by addition of 20% glycerol or 50 µM dinitrophenol we observed almost the same binding as to isolated outer membrane, which was characterised by a hyperbolic saturation curve and no activation of the enzyme. The results indicated two aspects: i) that additional contacts may have been restored in the presence of dextran and ii) that the function of contacts in case of hexokinase was to increase the binding affinity to the membrane surface and to activate the enzyme.

Structure of the hexokinase porin complex
Supposing that hexokinase bound to the pore protein in the contact sites, the cooperativity of the binding in the contact sites and the activation of the enzyme suggested the formation of oligomers in the complex. In support of this assumption Xie and Wilson, by cross-linking experiments, observed that hexokinase I formed tetramers upon binding to mitochondrial membranes [32]. We therefore screened for conditions of hexokinase porin interactions which caused activation of the enzyme.

Hexokinase I was isolated according to Wilson [33] and incubated with porin prepared from liver mitochondria in lauryl dimethyl amino oxide (LDAO) as described by DePinto [34]. The enzyme was activated 5–10 fold under conditions such as learned from binding of hexokinase to intact mitochondria: pH-(6.5), dependence on Mg^{2+} (10 mM). The activation of the enzyme by porin was inhibited by glucose-6-P (5 mM). Hexokinase was not activated (or did not interact?) with porin prepared in Triton X-100. This suggested that the porin N-terminus must be available for interaction with hexokinase, which agreed with the fact that antibodies against the porin N-terminus suppressed the activation of hexokinase by porin in LDAO (Wicker *et al.* unpublished). The N-terminus was found to be susceptible to proteases in porin prepared with LDAO in contrast to preparations of porin in Triton X-100 [35]. In intact mitochondria the porin N-terminus was exposed at the outer surface of mitochondria [35]. Analysis of the hexokinase/porin mixture by column chromatography [36] revealed activity of porin and hexokinase in a protein fraction of 480 kDa, suggesting the formation of a complex composed of a hexokinase tetramer and presumably porin dimer. A dimer of free hexokinase (170 kDa) and hexokinase fragments (76 kDa) as well as porin (30 kDa) were additionally eluted from the column. As indicated by SDS-gel electrophoresis and immuno blots with antibodies against hexokinase I and the N-terminal end of porin [these were a gift of Dr. Thinnes, Göttingen] porin and hexokinase were present in the 480 kDa fraction from the column. We were able to reconstitute the pore in the complex with hexokinase in artificial membranes, and to measure the voltage dependence of the conductance [36]. It was observed that the pore was less voltage sensitive, suggesting that the interaction with hexokinase might physiologically allow better permeation of adenine nucleotides (see below).

Effect of macromolecules on the attachment of mitochondrial surface components
The observation of a higher hexokinase affinity to contact sites suggested that more hexokinase might remain bound when the structure of the outer compartment would be preserved by the presence of dextran during isolation. We thus isolated liver mitochondria in a medium containing 140 mM KCl, 10 mM Hepes pH 7.4 and 15% dextran 20 and for comparison in 250 mM sucrose/10 mM Hepes. The specific activity of hexokinase in the mitochondrial fraction was 2.5 times higher compared to the control mitochondria (Table 2). Supposing that contacts were responsible for binding of hexokinase with higher affinity, these results suggested that isolation of mitochondria in the presence of macromolecules

might preserve contacts sites, and thus structures of the outer mitochondrial compartment. In general these results show that substitution of macromolecules during mitochondrial isolation not only preserves the outer mitochondrial compartment but in addition may have important consequences for the attachment of all kinds of regulatory proteins to the mitochondrial surface.

Function of the hexokinase porin complex in energy metabolism

The induction of contact sites by dextran and the influence on activation of hexokinase rose the question of the function of the hexokinase porin complexes. Besides of the activation of the enzyme we assumed a function in regulation of energy metabolism. Bessman and co-workers have produced evidence that the mitochondrial hexokinase [3, 4] preferentially utilises intramitochondrially generated ATP. They have also shown that this preference decreased with increasing total ATP concentrations. The same was observed in reticulocyte [37] and brain [38, 39] mitochondria. These experiments were performed in the absence of dextran. Addition of dextran to brain mitochondria, however, resulted in changes of the kinetic properties of the originally bound enzyme, which pointed to a functional coupling with the AdN-translocator [14, Fig. 6]. In agreement with Bessman and co-workers [3, 4] we noted a significantly lower Km for ADP (ATP) when the enzyme was utilising internal ATP as compared to external ATP (Fig. 6B). As observed by others [38, 39], the Vmax with internal ATP was about half of that with external ATP (compare Fig. 6A with 6B). The addition of dextran had no significant effect on these parameters, suggesting that contact site formation did not improve the supply of hexokinase

with internal ATP [14, Fig. 6B]. On the other hand, when we studied the stimulation of the oxidative phosphorylation by the ADP formed via the activity of originally bound hexokinase, we noticed a significant decrease in Km and a duplication of the catalytic efficiency (Vmax/Km) in the presence of dextran (Fig. 6C). The Vmax of hexokinase in this experiment (calculated from the O2 consumption and the ADP/O ratio) was almost that determined with external ATP (Fig. 6A) but the Km was decreased by dextran to the values observed with internal ATP (Fig. 6B). Thus the induction of contacts by dextran appeared to improve the saturation of the adenine nucleotide translocator with ADP (in the presence of increasing concentrations of external ATP) because of facilitated communication with hexokinase. On the whole the experiments emphasise the importance of contacts in the stimulation of the oxidative phosphorylation via peripheral kinases.

Based on their results, Bessman and co-workers [3, 4, 40], about 25 years ago, postulated a direct coupling between the adenylate translocator and hexokinase. It was subsequently shown that hexokinase specifically binds to the outer membrane pore [1, 2]. However, direct evidence of coupling between porin and the adenylate translocator has not yet been provided. We found that the relation of the translocator to the crista membrane marker cytochrome oxidase was 3 times higher in the isolated contact site fraction from liver mitochondria compared to the corresponding inner membrane, this was, however, not observed in subfractions of kidney mitochondria [41]. Evidence for a direct interaction between the translocator and porin came recently from the characterisation of the isolated mitochondrial benzodiazepine receptor from kidney mitochondria [42]. In these

Table 2. Effect of dextran addition during isolation of rat liver mitochondria on the activity of mitochondrial bound hexokinase

	HK mU/mg		G-6-Pase mU/mg		SDH U/mg		MAO mU/mg	
	Homog	Mitoch	Homog	Mitoch	Homog	Mitoch	Homog	Mitoch
Dextran	4.09	9.83	1.02	0.07	0.09	0.24	1.14	3.33
	± 2.07	± 5.01	± 0.39	± 0.03	± 0.05	± 0.13	± 0.55	± 1.21
Control	3.38	3.83	1.24	0.08	0.12	0.31	0.90	3.74
	± 1.65	± 1.65	± 0.06	± 0.01	± 0.07	± 0.12	± 0.34	± 1.66

The specific activity of hexokinase is compared to that of marker enzymes for endoplasmic reticulum ER (G-6-Pase), inner- (SDH) and outer mitochondrial membrane (MAO) in the homogenate (= Homog) and mitochondrial fraction (= Mitoch). While the specific activity increase of marker enzymes of inner and outer membrane in the mitochondrial fraction was comparable (2–2.5 fold) in control and dextran containing medium, the specific hexokinase activity in the mitochondrial fraction was 2.5 times higher in the presence of dextran compared to the control. As judged from the G-P-ase activity, the contamination by ER was comparable in both media. *Methods:* Mitochondria were isolated from a Teflon potter homogenate by fractional centrifugation in either 0.25 M sucrose, 10 mM Hepes pH 7.4 (= Control) or 140 mM KCl, 15% dextran 20 and 10 mM Hepes pH 7.4 (= Dextran). Mean of 8 experiments.

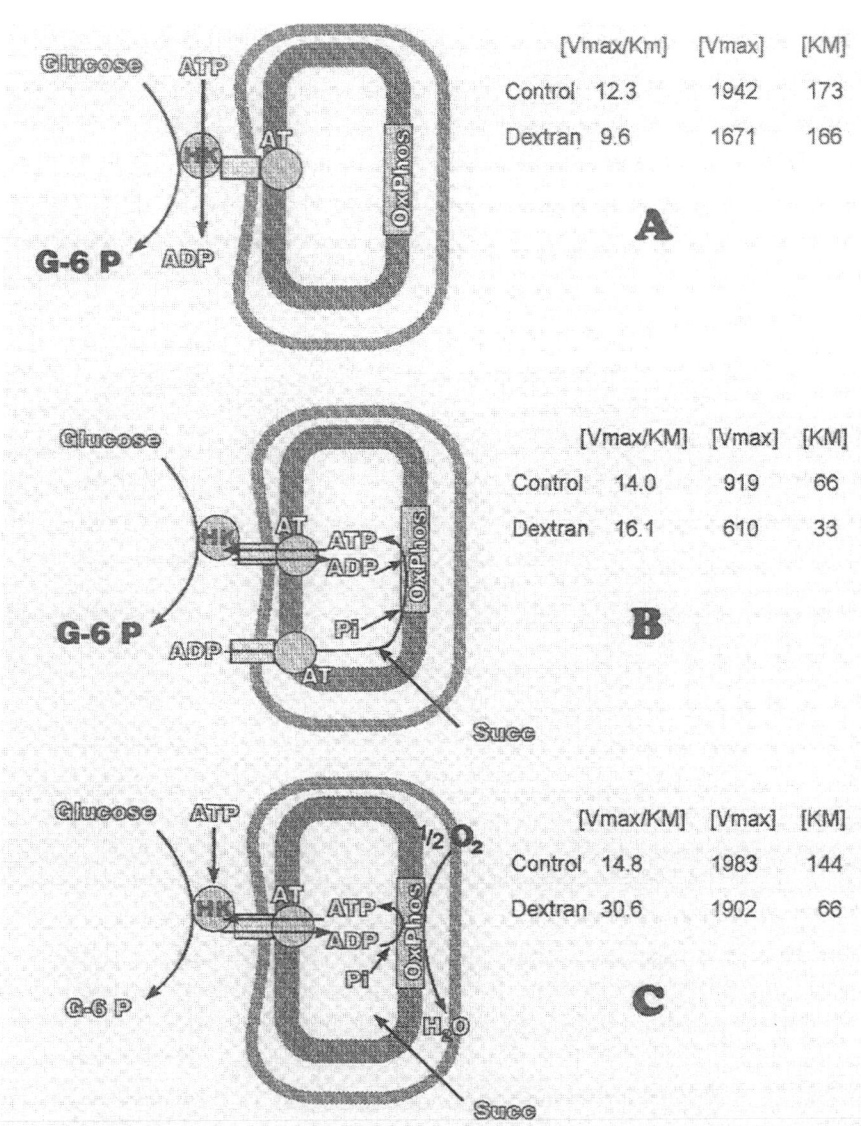

Fig. 6. Kinetic analysis of bound hexokinase in isolated mitochondria from rat brain. The experiments show the kinetic advantage of a functional coupling between hexokinase and adenylate translocator in the contact sites. When contacts were induced by dextran in experiment C, the velocity of hexokinase was as high as in A (with external ATP) while the Km was comparable to that observed in B with ATP produced by oxidative phosphorylation. *Methods:* The activity of originally bound hexokinase was analysed in 0.25 M sucrose/10 mM Hepes medium in the presence 2 mM glucose and 10% dextran 70 as indicated. The velocity of the activity was determined by registration of glucose-6-phosphate (G-6 P) formation by optical test [54] in experiment A and B, while the stimulation of state 3 respiration ($1/2 O_2$) was measured by oxygen electrode in experiment C. In experiment B ATP was formed via activity of oxidative phosphorylation in the presence of 5 mM succinate (Succ) and 5 mM phosphate (Pi) and 2 mM di-adenosine 5′-pentaphosphate. In experiment C the ADP to stimulate the oxidative phosphorylation was formed by the activity of hexokinase in the presence of 5 mM succinate and 5 mM phosphate. Hexokinase activity was calculated from the ADP/O ratio and the oxidation rate per min and ml mitochondrial suspension. The kinetic data shown in the figure were taken from [14] and are mean of 3 experiments.

experiments a protein complex of Mr 72 kDa was purified. On SDS-PAGE the mitochondrial benzodiazepine receptor appeared to be composed of three polypeptides which were identified as porin and adenylate translocator and a protein of Mr 18 kDa with so far unknown function which bound the benzodiazepine and was purified and sequenced by Krueger *et al.* [43].

Direct effects of macromolecules on the outer membrane pore

The compartmentation of adenine nucleotides in the mitochondrial inter-membrane space has been observed by several authors but the reason for this was still unclear. From radioisotopic measurements Viitanen *et al.*

concluded that the outer membrane represents a diffusion barrier for creatine kinase since compartmentation effects were reduced by removing the outer membrane [44], whereas we found in digitonin treated mitochondria that the enzyme was still not free accessible to negatively charged substances [13]. Brooks *et al.* came to the same conclusion [45]. These findings pointed to a somehow restricted AdN exchange through the outer membrane pore. The pore protein in the outer membrane is known to be voltage dependent [46]. The conductance of the pore, reconstituted in artificial membranes, decreases to about 50% at a voltage above 30 mV. The pore in this low conductance state was found to be cation selective [47]. It was assumed that the inner membrane potential might affect the pore in the contact sites, where a capacitive coupling between the two envelope membranes appeared possible [47]. However, Zimmerberg and Parsegian [48] observed that the presence of macromolecules increased the voltage sensitivity of the reconstituted pore. In agreement with these results, we demonstrated recently that addition of 10% of different dextrans of Mr between 20 kDa and 500 kDa led to pores in the low conductance state already at 10 mV transmembrane potential [15]. In view of these results, we assume that physiologically, besides the effect of cytosolic proteins on the structural organisation of the outer compartment, also a direct effect might occur on the permeability of AdN through all outer membrane pores.

Effects of macromolecules on the exchange of adenine nucleotides in isolated mitochondria
The functional consequences of this limitation were recently demonstrated by the observation that addition of 10% dextran to conventionally isolated liver mitochondria affected the export of ADP as well as the import into the inter-membrane space as is described in [15]. The ADP formed in the inter-membrane space by adenylate kinase was not readily accessible to extramitochondrial pyruvate kinase, and vice versa the adenylate kinase originally located in the inter-membrane space was less accessible to external ADP. The reduction of ADP diffusion correlated with the concentration of added dextrans of different Mr [15].

Conclusions

The important role of the outer mitochondrial compartment in regulation of the mitochondrial energy metabolism is based on the following aspects: i) the outer membrane pore is physiologically not freely permeable for adenine nucleotides so that the space between the two envelope membranes must be regarded as a separate compartment, ii) the outer compartment contains a number of different molecules located at the surface of the membranes, integrated into the membranes and between the membranes which form transient, functional complexes in the presence of macromolecules. Examples of the dynamics of these complexes are the specific binding of hexokinase to the contact sites where the enzyme associates as a tetramer [36] and the specific formation of creatine kinase octamers, consisting of 4 dimers as active building units, in these sites [13]. The functional advantage of these complexes presumably composed of the kinases, porin, and adenylate translocator would be molecular channelling of ADP into the inner compartment in order to keep a high phosphorylation potential in the cytosol also while ATP turnover is high. In this way, the mitochondrial energy metabolism can be activated (without general ADP increase) during rapid ATP turnover in muscle and brain by an interplay between cytosolic and mitochondrial creatine kinases supplying the intramitochondrial ADP and by bound hexokinase providing mitochondrial substrates via glucose phosphorylation (Fig. 7). On the other hand, also at rest, the activation of the mitochondrial metabolism via the kinases would be advantageous. Because ion pumps such as Ca^{2+} ATPase in the endoplasmic or sarcoplasmic reticulum depend very critical on the available free energy [49], large fluctuations of the phosphorylation potential must be avoided. Thus, if increased blood glucose levels, followed by high activity of glycogen synthesis in liver and resting muscle, demand activation of oxidative phosphorylation, this may be performed via bound hexokinase and nucleoside-diphosphate kinase (Fig. 7) which channel the ADP into the inner compartment avoiding aequilibration with the cytosolic AdN levels.

Correlated action of Ca^{2+} and ADP supply in regulation of energy metabolism

It has been proposed that Ca^{2+} ions serve as a signal to regulate oxidative phosphorylation on one hand by direct activation of mitochondrial ATP synthase [55] and on the other hand by activation of substrate supplying enzyme reactions (pyruvate-dehydrogenase, NAD-isocitrate dehydrogenase and 2-oxoglutarate dehydrogenase [56, 57]). This does not lower the importance of ADP supply by kinases discussed above in regulation of

Regulation of Mitochondrial Metabolism

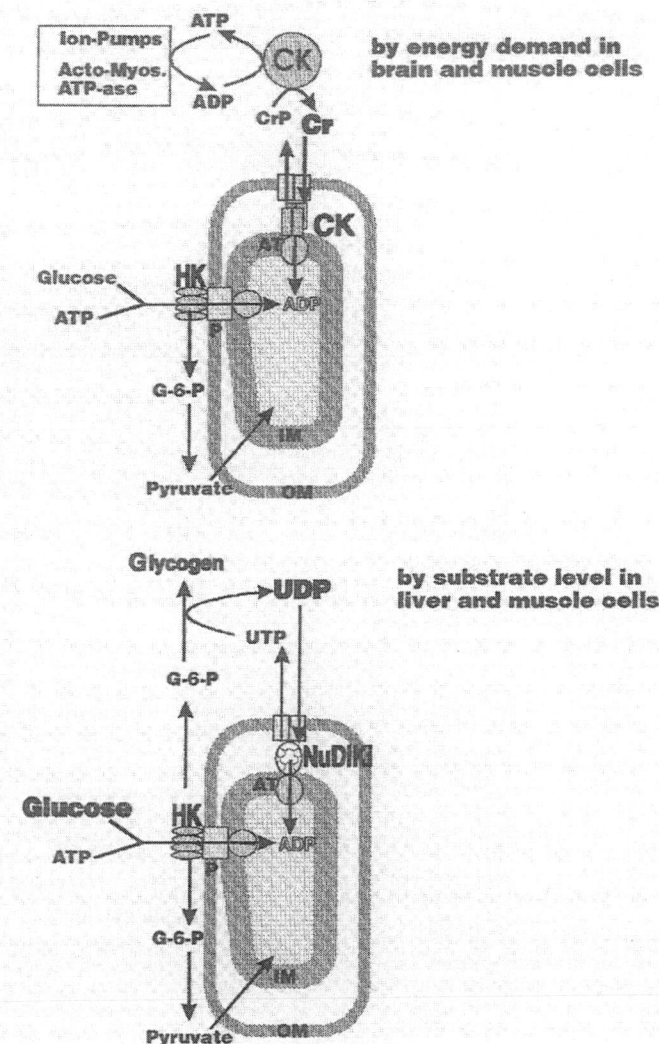

Fig. 7. Schematic representation of different ways to regulate the oxidative phosphorylation without changing cytosolic ATP and ADP levels. The intramitochondrial ADP is indirectly increased by the activity of peripheral kinases. A) in case of rapid ATP turnover upon stimulation of muscle or nerve cell activity, the level of creatine (Cr) increases and activates mitochondrial creatine kinase (CK) to produce intramitochondrial ADP. In parallel glucose uptake is increased by activation of hexokinase (HK). B) in case of liver cells or muscle cells at rest, an increased level of glucose and of the UDP, produced during glycogen synthesis, rises intramitochondrial ADP via bound hexokinase (HK) and nucleoside-diphosphate kinase (NuDiKi). In the contact sites, hexokinase and mitochondrial creatine kinase (which is a dimer in the active state) form tetramers and octamers, respectively.

mitochondrial activity but in contrast emphasises the function of the ADP producing reactions as a prerequisite of Ca^{2+} dependent activation. In view of this correlated action, it is worthwhile to mention that a Ca^{2+} binding protein was found to be concentrated in the contact sites of brain mitochondria [21] suggesting that Ca^{2+} might induce the formation ADP producing kinase complexes.

Acknowledgements

We are grateful to Dr. I. Riesinger and Dr. G. Wegmann for providing the EM figures. Furthermore, we thank Dr. F. Thinnes for the antibodies against the N-terminus of porin and Ph. Kaldis for the antibodies against mitochondrial creatine kinase. This work was supported by a postdoctoral grant of the Swiss Federal Institute of

Technology given to I.R., grants of the Deutsche Forschungsgemeinschaft (D.B.: Br773/3-2), the Swiss Foundation for Muscle Diseases (T.W.), and the Swiss National Science Foundation (T.W.: SNF 31-33907-92).

References

1. Fiek Ch, Benz R, Roos N, Brdiczka D: Evidence for identity between the hexokinase-binding protein and the mitochondrial porin in the outer membrane of rat liver mitochondria. Biochim Biophys Acta 688: 429–440, 1982

2. Lindén M, Gellerfors P, Nelson BD: Pore protein and the hexokinase-binding protein from the outer membrane of rat liver mitochondria and identical. FEBS Lett 141: 189–192, 1982

3. Gots RE, Gorin FA, Bessman SP: Kinetic enhancement of bound hexokinase activity by mitochondrial respiration. Biochem Biophys Res Commun 49: 1249–1255, 1972

4. Gots RE, Bessman SP: The functional compartmentation of mitochondrial hexokinase. Arch Biochem Biophys 163: 7–14, 1974

5. McCabe ERB: Human glycerol kinase deficiency: an inborn error of compartment metabolism. Biochem Med 30: 215–230, 1983

6. Östlund AK, Göhring U, Krause J, Brdiczka D: The binding of glycerol kinase to the outer membrane of rat liver mitochondria: Its importance in metabolic regulation. Biochem Med 30: 231–245, 1983

7. Kaneko M, Kurokawa M, Ishibashi S: Binding and function of mitochondrial glycerol kinase in comparison with those of mitochondrial hexokinase. Arch Biochem Biophys 237: 135–141, 1985

8. Bessman SP, Fonyo A: The possible role of the mitochondrial bound creatine kinase in regulation of mitochondrial metabolism. Biochem Biophys Res Commun 22: 597, 1966

9. Jacobus WE, Lehninger AL: Creatine kinase of rat heart mitochondria. Coupling of creatine phosphorylation to electron transport. J Biol Chem 248: 4803–4810, 1973

10. Brdiczka D: Review: Contact sites between mitochondrial envelope membranes. Structure and function in energy- and protein transfer. Biochim Biophys Acta 1071: 291–312, 1991

11. Malhotra SK, Van Harreveld A: Some structural features of mitochondria in tissues prepared by freeze-substitution. J Ultrastruct Res 12: 473–487, 1965

12. Rassow J, Guiard B, Wienhues U, Herzog V, Hartl F-U, Neupert W: Translocation arrest by reversible folding of a precursor protein imported into mitochondria. A means to quantitate translocation contact sites. J Cell Biol 109: 1421–1428, 1989

13. Kottke M, Adams V, Wallimann T, Nalam VK, Brdiczka D: Location and regulation of octameric mitochondrial creatine kinase in the contact sites. Biochim Biophys Acta 1061: 215–225, 1991

14. Wicker U, Bücheler K, Gellerich FN, Wagner M, Kapischke M, Brdiczka D: Effect of macromolecules on the structure of the mitochondrial inter-membrane space and the regulation of hexokinase. Biochim Biophys Acta 1142: 228–239, 1993

15. Gellerich FN, Wagner M, Kapischke M, Wicker U, Brdiczka D: Effect of macromolecules on the regulation of the mitochondrial outer membrane pore and the activity of adenylate kinase in the inter-membrane space. Biochim Biophys Acta 1142: 217–227, 1993

16. Knoll G, Brdiczka D: Changes in freeze-fracture mitochondrial membranes correlated to their energetic state. Biochim Biophys Acta 733: 102–110, 1983

17. Klug G, Krause J, Östlund AK, Knoll G, Brdiczka D: Alteration in liver mitochondrial function as a result of fasting and exhaustive exercise. Biochim Biophys Acta 764: 272–282, 1984

18. Brdiczka D, Riesinger I, Adams V, Bremm G: Function in metabolic regulation of contact sites between the mitochondrial boundary membranes 4th EBEC Short-Reports, p 200, Cambridge University Press, Cambridge 1986

19. Wojtczak L, Adams V, Brdiczka D: Effect of oleate on the apparent Km of monoamine oxidase and the amount of membrane-bound hexokinase in isolated rat hepatocytes: Further evidence for the controlling role of the surface charge in hexokinase binding. Mol Cell Biochem 79: 25–30, 1988

20. Weiler U, Riesinger I, Knoll G, Brdiczka D: The regulation of mitochondrial-bound hexokinases in the liver. Biochem Medicine 33: 223–235, 1985

21. Kottke M, Adams V, Riesinger I, Bremm G, Bosch W, Brdiczka D, Sandri G, Panfili E: Mitochondrial boundary membrane contact sites in brain: Points of hexokinase and creatine kinase location and of control of Ca^{2+} transport. Biochim Biophys Acta 395: 807–832, 1988

22. Adams V, Bosch W, Schlegel J, Wallimann T, Brdiczka D: Further characterization of contact sites from mitochondria of different tissues: topology of peripheral kinases. Biochim Biophys Acta 981: 213–225, 1989

23. Ohlendieck K, Riesinger I, Adams V, Krause J, Brdiczka D: Enrichment and biochemical characterization of boundary membrane contact sites in rat-liver mitochondria. Biochim Biophys Acta 860: 672–689, 1986

24. Schnyder T, Engel A, Lustig A, Wallimann T: Native mitochondrial creatine kinase forms octameric structures. II. Characterization of dimers and octamers by ultracentrifugation, direct mass measurements by scanning transmission electron microscopy, and image analysis of single mitochondrial creatine kinase octamers. J Biol Chem 263: 16954–16962, 1988

25. Rojo M, Hovius R, Demel RA, Nicolay K, Wallimann T: Mitochondrial creatine kinase mediates contact formation between mitochondrial membranes. J Biol Chem 266: 20290–20295, 1991

26. Biermans W, Bernaert I, De Blie M, Nijs B, Jacob W: Ultrastructural localization of creatine kinase activity in the contact sites between inner and outer mitochondrial membranes in rat myocardium. Biochim Biophys Acta 974: 74–80, 1989

27. Wegmann G, Huber R, Zanolla E, Eppenberger HM, Wallimann T: Differential expression and localization of brain-type and mitochondrial creatine kinase isoenzymes during development of the chicken retina: Mi-CK as a marker for differentiation of photoreceptor cells. Differentiation 46: 77–87, 1991

28. Hemmer W, Riesinger I, Wallimann T, Eppenberger HM, Quest AFG: Brain-type creatine kinase in photoreceptor cells outer segments: role of a phospho creatine circuit in outer segment energy metabolism and phototransduction. Exp Cell Res 1993 (in press)

29. Wyss M, Smeitink J, Wevers RA, Wallimann T: Mitochondrial creatine kinase: a key enzyme of aerobic metabolism. Biochim Biophys Acta 1102: 119–166, 1992

30. Ardail D, Lermé F, Louisot P: Further characterization of mitochondrial contact sites: effects of short-chain alcohols on membrane fluidity and activity. Biochem Biophys Res Commun 173: 878–885, 1990

31. Adams V, Bosch W, Hämmerle Th, Brdiczka D: Activation of low

Km hexokinases in purified hepatocytes by binding to mitochondria. Biochim Biophys Acta 932: 195–205, 1988

32. Xie G, Wilson JE: Tetrameric structure of mitochondrially bound rat brain hexokinase: A crosslinking study. Arch Biochem Biophys 276: 285–293, 1990

33. Wilson JE: Rapid purification of mitochondrial hexokinase from rat brain by a single affinity chromatography step on Affi-Gel blue Prep. Biochem 19: 13–21, 1989

34. DePinto V, Benz R, Palmieri F: Interaction of non-classical detergent with mitochondrial porin. A new purification procedure and characterization of the pore-forming unit. Eur J Biochem 183: 179–187, 1989

35. DePinto V, Prezioso G, Thinnes F, Link TA, Palmieri F: Peptide-specific antibodies and proteases as probes of the transmembrane topology of the bovine heart mitochondrial porin. Biochemistry 30: 10191–10200, 1991

36. Brdiczka D, Wicker U, Gellerich F: The function of the outer membrane pore in the regulation of peripheral kinases and energy metabolism. In: M Colombini, M Forte (eds) NATO ARW on: Molecular biology of mitochondrial transport systems. Springer Verlag, Berlin, Heidelberg 1994 (in press)

37. Schlame M, Gellerich FN, Augustin HW: Localization of hexokinase in mitochondria from rabbit reticulocytes studied by measurements of ^{32}P-fluxes. Acta Biol Med Germ 40: 617–623, 1981

38. Kabir F, Nelson BD: Hexokinase bound to rat brain mitochondria uses externally added ATP more efficiently than internally generated ATP. Biochim Biophys Acta 1057: 147–150, 1991

39. BeltrandelRio H, Wilson JE: Hexokinase of rat brain mitochondria: relative importance of adenylate kinase and oxidative phosphorylation as sources of substrate ATP, and interaction with intramitochondrial compartments of ATP and ADP. Arch Biochem Biophys 286: 138–194, 1991

40. Bessman SP: A molecular basis for the mechanism of insulin action. Am J Med 40: 740–749, 1966

41. Bücheler K, Adams V, Brdiczka D: Localization of the ATP/ADP translocator in the inner membrane and regulation of contact sites between mitochondrial envelope membranes by ADP. A study on freeze fractured isolated liver mitochondria. Biochim Biophys Acta 1061: 215–225, 1991

42. McEnery MW, Snowman AM, Trifiletti RR, Snyder H: Isolation of the mitochondrial benzodiazepine receptor: Association with the voltage-dependent anion channel and the adenine nucleotide carrier. Proc Natl Acad Sci 89: 3170–3174, 1992

43. Krueger KE, Mukhin AG, Michaluk L, Santi MR, Grayson DR, Guidotti A, Sprengel R, Werner P, Seeburg PH: Purification, cloning and expression of a peripheral-type benzodiazepine receptor. In: G Biggio, E Costa (eds) GABA and Benzodiazepine Receptor Subtypes. Raven Press, New York, 1990, pp 1–12

44. Erickson-Viitanen S, Viitanen P, Geiger PJ, Wang WC, Bessman SP: Compartmentation of mitochondrial creatine phosphokinase. J Biol Chem 257: 14395–14404, 1982

45. Brooks SPJ, Suelter CH: Compartmented coupling of chicken heart creatine kinase to the nucleotide translocase requires the outer membrane. Arch Biochem Biophys 257: 144–153, 1987

46. Colombini M: A candidate for the permeability pathway of the outer mitochondrial membrane. Nature 279: 643–645, 1979

47. Benz R, Kottke M, Brdiczka D: The cationically selective state of the mitochondrial outer membrane pore: a study with intact mitochondria and reconstituted mitochondrial porin. Biochim Biophys Acta 1022: 311–318, 1990

48. Zimmerberg J, Parsegian VA: Polymer inaccessible volume changes during opening and closing of a voltage-deplendent ionic channel. Nature 323: 36–39, 1986

49. Läuger P: Electrogenic ion pumps. Ca-pump from sarcoplasmic reticulum. Sinauer Inc. Publishers, Mass., USA, 1991, pp 226–251

50. Knoll G, Oebel G, Plattner H: A simple sandwich-cryogen-jet procedure with high cooling rates for cryofixation of biological materials in native state. Protoplasma 111: 161–176, 1982

51. Branton DS, Bullivant S, Gilula NB, Karnovsky MJ, Moor H: Muehlethaler K, Northcote DH, Packer L, Satir B, Speth V, Staehelin LA, Steere RL, Weinstein RS: Science 190: 54–56, 1975

52. Carlemalm E, Garavito EM, Villinger W: Advances in low temperature embedding for electron microscopy. Electron Microsc 2: 656–657, 1980

53. Carlemalm E, Garavito EM, Villinger W: Resin development for electron microscopy and an analysis of embedding at low temperature. J Microsc. 126: 123–144, 1982

54. Bücher Th, Luh W, Pette D: Hoppe-Seyler Thierfelder Handbuch der physiologisch- und pathologisch-chemischen Analyse, Bd VI/A. Springer, Berlin, 1964, pp 293–339

55. Harris DA, Featherstone J, Das AM: Regulation of the mitochondrial ATP synthase. Abstract 235.1/O XXXII IUPS Congress 1993, Glasgow

56. Denton and McCormack: On the role of the calcium transport cycle in heart and other mammalian mitochondria. FEBS Lett 119: 1–8, 1980

57. Hansford RG: Rev Physiol Biochem Pharmacol 102: 1, 1985

58. Kottke M, Wallimann Th, Brdiczka D: Dual electron microscopic localization of mitochondrial creatine kinase in brain mitochondria. Biochem Med and Metabol Biol (in press) 1994

Molecular and Cellular Biochemistry **133/134**: 85–104, 1994.
© 1994 *Kluwer Academic Publishers.*

II-2 The influence of the cytosolic oncotic pressure on the permeability of the mitochondrial outer membrane for ADP: implications for the kinetic properties of mitochondrial creatine kinase and for ADP channelling into the intermembrane space

Frank Norbert Gellerich,[1,2] Matthias Kapischke,[3] Wolfram Kunz,[3] Wolfram Neumann,[4] Andrey Kuznetsov,[5] Dieter Brdiczka[6] and Klaas Nicolay[1]

[1] *Dept. of in vivo NMR spectroscopy, Bijvoet Center for Biomolecular Research, Utrecht University, The Netherlands;* [2] *Permanent address: Department of Transplant Surgery, Clinical and Interdisciplinary Bioenergetics, University Hospital of Innsbruck, Austria;* [3] *Institut für Biochemie und* [4] *Klinik für Orthopädie der Medizinischen Fakultät der Universität Magdeburg, Germany;* [6] *Fakultät für Biologie, Universität Konstanz, Germany;* [5] *Laboratory of Bioenergetics, Cardiology Research Center, Moscow, Russia*

Summary

Cytosolic proteins as components of the physiological mitochondrial environment were substituted by dextrans added to media normally used for incubation of isolated mitochondria. Under these conditions the volume of the intermembrane space decreases and the contact sites between the both mitochondrial membranes increase drastically. These morphological changes are accompanied by a reduced permeability of the mitochondrial outer compartment for adenine nucleotides as it was shown by extensive kinetic studies of mitochondrial enzymes (oxidative phosphorylation, mi-creatine kinase, mi-adenylate kinase). The decreased permeability of the mitochondrial outer membrane causes increased rate dependent concentration gradients in the micromolar range for adenine nucleotides between the intermembrane space and the extramitochondrial space. Although all metabolites crossing the outer membrane exhibit the same concentration gradients, considerable compartmentations are detectable for ADP only due to its low extramitochondrial concentration. The consequences of ADP-compartmentation in the mitochondrial intermembrane space for ADP-channelling into the mitochondria are discussed. (Mol Cell Biochem **133/134**: 85–104, 1994)

Key words: mitochondria, creatine kinase, adenylate kinase, compartmentation, oncotic pressure, metabolic channelling

Address for offprints: F.N. Gellerich, Dept. in vivo NMR spectroscopy, Bijvoet Center for Biomolecular Research, University of Utrecht, Bolognalaan 50, NL-3584 CJ Utrecht, The Netherlands

Compartmentation of adenine nucleotides in the mitochondrial periphery

The mitochondrial outer membrane separates the inter-membrane space (i.m.s.) from the cytosol. The communication between both compartments is possible by the existence of porin pores which are present in the outer membrane. Estimated to range from 1.3 to 2×10^{-9} m [1] the radius of the pores is sufficient to allow the passage of molecules up a molecular weight of 6 kDa [2]. Small molecules such as adenine nucleotides, creatine and creatine phosphate can pass the outer membrane by passive diffusion through the porin pores. The mitochondrial outer membrane, however, is impermeable to proteins. Consequently, precursor proteins destinated for the mitochondria need a special transport machinery for crossing the outer membrane during import [3].

In the mitochondrial outer compartment, several ATP converting enzymes as creatine kinase (mi-CK), adenylate kinase (mi-AK), nucleoside diphosphate kinase, glycerol kinase and hexokinase are localized [4]. Mi-CK and mi-AK are localized within the i.m.s. While mi-CK is reversibly bound to the mitochondrial inner membrane [5] it is generally accepted that mi-AK is not bound to mitochondrial membranes, a property which allows that enzyme to be used as a marker for the i.m.s. [4]. The biological relevance of the mitochondrial localization of the above ATP splitting enzymes has been proposed in the past as an advantage in supplying these enzymes with mitochondrially formed ATP [6–8].

A new approach to this problem became possible through experiments with reconstituted systems [9–13] in which mitochondria and pyruvate kinase competed for ADP generated by kinases having different localizations with respect to the mitochondrial outer membrane. For mitochondria from heart [10, 11], liver [12] and brain [13] it was shown that the ADP supply to oxidative phosphorylation proceeded more effectively via mitochondrial creatine kinase or adenylate kinase compared to the ADP supply by extramitochondrially added enzymes such as yeast hexokinase. As a reason for this it was assumed that the mitochondrial outer membrane dynamically separates the i.m.s. from the extramitochondrial space [11, 16].

Reconstituted systems for detection of rate dependent concentration gradients between the mitochondrial intermembrane and the extramitochondrial space

To investigate whether or not the i.m.s. together with the extramitochondrial space form a homogenous pool for adenine nucleotides we developed three reconstituted systems consisting of (i) functionally intact mitochondria from different tissues plus AdN, P_i, substrates and (ii) pyruvate kinase plus PEP, both competing for ADP regenerated by ATP-utilising enzymes in varied localization with regard to the mitochondrial outer membrane (Fig. 1). If added or regenerated extramitochondrially by hexokinase ADP has to pass through the porin pores on its way to the AdN-translocator (HK system). If formed in the i.m.s. by mi-AK or mi-CK, ADP has to pass through the pores only on its route to pyruvate kinase (AK- or CK system). Obviously mi-CK and HK are only active if their substrates are added to the incubation. In contrast mi-AK permanently equilibrates the adenine nucleotides in the i.m.s. However, under stationary conditions there is no flux through this pathway due to the absence of extramitochondrial AMP converting enzymes (Fig. 1).

If the outer membrane were to present a barrier to ADP diffusion then, presumably, the competition between mitochondrial and extramitochondrial ADP phosphorylation would be expected to be influenced by the localization of ADP regenerating enzymes.

The adenylate kinase system

Using the adenylate kinase system (Fig. 1) in comparison to direct pulse additions of ADP into the extramitochondrial space it was at first shown for rat liver mitochondria [12] that ADP formed via mi-AK from AMP and ATP in the i.m.s. is preferentially used for oxidative phosphorylation. In these experiments equivalent additions of AMP and ADP brought about different responses of the system. Whereas the extramitochondrially added ADP was completely phosphorylated by pyruvate kinase, under the same conditions the ADP formed in the i.m.s. stimulated the oxidative phosphorylation.

It is well known that the specific adenylate kinase activity of liver mitochondria exceeds those of heart mitochondria by one order of magnitude [15]. Therefore it was interesting to know if the limited activity of mi-AK in heart mitochondria (about 0.3 U/mg) allows similar effects as in liver mitochondria. As shown in Fig. 2 the

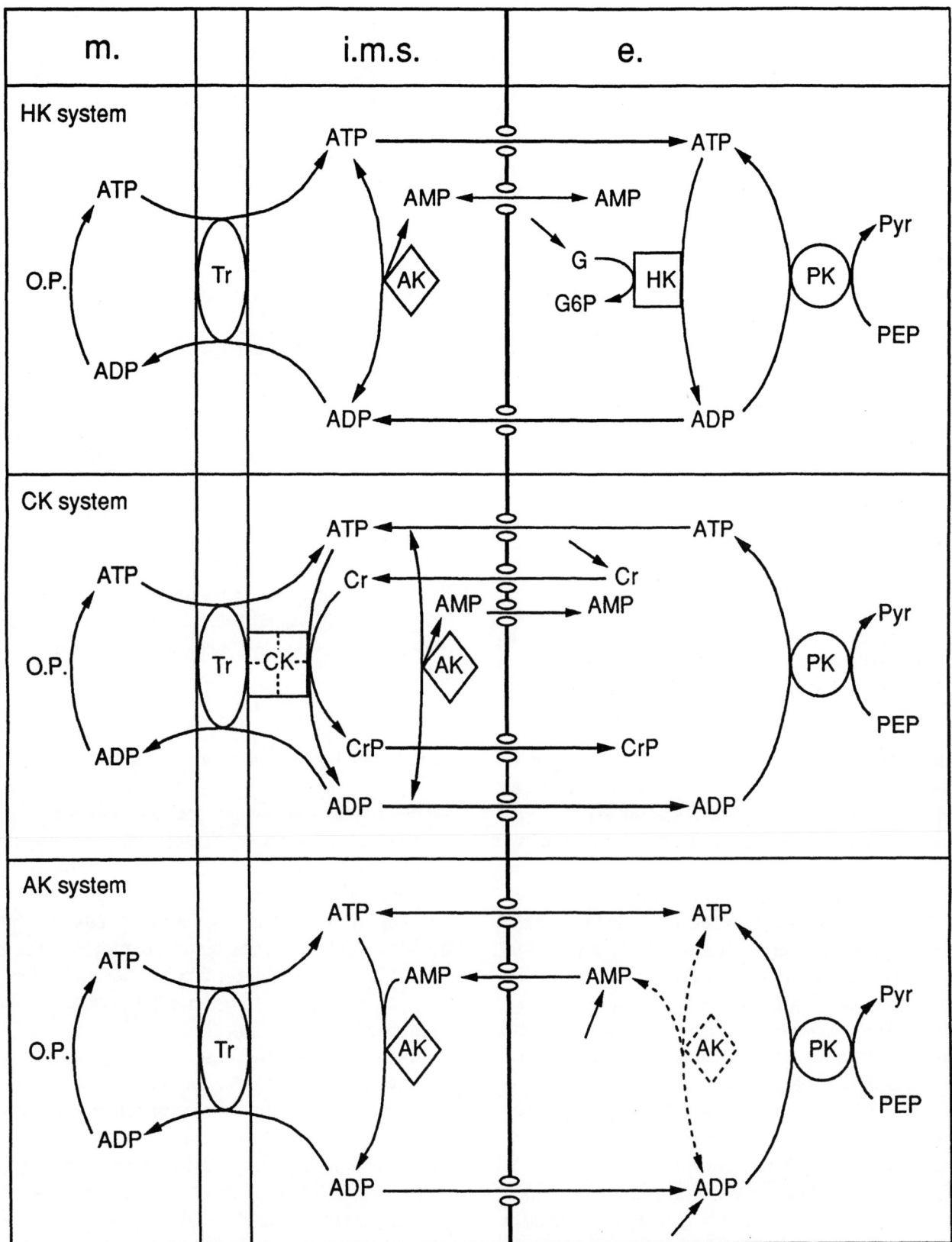

Fig. 1. Schematic representation of the various systems involved in investigating the AdN-compartmentation in the mitochondrial intermembrane space. Mitochondrial AdN-translocator (Tr) and pyruvate kinase (PK) compete for ADP formed by mitochondrial creatine kinase (CK), mitochondrial adenylate kinase (AK) or soluble yeast hexokinase (HK). The ATP-splitting enzymes HK (HK system), mi-CK (CK system) and mi-AK (AK-system) were activated by addtion of their substrates glucose, creatine or AMP (marked by diagonal arrows) in the presence of ATP inducing stimulation of respiration. By additions of pyruvate kinase, stationary intermediate rates of respiration can be adjusted. In the AK-system two modes are possible. Stationary mode: addition of AMP in high concentrations. Pulse mode: comparison of equivalent AMP and ADP additions in the micromolar range. In this mode we additionally investigated the effect of soluble extramitochondrial AK on the ADP compartmentation.

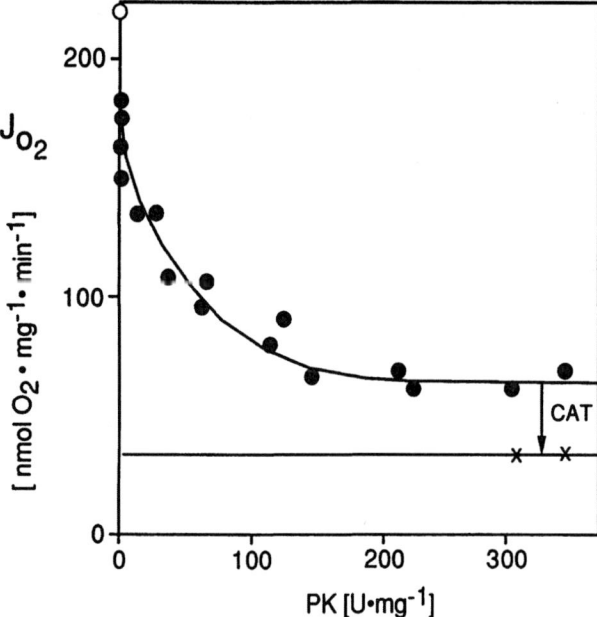

Fig. 2. Respiration rates of rat heart mitochondria stimulated by mi-AK, as a function of pyruvate kinase activity. Rat heart mitochondria (0.3 mg/ml) were incubated in sucrose medium [11] supplemented with 5 mM PEP, 5 mM ATP and PK as indicated. Rates of respiration were measured by means of a Clark-electrode and a custom made ratemeter. One minute after addition of 2 mM AMP respiratory rates were stationary for at least 5 min. After this time all added AMP was converted into ATP and the rate of respiration was the same as adjusted by addition of 10 μM carboxyatractyloside (CAT). The maximal mi-AK stimulated respiration was 75% of the state 3 respiration (O).

mitochondrial respiration was activated by addition of 2 mM AMP in the presence of 5 mM ATP. Under these conditions the respiratory rate was stationary for about 5 min. In further incubations increasing activities of pyruvate kinase were added thereby decreasing the respiratory rate. Interestingly even in the presence of excess pyruvate kinase activities it was impossible to reach the resting state of respiration indicating that a remarkable part of the ADP formed in the i.m.s. is not accessible for pyruvate kinase. It was shown earlier that pyruvate kinase is able to suppress the mitochondrial respiration which is stimulated by soluble yeast [14] or even by mitochondrial bound hexokinase [16] to resting state levels. In the adenylate kinase system however the resting state can only be adjusted by the inhibition of the AdN-translocator with carboxyatractyloside.

In an additional experiment shown in Fig. 3 we compared non stationary pulse additions of equivalent amounts of AMP and ADP to heart mitochondria in the presence of excess pyruvate kinase. Trace A shows the oxygen concentration in the oxygraph over the time. The first derivative (B) of this signal is directly proportional to the respiratory rate and the integrated peak in-

dicates the amount of oxygen consumed. The addition of ATP (4 mM) stimulated mitochondrial respiration due to ADP and AMP contaminations. The addition of excess pyruvate kinase adjusted the resting state since all extramitochondrial ADP is phosphorylated now by pyruvate kinase. Next, equivalent amounts of ADP and AMP were added to the incubations sequentially. Virtually all ADP added was phosphorylated by pyruvate kinase whereas ADP formed from AMP in the i.m.s. caused a marked oxygen consumption. When, additionally adenylate kinase was injected, then also AMP did not cause a stimulated respiration. In this case, adenylate kinase extramitochondrially converted all AMP into ADP which was rephosphorylated by pyruvate kinase.

Both experiments can be understood only if an effective diffusion barrier is assumed to exist between the intermembrane and the extramitochondrial space, resulting in concentration gradients for adenine nucleotides between both pools. Since concentration gradients are necessarily connected with fluxes and disappear with them, the term *dynamic compartmentation* has been proposed for diffusion-dependent differences in metabolite concentrations in subcellular compartments [11, 16].

The creatine kinase system

As described ten years ago especially mi-CK is able to stimulate the mitochondrial respiration in the presence of exceeding pyruvate kinase [10]. Due to the higher specific activity of mitochondrial creatine kinase (about 2 U/mg) in comparison to adenylate kinase (about 0.3 U/mg) and the specific binding of mi-CK to the mitochondrial inner membrane the compartmentation effects are more pronounced than those observed with adenylate kinase [17].

To obtain quantitative data concerning the compartmentation of adenine nucleotides in the i.m.s. of heart mitochondria we analysed heart mitochondrial respiration as a function of the pyruvate kinase activity added to the medium. Stationary respiratory rates were adjusted by addition of either creatine or glucose plus yeast hexokinase (Fig. 1). Single incubations were performed with increasing amounts of pyruvate kinase. In each sample the stationary rate of respiration was registered and the ADP concentrations were determined enzymatically after quenching samples as described previously [11]. As shown in Fig. 4 the rates of mitochondrial respi-

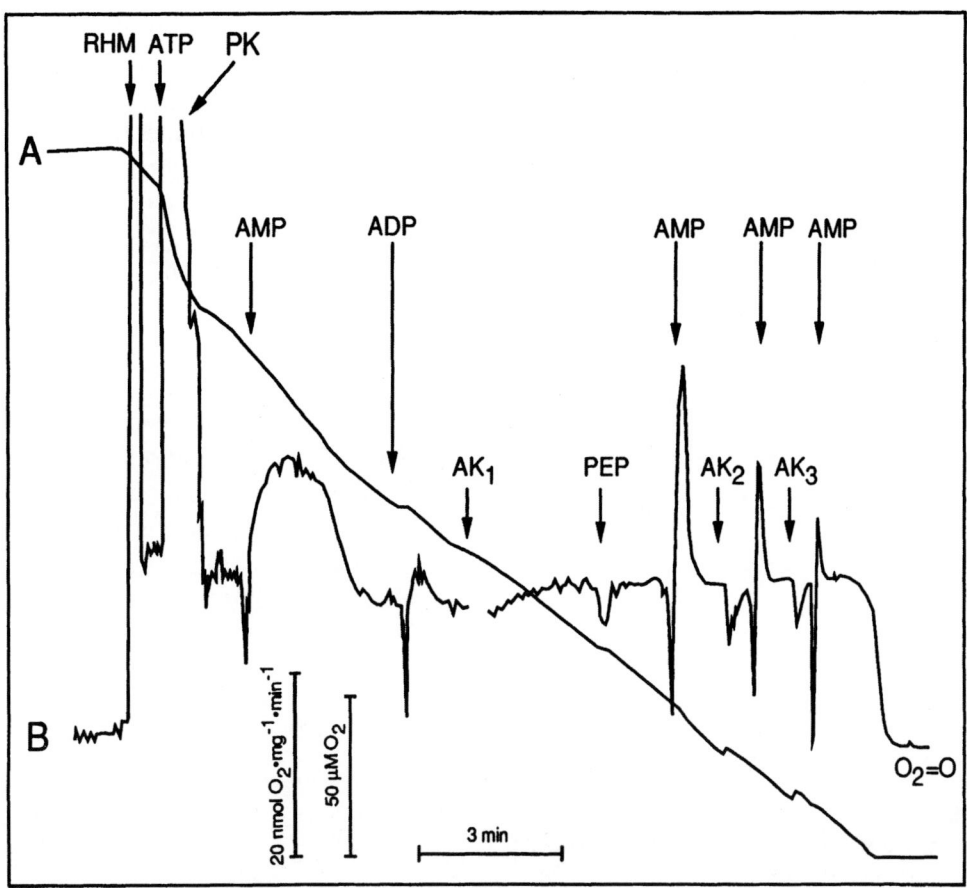

Fig. 3. Differential effects on oxygen consumption of equivalent additions of AMP and ADP to rat heart mitochondria and the prevention of AMP-induced stimulation of oxygen consumption by addition of soluble AK, in the presence of excess PK. Incubation of rat heart mitochondria (0.64 mg/ml) as in Fig. 2. Additions: ATP, 4 mM; PK, 121 U/mg; AMP, 595 nmol; ADP, 810 nmol (containing 188 nmol AMP); AK1, AK2 and AK3, 22, 55 and 88 U soluble AK/mg respectively in a final volume of 3 ml.

ration decreased as expected with increasing activities of pyruvate kinase. At sufficiently high PK activity, all ADP formed by yeast hexokinase was completely phosphorylated extramitochondrially leading to the induction of the resting state. The mi-CK stimulated respiration under these conditions was twice that of the resting state but could be diminished by addition of CAT. The bulk phase ADP concentration in both systems gave us further insight into the ADP compartmentation within the i.m.s. The most interesting points are marked by arrows in Fig. 4. The same ADP concentration in the bulk phase of both systems we only found at exceeding pyruvate kinase activity despite of a factor two difference in the respiratory rates (arrow 1). Since in the CK-system a higher respiration was detected, a higher ADP concentration should hold in the i.m.s. in comparison to the bulk phase. Arrow 2 marks conditions under which the same rate of respiration was measured in both systems while in this case the bulk phase ADP levels differed by a factor two. Since mitochondria phosphorylate at the same velocity in both systems, it is to be expected that in

the immediate vicinity of the AdN-translocator the same ADP concentration occurs.

Unfortunately, there is no possibility to directly detect the concentration of adenine nucleotides in the i.m.s. We therefore measured extramitochondrial AMP which is indicative of the ATP/ADP ratio in the i.m.s. where mi-AK metabolically connects AMP with ATP and ADP. AMP diffuses into the extramitochondrial space like ADP and ATP do, but unlike the latter two AMP cannot be metabolised there (Fig. 1). In experiments similar to that of Fig. 4 we determined in each incubation the adenine nucleotides ATP, ADP and AMP. The rates of respiration in both systems were plotted versus the bulk phase ATP/ADP ratio and versus the bulk phase AMP level as a probe for the ATP/ADP ratio in the i.m.s.

As shown in Fig. 5 at the same rate of respiration (50 nmol $O_2 mg^{-1} min^{-1}$), the extramitochondrial ATP/ADP ratio was 46 and 150 (arrows) in the HK and in the CK system, respectively. The AMP concentration was virtually the same in both systems (ca. 7 μM). Hence it

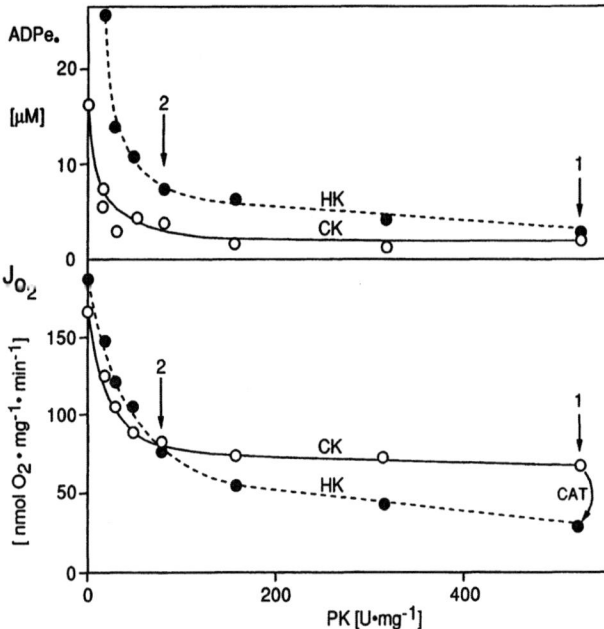

Fig. 4. Effect of the localization of ADP-regenerating enzymes on the bulk phase ADP concentrations and respiration rates of rat heart mitochondria. Rat heart mitochondria (0.28 mg/ml) were incubated in sucrose medium [11] containing additionally either 25 mM creatine or 10 mM glucose and 1.9 U HK/mg as well as PK as indicated. Ninety seconds after starting the reactions by addition of 1 mM ATP samples were quenched and the total ADP was determined enzymatically. Each point was one separate incubation. Arrow 1: 100% difference in the rates of respiration at similar ADP concentration. Arrow 2: vice versa. The addition of 10 μM CAT diminished the respiration of the CK-system to that of the HK-system.

was concluded that the ATP/ADP ratio in the i.m.s. was in effect identical in both systems. In other words, this experimental result clearly indicates that concentration gradients for ADP and ATP exist across the mitochondrial outer membrane.

Since we determined the rate of creatine phosphate formation (Fig. 5) in addition to the rate of respiration we were able to calculate the rates of AdN-diffusion through the mitochondrial outer membrane. The calculations were done using a relatively simple mathematical model as described previously [11, 16]. This model takes into account that in the creatine kinase system the ADP diffusion rate (v_{dADP}) through the pores must be equal to the difference between ADP production in the i.m.s. by mi-CK (v_{CK}) and the rate of mitochondrial phosphorylation (v_P). This part of ADP diffused from the mi-CK through the pores to the pyruvate kinase. The experimentally measured pyruvate formation however is higher than this ADP diffusion since there are usually contaminations with extramitochondrial ATPases [11].

$$v_{dADP} = v_{CK} - v_P \qquad (1)$$

In the hexokinase sytem, ADP diffuses in the opposite direction and v_{dADP} is equal to the rate of mitochondrial respiration ($v_{CK} = 0$ in Eq. 1). Under stationary conditions the mi-AK equilibrates the adenine nucleotides in the i.m.s. but does not influence the ADP fluxes. v_{CK} was experimentally determined from the formation of creatine phosphate. v_P was calculated from the measured respiration rates by adopting a simplified model that describes the rate of oxidative phosphorylation as a function of the extramitochondrial ATP/ADP ratio [18]. Then assuming that an identical rate of respiration in both systems was indicative of an identical ATP/ADP ratio in the vicinity of the AdN translocator, it was possible to calculate the ATP/ADP ratio in the i.m.s. A diffusion rate constant k_d was computed from the diffusion rates v_d and the different ATP/ADP ratios in both compartments, by using an equation similar to Fick's first law of diffusion:

$$k_d = \frac{v_{dADP}}{[ADP]_i - [ADP]_e} \qquad (2)$$

Using the experimental data from the experiment shown in Fig. 5, it was possible to calculate a diffusion rate constant $k_d = (8.7 \pm 4.7) \, 10^4 \, \mu l \cdot mg^{-1} min^{-1}$ [11,16]. This constant was employed together with Eq. 1 to estimate the concentration gradient between the AdN-translocator and the bulk phase which was 12.8 μM AdN at a maximal diffusion rate of 1222 nmoles $AdN mg^{-1} min^{-1}$ through the mitochondrial outer membrane.

If the mitochondrial outer membrane is creating a diffusion barrier then its partial removal should decrease the dynamic AdN compartmentation. As shown previously digitonin treatment of rat heart mitochondria decreases the compartmentation to 40% of the effects in intact mitochondria [19]. If the outer membrane is destroyed by the action of digitonin and looses its barrier function then the bulk phase should be representative of i.m.s. In suspensions of mitochondria as well as of mitoplasts maximal mi-CK rates were set by addition of 25 mM creatine whereafter the pyruvate kinase activity was varied and the respiratory rate was measured in dependence on the bulk phase ATP/ADP ratio. As shown in Fig. 6 we observed indeed a shift of the control characteristics of mitoplasts to lower ATP/ADP ratios in comparison to the curve of intact mitochondria. At a respiratory rate of 50 nmol $O_2 mg^{-1} min^{-1}$ the ATP/ADP ratios differed by a factor of about 2. This finding is in line with the main results of the experiments shown in Figs 4 and 5: mi-CK activity creates elevated ADP-concentrations in the i.m.s. in comparison to the bulk phase. If the outer

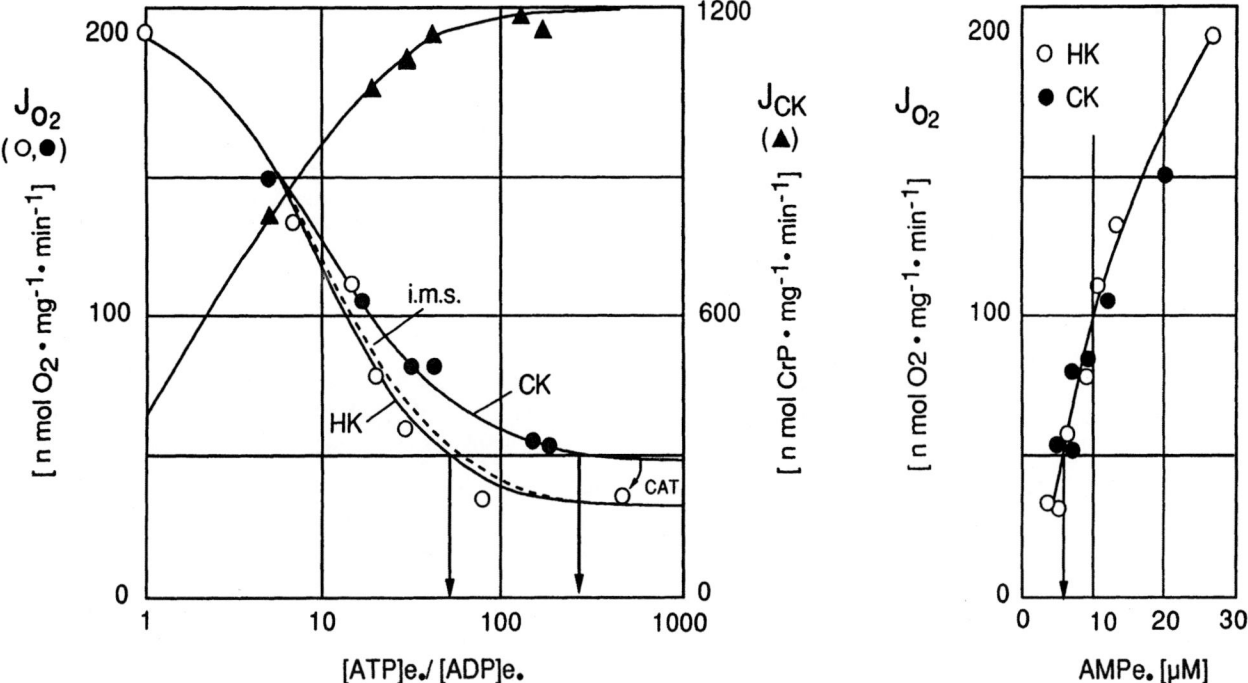

Fig. 5. Effect of the localization of ADP-regenerating enzymes on the dependence of mitochondrial oxidative phosphorylation on the extramitochondrial ATP/ADP ratio and AMP concentration. Incubation of rat heart mitochondria (0.14 mg/ml) and experimental protocol as described for Fig. 4 except that 2.2 U HK/mg was used for the HK-system. The concentrations of AdN and CrP were determined enzymatically. The rate of mi-CK was calculated while assuming stationary reaction velocity from the formation of creatine phosphate. Curve fitting of data points was accomplished by nonlinear regression as described in detail previously [11]. The estimated diffusion rate constant was $(8.7 \pm 4.7)10^4$ µlmin^{-1}mg^{-1}. Line i.m.s. represents the computed dependence of respiration on the ATP/ADP concentration ratio in the i.m.s. of both systems [11]. The dependence of the respiratory rates on bulk phase AMP-concentration which probes the ATP/ADP concentration ratio occurring in the i.m.s. is the same in both systems. The vertical arrows mark the extramitochondrial ATP/ADP ratios and AMP concentrations at the respiratory rate of 50 nmol O_2 min^{-1}mg^{-1}. The addition of 10 µM CAT diminished the rate of respiration in the CK incubations as indicated.

membrane is removed these elevated ADP concentrations hold in the entire bulk phase.

Estimation of the number of porin pores in the outer membrane of rat heart mitochondria

The rate constant for diffusion and the resulting concentration gradients for adenine nucleotides were calculated on the basis of experimental data only. No structural premise was used for the estimations. If, however, the limited number of pores in the outer membrane is assumed to restrict AdN diffusion between the two compartments it is possible to estimate the number of pores from the diffusion rate constant by employing Eq. 3 derived from Ficks' law of diffusion

$$k_d = \frac{D \cdot n \cdot A_{porin}}{dl} \qquad (3)$$

where D is the diffusion coefficient, n is the number of pores, A_{porin} is the cross-section of a single pore and dl is the diffusion distance across the outer membrane.

Assuming that (i) the AdN diffusion through the pores is comparable to that in an aqueous solution (D = $7.2 \ 10^{-9}$ m^2 min^{-1}, [20]), (ii) the cross section of a single pore is 6.3×10^{-18} m^2 as calculated from a diameter of 2 nm [1] and (iii) the diffusion distance is 7.5×10^{-9} m [21] we estimated the number of pores to be 1.0×10^{13} pores per mg mitochondrial protein. This value is very similar to that we obtained by direct determination of the specific mitochondrial porin concentration (21 ± 11 µg porin mg^{-1}) using gel electrophoresis. Assuming that a porin dimer forms one pore (60 kDa) we can estimate that 1 mg mitochondrial protein contains 2×10^{13} porin pores. Since 1 mg mitochondria contains 8×10^9 mitochondria [22] the area of porin pores for one mitochondrion ranges from 0.8 to 1.6×10^{-2} µm^2. According to Weibel *et al.* [22] the surface of one mitochondrion is 1.92 µm^2. From that it is possible to estimate that the porin pores keep open ca. 0.4 to 0.8% of the outer surface of the mitochondria.

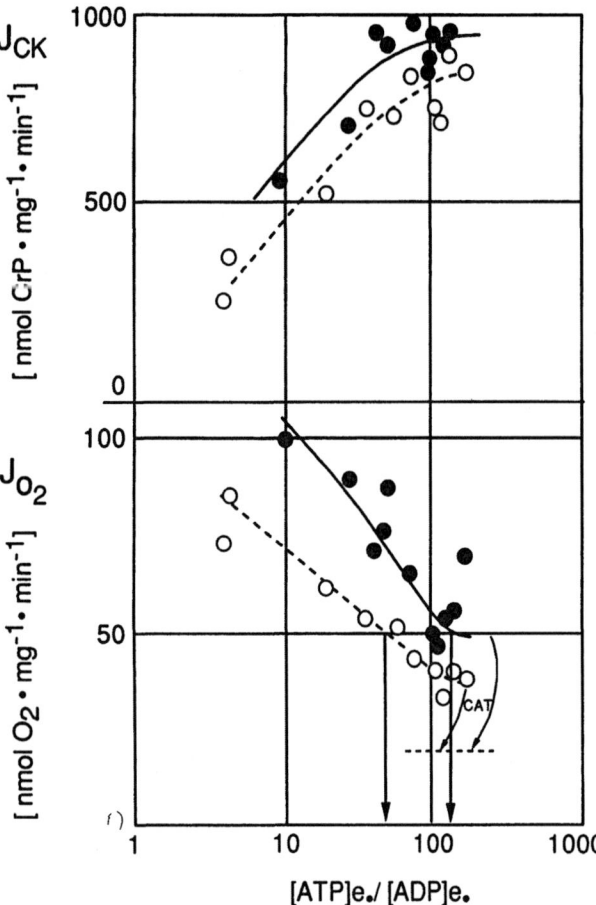

Fig. 6. Control of respiration of rat heart mitochondria and mitoplasts. Rat heart mitochondria (O, 0.45 mg/ml) and mitoplasts (●, 0.41 mg/ml) were incubated in sucrose medium [11] with 25 mM creatine (CK-system) and 1 mM ATP as described for Figs 4 and 5. The stationary rates of respiration and the rate of mi-CK were plotted versus the ATP/ADP concentration ratio. The addition of 10 μM CAT diminished the respiration of mitochondria and mitoplasts to the same rate. The vertical arrows mark the extramitochondrial ATP/ADP ratios at the same respiratory rate, as in Fig. 5.

*Functional versus dynamic compartmentation:
two aspects of one complex phenomenon*

The estimation of flux dependent concentration gradients for adenine nucleotides through the outer membrane of heart mitochondria from experimental flux measurements represents the first quantitative evidence for the existence of diffusion restriction in the outer mitochondrial compartment. Further qualitative evidence for a dynamic compartmentation of adenine nucleotides in the mitochondrial i.m.s. resulted from comparing experiments on mitochondria and mitoplasts using radio-chemical [23] and kinetic [24] measurements. Very recently Saks has presented evidence for the mitochondrial outer membrane to be an important diffusion barrier for ADP also in myocytes [25]. It is important to

mention that rate dependent diffusion gradients across the mitochondrial outer membrane occur also for the other metabolites of the creatine kinase reaction (ATP, Cr, CrP) and with the same extent as for ADP. However, whether or not a concentration gradient is connected with a dynamic compartmentation depends on the absolute concentration of that metabolite as previously shown by means of computer simulation [16]. In the dog heart, the cytosolic concentrations of ATP, CrP, Cr and ADP were 5, 9, 11 mM and 56 μM, respectively [26]. Gradients of 20 μM are equivalent to 0.004%, 0.002%, 0.0018% and 35% of the total concentrations. These numbers show that, despite the possible existence of concentration gradients for all metabolites, a remarkable compartmentation occurs only if the metabolite concentration and the gradient are in the same order of magnitude. These considerations lead to the conclusion that the ADP compartmentation is most important to regulation of the cellular energy metabolism.

On the other hand there is the concept of *functional coupling* favouring the compartmentation to be caused by interactions between the mitochondrial creatine kinase and the AdN-translocator [27], neglecting a role for the mitochondrial outer membrane [28] in compartmentation. To assess the relative importance of the two mechanisms postulated, we reinvestigated the influence of the binding of mi-CK to the inner membrane on AdN-compartmentation. This was done by comparing the compartmentation of AdN in rat heart mitochondria incubated in either a sucrose medium or in media containing KCl which releases mi-CK from the mitochondrial inner membrane [28, 29]. One method to investigate the functional coupling between the mi-CK and the AdN-translocator is to determine the K_{AdN} of mitochondrial oxidative phosphorylation in the presence of 25 mM creatine [28]. Table 1 shows that the release of mi-CK from the inner membrane has no effect on the K_{ADP} of oxidative phosphorylation demonstrating that the medium does not affect the oxidative phosphorylation itself. On the other hand under released conditions, the K_{AdN} increased from 32 μM AdN to 213 μM AdN indicating indeed a reduction of functional coupling between mi-CK and oxidative phosphorylation. These results completely confirm similar data published recently by the group of Saks [28]. However, mi-CK even if released from the mitochondrial inner membrane remains in the i.m.s. and should therefore (similar as mi-AK) at sufficiently high fluxes cause dynamic compartmentation effects. To verify this, we measured the AdN-compartmentation in the i.m.s. of heart mitochondria in

*Table 1.*App. K_m-values for AdN in regulation of mitochondrial respiration in different media

	app. K_{AdN}	
	ADP [μM]	ATP [μM]
Sucrose medium	26 ± 7	32 ± 10
KCl medium	20 ± 3	213 ± 29

Mitochondrial respiration of rat heart mitochondria with 10 mM glutamate and 2 mM malate as substrates was measured with a CYCLO-BIOS-Oxygraph. ADP titrations were performed in the presence of 10 U yeast hexokinase/mg plus 1 mM glucose. ATP titrations were performed in the presence of 25 mM creatine without hexokinase. Kinetic constants were calculated by means of nonlinear regression. Data are mean ± S.D. of 3 independent experiments.

KCl-medium. As shown previously [19] we could detect compartmentation effects also under these conditions. These results were confirmed in further experiments shown in Table 2. Here we investigated the effect of the binding state of mi-CK on the dynamic compartmentation at different AdN concentrations. The difference between the CK/PK and the CAT state was maximal (17.2 nmol O_2mg^{-1}min^{-1} = 100%) in sucrose medium in the presence of 2.3 mM AdN. At low AdN-concentration the compartmentation effect was lower (due to the lower V_{CK}) but much more pronounced in sucrose medium (44%) than in the KCl-medium (19%). These results again clearly demonstrate that even under conditions of released mi-CK, ADP is dynamically compartmentalized within the i.m.s. However, these effect are smaller in KCl medium in comparison to the sucrose medium demonstrating the existence of a functional coupling which is dependent on the binding of mi-CK to the mitochondrial inner membrane. It is clear that the binding of mi-CK to the inner membrane allows the smallest

possible distance between the mi-CK and the AdN-translocator.

The kinetic advantage of a small distance between different enzymes of one metabolic sequence was strikingly shown by Fossel and Hoefeler, who immobilised creatine kinase together with hexokinase or both individually on Sepharose beads [30]. They found that when reducing the distance between both enzymes from 0.1 mm to 10 nm the flux through the linked enzyme pair was much higher and much less dependent on the concentration of the intermediate metabolites. Even at adenine nucleotide concentrations far below the K_{ATP} of hexokinase substantial amounts of glucose-6-phosphate were produced when the enzymes were near but not when they were distant. Similar results were found by Mosbach and Mattiasson for a sequence of three enzymes bound to an artificial membrane [31]. The reduced lag phase and the increased fluxes at low substrate concentrations in the bulk phase can be explained by concentration gradients within unstirred layers at the surfaces of the membranes.

All results presented in this chapter provide evidence for the compartmentation of adenine nucleotides in the mitochondrial i.m.s. Our data allow to conclude that it is possible to adjust in the i.m.s. phosphorylation potentials differing from those in the cytosol. The difference between the phosphorylation potential in the extramitochondrial and the i.m.s. is rate dependent and therefore dynamic. These differences decrease with decreasing rates and when destroying the outer mitochondrial membrane. Our data also seem to imply that the distribution of AdN in the mitochondrial i.m.s. is non-homogenous. The finding that the released mi-CK (i) requires higher ATP concentrations for similar stimulation of oxidative phosphorylation and (ii) allows only diminished compartmentation effects suggests a microinhomoge-

*Table 2.*Influence of AdN concentration on AdN compartmentation in the intermembrane space of rat heart mitochondria in different media

	Rates of respiration [nmol O_2/mg/min]					
	PK state	CK/PK state I	CK/PK state II	CAT state	CK/PK I	CAT II
Sucrose	15.6 ± 3.0	23.1 ± 2.9	32.7 ± 2.3	15.5 ± 3.5	7.6 [44%]	17.2 [100%]
KCl	18.7 ± 2.9	21.8 ± 2.0	28.9 ± 3.3	18.5 ± 2.8	3.3 [19%]	10.4 [60%]

Rat heart mitochondria (0.23–0.58 mg/ml) were incubated in sucrose or KCl media with 4 mM glutamate and 2 mM malate as substrates. Rates of respiration were measured after addition of 195 U PK/mg in the presence of 0.33 mM AdN (PK state), after addition of 33 mM creatine (CK/PK state I), after addition of 1.7 mM AdN (CK/PK state II) and after addition of 20 μM CAT (CAT state). Data represent means of four independent experiments ± S.D. The difference between the CK/PK-state II and the CAT state in sucrose medium was assumed to be 100% (maximal compartmentation effect).

neity in the i.m.s. This might be caused by unstirred layer effects and enzyme enzyme interactions as assumed to exist between the AdN-translocator and the mi-CK [32, 33]. It is obvious that functional relations between the various components, as well as morphological and dynamic aspects together are responsible for such a complex phenomenon as the AdN-compartmentation in the i.m.s.

Influence of cytosolic macromolecules on the permeability of the mitochondrial outer membrane

In the intact cell, mitochondria are embedded in a 30% protein solution [34, 35] while isotonic media are used without macromolecules to isolate and investigate mitochondria. Because of the missing oncotic pressure, mitochondria exhibit a large i.m.s. in these media compared to mitochondria *in vivo* [36, 37]. The artificially enlarged mitochondrial i.m.s. can be reduced by adding macromolecules like polyvinyl pyrrolidone, ficol or albumin to isolation and incubation media [36–38].

Effect of oncotic pressure on frequency of boundary membrane contact sites

Contact sites between the mitochondrial membranes have been described as specific structures which are important to the uptake of mitochondrial precursor proteins [3] and the mitochondrial energy transfer [4]. To investigate whether or not the volume reduction of the mitochondrial i.m.s. is accompanied by an increase in the number of contact sites between the two envelope membranes we performed freeze fracture experiments with rat liver mitochondria and incubated them in 10% dextran 70 (i.e. having an average molecular mass of 70 kDa) under conditions of active and resting respiration. The presence of dextran increased the frequency of contact sites to 156% (active state) and 253% (resting state) of the value in the absence of dextran [38]. These findings have important implications for the investigation of the regulatory importance of the i.m.s. If the macromolecules increase the contact sites between both mitochondrial membranes to such an extent, then this might dramatically affect the structural and functional relationships between the components of the i.m.s. In a compressed i.m.s. the free water content should be re-

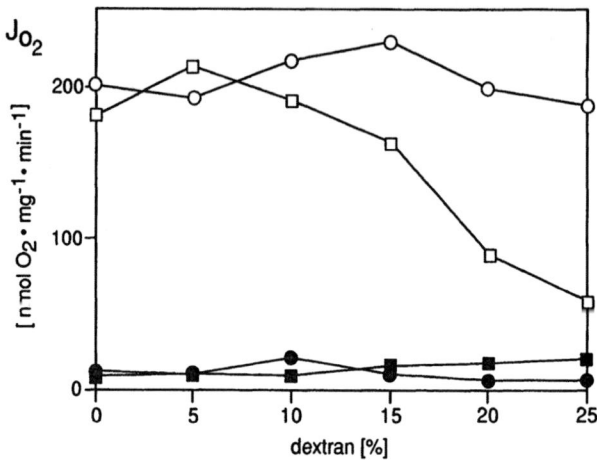

Fig. 7. Effect of dextran on functional properties of rat heart mitochondria in sucrose and KCl media. Rat heart mitocondria were incubated in sucrose- (□, ■) and KCl- (○, ●) medium, both containing increasing concentrations of dextran as indicated. The active rates of respiration (○, □) were adjusted by addition of 2 mM ADP whereas the CAT state of respiration (●, ■) was induced by addition of 20 μM CAT.

duced and the relative protein concentration should be increased. We can expect that under these conditions direct and long-lived interactions occur between the enzymes as observed in the mitochondrial matrix space [39]. This initiated the question whether or not the diffusion of adenine nucleotides into and from the mitochondrial i.m.s. might be changed by addition of macromolecules for mimicking the *in vivo* cytosolic oncotic pressure.

Effect of macromolecules on functional properties of mitochondria and kinetic properties of soluble enzymes

Before analyzing the effect of macromolecules on exchange of adenine nucleotides across the mitochondrial outer membrane we studied their influence on basic properties of isolated mitochondria. As previously shown the addition of 10% (w/v) bovine serum albumin or dextrans did not affect functional properties of rat liver mitochondria, e.g. active and resting rates of respiration [38, 40]. Similar experiments were performed with rat heart mitochondria. Figure 7 shows the effect of dextran 20 on state 3 respiration and on respiration in the presence of carboxyatractyloside (CAT) for rat heart mitochondria incubated in a sucrose and in a KCl medium. The addition of 10% dextran did not have any influence on both states in both media. Higher dextran concentrations however decreased the active respiration in sucrose medium while leaving this unchanged in the KCl medium. In order to keep the mi-CK bound to the mitochondrial inner membrane we decided to use the sucrose medium supplemented with 10% dextran for our further experiments.

In addition we investigated the influence of 10% dextran 70 on the kinetic properties of soluble enzymes as hexokinase (from yeast), pyruvate kinase (from muscle) as well as mi-CK and mi-AK released from rat heart and liver mitochondria, respectively [40]. Neither for pyruvate kinase (control: K_{ADP} = 372 ± 29 µM ADP, V_{max} = 223 ± 33 U/mg, 10% dextran: K_{ADP} = 378 ± 20 µM ADP, V_{max} = 216 ± 29 U/mg); mi-AK (control: 168 ± 9 µM ADP, V_{max} = 0.90 ± 0.02 U/mg, 10% dextran: K_{ADP} = 172 ± 9 µM; V_{max} = 0.9 ± 0.02 U/mg); yeast HK (K_{ATP} = 306 ± 15 µM ATP, V_{max} = 280 ± 4 U/mg; 10% dextran: K_{ATP} = 312 ± 13 µM ATP, V_{max} 282 ±3 U/mg) nor for mi-CK (control: K_{ATP} = 181 ±19 µM ATP, V_{max} = 1.78 ±0.04 U/mg, 10% dextran: K_{ATP} 180 ±16 µM ATP, V_{max} = 1.77 ±0.03 U/mg) the addition of 10% dextran 70 to the assay had an effect on the kinetic properties.

Effects of dextran on kinetic properties of mi-CK

To investigate whether or not dextrans have an influence on the permeability of the mitochondrial outer membrane and/or diffusion in the i.m.s., detailed kinetic investigations of enzymes localized in the i.m.s. are necessary. We performed kinetic studies of mi-CK in rat heart mitochondria by means of spectrophotometry. Since the oxidative phosphorylation was completely inhibited by rotenone, antimycin A and carboxyatractyloside the flux of adenine nucleotides through the outer membrane could be deduced from the rate of mi-CK probed by the indicator enzymes. All measurements were done in sucrose medium in the absence and in the presence of 10% dextran 70. In the experiment shown in Fig. 8 we additionally measured at 30% dextran. Obviously the K_{ATP} increases with increasing dextran concentrations from 332 µM (control) over 525 µM ATP (10% dextran) to 641 µM ATP (30% dextran) whereas the V_{CK} decreases from 1.98 U/mg over 1.71 U/mg to 1.08 U/mg. These results are clear indications for a decreased permeability of the outer membrane for adenine nucleotides. Since dextran does not influence the kinetic properties of soluble, released mi-CK the increased K_{ATP} could be explained by an increased concentration gradient across the outer membrane. If this is true, then after removal of the outer membrane by digitonin the dextran effect on mi-CK should disappear. As shown in the lower part of Fig. 8 this is indeed the case. It appeared, that more than

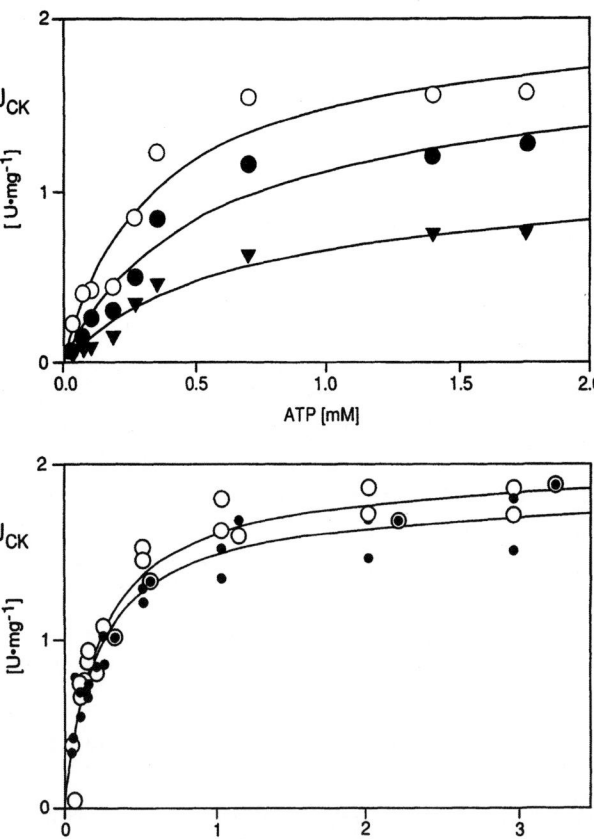

Fig. 8. Effect of dextran on kinetic properties of mi-CK in isolated rat heart mitochondria and in mitoplasts. Mi-CK was measured in isolated rat heart mitochondria (5 µg/ml, at the top) and in mitoplasts (6 µg/ml, at the bottom) incubated in sucrose medium by means of the optical test with PK and LDH as indicator enzymes in the presence of 25 mM creatine and increasing ATP concentrations as indicated. To inhibit the mitochondrial functions 2.5 µM rotenone, 1 µM antimycine A and 10 µM CAT were added. The rate of mi-CK was calculated as difference to controls without creatine. As a typical example we measured with the highest ATP concentration used (1.8 mM ATP) 2.36 U/mg with creatine and 0.86 U/mg without creatine = 1.49 U mi-CK/mg. Kinetic constants were calculated by means of nonlinear regression. Intact mitochondria: Control (O): K_{ATP} = 332 ± 91 µM, V_{CK} = 1.98 ± 0.2 U/mg; 10% dextran 70 (●): K_{ATP} = 525 ± 150 µM, V_{CK} = 1.71 ± 0.2 U/mg; 30% dextran 70 (▼): K_{ATP} = 641 ± 160 µM, V_{CK} = 1.06 ± 0.1 U/mg. Mitoplasts: Control (O): K_{ATP} = 230 ± 19 µM, V_{CK} 1.9 ± 0.05 U/mg; 10% dextran 70 (●): K_{ATP} = 238 ± 28 µM, V_{CK} = 1.82 ± 0.06 U/mg. By centrifugation of the probe it was evaluated that more than 75% of total mi-CK remained bound to the mitoplasts upon digitonin treatment.

75% of mi-CK remained bound under these conditions. Neither the K_{ATP} nor the V_{CK} of mi-CK was changed upon addition of dextran. These results very clearly show that one prerequisite for the dextran effect is an intact outer mitochondrial membrane. Moreover, even in the absence of dextran the K_{ATP} of mi-CK in mitoplasts (230 µM ATP) was somewhat lower than in intact mitochondria (332 µM ATP) suggesting that the mito-

Table 3. Temperature dependence of the effect of dextran on K_{ATP} of compartmentalized and soluble mi-CK

Temperature [° C]	K_{ATP} [µM] ATP		
	Control	10% Dextran	Difference
20	240 ± 74	365 ± 112	125
25	284 + 51	435 + 123	151
30	341 ± 50	553 ± 65	212
35	443 ± 81	689 ± 70	246

Kinetic constants were determined and calculated as described for Fig. 8 at temperatures as indicated.

Table 4. Complete kinetic analysis of the dextran effect on compartmentalized mi-CK

Parameter	Control	10% Dextran
K_a [µM] ATP	192 ± 20	234 ± 40
K_{ia} [µM] ATP	666 ± 92 (*)	1222 ± 185 (*)
K_b [mM] Creatine	11.4 ± 0.9 (*)	19.6 ± 2.0 (*)
K_p [µM] ADP	79 ± 7 (*)	142 ± 11 (*)
K_{ip} [µM] ADP	411 ± 100	381 ± 99
K_q [mM] CrP	1.12 ± 0.2	1.6 ± 0.2
V_{max} [U/mg] →	2.9 ± 0.1	3.1 ± 0.1
V_{max} [U/mg] ←	3.6 ± 0.1	3.7 ± 0.09

Calculation of kinetic parameters by means of the nonlinear regression programme BMDP. *, constants are significantly different (99.0%).

chondrial outer membrane also acts as a diffusion barrier in the absence of dextrans. This result completely confirms results obtained for chicken heart mitochondria by Brooks and Suelter [24].

If dextrans increase the rate dependent concentration gradients across the outer membrane as indicated by the increased Michaelis constants, then the dextran effect should increase with increasing rates of the mi-CK. To proof that, we determined the K_{ATP} of mi-CK in intact mitochondria and in a released, soluble state with and without dextran as a function of temperature. As shown in Table 3 in both the compartmentalized and the homogenous system, the K_{ATP} increased with rising temperatures. In accordance with the data mentioned above dextran has no influence on the K_{ATP} of the soluble mi-CK but at all temperatures we observed a dextran-dependent increase in the K_{ATP} for the compartmentalized mi-CK. The difference between the constants determined in the presence and in the absence of dextran increased from 125 µM ATP at 20° C to 246 µM at 35° C which fits with the increased rate of the mi-CK under these conditions.

The kinetic experiments described above were performed in the presence of 25 mM creatine. Under this condition we always observed a dextran decreased rate (V_{CK}) of the compartmentalized mi-CK (see Fig. 8). To determine the V_{max} of the mi-CK we performed a complete kinetic analysis and varied the creatine concentration between 6.25 and 100 mM in addition to a variation of the ATP levels. Further the backward reaction was measured with creatine phosphate in varied concentration. As shown in Table 4 dextran did not influence the V_{max} in both directions of mi-CK but significantly increased the dissociation constant (K_{ia}) for ATP as well as the Michaelis constants for creatine (K_b) and ADP (K_p). On the other hand the Michaelis constants for ATP (K_a)

and the dissociation constant for ADP (K_{ip}) were not significantly changed.

Effect of dextran on the K_{ADP} of oxidative phosphorylation

The above investigations of the dextran effects on the kinetic properties of mi-CK were performed with inhibited mitochondria. Since similar effects were found for mi-AK of rat liver mitochondria [40] it can be assumed that dextrans generally affect the exchange of adenine nucleotides between the i.m.s. and the cytosol. In this case dextrans should also increase the K_{ADP} of the oxidative phosphorylation. To proof this we titrated the rate of respiration of rat heart mitochondria in KCl medium and in sucrose medium with ADP. At least 24 U yeast hexokinase/mg mitochondrial protein was added to regenerate the ADP effectively. The results are shown in Fig. 9. In both media the K_{ADP} increased from ca 20 µM ADP (control) with increasing additions of dextran over 50 µM (15% dextran) up to 90 µM ADP at 25% dextran. At four independent experiments the K_{ADP} as measured in KCl-medium increased from 20 ±5 µM ADP in the absence to 144 ±36 µM ADP in the presence of 25% dextran. In the sucrose medium only additions up to 15% dextran were tested because at higher concentrations of dextran the active rate of respiration drastically decreased (Fig. 7). Similar observations were made for rat liver mitochondria [41]. In addition, we could show in detailed experiments that dextrans increased the AdN-compartmentation in the i.m.s. of rat liver mitochondria. With increasing dextran level the percentage of ADP as formed by mi-AK that was used for oxidative phosphorylation increased substan-

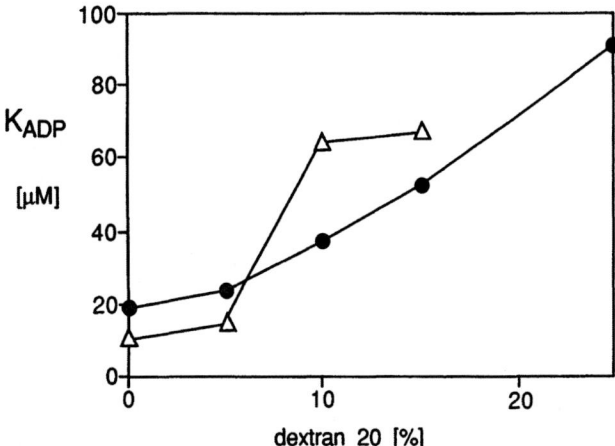

Fig. 9. Effect of dextran on app. K_{ADP} of oxidative phosphorylation. Rates of respiration of rat heart mitochondria were measured in sucrose (△) or KCl medium (●) with 10 mM glutamate and 2 mM malate as substrates and in the additional presence of 24 U HK/mg plus 1 mM glucose for an effective ADP regeneration.

tially [40]. In this type of experiments it was established that bovine serum albumin has the same effects as dextrans [40]. Therefore we concluded that dextrans are indeed appropriate to simulate the action of cytosolic proteins on the exchange of adenine nucleotides between the mitochondria and the cytosol in experiments with isolated mitochondria.

Analyzing the dextran effects on compartmentalized reactions we have to take into account that there are at least two steps determinating the overall flux through the investigated system. The first step is the compartmentalized enzyme and its control depends on its kinetic properties. The second step is the metabolite exchange between the compartment and the bulk phase. Since dextran affects the overall process, but not the kinetic properties of soluble enzymes we can conclude that the overall changes in kinetic properties of compartmentalized reactions are caused by changes in the transport processes. Therefore these results very clearly demonstrate that the understanding of *in vivo* regulation of oxidative phosphorylation requires information on intracellular compartments and the metabolite exchange between them.

Effects of dextrans on the electrical conductivity of reconstituted porin

As mentioned above, metabolites, which cross the mitochondrial outer membrane have to pass through the porin pores. It was previously shown by Zimmerberg and Persegian [42] that macromolecules reduce the electrical conductivity of porin pores inserted into black membranes. Since we observed that the dextran effects on rat liver mitochondria did not correlate with the molar concentration of dextrans but were dependent on the dextran content (w/v) [40] we used this typical behavior to test if the porin pores are included in the dextran effect. For that we isolated porin from rat liver mitochondria and investigated the effect of dextrans on the electrical conductivity of these pores by insertion into black membranes. Experiments were performed with small amounts of the isolated and purified pore protein added to both sides of a diphytanoyl lecithin bilayer membrane. Dextrans of different molecular weights (20, 70 or 500 kDa) were present in a concentration of either 10% (w/v) or 0.5 mM. The current at 10 mV membrane potential increased stepwise due to the insertions of single pores into the membrane. Under each condition we measured the changing of conductivity of the insertion of about 100 single pores. The relative number of insertions (observations) were plotted over the conductivity in nS. In agreement with the findings of Zimmerberg and Persegian [42], we observed a reduction in the electrical conductivity in the presence of dextrans (Fig. 10). In the absence of dextran most of the incorporated pores (35%) have a conductance of about 5 nS and a smaller group of 20% produced a conductance step of 2 nS. The addition of 10% of the different dextrans caused a shift towards a lower amplitude of 1–2 nS. Under these conditions less than 5% of the incorporated pores had a high conductivity of 5 nS. When 0.5 mM of the dextrans was added, the frequency of higher current steps decreased with the increase of molecular weight of the dextrans (Fig. 10, Table 5). In further experiments we measured the effect of dextrans on the current-voltage relationship of porin pores in black membranes. As commonly observed, the conductance of the pores incorporated into membranes decreased at a voltage above 30 mV [40]. In the presence of 10% dextrans 20, 70 and 500 as well as of 0.5 mM dextran 500 a significant increase in voltage sensitivity was observed, while this effect was less pronounced when adding dextran 20 and 70 at a concentration of 0.5 mM [40]. It was shown by Zimmerberg and Persegian that the addition of macromolecules to reconstituted porin pores caused changes in the pore structure to a low conductance state [42]. Our results are completely consistent with these observations. In addition, we found that these effects did not correlate with the dextran concentration but rather were dependent on the mass (w/v) of the macromolecules added regardless of their molecular weight.

98

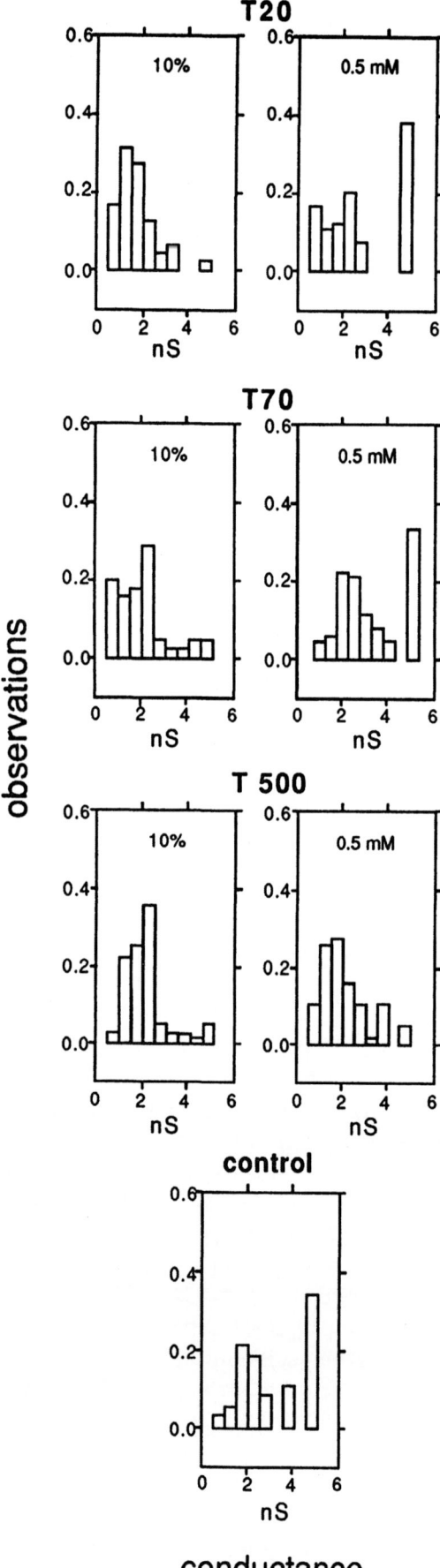

conductance

<-

Fig. 10. Effect of different dextrans on the conductivity of porin pores inserted into 'black' membranes. Histograms of conductance fluctuations of pores in membranes formed of diphytaloyl phosphatitylcholine/n-decane in 1 M KCl solution obtained as described previously [40]. Two ng rat liver porin/ml was used and a voltage of 10 mV was applied. The histograms were obtained from 50 to 110 single insertions into the membrane. The percentage of pores in the highest conductivity state (5 nS) of each histogram is presented in Table 5.

To obtain further evidence for the hypothesis that the porin pore is a central mediator of the dextran effect we measured the effect of different dextrans at 10% or 0.5 mM on the kinetic properties of compartmentalized mi-CK. As shown in Table 5 the effect of the various dextrans on mi-CK is practically the same at 10% (w/v) but very different when using them at 0.5 mM. The kinetic constants are compared with the effects of dextrans on the conductivity of porin pores. Obviously the high conductivity state of the porin pores (more than 30% pores at high conductivity of 5 nS) in black membranes as observed in the absence of dextran fits with a high permeability state of the outer mitochondrial membrane (low K_{ATP} of mi-CK) and vice versa. Although both sets of data were obtained by different experimental approaches the similar dependency of the measured parameters on the different dextrans support the conclusion that the dextrans keep the porin pores in a low permeability state thereby increasing the diffusion barrier between the i.m.s. and the extramitochondrial space.

Similar results were seen when testing different dextrans for their effects on the permeability of the outer membrane of rat liver mitochondria. In these experiments the reduction of the permeability of the outer mi-

Table 5. Effect of different dextrans on the conductivity of porin pores inserted into 'black membranes' as well as on kinetic constants of mitochondrial compartmentalized mi-CK

Addition		K_{ATP} [μM]	V_{CK} [U/mg]	Porin in a high conductivity state [%]
Control		255 ± 38	1.30 ± 0.06	32
Dextran 20	10%	480 ± 63	1.10 ± 0.05	2.5
	0.5 mM	293 ± 80	1.43 ± 0.1	37
Dextran M70	10%	593 ± 55	1.20 ± 0.05	5
	0.5 mM	316 ± 100	1.42 ± 0.1	32
Dextran M 500	10%	503 ± 69	1.10 ± 0.09	5
	0.5 mM	680 ± 100	0.70 ± 0.09	5

Kinetic constants of mi-CK were estimated as shown in Fig. 8. The percentage of porin pores in a high conductivity state (5 nS) were derived from the histograms shown in Fig. 10.

tochondrial membrane was deduced from the observed increased use of ADP, formed by mi-AK, for oxidative phosphorylation [40].

The mechanism of dextran effects on the permeability of the mitochondrial outer membrane

Several factors must be considered to explain the observed reduction of the metabolite exchange between the i.m.s. and the extramitochondrial compartment as brought about by an increased oncotic pressure (Fig. 11): (i) due to the increased number of contact sites [38] and the reduced volume of the i.m.s. [36–38], the resistance of lateral diffusion within the i.m.s. could have been reinforced, (ii) the permeability of the porin pore might have been reduced by oncotic stress [40, 42], (iii) the diffusion through the homogenous bulk phase to the pores could be hindered by the presence of macromolecules. The last possibility can be largely ruled out since dextran did not affect the kinetics of mi-CK in mitoplasts (see Fig. 8).

From the data presented above it can be concluded that the porin pores are involved in the dextran effects. However, the conductivity measurements of pores inserted into black membranes require non physiological conditions (1 M KCl). Therefore experiments are under way to elucidate further the contribution of the porin pores and the i.m.s. to the dextran effects.

Metabolite channelling into the mitochondrial intermembrane space

In the resting muscle the cytoplasmic free ADP concentration ranges from 1–60 µM [43–45]. These low ADP levels are advantageous to the thermodynamic efficiency of ATP hydrolysis [46]. They avoid the product inhibition of ATPases (as myosin, Ca-ATPase or K/Na-ATPase) and a net loss of adenine nucleotides via adenylate kinase and AMP desaminase [47]. Moreover, the low cytosolic ADP levels prevent the stimulation of glycolytic ATP production [48]. Up to now, the direct measurement of cytosolic ADP activity in the intact muscle is not possible. Therefore, it is usually calculated from ATP/CrP ratios determined by means of ^{31}P n.m.r. spectroscopy assuming creatine kinase equilibrium [49]. Under some conditions the calculated ADP levels in the heart muscle remain rather constant even if the load changes by a factor of 5 [26]. From these results, it was concluded

that the intramitochondrial calcium and the reducing equivalent supply to the mitochondrial respiratory chain are potential regulators in the control of oxidative phosphorylation [26, 50]. However even if the effector strength of calcium and reducing equivalents on the oxidative phosphorylation should really exceed that of ADP, then one serious problem remains; how is it possible to increase the ADP diffusion rate into the mitochondria 5 times without a changing in the ADP levels? This apparent discrepancy points to the intracellular compartmentation of ADP.

Obviously it seems inappropriate to assume that the i.m.s. together with the extramitochondrial space forms a homogenous pool for ADP. Recent work indicates that metabolites are micro-compartmentalized even without separation by membranes [51, 52]. We were able to demonstrate that at least in the i.m.s. ADP occurs compartmentalized. From the competition experiments performed with isolated mitochondria and pyruvate kinase (Figs 2–5) it became clear that the transport of the extramitochondrial formed ADP into the mitochondrial i.m.s. is a key element in cellular bioenergetics. Mimicking the intracellular conditions by addition of dextran to the incubation medium we could show, that the ADP diffusion into the i.m.s. is hindered by the macromolecules in comparison to incubations without dextran. To overcome this diffusion barrier higher concentration gradients are necessary than those determined for the absence of dextrans. However, the extent of dynamic compartmentation of a metabolite is limited by its total concentration. Due to its low cytosolic free concentration and the high K_{ADP} of oxidative phosphorylation the direct diffusion of ADP into the i.m.s. seems to be strongly diminished. The creatine diffusion however should not be limited since creatine is present in concentration higher than 10 mM allowing the existence of sufficiently high concentration gradients without having significant kinetic consequences. This is one basis for the metabolite channelling.

The general mechanism of such a channelling as based on proposals by Wittenberg [53] and Meyer et al. [54] is shown in the upper part of Fig. 12. Metabolite A is reversibly transformed into metabolite B by an enzyme localized in both i.m.s. and cytosol. Both A and B diffuse into the i.m.s. increasing the transport rate of A into the i.m.s.

In contrast to Wittenberg [53] and our previous publications [11, 16] we now avoid the term facilitated diffusion, since no protein in these shuttles is involved in the direct transport like carrier proteins in the glucose up-

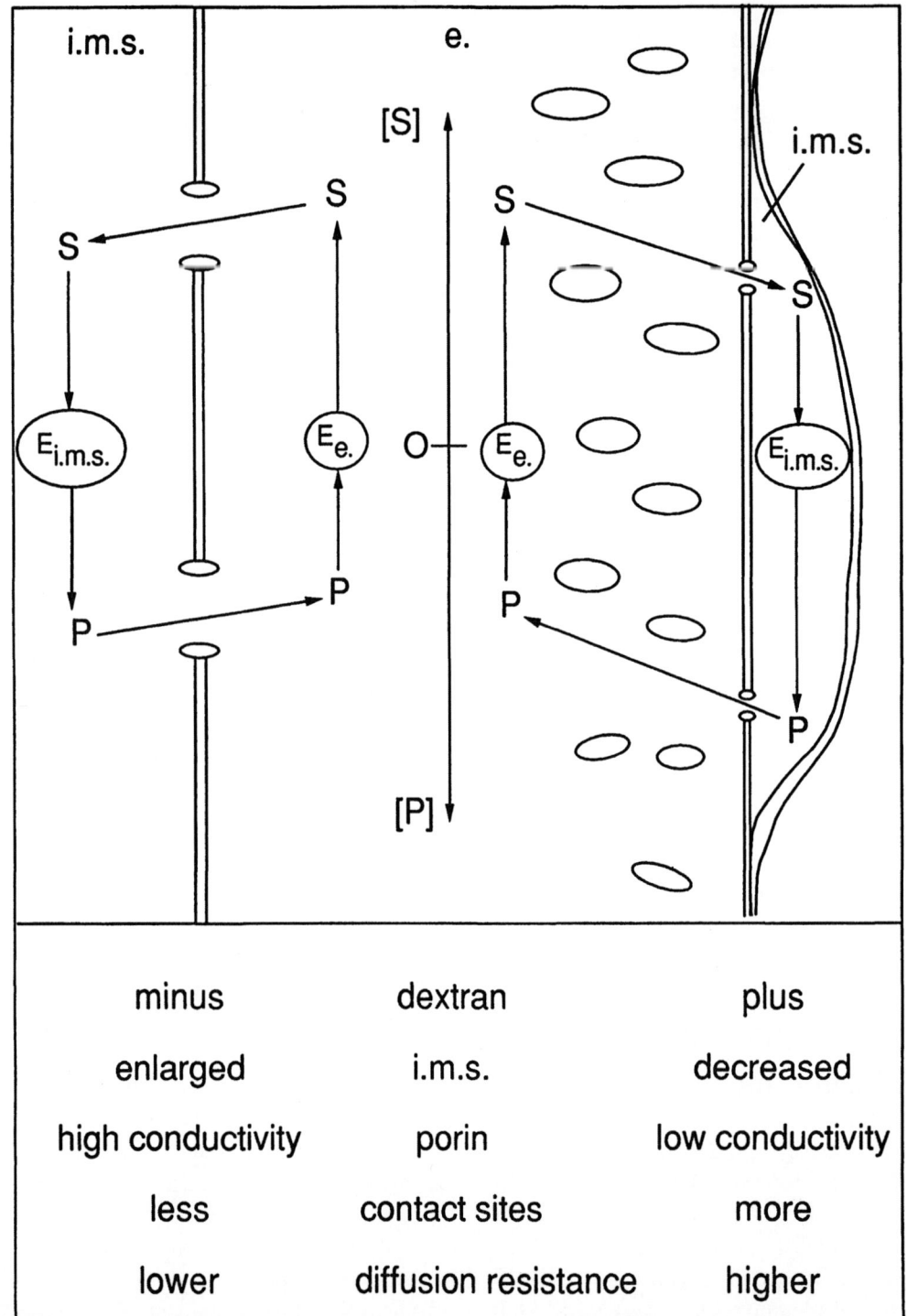

*Fig. 11.*Scheme representing the possible mechanisms of the action of macromolecules on the exchange of adenine nucleotides across the outer mitochondrial membrane. Enzymes localized within the intermembrane space ($E_{i.m.s.}$) are dynamically separated from those in the extramitochondrial space (E_e). Macromolecules reduce the volume of the i.m.s. [36–38], increase the number of contact sites [38] and keep the porin pores in a more closed state [41, 42]. These morphologically detectable changes are accompanied by an increased diffusion resistance through the outer membrane thereby requiring increased concentration gradients between the i.m.s. and the extramitochondrial space to achieve similar fluxes. In the i.m.s. these gradients cause decreased substrate and increased product concentrations in comparison to the extramitochondrial phase.

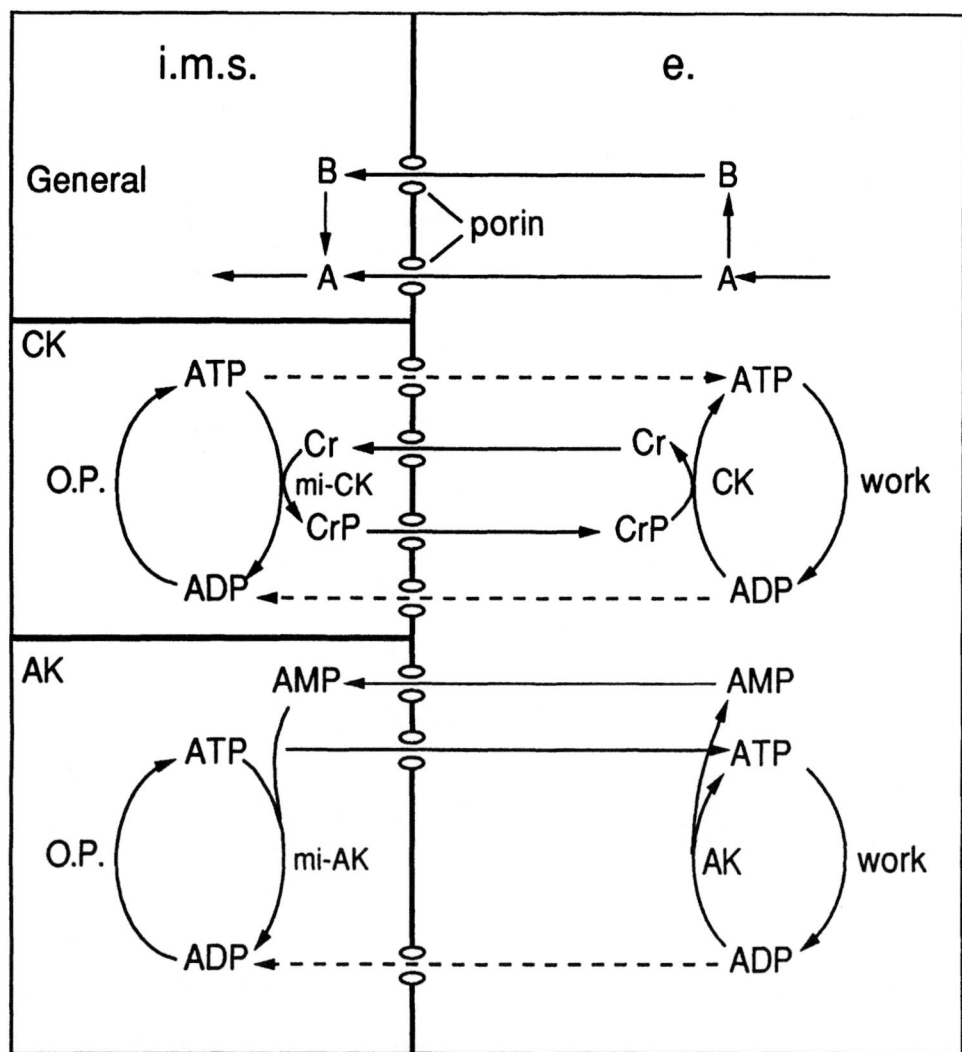

*Fig. 12.*Scheme representing the mechanism of ADP transport into the mitochondrial intermembrane space by means of a metabolite shuttle. General: General mechanism according to Meyer *et al.* [54]. A, B, metabolites. CK: CrP-shuttle; CrP, creatine phosphate; Cr, creatine; O.P., oxidative phosphorylation; mi-CK and CK, mitochondrial CK and cytosolic CK, respectively. AK: adenylate kinase shuttle; mi-AK and AK, mitochondrial AK and cytosolic AK, respectively. Dashed arrows: transport routes of minor importance at low substrate concentrations.

take across cell membranes or myoglobin in the intracellular oxygen transport. In contrast to other well investigated metabolite shuttles such as the malate/aspartate shuttle (transporting metabolites through the mitochondrial inner membrane) metabolite shuttles into different metabolic compartments are considered here. These compartments need not be separated from each other by membranes. Since the metabolites involved can diffuse through the porin pores, they do not require specific translocator proteins. In the lower part of Fig. 12 two shuttles are shown which can be regarded as special forms of the general mechanism. One is the widely accepted creatine phosphate shuttle (for a recent review see 45).

As a second shuttle, an adenylate kinase shuttle is proposed acting in a similar way to that of the creatine phos-

phate shuttle because mitochondrial and cytosolic adenylate kinase isoenzymes are compartmentalized similar to that of the creatine kinase isoenzymes. In this shuttle AMP carries ADP equivalents (like creatine) into mitochondria: AMP formed in the cytosol from ADP via cytosolic adenylate kinase diffuses into the i.m.s. forming ADP there via mitochondrial adenylate kinase. The adenylate kinase shuttle was originally proposed by Bessman [55]. It may operate in tissues with sufficiently high adenylate kinase activities such as liver, bovine spermatozoa [56] or certain muscles at least under conditions of elevated AMP concentrations. Recently it was shown for rat diaphragm muscle that the AK mediated transport of ADP-equivalents into the mitochondria is of minor importance. In this muscle the adenylate kinase rather realizes a close functional coupling of myosin AT-

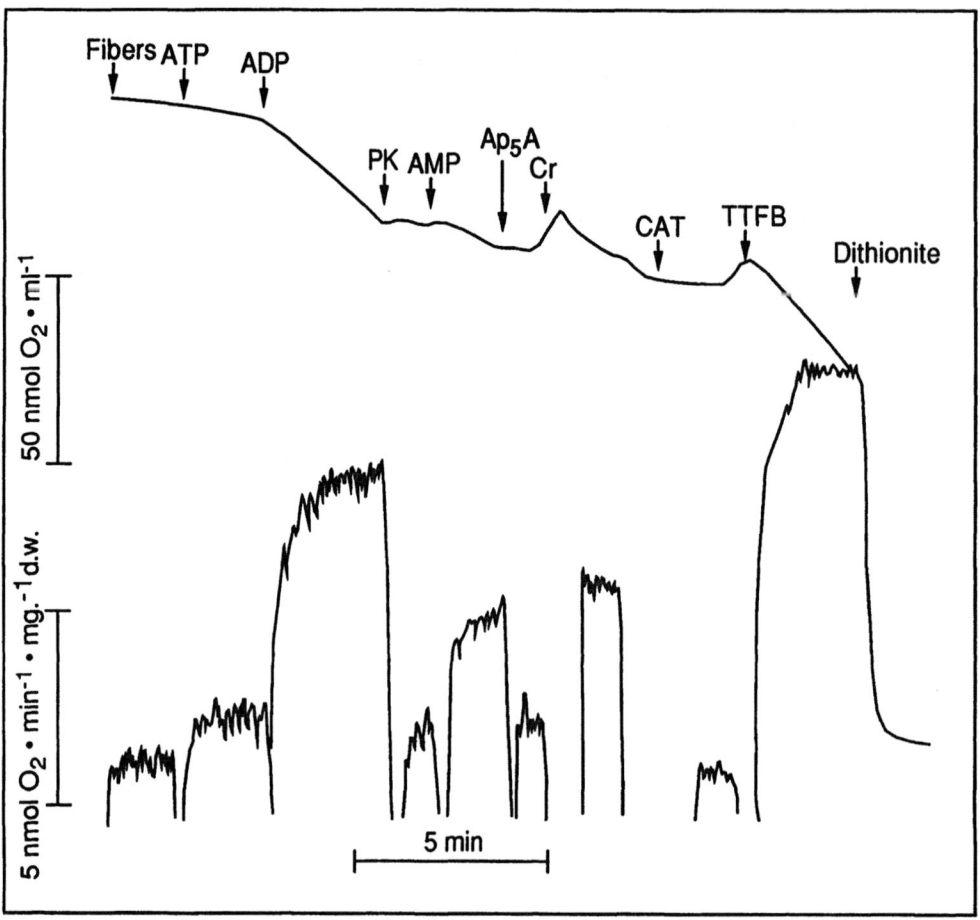

Fig. 13. Stimulation of the respiration of human skinned muscle fibers via mi-AK or mi-CK in the presence of excess PK. Preparation and skinning of muscle fibers with saponin as described previously [59]. Muscle probes were obtained from orthopaedic patients. The respiration was measured in a medium consisting of 75 mM mannitol, 25 mM sucrose, 100 mM KCl, 10 mM KH_2PO_4, 5 mM $MgCl_2$ and 20 mM Tris (pH = 7.4) in a CYCLOBIOS-Oxygraph. Additions: Fiber 3.9 mg dry weight, ATP, 2 mM; ADP, 1 mM; PK, 80 U/mg; AMP, 2 mM; AP_5A, 0.5 mM; Creatine; 25 mM; CAT, 10 μM; TTFB, 10 μM; Dithionite, a few mg $Na_2S_2O_4$.

Pase with the glycolytic ATP production [58]. However, for ejaculated bovine sperms (which contain no creatine kinases) it was shown that an adenylate kinase shuttle contributes to the ADP transport into the mitochondria [56].

In order to evaluate in which mammalian tissues an adenylate kinase shuttle could contribute to the ADP transport into mitochondria we analysed the stimulation of mitochondrial respiration by AMP (in the presence of ATP). The highest specific activity of mi-AK present in liver correlates well with the maximal stimulation of state 3 respiration of rat liver mitochondria by AMP. Under the same conditions however brain mitochondria exhibit only a negligible stimulation due to their very low activities of mi-AK. Interestingly we found for several human skeletal muscles a complete AMP stimulation of mitochondrial oxidative phosphorylation. This mi-AK stimulated respiration is at least comparable to the mi-CK stimulated respiration or even higher (Gel-

lerich FN, unpublished results). Moreover, it could be shown for human skinned muscle fibers in which the ADP stimulated respiration in suppressed by excess PK that AMP is able to stimulate the mitochondrial respiration in an extent similar as creatine does (Fig. 13). These data indicate the possibility that in the human skeletal muscle the adenylate kinase shuttle could contribute in addition to the creatine phosphate shuttle to the ADP transport into the mitochondria.

A precondition for the function of a metabolite shuttle is that the conversion between substrates A and B is directed differently in both compartments. Despite of the fact that mitochondrial and cytosolic isoenzymes involved in these shuttles have different kinetic properties [58], they can from thermodynamic reasons work in different directions only if the metabolite pattern is sufficiently different in both compartments. The experimentally determined dynamic ADP concentration gradients seem to be adequate to account for different directions of mitochondrial and extramitochondrial creatine kinase in the presence of virtually unchanged concentrations of ATP, Cr and CrP. Considering the fact that the rate dependent concentration gradients of ADP across the mitochondrial outer membrane were determined in the absence of macromolecules it is to be expected that still larger concentration gradients exist at high physiological oncotic pressure. This would imply that in the intact cell the preconditions for ADP metabolite shuttles into the i.m.s. are fulfilled.

Acknowledgements

F.N.G. thanks the Netherlands Organization for Scientific Research (NWO) for supplying a visitors grant allowing to continue the experimental work on this topic in Utrecht and to write this paper. Part of this work was supported by the DFG (Ge 663/1-1; Ge 663/3-1; Br 773/3-2). The authors thank Prof. R. Bohnensack and Dr. M. Schlame for computer simulation of the results shown in Fig. 5 and the estimation of the diffusion rate constant as well as Prof. W. Kunz for stimulating discussions.

References

1. Colombini M, Yeung CL, Tung J, König T: The mitochondrial outer membrane channel, VDAC, is regulated by a synthetic polyanion. Biochim Biophys Acta 905: 279–286, 1987
2. Zalmann LS, Nikaido H, Kagawa Y: Mitochondrial outer membrane contains a protein producing nonspecific diffusion channels. J Biol Chem 255: 1771–1774, 1980
3. Glick BS, Beasley EM, Schatz G: Protein sorting in mitochondria. TIBS 17: 453–459, 1992
4. Brdiczka D: Contact sites between mitochondrial envelope membranes. Structure and function in energy- and protein-transfer. Biochim Biophys Acta 1071: 291–312, 1991
5. Vial C, Godinot C, Gautheron D: Membranes creatine kinase (E.C. 2.7.3.2.) in pig heart mitochondria. Properties and role in phosphate potential regulation. Biochimie 54: 843–852, 1972
6. Bessman SP, Gots RE: The hexokinase acceptor theory of insulin action – Hormone control of functional compartmentation. Life Sci 16: 1215–1225, 1975
7. Saks VA, Chernousova GB, Gukovski DE, Smirnov VN, Chazov EI: Studies of energy transport in heart cells. Mitochondrial isoenzyme of creatine phosphokinase, kinetic properties and regulatory action of Mg^{2+} ions. Eur J Biochem 57: 237–2290, 1975
8. Gellerich FN, Augustin HW: Studies on the functional significance of mitochondrial bound hexokinase in rabbit reticulocytes. Acta Biol Med Germ 36: 571–577, 1977
9. Gosalvez M, Perez-Garcia J, Weinhouse S: Competition for ADP between pyruvate kinase and mitochondrial oxidative phospho-

rylation as a control mechanism in glycolysis. Eur J Biochem 46: 133–140, 1974
10. Gellerich FN, Saks VA: Control of heart mitochondrial oxygen consumption by creatine kinase: The importance of enzyme localization. Biochem Biophys Res Commun 1105: 1473–1481, 1982
11. Gellerich FN, Schlame M, Bohnensack R, Kunz W: Dynamic compartmentation of adenine nucleotides in the mitochondrial intermembrane space of rat heart mitochondria. Biochim Biophys Acta 722: 381–391, 1987
12. Gellerich FN: The role of adenylate kinase in dynamic compartmentation of adenine nucleotides in the mitochondrial intermembrane space. FEBS Lett 297: 55–58, 1992
13. Kottke M, Adams V, Wallimann T, Nalam VK, Brdiczka D: Localization and regulation of octameric mitochondrial creatine kinase in the contact sites. Biochim Biophys Acta 1061: 215–225, 1991
14. Gellerich FN, Bohnensack R, Kunz W: Control of mitochondrial respiration: The contribution of the adenine nucleotide translocator depends on the ATP and ADP consuming enzymes. Biochim Biophys Acta 722: 381–391, 1983
15. Watanabe K, Itakura T, Kubo S: Distribution of adenylate kinase isozymes in porcine tissues and their subcellular localization. J Biochem 85: 799–805, 1979
16. Gellerich FN, Bohnensack R, Kunz W: Role of the mitochondrial outer membrane in dynamic compartmentation of adenine nucleotides. In: A Azzi, KA Nalecz, MJ Nalecz, L Wojtczak (eds) The Anion Carriers of the Mitochondrial Membranes. Springer Verlag, Berlin Heidelberg. 1989, pp 349–359
17. Gellerich FN, Kunz W: Cause and consequences of dynamic compartmentation of adenine nucleotides in the mitochondrial intermembrane space in respect to exchange of energy rich phosphates between cytosol and mitochondria. Biomed Biochim Acta 46: 545–548, 1987
18. Bohnensack R: Rate law of mitochondrial respiration versus extramitochondrial ATP/ADP ratio. Biomed Biochim Acta 43: 403–411, 1984
19. Gellerich FN, Khuchua ZA, Kuznetsov A: Influence of the mitochondrial outer membrane and the binding of creatine kinase to the mitochondrial inner membrane on the compartmentation of adenine nucleotides in the intermembrane space of rat heart mitochondria. Biochim Biophys Acta 1140: 327–334, 1993
20. Kushmerick MJ, Podolsky RJ: Ionic mobility in muscle cells. Science 166: 1297–1298, 1969
21. Roos M, Benz R, Brdiczka D: Identification and characterization of the pore forming protein in the outer membrane of rat liver mitochondria. Biochim Biophys Acta 686: 204–214, 1982
22. Schwerzman K, Cruz-Orive LM, Eggman R, Sänger A, Weibel ER: Molecular architecture of the inner membrane of mitochondria from rat liver: A combined biochemical and stereological study. J Cell Biol 102: 97–103, 1986
23. Bessman SP, Carpenter CL: The creatine-creatine phosphate energy shuttle. Ann Rev Biochem 54: 831–865, 1985
24. Brooks SPJ, Sueter CH: Compartmentated coupling of chicken heart mitochondrial creatine kinase to the nucleotide translocase requires the outer mitochondrial membrane. Arch Biochem Biophys 267: 13–22, 1987
25. Saks VA, Vasiléva E, Belikova YuO, Kuznetsov AV, Lyapina S, Petrova L, Perov NA: Retarded diffusion of ADP in cardiomyocytes: possible role of mitochondrial outer membrane and creatine kinase in cellular regulation of oxidative phosphorylation. Biochim Biophys Acta 1144: 134–148, 1993

104

26. Heineman FW, Balaban RS: Control of mitochondrial respiration in the heart *in vivo*. Annu Rev Physiol 52: 523–542, 1990

27. Saks VA, Kupriyanov VV, Elizarova GV, Jacobus WE: Studies of energy transport in heart cells. The importance of creatine kinase localization for the coupling of mitochondrial phosphorylcreatine production to oxidative phosphorylation. J Biol Chem 255: 755–763, 1983

28. Kuznetsov VA, Khuchua ZA, Vassileva EV, Medvedeva NV, Saks VA: Heart mitochondrial creatine kinase revisited: The outer mitochondrial membrane is not important for coupling of phosphocreatine production to oxidative phosphorylation. Arch Biochem Biophys 268: 176–190, 1989

29. Wenger WC, Murphy MP, Brierley GP, Altshuld RA: Effect of ionic strength and sulfhydryl reagents on the binding of creatine phosphokinase to heart mitochondrial inner membranes. J Bioenerg Biomembr 17: 295–303, 1986

30. Fossel ET, Hoefeler H: A synthetic functional metabolic compartment. The role of propinquity in a linked pair of immobilized enzymes. Eur J Biochem 170: 165–171, 1987

31. Mattiasson B, Mosbach K: Studies on a matrix-bound three-enzyme system. Biochim Biophys Acta 235: 253–257, 1971

32. Hall N, DeLuca M: The effect of inorganic phosphate on creatine kinase in respiring rat heart mitochondria. Arch Biochem Biophys 229: 477–482, 1984

33. Saks VA, Khuchua ZA, Kuznetsov AV: Specific inhibition of ATP-ADP translocase in cardiac mitoplasts by antibodies against mitochondrial creatine kinase. Biochim Biophys Acta 891: 138–144, 1987

34. Woijceszyn JW, Wu ES, Jacobsen KA: Diffusion of injected macromolecules within the cytoplasm of living cells. Proc Natl Acad Sci USA 78: 4407–4410, 1981

35. Cameron IL, Fullerton GD: A model to explain the osmotic behavior of hemoglobin and serum albumin. Biochem Cell Biol 68: 894–898, 1990

36. Bakeeva LE, Chentsov YS, Jasaitis AA, Skulachev VP: The effect of oncotic pressure on heart muscle mitochondria. Biochim Biophys Acta 275: 319–332, 1972

37. Wrogemann K, Nylen EG, Adamson I, Pande SV: Functional studies on *in situ*-like mitochondria isolated in the presence of polyvinyl pyrrolidon. Biochim Biophys Acta 806: 1–8, 1986

38. Wicker U, Bücheler K, Gellerich FN, Wagner M, Kapischke M, Brdiczka D: Effect of macromolecules on the structure of the mitochondrial inter-membrane space and the regulation of hexokinase. Biochim Biophys Acta 1142: 228–239, 1993

39. Srere PA: Complexes of sequential enzymes. Annu Rev Biochem 56: 89–124, 1987

40. Gellerich FN, Wagner M, Kapischke M, Wicker U, Brdiczka D: Effect of macromolecules on the regulation of the mitochondrial outer membrane pore and the activity of adenylate kinase in the inter-membrane space. Biochim Biophys Acta 1142: 217–227, 1993

41. Gellerich FN, Kapischke M, Wagner M, Brdiczka D: Influence of macromolecules on the permeability of porin pores and dynamic compartmentation of adenine nucleotides in the mitochondrial intermembrane space. In: M Colombini, M Forte (eds) Proceedings on the NATO advanced research workshop (ARW) on: Molecular biology of mitochondrial transport systems. Springer Verlag, Berlin Heidelberg, 1993

42. Zimmerberg J, Persegian VA: Polymer inaccessible volume changes during opening and closing of a voltage-dependent ionic channel. Nature 323: 36–39, 1986

43. Bittl JA, Ingwall JS: Reaction rates of creatine kinase and ATP synthesis in the isolated rat heart. A 31P NMR magnetization transfer study. J Biol Chem 260: 3512–3517, 1985

44. Meyer RA, Brown TR, Kushmerick MJ: Phosphorus nuclear magnetic resonance of fast- and slow-twich muscle. Am J Physiol 248: C279–C287, 1985

45. Wallimann T, Wyss M, Brdiczka D, Nicolay K, Eppenberger HM: Intracellular compartmentation, structure and function of creatine kinase isoenzymes in tissues with high and fluctuating energy demands: the 'phosphocreatine circuit' for cellular energy homeostasis. Biochem J 281: 21–40, 1992

46. Stucki JW: The thermodynamic-buffer enzymes. Eur J Biochem 109: 257–267, 1980

47. Iyengar MR, Fluellen CE, Iyengar CL: Creatine kinase from the bovine myometrium: purification and characterization. J Muscle Res Cell Motil 3: 231–246, 1982

48. Kupriyanov VV, Seppet EK, Emilin IV, Saks VA: Phosphocreatine production coupled to the glycolytic reactions in the cytosol of cardiac cells. Biochim Biophys Acta 592: 197–210, 1980

49. Ackermann JH, Grove TH, Wong GG, Gadian DG, Radda GK: Mapping of metabolites in whole animals by ^{31}P NMR using surface coils. Nature, London 283: 167–170, 1980

50. Denton RM, McCormack JG: On the role of the calcium transport cycle in heart and other mammalian mitochondria. FEBS Lett 119: 1–8, 1980

51. Jones DP: Intracellular diffusion gradients of O_2 and ATP. Am J Phys 250: C663–C675, 1986

52. Miller DS, Horowitz SB: Intracellular compartmentalization of adenosine triphosphate. J Biol Chem 261: 13911–13916, 1986

53. Wittenberg JB: Myoglobin-facilitated oxygen diffusion: Role of myoglobin in oxygen entry in the muscle. Phys Rev 50: 559–636, 1970

54. Meyer RA, Sweeny L, Kushmerick MJ: A simple analysis of the 'phosphocreatine shuttle'. Am J Physiol 246: C365–C377, 1984

55. Bessman SP, Carpenter CL: The creatine-creatine phosphate energy shuttle. Annu Rev Biochem 54: 831–862, 1985

56. Schoff PK, Cheetham J, Lardy HA: Adenylate kinase activity in ejaculated bovine sperm flagella. J Biol Chem 264: 6086–6091, 1989

57. Zeleznikar RJ, Heyman RA, Graeff RM, Waslseth TF, Davis SM, Butz EA, Goldberg ND: Evidence for compartmentalized adenylate kinase catalyzis serving a high energy phosphoryl transfer function in rat skeletal muscle. J Biol Chem 265: 300–311, 1990

58. Bittl JA, DeLayre J, Ingwall JS: Rate equation for creatine kinase predicts the *in vivo* reaction velocity: ^{31}P NMR surface coil studies in brain, heart and skeletal muscle of the living rat. Biochem ioltry 26: 6083–6090, 1987

59. Kunz WS, Kuznetsov AV, Schulze W, Eichhorn K, Schild L, Striggow F, Bohnensack R, Neuhof S, Grashoff H, Neumann HW, Gellerich FN: Functional characterization of mitochondrial oxidative phosphorylation in saponin-skinned human muscle fibers. Biochim Biophys Acta 1144: 46–53, 1993

Molecular and Cellular Biochemistry **133/134**: 105–113, 1994.
© 1994 *Kluwer Academic Publishers.*

II-3 Influence of mitochondrial creatine kinase on the mitochondrial/extramitochondrial distribution of high energy phosphates in muscle tissue: evidence for a leak in the creatine shuttle

Sibylle Soboll,[1] Annette Conrad[1] and Siegbert Hebisch[2]

[1] *Institut für Physiologische Chemie I, Universität Düsseldorf, 40225 Düsseldorf, Universitätsstraße 1, Germany;* [2] *Institut für Herz- und Kreislaufforschung, Bayer AG, 42096 Wuppertal, Aprather Weg, Germany*

Abstract

The influence of mitochondrial creatine kinase on subcellular high energy systems has been investigated using isolated rat heart mitochondria, mitoplasts and intact heart and skeletal muscle tissue.

In isolated mitochondria, the creatine kinase is functionally coupled to oxidative phosphorylation at active respiratory chain, so that it catalyses the formation of creatine phosphate against its thermodynamic equilibrium. Therefore the mass action ratio is shifted from the equilibrium ratio to lower values. At inhibited respiration, it is close to the equilibrium value, irrespective of the mechanism of the inhibition. The same results were obtained for mitoplasts under conditions where the mitochondrial creatine kinase is still associated with the inner membrane.

In intact tissue increasing amounts of creatine phosphate are found in the mitochondrial compartment when respiration and/or muscle work are increased. It is suggested that at high rates of oxidative phosphorylation creatine phosphate is accumulated in the intermembrane space due to the high activity of mitochondrial creatine kinase and the restricted permeability of reactants into the extramitochondrial space. A certain amount of this creatine phosphate 'leaks' into the mitochondrial matrix.

This leak is confirmed in isolated rat heart mitochondria where creatine phosphate is taken up when it is generated by the mitochondrial creatine kinase reaction. At inhibited creatine kinase, external creatine phosphate is not taken up. Likewise, mitoplasts only take up creatine phosphate when creatine kinase is still associated with the inner membrane. Both findings indicate that uptake is dependent on the functional active creatine kinase coupled to oxidative phosphorylation.

Creatine phosphate uptake into mitochondria is inhibited with carboxyatractyloside. This suggests a possible role of the mitochondrial adenine nucleotide translocase in creatine phosphate uptake.

Taken together, our findings are in agreement with the proposal that creatine kinase operates in the intermembrane space as a functional unit with the adenine nucleotide translocase in the inner membrane for optimal transfer of energy from the electron transport chain to extramitochondrial ATP-consuming reactions. (Mol Cell Biochem **133/134**: 105–113, 1994)

Key words: subcellular creatine concentrations, creatine shuttle, adenine nucleotide translocase

Adress for offprints: S. Soboll, Institut für Physiologische Chemie I, Universität Düsseldorf, 40225 Düsseldorf, Universitätsstraße 1, Germany

106

Introduction

The mitochondrial and extramitochondrial creatine systems are the constituents of a creatine shuttle in heart and skeletal muscle (for a recent review see [1]). The main function of this pathway is the effective channelling of mitochondrial ATP across the intermembrane space to the myofibrils where it is consumed in the contraction cycle. The transport of energy in the cytosol is mediated by the creatine pool consisting of creatine (Cr) and creatine phosphate (CP) which are present there in high concentrations and converted via the local creatine kinases. By this way the cytosolic phosphorylation potential can be kept in a reasonable high range which is of basic importance for the functioning of muscle tissue.

The presumption for effective transport of mitochondrial ATP to the site of consumption via the creatine shuttle is that the mitochondrial creatine kinase (MiCK) located at the outer site of the inner mitochondrial membrane is shifted away from equilibrium to synthesise CP by a close coupling to oxidative phosphorylation [2, 3].

Three main hypotheses have been raised for the mechanism of coupling of mitochondrial creatine kinase to oxidative phosphorylation:

(i) coupling by compartmentation of adenine nucleotides within the inner membrane space [4]. In principle, it would be sufficient to raise the concentration of ATP and ADP in the intermembrane space of mitochondria simply by a diffusion barrier (i.e. the outer mitochondrial membrane) so that ADP could be rapidly removed into the matrix space by the adenine nucleotide translocase in exchange for mitochondrial ATP. By this way, at a high rate of oxidative phosphorylation and adenine nucleotide transport, MiCK would be forced to synthesise CP.

(ii) the same result can be obtained by a close association of the mitochondrial adenine nucleotide translocase with the MiCK as suggested by [5–7].

(iii) a perfect coupling of both processes is achieved by a 'functional coupling unit' in the contact sites between inner and outer mitochondrial membrane [8] consisting of mitochondrial adenine nucleotide translocase and creatine kinase in the inner membrane and porin in the outer membrane. It is the octamer form of the MiCK which is supposed to form this complex in the contact sites [9].

For all three models evidence has been presented so that one might suppose that *in vivo* all three mechanisms are in operation and every one might predominate in a special metabolic condition.

As a consequence of the limited diffusion of the reactants of the mitochondrial creatine kinase out of the intermembrane space, CP should accumulate there when the activity of the creatine shuttle is high. The activity of the creatine shuttle is difficult to estimate, since it is only operative in the intact cell and there flux rates can only be measured indirectly. It is especially difficult regarding creatine kinase since there two systems operate in the opposite direction, the mitochondrial enzyme towards CP-synthesis, whereas the extramitochondrial systems regenerate ATP. Mitochondrial CP, i.e. CP from the intermembrane space, could be an indicator for the activity of the creatine shuttle. In this study evidence is presented showing that CP from the intermembrane space 'leaks' into the matrix and that matrix CP is therefore a good indicator for the operation of the creatine shuttle.

Materials and methods

Materials

All enzymes and coenzymes were either from Boehringer (Mannheim, Germany) or from Sigma Chemie (München, Germany), chemicals were from Merck (Darmstadt, Germany) and were all of highest purity available.

Isolated perfusion of guinea pig hearts

Guinea-pigs of either sex with a body weight of 200–300 g, fed with a standard diet (Altromin, Lage, Germany) and water ad libitum, were killed and the hearts perfused in a Langendorff [10] mode as described in [3].

Experiments were started after an equilibration period of 1 h. KCl, atractyloside and isoprenaline were infused into the aortic cannula with a flow of 0.1 ml/min. Oxygen uptake was measured with a Clarke-type electrode in the effluent perfusate. The experiment was terminated by freeze-clamp. At this time all experimental protocols had reached a steady state.

Stimulation of rat gastrocnemius muscle [11]

In each experiment the left gastrocnemius muscle of the anaesthetised rat was stimulated isotonically via n. ischiadicus with a frequency of 5 Hz, pulse width of 1 ms and a variable amplitude of 1–3 V. The contralateral muscle was taken as control. The muscles were freeze-clamped *in situ* after 0.5, 1.0, 3.0 and 5 min of stimulation.

Determination of mitochondrial and extramitochondrial metabolite concentrations

Fractionation of tissue in non-aqueous solvents [12]

Left ventricular tissue or gastrocnemius muscle tissue was ground in liquid nitrogen and lyophilised at $-40°$ C. The dry tissue powder was sonicated in heptane/CCl_4, density 1.23 g/ml, in 5 s intervals for 5 min under continuous cooling in heptane/solid CO_2; the homogenate was then fractionated by density gradient centrifugation in heptane/CCL_4, the densities ranging from 1.29–1.38 g/ml. Centrifugation ($16500 \times$ g for 3 h) in a swing out rotor yielded 8 fractions each containing different proportions of mitochondrial and cytosolic protein. In each fraction the specific activities of marker enzymes for cytosol (phosphoglycerate kinase) and mitochondria (citrate synthase), as well as the contents of high energy phosphates and creatine were determined by enzymatic analysis [13] and mitochondrial and extramitochondrial contents were obtained by extrapolation from the contents in the fractions to pure mitochondrial and extramitochondrial fractions.

Isolation and incubation of rat heart mitochondria

Isolation of mitochondria and mitoplasts

Heart mitochondria from rats were isolated essentially according to [14] in a medium containing 0.25 M sucrose, 10 mM Tris, 5 mM KH_2PO_4, 20 mM KCl, 0.2 mM EDTA, pH 7.2, using 1 mg trypsin for 3 hearts. The mitochondria had respiratory ratios of 8–15 in the presence of 2.5 mM glutamate and 2 mM malate and in the absence of magnesium. In the presence of 5 mM $MgCl_2$ the respiratory control ratios were 6–8. The rate of respiration was about 200 nmol O_2/min $\times$ mg protein at 37° C.

Mitoplasts were prepared from mitochondria by incubating the mitochondrial pellet in 20 mM TES (2-trishydroxymethylaminomethane sulfonic acid) for 5 to 60 min (modified from [2]). Incubation of mitoplasts was performed as above with special additions or treatments indicated in tables and figures. Where 125 mM KCl were added, the osmolarity was maintained by decreasing the sucrose concentration.

Determination of the mass action ratio of mitochondrial creatine kinase

About 1 mg mitochondrial protein/ml was incubated in the medium described above with the addition of 5 mM $MgCl_2$, 1 mM ATP, 10 mM Cr, 2.5 mM glutamate and 2 mM L-malate for 20–60 min. CP, Cr, ATP and ADP were measured in perchloric acid extracts of the suspension using enzymatic analysis [13]. The mass action ratio of creatine kinase, Kapp = [ATP][Cr]/[ADP][CP], was calculated as described in [3].

Determination of CP-uptake into rat heart mitochondria by silicone oil centrifugation [16]

Mitochondria were incubated for 2 min as described above with substrate additions as indicated in tables and figures. The osmolarity was varied by varying sucrose and choline chloride concentrations in the medium. 200 µl of the suspension was then separated into mitochondria and supernatant by centrifugation through silicone oil into perchloric acid. Carry over of medium into the pellet was determined in parallel samples, containing ^{14}C-sucrose, from the amount of radioactivity found in the pellet [17]. CP was determined [13] in perchloric acid extracts of pellet and supernatant, respectively.

Results and discussion

Coupling of mitochondrial creatine kinase and oxidative phosphorylation

In 1985 Saks [2] showed that the mass action ratio of mitochondrial creatine kinase is shifted towards CP synthesis, i.e. towards a lower value compared to the equilibrium ratio, at active respiration of the mitochondria. This is confirmed by our study (Fig. 1A): when oxidative phosphorylation is inhibited by different ways, i.e. by inhibiting mitochondrial adenine nucleotide translocase, mitochondrial ATP-synthase or in the absence of oxygen, the mass action ratio is restored to the equilibrium ratio, which, using our incubation conditions, is 122 [3].

In a second series of experiments we wanted to know, whether other factors, besides active oxidative phosphorylation, influence the mass action ratio of mito-

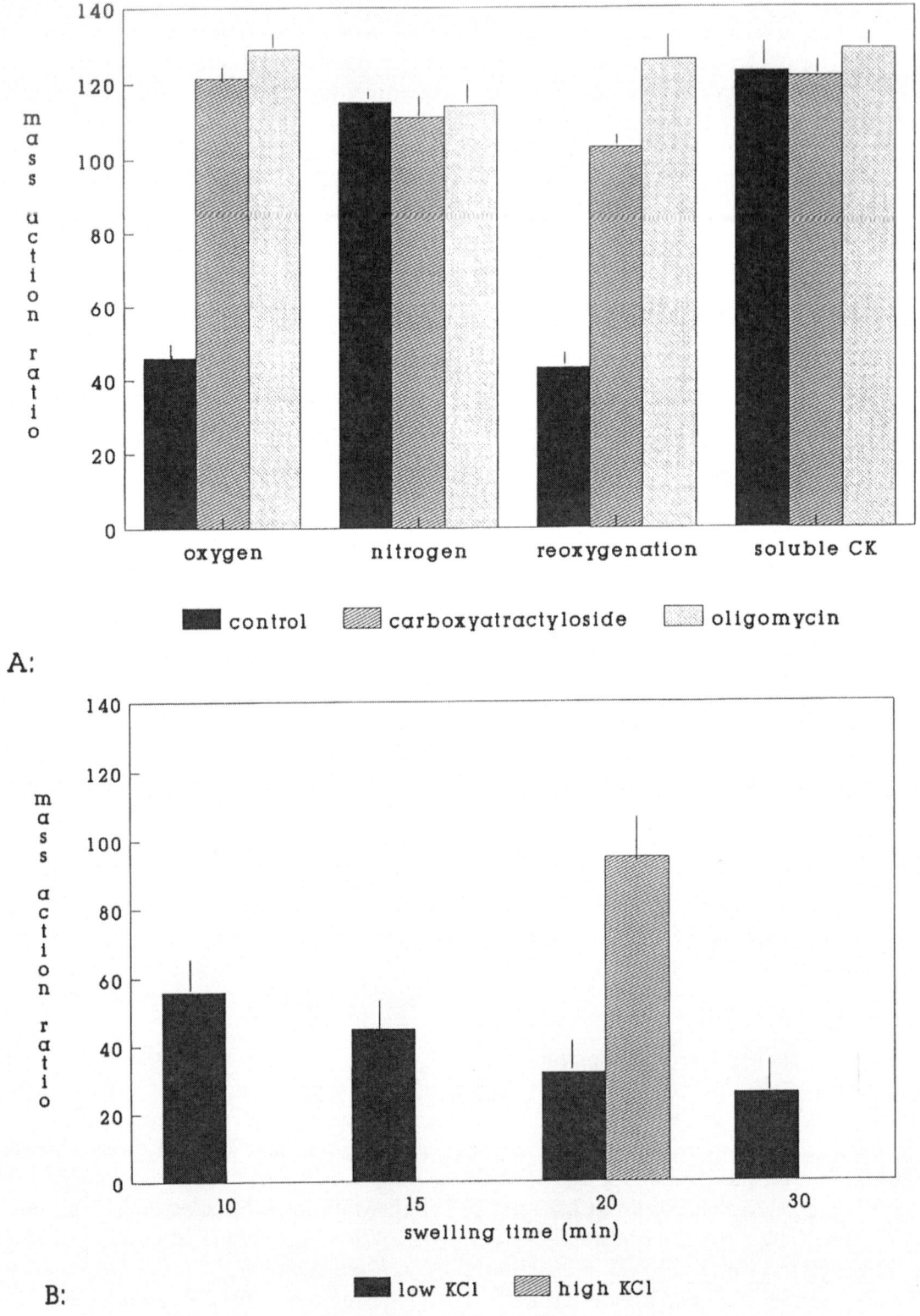

Fig. 1. Dependence of the mass action ratio of mitochondrial creatine kinase on coupling to oxidative phosphorylation in isolated rat heart mitochondria. Incubation conditions see Methods; n = 4 ± SEM; A: isolated mitochondria; 2 μM carboxy-atractyloside; 20 μM oligomycin; from [3]. B: isolated mitoplasts; swelling time is time where mitochondria were subjected to hypoosmolaric incubation; high KCl = 125 mM KCl.

chondrial creatine kinase. Thus is was shown by [4] that microcompartmentation of high energy phosphates in the intermembrane space regulates the supply of these compounds towards MiCK and thus CP synthesis. Therefore Kapp was determined in incubations using mitoplasts prepared from mitochondria by incubation in 10 mM TES. By this treatment the outer mitochondrial membrane is removed from the mitochondria but MiCK is only dissociated by about 15% (not shown) from the inner mitochondrial membrane. Kapp was shifted far from equilibrium (Fig. 1B) and the effect was more pronounced when the mitochondria were treated with hypoosmolaric TES-medium for a longer time. This increase in the shift of the mass action ratio time with swelling time indicates that the higher rate of CP-synthesis is dependent on the extent of removal of the outer mitochondrial membrane. When measuring the specific activity of MiCK (i.e. the rate of CP-synthesis) in mitoplasts normalised for the specific activity of citrate synthase, a matrix enzyme (Table 1), it is nearly twice as high in mitoplasts than in mitochondria already after 5 min of hypoosmolaric incubation. Thus under the experimental conditions used here the outer mitochondrial membrane does not contribute to the coupling of the mitochondrial creatine kinase reaction to oxidative phosphorylation but rather operates as a permeability barrier for ATP and Cr from the incubation medium to MiCK. On the other hand, if the MiCK is released from the inner membrane by additional treatment with 125 mM KCl for 20 min [15], the mass action ratio is close to the equilibrium ratio (95 versus 122), i.e. the coupling of the MiCK reaction to oxidative phosphorylation is abolished. Under this condition 34% of the total MiCK activity was released from the mitoplasts. It is concluded that the functional coupling of the mitochondrial creatine kinase reaction to oxidative phosphorylation is realised by the association of the MiCK with the

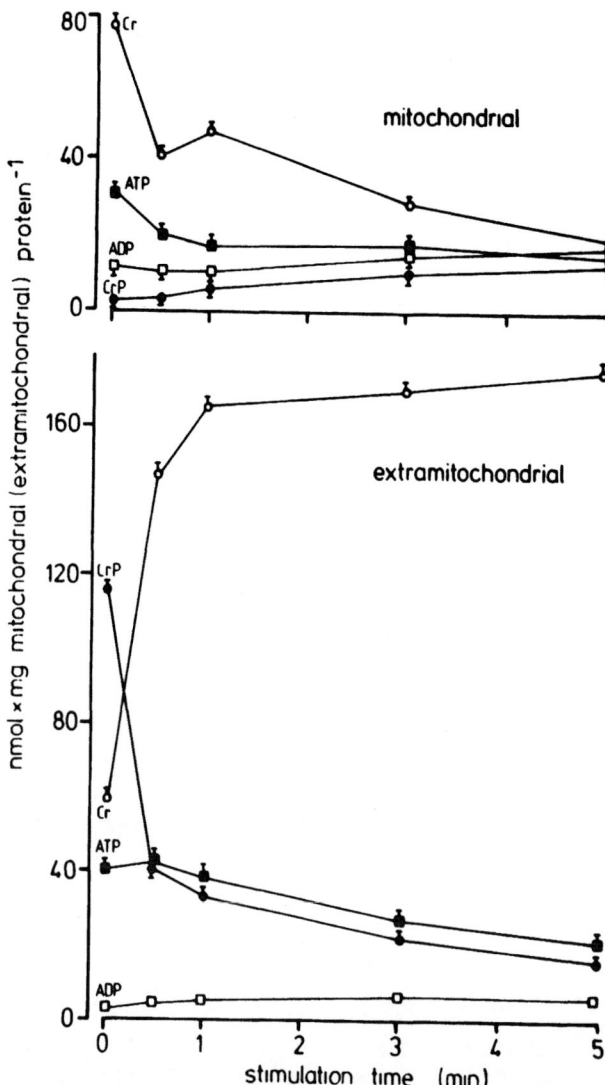

Fig. 2. Changes in mitochondrial and extramitochondrial high energy phosphates during isotonic stimulation of rat gastrocnemius muscle. From [11], mean from 5–11 determinations from 2–3 pools, each containing the tissue of 6–10 muscles ± SEM.

Table 1. Activity of creatine kinase normalised for activity of citrate synthase in rat heart mitochondria and mitoplasts

	CK/CS
Mitochondria	1.9 ± 0.2
Mitoplasts (5 min)	3.6 ± 0.3
(20 min)	3.5 ± 0.7
(30 min)	3.6 ± 0.1

CK: nmol CP produced/mg protein × min by creatine kinase; CS: mU/mg catalysed by citrate synthase; incubation medium as described in methods; mitoplasts were prepared from mitochondria by swelling in 10 mM TES for the time indicated. n = 4 ± SEM.

inner mitochondrial membrane and the activity of the electron transport chain, ATP-synthase and adenine nucleotide translocase. The outer mitochondrial membrane is an additional permeability barrier between extramitochondrial and inner membrane space probably with regulatory functions for the MiCK in the intact cell.

Activity of the creatine shuttle in the intact muscle tissue
When we studied the effect of isotonic stimulation of the mitochondrial/cytosolic distribution of high energy phosphates in rat gastrocnemius muscle [11] we not only observed the expected decrease in cytosolic CP and increase of cytosolic Cr with stimulation time (Fig. 2) but inverse changes of these compounds in the mitochondrial compartment. These findings were surprising, first-

Table 2. Mitochondrial concentrations of ATP, ADP, CP and Cr [mmol/l] in isolated perfused guinea pig hearts

	ATP	ADP	CP	CR	CP/Cr
Control, 10 mM glucose	16.5 ± 2.9	8.8 ± 1.4	2.5 ± 1.3	27.4 ± 3.8	0.13
40 µM atractyloside 40 mM KCl	32.6 ± 3.4	8.7 ± 1.3	0.4 ± 2.3	24.5 ± 3.4	0.04
Ischemia 30 min	18.8 ± 2.4	6.0 ± 0.8	0.9 ± 0.7	35.1 ± 3.5	0.04
Reperfusion 30 min	25.0 ± 4.1	6.4 ± 1.4	15.0 ± 2.9	39.0 ± 6.5	0.48
Isoprenaline 5 µg/l	21.2 ± 2.6	5.7 ± 1.3	7.8 ± 1.1	19.3 ± 3.4	0.51

n = 7–14 ± SEM; from [3].

ly, since we did not expect to detect mitochondrial CP at all, secondly mitochondrial CP rose from approximately zero in resting muscle to about 15 nmol/mg of mitochondrial protein with increasing stimulation time. It should be pointed out that with the fractionation method used here, the mitochondrial space is the space labelled by citrate synthase, i.e. mainly the matrix space. Also, in agreement with earlier measurements in heart and heart fibroblastoid cells [17, 18] a considerable amount of creatine is found in the mitochondria. This concurs with earlier results obtained with isolated mitochondria showing that the mitochondria are permeable for creatine [19].

Apart from the question, how CP reaches the matrix space, which will be dealt with below, we wanted to know whether matrix CP is related to the oxidative activity of muscle tissue. Since it has been shown that mitochondrial creatine kinase is closely coupled to oxidative phosphorylation, oxidative activity should correlate with the activity of the creatine shuttle in the intact muscle tissue. We therefore designed experimental conditions in Langedorff-perfused heart with low and high rates of oxygen consumption (Table 2). Oxygen consumption is very low compared to control in KCl-arrested, atractyloside inhibited hearts (2 µmol/min versus 5.5 µmol/min, expressed per total heart). In the presence of atractyloside, a complete inhibition of the creatine shuttle is to be expected due to the inhibition of adenine nucleotide translocase. Oxygen consumption is close to zero in ischemia, high with isoprenaline (7.2 µmol/min) and transiently high during reperfusion [3]. The changes in mitochondrial CP and the mitochondrial CP/Cr ratios closely corresponded to the oxidative activity of the hearts. Mitochondrial CP/Cr ratios were highest with isoprenaline and after 30 min of reperfusion but lowest after 30 min of ischemia and in atractyloside inhibited hearts. CP was highest after 30 min of reperfusion.

Therefore, these *in vivo* results are in agreement with the proposal that during phases of high extramitochondrial energy demand and therefore high activity of the creatine shuttle, CP accumulates in the mitochondria, since it is synthesised at high rates by the mitochondrial creatine kinase and since its permeability to the extramitochondrial space is restricted.

Table 3. Relationship between CP uptake into isolated rat heart mitochondria and the functional state of mitochondrial CK

Creatine phosphate	Supernatant [µM]	Mitochondria [nmol/mg protein]
Mitochondria		
1 mM ATP, 10 mM Cr	647 ± 54	3.3 ± 0.2
+ 10 µM carboxyatractyloside	320 ± 37	0.29 ± 0.13
+ 1 µM valinomycin + 125 mM KCl	141 ± 24	2.6 ± 0.28
+ 2 mM phosphoenolpyruvate and 7 U pyruvate kinase	1124 ± 179	n.d.
2 mM ADP, 0,5 mM CP	471 ± 44	n.d.
1 mM CP	1014 ± 117	n.d.
2 mM CP	2234 ± 266	n.d.
Mitoplasts		
1 mM ATP, 10 mM Cr	381 ± 30	3.7 ± 0.30
1 mM ATP, 10 mM Cr, 125 mM KCl*	482 ± 65	n.d.

n = 4–10 ± SEM; n.d. not detectable; incubation conditions see Methods; substrates as indicated; * incubated for 20 min with 125 mM KCl.

Creatine phosphate uptake into isolated mitochondria

In order to get more insight to the way how CP enters the mitochondria, CP-uptake was studied in isolated rat heart mitochondria. In the presence of 1 mM ATP and 10 mM creatine about 3 nmol CP/mg mitochondrial protein was found in the mitochondrial pellet after correction for extramitochondrial medium (Table 3). To be sure that this is not due to CP-accumulation in the intermembrane space, mitochondria were exposed to media of different osmolarity. Hypoosmolaric treatment did not only increase the matrix volume but also the amount of CP associated with the mitochondria showing that it is taken up into the matrix space (Fig. 3).

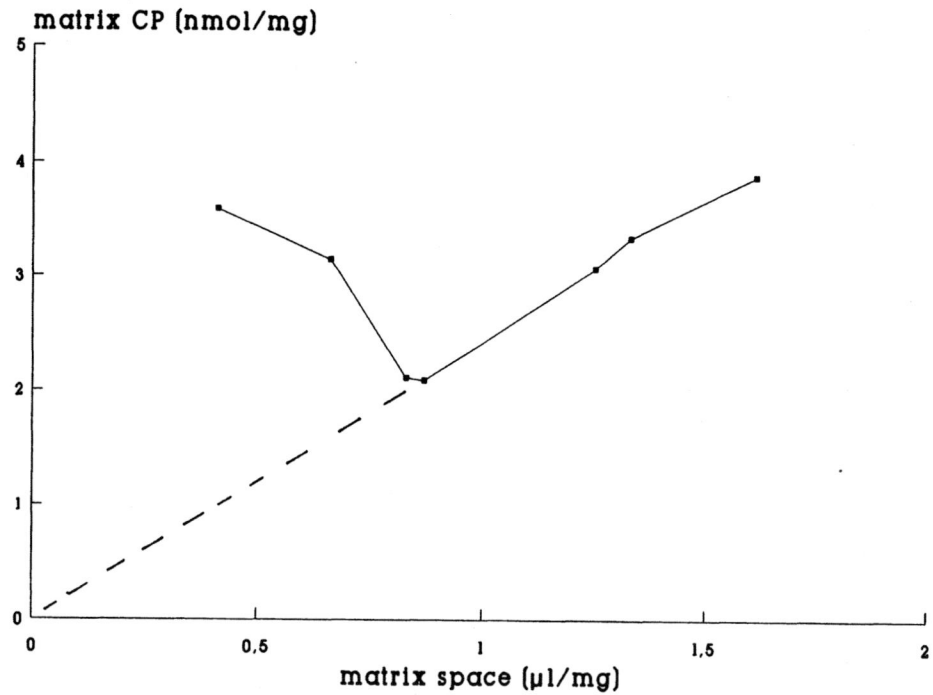

Fig. 3. Influence of hypoosmolaric incubation on matrix creatine phosphate in isolated rat heart mitochondria. Osmolarity varied from 70–1000 mosmol. The matrix space was determined by subtracting the ^{14}C-sucrose space of the pellet from the ^{3}H$_2$O-space as described in [17]. Each point is the average of 3 determinations.

If CP would have been only accumulated in the intermembrane space it should decrease upon swelling and not increase. On the other hand, when mitochondria were exposed to hyperosmolaric media, CP associated with the mitochondria increased due to an increase in unspecific permeability.

Since no transport system for CP has been described in the mitochondria, we examined several conditions which might influence CP-transport into the mitochondrial matrix. They are summarised in Table 3: Externally added CP was not taken up into the mitochondria at conditions where mitochondrial CP-synthesis was not possible. This is a further argument against a mere accumulation in the intermembrane space, since externally added CP should be able to permeate in this compartment, at least to a certain extent. Therefore, it is concluded that active MiCK is essential for mitochondrial CP-uptake. This finding is supported by another experiment where the addition of 7 units of pyruvate kinase + 2 mM phosphoenolpyruvate were added as external ATP-regenerating system. CP-accumulation in the mitochondria was completely inhibited.

The removal of the outer mitochondrial membrane does not inhibit CP-uptake (Table 3, mitoplasts). However, dissociation of creatine kinase from the inner membrane by treatment of mitoplasts with 125 mM KCl completely abolishes CP-uptake.

Taken together, these experiments show that a close association of MiCK in a functional compartment as suggested by [8, 15] is a presumption for CP-transport into the mitochondria. Thus, CP synthesised there could be picked up by the mitochondrial adenine nucleotide translocase located close to MiCK and taken into the mitochondria. Several authors have shown that other phosphorylated compounds like phosphoenolpyruvate [21, 22] or 2-phosphoglycerate [23] are transported in liver mitochondria by this system. When comparing phosphoenolpyruvate with CP (Fig. 4) the structural similarity of both compounds is apparent. If the mitochondrial adenine nucleotide translocase would be responsible for CP-uptake, it should be inhibited by carboxyatractyloside. 10 µM of the inhibitor completely abolished mitochondrial CP accumulation although a considerable amount of CP was formed by MiCK (the CP-concentration in the supernatant was 50% of the control, Table 3). Further, valinomycin + potassium which suppressed the CP-formation by MiCK to a higher extent, inhibited mitochondrial CP-accumulation only slightly. Therefore the abolition of CP-accumulation in the mitochondria found with carboxyatractyloside, is a strong indication for the involvement of the mitochondrial adenine nucleotide translocase in mitochondrial CP-accumulation.

As far as we know, mitochondrial Cr and CP have no

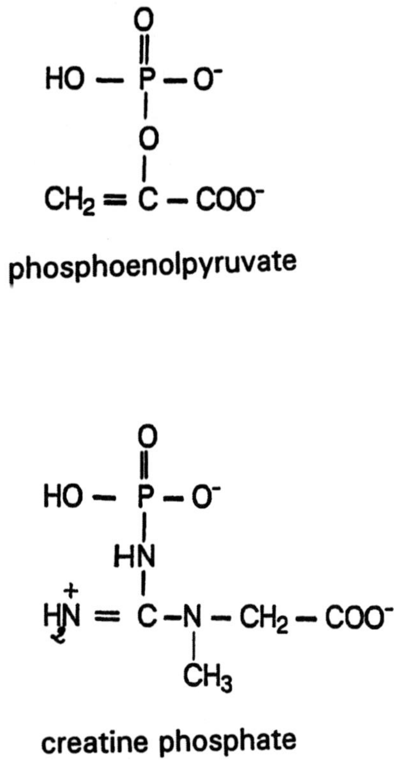

phosphoenolpyruvate

creatine phosphate

Fig. 4. Structural similarity between phosphoenolpyruvate and creatine phosphate.

physiological or biochemical function within the matrix. According to our finding it represents a leak in the creatine shuttle due to the unspecifity of the mitochondrial adenine nucleotide translocase and the microcompartmentation of high energy phosphates within the intermembrane space. It is still unclear, how CP is transported, i.e. whether transport is electroneutral, or whether it is electrogenic and thus dissipates energy, and which is the counter ion. Since valinomycin + K$^+$ only slightly inhibited CP-uptake, an electrogenic uptake appears not to be probable. Further experiments have to be designed to clarify this question.

Acknowledgement

This work was supported by a grant of the Deutsche Forschungsgemeinschaft, SFP 242, Projekt E3 and by Fa Bayer, Wuppertal.

References

1. Bessman SP: The creatine-creatine phosphate energy shuttle. Ann Rev Biochem 54: 831–862, 1985
2. Saks VY, Kuznetsov AV, Kupriyanov VV, Miceli MV, Jacobus WE: Creatine kinase of rat heart mitochondria: the demonstration of functional coupling to oxidative phosphorylation in an inner membrane-matrix preparation. J Biol Chem 260: 7757–7764, 1985
3. Soboll S, Conrad A, Keller M, Hebisch S: The role of the mitochondrial creatine kinase system for myocardial function during ischemia and reperfusion. Biochim Biophys Acta 1100: 27–32, 1992
4. Gellerich FN, Schlame M, Bohnensack R, Kunz W: Dynamic compartmentation of adenine nucleotides in the mitochondrial intermembrane space of rat-heart mitochondria. Biochim Biophys Acta 890: 117–126, 1987
5. Moreadith RW, Jacobus WE: Creatine kinase of heart mitochondria: functional coupling of ADP transfer to the adenine nucleotide translocase. J Biol Chem 257: 899–905, 1982
6. Saks VA, Kupriyanov VV, Elizarova GV, Jacobus WE: Studies of energy transport in heart cells: the importance of creatine kinase localisation for the coupling of mitochondrial phosphorylcreatine production to oxidative phosphorylation. J Biol Chem 255: 755–763, 1980
7. Barbour RL, Ribaudo J, Chan SHP: Effect of creatine kinase activity on mitochondrial ATP/ADP transport. J Biol Chem 259: 8246–8251, 1984
8. Brdiczka D, Bücheler K, Kooke M, Adams V, Nalam VK: Characterization and metabolic function of mitochondrial contact sites. Biochim Biophys Acta 1018: 234–238, 1990
9. Marcillat O, Goldschmidt D, Eichenberger D, Vial C: Only one of the two interconvertible forms of mitochondrial creatine kinase binds to heart mitoplasts. Biochim Biophys Acta 890: 233–241, 1987
10. Langendorff O: Untersuchungen am überlebenden Säugetierherzen. Arch Ges Physiol 61: 291–332, 1895
11. Hebisch S, Sies H, Soboll S: Function dependent changes in the subcellular distribution of high energy phosphates in fast and slow rat skeletal muscles. Pflüger's Arch 406: 20–24, 1986
12. Soboll S, Elbers R, Heldt HW: Metabolite measurements in mitochondria and in the extramitochondrial compartment by fractionation of freeze-stopped liver tissue in nonaqueous media. Methods Enzymol XVI: 201–206, 1979
13. Bergmeyer HU: Methoden der enzymatischen Analyse. 2. Auflage. Verlag Chemie, Weinheim, Germany, 1970
14. Jacobus WE, Saks V: Creatine kinase of heart mitochondria: changes in its kinetic properties induced by coupling to oxidative phosphorylation. Arch Biochem Biophys 219: 167–178, 1982
15. Wyss M, Smeitink J, Wevers RA, Wallimann Th: Mitochondrial creatine kinase: a key enzyme of aerobic energy metabolism. Biochim Biophys Acta 1102: 119–166, 1992
16. Pfaff E, Klingenberg M: Adenine nucleotide translocation of mitochondria – Specifity and control. Eur J Biochem 6: 66–79, 1968
17. Pfaff E, Klingenberg M, Ritt E, Vogell W: Korrelation des unspezifisch permeablen mitochondrialen Raumes mit dem Intermembran Raum. Eur J Biochem 6: 222–232, 1968
18. Soboll S, Bünger R: Compartmentation of adenine nucleotides in

the isolated working guinea pig. Heart. Hoppe-Seyler's Z Physiol Chem 362: 125–132, 1981

19. Soboll S, Werdan K, Bozsik M, Müller M, Erdmann E, Heldt HW: Distribution of metabolites between mitochondria and cytosol of cultered fibroblastoid rat heart cells. FEBS Lett 100: 125–128, 1979

20. Altschuld RA, Nerola AJ, Brierley GP: The permeability of heart mitochondria to creatine. J Mol Cell Card 7: 451–462, 1975

21. Shug A, Shrago E: Inhibition of phosphoenolpyruvate transport via the tricarboxylate and adenine nucleotide carrier systems of rat liver mitochondria. Biochem Biophys Res Commun 53: 659–665, 1973

22. Bryla J, Dzik JM: Phosphoenolpyruvate efflux from kidney cortex mitochondria of rabbit. Biochim Biophys Acta 638: 250–256, 1981

23. Grivell AR, Halls HJ, Berry MN: Role of mitochondria in hepatic fructose metabolism. Biochim Biophys Acta 1059: 45–54, 1991

Molecular and Cellular Biochemistry **133/134**: 115–123, 1994.
© 1994 *Kluwer Academic Publishers.*

II-4 The structure of mitochondrial creatine kinase and its membrane binding properties

Thomas Schnyder, Manuel Rojo,[1] Rolf Furter and
Theo Wallimann
Institute for Cell Biology, ETH-Hönggerberg, CH-8093 Zürich, Switzerland and [1] Department of Biochemistry, University of Geneva, Science II, 30, quai Ernest-Ansermet, CH-1211 Geneva, Switzerland

Abstract

The biochemical and biophysical characterization of the mitochondrial creatine kinase (Mi-CK) from chicken cardiac muscle is reviewed with emphasis on the structure of the octameric oligomer by electron microscopy and on its membrane binding properties. Information about shape, molecular symmetry and dimensions of the Mi-CK octamer, as obtained by different sample preparation techniques in combination with image processing methods, are compared. The organization of the four dimeric subunits into the Mi-CK complex as apparent in the end-on projections is discussed and the consistently observed high binding affinity of the four-fold symmetric end-on faces towards many support films and towards each other is outlined. A study on the oligomeric state of the enzyme in solution and in intact mitochondria, using chemical crosslinking reagents, is presented together with the results of a search for a possible linkage of Mi-CK with the adenine nucleotide translocator (ANT). The nature of Mi-CK binding to model membranes, demonstrating that rather the octameric than the dimeric subspecies is involved in lipid interaction and membrane contact formation, is resumed and put into relation to our structural observations. The findings are discussed in light of a possible *in vivo* function of the Mi-CK octamer bridging the gap between outer and inner mitochondrial membranes at the contact sites. (Mol Cell Biochem **133/134**: 115–123, 1994)

Key words: mitochondrial creatine kinase, electron microscopy, crosslinking, membrane binding

Introduction

The isoenzymes of creatine kinases (CK) catalyze the reversible transfer of a phosphoryl group from ATP to creatine and are expressed in a temporarily regulated manner in tissues specified by high energy turnover, e.g., brain, retina, spermatozoa, heart and skeletal muscle (reviewed in [1]). Sequence comparison and circular dichroism studies of the four CK isoforms identified so far show high sequence homology and secondary structure conservation [2]. In order to understand the role of CK isoenzymes *in vivo*, their isoform-specific subcellular localization is of particular interest; the cytosolic M- and B-isoforms are soluble and partly associated with structures at sites of energy consumption, such as myofibrils, sarcoplasmic reticulum and plasma membranes. In contrast, the two mitochondrial isoforms, the sarcomeric Mi_b-CK and the ubiquitous Mi_a-CK, are restricted to the mitochondria, the site of energy production through oxidative phosphorylation.

The spatial sequestering of cytosolic and mitochondrial CK isoenzymes led to the postulation of a phospho-

Address for offprints: T. Schnyder, Institute for Cell Biology, ETH-Hönggerberg, CH-8093 Zürich, Switzerland

116

creatine/creatine circuit, proposing that creatine (Cr) is phosphorylated to phosphocreatine (PCr) in the mitochondrial intermembrane space by Mi-CK which then is transported across the outer membrane. Liberated into the cytosol, PCr diffuses to sites of energy consumption where the cytosolic isoenzymes make use of it to regenerate *in situ* ATP for cellular processes. The creatine so generated diffuses back to the mitochondria where it is transported across the outer membrane to serve again as a substrate for Mi-CK.

Several lines of evidence concerning the intracellular compartmentation and functional coupling of CK isoenzymes to ATP-producing and ATP-requiring processes support the existence of such a PCr/Cr circuit (reviewed in [1]). This article reviews and discusses findings on the unique structure and membrane binding property of the Mi-CK isoenzymes which suggest to play a crucial role in the PCr/Cr circuit.

Localization, isolation and biochemical characterization of chicken Mi-CK

Mi-CK [3] is restricted to the mitochondrial intermembrane compartment, where it binds to the inner mitochondrial membrane [4]. In respiring mitochondria, Mi-CK preferentially utilizes ATP, that is synthesized in the mitochondrial matrix for the phosphorylation of extra-mitochondrial Cr [5]. Mi-CK activity is therefore functionally coupled to oxidative phosphorylation. Such a coupling involves two additional components, the adenine nucleotide translocator (ANT) in the inner mitochondrial membrane and the outer mitochondrial membrane porin VDAC (reviewed in [6]). The mechanism underlying the functional coupling for the transport of CK substrates and products remains to be established.

Isolation of chicken mitochondrial creatine kinase from heart tissue yielded enzyme preparations consisting of two oligomeric species [7]. Gel permeation chromatography [7], hydrodynamic measurements by ultra-centrifugation, electron optical investigations by scanning transmission electron microscopy [8] and radiation inactivation measurements [9] showed that the majority of Mi-CK exist as octameric species while a minor species is dimeric. This oligomerization pattern has also been shown to occur for Mi-CK purified from chicken brain [10] and was later corroborated in several tissues of many vertebrates (reviewed in [11]).

The octamer to dimer conversion can be reversibly modulated; dimerization is favored at low protein concentration [7], high pH and by the presence of substrates [12]. Interestingly, Mi-CK isoenzymes isolated from different organisms and tissues show different octamer stability [10]. Monomerization of Mi-CK only occurred in high concentrations of guanidine hydrochloride as judged from ultra-centrifugation experiments [8]. Hybridization experiments with the two Mi-CK isoforms (Mi$_b$-CK and Mi$_a$-CK) revealed hetero-octamers consisting of intact Mi$_a$-CK and Mi$_b$-CK homo-dimers, but no hetero-dimers could be detected [11]. Therefore, the dimer can be considered as the basic building block of the octamer, which is the oligomeric form of Mi-CK that is energetically favored in solution. This is also indicated by the observation that Mi$_b$-CK heterologously expressed in *E. coli* formed octameric molecules as well [13]. For octamer assembly, ionic interactions of the very *N*-terminal sequence [14] as well as hydrophobic interactions involving a Mi-CK-specific tryptophan residue are relevant [15]. The findings that extraction of Mi-CK from mitochondria yielded mostly octameric molecules, that rebinding of dimeric Mi-CK to mitoplasts lead to partial re-octamerization [16] and that the dimeric molecule was only favored in a very dilute aqueous solution of pure Mi-CK, argue for the octameric species to be predominant *in vivo*. However, whether the octamer-dimer conversion observed *in vitro* may have a regulatory role of Mi-CK function and energy supply is still a matter of debate.

In order to better understand the structure-function relationship of the mitochondrial creatine kinase, its role in energy metabolism and in the postulated PCr/Cr circuit, over the past years the Mi-CK octamer has become a subject of extensive electron optical investigation.

Structure of the Mi-CK octamer by electron microscopy

A broad spectrum of sample preparation techniques for electron microscopy was applied in combination with methods for image reconstruction to study the Mi-CK octamer [8,17]. Negatively stained octamers are square-shaped projections of about 10 nm in side-length with peripheral and central stain accumulation. The laterally accumulated stain protrudes less densely packed protein regions, giving rise to the characteristic pinwheel-like appearance of the octamers top/bottom projection (Fig. 1A). The central stain spot with a diameter of

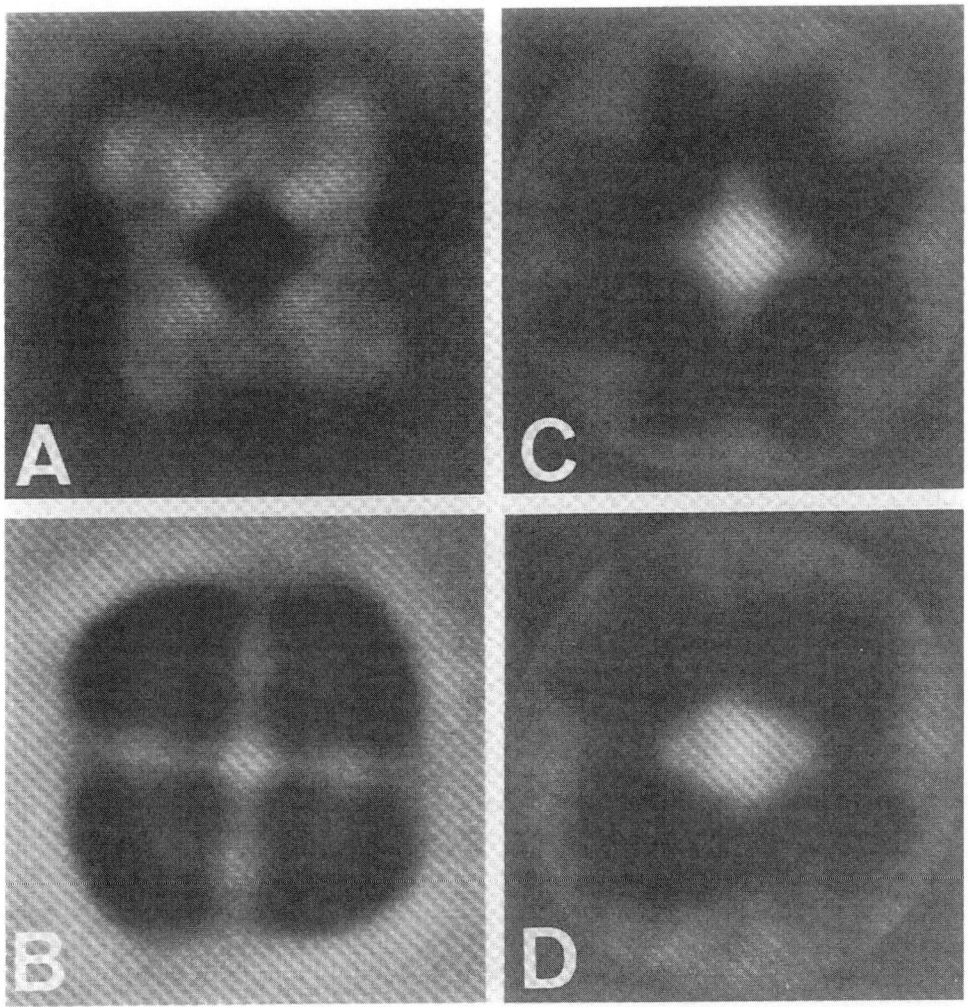

Fig. 1. Comparison of averaged images of Mi-CK octamers obtained by different sample preparation techniques: end-on views of negatively stained (A), rotary shadowed (B), and ice-embedded Mi-CK octamers (C) [22]. The most obvious feature in common is the four-fold symmetry axis through the center of the molecule. Using the bare-grid technique for cryo-electron microscopy the side-on view could be visualized (D). The octamer with the large water filled central cavity has molecule dimensions of about $10 \times 10 \times 8.4$ nm.

2.5 nm is indicative of a cavity or depression and is the origin of the octamer's four-fold symmetry axis.

To circumvent problems with the interpretation of images often caused by the negative stain method, the octamer's surface topography was replicated by evaporation of heavy metal at low object temperature [18]. Unidirectional shadowing led to the perspective impression of the surface of the octamer with its four hill-like quadrants. From the length of the molecules' shadow a quasi cubic three-dimensional shape of the octamer could be deduced [17]. The averaged image of rotary shadowed particles (Fig. 1B) shows strong contrast at the periphery due to metal accumulation at the vertical faces of the molecule. Symmetry and dimensions are in correspondence to negatively stained molecules. The cross of low contrast in the center of the octamer surface, *viz.* crevic-

es that are shielded from heavy metal deposition, subdivides the molecule into four quadrants. Per quadrant two maxima of contrast are observed, one located near the center and another at the corner of the molecule. These findings strongly suggest an arrangement of the long dimer axis being parallel to the four-fold axis of the octamer. This would imply that the molecule exhibits vertical or side-faces which are different from the top/bottom faces visualized so far (Fig. 1A and B). However, visual inspection of well-preserved particles and multivariate statistical analysis [19] applied to more than thousand particles failed to reveal other views of the molecule [20].

Only in accidentally created linear Mi-CK filaments the octamer's side-view could be finally visualized [17]. In this case, top and bottom faces became associated in

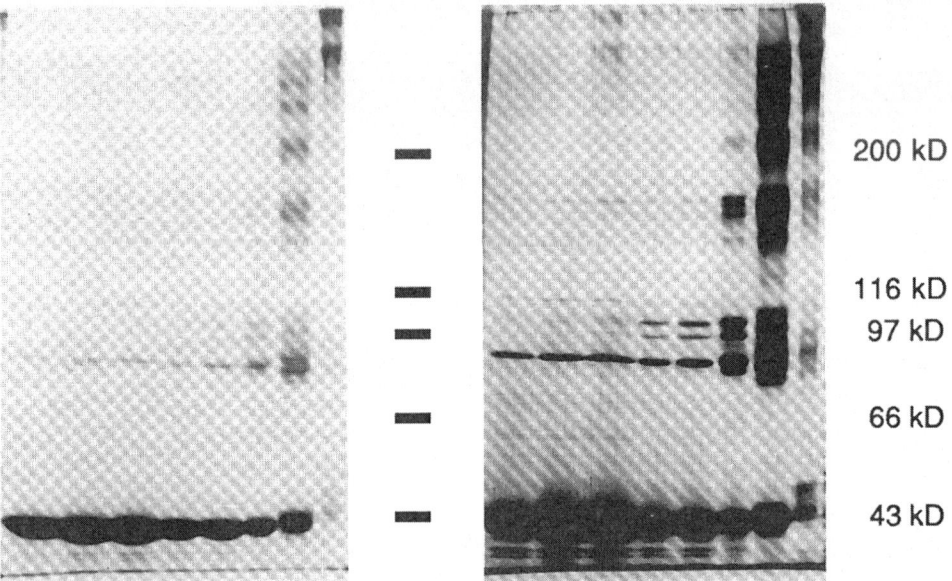

Fig 2 Mi-CK treated with a lysine-specific crosslinking agent and analyzed by SDS-PAGE with Coomassie Blue (left) and silver staining (right) About 1 mg/ml soluble Mi-CK octamers were incubated with dimethyl suberimidate (DMS) in a concentration range from 1 μM to 50 mM (from left to right) and run on a gradient gel [24] At 25 mM DMS about 18 crosslink products differing in their molecular weight are seen to be clustered in eight categories representing the oligomeric states of Mi-CK from monomer to octamer (second lane from the right in both pictures) At a concentration of 50 mM DMS, intermolecular crosslinkages (collision complexes) were formed, which were to large to enter the stacking gel (first lane from the right in both pictures) Molecular weight standards (middle) are defined on the right

solution by forming an alternate stacking of the octamers. The two-fold symmetric side-view of the octamer in the filament, showed the banana-shaped dimers arranged with their long axes (8.4 nm) parallel to the filament axis [17].

The fact that octamers can form linear filaments in solution and that only one view of the molecule (top/bottom projection) could be observed, suggest that the top and bottom faces of Mi-CK must have a strong tendency to interact with supporting films and, under certain conditions, with each other.

To avoid such adsorption constraints, Mi-CK octamers were vitrified in a thin ice film [21] and imaged in an unstained frozen-hydrated state. With the cryo-technique a projection of the molecule's volume is observed similar to negative staining, but specimen flattening and stain accumulation artefacts are circumvented. Although the top/bottom projection ('end-on view') again was predominant, several views of randomly oriented molecules also could be distinguished. The averaged motif of the end-on view (Fig. 1C) shows the expected molecular dimensions and symmetry [22]. The four protein densities organized around the water filled hole are bi-lobed, each lobe most probably representing a part of the Mi-CK monomer.

The averaged side-view obtained from frozen hydrat-

ed octamers (Fig. 1D) must be treated as a preliminary result since the number of molecules taken for image processing was limited by the low and random yield of molecules differing from the end-view orientation and the low signal-to-noise ratio of ice-embedded molecules. Nevertheless, molecular dimensions (8.4 × 10 nm) and the expected two-fold symmetry correspond to the motif found in filaments. This infers that in Mi-CK octamers top and bottom faces are structurally identical.

Chemical crosslinking of Mi-CK octamers

In vivo, Mi-CK molecules are restricted to the narrow mitochondrial intermembrane compartment and are bound to the inner mitochondrial membrane [23]. In order to investigate the oligomeric nature of the membrane bound enzyme, the technique of chemical crosslinking was applied. In a first step, different types of crosslinking agents were tested with purified Mi-CK octamers in solution, and the crosslink products were analyzed by SDS-PAGE. Incubation with a lysine-specific crosslinking agent revealed around 18 high molecular weight bands clustered into eight categories (Fig. 2) representing intramolecular crosslinking products of the monomeric up to the octameric subspecies [24]. These results corroborate the octameric structure of Mi-CK and are in agreement with similar crosslinking studies from other authors [25, 26].

In a second step, the reactivity of crosslinkers was tested in buffer compositions where Mi-CK was shown to be bound to the inner mitochondrial membrane [24]. After having found suitable buffer milieu and concentration range for several types of crosslinker, they were directly applied to mitochondria and mitoplasts and the crosslink products analyzed. In another experiment, pure Mi-CK was first labelled by photo-activable heterobifunctional crosslinkers and rebound to mitoplasts, that have been depleted of Mi-CK before. After removal of excess labelling agents, the solution of mitoplasts with bound, modified Mi-CK was illuminated by UV-light to induce photoactivable crosslinking. A comparison of all experiments revealed SDS-gel banding patterns of Mi-CK crosslinking products which corresponded to that of crosslinked pure Mi-CK octamers. This infers that rather the octameric than the dimeric species of Mi-CK is located at the mitochondrial inner membrane, which is in correspondence to the results obtained by radiation inactivation [9] and gel chromatography [16].

Search for heterologous crosslinking products

Based on the experimental demonstration of a functional coupling of Mi-CK activity to oxidative phosphorylation [4, 27], a direct physical interaction of membrane bound Mi-CK with the adenine nucleotide translocator (ANT) (see e.g. [28]) as well as with the mitochondrial outer membrane pore [29] has been postulated. In order to investigate the putative physical interaction between Mi-CK and ANT, mitochondria were treated with crosslinking agents. The crosslinking products were separated on SDS-PAGE and analyzed with specific antibody staining. Under reaction conditions where intramolecular crosslinking of Mi-CK was observed, no heterologous crosslinking product between Mi-CK and ANT could be demonstrated by immuno-blotting with anti-Mi-CK and anti-ANT antibodies [24]. With increase of the crosslinking reagent concentration, the ANT-specific immuno-signal became broader and weaker, shifted progressively toward higher apparent molecular weights, without forming any discrete crosslinking products [24].

In another series of experiments, freshly prepared mitochondria were treated with several types of crosslinking agents and were then differentially extracted. After the crosslinking reaction, mitochondria were hypotonically shocked and mitochondrial proteins extracted, first by phosphate and then by urea treatment. All fractions, including the residual membrane pellets after urea extraction, were separated on SDS-PAGE for immunochemical detection. Mi-CK could be found in the phosphate extract only. Depending on the type of chemical agent, different discrete high molecular weight bands of Mi-CK were observed. The crosslinking products co-migrated with those obtained from crosslinking of soluble, pure Mi-CK octamers. Immuno staining against ANT could be exclusively found in the membrane pellet fraction; at lower crosslinker concentration a signal corresponding in size to monomeric ANT and at higher concentration a high molecular smear was usually found. These results show that all defined Mi-CK crosslinking products were of intramolecular character. Neither an oligomeric state of ANT nor a linkage between Mi-CK and ANT or another protein could be demonstrated under the conditions chosen. These findings have been discussed in light of the difficulties to crosslink a soluble protein to a membrane protein elsewhere [24].

The interaction of Mi-CK with membranes

The inability to demonstrate a physical interaction of Mi-CK to membrane proteins suggests that lipids of the bilayer might act as the binding targets for Mi-CK on mitochondrial membranes. Therefore, the interaction of Mi-CK with membranes and lipid monolayers spread at the air/water interface was studied [34]. It was shown that Mi-CK induced an increase in surface pressure when interacting with lipid films, a property shared by many lipid binding proteins. The increase in surface pressure was dependent on the initial pressure and on the protein concentration in the subphase. For octameric Mi-CK, the induction of pressure increase ceased at an initial surface pressure of 30–32 mN/m (i.e. critical surface pressure). Since this is the equivalent surface pressure of lipid bilayers [31], this suggests that the cationic Mi-CK molecules may mainly interact with the polar head groups of the lipids. Dimeric Mi-CK showed a lower affinity to lipids and a lower critical surface pressure than the octameric species, indicating that the octamer of Mi-CK might be the species that preferentially interacts with mitochondrial membranes *in vivo* as well.

120

Mi-CK interacted in a similar manner with spread inner and outer mitochondrial membranes, as well as with those formed with lipids from microsomal membranes [34]. These results favour the hypothesis that the binding of Mi-CK to membranes is mediated by lipids. This explains the observation that Mi-CK-depleted heart mitoplasts rebind exogenous Mi-CK in a two to 16-fold excess of the endogenous amount. The same was found for liver mitoplasts, which are free of endogenous Mi-CK, but have a lipid composition similar to heart mitoplasts [32].

The interaction of Mi-CK with pure phosphatidylcholine monolayers resulted in a lower surface pressure increase than with monolayers of cellular lipid extracts. The increase in surface pressure induced by Mi-CK augmented with increasing content of anionic phospholipid in the phosphatidylcholine monolayer with the effect being more prominent for cardiolipin than for phosphatidyl-serine or -inositol. These findings sustain the notion, that the cationic Mi-CK interact solely with the polar head groups of lipid molecules in an electrostatic way. This is also in agreement with earlier suggestions based on the fact that Mi-CK desorption from the mitoplasts is strongly dependent on both, the ionic strength [33] and the pH of the extraction buffer [16].

The ability of Mi-CK to mediate intermembrane adhesion

Since Mi-CK was shown to be able to interact in a similar manner with inner and outer mitochondrial membranes, the question was raised, whether Mi-CK can bind two membrane interfaces simultaneously [30]. For this, Mi-CK was bound to a monolayer and the subphase was washed with buffer in order to remove unbound protein. Then, radioactive labelled vesicles were injected in the subphase. A subsequent raise in surface radioactivity showed that Mi-CK was able to interact simultaneously with inner and outer mitochondrial membranes. Mi-CK mediated intermembrane adhesion of whole and pure lipid membranes, whereas other cationic enzymes of the intermembrane space (cytochrome c and adenylate kinase) or the cytosolic isoenzymes of CK failed to induce contact formation. The two oligomeric forms of Mi-CK differed substantially in their ability to mediate membrane contact formation, the octamer being about three fold more potent. Regarding the basic nature of Mi-CK, control experiments with poly-L-lysine peptides showed an induction of intermembrane

binding as well as membrane fusion. Interestingly, the extent of contact formation mediated by poly-L-lysines was lower than that of octameric Mi-CK [30].

Significance for *in vivo* function of Mi-CK

Cube-like Mi-CK octamers visualized by electron microscope have indistinguishable top and bottom faces, which were able to adsorb to several kinds of support films [17]. In contrast to the sticky top/bottom faces (end-on views), the four side faces could not be visualized as long as an adsorption matrix was present. They became, however, manifest when top/bottom faces of the octamers got associated to form linear filaments [17]. Here, we could directly demonstrate the second view (side-on view) of the Mi-CK octamer by vitrifying single molecules in ice.

Mi-CK octamers have been shown to interact with model as well as purified mitochondrial membranes [34] and to be involved in intermembrane adhesion [30]. Since the interacting faces of the octamer must have identical binding properties, the top and bottom faces are suggested to mediate the membrane contact formation *in vitro*. It is tentative that *in vivo* Mi-CK interacts in a similar way with opposed mitochondrial membranes. This view is supported by the narrow gap observed between all membranes in mitochondria [35]. Therefore, the Mi-CK octamer is expected to simultaneously interact by its top and bottom faces with inner and outer membrane and so functioning as a mitochondrial membrane 'gap junction'.

Even though we could not show a direct interaction between Mi-CK and any other protein, the above hypothesis is in line with those studies which have described a functional coupling of Mi-CK to the ANT (see e.g. [28]) and to the mitochondrial porin [29]. Together with the presented low resolution structure, the membrane properties and the ultrastructural location of Mi-CK, we suggest that Mi-CK is able to form a multienzyme complex mediated by phospholipid interactions. This complex would consist of the ANT in the inner membrane, Mi-CK as the gap bridging protein, and porin in the outer membrane (Fig. 3).

A reversible formation of a multienzyme complex would be an appropriate mechanism for coupling Mi-CK catalysis to oxidative phosphorylation and the delivery of PCr to the cytosol (Fig. 3). Possible ways to achieve the assembly and disassembly of the multien-

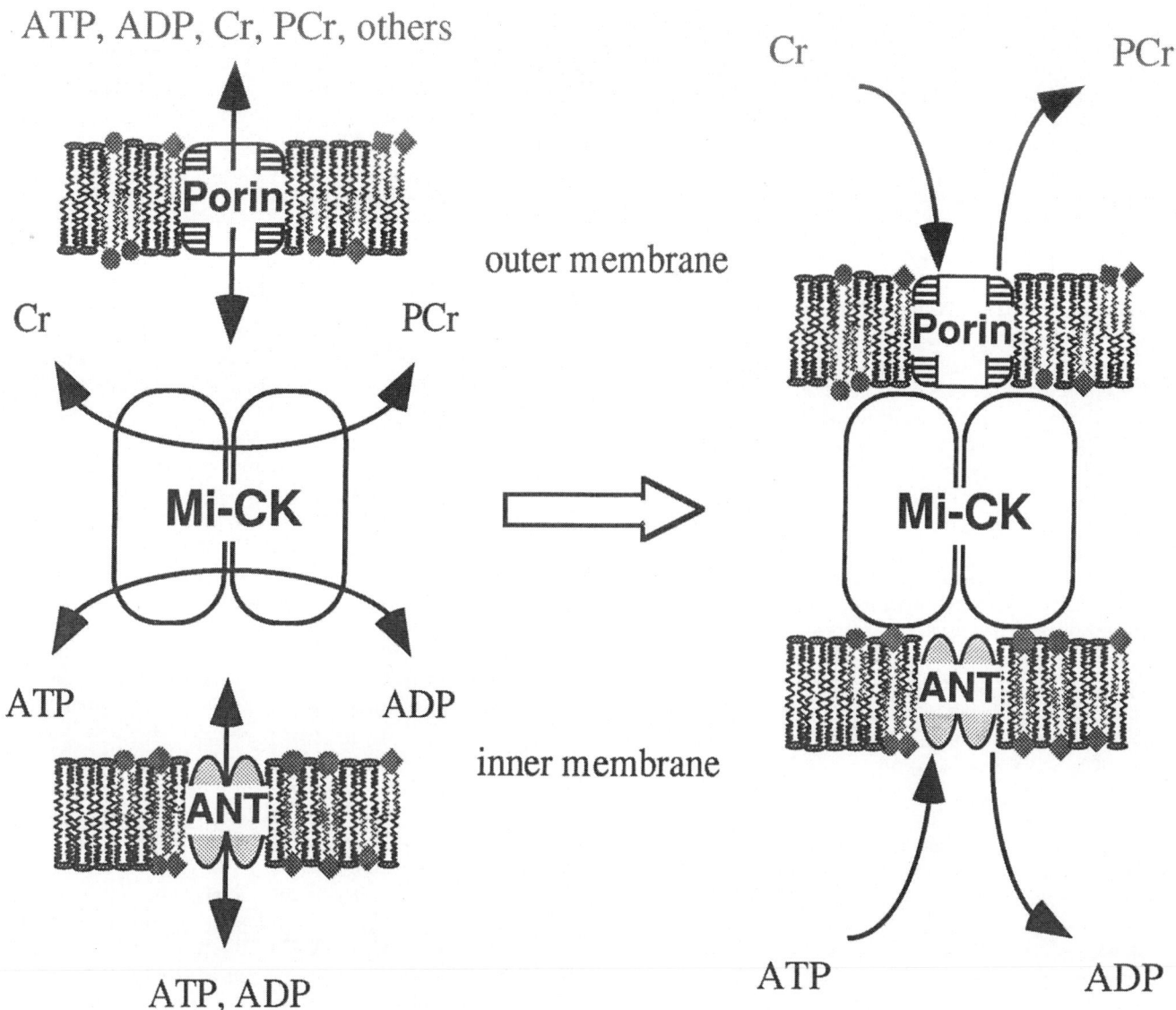

ATP, ADP, Cr, PCr, others

outer membrane

inner membrane

ATP, ADP

Fig 3 Model for a multienzyme complex where the creatine kinase reaction is coupled to oxidative phosphorylation (according to [37]) The enzymes involved in this functional coupling, Mi-CK, ANT and porin, all catalyze reversible reactions The creatine kinase reaction utilizes ATP synthesized by oxidative phosphorylation in the mitochondrial matrix and creatine (Cr) from the cytosolic compartment and shuttles phospho-creatine (PCr) back to the cytosol, whereas ADP gets imported into the matrix (left drawing) An intimate coupling is maintained by a simultaneous interaction of the Mi-CK octamer with the mitochondrial inner and outer membrane bringing the enzyme in close contact to ANT and porin The postulated multienzyme complex (right drawing) would ensure a synchronous and vectorial action of the enzymes involved and would increase the thermodynamic efficiency of mitochondrial high energy phosphate synthesis and transport

zyme complex could be a shift in the dimer/octamer ratio of Mi-CK [16] resulting in altered membrane binding properties (see above), or changes in membrane potential or membrane structure in dependence of the metabolic state, which were demonstrated to occur in mitochondria (see e.g. [36]). Experimental approaches to detect short-lived protein-lipid and protein-protein interactions will be useful to address questions concerning the postulated multienzyme complex.

Acknowledgements

We thank Dr. H. Winkler for excellent image processing work and Prof. Dr. M. Klingenberg for providing us with anti-ANT antibodies. This work was supported by a ETH graduate training program (for T.S. and M.R.) and Grant No. 31-33907.92 from the Swiss National Science Foundation (for T.W. and R.F.).

References

1. Wallimann T, Wyss M, Brdiczka D, Nicolay K, Eppenberger HM: Intracellular compartmentation structure and function of creatine kinase isoenzymes in tissues with high and fluctuating energy demands; the 'phosphocreatine circuit' for cellular energy homeostasis. Biochem J 281: 21–40, 1992

2. Mühlebach SM, Gross M, Wirz T, Wallimann T, Perriard J-C, Wyss M: Sequence homology and structure predictions of the creatine kinase isoenzymes. Mol Cell Biochem 133/134: 245–262, 1994

3. Jacobs H, Heldt HW, Klingenberg M: High activity of creatine kinase in mitochondria from muscle and brain and evidence for a separate mitochondrial isoenzyme of creatine kinase. Biochem Biophys Res Commun 16: 516–521, 1964

4. Jacobus WE, Lehninger AL: Creatine kinase of rat heart mitochondria. Coupling of creatine phosphorylation by to electron transport. J Biol Chem 248: 4803–4810, 1973

5. Saks VA, Kuznetsov AV, Kupriyanov VV, Miceli MV, Jacobus WE: Creatine kinase of rat mitochondria. The demonstration of functional coupling to oxidative phosphorylation in an inner membrane-matrix preparation. J Biol Chem 260: 7757–7764, 1985

6. Brdiczka D: Contact sites between mitochondrial envelope membranes. Structure and function in energy- and protein transfer. Biochim Biophys Acta 1071: 291–312, 1991

7. Schlegel J, Zurbriggen B, Wegmann G, Wyss M, Eppenberger HM, Wallimann T: Native mitochondrial creatine kinase (Mi-CK) forms octameric structures. I. Isolation of two interconvertible Mi-CK-forms; Dimeric and octameric Mi-CK. Characterization localization and structure-function relationship. J Biol Chem 263: 16942–16953, 1988

8. Schnyder T, Engel A, Lustig A, Wallimann T: Native mitochondrial creatine kinase (Mi-CK) forms octameric structures. II. Characterization of dimers and octamers by ultracentrifugation, direct mass measurement by STEM and image analysis of single Mi-CK octamers. J Biol Chem 263: 6954–6962, 1988

9. Quemeneur E, Eichenberger D, Goldschmidt D, Vial C, Beauregard G, Potier M: The radiation inactivation method provides evidence that the membrane-bound mitochondrial creatine kinase is an oligomer. Biochem Biophys Res Commun 153: 1310–1314, 1988

10. Wyss M, Schlegel J, James P, Eppenberger HM, Wallimann T: Mitochondrial creatine kinase from chicken brain; Purification, biophysical characterization and generation of heterodimeric and heterooctameric molecules with subunits of other creatine kinase isoenzymes. J Biol Chem 265: 15900–15908, 1990

11. Wyss M, Smeitink J, Wevers RA, Wallimann T: Mitochondrial creatine kinase; a key enzyme of aerobic energy metabolism. Biochim Biophys Acta 1102: 119–166, 1992

12. Marcillat O, Goldschmitt D, Eichenberger D, Vial C: Only one of the two interconvertible forms of mitochondrial creatine kinase binds to mitoplasts. Biochim Biophys Acta 890: 233–241, 1987

13. Furter R, Kaldis Ph, Furter-Graves EM, Schnyder T, Eppenberger HM, Wallimann T: Expression of active octameric chicken cardiac mitochondrial creatine kinase in *Escherichia coli*. Biochem J 288: 771–775, 1992

14. Kaldis Ph, Furter R, Wallimann T: The N-terminal heptapeptide of mitochondrial creatine kinase is important for octamerization. Biochemistry 33: 952–959, 1994

15. Gross M, Wallimann T: Kinetics of assembly and dissociation of the mitochondrial creatine kinase octamer. A fluorescence study. Biochemistry 32: 13933–13940, 1994

16. Schlegel J, Wyss M, Eppenberger HM, Wallimann T: Functional studies with the octameric and dimeric form of mitochondrial creatine kinase. J Biol Chem 265: 9221–9227, 1990

17. Schnyder T, Gross H, Winkler H, Eppenberger HM, Wallimann T: Structure of the mitochondrial creatine kinase octamer; High-resolution shadowing and image averaging of single molecules and formation of linear filaments under specific staining conditions. J Cell Biol 112: 95–101, 1991

18. Gross H, Müller T, Wildhaber I, Winkler H: High resolution metal replication quantified by image processing of periodic test specimens. Ultramicroscopy 16: 287–304, 1985

19. Van Heel M: Classification of very large electron microscopical image data sets. Optik 82: 114–126, 1989

20. Winkler H, Gross H, Schnyder T, Kunath W: Circular harmonic averaging of rotary-shadowed and negatively stained creatine kinase macromolecules. J Electron Microsc Techn 18: 135–141, 1991

21. Dubochet J, Adrian M, Chang J-J, Homo J-C, Lepault J, McDowall AW, Schultz P: Cryo-electron microscopy of the vitrified specimen. Q Rev Biophys 21: 129–228, 1988

22. Winkler H, Schnyder T, Lücken U: Electron microscopic comparison of frozen-hydrated with negatively stained and rotary-shadowed creatine kinase single. In: Proceedings of the XIIth International Congress for Electron Microscopy, San Francisco Press Inc, 1990, pp 262–263

23. Scholte HR, Weijers PJ, Wit-Peeters EM: The localization of creatine kinase in heart and its use for the determination of the sideness of submitochondrial particles. Biochim Biophys Acta 291: 764–773, 1973

24. Schnyder T: Structure of the mitochondrial creatine kinase octamer by electron microscopy, protein crystallography and chemical crosslinking. Ph.D. Thesis No. 9250, Eidgenössische Technische Hochschule (ETH), Zürich Switzerland, 1990

25. Font B, Eichenberger D, Goldschmidt D, Vial C: Interaction of creatine kinase and hexokinase with the mitochondrial membranes and self-association of creatine kinase; crosslinking studies. Mol Cell Biochem 78: 131–140, 1987

26. Lipskaya TYu, Trofimova ME: Study on heart mitochondrial creatine kinase using a cross-linking bifunctional reagent. I. The binding involves the octameric form of the enzyme. Biochem Int 18: 1029–1039, 1989

27. Bessman SP, Fonyo A: The possible role of the mitochondrial bound creatine kinase in regulation of mitochondrial respiration. Biochem Biophys Res Commun 22: 597–602, 1966

28. Barbour RL, Ribaudo J, Chan SHP: Effect of creatine kinase activity on mitochondrial ADP/ATP transport. Evidence for a functional interaction. J Biol Chem 259: 8246–8251, 1984

29. Kottke M, Adams V, Wallimann T, Nalam KV, Brdiczka D: Location and regulation of the octameric mitochondrial creatine kinase in the contact-sites. Biochim Biophys Acta 1061: 215–225, 1991

30. Rojo M, Hovius R, Demel RA, Nicolay K, Wallimann T: Mitochondrial creatine kinase mediates contact formation between mitochondrial membranes. J Biol Chem 266: 20290–20295, 1991b

31. Seelig A: Local anesthetics and pressure; a comparison of dibucaine binding to monolayers and bilayers. Biochim Biophys Acta 899: 196–204, 1987

32. Hovious R, Lambrechts H, Nicolay K, de Kruijff B: Improved methods to isolate and subfractionate rat liver mitochondria. Lipid composition of the inner and outer membrane. Biochim Biophys Acta 1021: 217–226, 1990

33. Brooks SPJ, Suelter CH: Association of chicken mitochondrial creatine kinase with the inner mitochondrial membrane. Arch Biochem Biophys 253: 122–132, 1987

34. Rojo M, Hovius R, Demel R, Wallimann T, Eppenberger HM, Nicolay K: Interaction of mitochondrial creatine kinase with model membranes. A monolayer study. FEBS Lett 281: 123–129, 1991a

35. Sjöstrand FS: The structure of mitochondrial membranes; a new concept. J Ultrastruct Res 64: 217–245, 1978

36. Bücheler K, Adams V, Brdiczka D: Localization of the ATP/ADP translocator in the inner membrane and regulation of contact sites between mitochondrial envelope membranes by ADP. A study on freeze-fractured isolated liver mitochondria. Biochim Biophys Acta 1056: 233–242, 1991

37. Rojo M; Studies on the interaction of mitochondrial creatine kinase with membranes and structural studies on the adenine nucleotide translocator. Ph.D. Thesis No. 10042, Eidgenössische Technische Hochschule (ETH), Zürich Switzerland, 1993

Molecular and Cellular Biochemistry **133/134**: 125–144, 1994.

II-5 Myofibrillar creatine kinase and cardiac contraction

Reneé Ventura-Clapier, Vladimir Veksler[1] and Jacqueline A. Hoerter

CJF 92-11 INSERM, Université Paris-Sud, 92296 Châtenay-Malabry, France; [1] *Laboratory of Experimental Cardiac Pathology, Cardiology Research Center, 3, Cherepkovskaya Street 15A, Moscow 121552, Russia*

Abstract

This article is a review on the organization and function of myofibrillar creatine kinase in striated muscle. The first part describes myofibrillar creatine kinase as an integral structural part of the complex organization of myofibrils in striated muscle. The second part considers the intrinsic biochemical and mechanical properties of myofibrils and the functional coupling between myofibrillar CK and myosin ATPase. Skinned fiber studies have been developed to evidence this functional coupling and the consequences for cardiac contraction. The data show that creatine kinase in myofibrils is effective enough to sustain normal tension and relaxation, normal Ca sensitivity and kinetic characteristics. Moreover, the results suggest that myofibrillar creatine kinase is essential in maintaining adequate ATP/ADP ratio in the vicinity of myosin ATPase active site to prevent dysfunctioning of this enzyme. Implications for the physiology and physiopathology of cardiac muscle are discussed. (Mol Cell Biochem **133/134:** 125–144, 1994)

Key words: skinned fibers, sarcomere, functional coupling, cardiac hypoxia, myosin ATPase, crossbridges, cardiac mechanics, ischemic contracture

Introduction

The history of the discovery of creatine kinase role in muscle cells is intimately connected to the understanding of muscle contraction. At first, lactic acid formation was thought to be the primary energy-producing reaction. In 1927, Fiske and Subbarow [1] discovered the labile phosphorus of muscle, phosphocreatine (PCr), and showed that it decreased with contraction and was restored during recovery. The overthrow of the lactic acid theory was brought about by Lundsgaard in 1930 [2] when he showed that a muscle poisoned with iodoacetic acid was capable of performing many contractions without producing lactic acid. The relation between contraction and PCr content was the basis for the 'phosphagen era'. This was the start of increasing interest for phosphocreatine, until ATP was proposed as the phosphate donor. In fact, despite numerous attempts, it had been impossible to observe a decrease in ATP during a contraction cycle, until Cain and Davies in 1962 [3] by inhibiting creatine kinase with dinitrofluorobenzene, established that the primary reaction of muscle contraction was the conversion of ATP to ADP + Pi. This was the reason for the decreasing interest of muscle physiologists for creatine kinase and creatine phosphate in muscle contraction. PCr has been considered as a secondary energy reservoir to maintain sarcoplasmic ATP during pe-

Address for offprints: R. ventura-Clapier, CJF92-11 INSERM, Université Paris-Sud, 92 296 Châtenay-Malabry, France.

126

riods of excessive demand and such consideration is still widespread nowadays.

Myofibrillar creatine kinase as an integral part of the contractile unit

Contractile activity is the acceptor of energy synthesis and transfer in muscle cells. It is linked to two main events: ionic changes and particularly in calcium concentration, as the source of activation, and actomyosin ATPase activity as the chemio-mechanical transduction system. In the process of excitation-contraction coupling in cardiac cells (see Fig. 1 in chapter I-1), membrane depolarization triggers a calcium current. Calcium is then liberated from the sarcoplasmic reticulum mainly through calcium-induced calcium release mechanism [4]. These changes in intracellular calcium concentration lead to activation of actomyosin ATPase reaction and development of contraction. The relaxation process necessitates the active pumping of calcium by Ca-ATPases of sarcoplasmic reticulum and sarcolemma and the recovery of ionic concentrations to the basal level through different transmembrane exchangers and the Na-K ATPase of the sarcolemma. The main energy consuming reactions in muscle cells are associated with excitation-contraction coupling.

Creatine kinase is associated with the different sites of excitation-contraction coupling in muscle [5–7]. The M-isoform was identified in isolated fractions of sarcoplasmic reticulum and shown to be coupled with calcium transport and ATPase activity [8–10]. In sarcolemma, creatine kinase was identified in membrane vesicles and functional coupling between Na-K-ATPase and creatine kinase has been evidenced [11, 12]. However, the most important of these energy consuming reactions takes place in the myofibrillar compartment. The energetic correlates of muscle contraction have long been neglected with the idea that the high energy phosphate reserves (in the form of ATP and PCr) are largely in excess and prevent any substrate limitation for muscular contraction. In fact, in heart muscle, the safety margin is narrow and the regulations more complex and not simply quantitative. When taking the flux of energy inside the cell compared to the actual high energy phosphate concentration, it is clear that this would provide not more than a few seconds of activity, in the absence of oxidative phosphorylation [13]. This is a really narrow safety margin, and considering the higher content of PCr

in skeletal muscle, this rapidly points out the specificity of energy transfer in cardiac muscle.

The actual accomplishment of a long evolutionary process leads to a complex organization of intracellular functions highly hierarchised and interrelated. Examination of the end product and evolutionary as well as development processes leading to adult mammalian cardiac cells, is of great interest for the understanding of living processes. In this review, we will mainly focus on the myofibrillar compartment, its functioning and the role that myofibrillar creatine kinase plays in cardiac function.

Organization of myofilaments

The myofibrillar compartment in mammalian adult cardiac cells shows minimal species variability in volume density. It amounts to about 60% of the mammalian cardiac cell volume [14]. Together with a mitochondrial compartment within the range of 22 to 37% of volume density, and a sarcoplasmic reticulum compartment of 3–5%, this emphasizes the highly organized structure of adult mammalian cardiac cell.

The myofibrillar compartment represents a three-dimensional arrangement of fibrillar proteins. In striated muscles, the contractile proteins are organized in myofilaments made of a series of contractile units: the sarcomere (Fig. 1). Each sarcomere is a repeat of interdigitating thick and thin filaments. Contraction is the consequence of interactions between thick and thin filaments. The thick filament forms the A-band and is constituted of myosin molecules bearing the ATPase activity on their two globular heads protruding from the filament. The thin filament or regulatory filament is constituted of two strands of polymerized actin and two strands of tropomyosin. This filament also contains the troponin unit formed of three different subunits C (calcium-binding component), I (inhibitory component), T (tropomyosin-binding component) which represent the calcium sensor and are organized in a repeat. Thus the thick filament can be regarded as the location of the energy transduction system while the thin filament is the regulatory structure or calcium sensor and plays the role of modulation of contractile amplitude. The myosin molecules in the sarcomere are linked together in their center at the level of the M-line. Other proteins like α-actinin or desmin form the Z-line which joins the actin filaments. One sarcomere is delimited by two Z-lines and contraction takes place by the sliding of thin filaments

Compartmentation of adenine nucleotides

in myofibrils

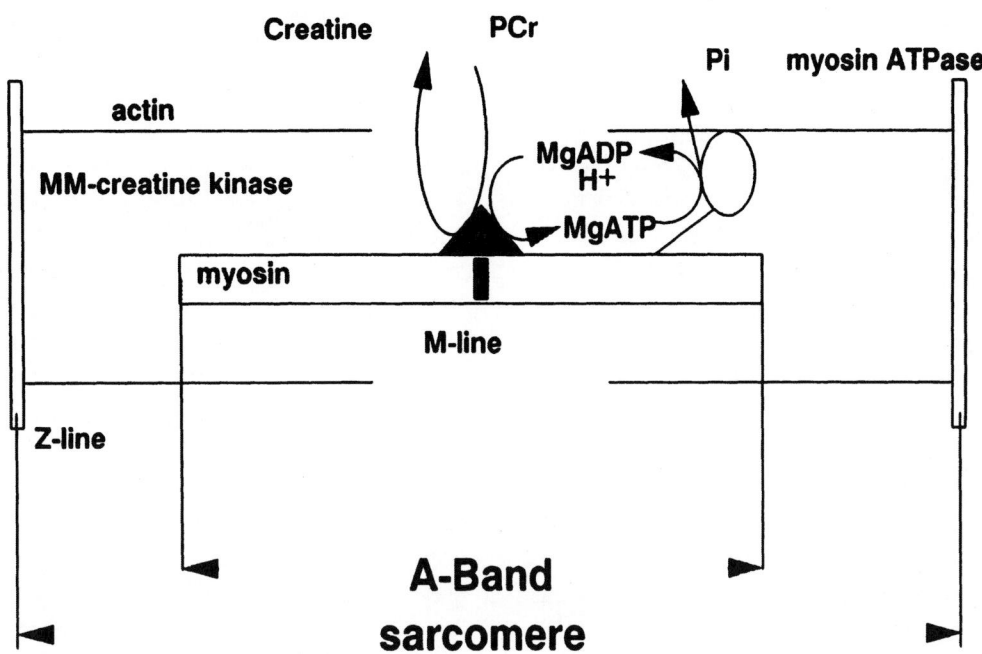

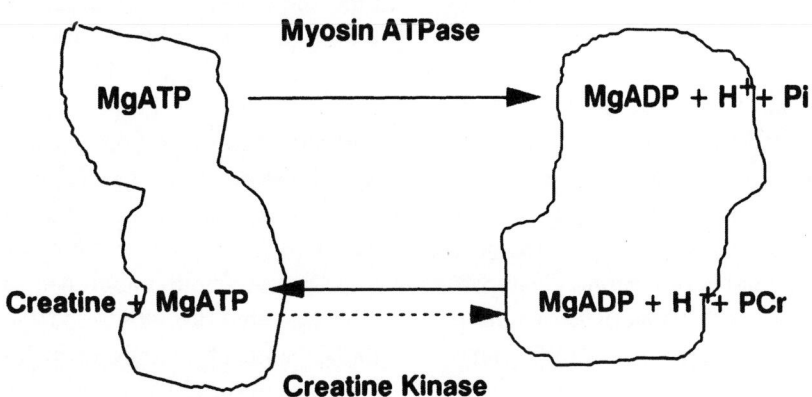

Fig. 1. Compartmentation of adenine nucleotides in myofibrils.

between thick filaments due to the cyclic attachment of myosin heads to actin. The M-line forms the center of the sarcomere.

This complex proteic structure entails peculiar properties. Given the high amount of proteins in mammalian cells, and the density of intracellular structures, the volume occupied by proteins should more closely approxi-mate crystals than dilute solutions [15]. Accumulating evidences suggest that a large fraction of total cell water exhibits solvent properties that differ from those of pure water [16]. The extent to which cellular compartments are concentrated solutions of macromolecules, metabolites, ions and either solutes is a matter of importance in the relationship between structure and function [16].

High protein concentration and the presence of hydration or organized water will confer unique enzymatic properties and capabilities of assembly within the intracellular medium. This dense protein assembly represents a crystalline structure in which water molecules are organized. It is widely agreed that each filament possesses a net charge on its surface and is surrounded by a diffuse layer of counter-ions. Morel suggests that myofilament lattice is a succession of water-protein interfaces, the interfilamentary water having properties comparable to those of vicinal water found near solid surfaces and, therefore, quite different from those of bulk water. This may be the reason for the high stability of the filament lattice [17]. The fixed charge density of myofilaments, mainly due to aminoacids is estimated to be − 42 meq/l cytoplasmic fluid. These negative fixed charges in myofibrils will exclude anions while concentrating cations [18]. It will additionally create a more acidic environment. A nonuniform distribution of potential was observed in skinned fibers with more negative potential in the A-band than in the I-band during the rigor state and changing during relaxation [19]. These microheterogeneities in charges and potential participate in compartmentation of charged molecules in the myofibrils. Maughan et Godt [18] however calculated that these microheterogeneities cannot account for compartmentation of PCr in the I-band observed by Hill [20]. Thus, the sarcomeric structure can be viewed as a supramolecular assembly devoted to a specific function.

Localisation of creatine kinase in myofilaments

The localization and function of M-line bound creatine kinase has been reviewed a few years ago [21]. We will briefly summarize the main points. Ottaway [22] showed in 1967 that a significant amount of creatine kinase was strongly bound to myofibrillar structure of heart muscle. In 1973, creatine kinase has been recognized by Turner *et al.* [23] as an integral part of the sarcomere organization, binding specifically to the M-line structure. Physicochemical studies indicated that a dimer of creatine kinase can be isolated from the M-line and can associate with the rod portion of the myosin molecule [24]. Strong evidences that the M-form of creatine kinase (M-CK) is an integral element in the M-band structure have been found by the use of immunological and electron microscopy techniques [6, 7]. M-CK has been identified by immunofluorescence as one of the constituent elements of the M-line [21]. In chicken skeletal muscle, myofibrils contain approximately 0.8 IU of creatine kinase activity per milligram dry weight representing 5% of the total cellular CK [6]. In chicken heart, lacking the M-line and containing almost exclusively the BB-isoenzyme, the B-form of creatine kinase was identified in the Z-line where it represents 2% of total CK activity [7]. However, adult mammalian cardiac cells contain a well developed M-band structure as well as M-line-bound CK like mammalian skeletal muscle [7, 25, 26].

M-creatine kinase is the product of one gene. The MM-form of CK is specifically associated with the M-line while B-CK is generally found at the Z-line. When mRNA or cDNA for M-CK are injected in cultured chicken heart cells which do not express M-CK, only the M-CK product display a pattern of interaction in the A-band region, and associates in an isoprotein-specific manner with the M-band of myofibrils [27]; it was determined using chimeric cDNA constructs between M and B-CK that the head COOH-terminal region of the molecule was responsible for the isoprotein specific interaction with the cytoarchitecture. Creatine kinase thus plays a role in the formation of the M-line and the structuration of the A-band in the sarcomere.

A controversy still exists concerning the exact location of M-CK within the myofilaments [6, 7, 21]. Indeed, selective M-line immunofluorescence has been obtained after extensive washing of the preparations to eliminate cytosolic and loosely bound M-CK. More recent studies have investigated, in more details, the localisation of cytosolic M-CK in intact sarcomeric muscles. In 1989, Otsu and his collaborators [28], studied the localization of M-CK in canine myocardium by direct immunoperoxidase method. These authors concluded that M-CK was present in the whole A-band and the Z-line. They found it mainly associated with the thick filament with only a small amount being in the M-line. Using a similar experimental approach, Robert and colleagues [29] suggested that M-CK in frog muscle is localized at the A-band. This would fit with *in vitro* association of creatine kinase with the myosin heads subfragment I described by Yagi and Mase [30] and Botts and Stone [31] in *in vitro* studies. However, Wegmann and collaborators [26] pointed out the possible diffusion and trapping of stain products using this indirect technique. They presented new data on the localization of M-CK in *in situ* cryosections of mammalian cardiac muscle and chicken skeletal muscle, showing that M-CK was localized at the M-line as well as the sarcoplasmic reticulum. Surprisingly, it was also highly compartmentalized and mainly confined to the I-band (actin filament), in the region of no-overlap between thick and thin filament. They gave evidence for a co-localization of adenylate kinase and aldolase with creatine kinase in this region. Other enzymes associate in a more or less loosely fashion to myofibrillar proteins. In particular, glycolytic enzymes and myokinase are associated with the actin filament [32, see chapter II-6]. AMP deaminase can reversibly bind to myosin during intensive work [33].

The acto-myosin overlap region is a region of dense protein concentration, creating a specific environment with crystalized water, unstirred layers and charge density. This environment makes this compartment less favorable for free diffusion of large proteins. Wegmann et al. [26] concluded from their observations that the acto-myosin overlap region of the A-band acts as a molecular sieve, excluding to a large extent large molecules (for illustration see their Fig. 8).

Association of creatine kinase with myofibrils

In mammalian cardiac as in chicken skeletal muscle, M-creatine kinase is associated with myofibrillar structures. Two types of associations have been described: a very weak binding, closer to adsorption on the actin filament and a stronger binding, observed after intensive washing within the M-line [26]. Binding studies in washed Triton X-100 skinned fibers and isolated myofibrils of rat heart were performed to clarify the nature of the interaction of creatine kinase with cardiac myofibrils. CK activity was mostly due to reversible binding of the enzyme with an apparent Km value of 2 μM; due to a high concentration of creatine kinase in cardiac cells (20 μM), these sites should be saturated in *in vivo* conditions [34].

By comparing the published data of myofibrillar CK activity (Table 1) it appears that specific activity of M-CK in myofibrils is constant. In extensively washed myofibrils from chicken pectoralis muscle, the specific activity of creatine kinase in myofibrils is 0.79 IU per mg of protein [21]. Since the total activity of creatine kinase in this fast skeletal muscle is about 2200 IU/g wet weight, a

value of 3–5% of total CK is compartmentalized in myofibrils in skeletal muscle. According to Wallimann et al. [7], it can be estimated that chicken heart myofibrils contain 0.07 IU/mg protein of BB-CK, amounting to about 2% of total CK activity. A value of 1 IU/mg of myofibrillar proteins was reported in cardiac myofibrils isolated in an EGTA containing buffer [35] representing more than 20% of total CK activity in rat heart [34]. This difference in the relative amount of bound CK simply reflects the lower total CK activity in these two muscles (2200 IU/gww in fast skeletal muscle versus 400 IU/gww in rat heart). It points to the difference in the organization of the creatine kinase system in skeletal *versus* cardiac muscle where the main part of CK is compartmentalized within the cell. Indeed, the specific amount of MM-creatine kinase in myofibrils is not different in mammalian heart (3 mg/g) [34, 35] and chicken skeletal muscle (3 mg/g [21], in accordance with the proposed structural role of MM-CK in the M-line [21]. The molar ratio of creatine kinase to myosin has been estimated and vary from 1/40 CK/myosin for skeletal muscle [21] to 1/10 for heart muscle [34]. These calculations are based on many approximations and the discrepancy may not be of real significance. Anyway, it unequivocally excludes that, *in vivo*, one CK molecule would be associated with one myosin molecule as was shown *in vitro* [30, 31]. One can thus conclude that the interaction between M-CK and myosin ATPase is not an enzyme-enzyme interaction and would be more directly influenced by the specificity of the intramyofibrillar milieu.

Myofibrillar creatine kinase as an actor of contractility

Considering the highly organized ultrastructural characteristics of mammalian cardiac cells, it confers unique properties to these intracellular structures and to each individual reaction within each compartment. Each reaction should thus be considered inside its own environment, and in connection with the peculiar function assigned to each compartment.

Table 1. Myofibrillar ATPase and creatine kinase activities.

	total CK IU/g ww	myosin ATPase IU/min/mg prot	CKmyof IU/min/mg prot	CKmyof/total CK %	CK amount mg/g	refs
chicken fast	2,200	0.35	0.79	5	3	[6, 21]
rat heart	400	0.13	1	20	3	[34, 35]
chicken heart	730	0.08	0.02	2	1	[7]

Sarcomere function. Chemio-mechanical transduction in striated muscles

To understand the possible role of creatine kinase in modulating striated muscle contraction, it is important to know at which steps of the crossbridge cycle, hydrolysis of ATP and release of reaction products occur. Muscle contraction occurs when the two sets of interdigitating filaments slide past each other. The widely accepted theory proposed by A. Huxley [36] is the cross bridge theory of muscle contraction. Muscle contraction or shortening occurs as a result of cyclic interactions between thick and thin filaments and cross bridge extension from myosin to actin filament with the consumption of chemical energy liberated by ATP hydrolysis in the presence of calcium. Troponin C moiety in the regulatory complex has the key role in setting the activation characteristics of the contractile proteins [37]. Much efforts have been made to connect the mechanical changes of muscle fibers with the kinetics of ATP hydrolysis by the actomyosin complex [38]. The current theory is that cross-bridges can be classified into two types, i.e., a weakly-bound state where myosin binds ATP or rather its products ADP·Pi and a strongly bound state with ADP or without bound nucleotides (rigor bonds) [39]. The transition between these two states is coupled to the release of Pi from the complex and leads to the generation of force [40].

During a cross bridge or actomyosin cycle (Fig. 2), four major steps may be described. 1) At the low calcium concentration prevailing in diastole (about 200 nM), ATP is bound to myosin and hydrolyzed while the products (ADP and Pi) are not dissociated. The interaction or strong binding of actin and myosin is inhibited and crossbridges are detached. 2) Upon calcium entry and subsequent binding to TNC, thin filaments are activated and actomyosin complexes are formed, *i.e.* cross bridges are attached. 3) This step activates the release of the products of ATP hydrolysis and leads to the formation of rigor-bonds with conversion of chemical energy into mechanical work. 4) Dissociation of actin and myosin after fixation of a new ATP molecule on myosin and relaxation of the muscle. If Ca is still present a new cycle will be initiated. ATP binding and hydrolysis are considered to be fast and non-limiting processes [41]. The Km for MgATP is in the order of 10 µM in myofibrils in the presence of an ATP regenerating system [42]. The predominant steady-state intermediates of relaxed muscle is M·ADP·PI. The release of products is activated 100 times by the binding of actin to myosin [43]. Pi release is

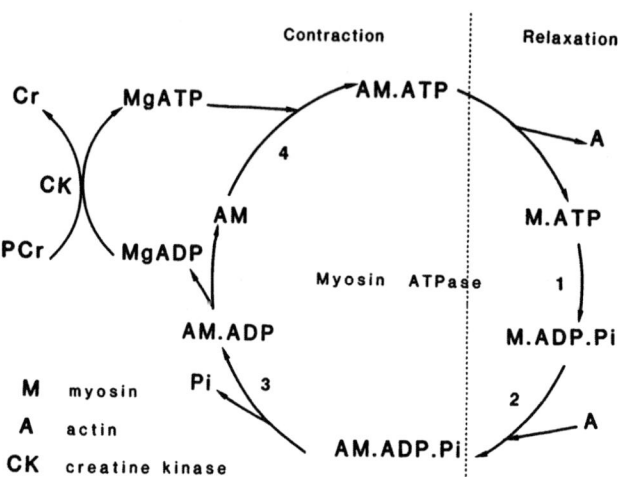

Fig. 2. Cross bridge cycle and creatine kinase in striated muscles. A, actin; M, myosin; AM, actomyosin complex. In step 1, ATP is bound to myosin and hydrolysed. The binding of activated actin (2) increases the rate of products detachment and the chemiomechanical coupling which takes place during the Pi release step 3. Detachment of ADP is thought to limit the crossbridge cycling rate and the rebinding of ATP and further detachment of actin (step 4). The CK reaction acts at the rate limiting step *i.e.* ADP detachment, rephosphorylation and rebinding of MgATP.

associated with the force generating step which will be dependent on the mechanical constraints applied to the muscle. ADP release appears to limit the rate of cross bridge detachment, and maximum shortening velocity [44, 45]. The kinetic steps are influenced by the strain in the crossbridges [38] which explains that the cycling rate is slower in non shortening than in shortening muscles [46]. In isometrically contracting cardiac skinned preparation, the release of products is slower than for isolated cardiac proteins or myofibrils and the slow step is likely to be the ADP release from the actomyosin·ADP complex [47]. The dissociation or inhibitory constants of MgADP on cardiac actomyosin or myofibrils are comparable and in the range of 10 µM while being lower than the corresponding values for skeletal muscle (170 µM) [44, 48–50]; MgADP acts as a pure competitive inhibitor over a large range of concentration [51]. Thus, the intramyofibrillar concentration of MgADP would play a key role in determining ATPase activity, crossbridge cycling and contractile force. These kinetic considerations allow to speculate that any ADP regenerating enzyme in the myofibrillar compartment, able to control ADP concentration near the myosin active site, would be of great importance in influencing crossbridge cycling rate and contractile activity. This can be the main role of M-form of CK associated with myofibrils.

During normal cardiac activity, it can be calculated that 1 mM·s^{-1} of ATP (calculated per liter of cell water) is utilized, the main part (60–80%) being due to cross-bridge cycling. Thus, a similar amount of MgADP is to be produced in myofibrils. In the relaxed state, myosin is saturated with ADP and if the concentration of myosin active site in the cell is 150 µM [47], a similar concentration of ADP is associated with myosin [52]. It is impossible to directly measure free ADP concentration in muscle due to the high amount of ADP associated with actin and the subthreshold concentration of free ADP which makes it invisible by NMR technique. However MgADP concentration are deduced from the equilibrium reaction of creatine kinase and myokinase and should be of the order of 50 µM in cardiac tissue [31], *i.e.* in the inhibitory range of myosin ATPase as discussed above. On the other hand, this also excludes that ADP itself would be the direct signal from myofibrils to mitochondria for adequation of energy production to utilisation.

Functional coupling of creatine kinase and myosin ATPase

A series of different experiments have led to the conclusion that myosin ATPase and creatine kinase were functionally coupled in myofibrils. By this statement we mean that a small pool of adenine nucleotides (ATP and ADP as substrate and product) is rapidly turning over in the ATPase and the creatine kinase reactions, at a rate higher than their diffusion rate in the highly organized structure, and in this way maintaining a high ATP/ADP ratio inside the sarcomeres. This concept has evolved from a variety of experimental conditions and observations.

Biochemical evidences
In a reconstituted system consisting of myosin ATPase and CK, Yagi and Mase [30] showed that the Michaelis-Menten constant of the ATPase for ATP was 1 or 2 orders of magnitude lower when creatine kinase was associated with myosin shifting from 300 to 7 µM; they proposed that in the two-enzyme system, the substrate concentration near the ATPase active site would be much higher than the mean concentration in the reaction mixture [53]. In this coupled reaction, no difference was observed whether ATP or ADP were present and the rate limiting step seemed to be at the ATP hydrolysis step. Creatine kinase extracted from myofibrils is kinetically close to mitochondrial creatine kinase [54]; Km for ADP is equal to 0.08 mM and for PCr 1.7 mM and the ATP

formation reaction is kinetically preferable. In fact, the kinetic properties of myofibrillar creatine kinase make it favorable to interact with myosin ATPase. Saks *et al.* [54] showed that in the coupled reaction, CK binds ADP produced by the ATPase with a high affinity (80 µM); on the other hand, there is a six-fold difference in K_M for ATP in the CK reaction (0.95 mM) and in the ATPase reaction (0.16 mM), such that the ATP formed will be rapidly redistributed from CK to myosin to start a new cycle [54]. Bessman *et al.* [55] showed by radioactive labelling of Pi that there is preferential liberation of Pi by ATPase from ATP generated by creatine kinase than from bulk ATP. Additional evidence in favor of a close functional relationship between the two enzymes came from ^{31}P-labeled ATP experiments indicating that the ATP formed by PCr through the CK reaction can reach the ATPase active site more readily than labeled ATP from the medium [55]. This close interaction and preferential access of products of ATPase reaction to myofibrillar CK and vice versa were identified as functional coupling.

Further characterization of the functional coupling was obtained using a competitive enzyme system for the ADP produced by the creatine kinase reaction [12]. In these experiments soluble enzymes were compared to myofibrils bearing both ATPase and CK activities. The competitive enzyme was pyruvate kinase in the presence of phosphoenolpyruvate. Although the production of creatine and pyruvate in the soluble enzyme system depends directly on the respective activities of both enzymes, in the myofibrillar preparation, even a PK/CK ratio of 100 was not enough to inhibit the production of creatine [12], showing preferential access of ADP to creatine kinase reaction due to immobilization of the two enzymes.

Using the pH-stat assay system to monitor the net consumption of protons by the CK reaction in the coupled myofibrillar CK and ATPase system, Wallimann and colleagues [21, 56] showed that endogenous M-bound CK was sufficient to regenerate all ATP hydrolyzed by the myofibrillar actin- activated myosin ATPase. These experiments additionally showed that CK reaction in myofibrils led to local alkalinization despite production of protons by the ATPase reaction. This effect may also be of significant importance in the physiology of muscle contraction. Indeed, these authors showed that the optimum pH for the CK reaction is between 6.5 and 6.6 while the optimum pH for the MgATPase reaction in chicken fast skeletal muscle is between 7.5 and 7.8. The close functional association of the two enzyme will also avoid inhibition of the ATPase by proton production and their difference in pH sensitivity will preserve within a physiological pH range (6.6–7.5) the high matching and efficiency of the two reactions to avoid any influence of local proton production on myosin ATPase reaction. CK reaction can thus be viewed as a local pH buffering system. In this respect, it is noteworthy for muscle function and for integration between energy consumption and utilization, to consider that the CK reaction, due to its functional coupling with translocase in mitochondria (see corresponding chapters in this book), will function in the direction of proton production and restitution just where they will be able to enter oxidative phosphorylation pathway. Thus, in addition to high energy phosphate transport, CK reaction can be viewed as a proton transport system from sites of production i.e. the ATPases to sites of utilization (oxidative phosphorylation in oxidative muscles).

Other kinetic considerations came from experiments performed on myosin ATPase activity. In isolated cardiac myofibrils, the apparent Km for MgATP was observed to be 80 µM in the absence of PCr and was shifted to 14 µM in its presence; exogenous creatine kinase or pyruvate kinase as ATP-regenerating systems do not produce such a high shift in Km, showing that the localization of creatine kinase inside the myofibrils is of primary importance for maximal efficacy [57]. On the other hand, although kinetic properties of myosin ATPase seem to be greatly affected in the myofibrillar structure, the binding of creatine kinase by itself does not change its kinetic properties [58].

How can we explain this functional coupling? Compartmentation can be established by semi-permeable limiting membranes, multi-enzyme complexes with covalent substrate transport, enzyme-enzyme proximity effects without direct substrate transport, exclusive effects imposed by structured water, or the establishment of Nernst unstirred layers [59, 60]. For myofibrillar compartmentation, a direct interaction between ATPase site and creatine kinase seems to be excluded for different reasons: first the stoichiometry gives at most a 1/10 ratio between CK and ATPase. Additionally, careful morphological data as well as steric considerations indicate that the myosin head region is devoid of creatine kinase activity [26]. The role of proximity or propinquity in substrate channeling was described for hexokinase and creatine kinase co-immobilized on sepharose beads; when the two enzymes were separated by less than 10 nm, they comprised a functional compartment in which hexokinase had preferential access to ATP formed by CK [61].

However CK and myosin heads are distant by more than 10 nm. On the other hand, the co-localization on a surface of enzymes which catalyze consecutive reactions, may increase the efficiency of their kinetic coupling. This vicinity is able to favor the channelling of intermediate metabolites. Such a channelling was observed in Triton-X 100 lyzed frog heart cells: ATP is channelled from creatine kinase to the ATPase site of myosin and do not diffuse into the bulk solution [62]. A modelling of coupled kinetics could be obtained by co-immobilization of creatine kinase and myosin on an artificial membrane and Arrio-Dupont et al. [63] described similar ADP channelling from ATPase to creatine kinase. The channelling of substrates between co-immobilized enzymes is due to the presence of an unstirred layer near the active surface. Thus unstirred layer represents a microcompartment in which the local concentrations of the intermediate substrates and products are different from those in the bulk solution, creating more favorable operating conditions for the second enzyme although the two enzymes are not in close proximity.

Mechanical evidences
The subsequent question is how this functional coupling interferes with contractile activity and crossbridge cycle. As early as in 1953, Bozler [64] showed that PCr intensifies the action of ATP to a surprising degree, increases the tension of contracting fibers and accelerates relaxation, and concluded that contraction requires only a small amount of adenine nucleotides firmly bound to the contractile elements. Similar results were reported by Perry in 1954 [65] who found that a given low concentration of ATP, maintained by the creatine phosphokinase system, was much more effective in inducing contraction than the same concentration of ATP alone. In 1962, Yagi and Mase [30] have shown *in vitro* formation of a two enzyme complex between creatine kinase and myosin ATPase with characteristics of functional binding between these two enzymes. Savabi et al. [66], demonstrated in glycerinated psoas fibers that in the presence of small amount of ADP, 10 mM PCr produced much faster and stronger contraction and faster and more complete relaxation than 10 mM ATP. McClellan et al. [67] using hyperpermeable cells described two important features of energy transfer in cardiac cells. First, adenine nucleotides are bound to the contractile system in resting muscle; 2) these bound nucleotides can be rephosphorylated in the immediate vicinity of the contractile system by phosphate transfer from PCr.

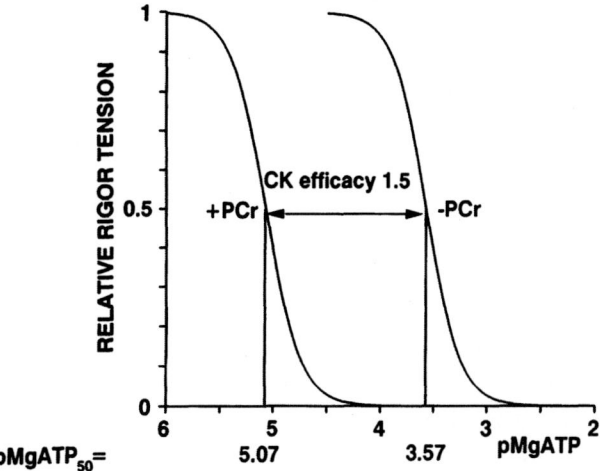

Dependence of rigor tension development on MgATP

Fig. 3. Dependence of rigor tension development on MgATP. Triton X-100 skinned fibers are successively bathed in very low calcium (pCa 9) solutions of decreasing MgATP in the presence or in the absence of 12 mM PCr. Rigor tension is plotted as a function of pMgATP using the Hill equation. pMgATP$_{50}$ the pMgATP for half maximal rigor tension development is calculated using linearization of the Hill equation. CK efficacy is the difference between pMgATP$_{50}$ values in the absence and in the presence of PCr. The presence of PCr and the endogenous CK activity induce a drop in MgATP needed to induce rigor tension.

Functional coupling studies using skinned fibers
In view of the importance of intracellular architecture and protein organization, we have chosen to study the contractile properties of cardiac myofibrils using skinned fibers. This technique of chemical solubilization of cellular membranes with detergents, allows to control ionic and metabolic environment around myofibrils, while keeping the complex cell architecture. Detergents such as Triton X-100 induce complete disruption of all cellular membranes and result in removal of cytosolic and membrane associated fraction of proteins. Skinned fiber bundles can be attached to a force transducer to measure Ca activated maximal force, Ca sensitivity of myofilaments and rigor tension (*i.e.* tension developed at low calcium and low ATP concentrations). Responses of the fibers to quick length changes probe stiffness and rate of cross-bridge turnover [68].

ATP sensitivity. When myosin is deprived of MgATP, it strongly interacts with actin, independently of calcium ions, leading to rigor tension. It is thus possible to observe an increase in contractile force when MgATP concentration is stepwise decreased in the absence of calcium ions (Fig. 3). In this case crossbridges exhibit no or very slow turnover. Reuben *et al.* [69] observed in crayfish muscle that the resulting relation for force versus MgATP in the absence of calcium was a bell-shaped

curve, such as frequently obtained in studies of the kinetics of enzymes which manifest substrate inhibition. Bremel and Weber [70] further demonstrated that myosin free of bound nucleotides has a high affinity for actin molecules, even in the presence of calcium free troponin. Thus, rigor tension development seems to be associated with an increase in myosin ATPase activity linked to the presence of rigor cross-bridges. The study of rate and extent of rigor tension relaxation is a good measure of the actual MgATP concentration in the immediate vicinity of the myosin ATPase substrate binding sites. In 1984 [71] and 1985 [72], we successively described this kind of observations. In EGTA treated fibers, Veksler *et al.* [71] observed that the relaxation of rigor tension is more sensitive to ATP in the presence than in the absence of PCr, producing a shift in the apparent Km for relaxation towards lower MgATP concentration (Fig. 3). The given interpretation was that, due to the close proximity between CK and ATPase, ADP was rapidly rephosphorylated creating a higher ATP concentration close to the active site of the ATPase. It was further shown that this effect was inhibited by the addition of FDNB, an inhibitor of CK and that an external MgATP regenerating system consisting of pyruvate kinase and phosphoenolpyruvate was not as efficient as endogenous CK in relaxing rigor tension [72]. Half maximal effects were obtained for 2 mM PCr and 14 µM ADP [71, 72], *i.e.* values very close to the K$_M$ values of M-CK for its substrates [54]. The pMgATP value for half maximal relaxation of rigor tension (pMgATP$_{50}$) was further used as an index of creatine kinase efficacy. This observation means that the apparent Km for relaxation of rigor tension is shifted from 300 µM MgATP to 10 µM. This amounts approximately to 1.5 log units (Fig. 3). This index proved himself to be very useful for the estimation of the presence, function and possible alterations of myofibrillar creatine kinase. It was found relatively constant among adult mammalian myocardium (see Table 2), decreasing in some cases of cardiomyopathies [73] and increasing during development [74] (see chapters V-1 and V-2). This indicates that the actual MgATP concentration in the vicinity of myosin ATPase is higher than in the bulk solution. But it may also mean that MgADP concentration is lower in the vicinity of the myosin active site, or both. In an attempt to understand the influence of metabolic products on contractile activity, we recently investigated the influence of protons, ADP and Pi on rigor tension development [75]; the dependence of rigor tension development on MgATP decrease is lowered by protons and greatly enhanced by

134

Table 2. Creatine Kinase efficacy in the heart of different species.

	Creatine Kinase Efficacy Log units	type	Refs
Rat	1.58	M	[73]
Rabbit	1.92	M	[74]
Guinea-pig	1.79	M	[114]
Hamster	1.69	M	[73]
Ferret	1.66	M	unpublished
Chicken	0.28	B	unpublished
Pigeon	0.3	B	unpublished
Frog	0.17	'M'	unpublished

CK efficacy is the difference in pMgATP concentration needed to induce half rigor tension development in the presence and in the absence of phosphocreatine (see Fig. 3).

MgADP. Indeed, addition of 250 µM MgADP further shifts the apparent Km for MgATP to higher MgATP concentrations (550 µM). This effect persists even in the presence of both Pi and acidosis (Fig. 4). Thus rigor tension development appears to be very sensitive to ADP concentration. This effect of MgADP is completely removed in the presence of PCr since no rigor tension has been observed even in the complete absence of MgATP showing the high efficacy of endogenous CK to remove intramyofibrillar ADP.

One possible problem in interpreting these results is the use of multiple cell preparations. We should therefore consider the intercellular diffusion problems inherent to cardiac skinned fibers. It can be argued that diffu-

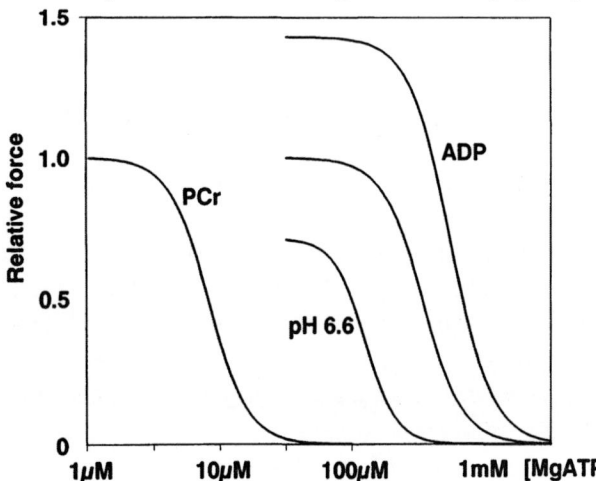

Dependence of relative rigor tension on [MgATP]

Fig. 4. Effects of MgADP and protons on rigor tension. In the absence of PCr, MgADP strongly increase rigor tension and shift the threshold towards higher MgATP concentration, while protons induce opposite effects. In the presence of PCr, MgADP prevented the development of rigor tension whatever the MgATP concentration.

sional limitations due to the size of the preparations and MgATP consumption inside skinned fibers would produce an artifactual shift in the apparent Km for MgATP towards higher MgATP concentrations. Although such a problem does exist, we think that it does not interfere to a high extent with the results reported below for the following reasons. We have reported that CK efficacy was not dependent on the diameter of the preparations between 1000 and 50 µM [68] in the presence or absence of PCr. Indeed, Cooke and Pate [51], by modelling diffusion limitation inside a fiber, suggested that the gradients in nucleotide concentration across the fibre did not greatly influence the data. If simple intercellular diffusion would be observed for MgATP, a similar observation i.e. shift in apparent Km should be observed with PCr since these two molecules do not have large differences in their diffusion coefficient [76]. Indeed, the apparent Km for PCr [71, 72] is very close to the real Km of creatine kinase for PCr (2 mM) [77], showing no limitation of diffusion for this substrate inside the preparations in rigor conditions. Furthermore, the enzymatic coupling between creatine kinase and myosin ATPase has been evidenced biochemically in skinned fibers as well as in isolated myofibrils where the diffusion distances are restricted to the size of myofibrils (few microns) [34, 35]. Using saponin-permeabilized myocytes, Nichols and Lederer [78] showed that the MgATP dependence of shortening velocity had a threshold of 1 mM and a K1/2 of 100 µM MgATP; these values were shifted respectively to 100 µM and 10 µM in the presence of PCr. Recently, using isolated skinned cardiomyocytes, where intercellular diffusion distances have been abolished, we showed that the threshold for rigor tension development in isolated cells was close to the value determined using skinned fibers; interestingly, endogenous creatine kinase system also shifted the threshold for rigor tension development (Fig. 5). This suggests that the problem of concentration gradients of nucleotides is an intracellular or an intramyofibrillar phenomenon. However, such a limitation of MgATP diffusion is still surprising. In the presence of a MgATP regenerating system, the Km for MgATP of actomyosin is close to 10 µM, the value we observed for relaxation of rigor tension, but far from the value obtained with bulk ATP which is in the order of 300 µM. To explain this difference we should infer a high hindrance to ATP diffusion inside the myofibrillar space. Interestingly, Yagi and Mase using isolated myosin showed that the Km for MgATP was shifted from 300 µM to 7 µM in the two enzyme system [30]. Our data are also consistent with their

Rigor tension at various values of pMgATP

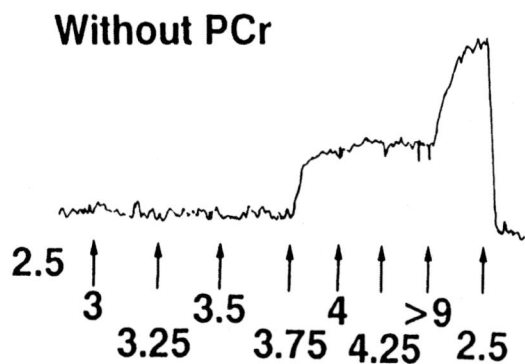

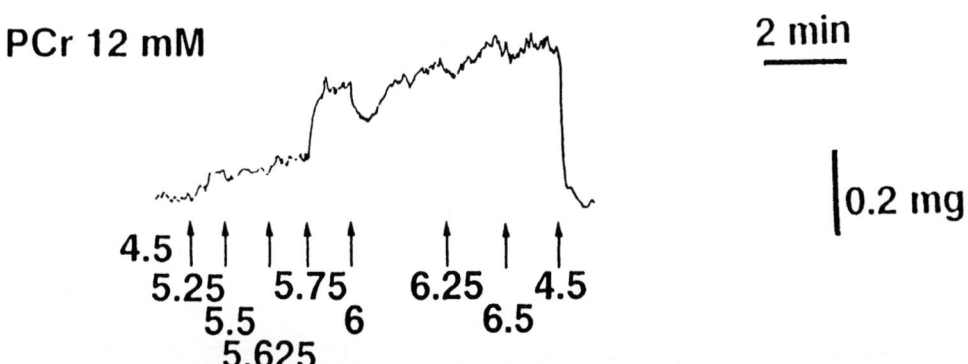

Fig. 5. Rigor tension development in low calcium (pCa7) at various values of pMgATP in an isolated skinned cardiomyocyte. Solution changes were done by rapid superfusion of the cardiomyocyte. The threshold for rigor tension development was pMgATP 3.75 in the absence and 5.25 in the presence of PCr showing compartmentation of adenine nucleotides in the myofibrillar compartment.

conclusion that ATP hydrolysis by myosin is the limiting step in the coupled reaction as rigor tension development in the presence of PCr seems to be limited by the Km of myosin ATPase for MgATP rather than by the CK reaction itself.

A more plausible conclusion is that accumulation of ADP rather than ATP consumption is the reason for rigor tension development. Krause and Jacobus [57] described a non-competitive inhibitory effect of MgADP in the millimolar range and therefore concluded that K_{iADP} was far too high to have much influence on the ATPase. However, this uncompetitive inhibition is apparently different from the pure competitive inhibition described in other studies (see discussion above). Considering that myosin ATPase will produce an amount of ADP equal to the amount of ATP consumed and that Ki

for competitive inhibition for MgADP is around 15 μM in cardiac muscle, this suggests that myosin ATPase will be in fact more inhibited by MgADP accumulation. Thus a major role of bound creatine kinase would be to maintain a low ADP concentration. The apparent high Km for ATP observed on the absence of ATP regenerating system thus seems to arise from the competitive action of low ATP and high ADP at the myosin ATPase active site.

b- Active tension development and role of myofibrillar creatine kinase. Knowing the importance of creatine kinase for adenine nucleotide compartmentation in myofibrils, it was of interest to investigate the influence of this functional coupling between myosin ATPase and creatine kinase during activation of contractile machin-

136

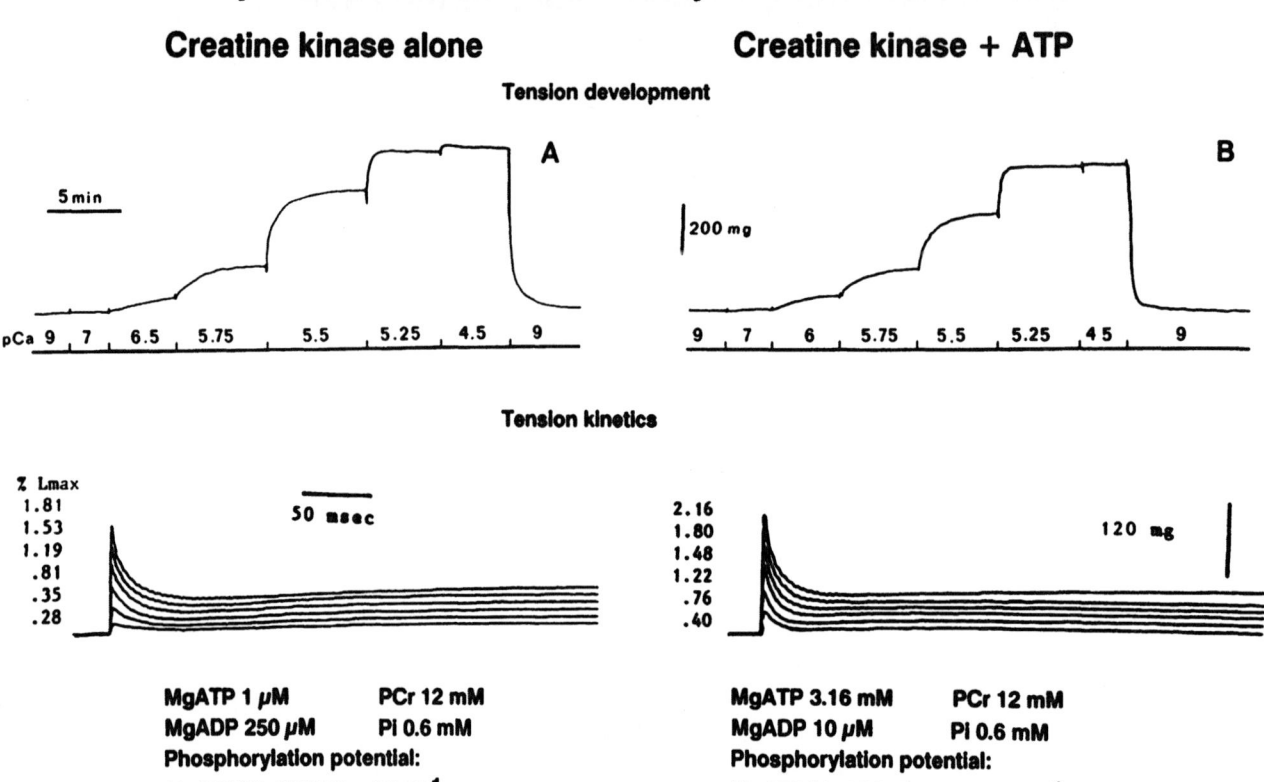

Role of myofibrillar creatine kinase. Study in skinned cardiac fibers.

Creatine kinase alone   Creatine kinase + ATP

Tension development

Tension kinetics

MgATP 1 μM   PCr 12 mM
MgADP 250 μM   Pi 0.6 mM
Phosphorylation potential:
MgATP/MgADP*Pi= 6.7 M⁻¹

MgATP 3.16 mM   PCr 12 mM
MgADP 10 μM   Pi 0.6 mM
Phosphorylation potential:
MgATP/MgADP*Pi= 530 000 M⁻¹

Fig. 6. Role of myofibrillar creatine kinase. Study in skinned cardiac fibers. Experiments were performed either at very low MgATP (1 μM) and high MgADP (250 μM) and PCr (12 mM) to check endogenous creatine kinase efficacy (lane A) or in the presence of high MgATP (3.16 mM) and PCr (12 mM) for control (Lane B). In the upper part, force was induced by stepwise increasing calcium concentration. In the middle part, tension kinetics in response to quick stretches from 0.28 to 2.16% muscle length are shown. In the lower part, calculations of the phosphorylation potential in these two conditions are given.

ery by calcium ions. Again this is made possible using skinned fibers at different calcium concentration and investigating Ca sensitivity of tension development. Additionally, kinetic behavior of actomyosin interactions is of primary importance in a muscle exhibiting cyclic activity.

The first question was the real capacity of endogenous creatine kinase to fulfil the requirements of myofibrillar activity. Using skinned fibers which, as we have seen, still contain their endogenous, firmly bound creatine kinase, it is possible to investigate the intrinsic mechanical properties when PCr and ADP are the only substrates or when MgATP is present with PCr [34]. This second condition approximates the *in vivo* situation. The first condition checks the effectiveness of creatine kinase to adequately fulfil ATPase requirements. The results are summarized in Fig. 6. Several parameters characterize myofibrillar properties. The upper part of the figure shows the response of cardiac fibers to increasing calcium concentration from 1 nM to 31.6 μM. Maximal force developed by skinned fibers at saturating calcium con-

centration represents the contractile or inotropic reserve of myocardial tissue. It is clearly observed that both calcium sensitivity and inotropic response are normal when creatine kinase alone provides the energy for contraction through local ADP regeneration. Another key parameter of cardiac contractility is the kinetic characteristics. They can be assessed by the tension responses to length changes which are shown in the middle part of Fig. 6. This provides an estimate of the speed of interaction between actin and myosin and again it can be seen that the kinetic properties are identical whether or not ATP is present. This is in line with biochemical data showing that creatine kinase activity is in excess of myosin ATPase activity in skeletal muscle and mammalian cardiac muscle and can rephosphorylate virtually all the ADP produced at maximal ATPase rate by creatine kinase [12, 56]. This shows that creatine kinase is a very powerful energy supplier for cardiac contraction. It also shows that cardiac contraction can work independently of cytosolic ATP concentration, and rely on local intracompartmented substrates and products. This does not mean that cytosolic ATP cannot reach the myosin AT-Pase, this just suggests that we must be careful in correlating contractile activity with total concentrations of adenine nucleotides in cardiac tissue.

By a matter of consequences, the same holds true when considering phosphorylation potential as a regulatory factor of cardiac contraction [79, 80]. This has been the matter of long debate to envisage enzymatic regulation through kinetic or thermodynamic considerations and tentative conciliations of the approaches in what concerns creatine kinase are under studies [81]. However both approaches rely on true concentrations of products and substrates of enzymatic reactions. In addition to the direct effects of MgADP and Pi on the crossbridge cycle [82], myosin ATPase reaction and chemiomechanical coupling may thus depend on the free energy of ATP hydrolysis and on the ratio ATP/ADPxPi. Calculations of the phosphorylation potential in the medium, taking the contaminating Pi to be 0.6 mM [83] and contaminating MgADP to be 10 μM, give two values differing by a factor of 8×10^4, for identical mechanical properties showing that compartmentation of phosphorylation potential takes place inside myfibrillar compartment (lower part in Fig. 6). The phosphocreatine-creatine kinase system could be a necessary mechanism in excitable tissues in order to keep a high level of free energy of ATP hydrolysis together with a high efficiency of energy transduction due to the low rate of ADP diffusion [80].

Another kind of experiments is to see what could be the effects of a decrease in myofibrillar CK activity on force development. If creatine kinase is needed under physiological conditions (to ensure optimal activity of myofibrillar ATPase), the decrease in PCr, by limiting the rephosphorylation of MgADP, may have an effect on the contractile performances of cardiac muscle. Decrease of PCr from 30 to 10 mM slightly increases maximal force while increase in creatine from 0 to 30 mM has no effect [84]. On the other hand, further lowering in PCr from 12 mM to zero decreases velocity of shortening and increases maximal force and Ca sensitivity which can be explained by the slowing of crossbridge detachment [68, 85, 86]; these effects are similar to the effects of ADP. As a result of the difference in kinetic constants (Km for ATP 15 μM and Ki for ADP 10 μM) and cellular concentrations of ATP and ADP (respectively 8 mM and 50 μM compared to their kinetic contants): it is likely that ADP removal would be the most critical point for optimal contractility. The apparent discrepancy between a decrease in energy supply and rise in con-

tractile strength lies in the role of creatine kinase in maintaining a high ATP/ADP ratio in myofibrils and in the role of ATP and ADP in the crossbridge cycle. As discussed previously, MgADP accumulation or MgATP shortage will lead to a slowing of the detachment rate constant of crossbridges and an increase in their number. This is accompanied by a slowing of crossbridge kinetics and an overall alteration of contractile activity [51, 68, 87].

The net cellular consequence of energy splitting in a cell is PCr- - -> Pi + Cr as a consequence of creatine kinase reaction. At constant PCr concentration, inorganic phosphate has been shown to markedly influence contractile force; it decreases both maximal force and calcium sensitivity [83, 84, 86, 88, 89] with a half maximal effect of 2 mM Pi, in the physiological concentration range. However, an increase in inorganic phosphate is always paralleled by a decrease in PCr. We have shown the conflicting interactions of these two phenomenons at the level of calcium sensitivity and mechanical properties [89]. In a simple way, it can be observed that the negative inotropic effects of phosphate ions are weakened by the decrease in PCr and the intramyofibrillar changes in adenine nucleotides, induced by inhibition of the creatine kinase reaction.

Pathophysiological implications

To understand the influence of CK reaction on twitch tension development and contractile activity, models of high energy phosphate depletion are of great interest.

Hypoxic or ischaemic contractile failure

The results aforementioned add further insight into ischemic contractile failure. Ischemic contractile failure is one of the most frequent event in cardiac pathophysiology. It is usually caused by partial or total occlusion of coronary arteries and results in reduced blood flow which deprives cardiac cells of oxygen and substrates with accumulation of metabolic products. The effects of ischemia or hypoxia on contractile performances can be summarized as follows. When a heart or a cardiac muscle preparation is subjected to a decrease in oxygen supply, there is a rapid fall in force production which is completed within 1 to 5 min, followed by a delayed rise in diastolic tension which increases in amplitude with time. Despite extensive description, the fundamental basis for

138

ischemic contractile failure is still under debate. Contractile force is regulated by interaction of calcium ions with troponin C and amplitude and kinetics of transitory as well as steady state changes in calcium concentration are crucial in determining contractile force. However, the relation between calcium and force is not unique. Contraction needs also a well controlled and high energy flux. Although energy supply to myosin ATPase cannot be considered as a physiological regulatory parameter of cardiac contraction, the status of metabolic substrates and products becomes crucial during energy shortage. In ischemia or hypoxia, the initial reduction of work output occurs with little change in ATP content, an observation at the origin of the concept of PCr shuttle [90]. The main changes in high energy phosphates are PCr decline and Pi and calculated ADP accumulations. As a result of creatine kinase reaction, ATP decrease is delayed. ATP splitting and activation of glycolysis and glycogenolysis lead to proton accumulation. Disorders in calcium homeostasis, *i.e.* decrease in calcium current and calcium transient, increase in resting calcium concentration are usually described. The participation of these different events to ischemic contractile failure has been reviewed in details [91, 92]. A clear dissociation between calcium entry inside the cells assessed by electrophysiological studies and twitch force development has been observed during metabolic inhibition [93]. The same conclusion can be drawn from the analysis of the calcium transient using bioluminescent probes to monitor intracellular calcium changes [94]. Increased proton concentration induces both a decrease in maximal force and calcium sensitivity together with a decrease in mechanochemical transduction efficiency [86, 88, 95, 96]. In normal conditions, proton production by myosin ATPase will be buffered by the creatine kinase reaction. After PCr exhaustion, acidification in myofibrillar space will inhibit myosin ATPase and greatly lower contractile force. Intracellular acidification does not systematically take place and may occur as a consequence of extracellular accumulation of metabolites during ischemia or glycolytic production of protons and lactate when glycolysis is stimulated. But, it cannot be the only mechanism of contractile failure since in addition, during the early contractile failure, a slight alkalinization can be observed which can be attributed to PCr consumption and net protons consumption by creatine kinase reaction. Herzig and Ruegg [97] were the first to propose that the increase in Pi may be involved in the fall of tension induced by hypoxia. Since then, an extensive number of studies has been devoted to the effects of inorganic

phosphate on tension development and cross bridge kinetics. The influence of these metabolites on contractile dysfunction has already been considered [92] with the main emphasis being put to the role of Pi ions. Inorganic phosphate induces a decrease in maximal contractile capacities of myofilaments together with a decrease in Ca sensitivity (see discussion above). During hypoxia, a large increase in inorganic phosphate occurs, equal to the sum of PCr and ATP decreases. Thus inorganic phosphate seems to be the best candidate for explaining contractile failure. It should be noted however that, based on skinned fiber experiments, only MgADP increase or MgATP decrease may explain decreased kinetics of contraction. Indeed, while protons do not affect crossbridge kinetics, Pi increases it while clear crossbridge slowing was observed with PCr decrease and/or MgADP increase [68, 96].

However, in light of the role of creatine kinase in compartmentation of nucleotides and protons inside myofibrils, additional factors should be considered. Inhibition of creatine kinase activity both by a decrease in creatine phosphate in the mM range and Pi inhibition, will lead to local ATP depletion and accumulation of ADP and protons in addition to inorganic phosphate. From the results we obtained of the combined effects of PCr depletion and Pi accumulation [89], antagonistic effects have been observed resulting mainly in a decreased influence of phosphate ions on maximal force and calcium sensitivity. At the beginning of tension fall, PCr level is still in a range exceeding the Km of creatine kinase and will little influence the force, while Pi effect will be predominant. As hydrolysis of PCr proceeds, Pi effect will level off while PCr will reach a concentration range in which it will decrease the number of actively cycling crossbridges and participate in the fall of twitch tension.

Other models of energy depletion

Studies of cardiac contraction in relation to high energy phosphate metabolism using NMR technique have shown a nonlinear relationship between developed force and ATP or PCr content in models of depletion in which energy production was not inhibited [98, 99]. These experiments pointed out several important observations. Hoerter *et al.* [99] used the capacity of cardiac cells to phosphorylate 2-deoxy-glucose into 2DG 6-phosphate which accumulates in the cell. This has several consequences. First, inorganic phosphate is trapped and does not accumulate in the cells since it is not metabolized further. This is a unique model to study *in vivo* the effects of high energy phosphate depletion without Pi accumulation. Second, PCr and subsequently ATP are decreased due to lack of Pi for rephosphorylation despite normal oxidative function. Thirdly, adenine moieties progressively disappear from the cells as a consequence of myokinase and deaminase activities. Examination of heart function and energy-rich phosphate concentrations lead to several conclusions. The main observation is that as 2DG 6-phosphate accumulates, PCr level decreases with no or only small changes in ATP, in contractile force and Pi. As phosphorylation of 2-DG proceeds, ATP decreases and ATP and PCr levels as low as respectively 10 and 15% could be obtained. Although during metabolic inhibition such low ATP and PCr levels result in cessation of systolic activities and rigor type contracture, in the 2-DG model contractility is only reduced by 35% in isovolumic conditions. It is remarkable that β-stimulation still induces normal increase in contractility confirming the minor importance of the reservoir function of PCr in myocardium when energy fluxes are not inhibited. Although at first surprising, a high and sustained steady state contractility at such low ATP and PCr content is compatible with the kinetic data of isolated organelles. Indeed, PCr level (2–3 mM) is still above the Km of creatine kinase and is not expected to profoundly affect force development, whatever the ATP content. In models of chronic PCr depletion, it was further shown that, when PCr was decreased within this range, the main alterations in mechanical activity appeared to be impaired relaxation and increase in end diastolic pressure which reduced ventricular filling [100, 101]. This can be explained by ADP accumulation as a result of PCr drop.

One limitation of the skinned fiber technique is that it gives access to steady state variations in force and not to transient changes. Viewed within the cardiac cycle, during the hypoxic contractile failure with constant calcium transient [94], the development of force will depend on the relative influence of the kinetics of calcium transient and force development. In view of the respective time course of calcium transient and force development, this last event always lies behind the calcium transient, evidencing a rate limiting step of crossbridges in tension development. Within the short duration of the calcium transient, the amplitude of force development will be limited by crossbridge kinetics. Thus, inorganic phosphate, favoring higher crossbridge kinetics may paradoxically induce a positive inotropic effect while the decrease in ATP/ADP ratio will slow the detachment rate

and the overall crossbridge cycle, inducing by itself a decrease in developed tension despite increased maximal force and calcium sensitivity. Thus it can be inferred that not only phosphate ions, but more widely, metabolites changes inside the myofibrillar structure will participate in the overall alterations in twitch tension during metabolic inhibition.

This points out that total metabolite concentration is not an unequivocal reflection of cellular activities and that local metabolite concentrations are more directly correlated to cardiac function and energy production. Energetic fluxes more closely reflect the energetic status of the compartments than total concentrations. A major consequence of adenine nucleotide compartmentation in cardiac tissue is the fact that the decrease in high energy turnover, rather than their tissue contents, may be a metabolic basis for ischemic contractile failure. A parallel decrease in CK flux and heart contractility has been observed *in vivo* during graded ischemia or hypoxia at high energy phosphate contents [102, 103]. The reservoir function of PCr in cardiac muscle appears of minor importance due to efficient integration between ATP production and consumption through the CK pathway. We may infer that this reservoir could be used during local and short hypoxic episodes to avoid contractile failure. As discussed in this book by Valdur Saks, PCr and ATP concentrations in cells represent the consequence of the energetic state rather than its regulatory factor (see chapter III-1).

Ischemic contracture

Another consequence of ischemia is the gradual rise in diastolic stiffness and force, the 'ischemic contracture'. Despite extensive description, the fundamental basis of the development of ischemic 'contracture' is still under debate. An alteration of the contractile proteins themselves does not seem to be involved since no evidence of myofibrillar damage could be shown following severe ischemia and reperfusion [104, 105, but see 106]. This rise in tension can be attributed to a rise in internal calcium leading to actively cycling crossbridges or to the formation of rigor crossbridges following local ATP depletion. Arguments have been accumulated in favour of the second hypothesis. Indeed, ischemic or hypoxic contrature may develop even in the absence of any rise in internal calcium and many authors have pointed to a dissociation between the rise in internal calcium and rise in resting tension (for review see [91, 92]). Recent data obtained using NMR to measure Ca_i and ATP in the same preparations indicate that ATP depletion correlates much better than changes in Ca_i with the appearance of ischemic contracture, and appears to be its primary cause [107]. Alternatively evidences have been brought about concerning the rigor origin of this increase in tension. In frog as well as in rat heart the rise in resting tension following hypoxia or metabolic inhibition is not related to an increase in the activation of the contractile proteins by myoplasmic calcium but is rather mediated by formation of rigor bridges suggesting a low ATP availability at the myofilaments [108]. More recently, Leijendekker et al. [109] showed that the development of unstimulated force during severe hypoxia in rat trabeculae was completely due to the formation of rigor links while Ca-dependent crossbridges can contribute to the rise in force during less severe hypoxia.

The exact relationship between ATP depletion and rise in rigor tension is not fully understood. Ischemic contracture develops when PCr has been exhausted. Impairment of the PCr shuttle may be one of the numerous factors responsible for contractile abnormalities in failing myocardium. Hearse et al. [110] evaluated the threshold for ischemic contracture development to be 12–15 μmol/g dw (i.e. 2 mM) ATP with very low PCr concentration. In this case, myofibrillar creatine kinase is not anymore able to actively rephosphorylate ADP produced by myosin ATPase. Left ventricular chamber stiffness increases when myocardial O_2 demand exceeds supply; this can be associated with a substantial prolongation of the time constant of left ventricular relaxation [111]. The diastolic properties of the left ventricle are important determinants of cardiac function. An increase in diastolic tension will decrease cell perfusion, impede relaxation and decrease the ventricular filling and cardiac output. Decreased ventricular filling, as well as increased diastolic force and stiffness are common features of cardiomyopathies of different etiologies [113, 100]. When PCr shuttle is decreased by feeding animals with a creatine analog, the main consequences are the impairment of diastolic function and a steeper rise in stiffness at increased afterloads in association with increased energy breakdown [100]. A similar combination of decreased PCr pool and increased ventricular pressure was observed in hereditary cardiomyopathy of the hamster [112]. Interestingly, in a model of creatine kinase inhibition, similar functional alterations were described [114]. These observations can, for a part, be explained by alterations in intracellular calcium handling. Accumulation of ADP in the myofibrillar space de-

creasing cross bridge detachment rate, can also be responsible for a slowed and incomplete relaxation. We thus propose that 'ischemic contracture' and some diastolic dysfunctions in heart failure may be the result of ATP depletion and ADP accumulation inside myofibrils, following PCr depletion.

In the severely ischemic zones, increase in diastolic tension and stiffness occurs at a time when systolic pressure is already abolished or greatly impaired. Under these circumstances, rigor tension development in the absence of PCr may appear as a protective mechanism for the myocardium. In this case, it allows to sustain tension at a low energy cost. This maintained tension may be of vital importance for ischemic cardiac cells by preventing cell overstretching and membrane disruption. Furthermore, for the whole heart, the development of rigor tension in ischemic zone, may help to avoid local dilation and development of aneurism; it will also preserve mechanical coupling between adjacent contracting regions. When energy production is impaired, accumulation of ATPase products due to creatine kinase inhibition can be viewed as a 'switch-off' of energy expenditure for contraction when PCr is exhausted in order to preserve vital cell functions. In this sense PCr/CK system behave in the cell as a specialized energy pathway for contractile activity. The localization of the different isoenzymes serves such a purpose.

Conclusions

The present structure of the myofilaments is the result of a long evolutionary process resulting in a very fine protein assembly consisting of a main enzyme, myosin ATPase, the mechanical transduction system, the activating system, – the thin filament and the providing system – creatine kinase. This protein assembly is functionally efficient and structurally elegant. Myofibrillar creatine kinase can be considered as the energetic provider of the myofibrillar compartment. It helps to keep favorable ATP, ADP, and proton concentrations, it allows myosin ATPase to be independent of substrate and product fluctuations in order to be finely regulated by calcium ions and other signalling pathways. Together with Pi, PCr and creatine serves as metabolic signalling between energy producing and utilizing processes.

Acknowledgements

We wish to thank all our friends and colleagues, from Moscou, Orsay, Châtenay-Malabry, Tallinn, Oujda, Nantes who happened to be taken aboard. We thank 'Institut National de la Santé et de la Recherche Médicale' and 'Fondation pour la Recherche Médicale' for their support in this work.

References

1. Fiske CH, Subbarow Y: The nature of the 'inorganic phosphate' in voluntary muscle. Science 65: 401–403, 1927
2. Lundsgaard E: Untersuchungen under Muskelkontraktionen ohne Milchsaurebildung. Biochem Z 217: 162–177, 1930
3. Cain DF, Davis RE: Breakdown of adenosine triphosphate during a single contraction of working muscle. Biochem Biophys Res Commun 8: 361–366, 1962
4. Fabiato A, Fabiato F: Contractions induced by a calcium-triggered release of calcium from the sarcoplasmic reticulum of single skinned cardiac cells. J Physiol 249: 469–495, 1975
5. Sharov VG, Saks VA, Smirnov VN, Chazov EI: An electron microscopic histochemical investigation of the localization of creatine phosphokinase in heart cells. Biochim Biophys Acta 468: 495–501, 1977
6. Wallimann T, Turner DC, Eppenberger HM: Localization of creatine kinase isoenzymes in myofibrils. I. Chicken skeletal muscle. J Cell Biol 75: 297–317, 1977
7. Wallimann T, Turner DC, Eppenberger HM: Localisation of creatine kinase in myofibrils. II. Chicken heart muscle. J Cell Biol 75: 318–325, 1977
8. Levitskii DO, Levchenko TS, Saks VA, Sharov VG, Smirnov VN: Functional coupling between Ca²⁺-ATPase and creatine phosphokinase in sarcoplasmic reticulum of myocardium. Biokhimia 42: 1766–1773, 1977
9. Rossi AM, Eppenberger HM, Volpe P, Cotrufo R, Wallimann T: Muscle-type MM creatine kinase is specifically bound to sarcoplasmic reticulum and can support Ca-2+ uptake and regulate local ATP/ADP ratios. J Biol Chem 265: 5258–5266, 1990
10. Korge P, Byrd SK, Campbell KB: Functional coupling between sarcoplasmic-reticulum-bound creatine kinase and Ca²⁺-ATPase. Eur J Biochem 213: 973–980, 1993
11. Grosse R, Spitzer E, Kupriyanov VV, Saks VA, Repke KR: Coordinate interplay between (Na+/K+)-ATPase and creatine phosphokinase optimizes (Na+/K+)-antiport across the membrane of vesicles formed from the plasma membrane of cardiac muscle cell. Biochem Biophys Acta 603: 142–156, 1980
12. Saks VA, Ventura-Clapier R, Huchua ZA, Preobrazhensky AN, Emelin IV: Creatine kinase in regulation of heart function and metabolism I. Further evidence for compartmentation of adenine nucleotides in cardiac myofibrillar and sarcolemmal coupled ATPase-creatine kinase systems. Biochim Biophys Acta 803: 254–264, 1984
13. Jacobus WE: Respiratory control and the integration of heart high-energy phosphate metabolism by mitochondrial creatine kinase. Ann Rev Physiol 47: 707–726, 1985
14. Barth E, Stammler G, Speiser B, Schaper J: Ultrastructural quantitation of mitochondria and myofilaments in cardiac muscle from 10 different animal species including man. J Mol Cell Cardiol 24: 669–681, 1992
15. Fulton AB: How crowded is the cytoplasm? Cell 30: 345–347, 1982
16. Clegg JS: Properties and metabolism of the aqueous cytoplasm and its boundaries. Am J Physiol 246: R133–R151, 1984
17. Morel JE: Discussion on the state of water in the myofilament lattice and other biological systems, based on the fact that the usual concepts of colloid stability can not explain the stability of the myofilament lattice. J Theor Biol 112: 847–858, 1985
18. Maughan DW, Godt RE: Equilibrium distribution of ions in a muscle fiber. Biophys J 56: 717–722, 1989
19. Bartels EM, Elliott GF: Donnan potentials from the A- and I-bands of glycerinated and chemically skinned muscles, relaxed and in rigor. Biophys J 48: 61–76, 1985
20. Hill DK: The location of creatine phosphate in frog's striated muscle. J Physiol 164: 31–50, 1962
21. Wallimann T, Eppenberger HM: Localization and function of M-line-bound creatine kinase. M-band model and creatine phosphate shuttle. In: JW Shay (ed.) Cell and Muscle Motility: Plenum Publishing Corp, New-York, vol 6, 1985, pp 239–285
22. Ottaway JH: Evidence for binding of cytoplasmic creatine kinase to structural elements in heart muscle. Nature 215: 521–522, 1967
23. Turner DC, Wallimann T, Eppenberger HM: A protein that binds specifically to the M-line of skeltal muscle is identified as the muscle form of creatine kinase. Proc Nat Acad Sci 70: 702–705, 1973
24. Mani RS, Kay CM: Physicochemical studies on the creatine kinase M-line potein and its interaction with myosin and myosin fragments. Biochem Biophys Acta 453: 391–399, 1976
25. Carlsson E, Kjorell U, Thornell L-E: Differentiation of the myofibrils and the intermediate filament system during postnatal development of the rat heart. Eur J Cell Biol 27: 62–73, 1982
26. Wegmann G, Zanolla E, Eppenberger HM, Wallimann T: *In situ* compartmentation of creatine kinase in intact sarcomeric muscle: the acto-myosin overlap zone as a molecular sieve. J Muscle Res Cell Motility 13: 420–435, 1992
27. Schafer BW, Perriard JC: Intracellular targeting of isoproteins in muscle cytoarchitecture. J Cell Biol 106: 1161–1170, 1988
28. Otsu N, Hirata M, Tuboi S, Miyazawa K: Immunochemical localization of creatine kinase M in canine myocardial cells: most creatine kinase M is distributed in the A band. J Histochem Cytochem 37: 1465–1470, 1989
29. Robert J, Barandun B, Kobel HR: A Xenopus laevis creatine kinase isoenzyme (CK-III/III) expressed preferentially in larval striated muscle: cDNA sequence, developmental expression and subcellular immunolocalization. Genet Res Camb 58: 35–40, 1991
30. Yagi K, Mase R: Coupled reaction of creatine kinase and myosin A-adenosine triphosphate. J Biol Chem 237: 397–340, 1962
31. Botts J, Stone M: Kinetics of coupled enzymes: CK and myosin A. Biochemistry 7: 2688–2696, 1968
32. Masters CJ: Interactions of enzymes and subcellular structure. CRC Crit Rev Biochem 11: 105–143, 1981
33. Rundell KW, Tullson PC, Terjung RL: AMP deaminase binding

142

in contracting rat skeletal muscle. Am J Physiol 263: C287–C293, 1992

34. Ventura-Clapier R, Saks VA, Vassort G, Lauer C, Elizarova G: Reversible MM creatine kinase binding to cardiac myofibrils. Amer J Physiol 253: C444–V455, 1987

35. Elizarova GV, Sukhanov AA, Saks VA: Myofibrillar creatine kinase: reversible binding with contractile proteins, stoichiometric ratio with myosin, and functional significance. Biokhimiya 52: 667–675, 1987

36. Huxley AF: Muscle structure and theories of contraction. Progress in Biophys 7: 255–318, 1957

37. Holroyde MJ, Robertson SP, Johnson JD, Solaro RJ, Potter JD: The calcium and magnesium binding sites on cardiac troponin and their role in the regulation of myofibrillar adenosine triphosphatase. J Biol Chem 255: 11688–11693, 1980

38. Eisenberg E, Hill TL: Muscle contraction and free energy transduction in biological systems. Science 277: 999–1006, 1985

39. Stein LA, Schwartz R, Chock PB, Eisenberg E: Mechanism of actomyosin adenosine triphosphate. Evidence that adenosine 5′-triphosphate hydrolysis can occur without dissociation of the actomyosin complex. Biochemistry 18: 3895–3909, 1979

40. Hibberd MG, Dantzig JA, Trentham DR, Goldman YE: Phosphate release and force generation in skeletal muscle fibers. Science 228: 1317–1319, 1985

41. Goldman YE, Hibberd MG, Trentham DR: Inhibition of active contraction by photogeneration of adenosine-5′-triphosphate in rabbit psoas muscle fibers. J Physiol 354: 605–624, 1984

42. Glyn H, Sleep J: Dependence of adenosine triphosphatase activity of rabbit psoas muscle fibres and myofibrils on substrate concentration. J Physiol 365: 259–276

43. Eisenberg E, Moss C: The adenosine triphosphatase activity of acto-heavy meromyosin. A kinetic analysis of actin activation. Biochemistry 7: 1486–1489, 1968

44. Siemankowski RF, Wiseman MO, White HD: ADP dissociation from actomyosin subfragment-1 is sufficiently slow to limit the unloaded shortening velocity in vertebrate muscle. Proc Nat Acad Sci 82: 658–662, 1985

45. Dantzig JA, Hibberd MG, Trentham DR, Goldman YE: Crossbridge kinetics in the presence of MgADP investigated by photolysis of caged ATP in rabbit psoas muscle fibres. J Physiol 432: 639–680, 1991

46. Curtin NA, Gilbert C, Kretzschmar KM, Wilkie DR: The effect of the performance of work on total energy output and metabolism during muscular contraction. J Physiol 238: 455–472, 1974

47. Barsotti RJ, Ferenczi MA: Kinetics of ATP hydrolysis of tension production in skinned cardiac muscle of the guinea pig. J Biol Chem 263: 16750–16756, 1988

48. Johnson RE, Adams PH: ADP binds similarly to rigor muscle myofibrils and to actomyosin-subfragment one. FEBS Let 174: 11–14, 1984

49. Siemankowski RF, White HD: Kinetics of the interaction between actin, ADP and cardiac myosin S1. J Biol Chem 259: 5045–5053, 1984

50. Sleep J, Glyn H: Inhibition of myofibrillar and actomyosin subfragment 1 adenosinetriphosphatase by adenosine 5′-diphosphate, pyrophosphate, and adenyl-5′-yl imidodiphosphate Biochemistry 25: 1149–1153, 1986

51. Cooke R, Pate E: The effects of ADP and phosphate on the contraction of muscle fibres. Biophys J 48: 789–798, 1985

52. Hebisch S, Sies H, Soboll S: Function dependent changes in the subcellular distribution of high energy phosphates in fast and slow rat skeletal muscles. Pflugers Arch 406: 20–24, 1986

53. Yagi K, Mase R: Possible compartmentation of adenine nucleotides in a coupled reaction system composed of F-actomyosin-adenosine-triphosphatase and creatine kinase. In: S Ebashi, F Oosawa, T Sekine, Y Tonomura (eds.) Molecular Biology of Muscular Contraction. Elsevier, Amsterdam, 1965, pp 109–123

54. Saks VA, Chernousova GB, Vetter R, Smirnov VN, Chazov EI: Kinetic properties and the functional role of particulate MM-Isoenzyme of creatine phosphokinase bound to heart muscle myofibrils. FEBS Lett 62: 293–296, 1976

55. Bessman SP, Yang WC, Geiger PJ, Erickson-Viitanen S: Intimate coupling of creatine phosphokinase and myofibrillar adenosine triphosphatase. Biochem Biophys Res Commun 96: 1414–1420, 1980

56. Wallimann T, Schlosser T, Eppenberger HM: Function of M-line bound creatine kinase as intramyofibrillar ATP regenerator at the receiving end of the phosphorylcreatine shuttle in muscle. J Biol Chem 259: 5238–5246, 1984

57. Krause SM, Jacobus WE: Specific enhancement of the cardiac myofibrillar ATPase by bound creatine kinase. J Biol Chem 267: 2480–2486, 1992

58. Dowell RTn, Fu MC: Cardiac myofibrillar creatine kinase Km is not influenced by contractile protein binding. Life Sci 50: 1551–1559, 1992

59. Mosbach K: Immobilized coenzymes in general ligand affinity chromatography and their use as active coenzymes. Adv Enzymol 46: 205–278, 1978

60. Mosbach K, Mattiason B: Immobilized model systems of enzyme sequences. Curr Top Cell Regul 14: 197–262, 1978

61. Fossel ET, Hoefeler H: A synthetic functional compartment. The role of propinquity in a linked pair of immobilized enzymes. Eur J Biochem 170: 165–171, 1987

62. Arrio-Dupont M: An example of substrate channeling between co-immobilized enzymes. Coupled activity of myosin ATPase and creatine kinase bound to frog heart myofilaments. FEBS Lett 240: 181–185, 1988

63. Arrio-Dupont M, Bechet JJ, d'Albis A: A model system of coupled activity of co-immobilized creatine kinase and myosin. Eur J Biochem 207: 951–955, 1992

64. Bozler E: The role of phosphocreatine and adenosine-triphosphate in muscular contraction. J Gen Physiol 37: 63–70, 1953

65. Perry SV: Creatine phosphokinase and the enzymic and contractile properties of the isolated myofibrils. Biochem J 57: 427–431, 1954

66. Savabi F, Geiger PJ, Bessman JP: Kinetic properties and functional role of creatine phosphokinase in glycerinated muscle fibers further evidence for compartmentation. Biochem Biophys Res Commun 114: 785–790, 1983

67. McClellan G, Weisberg A, Winegrad S: Energy transport from mitochondria to myofibrils by a creatine phosphate shuttle in cardiac cells. Amer J Physiol 245: 423–427, 1983

68. Ventura-Clapier R, Mekhfi H, Vassort G: Role of creatine kinase in force development in chemically skinned rat cardiac muscle. J Gen Physiol 89: 815–837, 1987

69. Reuben JP, Brandt PW, Berman M, Grundfest H: Regulation of tension in the skinned crayfish muscle fiber. J Gen Physiol 57: 385–407, 1971

70. Bremel RD, Weber AM: Cooperation within actin filament in vertebrate skeletal muscle. Nature 238: 97–101, 1972

71. Veksler VI, Kapelko VI: Creatine kinase in regulation of heart function and metabolism. II. The effect of phosphocreatine on the rigor tension of EGTA-treated rat myocardial fibers. Biochim Biophys Acta 803: 265–270, 1984

72. Ventura-Clapier R, Vassort G: Role of myofibrillar creatine kinase in the relaxation of rigor tension in skinned cardiac muscle. Pflügers Arch 404: 157–161, 1985

73. Veksler VI, Ventura-Clapier R, Lechene P, Vassort G: Functional state of myofibrils mitochondria and bound creatine kinase in skinned ventricular fibers of cardiomyopathic hamsters. J Mol Cell Cardiol 20: 329–342, 1988

74. Hoerter J, Kuznetsov A, Ventura-Clapier R: Functional development of the creatine kinase system in perinatal rabbit heart. Circ Res 69: 665–676, 1991

75. Ventura-Clapier R, Veksler VI: Myocardial ischemic contracture: metabolites affect rigor tension and stiffness. Circ Res (in press)

76. Yoshizaki K, Nishikawa H, Watari H: Diffusivities of creatine phosphate and ATP in an aqueous solution studied by pulsed field gradient ^{31}P NMR. Jap J Physiol 37: 923–928, 1987

77. Saks VA, Seppet EK, Lyulina NV: Comparative investigation of the role of creatine phosphokinase isoenzymes in energy metabolism of skeletal muscles and myocardium. Biokhimiya 42: 579–588, 1977

78. Nichols CG, Lederer WJ: The role of ATP in energy-deprivation contractures in unloaded rat ventricular myocytes. Canadian J Physiol Pharmacol 68: 183–194, 1990

79. Kammermeier H, Schmidt P, Jungling E: Free energy change of ATP hydrolysis: a causal factor of early hypoxic failure of the myocardium? J Mol Cell Cardiol 14: 267–277, 1982

80. Kammermeier H: Why do cells need phosphocreatine and a phosphocreatine shuttle. J Mol Cell Cardiol 19: 115–118, 1987

81. Gnaiger E, Jacobus WE: Adenine nucleotide thermodynamic or kinetic control of mitochondrial oxidative phosphorylation: a reconciliation via Einstein's diffusion equation. Biophys J 55: 568a, 1989

82. Kentish JC, Allen DG: Is force production in the myocardium directly dependent upon the free energy change of ATP hydrolysis? J Mol Cell Cardiol 18: 879–882, 1986

83. Kentish JC: Combined Inhibitory Actions of Acidosis and Phosphate on Maximum Force Production in Rat Skinned Cardiac Muscle. Pflügers Arch 419: 310–318, 1991

84. Kentish JC: The effects of inorganic phosphate and creatine phosphate on force production in skinned muscles from rat ventricle. J Physiol 370: 585–604, 1986

85. Maughan DW, Low ES, Alpert NR: Isometric force development, isotonic shortening and elasticity measurements from Ca2+-activated ventricular muscle of the guinea-pig. J Gen Physiol 71: 431–451, 1978

86. Godt RE, Nosek TN: Changes of intracellular milieu with fatigue or hypoxia depress contraction of skinned rabbit skeletal and cardiac muscle. J Physiol 412: 155–180, 1989

87. Hoar PE, Mahoney CW, Kerrick WG: MgADP$^-$ increases maximum tension and Ca^{2+} sensitivity in skinned rabbit soleus fibers. Pflügers Arch 410: 30–36, 1987

88. Kentish JC, Palmer S: The influence of pH, phosphate, and ionic strength on contraction in skinned cardiac muscle. In: Modulation of Cardiac calcium sensitivity. JA Lee, DG Allen (eds.) Oxford University Press 1993, p 67–88

89. Mekhfi H, Ventura-Clapier R: Dependence upon high-energy phosphates of the effects of inorganic phosphate on contractile properties in chemically skinned rat cardiac fibers. Pflügers Arch 411: 378–385

90. Gudbjarnason S, Mathes P, Ravens KG: Functional compartmentation of ATP and creatine phosphate in heart muscle. J Mol Cell Cardiol 1: 325–339, 1970

91. Allen DG, Orchard CH: Myocardial contractile function during ischemia and hypoxia. Circ Res 60: 153–168, 1987

92. Lee JA. Allen DG: Mechanisms of acute ischemic contractile failure of the heart. Role of intracellular calcium. J Clin Invest 88: 361–367, 1991

93. Ventura-Clapier R, Vassort G: Electrical and mechanical activities of frog heart during energetic deficiency. J Muscle Res Cell Mot 1: 429–444, 1980

94. Allen DG, Orchard CH: Intracellular calcium concentration during hypoxia and metabolic inhibition in mammalian ventricular muscle. J Physiol 339: 107–122, 1983

95. Fabiato A, Fabiato F: Effects of pH on the myofilaments and the sarcoplasmic reticulum of skinned cells from cardiac and skeletal muscles. J Physiol 276: 233–255, 1978

96. Ventura-Clapier R, Mayoux E, Coutry N, Lechene P, Marotte F: Effects of acidosis and alkalosis on mechanical properties of skinned fibers from control and hypertrophied hearts. J Mol Cell Cardiol 24, S50, 1992

97. Herzig JW, Ruegg JC: Myocardial cross-bridge activity and its regulation by Ca^{2+}, phosphate and stretch. In: G Riecker, A Weber, J Goodwin (eds.) Myocardial failure. Springer, Berlin Heidelberg, New-York, 1977, pp 41–51

98. Kupriyanov VV, Lakomkin VL, Kapelko VI, Steinschneider AY, Ruuge EK, Saks VA: Dissociation of adenosine triphosphate levels and contractile function in isovolumic hearts perfused with 2-deoxyglycose. J Mol Cell Cardiol 19: 729–740, 1987

99. Hoerter JA, Lauer C, Vassort C, Gueron M: Sustained function of normoxic hearts depleted in ATP and phosphocreatine: a 31P-NMR study. Amer J Physiol 255: C192–C201, 1988

100. Kapelko VI, Kupriyanov VV, Novikova NA, Lakomkin VL, Steinschneider AYa, Zueva MY, Veksler VI, Saks VA: The cardiac contractile failure induced by chronic creatine and phosphocreatine deficiency. J Mol Cell Cardiol 20: 465–479, 1988

101. Zweier JL, Jacobus WE, Korecky B, Brandejs-Barry Y: Bioenergetic consequences of cardiac phosphocreatine depletion induced by creatine analogue feeding. J Biol Chem 266: 20296–20304, 1991

102. Bittl JA, Balschi JA, Ingwall JS: Contractile failure and high energy phosphate turnover during hypoxia: 31P-NMR surface coil studies in living heart. Circ Res 60: 871–878, 1987

103. Neubauer S, Hamman BR, Perry SB, Bittl JA, Ingwall JS: Velocity of the creatine kinase reaction decreases in post ischemic myocardium: A 31P-NMR magnetization transfer study of the isolated ferret heart. Circ Res 63: 1–15, 1988

104. Ventura-Clapier R, Veksler VK, Elizarova GV, Mekhfi H, Levitskaya EL, Saks VA: Contractile properties and creatine kinase activity of myofilaments following ischemia and reperfusion of the rat heart. Biochem Med Metab Biol 38: 300–310, 1987

105. Krauss SM: Effect of global myocardial stunning on Ca2+-sensitive myofibrillar ATPase activity and creatine kinase kinetics. Am J Physiol 259: H813–H819, 1990

106. Greenfield RA, Swain JL: Disruption of myofibrillar energy use: dual mechanisms that may contribute to postischemic dysfunction in stunned myocardium. Circ Res 60: 283–289, 1987

144

107. Koretsune Y, Marban E: Mechanism of ischemic contracture in ferrets hearts: relative roles of Ca_i elevation and ATP depletion. Am J Physiol 258: H9–H16, 1990

108. Ventura-Clapier R, Vassort G: Rigor tension during metabolic and ionic rises in resting tension in rat heart. J Mol Cell Cardiol 13: 551–561, 1981

109. Leijendekker WJ, Gao WD, ter Keurs HEDJ: Unstimulated force during hypoxia of rat cardiac muscle – stiffness and calcium dependence. Amer J Physiol 258: H861–H869, 1990

110. Hearse DJ, Garlick PB, Humphrey SM: Ischemic contracture of the myocardium: mechanisms and prevention. Am J Cardiol 39: 986–993, 1977

111. Serizawa T, Vogel WM, Apstein K, Grossman W: Comparison of acute alterations in left ventricular relaxation and diastolic chamber stiffness induced by hypoxia and ischemia. Role of myocardial oxygen supply-demand imbalance. J Clin Invest 68: 91–102, 1981

112. Kapelko VI, Parmley WW, Wu S, Stone RD, Jasmin G, Wilkman-Coffelt J: Increased left ventricular diastolic stiffness in the early phase of hereditary cardiomyopathy. Am Heart J 116: 765–770, 1988

113. Kapelko VI, Popovich MT, Veksler VI, Ventura-Clapier R, Khuchua ZA, Saks VA: Subcellular basis for increased diastolic stiffness in experimental cardiomyopathies. In: M Nagano, N Takeda, NS Dhalla: The Cardiomyopathic Heart, Raven Press Ltd, New-York, 1994, pp 185–195

114. Kupriyanov VV, Lakomkin VL, Korchazhkina OV, Steinschneider AYa, Kapelko VI, Saks VA: Control of cardiac energy turnover by cytoplasmic phosphates: ^{31}P-NMR study. Am J Physiol Supp (Oct) 261: 45–63, 1991.

Molecular and Cellular Biochemistry **133/134**: 145–152, 1994.
© 1994 *Kluwer Academic Publishers.*

II-6 Interaction of creatine kinase and adenylate kinase systems in muscle cells

Fatemeh Savabi

University of Southern California, School of Medicine, Department of Pharmacology and Nutrition, 2025 Zonal Ave.
Los Angeles, CA 90033, U.S.A.

Abstract

Elsewhere in this book the important role of creatine kinase and its metabolites in high energy phosphate metabolism and transport in muscle cells has been reviewed. The emphasis of this review article is mainly on the compartmentalized catalytic activity of adenylate kinase in relation to creatine kinase isoenzymes, and other enzymes of energy production and utilization processes in muscle cells. At present the role of adenylate kinase is considered simply to equilibrate the stores of adenine nucleotides. Recent studies by us and others, however, suggest an entirely new view of the metabolic importance of adenylate kinase in muscle function. This view offers a closer interaction between adenylate kinase and creatine kinase, in the process of energy production (at mitochondrial and glycolytic sites), and energy utilization (at myofibrillar sites and perhaps other sites such as sarcoplasmic reticular, sarcolemmal membrane, etc.), thus being an integral part of the high energy phosphate transport system.

This review article opens up the opportunity to further examine the metabolism of adenine nucleotides and their fluxes through the adenylate kinase system in intact muscle cells. Using an intact system, having a preserved integrity of their compartmentalized enzymes and substrates, is essential in clarifying the exact role of adenylate kinase in high energy phosphate metabolism in muscle cells. (Mol Cell Biochem **133/134:** 145–152, 1994)

Key words: Myokinase, high energy phosphate metabolism, heart, compartmentation, glycolysis

Introduction

The primary function of muscle is contraction which utilizes a substantial amount of cellular energy. Optimal function of muscle cells, therefore, warrants an immediate and efficient access to the available energy and a precise coupling of energy production to utilization. There has been a continued interest and a great deal of progress made in our understanding of the above concept. It is well accepted that the hydrolysis of ATP provides the free energy for all cell functions including contraction [1–3]. In spite of such a great rate of ATP utilization in the process of contraction-relaxation, the coordination of ATP synthesis with its utilization is so efficient that no change in the ATP pool can be detected [4, 5]. Creatine kinase (CK) and adenylate kinase (AK) have long been thought to play a crucial role in this process.

Back in 1950's Lonard [6] and Bendall [7, 8] reported that the high energy phosphate transferring enzymes, CK and AK, were relaxing factors for muscle cells. This concept raised the question concerning the properties of theze enzymes and the mechanism by which the transfer of so called high energy phosphate is effected. The important role of CK isoenzyme and its metabolites (cre-

Address for offprints: F. Savabi, University of Southern California, School of Medicine, Department of Pharmacology and Nutrition, KAM-110, 2025 Zonal Ave. Los Angeles, CA 90033, USA

Table 1. Adenylate kinase activity and its isoenzyme distribution in skeletal and heart muscle subfragments from various species.

Species	Tissue	Total	Unite	Cyto.(CK1)	Mito.(CK2)	Reference
Bovine	Heart	6.42	IU/mg protein	$\sim$90% of total	$\sim$10% of total	19
Porcine	Heart	90.2	IU/g of tissue	91.4% of total	9.5% of total	20
Porcine	Muscle	276.6	IU/g of tissue	98.7% of total	1.9% of total	20
Rabbit	Muscle (R)	$\sim$40	IU/g fresh tissue	–	–	44
Rabbit	Muscle (W)	$\sim$225	IU/g fresh tissue	–	–	44
Pigeon	Muscle (R)	$\sim$110	IU/g fresh tissue	–	–	44
Pigeon	Muscle (W)	$\sim$200	IU/g fresh tissue	–	–	44
Rat	Heart (A)	1.32	IU/mg protein	–	0.14 IU/mg prot.	45
Rat	Heart (V)	0.84	IU/mg protein	–	0.2 IU/mg prot.	46
Rat	Heart (A)	0.85	IU/mg protein	–	–	46
Rat	Muscle	3.13	IU/mg protein	–	–	46
Rat	Heart (A)	1.65	IU/mg protein	–	–	*
Rat	Heart (V)	2.15	IU/mg protein	–	2.54 IU/mg prot.	*
Human	Muscle	3.00	IU/mg protein	92% of total	–	16

Abbreviations used are: R = red, W = white, A = atria, V = ventricle, * = our unpublished data

atine phosphate and creatine) in facilitating energy distribution and responding to energy demand, called the phosphocreatine energy shuttle, has been the subject of interest and extensive investigation for the past several decades [reviewed in ref. 9–12]. The role of AK, however, has not yet been extensively investigated. The emphasis of this article is primarily on this subject.

Muscle adenylate kinase

Muscle adenylate kinase (AK), usually called myokinase (EC2.7.4.3), is an enzyme which reversibly catalyzes the following equilibrium reaction, 2ADP- -< ATP + AMP, thereby influencing the relative pool size of the three adenine nucleotides. The presence of two main adenylate kinase isoenzymes, basic (cytosolic, AK1) and acidic (mitochondrial, AK2) have been reported in many tissues including heart and skeletal muscle cells [13–15]. The cytosolic AK (AK1) is polymorphic in man (AK1*1 and AK1*2) and AK1*1 accounts for 95% of the AK1 isoenzyme [16]. Part of the cytosolic AK is bound to myofibrils and is called myofibrillar AK [17–19]. The mitochondrial AK is located between the mitochondrial membranes with fatty acid-CoA ligase [14, 19]. The extent of mitochondrial AK activity depends very much on the method of preparation, the tissue, and the species. It could be varied from as high as 40% in intact crude mitochondrial preparation obtained from hypo-osmatic bovine cardiac homogenates down to 0.3% of the total activity in bovine aorta smooth muscle cells [19]. It is usually found to be around 10% and the

rest accounted as cytosolic isoenzyme. Table 1 lists values for AK activity and its subfractional distribution in heart and muscle cells from different species. It has been found that the mitochondrial AK has a lower Km and a different pH profile for AMP and ATP than cytosolic AK [20]. There are indications that cytoplasmic and mitochondrial AK may catalyze the reactions in different directions [19]. For example the mitochondrial AK, having a lower Km for AMP and ATP, may mostly catalyze the reverse direction to regulate mitochondrial respiration as do mitochondrial creatine CK (CKm) and hexokinase [21]. Thus muscle AK, by being bound to subcellular structures that are involved in energy production and utilization, promotes the idea of having an important role in cellular high energy phosphate metabolism and transport.

Adenylate kinase and the phosphocreatine energy shuttle

The general concept of the role of AK is that AK is at or near equilibrium in muscle cells, and is regulated by the cellular level of AMP and ADP. Recent data, however, suggest the concept of compartmentation of AK, interacting with compartmentalized CK isoenzymes and being an integral part of the high energy phosphate transfer system. There seems to be a direct interaction between AK and CK at both the mitochondrial and myofibrillar ends of the phosphocreatine energy shuttle [9]. This close interaction between AK and CK became apparent when quantitative experiments were conducted

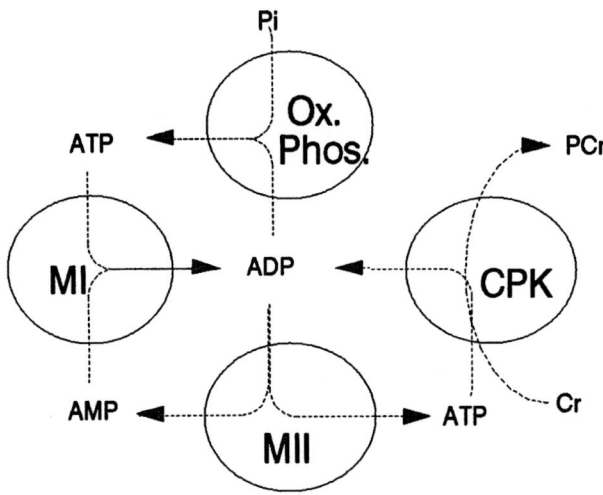

Fig. 1. A possible tetrahedral arrangement of three enzymes, one molecule of creatine kinase (CPK) and two molecules of adenylate kinase (AK) around the site of mitochondrial oxidative phosphorylation (Ox. Phos.), sharing a common pool of ADP and producing the final high energy phosphate as phosphocreatine (PCr).

to test the degree of coupling of the mitochondrial and myofibrillar bound creatine kinase isoenzymes to the localized energy producing (mitochondrial oxidative phosphorylation) and utilizing (myosin ATPase) enzymes. For example, tracer studies of the mitochondrial synthesis of creatine phosphate (CP) from nascent ATP, in the presence of added [γ-^{32}P]ATP (incubated for 5–10 sec.) resulted in CP that had lower specific activity (about 1/3) than that of γ-phosphate of ATP [22]. This meant that not all the CP synthesized during this period could have come from the γ-phosphate of ATP. Incubation with either an inhibitor of oxidative phosphorylation, CCCH, or with atractyloside, an inhibitor of nucleotide transport, permitted the formation of approximately 1/3 of the control amount of CP, but with the same specific activity as the γ-phosphate of ATP [21, 22]. It was concluded that 1/3 of the CP formed was derived from exogenous ATP and 2/3 from some other sources including the β-phosphate of ATP. Incubation with β-γ-labeled ATP resulted in the formation of CP with a similar specific activity to the phosphate label of ATP. These types of studies demonstrated the close interaction of mitochondrial oxidative phosphorylation with mitochondrial bound CK and AK. It was suggested that in order for AK to be involved in this system, there must be two molecules of AK attached to each other, one of them bound to an oxidative phosphorylation site and the other one to a molecule of CK (Fig. 1). This would provide two portals of entry of ATP and one of AMP and an exit portal of creatine phosphate. This tetrahedral arrangement, clustered around the translocase pro-

vides a prompt and efficient removal of ATP and introduction of ADP to stimulate mitochondrial respiration.

A similar type of arrangement has been proposed for the energy utilizing site of the phosphocreatine energy shuttle. The location of AK near the myofibrillar ATPase site was proposed by Tice & Barnett [18], when in their study on histochemical distribution of myofibrillar ATPase, using lead precipitation of liberated phosphate, they showed that both ATP and to a lesser degree ADP caused the formation of free phosphate throughout the M-band, as well as the Z-line. The compartmentalized catalytic activity of AK at the myofibrillar end of the phosphocreatine energy shuttle was demonstrated in our laboratories [23], when the contractile function of glycerinated rabbit psoas muscle fibers, containing native CK, AK, and ATPase activities, was studied. In this investigation, the isometric contraction and relaxation responses of these glycerinated fibers to either ADP, ATP + CP, or ATP, in the presence and absence of AP5A (P1, P5-di-adenosine 5'pentaphosphate) an adenylate kinase inhibitor, were compared. This study not only emphasized the important role of the myofibrillar bound CK [23, 24] but also the role of AK in the process of muscle contractile function. In the presence of adenylate kinase inhibitor, the ability of the fibers to stay at the contracted state was increased and the their ability to relax was diminished [23]. Since ADP inhibits and ATP promotes relaxation, this study supported the notion that not only the myofibrillar CK, but also the myofibrillar AK has a crucial role in the removal of the contraction-produced ADP and its rephosphorylation to ATP at or near the site of myosin ATPase. It was concluded that a coordinated participation of the three myofibrillar enzymes, ATPase, CK, and AK, and the presence of small amount of compartmentalized adenine nucleotide, are necessary for the optimal contractile function of muscle. On this basis, a tetrahedral model, analogous to that proposed for mitochondrial site, was proposed for myofibrillar site [23] by substituting the contractile site for the oxidative phosphorylation site (Fig. 2). According to this model: 1) the compartmentalized ADP pool, although very small, would provide enough substrate for CK to prime the CP-induced contraction; 2) the contraction-produced ADP would be efficiently and quickly rephosphorylated, preventing its accumulation at the site of contraction.

In addition, there are other findings that support the above notion. It has been shown that the ADP generated in the mitochondrial membrane, by either mitochondrial bound CK or AK, is more effective in stimulating

148

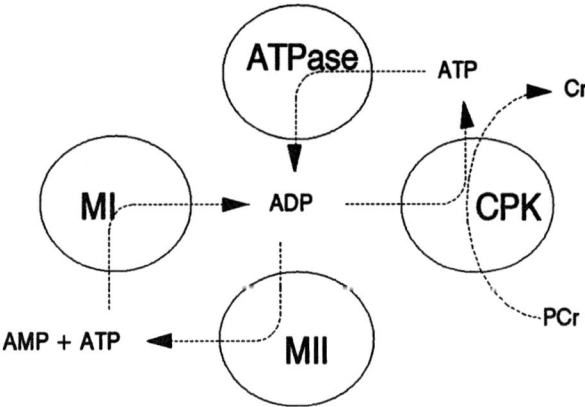

Fig. 2. The hypothesized tetrahedral arrangement of three enzymes, one molecule of CK and two molecules of AK around the energy utilizing sites such as that of myosin ATPase, showing a common pool of ADP and the delivery of energy as ATP from PCr.

the state 3 respiration in isolated mitochondria than added ADP [9, 10, 25]. In an attempt to synthesize ^{33}P labeled creatine phosphate with a high specific activity by isolated mitochondria, AMP was found to be a better substrate than ADP [26].

A more recent study by Goldberg's group [27] provides further and more concrete evidence for the compartmentalized role of AK in the high energy phosphate transfer system in muscle. Previous isotope studies in this area were done with ^{32}P, and the kinetic analysis of the results was criticized due to the slow rate of the orthophosphate transport into the cells, and the likelihood of its non-uniform incorporation into metabolic compartments. This problem was taken care of when Zeleznikar *et al.* [27] used ^{18}O-H$_2$O to study the rate of AK catalyzed phosphoryl transfer by determining the incorporation of ^{18}O-labeled β-phosphate into ADP and ATP in intact rat diaphragm. This study revealed that the AK activity, normally operating at a very low level (1/1000 of its Vmax) in resting muscle cells (cell free system), perhaps due to restricted levels of ADP, can increase in size with increase in contractile frequency. This indicates that active AK metabolic compartments, which are small in number in resting muscle, can increase in size or number in contracting muscle. It is likely that ADP generated via the contraction process (myosin ATPase and/ or calcium ATPase of sarcoplasmic reticulum) is the driving force of this metabolic flux. The enhanced rate of AK was found to be enough to process all the cellular ATP and ADP in approximately one minute, while the levels of ATP, ADP, and AMP remained near their basal steady state. This is an indication that the AK-catalyzed reaction, instead of catalyzing or salvaging ATP, transfers energy rich β-phosphate from one adenine to an-

other to produce a rapid equilibrium between ADP and ATP, even under increased rate of ATP hydrolysis. Moreover, from the metabolic behavior of [β-^{18}O]ATP, it became apparent that a compartmentalized adenine nucleotide pool exists for this action of AK. A secondary, much slower and frequency-independent rate of [β-^{18}O]ATP, emerged after the appearance of the rapid and frequency-dependent rate of [β-^{18}O]ATP and [β-^{18}O]ADP.

The above conclusion reached by these investigators is consistent with the metabolic scheme of interaction of two molecules of AK and one molecule of creatine kinase, with adenine nucleotide translocase at the mitochondrial energy producing site, and with myosin ATPase at the energy utilizing site of the phosphocreatine energy shuttle. Since the amount of β-phosphate transfer, compared with that of γ-phosphate transfer, was found to be very small in resting muscle (2–3%), and was increased 10-fold in contracting muscle [27], they concluded that the β-phosphate transfer system could not play a major role in transferring mitochondrially generated ATP as CK system does, but it is perhaps more important in facilitating the utilization of high energy phosphate at the contractile site. However, it should be kept in mind that an increase in contractility would be associated with an increase in the rate of oxidative phosphorylation that could result in a higher rate of adenine nucleotides turnover at the mitochondrial compartment (Fig. 1). This could be partly responsible for the higher rate of β-phosphate transfer observed under increased contractility [27]. It is apparent that further studies are necessary to clarify this question.

Adenylate kinase and the glycolytic metabolism of high energy phosphate

The role of AK-catalyzed transfer of high energy β-phosphate could be even more important under anaerobic conditions. Similar to contraction, oxygen deprivation can also increase the activity of AK due to a decrease in mitochondrial ATP synthesis and thus a build up in ADP pool. In fact, Zeleznikar *et al.* [27] have demonstrated that the velocity of the AK can be increased 35-fold by oxygen deprivation in intact rat diaphragm. The rate of increase in AK-catalyzed β-phosphate transfer coincided with the enhanced glycolytic flux, while the steady state levels of cellular ATP were maintained during a short period (8 min.) of anoxia. The ratio of the amount of adenine nucleotide processed through the

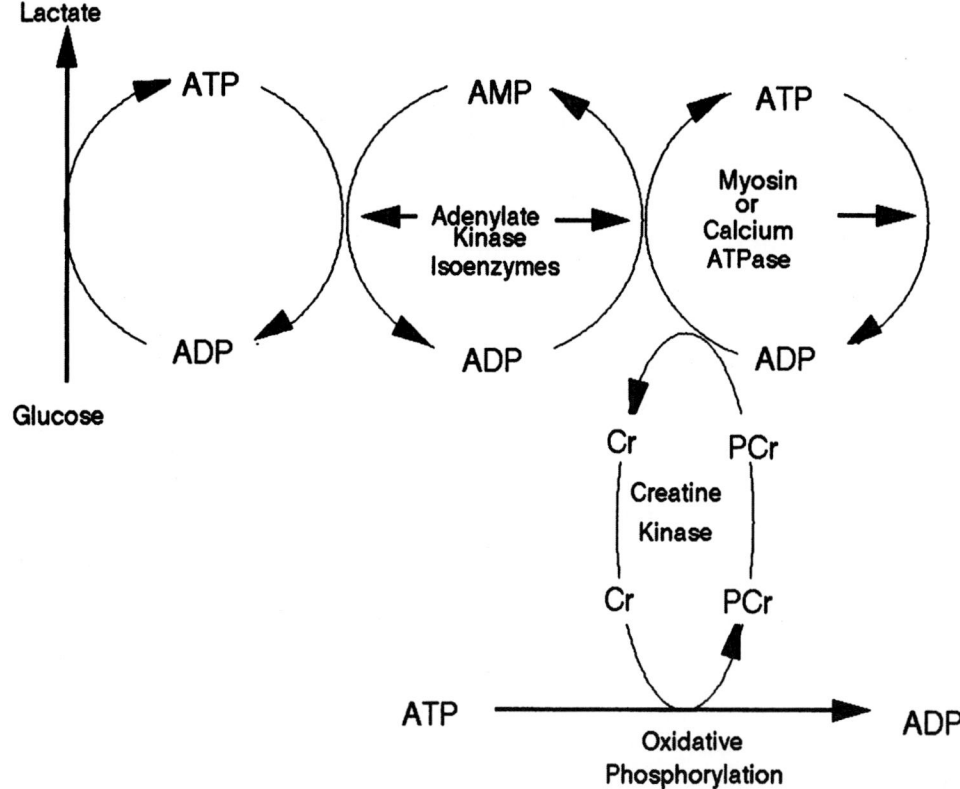

Fig. 3. A proposed scheme for adenylate kinase-catalyzed high energy phosphate transfer by Zeleznikar *et al.* [26]. This scheme proposes a mechanism by which isoenzymes of AK may serve to transfer high energy phosphate generated by glycolytic metabolism to ATP utilizing contractile components in muscle which operates in parallel with the phosphocreatine energy shuttle coupled to oxidative phosphorylation of mitochondria.

β-phosphate transfer pathway (AK) to that of γ-phosphate transfer in resting muscle, increased up to 40-fold under their short period of anaerobic condition. On the basis of finding that β-phosphate transfer plays an important role in oxygenated contracting muscle and in anaerobic resting and contracting muscle, they concluded that the AK-catalyzed high energy phosphate transfer is probably more important in mediating the transport of energy between glycolytic rather than mitochondrial site of energy production and the energy consuming sites, whereas, CK system would be involved at the mitochondrial site (Fig. 3). While this proposal might be correct, the possibility of the role of cytosolic creatine kinase in transferring energy from glycolytic site to the energy utilizing sites and thus their interaction with cytosolic AK, should also be considered. A functional coupling of CK with glycogenolysis and glycolysis is supported by several lines of evidence [11, 12, 28]. In muscle cells, the glycolytic enzymes, forming a so-called 'glycolytic complex' [29], are located at the I-band where they are loosely associated with the thin filaments [30, 31]. Most of the soluble MM-CK is also found at this location [11, 32] and is eluted from skinned muscle fibers along with the glycolytic enzymes. In addition the amount of

soluble MM-CK correlates well with the glycolytic potential of a muscle [33]. Functional coupling of glycolysis and phosphocreatine utilization has also been demonstrated by 31P-NMR [28, 34]. Therefore, with evidence supporting the functional coupling between the cytosolic CK and glycolysis and also between glycolysis and AK, a third form a tetrahedral complex could be also proposed for the interaction among glycolytic enzymes complex and the cytosolic CK and AK in muscle. This complex would be similar to that of Fig. 1, where glycolytic site of energy production displaces the mitochondrial oxidative phosphorylation, and cytosolic CPK and AK displace the mitochondrial bound CPK and AK sites.

The process of AK-catalyzed β-phosphate transfer under an anaerobic condition could be both beneficial and harmful to muscle function and muscle cell integrity. While during the short period of anoxia, AK along with CK, can maintain the ATP level and furnish the contractile apparatus with ATP, a longer period of anoxia can result in loss of adenine nucleotide pool and thus can impede the recovery of the contractile function during reoxygenation. It has been shown that the extent of recovery of myocardial contractile function after a period

150

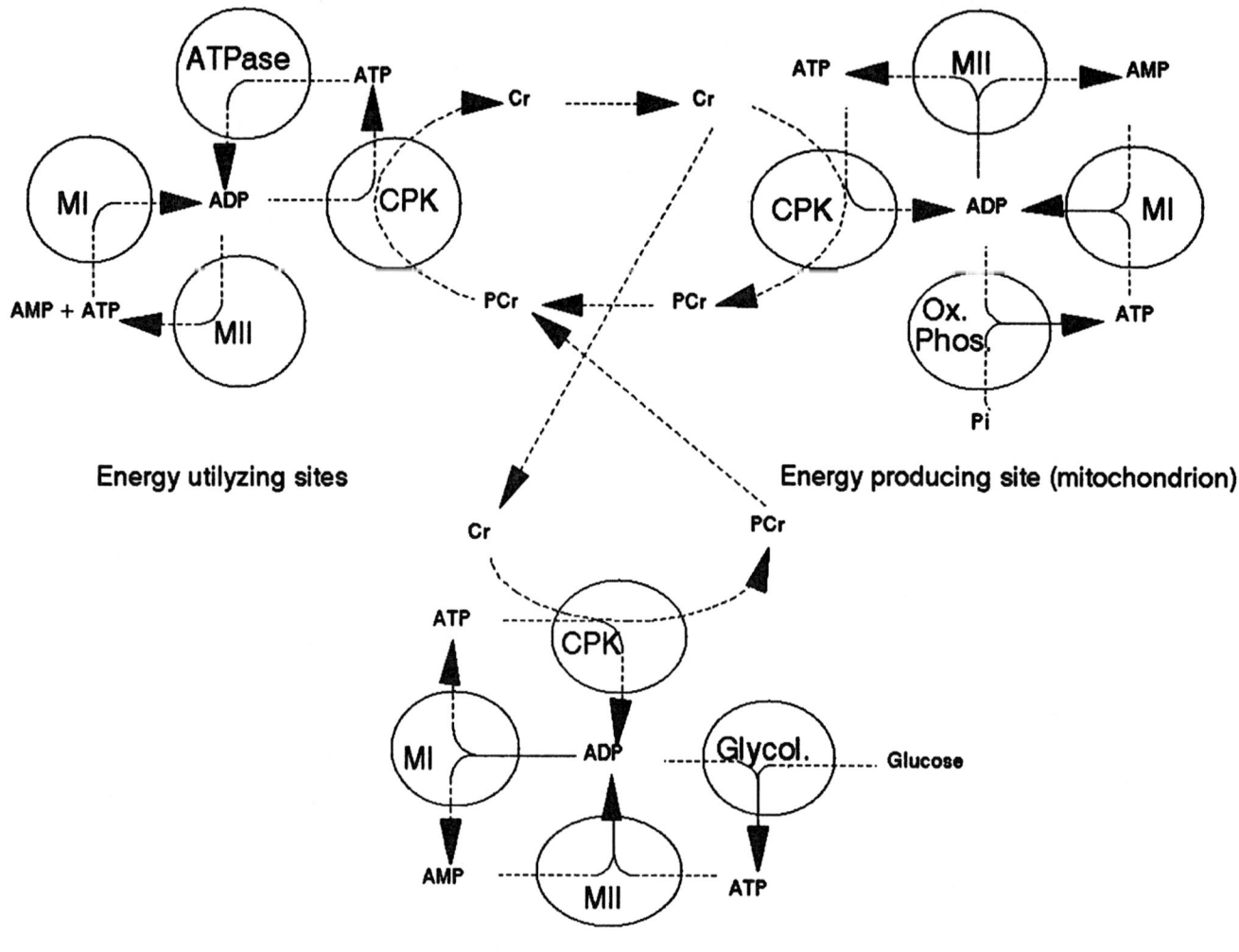

Energy utilyzing sites

Energy producing site (mitochondrion)

Energy producing site (glycolytic)

Fig. 4. A proposed scheme for the compartmentalized role of AK in catalyzing the high energy phosphate transfer at energy producing (glycolytic and oxidative phosphorylation of mitochondria) and energy utilizing sites (myosin ATPase, sarcoplasmic reticular Ca^{2+}-ATPase, sarcolemmal Na^{+}-K^{+}-ATPase, etc.) in muscle cells.

of anoxia is well correlated with the size of total adenine nucleotides [35, 36]. The action of AK under anaerobic conditions could result in accumulation of AMP. AMP is then deaminated by AMP deaminase [37–39] which is bound to myofibrils [40] at both ends of the A band [41] into IMP. AMP and IMP are then dephosphorylated by 5'-nucleotidase located on the sarcolemma [42] to give adenosine and inosine which can leak out of the cell and lead to the loss of the adenine nucleotide pool [43]. This became apparent when we showed that if the action of myocardial AK was reduced during the anoxic period, the nucleotide pool was preserved and a greater and more complete recovery of the contractile function was possible after reoxygenation [36].

Based on the above evidence, a hypothetical model concerning the functional and structural coupling of muscle AK and CK isoenzymes, with other energy me-

tabolizing enzymes such as those of energy producing sites (mitochondrial oxidative phosphorylation and glycolytic enzymes) and energy utilizing sites (various AT-Pases such as that of myosin ATPase, sarcoplasmic reticulum-Ca^{2+}-ATPase, sarcolemmal membrane-Na^{+}-K^{+}-ATPase, etc.) is presented as Fig. 4. It is apparent that functional details of the role of muscle AK in energy metabolism requires more experimentation. Further studies on AK system should focus on its compartmentalized interaction with other related enzymes, in different intact organisms, under various conditions, using a multidisciplinary biochemical and immunohistochemical approach which has been applied in the study of CK isoenzymes and the phosphocreatine energy shuttle system.

References

1. Leninger LA: Principles of Biochemistry, Worth Publishers Inc, New York. 1982
2. Taylor EW: Chemistry of muscle contraction. Ann Rev Biochem 41: 577–616, 1972
3. Weber A, Murray JM: Molecular control mechanisms in muscle contraction. Physiol Rev 53: 612–673, 1973
4. Mommaerts WFHM: Energetics of muscle contraction. Physiol Rev 49: 427–508, 1969
5. Seraydarian MW: The correlation of creatine phosphate with muscle function. In: WE Jacobus, JS Ingwall (eds.) Heart Creatine Kinase. Williams and Wilkins, Baltimore, London. 1980, pp 82–90
6. Lonard L: Adenosine triphosphate-creatine transphosphorylase as relaxing factor of muscle. Nature 172: 1181–1183, 1953
7. Bendall JR: Myokinase as relaxing factor in muscle. Nature 173: 548–549, 1954
8. Bendall JR: The relaxing effect of myokinase on muscle fibers, its identity with 'Marsh' factor. Proc Roy Soc London, Series B, 142: 409–412, 1954
9. Bessman SP, Carpenter CL: The creatine-creatine phosphate energy shuttle. Ann Rev Biochem 54: 831–862, 1985
10. Saks VA, Rosenshtraukh LV, Smirnov VN, Chazov EI: Role of creatine phosphokinase in cellular function and metabolism. Can J Physiol Pharmacol 56: 691–706, 1978
11. Wegmann G, Zanolla E, Eppenberger HM, Wallimann T: *In situ* compartmentation of creatine kinase in intact sarcomeric muscle: the acto-myosin overlap zone as a molecular sieve. J Muscle Research and Cell motility 13: 420–435, 1992
12. Wallimann T, Wyss M, Brdiczka D, Nicolay K: Intracellular compartmentation, structure and function of creatine kinase isoenzymes in tissues with high and fluctuating energy demands: the 'phosphocreatine circuit' for cellular energy homeostasis. Biochem J 281: 21–40, 1992
13. Hamada M, Takenaka H, Fukumoto K, Fukamachi T, Yamaguchi T, Sumida M, Shiosaka T, Kurokava Y, Okuda H, Kuby SA: Structure and function of adenylate kinase isoenzymes in normal humans and muscular dystrophy patients. In: CL Markert, JG Scandolios (eds.) Isoenzymes: Current Topics in Biological and Medical Research. Liss, New York, 1987, 16: pp 81–99
14. Frank R, Trosin M, Tomasselli A, Schuls GE, Schirmer RH: Mitochondrial adenylate kinase (AK2) from bovine heart. Eur J Biochem 141: 629–636, 1984
15. Russell PJ, Horenstein JM, Goins L, Jones D, Laver M: Adenylate kinase in human tissues. J Biol Chem 249: 1874–1879, 1974
16. Luz CM, Konig I, Schirmer RH, Frank R: Human cytosolic adenylate kinase allelozymes; purification and characterization. Biochem. Biophys. Acta 1038: 80–84, 1990
17. Kityarakara A, Harman JW: The cytological distribution of pigeon skeletal muscle of enzymes acting on phosphorylated nucleotides. J Exp Med 97: 553–572, 1963
18. Tice LW, Barnett RJ: Fine structural localization of adenosinetriphosphatase activity in heart muscle myofibrils. J Cell Biol 15: 401–415, 1962
19. Walker EJ, Dow JW: Location and properties of two isoenzymes of cardiac adenylate kinase. Biochem J 203: 361–369, 1982
20. Watanabe K, Itakura T, Kubo S: Distribution of adenylate kinase isoenzymes in porcine tissues and their subcellular localization. J Biochem 85: 799–805, 1979
21. Bessman SP, Geiger PJ: Compartmentation of hexokinase and creatine phosphokinase, cellular regulation and insulin action. Curr Top Cell Reg 16: 55–86, 1980
22. Yang WCT, Geiger PJ, Bessman SP, Borrebaek B: Formation of creatine phosphate from creatine and ^{32}P-labeled ATP by isolated rabbit heart mitochondria. Biochem Biophys Res Commun 76: 882–887, 1977
23. Savabi F, Geiger PJ, Bessman SP: Myokinase and contractile function of glycerinated muscle fibers. Biochem Med & Metab Biol 35: 227–238, 1986
24. Savabi F, Geiger PJ, Bessman SP: The myofibrillar end of the creatine phosphate energy shuttle. Am J Physiol 247: C424–C432, 1984
25. Gellerich FN, Kunz W: Cause and consequences of dynamic compartmentation of adenine nucleotides in the mitochondrial intermembrane space in respect to exchange of energy rich phosphates between cytosol and mitochondria. Biomed Biochim Acta 46: S545–S548, 1987
26. Savabi F, Geiger PJ, Bessman SP: Three-step preparation and purification of phosphorus labeled creatine phosphate of high specific activity. Anal Biochem 137: 374–379, 1984
27. Zeleznikar RJ, Heyman RA, Graeff RM, Walseth TF, Dawis SM, Butz E, Goldberg D: Evidence for compartmentalized adenylate kinase catalysis serving a high energy phosphoryl transfer function in rat skeletal muscle. J Biol Chem 265: 300–311, 1990
28. Kuprianov VV, Seppet EK, Emelin IV, Saks VA: Phosphocreatine production coupled to the glycolytic reactions in the cytosol of cardiac cells. Biochem Biophys Acta 592: 197–210, 1980
29. Brooks SP, Storey KB: Reevaluation of the glycolytic enzymes with thin filament proteins. Can J Biochem 59: 494–499, 1981
30. Arnold H, Pette D: Binding of aldose and triosephosphate dehydrogenase to F-actin and modification of catalytic properties of aldose. Eur J Biochem 15: 360–366, 1979
31. Bronstein WW, Knull HR: Interaction of muscle glycolytic enzymes with thin filament proteins. Can J Biochem 59: 494–499, 1981
32. Wallimann T, Schnyder T, Schlegel J, Wyss M, Wegmann G, Rossi AM, Hemmer W, Eppenberger H, Quest AFG: Subcellular compartmentation of creatine kinase isoenzymes, regulation of CK and octameric structure of mitochondrial CK: important aspects of the phosphoryl-creatine circuit. Prog Clin Biol Res 315: 159–176, 1989
33. Wallimann T, Schlosser T, Eppenberger HM: Function of M-line-bound CK as intramyofibrillar ATP regenerator at the receiving end of the phosphorylcreatine energy shuttle in muscle. J Biol Chem 259: 5238–5246, 1984
34. Van Waarde A, Van Den Thillart G, Erkelens C, Addink A, Lugtenburg J: Functional coupling of glycolysis and phosphocreatine utilization in anoxic fish muscle. J Biol Chem 265: 914–923, 1990
35. Savabi F, Bessman SP: Recovery of isolated rat atrial function related to ATP under different anoxic conditions. Arch Biochem Biophys 248: 151–157, 1986
36. Savabi F, Bessman SP: Post-anoxic recovery of spontaneously beating isolated atria; pH related role of adenylate kinase. Biochem Med Metab Biol 35: 345–355, 1986
37. Kushmerick MJ, Davies RE: The chemical energetics of muscle contraction. II. The chemistry, efficiency and power of maximally

152

working sartorius muscles. Proc R Soc London Ser B 174: 315–353, 1969

38. Lowenstein JM: Ammonia production in muscle and other tissues: The purine nucleotide cycle. Physiol Rev 52: 382–414, 1972

39. Hamada M, Kuby SA: Studies on adenosine triphosphate transphosphorylases. Arch Biochem Biophys 190: 772–792, 1978

40. Ashby B, Frieden C, Bischoff R: Immunofluorescent and histochemical localization of AMP deaminase in skeletal muscle. J Cell Biol 81: 361–373, 1979

41. Cooper J, Trinick J: Binding and location of AMP deaminase in rabbit psoas muscle myofibrils. J Mol Biol 177: 137–284, 1984.

42. Bowditch J, Nigdikar S, Brown AK, Dow JW: 5'-nucleotidase activity of isolated mature rat cardiac myocytes. Biochem Biophys Act 845: 21–26, 1985

43. Jennings RB, Reimer KA, Hill ML, Mayer SE: Total ischemia in dog hearts, *in vitro* 1-comparison of high energy phosphate production, utilization, and depletion, and of adenine nucleotide catabolism in total ischemia *in vitro* vs. severe ischemia *in vivo*. Circ Res 49: 892–900, 1981

44. Raggi A, Ronca-Testeni, Ronca G: Muscle AMP aminohydrolase. II. Distribution of AMP aminohydrolase, myokinase activities in skeletal muscle. Biochem Biophys Acta 178: 619–622, 1969

45. Savabi F: Mitochondrial creatine kinase deficiency in diabetic rat heart. Biochem Biophys Res Commun 154: 469–475, 1988

46. Kirsch A, Savabi F: Effect of food restriction on the phosphocreatine energy shuttle components in rat heart. J Mol Cell Cardiol 24: 821–830, 1992.

PART III

CREATINE KINASES AND
METABOLIC INTEGRATION

Molecular and Cellular Biochemistry **133/134**: 155–192, 1994.

III-1 Metabolic compartmentation and substrate channelling in muscle cells

Role of coupled creatine kinases in in vivo *regulation of cellular respiration – a synthesis*

V.A. Saks,[1] Z.A. Khuchua, E.V. Vasilyeva, O.Yu Belikova and A.V. Kuznetsov

Group of Bioenergetics, Cardiology Research Center, Moscow, Russia; [1] *Present address: Laboratory of Bioenergetics, Institute of Chemical and Biological Physics, Ravala 10, EE0001 Tallinn, Estonia*

'It is easier to explain biochemistry in terms of transport than it is to explain transport in terms of biochemistry.'
P. Mitchell The Ninth Sir Hans Krebs Lecture, Dresden, July 2, 1978

Abstract

The published experimental data and existing concepts of cellular regulation of respiration are analyzed. Conventional, simplified considerations of regulatory mechanism by cytoplasmic ADP according to Michaelis-Menten kinetics or by derived parameters such as phosphate potential etc. do not explain relationships between oxygen consumption, workload and metabolic state of the cell. On the other hand, there are abundant data in literature showing microheterogeneity of cytoplasmic space in muscle cells, in particular with respect to ATP (and ADP) due to the structural organization of cell interior, existence of multienzyme complexes and structured water phase. Also very recent experimental data show that the intracellular diffusion of ADP is retarded in cardiomyocytes because of very low permeability of the mitochondrial outer membrane for adenine nucleotides *in vivo*. Most probably, permeability of the outer mitochondrial membrane porin channels is controlled in the cells *in vivo* by some intracellular factors which may be connected to cytoskeleton and lost during mitochondrial isolation. All these numerous data show convincingly that cellular metabolism cannot be understood if cell interior is considered as homogenous solution, and it is necessary to use the theories of organized metabolic systems and substrate – product channelling in multienzyme systems to understand metabolic regulation of respiration. One of these systems is the creatine kinase system, which channels high energy phosphates from mitochondria to sites of energy utilization. It is proposed that in muscle cells feed-back signal between contraction and mitochondrial respiration may be conducted by metabolic wave (propagation of oscillations of local concentration of ADP and creatine) through cytoplasmic equilibrium creatine and adenylate kinases and is amplified by coupled creatine kinase reaction in mitochondria. Mitochondrial creatine kinase has experimentally been shown to be a powerful amplifier of regulatory action of weak ADP fluxes due to its coupling to adenine nucleotide translocase. This phenomenon is also carefully analyzed. (Mol Cell Biochem **133/134**: 155–192, 1994)

Key words: myocytes, mitochondria adenine nucleotides, compartmentation, diffusion, creatine kinase, oxidative phosphorylation, regulation, metabolic wave, functional coupling

Address for offprints: V.A. Saks, Laboratory of Bioenergetics, Institute of Chemical and Biological Physics, Ravala 10, Tallinn EE 0001, Estonia

Introduction: the problem of regulation of respiration *in vivo*

In the field of cell physiology one of the most discussed and controversial topics is cellular and molecular mechanism of regulation of respiration and mitochondrial oxidative phosphorylation [1–24]. Many current experimental approaches are described by Mary Osbakken in this issue (chapter I-2). The problem is not as yet solved and sometimes even seems to get farer from its solution. In this work we consider several new groups of experimental data and try to analyze some possibilities of the solution of the problem.

The basic discovery in this field was made by Lardy and Wellmann in 1952 and Chance and Williams in 1952–1955: they described the phenomenon of respiratory control by ADP for isolated mitochondria [25–27]. This phenomenon – manifold activation of respiration by ADP in the presence of inorganic phosphate due to strict control exerted by phosphorylation system (membrane potential) over respiratory chain – shows that high respiration rates may be observed only in the presence of ADP, in the State 3. During forty years of bioenergetic research oxidative phosphorylation of this ADP was main topic of studies and was elucidated to impressive extent including the discovery of its chemiosmotic mechanism by Mitchell and registration of membrane potentials by Lieberman and Skulachev, but probably it is not as yet completely solved [28–30]. The cellular mechanisms of its regulation are much less understood and the existing concepts controversial. In intact heart the oxygen uptake by cardiac tissue is strictly related to the work performance, as it was first shown in the pioneering works by Neely *et al.* in 1967 [31, 32] and then confirmed by Williamson group [33] and by many others [7–11, 34–45]. That means that, in general sense, the respiratory control is operative in the intact cells: increased ADP production in the ATPase reactions results in elevation of mitochondrial respiration rate. The ranges of these changes are wide: in the resting state the respiration rate is from 6 to 12–15 micromoles (µmoles) of oxygen per g of dry weight per min in rabbit and rat hearts, respectively [15, 38, 40, 41], and at maximal workloads it may reach the value of 40–55 µmoles of oxygen per min per g dry weight for Langendorff-perfused rat hearts [15, 38, 40, 41, 43] or even 84 – 110 micromoles of oxygen per g of dry weight per min for Langendorff-perfused hearts, stimulated with catecholamines [34, 41, 43, 44] and up to 116–160 micromoles oxygen per min per g

dry weight for the working heart [33, 39]. When recalculated per mg of mitochondrial protein (the content of mitochondrial protein in heart tissue is 65–70 mg per g of wet weight or 320–350 mg per g of dry weight [46, 47] the maximal respiration rate in the intact working heart is around 0.7–1.0 µg-atoms of oxygen per min per mg of mitochondrial protein at 37° C. For isolated mitochondria the State 3 respiration rate is usually in the range of 350–400 ng-atoms per min per mg at 30° C and close to 700–800 ng-atoms at 37° C in *in vitro* conditions [48]. Thus, we may assume that at maximal workloads mitochondria achieve the State 3 respiration rate (the maximal possible rate) in the intact cells *in vivo*. Similar conclusion was made by Geisen and Kammermeier [37], who, however reported the oxygen consumption rates as high as 2–4 µg-atoms per min per mg of mitochondrial protein both for *in vivo* and *in vitro* conditions, that seems to exceed the value obtained in physiological experiments by many authors [7–11, 15, 38–41, 43].

Thus, the basic characteristics of the energy metabolism – the strictly linear relationship between the rate of oxygen uptake and cardiac work – implies the existence of an extremely effective feedback between energy consumption and production. The unsolved question is: by which cellular mechanism is this signal passed by, in which way the information of increased ADP production by myofibrillar ATPases reaches the mitochondria?

The concepts of the mechanism of this feedback signal are dependent on the point of view on the organization of cellular metabolism. Here we can see the existence of two parallel ways of thinking which intercept very rarely (see interesting reviews on this topic by J. Glegg, ref. 49, 50).

1. First one (it was a first necessary approximation) was to consider the cell as a homogenous system in which low molecular weight substrates are distributed within different cellular compartments – cell cytoplasm and mitochondrial matrix – where their concentrations can be easily calculated on the basis of knowledge of their tissue content and cell volume characteristics [37, 51–58]. In the field of cardiac energetics this seems to be the dominating concept and all discussions are held in its framework [6–11, 19, 37–45, 52–58]. This classical approach is reviewed by M. Osbakken in this tissue. Two groups of data can be very clearly identified.

A. Very intensive excellent experimental work has been carried out by the groups of Wilson and Erecinska [34, 35] Williamson [33], Neely [31, 32] and Opie [36], by Hassinen *et al.* [14, 52, 53], Kammermeier *et al.* [37] and others on the isolated working or Langendorff-perfused hearts with subsequent enzymatic analysis of the metabolite contents in tissue extracts. In rather good concord with each other (with exception of a study by Neely *et al.* [32]), these works have shown that there are distinct metabolic changes connected to the work transitions: when the workload is increased and oxygen consumption rates elevated, phosphocreatine (PCr) content decreases and that of inorganic phosphate (Pi) and creatine (Cr)

Table 1. Metabolic characteristics of isolated hearts: effect of work transition.

Perfusion technique method of analysis and reference	Oxygen uptake (µmoles/ min/g dry weight)	Tissue contents (µmole/g dry w)				Fluxes (µmol/g dry w/s)	
		ATP	PCR	PCR/Cr	PCR/ATP	$F_{PCr-ATP}$	F_{ATP}
I GROUP							
Langendorff perfusion							
1. Freeze-clamp (32)							
l.w.	25	24,9	39,4				
h.w.	61	24,5	42,0				
2. Freeze-clamp (52)							
l.w.	11	26,6	49,8				
h.w.	26,8	25,4	42,1				
3. Freeze-clamp (34)							
l.w.	14	26,3	27,5	2,27			
h.w.	108	25,4	21,0	0,82			
4. NMR (40)							
l.w.	6,7				2,1		
h.w.	42,8				2,2		
5. NMR (42)							
h.w.	46,3	27,4			2,08	27	4,6
6. NMR (43) gluc							
l.w.	37,2	25,3	37,5				
h.w.	82,7	20,3	35,4				
NMR (43) palm. gluc.							
l.w.	50	24,4	48,3				
h.w.	92	19,8	43,0				
7. NMR (44)							
a.	5.9					10	
l.w.	25	25				18	1,49
m.w.	33,7					25	2,3
h.w.	45,8					30	4,5
8. Freeze-clamp (58) pyr.							
l.w.	25,3	22	41,6				
h.w.	111	15,3	33,9				
gluc. l.w.	25,3	21,2	23,1				
h.w.	111	17,5	29,9				
II GROUP							
Working heart							
9. Freeze-clamp (56)							
l.w.	20	22,7	34,8				
h.w.	46	21,1	27,8				
10. Freeze-clamp (33)							
l.w.	25	21,4		1,5			
h.w.	100	20,1		1,0			
11. NMR (39)							
l.w.	39,9				1,47		
h.w.	116				1,59		

a., l.w., m.w., h.w.: arrested, low work, medium work and high work state of heart, respectively (arbitrary units, used by authors). pyr.: pyruvate; gluc.: glucose; palm.: palmitate. Substrates are shown only if they all were used in the same work. Metabolic fluxes between PCr and ATP ($F_{PCr-ATP}$) and Pi and ATP (F_{ATP}), determined by NMR method.

increase at more or less constant level of ATP [14, 33–37, 52, 53]). Selective data of these groups are summarized in Table 1. Klingenberg, Wilson and Erecinska and Hassinen *et al.* proposed that there is an equilibrium state between cytoplasmic adenine nucleotides and mitochondrial pyridine nucleotides and cytochrome c, assuming very rapid translocation of adenine nucleotides by translocase across the inner mitochondrial membrane (and very rapid diffusion of adenine nucleotides through the outer mitochondrial membrane) [14, 34, 35, 51–53].

B. In recent years a new group of data emerged in the works with the use of phosphorus NMR. It is reported by rapidly increasing number of investigators that the changes in the workloads and correspondingly in the oxygen uptake are observed at a constant level of phosphocreatine, inorganic phosphate and ATP both in *in vitro* and *in vivo* experiments [6–11, 39–43]. This group includes recent work by Jeffrey and Malloy [39] who practically reproduced the earlier work by Williamson *et al.* [33] but with different results. In some cases the use of NMR gave results depending on the nature of substrates both for the heart [41, 43, 54] and skeletal muscle [57]. This result is not completely due to use of NMR since Aussedat *et al.* observed, by using the same method, clear changes in the PCr level at different loads due to different calcium concentrations [38]. On the other hand, in very recent detailed studies carried out in Kathrine F. Lanoue laboratory [58], classical freeze-clamp technique was used to investigate metabolic changes in Langendorff-perfused hearts in which also mitochondrial membrane electrical potential gradients were measured at different workloads. No decrease of phosphocreatine with elevation of workload was seen when glucose was used as a substrate, and when pyruvate was used as a substrate oxygen consumption rate was not different from that with glucose in spite of higher PCr levels [58]. Mitochondrial membrane potential slightly decreased with elevation of workload but calculated ADP content showed no correlation with oxygen uptake [58]. In general, this new group of data (some of which are also indicated in Table 1) resulted first in some kind of confusion and then in firm conclusion that there is no metabolic regulation of respiration. We may refer here to the summary of the Jeffrey and Malloy's work: 'Thus, metabolites of ATP synthesis are not normally involved in respiratory control, . . . the normal control of respiration in the heart at steady state cannot occur at the level of substrates.'[39]. Katz, Swain, Portman and Balaban were more cautious: 'These data suggest that some other

parameters or cooperativity effects involving the phosphate metabolites must play a role in the feedback between respiration and work in the heart muscle' [11]. On the basis of these data and of the discovery by Hansford and McCormac and Denton (see ref. 24 and 59 for review) of the regulation of the Krebs cycle dehydrogenases by calcium ions, an assumption of simultaneous regulation of contraction and mithochondrial respiration by calcium ions was developed, that was called 'stimulus-response metabolic coupling' [6–11, 22, 24, 39, 40, 60]. According to this assumption, activation of mitochondrial matrix dehydrogenases, in response to elevated cytoplasmic and subsequently matrix calcium increases the level of mitochondrial NADH that in turn increases the rate of electron transport and oxygen uptake [6, 11, 24, 39, 40, 60]. This is possible only if the ADP supply to mitochondria is not restricted [22, 59]. However, application of these ideas is seriously restricted by the experimental fact that perfusion of the heart with ruthenium red which inhibits mitochondrial uptake of calcium does not abolish the linear dependence of oxygen uptake rate on the heart work but only modifies the metabolic response to these manipulations [60–64]. In fact, ruthenium red was found even to protect mitochondria in ischemic hearts *in situ* by inhibiting calcium overload and ensure better restoration of cardiac energetics and function at the reperfusion [65]. That means that ruthenium red does not destroy feedback between contraction and respiration significantly. Also, no changes in NAD(P)H fluorescence were seen by Heineman and Balaban in response to alteration of afterload of perfused working rabbit heart [66]. A proposal of possible activation of ATP synthase by Ca ions [67] does not seem to explain the respiration regulation in the heart, as pointed out by Kathrine Lanoue group [58]. In addition to all this controversy, Moravec and Bond showed very recently [68] by using electron probe microanalysis that inotropic stimulation by isoproterenol that increased heart work and is known to increase oxygen uptake, did not increase calcium content in mitochondria. There is little doubt, however, that stimulatory effect of calcium on mitochondrial dehydrogenases modulates the response of mitochondrial respiration to work transitions, as it has been experimentally shown by Moreno-Sanchez [23] and very recently by Hak *et al.* [63].

As it was already mentioned above, interpretation of both these groups of data is based on the assumption (not clearly stated in papers but obvious from their design and based on 'conventional wisdom' [49, 69]) that mitochondrial respiration obeys the simple kinetics of regulation of the enzyme action in the homogenous aqueous solutions: the rate of respiration is thought to be dependent (but not often found) on cytoplasmic ADP concentration (more complicated derived versions are ATP/ADP ratio, phosphorylation potential etc., see the chapter by M. Osbakken in this issue) and in more simple cases the value of Km for ADP is attempted to be determined for the intact cells [11, 16, 19, 41, 43]. In all cases it is accepted a priori that the cell cytoplasm is a homogenous phase for substrates including ADP which concentration as well as phosphorylation potential may be calculated from creatine kinase equilibrium [19, 41, 71–749, and it is assumed that the physico-chemical properties of this phase are at least close to those of the aqueous diluted solutions; it is assumed that this cytoplasmic ADP has free access to the intermembrane space of mitochondria and thus to the adenine nucleotide translocase.

2. The second alternative concept of the cellular metabolism considers it as a result of functioning of organized myltienzyme systems, where the behaviour of the enzymes and even substrates differ remarkably from that in aqueous solution *in vitro*. This concept has been developed during several decades in the works of Welch, Sols and Marco, Masters *et al.*, Srere, Friedrich and Kelety, Ottaway and Mowbray, Kurganov, Glegg and by others [49, 50, 69, 75–89] and now it is a rather advanced and detailed theory (see the paper by Kholodenko, Cascante and Westerhoff in this issue and references 80–82 for review). This concept, however, has not as yet been intensively exploited for the analysis of the phenomenon of regulation of cellular respiration, may be because of the complexity of its use. Some attempts in this direction are being be made in this issue. According to the concept of organized metabolic systems, enzymes in the cell are controlled by their microenvironment, and by other components of the system by a mechanism of direct substrate and product chanelling, therefore the diffusion distances for intermediates and also enzymes are substantially decreased and the transient times of sequential enzymatic reactions reduced by precise structural organization of multienzyme complexes, and high reaction rates can be achieved in coupled systems independently from the average concentration of substrates or intermediates [75–89]. Correspondingly, the problem of intracellular regulation of respiration and cellular energetics in general may be analyzed in very different way: due to direct channelling of the substrates, the fluxes of metabolites and energy which are most important characteristics of metabolism are dissociated from the total content of metabolites but are dependent only on the rate of substrate supply and product removal between neighbouring enzymes. Accordingly, total tissue contents (or cytoplasmic concentrations) of high energy phosphates as integral parameters of the cell are of limited value for finding out the cause (via calculation of mean cytoplasmic ADP) of regulation of respiration and oxidative phosphorylation for which they are only the end products. These integral parameters indicate only the overall balance between energy production and utilization, without any direct feedback to the reaction rates. What is important is the structural organization of all these systems which determines the values of the metabolic fluxes. In favour of this concept are two important newly described phenomena: compartmentation of ATP and ADP in the cells including microcompartments in the cytoplasm due to its high degree of organization (structural heterogeneity) and the very low accessibility of the cytoplasmic ADP into the intermembrane space of mitochondria because of decreased permeability of the mitochondrial outer membrane for this substrate (see the following sections). This kind of analysis leads to the new concept of the feedback signal as a metabolic wave in the structured medium (cytoplasm) amplified by the coupled reactions in mitochondria. The experimental basis for these concepts will be considered in details in the following sections, but it is interesting to mention here that several very important ideas have been described in the literature long time ago but not recognized and by now are almost totally forgotten. Thus, this review is the synthesis of new and old data.

Compartmentation and microcompartmentation of substrates in heart cells

Cell structure and compartmentation in muscle cells

It is usual practice in the metabolic research that the contents of metabolites in heart are calculated per g of wet tissue, or to avoid the artifacts of cell oedema at perfusion, per g of dry weight. The cytoplasmic concentrations are calculated using the known values of extra- and intracellular water spaces determined by labelled compounds which do not penetrate through sarcolemma, and these values are respectively 1–2 ml for extracellular (seems to be variable and dependent of perfusion time) [19, 90] and 2.6–2.9 ml of intracellular water per g of dry weight [19, 53, 54, 56, 90, 92]. That is in good agreement with the earlier value of cytoplasmic water equal to 0.44–0.46 ml per g of wet weight [53, 91, 92] which is basically used by many authors [19, 41, 43, 54–56, 92]. This water space does not include water in mitochondrial matrix, that gives additional 15% of tissue wet weight; thus, it is generally accepted that intracellular water accounts for 60% of tissue wet weight of intact tissue [53, 58, 90, 92].

On the other hand, we have been given a perfect description of the ultrastructure of cardiac cells by several teams of researchers [93–97]. Quantitative morphometric measurements by Schaper *et al.* of cardiac cells in the heart of different animals including man have shown that the cytoplasmic volume makes up not more than 10–20% of cell volume, 50–60% being occupied by myofibrils and 25–35% by mitochondria [94, 95]. The 10–20% of cell volume occupied by the cytoplasma must accomodate the fraction of soluble (or easily solubilized) proteins, about 10% of total protein of the cell (our unpublished data), and the elements of cytomatrix, or cytoskeleton – microtubules, microfilaments, intermediate filaments and interlocking microtrabecular lattice [89]. If we compare these data with the value of 0.4–0.5 ml for cytoplasmic water per g of wet weight (see above), we must admit that most of this water should also come from the cellular structures: myofibrils, intermembrane space of mitochondria, nucleus etc.

Analyses of the state of water in the cells and cytoplasmic structures have been given by many authors [49, 50, 88, 99–101], for the purpose of this work it will be sufficient to refer to reviews by Glegg [49, 50, 88], Srere [78, 80], Fulton [98] and comprehensive work by Ling [99, 100]. Fulton has already in 1982 concluded that cytoplasmic proteins are most probably organized into liquid crystal-like structures where the protein concentration is about 50% of volume [98]. Similar analysis for the structure of mitochondrial matrix has been given by Srere with exactly the same conclusion [78]. Morel has analyzed the structure of water and its role in organization of myofibrillar space [79]. Taking into account three dimensional well developed structure of cytoskeleton network – microtubular lattice [88, 89] and protein content both in cytoplasm and in other cellular structures, it is clear that the surface area of proteins as well as membranes is very significant [75–79]. This surface influences the structure of water, which becomes structured in the interface area ('vicinal' water characterized by decreased entropy, as compared to the bulk water phase [79, 88, 102]. The estimations show that this structured water may make up to 30% of intracellular water, and

the properties of this phase such as solvent properties etc.) may very significantly differ from those of bulk water [79, 88, 98–103]. In addition to that Waterson has postulated that in the cells 'ordered water' may exist in a form of cubic clusters which may behave as independent entity [104]. In the microcompartments related to the multienzyme complexes this interface water may be the most abundant water phase and significantly change the physico-chemical characteristics of the reaction (such as the free energy change etc.) as compared to the bulk phase of cytoplasmic water, and may additionally increase the compartmentation (microcompartmentation) of substrates in this space [102, 103]. For example, it is not excluded at all that important metabolites may partition between different water phases [75, 88].

In conclusion, in the cytoplasm all proteins which have been thought to be soluble are most probably organized into liquid-crystal like structures and multienzyme complexes and even the water phase is not homogenous, that gives rise to multiple microcompartments in the cells.

For further analysis, it is important to give precise definitions of different cellular compartments.

In the literature the following clear definitions of cellular compartments have been given, in particular by Friedrich [103] and Kempner [105] which will be used in this work.

Intracellular metabolic compartment has been defined as 'a subcellular region of biochemical reactions kinetically isolated from the rest of cellular processes' [105]. Several different cellular compartments may exist [103].

a) Macrocompartments are the cellular organelles and cytoplasm which are large relative to the molecular dimension;

b) microcompartments are multiple compartments which dimensions are of the order of the size of metabolites; the examples for these microcompartments are channels in multienzyme clusters; Schoolwerth and LaNoue [106] defined microcompartmentation as a 'situation in which a metabolite appears to move from enzyme to enzyme for which it is a substrate, without mixing with the total metabolite pool in the same macrocompartment'. This is also a definition for metabolite channelling;

c) dynamic compartments: in contrast to static compartments which are durable, the dynamic compartments are transient and they form and decompose rapidly as a result of association of different enzymes with each other or with membrane surfaces, and the relative metabolite pools may undergo rapid dynamic changes [103].

Compartmentation of ATP(ADP) in muscle cells

In all types of mammalian cells the adenine nucleotides are devided into intra- and extramitochondrial pools (macrocompartments) interconnected via the adenine nucleotide translocase within the inner mitochondrial membrane. Structure of this translocase and its mode of action is in details described by Vignais and Klingenberg groups [107–110]. By selectively dissolving first the sarcolemma of isolated cardiomyocytes by digitonin and then the intracellular membranes by Triton, Geisbuhler and Altschuld et al. have quantitatively estimated the sizes of these pools and found the following distribution: total adenine nucleotide pool (95% ATP) was 23 nmoles per mg of cell protein, of this amount 17 nmoles (74%) was found in cytoplasm, 5 nmoles (21%) in mitochondria and 1.3 nmoles of ADP (5% of total adenine nucleotides) was bound to the myofibrils [111]. These data are in good accordance with other estimations [33, 53, 113, 114]. Similar distribution of cellular ATP has also been found for other types of cells [111–113].

However, besides such a clear macrocompartmentation, there seems to be a very interesting phenomenon of extramitochondrial adenine nucleotide compartmentation in muscle cells. The evidence for that kind of compartmentation has been slowly accumulating and is basically functional, but it has already been discussed during more than thirty years.

Probably, first indications for compartmentation of cytoplasmic ATP can be found in the classical work by Infante and Davies when they solved the controversy about the roles of phosphocreatine and ATP in skeletal muscle contraction: they showed that in muscles in which the creatine kinase reaction was poisoned by fluorodinitrobenzene (FDNB), the contraction was related to the energetically equivalent decrease of ATP [115]. However, the muscle was able to respond to stimulation only 3–4 times and then lost its ability to contract, in contrast to the normal noninhibited muscle which contracted about 75–100 times up to the exhaustion of phosphocreatine [115, 116]. However, after 10–15 min of the resting period the inhibited muscle recovered and regained its ability to contract. It was concluded that this time was necessary for diffusion of new amount of ATP from cytoplasm to myofibrils and replacement of ADP which

cannot be rephosphorylated in myofibrils since creatine kinase was inhibited [115]. Some years later, a similar work was performed by Gercken and Schlette who used the same inhibitor of creatine kinase in studies of isolated perfused heart and found that heart contraction stopped when only about 10% of cellular ATP was used (ATP resynthesis in oxidative phosphorylation was also inhibited) [117]. The authors concluded that this 10% of cellular ATP represent a small pool of ATP with high turnover rate which is immediately accessible for contraction, and this pool is connected with other ATP pools via the creatine kinase reactions [117]. This important work, however, is not widely cited, may be because of the non-specificity of the inhibitory action of FDNB which may covalently bind SH and amino-groups of many enzymes and proteins including those involved in the excitation-contraction coupling. Much more references have been given to the work by Gudbjarnason et al. [118]. These authors were among the first to study in details the metabolic changes in the isolated heart in acute ischemia. They found that heart contraction decreased in parallel to decrease of phosphocreatine content and stopped when only 10% of cellular total ATP was used up. Their conclusion was identical to that made by Gercken and Schlette (see above) [117, 118]. More rapid decline of phosphocreatine at almost constant ATP level is not surprising since it is the consequence of the thermodynamics of the creatine kinase reaction and in concord with the equilibrium constant value K_{eq} = 166 in direction of ATP synthesis [72] but the surprising thing was rapid decline of contraction when almost all cellular ATP was present [118].

The importance of only a small pool of ATP for contraction was confirmed in the studies with the opposite approach when the major part of cytoplasmic ATP in isolated perfused rat hearts was depleted by hypoxic period of pre-perfusion in the experiments by Neely and Grotyohan [119] or when the cytoplasmic ATP was trapped in experiments by Hoerter et al. [120] and then by others [121] by deoxyglucose – it is phosphorylated but not converted further, but the adenine nucleotides are depleted since increased cellular ADP is first degraded by myokinase into AMP and then by 5 – nucleotidase and deaminases into inosine or adenosine which easily leave the cells [119, 120]. In all cases when the contractility and work performance were studied again under aerobic conditions or with glucose as substrate they were not changed even if the total ATP content was decreased by 70% (to 30% of its original value) – that means that when aerobic production of ATP in mitochondria was restored and phosphocreatine pathway of energy transport well operating, only small pools of ATP in cytoplasm which remained intact during this treatment were necessary to maintain high energy fluxes and the contractile function of the heart [119–121]. This is in concord with the finding by Gudbjarnason et al. that in the normoxic area of the heart with regional infarction cardiac contraction is normal at ATP content as low as 1.5–2 µmoles per g w.w. [118] and with the results of systematical studies by Buckberg group of biochemical changes in hearts during intraoperational ischemia in cardiac surgery when they achieved a good recovery of heart function at very low ATP content by using perfusion with cardioplegic solution to remove the catabolites [122].

These important works established firmly that the total tissue ATP content is in fact totally dissociated from the cardiac contractile function and energy fluxes [117–122].

Most directly the existence of cytoplasmic compartmentation of adenine nucleotides was shown first by McLellan and Vinegrad [123]. They studied the ATP-dependent rigor state release in chemically skinned cardiac fibers with well preserved mitochondria and found that there is an ADP pool in myofibrils which can be easily rephosphorylated into ATP by phospocreatine-myofibrillar creatine kinase but which is not accessible for mitochondria [123].

Independent line of further support for the idea of compartmentation of adenine nucleotides in cytoplasm came from the studies of substrate-dependence of the effect of ischemia (hypoxia) of different functions of cardiac cells. Most of these studies have been exhaustively analyzed in several reviews by Opie [124, 125]. In short translation, this group of data shows the following. In low flow ischemia the development of ischemic contracture of isolated perfused rat heart was delayed when glycolysis was active in hearts with inhibited acetate oxidation in mitochondria, and accelerated when glycolysis was inhibited and acetate used as substrate. The protective effect was directly dependent on the rate of glycolysis and not related to the total ATP content which was even lower in glucose-perfused hearts [126–128]. Kingsley et al. [129] studied the metabolic changes in heart during contracture development and found that the onset of contracture coincided with cessation of glycolysis as seen by the pH leveling off. Also, it was found that inhibition of glycolysis in the presence of pyruvate results in much worse functional recovery at the reperfusion than in control with non-inhibited glycolysis [130]

and in increased release of lactate dehydrogenase [126]. Weiss and Hiltbrand [131] showed that inhibition of oxidative phosphorylation in perfused rabbit septum caused a significant reduction of contractile activity but only a small increase of intracellular potassium; inhibition of glycolysis had an opposite effect. In development of this work, Weiss and Lamp have directly demonstrated by using patch-clamp technique in studies of permeabilized ventricular myocytes that glycolysis (and glycolytic substrates for ATP-producing steps) were more effective than oxidative phosphorylation in preventing ATP-sensitive channels for opening [132]. All these data were taken to show that 'glycolytic ATP' is a preferable substrate for membrane-linked cellular functions and prevention of contracture [124–132]. (Delayed ischemic contracture is primarily due to decreased ATP deficiency in the myofibrillar space and prevention of rigor complex formation, in some part it may be also due to prevention of Ca-overload resulting from increased intracellular Na and decreased Ca sequestering [133]).

Concerning the term 'glycolytic ATP', we may need to clarify the definition. Most probably, these are the pools (microcompartments) of ATP, spatially related to the myofibrillar and subsarcolemmal compartments and immediately accesible for corresponding ATPases and most rapidly replenished by the glycolytic complex located in the same area of the cell. Indeed, the connection of the glycolytic system to the cell membrane has been well described for erythrocytes [134], for smooth muscle cells (see Ishida, Paul and Wallimann in this volume), for sarcoplasmic reticulum and skeletal muscle triads [135]. The experiments of Weiss and Lamp demonstrate also the connection between key glycolytic enzymes and sarcolemmal membrane or adjacent cytoskeleton in the heart [132]. These membrane-bound glycolytic enzymes form a real microcompartment of ATP which is kinetically different from soluble ATP and does not exchange rapidly with the latter: this ATP was not accessible for added hexokinase and deoxyglucose, in contrast to ATP produced in mitochondria [132, 134, 135]. Recently, Vaughan et al. have shown by extraction of permeabilized cells that there is also a whole glycolytic complex directly connected by myofibrils [136] in accordance with earlier data by Masters et al. [77]. Kurganov has studied structural organization of glycolytic enzymes into one complex, which may contain tetrameric molecule of 6-phosphofructokinase and two molecules of each of other glycolytic enzymes but not hexokinase [83–86]. Through 6-phosphofructokinase this complex may be attached to the thin F-actin filaments in myofibrils, or to membrane proteins, such as 3-band protein of erythrocytes [83–86]. The complex may have a dynamic structure and dynamic equilibrium may exist between the complex and free enzymes. Structural connection of glycolytic complex to the sarcolemmal and myofibrillar ATP pools (multiple microcompartments) allow effective supply of ATP to these structures under critical conditions when mitochondrial energy production is ceased, but this complex has a very limited functional capacity and cannot maintain the cardiac contraction to the full extent under conditions of high energy requirements [124, 125]. Also, under normal aerobic conditions these microcompartments of ATP can be easily replenished at the expense of phosphocreatine due to both cytoplasmic creatine kinase and myofibrillar and sarcolemmal coupled creatine kinases [48, 137–139]. It is very well established that in heart and skeletal muscle MM creatine kinase is structurally bound to sarcolemma, sarcoplasmic reticulum, and located mostly in I-band space in myofibrils and functionally coupled to respective ATPases [137–139]. In fact, already in 1932 Clark had shown that aerobic heart contraction was not dependent on glycolysis, in contrast to the behavior of the anoxic heart [140]; very recently similar effect was described by Jeremy et al. (see Fig. 1 in ref 130) and in fact is well known.

In development of these ideas, the existence of rather deep cytoplasmic concentration gradients of ATP and other substances, including oxygen and sodium have been described for different types of cells [141, 142].

Jones et al. have studied the activities of ATP requiring enzymes with different subcellular localization in isolated hepatocytes in which average cellular ATP concentrations were varied. They have found that the activity of cytoplasmic ATP-sulfurylase varied linearly with the cellular ATP concentration but the plasma membrane Na,K-ATPase activity measured by Rb uptake was already zero at 40% of cellular ATP. These results were taken to indicate that ATP-utilizing enzymes located in the plasma membrane are exposed to a lower ATP concentration than are the enzymes in the cytosolic fluid surrounding the mitochondria due to heterogeneity of ATP distribution in the cytoplasm [141, 142]. Similar experimental data were obtained on cardiac cells in numerous studies of ATP-dependent (inhibited) potassium channels which are open in ischemic heart at much higher ATP levels (about 60% of initial normoxic level, around 2 μmoles per g w.w., or 4 mM in 'cytoplasmic water') than ATP level necessary for inhibition of these channels in inside-out isolated patches obtained from cardiomyocytes where the apparent Km was about 25–100 μM [143–146]. This result is in concord with the concept of subcellular compartmentation of ATP and existence of its concentration gradients, as it has been well argued in the work of Nicols and Lederer [146]. By comparison of inside-out patches and permeabilized cardiomyocytes they found that in the latter structures apparent Km for ATP for ATP-dependent potassium channels was increased 5 times as compared to patches,

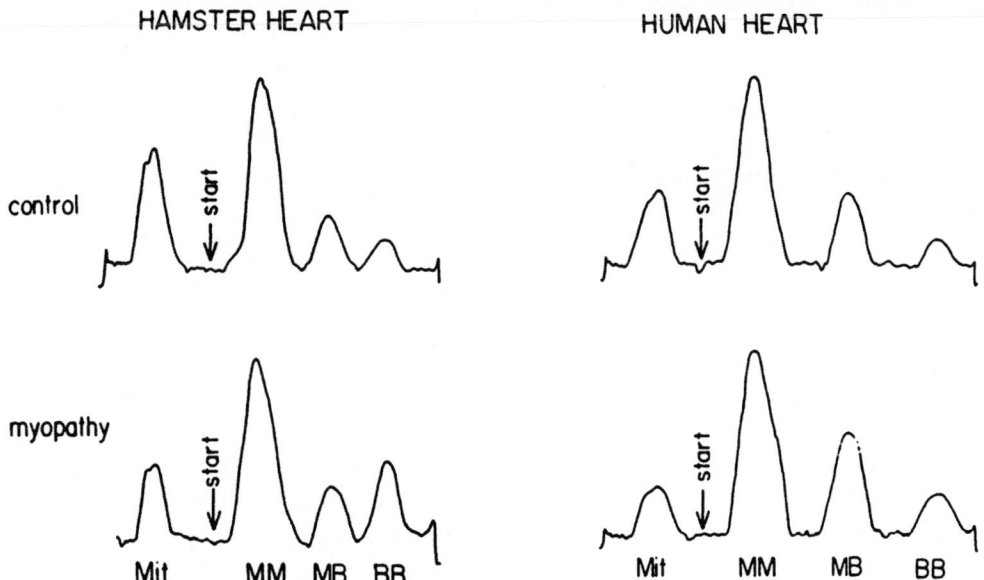

Fig. 1. Isoenzyme patterns of creatine kinase visualized after gel electrophoresis by activity staining. Hamsters: control-disease free golden hamsters; myopathy, cardiomyopathic Syrian hamsters, line CHF 146. Human heart: control was taken from a donor heart prepared for translation; myopathy, heart of 37 yr-old male recipient with diagnosis of dilated cardiomyopathy, functional class III. Mit.-mitochondrial isoenzyme; MM, MB and BB – respective creatine kinase isoenzymes. (From Ref. 182 with permission).

164

this being a result of diffusion problems and ADP formation inside the cells. However, this Km value was decreased by 10 mM phosphocreatine, which supports the production of ATP in the creatine kinase reaction from ADP at the sites of its formation; phosphocreatine also attenuated the K channel activity even in the absence of added ATP [146]. Most probably the subsarcolemmal and myofibrillar pools (microcompartments) of ATP (ADP) which are very slowly (if at all) exchanging with cytoplasmic ATP in bulk water phase are accessible not only to the glycolytic system, but also to the particulate creatine kinases [48, 147] which also may be built into the same microcompartments; that conforms to the definition given for 'glycolytic' ATP (see above). In connection to this problem, very interesting information is given by Muhlbach *et al.* in this issue: it appears that there are significant structural similarities between phosphoglycerate kinase and creatine (guanidino) kinases – it may be that that reflects their link to some common structural elements of the cell. In the already cited work Kingsley *et al.* [129] observed the onset of ischemic contracture at different high ATP levels and also concluded that ATP in cytosolic bulk (water) phase is not able to diffuse rapidly into myofibrillar compartment and replace ADP in it. This kind of evidence is increasingly accumulating [124].

We may also recall that already in 1964 Hodgkin described the preference of a phosphagen, argininephosphate, over ATP in supporting sodium efflux from the giant axon cells and already then, ahead of time, supposed the existence of membrane-connected microcompartment of adenine nucleotides related to the phosphagen pathway for intracellular energy transfer [148].

Concerning the intracellular compartmentation of adenine nucleotides, the interesting question is if it can be directly studied by phosphorus NMR method. According to several publications, comparison to the total tissue contents of ATP measured by this method and by biochemical analyses (or HPLC method) shows that there is practically no difference between these two groups of results in the normoxic heart for ATP, but some amount of ATP became NMR-invisible after ischemia [55, 92, 149, 150, but see also ref. 151 for different point of view]. However, the question can not be solved easily in full extent by this method because of rather low sensitivity of NMR method and its inability to differentiate between different exchangable compartments [6, 9, 152–156]; also, the degree of immobilization of ATP in different macro- and microcompartments is not known and therefore the interpretation of the NMR data of to-

tal ATP contents for answering this question is not clear [154–159]. Much better solution of the problem can be obtained if the magnetization transfer is measured [152–156, 159]. As a very good example we may refer to the work of Ingwall group who used magnetization transfer between phosphocreatine and ATP to determine the fluxes through creatine kinase reaction in the perfused hearts and found the experimental evidence for the fraction of ATP with NMR properties different from those of the soluble cytoplasmic ATP [154, 156]. To explain the double-exponential behavior of the magnetization transfer data, they have developed a mathematical model to describe the NMR-invisible ATP pool which in major part seems to be located in mitochondria, but localization of some of it in myofibrils is not excluded [156]. This model was used for discrimination of energy fluxes through mitochondrial creatine kinase from total flux, and it was found that the ATP synthesis rate was equal to the flux through mitochondrial creatine kinase, in good accordance with the *in vitro* data [154, see also below]. Thus, the concept of ATP compartmentation allowed the authors to develop a new effective tool of research by using NMR magnetization transfer method.

Finally, in a very recent publication Carmeliet and Niggli and Lipp have reviewed evidence for existence of calcium and sodium concentration gradients in the cardiomyocytes close to the sarcolemmal membrane, in so-called 'fuzzi subsarcolemmal space' or subcellular restricted spaces [160, 161]. The existence of these gradients was confirmed directly by X-ray microprobe analysis [162]. For calcium ions the diffusion restrictions have been known for long time and are considered to be the result of multiple binding sites [163].

The problem of intracellular compartmentation of metabolites is directly related to the problem of their intracellular diffusion. For phosphorus compounds (with exception for ADP) the intracellular diffusion coefficients have been determined by an isotopic method for skeletal muscle [163] and by a NMR pulse gradient method for perfused heart [164, 165]. In both cases the diffusion coefficients in the cells have been found to be only two times less than their values in diluted water solution [163–165]. However, if there are functionally important ATP pools (microcompartments) which do not exchange rapidly with cytoplasmic ATP in bulk water phase, these diffusion studies probably concern only the latter phase in the cells. However, this ATP compartment is not very important for the cell function (as mentioned above, it can be removed without too much changes in heart function). Interestingly enough, very similar situation has been found for intracellular sodium which may build up high subsarcolemmal gradients in spite of high experimentally determined diffusion coefficient [160].

This group of data makes much weaker the often used arguments that the diffusion length for creatine kinase substrates including ATP and ADP calculated on the basis of enzyme turnover time and *in vivo* diffusion coefficients is larger than distance between mitochondria and energy consuming structures [166, 167]; this argument will look even much weaker if we consider the new data of limited diffusion (permeability) of ADP through outer mitochondrial membrane (see next sections).

In summary, there are abundant experimental data showing the microheterogeneity of cytoplasmic space in muscle cell, in particular with respect to ATP (and ADP) due to the structural organization of the cell interior. Most probably, there exist multiple microcompartments of ATP(ADP) in cytoplasm as the result of formation of multienzyme complexes and liquid-crystal like structure of cytoplasm and because of existence of significant structured cell water phase. These microcompartments have been shown to exist in subsarcolemmal and myofibrillar space, but it is not excluded that the dynamic microcompartments exist also due to transient association of cytoplasmic enzymes including those of glycolytic system and creatine kinase [81–86, 103, 132, 134, 136]. One of important consequences of this is that the exchange (turnover) of ATP and ADP molecules within the microcompartments between adjacent enzymes may be much faster than their diffusion within the cell volume – this may have an important consequences for the metabolic feedback signal in the cardiac cells (see section IV).

Functional role of coupled creatine kinases in the cells

Heterogeneity of the creatine kinase system: intracellular distribution of isozymes in muscle cells

The 'classical', simplest point of view on the cellular metabolism as an homogenous system (see above) has long time been satisfied with an assumption that the creatine kinase reaction is in equilibrium and determines the cytoplasmic ADP concentration in muscle cells, according to the equilibrium constant and cytoplasmic concentrations of creatine, phosphocreatine, ATP and intracellular pH [72, 168]. Since all creatine kinase isoenzymes have very similar kinetic characteristics [169, 170], this equilibrium function requires only one of the isoenzymes in sufficiently high activity to be present in cells. Fractional extraction studies have shown that it is in fact true for the cytoplasmic compartment which in the skeletal muscle contains mostly MM isoenzyme; in the heart it contains half of cellular content of MM and additionally BB and MB isoenzymes in significantly lower activities; their relative content may change in pathological states [171–173]. However, the reality is that cellular creatine kinase activity is not limited to cytoplasmic compartment: another half of the cellular content of MM isoenzyme is structurally bound to myofibrils and cellular membranes – to the energy utilizing structures [171–175]. Most important for the regulation of cellular respiration is the existence of the specific mitochondrial creatine kinase isoenzyme which was discovered in 1964 by the Klingenberg group [176]; it has been shown to be encoded by two different nuclear genes (see the chapters by Payne and Strauss, Muhlbach *et al.* in this issue) and is localized exclusively on the inner mitochondrial membrane attached to its outer surface [177]. Figure 1 shows typical isoenzyme spectrum of total fraction of creatine kinase from hamster and human heart, both in normal tissue and cardiomyopathy. The creatine kinase isoenzyme spectrum changes very significantly in pathological states and also during pre- and postnatal development (see the chapters by Hoerter and Ventura-Clapier, Muhlbach *et al.* and Strauss and Payne in this issue). According to the results of quantitative electrophoretic analysis, in human heart as well as in rat, dog, guinea pig and hamsters cardiac muscle mitochondrial isoenzyme represents 21–45% of total cellular activity of creatine kinase [171, 172]. (The content of mitochondrial creatine kinase in hearts equal to 6–10% of total [178] may be some underestimation, probably because of very quick oxidative inactivation during electrophoresis). It is in slow skeletal muscle that it makes up about 6% of total activity because of three times higher activity of MM isoenzyme in cytoplasm [173]. Different relative distribution of creatine kinase isoenzymes in the heart and skeletal muscle is related to different metabolic patterns, intracellular volume characteristics and results in different metabolic changes of high energy phosphates during contraction [57, 173–175]. In skeletal muscle both glycolytic and creatine kinase activities are several times higher than in heart and during contraction PCr changes very significantly because of its slower production in mitochondria, just because of their lower amount in the cells [70, 173–175]. However, the mitochondrial creatine

Table 2. Characteristics of creatine kinase and related enzymes in subcellular fractions.

Source	Creatine kinase		ANT nmoles/mg prot.	cyt.aa3 nmoles/mg prot.	Ref.
	IU/mg prot.	nmoles/mg prot.			
A. Mitochondria:					
1. Heart			2.1 – 3.0 (90 µg/mg)	0.3 – 0.45	184, 62
Rat	4.5 ± 0.5	1.04 ± 0.1	2.54 ± 0.1	0.445 ± 0.09	184
Rabbit	3.0 ± 0.5	0.86 ± 0.1	1.94 ± 0.08	0.353 ± 0.09	184
Dog	4.1 ± 0.3	0.98 ± 0.08	2.46 ± 0.08	0.44 ± 0.05	184
Chicken	0.86 ± 0.2	0.92 – 1.1	2.0 – 2.6	0.354 ± 0.006	184, 263
2. Skeletal muscle (soleus)	2.7 ± 0.5	–	–	–	173, 174, 264
3. Smooth muscle (uterus)	0.2 – 0.8	–	–	–	265
B. Sarcoplasmic reticulum:					
1. Heart	0.63 ± 0.08	–	–	–	266
2. Skeletal muscle	2.05 ± 0.35 (37° C)	–	–	–	267
C. Myofibrils					
1. Rat heart	0.41 ± 0.63	–	–	–	268
	0.99 ± 0.12				224
D. Sarcolemma:					
1. Rat heart	0.64 ± 0.14				269

Reaction rates are determined in direction of ATP formation under standard conditions.

kinase content in skeletal muscle increases during endurance training [179, 180]. In heart, the mitochondrial creatine kinase drops very significantly in cardiomyopathy, when the B subunit is elevated [181, 183]. Specific activities of creatine kinase in cardiac and skeletal muscle mitochondria are the same, as it is indicated in Table 2 which summarizes the characteristics of mitochondrial isoenzyme of creatine kinase in the cell and in isolated mitochondria. The specific activities of MM creatine kinase in other subcellular structures in the heart are also indicated in Table 2.

Molar content of creatine kinase in mitochondria has been studied in our laboratory by using different experimental methods, including immunochemical, specific SH group determination and oxy-[³H] ADP method [183–185]. The results agree with each other and show that the molar content of creatine kinase monomer in the heart mitochondria is in the range of 0.7–1.0 nmoles per mg of mitochondrial protein. Since the molecular mass of a monomer is 42 kD [183], this amount corresponds to creatine kinase protein content of 29–42 µg per mg of mitochondrial protein, or 2.9–4.2% of the latter. If we take the specific activity of purified mitochondrial creatine kinase to be 100 IU/mg (our unpublished results), we get the specific activity of creatine kinase in mitochondria in the range of 2.8–4 IU/mg that fits the data of Table 2.

The localization of creatine kinase in mitochondria has been studied in details in the earlier work by Scholte in 1974 [177], by electron microscopic histochemical and immunochemical investigations by Baba *et al.* [186], Sharov *et al.* [147] and Wallimann group [187] and in Carafoli's laboratory [188, 189]. All these studies, very consistently with each other, show homogenous distribution of the creatine kinase along the cristae membrane (for an exception see 190). After sonication of mitochondria to produce inside-out submitochondrial particles and remove the outer membrane, the creatine kinase activity became latent and could be revealed after solubilization of mitochondrial membrane [177]. On the basis of these data Scholte proposed to use creatine kinase activity for determination of sidedness of the mitochondrial membrane as a marker enzyme of the outer surface of inner mitochondrial membrane [177]. Cheneval and Carafoli *et al.* have shown that positively charged creatine kinase is fixed to negatively charged cardiolipin moiety in the inner mitochondrial membrane with participation of positively charged Arg 19, Lys 20 and His 21 residues in the enzyme molecule [189]. The same cardiolipin moiety surrounds also the ATP-ADP translocase

[191], connecting these proteins together into one complex.

Functional coupling of mitochondrial creatine kinase to ATP-ADP translocase and mitochondrial oxidative phosphorylation: review of evidence

The first study of the functional role of mitochondrial creatine kinase was published by Bessmann and Fonyo in 1966 in which they showed the ability of creatine to exert effective acceptor control of respiration of mitochondria due to production of ADP from mitochondrial ATP [192]. This work was followed by studies by Vial *et al.* in 1972 [193], Jacobus and Lehninger in 1973 [194] and Saks *et al.* in 1974 [172], all describing the functional role of mitochondrial creatine kinase in production of phosphocreatine (PCr) from mitochondrial ATP with overall PCr/O = ADP/O close to 3.0, and the common conclusion was that in fact not ATP but PCr is the end-product of mitochondrial reactions of high energy phosphate production under physiological conditions – in particular, in the presence of creatine. The mechanism of this effective high energy phosphate conversion process was studied in more details in our laboratory in 1975 by kinetic methods using also mathematical modelling [169]. The conclusion in this work was that the mitochondrial ATP is much more effectively (preferentially) used for PCr synthesis by mitochondrial creatine kinase than ATP added into the surrounding medium [169, 195]. It was proposed for explanation of this phenomenon that there exists a tight functional coupling between ATP-ADP translocase and creatine kinase with direct channelling of ATP to creatine kinase in exchange for ADP produced from ATP after creatine phosphorylation [169, 195]. This mechanism compensates for unfavourable thermodynamic parameters of the creatine kinase reaction for phosphocreatine production [169]. Also, the kinetic properties of the mitochondrial creatine kinase reaction do not differ from those of other isoenzymes and are not in favor of phosphocreatine production [169, 179]. It was supposed that in mitochondria these unfavourable properties of the enzyme are superimposed by its interaction with ATP-ADP translocase which controls the creatine kinase microenvironment by creating very high local ATP and low ADP concentrations, and in this way increases the free energy (just by changing the concentrations) of ATP system in the microcompartment around creatine kinase to a level much higher than necessary for PCr production. An extreme case

might be that there is direct exchange (channelling) of ATP and ADP molecules that makes it not necessary to consider the energetics in microcompartment; in both cases the role of ATP and ADP is reduced to that of necessary cofactors in the overall reaction of PCr production coupled to the oxidative phosphorylation. Kinetically, by saturating creatine kinase by a substrate ATP and removing the product ADP, translocase drives the creatine kinase reaction unidirectionally towards phosphocreatine production [169, 195]. On the other hand, when creatine kinase reaction is running under steady state conditions, it quickly utilizes ATP supplied by translocase and converts it into ADP, releasing the translocase from any kind of possible inhibition by high concentrations of ATP. This was the hypothesis in 1975 to explain the unexpectedly high rates of phosphocreatine production in heart mitochondria [169]. In further studies the proposed tight functional coupling between ATP-ADP translocase and creatine kinase and a preferential use of mitochondrial ATP for PCr production was experimentally confirmed in many laboratories by radioactive isotopic methods [196–198], by competetive enzyme methods using pyruvate kinase to trap ADP [199, 200] and by detailed biochemical and kinetic studies [201–209]. These works have already been reviewed by several authors [208, 210–212]. This direction of studies resulted, in particular, in developing the mathematical model for PCr production based on the direct substrate (ATP and ADP) channelling [ref. 213, see also the chapter VI-2 by Aliev and Saks in this issue]. Given below are only several selected results of these experimental works which are important for understanding the possible mechanism of regulation of respiration *in vivo*.

Figure 2 reproduces the results obtained in studies of isolated heart mitochondria. It shows the dependence of the rate of phosphocreatine production on the MgATP concentration in the absence and in the presence of the oxidative phosphorylation. In the former case, the dependence was rather low and gave an apparent Km for ATP around 0.3 mM, in accordance with the kinetic parameters of the enzyme [204]. However, when oxidative phosphorylation was an additional source of ATP in the medium, the reaction kinetics was significantly changed: at any concentration of ATP the reaction rate was significantly enhanced and the apparent Km value was decreased by an order of magnitude, to 14–40 μM [169, 202, 204]. The value of Vmax was exactly the same in both cases and equal to 1,0 μmole per min per mg of protein, that is equal to the maximal rate of ATP production in oxidative phosphorylation under these conditions [48,

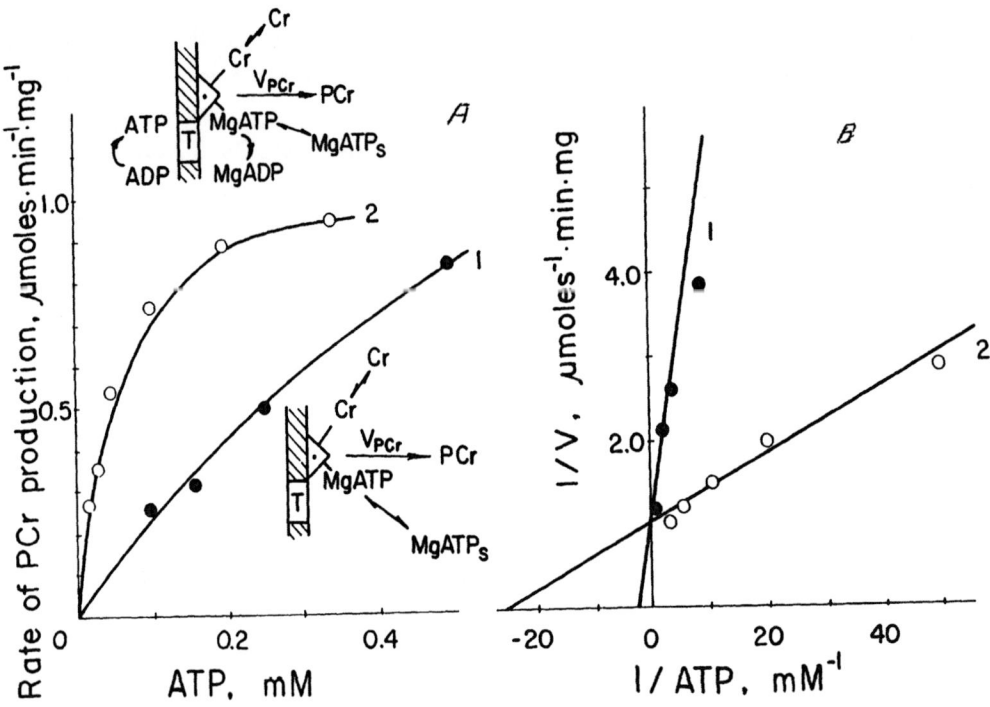

Fig. 2. (A) The dependence of the rate of phosphocreatine production in the creatine kinase reaction in isolated rat heart mitochondria on the ATP concentration in the medium in the absence (curve 1) and in the presence (curve 2) of oxidative phosphorylation. Creatine concentration was 25 mM, oxidative phosphorylation (curve 1) was inhibited by oligomycin (5 µg/mg of protein). Other experimental conditions as described in ref. 48. (From Ref. 48 with permission). (B) Linearization of data from Fig. 2A in double-reciprocal plots. Numbers as in Fig. 2A.

202]. In more details this kinetic effect is considered in several publications [202–208], but the data in Fig. 2 demonstrate very clearly the amplifying effect of coupling of oxidative phosphorylation and ATP-ADP translocation on the mitochondrial creatine kinase reaction of phosphocreatine production: due to rapid turnover of adenine nucleotides in these coupled reactions without significant release into the medium, high reaction rates are achieved at very low ATP concentration in the medium [48]. Very similar results were reported by Kim and Lee [209] who achieved rather significant rates of phosphocreatine synthesis under conditions of oxidative phosphorylation in pig mitochondria even without any added adenine nucleotides. It will be shown in the next section that due to this tight functional coupling which results in high turnover of ATP and ADP in mitochondria (Fig. 2), the unidirectional creatine kinase reaction may exert, on the other hand, very strong amplyfing effect on oxidative phosphorylation at very low ADP concentrations (fluxes) *in vivo.*

One of the most interesting questions regarding this effect is that of its structural basis. One of the possible solutions of this question is given by Walliman *et al.* [187, 210, 211, see the chapter by Schneyder *et al.* in this issue] who has found evidence for an assumption that the octameric form of creatine kinase is attached to the tetrameric form of translocase in the membrane-bound multienzyme complex probably involving also the special porin channel (see below) in the outer membrane. In this complex ATP is directly entering the channel inside the octameric creatine kinase molecule with active sites inside this channel, and then in the presence of creatine there is no way for ATP to escape its conversion into ADP (which goes back to translocase) and to phosphocreatine which leaves the complex for cytoplasm. The only problem with this beautiful picture is that it requires the stoichiometric ratio of creatine kinase to translocase to be at least 8:4 = 2:1. The translocase makes up about 9–10% of mitochondrial protein [107–110], against 3–4% accounted for by creatine kinase (Table 2 shows, see above). Since the molecular masses of creatine kinase and translocase monomers are close and equal to 40–42 and 38, 87 kD, correspondingly [107–110, 184, 210, 211], their molar ratio is most probably 1:2 in heart mitochondria. In fact Kuznetsov *et al.* have found earlier by titration with oxy-[³H]ADP that number of creatine kinase active sites (located on each monomer) is equal to number of atractyloside – sensitive ADP binding sites of translocase in mitochondrial membrane, representative of number of dimers, since translocase carries one binding site per dimer [108, 110, 184].

At this point, we want to reproduce the results ob-

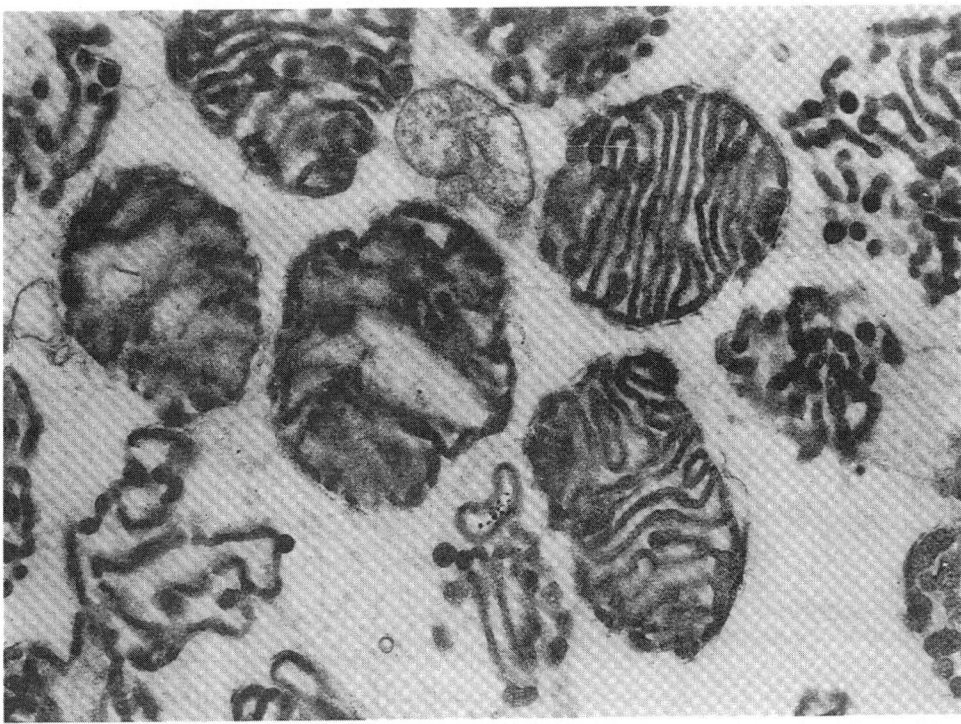

Fig. 3. Electron micrograph of the mitoplast preparation obtained from rat heart mitochondria by a digitonin procedure [206]. Note the high extent of outer membrane removal. Magnification 55 000 ×.

tained in our laboratory which, from our point of view, should be accounted for in any kind of speculations on this topic.

These results were obtained in the studies of mitoplasts – the mitochondrial inner membrane – matrix preparations after removal o membrane by digitonon treatment (120 μg per mg of mitochondrial protein for 8 min at 0° C [207]). The resulting membrane preparation is shown in Fig. 3, it is practically deprived of outer membrane but retains all creatine kinase activity which stays attached to the membrane even at the high ionic strength of physiological salt solution when the damaging effect of chloride ion is avoided and this creatine kinase is still functionally tightly coupled to oxidative phosphorylation via translocase [203, 206]. The complete removal of the outer membrane as a barrier is illustrated in Fig. 4 which shows that the antibodies against mitochondrial creatine kinase (molecular mass 150 kD) have a good access to the enzyme and inhibit it completely. The very important consequence of this inhibition is that it results in strong inhibition of oxidative phosphorylation of added ADP to similar extent as of creatine kinase (Fig. 4). Direct measurements with labelled [³H]ADP have shown that the step of ADP-ATP translocation is inhibited under these conditions [207]. It has also been shown that when the non-inhibitory anti-

bodies against mitochondrial creatine kinase were used which bind to the epitope far from the active site of the enzyme, the translocase was not inhibited at all [207]. Interpretation of these data is given by the scheme in Fig. 4C: the translocase in the membrane and creatine kinase at the membrane are structurally close to each other, the distance between them being less than molecular size of antibodies (about 16 nm [207]). One possible way of their interaction may also include frequent collisions due to lateral diffusion [214]. Taking into account the data by Cheneval and Carafoli [188, 189], one may suppose that most probably cardiolipin moiety in the inner mitochondrial membrane surrounding translocase acts to clue it with creatine kinase, but some additional factor of structural orientation in membrane may be necessary [207, 215]. Important information from this type of experiments is as following: all population of translocase is structurally closely related to the creatine kinase with rather precise orientation of active sites of enzyme and nucleotide binding site of the translocase. The best way to accommodate these data with the stoichiometric ratio 1:2 between creatine kinase and translocase monomers is to suppose that dimeric molecule of creatine kinase is structurally and functionally related to the tetramer of translocase [207, 216] (see Fig. 4D). If creatine kinase or some part of it is attached to the mem-

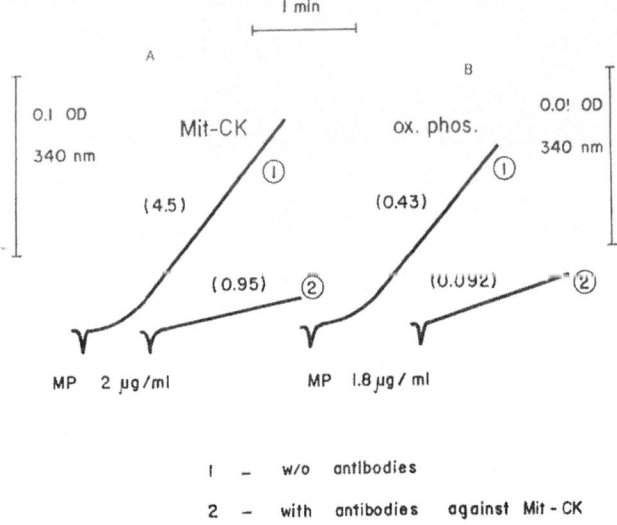

Fig. 4. (A) Recordings of the enzymatic activities of the rat heart mitochondrial inner membrane + matrix (mitoplast) preparation. The effect of antibodies (Ab) against mitochondrial creatine kinase (Mit.-CK). Antibodies against rat mitochondrial creatine kinase were produced in chicken. Activities were recorded spectrophotometrically by using a hexokinase-glucose 6-phosphate dehydrogenase coupled system [207]. A. Creatine kinase activity without (1) and with (2) antibodies in ratio 40:1 with respect to mitoplast protein. B. Oxidative phosphorylation, conditions are the same as in Fig. 4A. (From Ref. 207 with permission).
(C) Titration curve of creatine kinase reaction (solid line) and oxidative phosphorylation (dotted line) with chicken antibodies against rat heart mitochondrial creatine kinase. Experiments were performed as described in Fig. 4.
(C) Schematic presentation of the structure of inner membrane of heart mitochondria showing possible spatial relationship between adenine nucleotide translocase, T, and creatine kinase, CK. Adenine nucleotide translocase is associated with cardiolipin, which, in turn, as a glue, binds mitochondrial creatine kinase. Ab-CK-antibodies against mitochondrial creatine kinase. Possible association of dimeric creatine kinase with a tetramer of translocase is indicated on the basis of stoichiometric studies (see the text°).

brane in the octameric form (see Walliman *et al.* [187, 210, 211]), the inhibition curves shown in Fig. 4 may be explained only if we assume that some part of translocase population is inactive. In this case the inhibition of oxidative phosphorylation by creatine kinase antibodies after binding to the creatine kinase octamers is predictable. Thus, most probably the precise structure of the translocase-creatine kinase complex in heart, skeletal muscle and brain mitochondria awaits further work for its solution.

Independently from its solution, it is already clear that this complex forms the microcompartment of adenine nucleotides between translocase and creatine kinase which governs the behavior of both of them. In the case of octameric form of creatine kinase attached to the membrane this microcompartment is built up by the channel inside the octamer [210, 211]. In the case of dimer there should be direct close spatial orientation of creatine kinase active center and ATP,ADP-binding site of translocase (Fig. 4), resulting in direct exchange of ATP and ADP, the efficiency of such interaction being increased possibly due to frequent collisions of these proteins as a result of lateral diffusion, or including some kind of 'bottleneck' effect. As already referred to in several papers of this issue, Fossel and Hoefeler [212] have described a functional metabolic compartment created by covalent being of creatine kinase and hexokinase on Sepharose beads in which hexokinase has a preferential access to ATP produced by creatine kinase when the average distance between enzymes does not exceed 10 nm, most probably because of decreased diffusion distance for substrates. Interestingly enough, this size is comparable with the size of antibodies against mitochondrial creatine kinase – 16 nm, which block simulateneously creatine kinase and translocase after binding to the former (see Fig. 4). Thus, close spatial proximity of creatine kinase and translocase may be already enough to create a microcompartment of adenine nucleotides with effectively reduced diffusion distance and increased turnover at the outer surface of inner mitochondrial membrane. The existence of this microcompartment is also clearly seen from Fig. 6 of chapter II-2 by Gellerich *et al.*, this issue. Although not commented by authors, it is shown in that figure that the presence of excessive amount of pyruvate kinase decreased creatine kinase stimulated respiration of mitoplast preparation not more than 40%, that meaning that 60% of creatine kinase produced ADP is not accessible to pyruvate kinase in spite of the absence of the outer membrane. The pyruvate kinase system completely suppresses, however, hexokinase stimulated respiration [199, 200]. In addition to such a microcompartment of ATP (ADP) at the membrane, outer membrane may build up some dynamic, rate dependent compartment of ATP in the intermembrane space (see below and Gellerich *et al.*, this issue).

The existence of microcompartment on the mitochondrial membrane is seen from the data shown in Fig. 5. Figure 5 shows the results of the analysis of the composition of the medium surrounding rat heart mitoplasts respiring in the presence of creatine [206]. Initially the reaction mixture was made up to contain substrates and products in such concentrations that the creatine kinase mass action r = [PCr] × [ADP]/[Cr] × [ATP] exceeded the equilibrium constant (Keq = 1/166 = 0.006 [72, 168]) for the reaction; if the thermodynamics of the medium governs the whole system, further production of PCr should be impossible and the system should approach the equilibrium. However, what we see is just opposite: the mass action ratio went farer from the equilibrium constant value due to PCr production supported by oxidative phosphorylation via adenine nucleotide translocase, and only when oxidative phosphorylation was inhibited by addition of oligomycin, the reaction changed its direction and started to approach the equilibrium (Fig. 5). Thus, when coupled to oxidative phosphorylation, the mitochondrial creatine kinase reaction was driven out of equilibrium unidirectionally towards phosphocreatine production, the overall rate of reactions being close to Vmax for creatine kinase (all enzyme molecules are involved) [48]. These results were recently quantitatively reproduced by Soboll *et al.* (see chapter II-3) who also showed that if the membrane attached

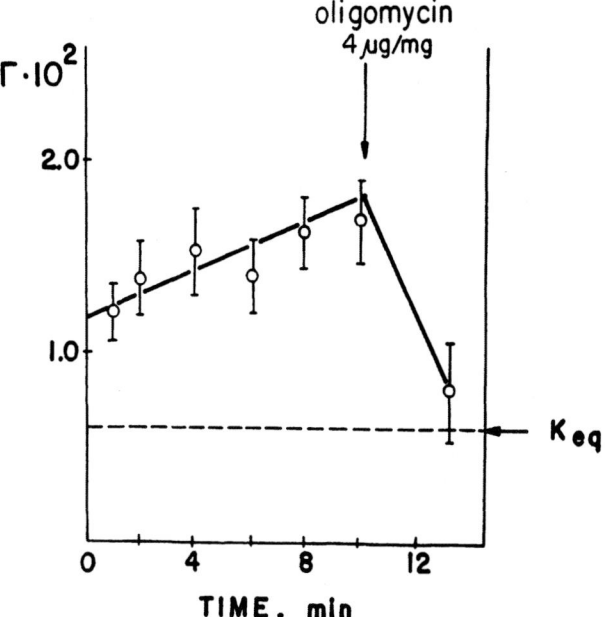

Fig. 5. Changes of the mass action ratio of the creatine kinase reaction under conditions of oxidative phosphorylation in rat heart mitoplast preparations. Initial composition of reaction mixture was: PCr 4 mM, 0,12 mM ATP, 0,05 mM ADP, 40 mM creatine. Keq = 1/166 = 0,006. (From ref. 206 with permission).

complex in intact mitochondria is dissociated in the presence of 20 mM phosphate, such a thermodynamic coupling disappears [217]. This effect is also quantitatively described by a mathematical model based on an assumption of direct substrate channeling between ADP-ATP translocase and creatine kinase (see the chapter VI-2 by Aliev & Saks, this issue). Earlier it was shown that kinetic coupling is lost in intact mitochondria in KCl medium when creatine kinase is detached from membrane but stay in the intermembrane space [203].

The conclusion from all of these works for the main problem which is under discussion in this study – regulation of cellular respiration – is that under experimental conditions simulating physiological situation in the cells (respiration in the presence of creatine) the thermodynamic parameters of the reaction medium do not have any influence and control over rate of coupled oxidative phosphorylation-creatine kinase, that means that they do not control the main process of phosphocreatine production. If we extrapolate this conclusion to the conditions *in vivo,* we may say that the processes in cytoplasmic compartment do not have any direct and simple thermodynamic control over mitochondrial respiration when it is coupled to the creatine kinase reaction. In the cells it obviously is coupled, and therefore the respiration is controlled by multienzyme complex at the inner mitochondrial membrane including creatine kinase, and

172

the value of direct supply (flux) of ADP available for translocase and thus for phosphorylation is determined inside this complex and by kinetic factors in cytoplasm, such as creatine concentration etc. We may propose that this complex is a receiver unit for the metabolic signal from cytoplasm given by changing creatine and ADP concentrations and transforming it into ADP flux available to translocase. The next section shows that it is most probably a true conclusion, the validity of which is increased by a decreased permeability of the outer mitochondrial membrane for ADP inside the intact cells.

Because of these coupled reactions, which control oxidative phosphorylation in the cells, a linear model of muscle respiration proposed by Ronald A. Meyer [73] in which the equilibrium creatine kinase is modelled as a capacitance seems to be only of limited value, lacking the description of above indicated coupled reactions which clearly disconnect system of oxidative phosphorylation from cytosolic phosphorylation potential. Cytosolic phosphorylation potential and work-energy cost transfer functions of Chance (see M. Osbakken, this issue) may have, however, some value for estimating the energy status of the cytoplasmic bulk phase indicating the general balance between processes of energy production and utilization. For better understanding of regulation of respiration in skeletal muscle it seems to be very worthwhile to come back to the work by Michael Mahler, who concluded from the analysis of monoexponential changes in oxygen consumption and PCr (and creatine) contents that mitochondrial creatine kinase is out of equilibrium in the cells *in vivo* [218]. Pseudo-first order of creatine kinase reaction *in vivo* [218] is easily explained by coupled reactions in mitochondria regulated by cytoplasmic creatine (see above). Probably, for more precise quantitative description of control of respiration on the basis of theoretical considerations given by Kholodenko *et al.*, chapter VI-1 will be helpful.

There is one physiological phenomenon not too often mentioned that seems to conform well to this conclusion made in experiments *in vitro* – this is the phosphocreatine overshoot phenomenon at the reperfusion after short period of ischemia [219–222]. In the best way this phenomenon was described by Flaherty *et al.* for heart [219] and this phenomenon is also seen in smooth muscle (see Ishida *et al.*, this issue). During ischemia ATP concent was decreased close to half of normal value, and it is known that ADP level is transitorily elevated [219–222]. Under these conditions, according to equilibrium expression PCr/Cr = ATP/(ADP·Keq), the PCr/Cr at reperfusion should not exceed 1/2 of its normal value – but

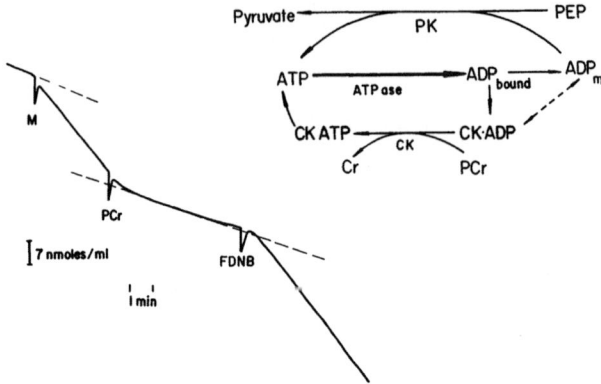

Fig. 6. Functional compartmentation of adenine nucleotides in myofilaments of skinned muscles due to particulate creatine kinase. MgADP release was recorded fluorimetrically by a coupled pyruvate kinase-phosphoenol pyruvate system. In right upper corner the scheme of reactions in the presence of phosphocreatine is shown. M-addition of skinned muscle fiber into the cell; FDNB-fluorodinitrobenzene.

it was found to get close to 200% of normal [219]. Recently it was found that during mild (low flow) ischemia simulating 'hibernating' myocardium phosphocreatine content rises spontaneously after rapid initial drop at the background of decreased ATP [223]. Both of these results are easily explained if we consider phosphocreatine synthesis to be unidirectional in mitochondria and geared to oxidative phosphorylation via translocase (see above). In this case these results simply mean that mitochondrial system of PCr production is better preserved or function at higher rate than the system of its utilization during ischemia.

Symmetrical conclusion comes out from the works in which the myofibrillar end of the creatine kinase and the regulation of myofibrillar contraction were studied. These works are reviewed by R. Ventura-Clapier, V. Veksler and J. Hoerter, chapter II-5, in this issue and they lead us to the conclusion that the cytoplasmic compartment does not control the events in the myofibrillar space either, which are controlled by the processes in the multienzyme complexes in this space. Very simple (easy exercise for students) experimental evidence for this conclusion is given in Fig. 6. This result can be obtained by using Triton treated myocardial fibers or isolated myofibrils with maximal activity of creatine kinase [224]. This Fig. 6 shows the spectrophotometric recording of the myofibrillar ATPase activity by a coupled enzyme system consisting of lactate dehydrogenase to reduce pyruvate at the expense of NADH (which concentration changes are recorded) and pyruvate kinase in high activity to trap ADP (see the scheme in upper right corner). In spite of the fact that the activity of pyruvate kinase exceeded that of endogenous creatine kinase

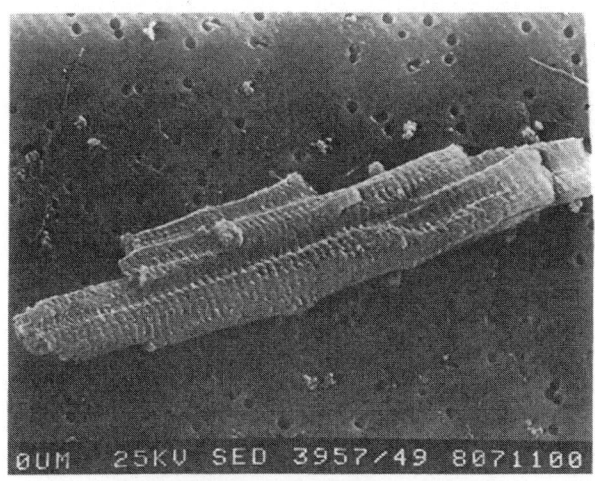

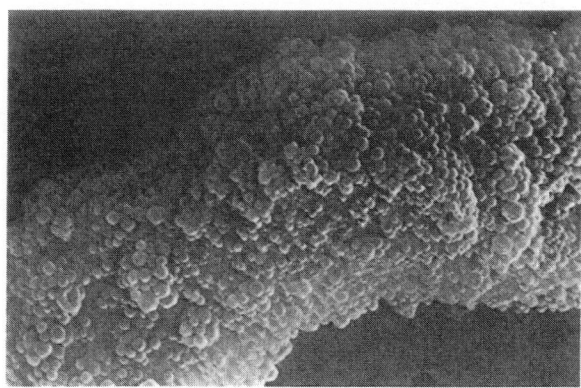

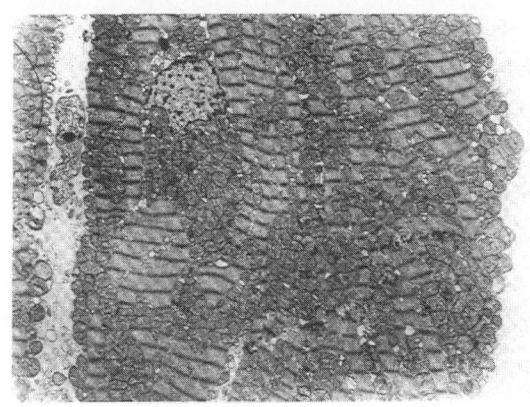

Fig. 7. (A) Scanning electron microscopy of the intact isolated rat cardiomyocytes, magnification 1100 ×. (B) The same as A, after treatment with saponin and hypercontracted in the presence of 1 mM calcium. Note complete dissolution of sarcolemma and appearance of subsarcolemmal mitochondria. 2500 ×. (C) Transmission electron micrograph of saponin-skinned cardiac fibers in physiological salt solution. Complete dissolution and vesicularization of sarcolemma (after saponin removal) is seen. Magnification 5280 ×. (Reproduced from Ref. 182 and 231 with permission).

about 20 times, simple addition of PCr to activate creatine kinase completely ceased the reaction of NADH oxidation due to ADP rephosphorylation inside the myofibrils [224]. When creatine kinase was inhibited by FDNB, ADP outflow was restored completely. This result means that ADP produced in contraction is immediately rephosphorylated within the myofibrillar complex including creatine kinase, and the metabolic result of these processes is formation of creatine.

This result is in very good concord with classical physiological data of direct connection of contraction with creatine production at the expense of PCr [118, 119, 225]. What is new and interesting is that for this classical phenomenon to occur the myofibrillar creatine kinase alone is sufficient because of its effective interaction with myofibrillar ATPase and the cytoplasmic creatine kinase is not necessary for that purpose.

Retarded diffusion of ADP in cardiomyocytes: outer mitochondrial membrane as a permeability barrier

The crucial question we always have to ask is: to which extent we may use the experimental data collected *in vitro* for explanation of cellular mechanisms in the cells *in vivo*? To answer this question, one needs information of the system behavior *in vivo*. One of the possible ways of investigations of the cellular systems *in vivo* is the use of very popular now but rather expensive NMR methods [3, 7–11, 15, 17, 18, 21, 38–45, 54, 55, 57, 60, 62, 130, 149–156, 164–166, 226, 227]. However, in some cases it is much more informative to use more simple methods such as permeabilized cell technique (obviously, the best way is their combined use). In the latter case isolated intact cardiomyocytes (Fig. 7A) are treated with low concentration of saponin under mild conditions (40 µg per ml, 30 min at 4° C) to dissolve very selectively the sarcolemma (Fig. 7B); under these conditions all soluble enzymes leave the cell but the subcellular structures stay intact [182, 228–233]. Similar procedure may be performed with small (3–5 mg) bundles of myocardial fibers without isolation of cardiomyocytes (Fig. 7C), that is very convenient in the studies of biopsy material taken from patients [182]. In this way we can get access to subcellular structures inside the cells in their more or less normal milieu and obtain necessary information of their

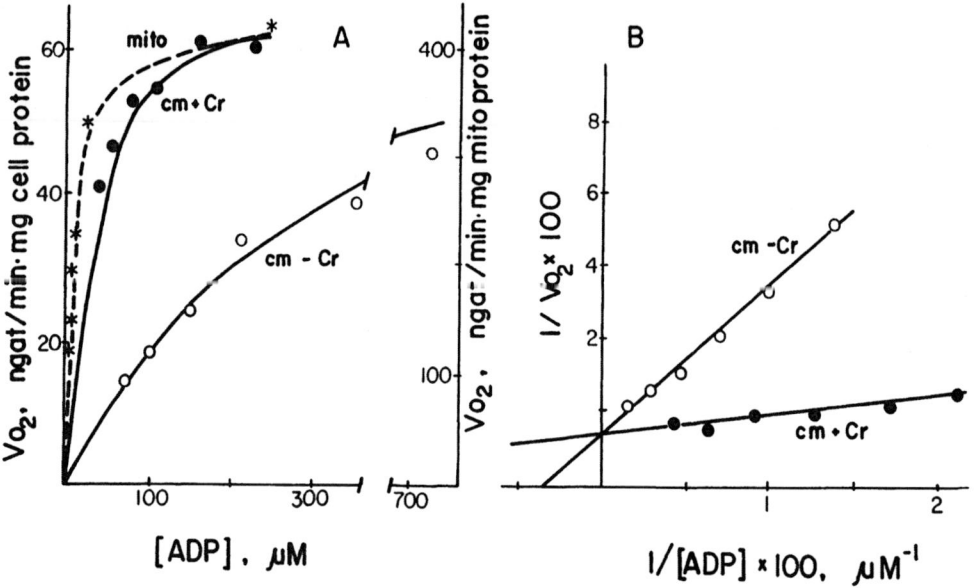

Fig. 8. (A) Dependences of the respiration rate of skinned cardiomyocytes – (left ordinate axis) and isolated rat heart mitochondria (right ordinate axis) on ADP concentration in medium. Dotted line – mitochondria (mito); solid lines: open circles – cardiomyocytes (cm) in the absence of creatine, closed circles – cardiomyocytes in the presence of 25 mM creatine. (B) Linearization of dependences for cardiomyocytes in double reciprocal plots. (Reproduced from Ref. 182 with permission).

behavior in this natural surrounding. One of the problems which may be investigated in this way is diffusion of substrates inside the cell. The use of this method has already shown the existence of some diffusion problems for ATP (see above) and its use has given new information about diffusion of ADP which cannot be studied by NMR method because of its low cellular content. The latter problem became apparent first in a study of oxygen consumption by permeabilized cardiomyocytes by Kummel [228] and was recently studied further in our laboratory in details resulting in some surprises [182, 229, 231]. The main results of these studies are shown in Fig. 8. This Figure shows the dependence of the respiration by isolated heart mitochondria (dotted line) and permeabilized cardiomyocytes (solid lines) on the concentration of ADP in the medium. As it was already described by Lardy and Chance *et al.* in 1953 [25–27], isolated mitochondria have very high affinity to ADP, the value of apparent Km is 17 µM (Fig. 8A). However, permeabilized cardiomyocytes from which cytoplasmic enzymes were released showed very slow dependence of respiration on the ADP concentration, and it was necessary to add millimolar concentration of ADP to achieve maximal respiration (Fig. 8A); the value of apparent Km was equal to about 300 µM (Fig. 8B). However, when these experiments were performed in the presence of creatine (25 mM), the sensitivity to ADP was increased again and the apparent Km value returned to 36 µM (Fig. 8A and B). Special experiments showed

that there was no difference for apparent Km for creatine between isolated mitochondria and cardiomyocytes (or fibers) [182, 229, 231]. Thus, these results show that there are significant specific problems for diffusion of ADP which is retarded in cardiac myocytes and these difficulties are overcome, or diminished in the presence of creatine. Exactly the same kinetic parameters were obtained in the experiments with skinned fibers [229, 233] – that means that the diffusion of ADP is retarded only inside the cells. The effect of creatine was no mystery taking into account the data shown in Fig. 2 and all the information we have collected in last two decades on the functional coupling between ATP-ADP translocase and creatine kinase: since in heart mitochondria oxidative phosphorylation is tightly coupled to the creatine kinase reaction, the activation of the latter in the presence of creatine increases the turnover of adenine nucleotides in the coupled reactions and by this means amplifies the response of mitochondria to small amount of ADP which may reach them at the low ADP concentration in the medium. The puzzling problem was: what makes the ADP diffusion so slow in the cardiomyocytes? The possible influence of mitochondrial clustering [142] was excluded, since in this case creatine should not activate the reaction, and this activation disappeared in KCl solution that showing that the intact structure of functionally coupled complex is important [233]. Two possible mechanisms were considered: intermediate binding of ADP to the myofibrillar structures or its

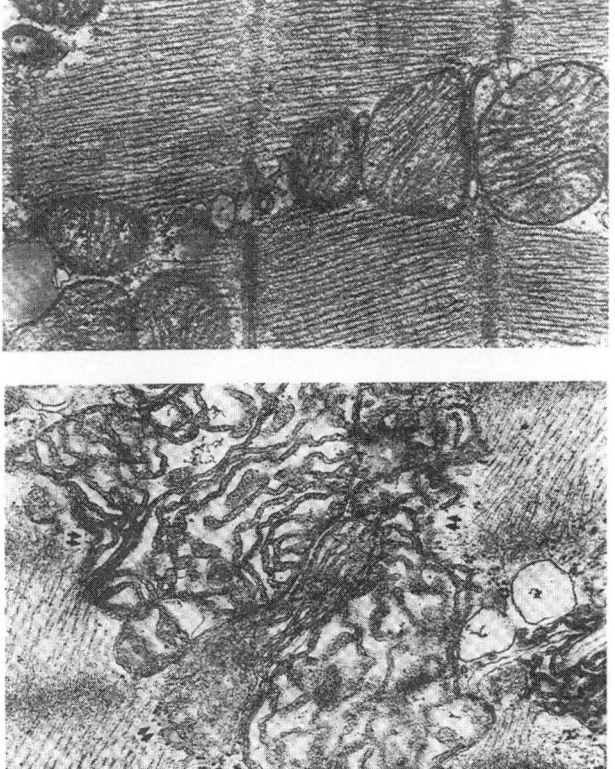

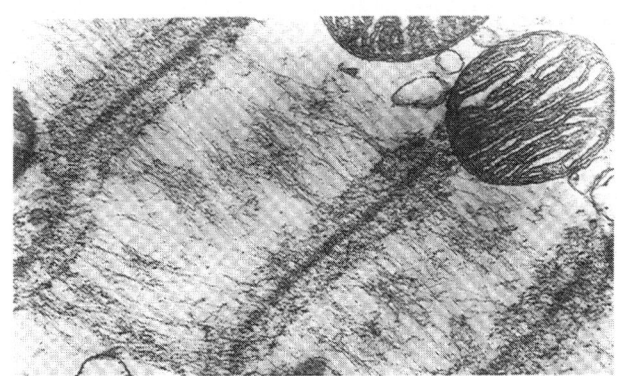

Fig. 9. Transmission electron micrographs of cardiac fibers. (A) Control, intact fiber. 40 000 ×. (B) The saponin-skinned fiber after treatment with 800 mM KCl solution – 'ghost fibers'. Note complete dissolution of thick filaments. 54 000 ×. (C) Skinned fibers after osmotic shock in 40 mOsM solution; 50 000 ×. Mitochondrial outer membrane rupture is obvious (double arrows). (Reproduced from Ref. 231 with permission).

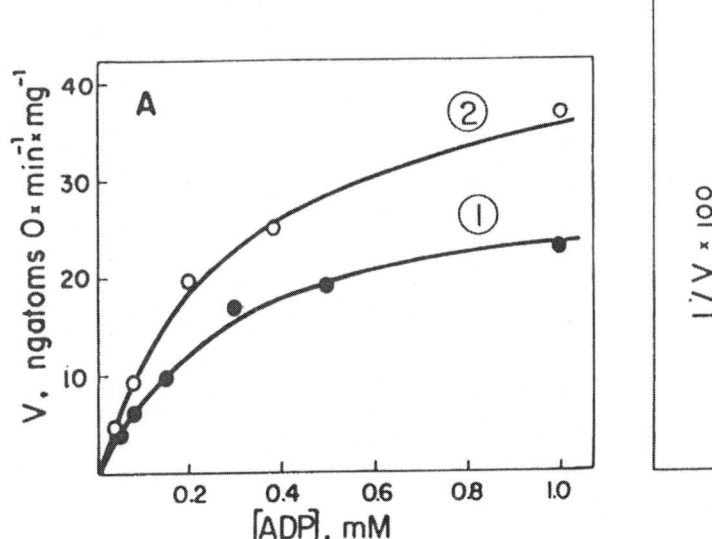

restricted penetration through the outer mitochondrial membrane [229, 233]. It was easy to discriminate between these two possible mechanisms since myosin thick filaments can be completely dissolved by incubation of skinned fibers for 30 min in 800 mM KCl solution

(Fig. 9B) and the outer mitochondrial membrane can be disrupted by hypoosmotic treatment (Fig. 9C). Complete destruction of myosin thick filaments did not change the kinetics of cellular respiration: before and after KCl treatment the ADP dependence of the rate of

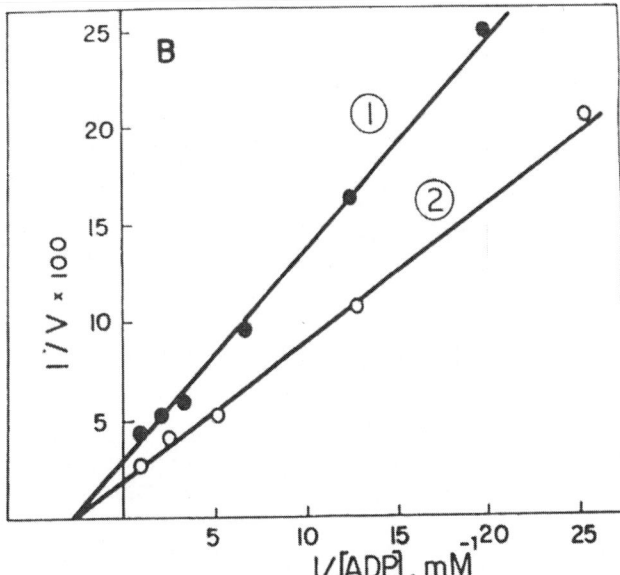

Fig. 10. (A) The dependences of respiration of skinned ghost fibers on external ADP concentration. 1-control, skinned fibers. 2-ghost fibers obtained after treatment of skinned fibers with 800 mM KCl solution. (B) Linearization of the dependences from Fig. 10A in double reciprocal plots. (Reproduced from ref 231 with permission).

oxygen consumption was the same (Fig. 10). Thus, myofibrillar structures are not a major barrier on the way of ADP to mitochondria.

The hypoosmotic (40 mOsM) treatment resulted in the appearance of the population of mitochondria with disrupted outer membrane (Fig. 9C), as seen also by increased sensitivity to the exogenous cytochrome c addition (cytochrome c test). This test showed that the population of mitochondria with the disrupted outer membrane made up about 50% of whole population (Fig. 11A). This Figure shows also that carboxyatractyloside inhibited the respiration to the basal level in these mitochondria – that means that the inner membrane was not disrupted and all respiration was controlled by ATP-ADP translocase in these experiments. The ADP dependence of cellular respiration of hypoosmotically treated skinned fibers showed biphasic kinetics: at low ADP concentration the rates of respiration were elevated and in double reciprocal plots two straight lines were obtained (Fig. 11B). These two affinities corresponded

to two different populations of mitochondria: one had apparent Km equal to 315 µM that obviously corresponded to the intact population; the new population, apparently that with disrupted outer membrane, had apparent Km for ADP 32 µM [233]. All values of kinetic constants obtained are summarized in the Table 3.

Two conclusions can be made from this observation: in the intact cardiomyocytes ADP diffusion is retarded mostly because of restricted permeability of the outer mitochondrial membrane for ADP; since isolated intact preparation of heart as well as liver mitochondria have very high affinities for ADP [25, 27, 233], there should be intracellular factor(s) controlling the permeability of the outer mitochondrial membrane for ADP which are lost during isolation of mitochondria. If these conclusions are valid, there will be a third one: to understand the phenomena of the regulation of mitochondrial activities *in vivo*, it is not sufficient to study the isolated mitochondria which obviously lack some important intracellular factors.

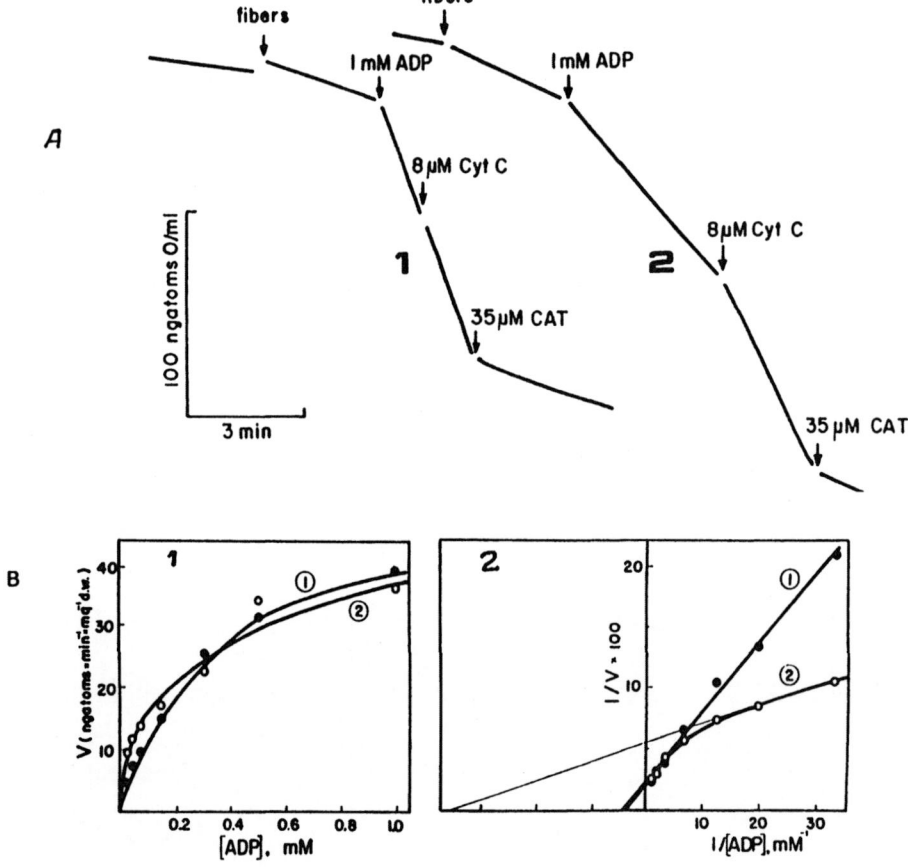

Fig. 11. (A) Recordings of respiration of skinned fibers before (1) and after (2) osmotic shock in 40 mOsM solution. The respiration rates were recorded in the medium containing 125 mM KCl to detach the cytochrome c from the inner membrane to test the intactness of outer membrane. At the points indicated with arrows, cytochrome c was added.
(B) 1-The dependences of respiration rates of skinned cardiac fibers before (curve 1) and afer osmotic shock (curve 2) on external ADP concentration; 2-Linearization of the dependences from 11. B1 in double reciprocal plots. (Reproduced from Ref. 231 with permission).

Table 3. Km values for ADP in regulation of respiration in different preparations at pH 7.2.

Preparation	Km app. for ADP (μM)		
	– creatine		+ creatine (25 mM)
1. Skinned cardiac fibers	297 ± 35		85 ± 5
2. Ghost cardiac fibers (without myosin)	349 ± 24		–
3. Skinned cardiac fibers with swollen mitochondria	**I**	**II**	–
	315 ± 23	32.3 ± 5	
4. Isolated mitochondria			
heart	17.6 ± 1.0		13.6 ± 4.4
liver	18.4		–

Probably, this conclusion will not make the investigators of mitochondria very happy, but we can give them good advise which will make life easier: knowing the kinetics of mitochondrial respiration in intact cells, it is worthwhile not to isolate mitochondria at all but to study the mitochondria behavior in the skinned fibers using small (5–7 mg) samples of tissue instead of several grams needed for isolation of mitochondria. That makes it possible to study the biopsy samples taken from human heart during angiography or from skeletal muscle and is especially important in studies of pathology when mitochondrial population becomes heterogenous (when isolated, nobody can tell from which population mitochondria came from, since about 90% of them go into debris during isolation). 7-years of practice of the use of the method of skinned cardiac fibers has shown its very high efficiency for studies of respiration *in situ* [182]. Precise protocols of these experiments have been published elsewhere [182, 229, 232, 233]. There are already very good examples of application of this methods for studies of heart and skeletal muscle in experiments and in patients [234–236].

In recent years the mitochondrial outer membrane has been in focus of many studies (see the chapter II-1 by Brdiczka and Wallimann in this issue), and now it is well understood that the high permeability of the outer membrane of isolated mitochondria to ADP is due to the existence of protein pores, or voltage-dependent anion selective channel, VDAC, with pore size around 2 nm [237–240]. The pore protein, porin, has been isolated from mitochondria of different sources and its structure and function (after incorporation into artificial phospholipid membranes) studied in great detail [239, 240]. These porin channels conduct anions better than cations, switch to low conductance at applied voltage above 10–20 mV and are easily regulated by several factors which may interact with a special sensor part in the channels or induce general conformation changes in the channel protein. It has been supposed that high oncotic pressure might be one of the factors significantly decreasing the channel permeability (see Gellerich *et al.* in this issue). Brdiczka and Wallimann have found that mitochondrial porin channels may be involved in formation of the dynamic contacts between the inner and outer mitochondrial mmebranes with participation of mitochondrial creatine kinase octamers [210]. Such a supramolecular complex of porin-creatine kinase-adenine nucleotide translocase assumes direct channelling of substrates and very efficient phosphocreatine production from cytoplasmic creatine and mitochondrial ATP [210, 211].

Thus, there are already sufficient data in the literature to suppose that the VDAC in the outer mitochondrial membrane may be of importance in regulation of metabolic fluxes between cytoplasm and intermembrane space of mitochondria. Our data show that in the intact cells the permeability of this porin channel is significantly reduced due to existence of some unknown intracellular factor. It is not excluded that this factor is some protein similar to that recently isolated by Liu and Colombini *et al.* [240] which is significantly lost during mitochondrial isolation. Schematically we have illustrated the connection of this unknown factor to the outer mitochondrial membrane *in vivo* in Fig. 12. Hypothetically, this factor may have some connection to cytoskeleton, that might be one of the reasons why we lose it during isolation. Search for this factor is one of interesting tasks of future studies.

Also, it is not excluded that the permeability of outer mitochondrial membrane to ADP has some connection to the configuration of mitochondrial membrane. The conclusion that mitochondria in the cells *in vivo* are different from those *in vitro* has already been made in 1978 by Sjostrand, but was left without much of attention [241]. Sjostrand concluded, on the basis of ultrastructural studies with carefully selected fixation technique to minimize the artifacts, that in intact cells the outer and inner mitochondrial membranes are not separated but form one proteinous-lipid phase with high viscosity [241] – using contemporary language, this is the continuous contact site. Accordingly, isolation of mitochondria leads to separation of the membranes, except at the contact sites, because of the use of artificial isolation media [241, see also Brdiczka and Wallimann, chapter II-1].

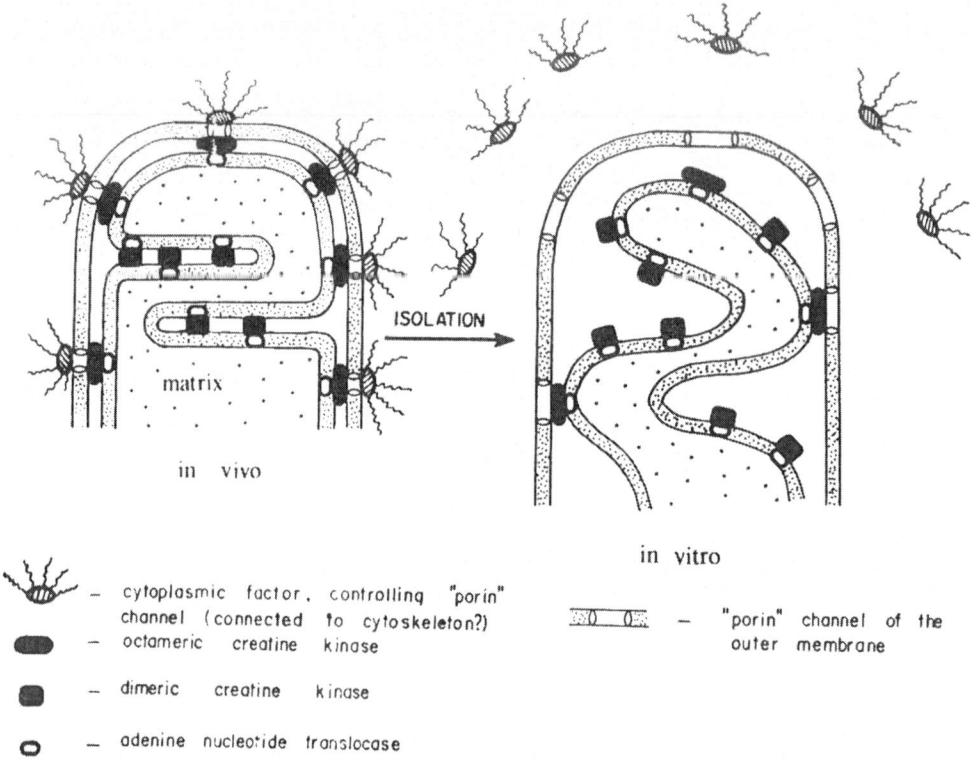

Fig. 12. Schematic illustration of the mitochondrial membrane systems *in vivo,* where the outer membrane porin channel permeability to ADP is controlled by some cytoplasmic factor, hypothetically connected to cytoskeleton. During isolation of mitochondria this factor is lost and the outer mitochondrial membrane becomes absolutely permeable for ADP (explanation see in test).

The schematic presentation of these ideas is reproduced from the work by Sjostrand in Fig. 13, because of its possible importance due to current high interest in the contact sites. The recent discoveries of the multienzyme complexes including creatine kinase and outer membrane porin in the space considered to be an intermembrane space [210, 211] take us back to these ideas of Sjostrand and emphasize the complexity of the molecular architecture of this intermembrane area where direct substrate channelling may become increasingly important due to increased viscosity. How much this structural contact between the membranes influences the permeability of the outer membrane to ADP and how much this permeability is controlled by other factors including proteins remains to be again a topic for further studies. However, from the available data one conclusion emerges very firmly – that the cytoplasmic ADP does not have easy access to translocase in the inner mitochondrial membrane because of the controlled, low permeability of the outer membrane for this substrate in the cardiac cells *in vivo.* Thus, in addition to the matrix pool of adenine nucleotides, there appears to be a second pool of ATP(ADP) between outer and inner membranes in mitochondria functionally (catalytically) ac-

tive inside the multienzyme complex translocase- creatine kinase-porin, that most probably represents the NMR-invisible ATP involved in phosphocreatine synthesis and mathematically modelled by Zahler *et al.* [154, 156].

It has been shown that in the low conductivity state the VDAC size is around 0.9 nm and it is not a permeability barrier for low molecular weight compounds such as creatine and phosphocreatine [238].

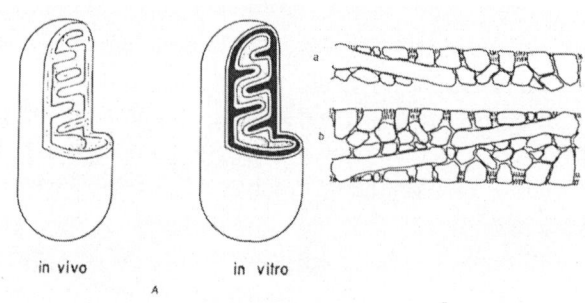

Fig. 13. Mitochondrial membrane structure *in vivo* and *in vitro* according to Sjostrand, redrawn on the basis from data from Ref. 241. *In vivo* (A) the outer and inner membranes of mitochondria form one continuous phase; during isolation the two membrane phase become separated, B shows the membrane configuration *in vitro.* On the right: a-structure of membrane; b-structure of two membrane system *in vivo*).

It is interesting to point out here that several authors have already concluded from analysis of the results of *in vivo* experiments that it is not always easy to apply experimental data of *in vitro* mitochondrial studies for interpretation of results of *in vivo* experiments [58, 66]. For example, Lanoue group has recently concluded that 'components present in the intact cells needed for regulation may be lost during mitochondrial preparation' [58]. Most probably, this is a correct conclusion.

Very low permeability of the outer mitochondrial membrane for ADP *in vivo* makes it necessary to modify many existing hypothesis and theories about the nature of feedback signal between cardiac work and cellular respiration.

Metabolic wave (propagation of local oscillations) in structured cytoplasm as a feedback signal between contraction and respiration

Estimations of free ADP concentration in cardiac cell cytoplasm have given for this parameter a value around 30–50 µM in normal heart [3, 6–11, 14–16, 33–37, 41, 43, 53, 55, 72]. As it has been pointed out by several investigators [7–11, 55] this ADP concentration exceeds Km for ADP for isolated mitochondria reported to be 17 µM [25–27, 231]; it we think that this Km is valid *in vivo* the calculated ADP should keep the respiration constantly close to maximal level. It is not so. If we use the value of Km for ADP found in this work for intact cardiomyocytes (300 µM), the cytoplasmic concentration of ADP around 50 µM gives the respiration rate not more than 1/12 of Vmax. Obviously, if ADP flux is by order of magnitude lower than required for maximal activation and we still believe firmly (as we do) that respiratory control by ADP exists, activation of Krebs cycle and respiratory chain by calcium will not be enough to achieve maximal respiration rates in response to increased workload – simply because in this case ADP supply will be rate-limiting factor. According to calculations of Katz *et al.* [11] to achieve the respiration rate experimentally found in epinephrine-stimulated hearts the levels of ADP and Pi should increase 4 times starting from Km value. In our case, ADP concentration should rise to 1,2 mM that will decrease phosphocreatine level in the equilibrium system 24 times, that means to about 1.5 µmoles per g dry weight – clearly, phosphorus NMR spectrum in this heart will be very different from that observed in experi-

ments with normal aerobic heart [6–11, 38–45, 58, 149–156]. Thus, the simple point of view of homogenous cytoplasm and regulation of respiration by cytoplasmic ADP leads us only into more and more difficulties.

To accomodate the necessity of high fluxes of ADP to translocase for maximal respiration with very low permeability of the outer membrane for ADP, we have to assume that this ADP is generated between outer and inner membranes, in so called intermembrane space, in response to some stimulus from cytoplasm. Such a system in mitochondria has already been described in great details here and in chapters by Brdiczka and Gellerich *et al.* – this is, without any doubt, a powerful amplifying device for adenine nucleotides represented by the complex ATP-ADP translocase-mitochondrial creatine kinase (octameric-dimeric) – outer membrane porin channel, directly connected, via translocase, to the system of oxidative phosphorylation in mitochondrial matrix and may also include other kinases. This complex can receive two kinds of signals – changes of ADP or/and creatine concentrations. The only question we have to solve now is to explain how these signals can come to mitochondria unnoticed by NMR people.

The possible answer was given in german language even before NMR method was applied for heart research, but again was not properly recognized most probably because it was also much ahead of time. In 1970 S. Nagle published two papers in the journal Klinische Wochenschrift, one of which had a title 'Die Bedeutung von Kreatinphosphat und Adenosintriphosphat im Hertzmuskel' [242, 243]. In these papers the author analyzed all experimental data available by that time in the literature on cardiac energy metabolism in normal and hypoxic heart including earlier work by Gercken and Schlette [117] and compared these data with calculations of ATP diffusion rates in cells [242]. He came to several interesting conclusions: ATP diffusion seems to be restricted in the cells and the main part on energy is transported in the cytoplasm via equilibrium creatine kinase reaction without direct physical diffusion of ATP, but in the way of facilitated diffusion (high energy phosphoryl group transfer via enzymatic equilibrium creatine kinase reaction) and via the same reaction the feedback signal, a change of ADP, goes back to mitochondria in the following way: in the contraction cycle ADP is released in myofibrillar space and this induces the cyclic shift in the creatine kinase reaction equilibrium in the next, adjacent point ('Raumelement') of cytoplasmic space, which leads to restoration of myofibrillar ATP but in this 'Raumelement' creatine is produced from

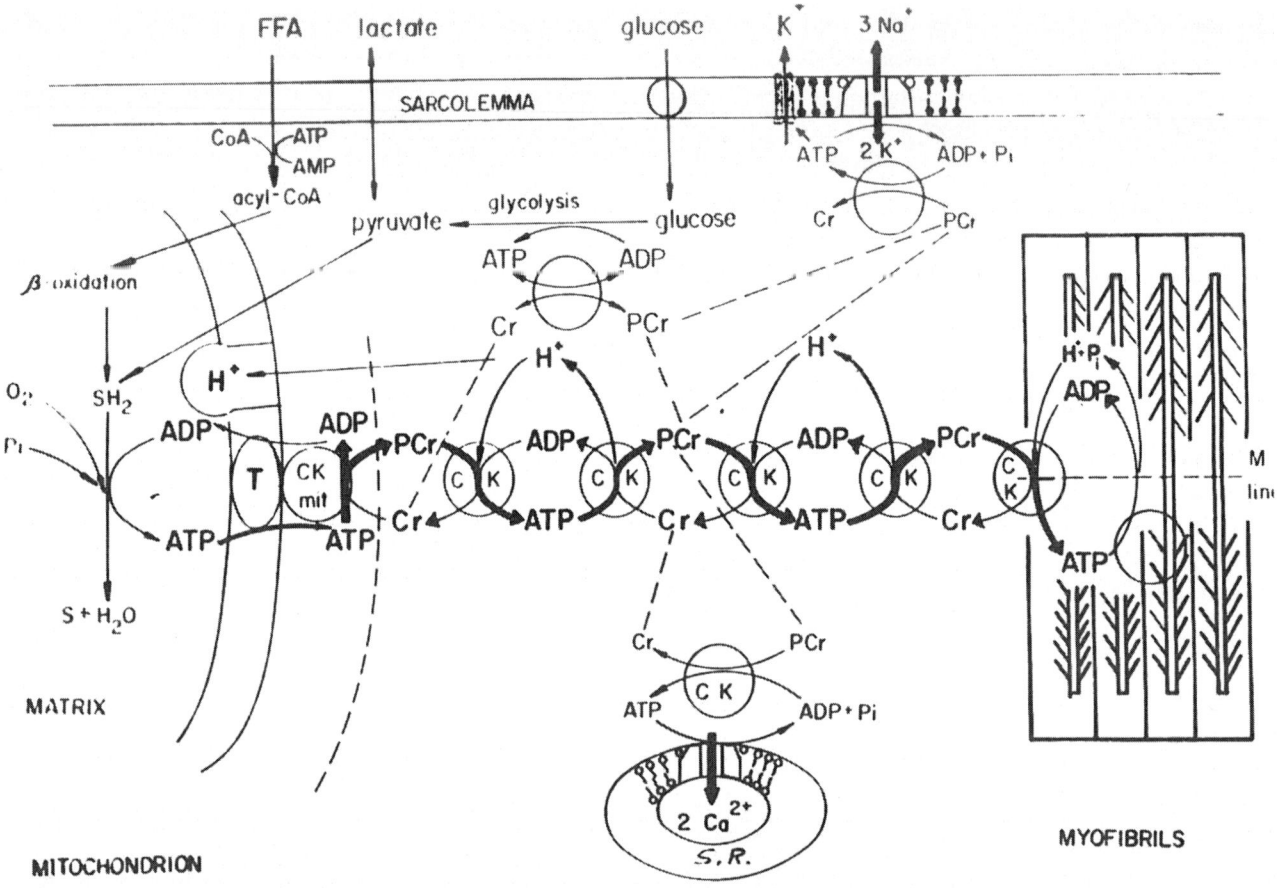

Fig. 14. Phosphocreatine pathway for intracellular energy transport in muscle cells. For explanation see the text. S.R. – sarcoplasmic reticulum.

phosphocreatine, that in turn will shift the equilibrium in the next point ('Raumelement'). In this way the shift of creatine kinase from equilibrium distributes, or propagates, from myofibrils to mitochondria in wave-like manner, 'somit pflanzt sich der Auslenkungeffekt gewissermassen wellenformig fort' [242], finally resulting in the local cyclic changes of ADP near mitochondria. Thus, already in 1970 Nagle formulated a concept of metabolic oscillation or wave as a feedback signal in muscle cells. Today we may add many new details to this description which seems to be basically true and worthwhile of very careful study.

The basic pathway for intracellular energy transport and correspondingly for feedback signal conduction in muscle, brain and several other types of cells (see a chapter by Wallimann in this issue) is the phosphocreatine pathway [13, 210, 211, 244, 245] ('circuit' according to Wallimann [210, 211] and 'shuttle' according to Bessmann [244, 245]), shown in details in Fig. 14. What was unknown to Nagle in that time was the existence of coupled kinases. Now we know that in mitochondria the creatine kinase reaction is driven by translocase unidi-

rectionally in the steady state in the direction of phosphocreatine synthesis. In myofibrils it is unidirectionally driven in the steady state by myosin ATPase in direction of phosphocreatine utilization, this is a kinetically favorable reaction and close spatial localization of these two proteins in the same cellular compartment makes their interaction very efficient (see the chapter by R. Ventura-Clapier and Veksler in this issue). Exactly the same is true for subcellular membranes – sarcoplasmic reticulum and sarcolemma [13, 137–139, 208, 210, 211, 244, 245]. And only in the cytoplasmic compartment the creatine kinase reaction is in equilibrium, according to all available data [72]. However, this equilibrium reaction most probably occurs in highly structured medium and for several components of reaction – adenine nucleotides – multiple microcompartments and deep concentration gradients may exist (see section II). We may assume that the total reaction equilibrium in cytoplasm is a result of multiple equilibrium states of cellular PCr-creatine pool (s) with ATP-ADP in microcompartments. In cytoplasm creatine kinase concentration (activity) is high [172, 173] and recent detailed immunohistochemical studies by

Wegman *et al.* [246] showed that it is mostly located in the l-band space of sarcomers. Glucolytic enzymes are located in the same area [77, 83–86, 246]. Thus, cytoplasmic creatine kinase may form dynamic complexes with other proteins (cytoskeleton, glycolytic enzymes etc.) resulting in formation of the dynamic microcompartments for ATP or ADP in the cytoplasmic space (see above and ref. 125, 145) between two neighbouring creatine kinase molecules [136, 246]. These dynamic cytoplasmic microcompartments of ATP may be representative of the 'Raumelements' of Nagle [242]. The chain of concentration changes within phosphocreatine pathway induced by ADP production in the contraction cycle in myofibrils is illustrated in Figs 15A and B, as we see it now. In concord with and in addition to the concept of Nagle Fig. 15 shows (see also Fig. 6) that already within the myofibrillar space local increase in ADP is quickly replaced by change in creatine concentration due to ADP rephosphorylation at the expense of phosphocreatine in coupled creatine kinase reaction – this is the first step of signal transformation. The next steps are sequential local cyclic changes of ADP and creatine concentrations due to phosphoryl group transfer in equilibrium cytoplasmic creatine kinase (facilitated diffusion, or vectorial ligand conduction). Since each next step brings the previous step back into equilibrium position and quenches the deviation from equilibrium at that step, only one single metabolic wave-alternating local changes in ADP and creatine concentrations – is propagating from myofibrils to mitochondria, as it is shown in Fig. 15B. The frequency of this signal is that of contraction (heart rate, for example), but the amplitude of signal may decrease in direction of wave propagation due to high mobility of creatine and phosphocreatine [242]. In cardiomyocytes sarcomer contractions are synchronized and the waves described are distributing in the space in all directions to all mitochondrial populations. Reaching mitochondria, the local change in creatine concentration stimulates coupled reactions of its rephosphorylation; if mitochondria receive a signal as local change in ADP concentration, this signal is manifold amplified by the coupled mitochondrial systems that again results in rapid phosphocreatine production and quenching of the signal.

Very important is that the metabolic signal transduction may occur with the rate exceeding the rate of ATP turnover in mitochondria and myofibrils by order of magnitude. The use of NMR magnetization transfer methods has given precise values for the rate of ATP synthesis (Pi-ATP) in hearts *in vivo*, that is in the range of 1–4 µmoles per g dry weight per second in dependence on workloads and is close to the rate of ATP synthesis calculated from rates of oxygen uptake [54, 130, 154, 156, 247–256], and ratio of Pi-ATP transfer to oxygen uptake is close to 3 in *in vivo* (if ATP production in glycolysis is accounted for [130]) as it is in isolated mitochondria. The total rate of creatine kinase reaction (ATP-PCr or PCr-ATP transfer) is, however 20–40 µmoles per g dry weight per second [25, 130, 154, 156, 247–256], it includes as a part the rate of unidirectional reaction in mitochondria or myofibrils that is equal to ATP turnover rates [156], and equilibrium creatine kinase reaction in cytoplasm which has a constant value about 20–25 µmoles per g dry weight per second [156, 247–256]. For some unknown reason, the high value of this equilibrium creatine kinase rate was often used in discussions as argument against the phosphocreatine pathway and was considered to show only the ATP buffering role of PCr [166, 250–252]. In the elegant ('simple') mathematical analysis of facilitated diffusion of ATP and PCr in cells Meyer *et al.* showed that in fact the opposite is true and in the equilibrium creatine kinase system the phosphocreatine flux is dominant, representing more than 99% of high energy phosphate flux [167]. Here, starting with the work by Nagle and taking into account the information of the cytoplasmic structure, we come to the conclusion that the equilibrium creatine kinase is a necessary step also for rapid signal transduction between two (or more) coupled creatine kinase systems in myofibrils (and at membranes) and in mitochondria, respectively. Thus, both facilitated diffusion of high energy phosphoryl group from mitochondria and feedback signal transduction to mitochondria are the results of the functioning of the same equilibrium creatine kinase in structured cytoplasm.

Simultaneously with such a signal transduction there should be the flux of inorganic phosphate from myofibrillar space to mitochondria to maintain the steady state rates of oxidative phosphorylation [17, 18]. Most probably this small molecule may diffuse rapidly in the bulk phase of cytoplasmic water and will be transported into mitochondria by a special carrier system [17–21].

Concerning the feedback signal transduction, it is possible that there is another similar system functioning in the muscle cells – the adenylate kinase system (see the chapters by Savabi and Gellerich *et al.* in this issue). At the background of high concentration of ATP this reaction ATP + AMP ↔ 2 ADP may contribute in the feedback signal which reaches mitochondria as local changes of ADP (or AMP) concentration. In the transgenic ani-

182

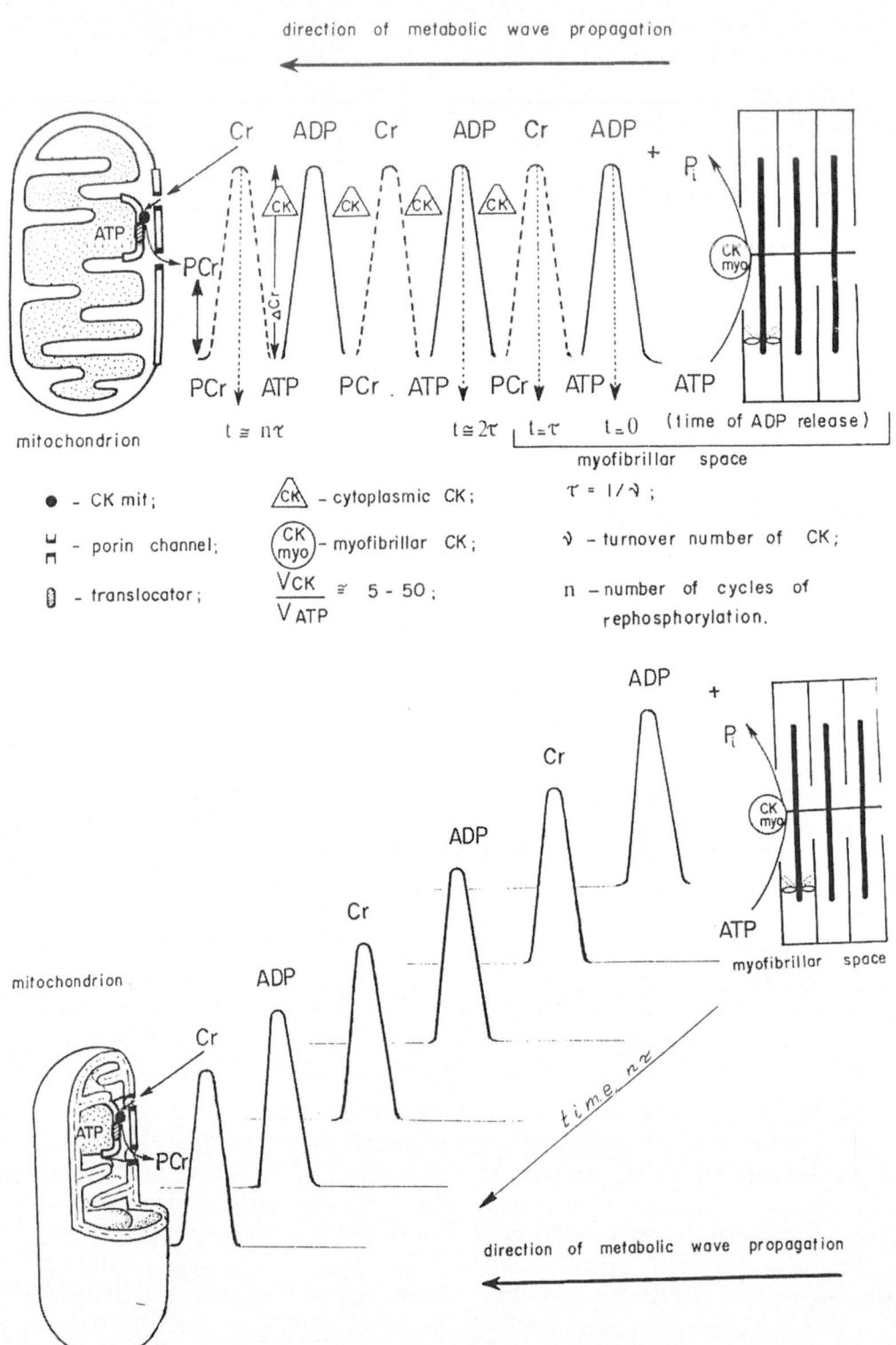

Fig. 15. (A) Consecutive ADP and creatine concentration changes in microcompartments around cytoplasmic creatine kinase as a result of shifts from equilibrium induced by cyclic ADP liberation during contraction.
(B) The resulting metabolic wave was a feedback signal from myofibrils to mitochondria. For explanation see the text.

mals with switched off cytoplasmic creatine kinase (see chapter IV-3) this adenylate kinase system may become more important for signal transduction. However, because of low permeability of the mitochondrial outer membrane for ADP, the presence of creatine is necessary for the signal amplification and for phosphocreatine production.

Such a concept of metabolic wave propagation in structured cytoplasmic medium which may seem new is in fact an extension of the Peter Mitchell theory of vectorial ligand conduction in catalytic domains of enzymes and catalytic carriers [253]. According to this theory (which formed a basis for chemiosmotic concept of oxidative phosphorylation), 'all enzymatic group transfer processes have not only scalar chemical transformation aspects, but also an intrinsic vectorial transport, or osmotic aspect which could be the cause of a macroscopic osmotic process if molecules of enzymes were appropriately plugged through the membrane or if they were organized in pairs of sequences' [253]. Macroscopic osmotic process – vectorial ligand movement – may be an effective system for intracellular signal transduction because of much higher sensitivity for regulation than an homogenous system, as very clearly and quantitatively described by Kholodenko, Cascante and Westerhoff in chapter VI-1 of this issue.

It is easy to show both theoretically and experimentally that the flux through equilibrium creatine kinase system does not depend on absolute concentrations of reactants until their low values [254]. Because of the amplifying effects of coupled reactions, they are also perfectly functioning at very low ATP concentrations. Thus, it becomes rather clear why the cardiac function is totally dissociated from the tissue ATP content [119–123]. The same is true for phosphocreatine content. Kapelko *et al.* [255] and Zweier *et al.* [256] have demonstrated convincingly that creatine and phosphocreatine contents should be depleted significantly by feeding animals the creatine structural analogues, but at total creatine lower than 20–30% of normal both cardiac work and oxygen uptake decrease rapidly because of decreased rate of PCr production and probably signal transmission. Some of these results are illustrated in Fig. 16 for working isolated rat heart. Thus, for the phosphocreatine pathway 20–30% of normal phosphocreatine level is sufficient to saturate coupled creatine kinases in energy consuming structures and maintain equilibrium reaction in cytoplasm for signal transduction, and the same per cent of creatine is sufficient to maintain maximal rates of coupled reactions in mitochondria.

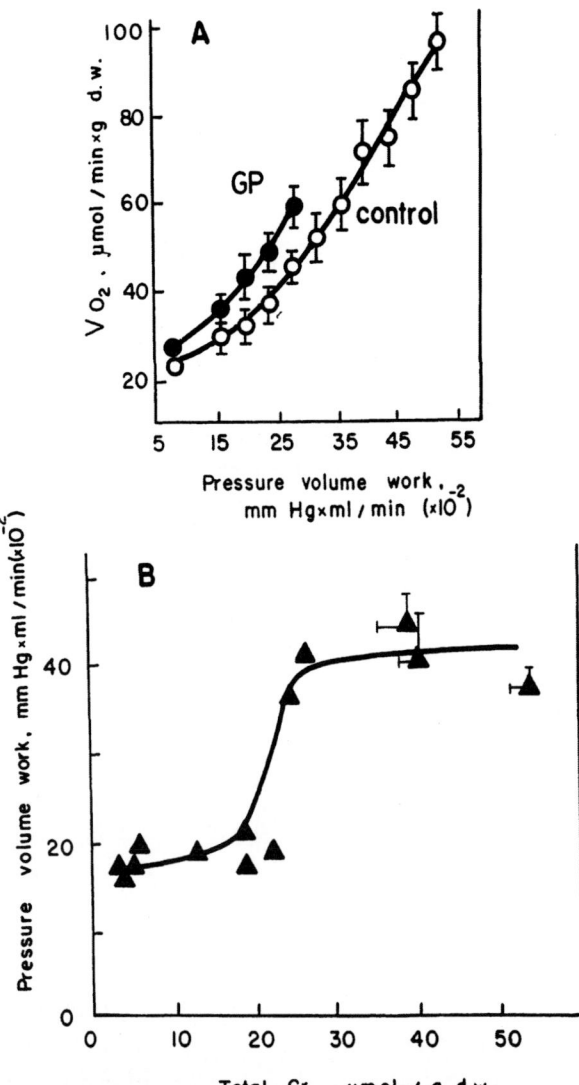

Fig. 16. (A) Dependences of cardiac oxygen consumption on pressure-volume work for control and guanidinopropionate (GP) fed rat hearts (from Ref. 255 with permission). (B) Cardiac work indices of working heart preparation as function of degree of creatine depletion by guanidino propionate (from Ref. 255 with permission).

Probably, high cellular steady state levels of ATP and phosphocreatine in normal cells have some additional physiological importance beside trivial energy store function [257, 258]. For phosphocreatine that may be the additional membrane stabilizing function in details described in recent years in pharmacological and clinical studies [257]. For high cellular concentration of ATP several non-catalytic functions can be also easily found [258].

Since the feedback signal transduction rate is higher than ATP turnover rate, the steady state PCr level depends on the rapidity of the mitochondrial response. If the rate of mitochondrial response to local changes of ADP and creatine concentrations – a signal – is high,

PCr may be synthesized so rapidly that no alterations in its content during work transitions could be seen. However, if the response rate is low and metabolic signals are 'accumulating' in mitochondria or around them in the cytoplasmic space as elevated creatine and ADP concentrations, steady state level of PCr will obviously decrease. This may well explain the controversy of experimental data described in the first section of this paper: the conflicting results may just reflect differences of mitochondrial response (different substrates, intracellular and matrix calcium etc.) to very quick metabolic signal.

Recently, direct method of measurement of mitochondrial response time in heart tissue was worked out and it was shown that this parameter is increased if calcium entry into mitochondria is blocked by ruthenium red and Krebs cycle activity decreased, and decreased by pyruvate which activates Krebs cycle [63, 259].

Some experimental support for the concept of metabolic wave as a regulatory signal may be found in the work by Wikman-Coffelt *et al.* who studied the metabolic changes in the cardiac cycle by using stimulator-triggered freeze clamp technique [260]. The authors found cyclic changes in energy-related metabolites in glucose-perfused hearts but not pyruvate-perfused isolated working hearts [260]. Since pyruvate decreases response time of mitochondria via pyruvate dehydrogenase activation [63, 259], in these hearts metabolic signals may be quickly quenched and changes not observed. In glucose-perfused hearts the metabolic changes may be observed within time intervals less than duration of contraction cycle because of the slower mitochondrial responses, and the steady state level of PCr is usually lower in glucose-perfused than in pyruvate-perfused hearts [54, 58, 260], in concord with the model predictions.

Thus, the metabolic signal between contraction and respiration most probably exists, but is too quick and not observable (in particular by low sensitivity NMR method) if mitochondrial responses are quick enough.

The problem of regulation of ATP turnover was recently analyzed also in a review by Hochachka and Matheson [261]. Application of the simple Michaelis-Menten enzyme kinetics for investigation of regulation of respiration in muscle cells was also found to be completely unsatisfactory. The authors assumed the existence of some enzyme activation (regulation of Vmax) mechanism in the cells *in vivo* which could explain the very wide range of the rates of ATP turnover [261]. The present work shows that this assumption is not necessary if we accept the concept of heterogeneity of cell cytoplasm and direct substrate-product channelling between structurally organized enzymes in the cells, leaving the point of view on cell interior as homogenous solution for history. Or otherwise, the assumption mentioned may be modified by saying that respiratory enzymes are activated by substrate-product fluxes.

Finally, we can mention that there is already a good historical analogy for our current troubles with respiration regulation *in vivo* – it was the controversy between *in vitro* and *in vivo* data on physiological roles of ATP and phosphocreatine in the cells, when in *in vivo* experiments ATP was never seen to change during contraction [118]. The reason was very simple – effective interaction between myofibrillar creatine kinase and myosin ATPase (see the excellent review by Mommaerts, ref. 225). Probably, we are now dealing with more sophisticated aspect of this general phenomenon – very high efficiency of intracellular interactions which simply keep generalized metabolic parameters (cytoplasmic PCr, ADP etc.) constant during work transitions.

Conclusions

1. Outer mitochondrial membrane has a very limited permeability for ADP *in vivo*. The coupled multienzyme system ATP-ADP translocase-mitochondrial creatine kinase – outer membrane porin may operate as a receiving unit amplifying weak ADP signals from cytoplasm due to increased turnover of adenine nucleotides and in this way control the oxidative phosphorylation.

2. Cell cytoplasm in myocytes has highly organized structure and contains multiple microcompartments for ATP (ADP) as a result of formation of multienzyme complexes, liquid crystal like protein structures and presence of structured water phase. Cytoplasmic microcompartments of ATP(ADP) are connected to creatine-phosphocreatine pools via equilibrium creatine kinase.

3. Feedback signal between contraction and the mitochondrial respiration may be conducted in cells by metabolic wave (propagation of oscillations of local concentrations) through cytoplasmic equilibrium creatine kinase (+ adenylate kinase).

4. The rate of metabolic wave propagation exceeds the rate of ATP turnover in contraction and oxidative phosphorylation 5–10 times and is not directly related to total ATP and phosphocreatine contents which depend on the rate of mitochondrial response to metabolic signal. These integral parameters are dissociated

from cardiac function until they reach low values (about 20–30% of normal cellular content).

Acknowledgements

This work was accomplished in large part due to generous assistance from INSERM, France and Joseph Fourier University, Grenoble, France, which provided excellent conditions for careful analysis of the problem. Very helpful discussions with Prof. Andre Rossi and Dr. Christiane Keriel are gratefully acknowledged. The authors wish also to thank all our former co-workers in Cardiology Center in Moscow who have participated in experimental part of this research.

References

1. Williamson JR: Mitochondrial function in heart. Annu Rev Physiol 41: 485–506, 1979
2. Mela-Riker L, Bukoski RD: Regulation of mitochondrial activity in cardiac cells. Annu Rev Physiol 47: 645–663, 1985
3. Ugurbil K, Kingsley-Hickman PB, Sano EI: P-NMR Studies of the kinetics and regulation of oxidative phosphorylation in the intact myocardium. Ann NY Acad Sci US 508: 265–286, 1987
4. McMillin JB, Payly DF: Control of mitochondrial respiration in muscle. J Mol Cell Biochem 81: 121–129, 1988
5. Brown C: Control of respiration and ATP synthesis in mammalian mitochondria and cells. Biochem J 284: 1–13, 1992
6. Heineman FW, Balaban RS: Control of mitochondrial respiration in the heart *in vivo*. Annu Rev Physiol 52: 523–542, 1990
7. Balaban RS, Kantor HL, Katz LA, Briggs RW: Relation between work and phosphate metabolites in the *in vivo* paced mammalian heart. Science 232: 1121–1123, 1986
8. Koretzky AP, Katz LA, Balaban RS: The mechanism of respiratory control in the *in vitro* heart. J Mol Cell Cardiol 21: 59–66, 1989
9. Balaban RS: Regulation of mitochondrial oxidative phosphorylation in mammalian cells. Am J Physiol 258: C377–389, 1990
10. Katz LA, Koretsky AP, Balaban RS: Respiratory control in the glucose perfused heart. A P-NMR and NADH fluorescence study. FEBS Lett, 221: 270–276, 1987
11. Katz LA, Swain JA, Portman MA, Balaban RS: Relation between phosphate metabolites and oxygen consumption of heart *in vivo*. Am J Physiol 256: H265–H274, 1989
12. Das AM, Harris DA: Reversible modulation of the mitochondrial ATP synthesis with energy demand in cultured rat cardiomyocytes. FEBS Lett 256: 97–100, 1989
13. Jacobus WE: Respiratory control and integration of heart high energy phosphate metabolism by mitochondrial creatine kinase. Annu Rev Physiol 47: 707–725, 1985
14. Hassinen IE: Mitochondrial respiratory control in myocardium. Biochim Biophys Acta 853: 135–151, 1986
15. Bittl J, Balschi JA, Ingwall JS: Effects of norepinephrine infusion on myocardial high energy phosphate content and turnover in the living heart. J Clin Invest 79: 1852–1859, 1987
16. Gibbs C: The cytplasmic phosphorylation potential. Its possible role in the control of myocardial respiration and cardiac contractility. J Mol Cell Cardiol 17: 727–731, 1985
17. Chance B, Leigh JS, Kent J, McCully K, Nioka S, Clark BJ, Mais JM, Graham T: Multiple controls of oxidative metabolism in living tissues as studied by phosphorus magnetic resonance. Proc Natl Acad Sci USA 83: 9458–9462, 1986
18. Chance B, Leigh JS, Clark BJ, Kent J, Nioka S, Smith D: Control of oxidative metabolism and oxygen delivery in human skeletal muscle: a steady analysis of the work-energy cost transfer function. Proc Natl Acad Sci 82: 8384–8388, 1985
19. Connett RJ, Honig CR: Regulation of VO2 in red muscles: do current biochemical hypothesis fit *in vivo* data? Am J Physiol 256: R898–R906, 1989
20. Osbakken M, Mitchell MP, Zheng D, Mayevsky A, Chance B: *In vivo* correlation of myocardial metabolism, perfusion and mechanical function during increased cardiac work. Cardiovasc Res 25: 749–756, 1991
21. Osbakken M, Ligeti L, Huddell J, Duska C, Ponomarenko I, Chance B: *In vivo* myocardial bioenergetics during acute volume and/or pressure loading in a canine model: a P-NMR study. Cardiology 76: 405–417, 1989
22. Moreno-Sanchez R, Hogue BA, Hansford RG: Influence of NAD-linked dehydrogenase activity on flux through oxidative phosphorylation. Biochem J 268: 421–428, 1990
23. Moreno-Sanchez R: Regulation of oxidative phosphorylation in mitochondria by external free Ca concentration. J Biol Chem 260: 4028–4034, 1985
24. McCormac JC, Halestrap AP, Denton RM: Role of calcium ions in regulation of mammalian intramitochondrial metabolism. Physiol Rev 70: 391–425, 1990
25. Lardy HA, Wellman H: Oxidative phosphorylation. Role of inorganic phosphate and acceptor systems in control of metabolic rates. J Biol Chem 195: 215–224, 1952
26. Chance B, Williams GR: Respiratory enzymes in oxidative phosphorylation. III. The steady state. J Biol Chem 217: 409–427, 1955
27. Chance B, Williams GR: The respiratory chain and oxidative phosphorylation. Adv Enzymol 17: 65–134, 1956
28. Skulachev VP: Membrane bioenergetics. Springer Verlag, Berlin 1–442, 1988
29. Nicolls D, Ferguson SJ: Bioenergetics 2. Academic Press, London-New York-Toronto, p. 1–247, 1992
30. Boyer PD: The binding change mechanism for ATP synthase. Some probabilities and possibilities. Biochim Biophys Acta 1140: 215–250, 1993
31. Neely JR, Liebermeister H, Battersby EJ, Morgan HE: Effect of pressure development on oxygen consumption by isolated rat heart. Am J Physiol 212: 804–814, 1967
32. Neely JR, Denton RM, England PJ, Randle PJ: The effects of increased heart work on the tricarboxylate cycle and its interactions with glycolysis. Biochem J; 128: 147–159, 1972
33. Williamson JR, Ford G, Illingworth J, Safer B: Coordination of citric acid cycle activity with electron transport flux. Circ Res 38, Suppl. I, I-39–I-51, 1976
34. Nishiki K, Erecinska M, Wilson D: Energy relationship between cytosolic metabolism and mitochondrial respiration in rat heart. Am J Physiol 234: C73–C81, 1978

186

35. Erecinska M, Wilson D: Regulation of cellular energy metabolism. J Membr Biol 70: 1–14, 1982
36. Opie LH, Mansford KRC, Oven P: Effects of increased heart work on glycolysis and adenine nucleotides in the perfused heart of normal and diabetic rats. Biochem J 124: 457–490, 1971
37. Geisen J, Kammermeier H: Relationship of phosphorylation potential and oxygen consumption in isolated perfused rat hearts. J Mol Cell Cardiol 12: 891–907, 1980
38. Aussedat J, Ray A, Lorter S, Reutenauer H, Grably S, Rossi A: Phosphorylated compounds and function in isolated hearts: a P-NMR study. Am J Physiol 260: H110–H117, 1991
39. Jeffrey FMH, Malloy CR: Respiratory control and substrate effects in the working rat heart. Biochem J 287: 117–123, 1992
40. Lewandowski ED: Nuclear magnetic resonance evaluation of metabolic and respiratory support of workload in intact hearts. Circ Res 70: 576–582, 1992
41. From AHL, Petein MA, Michursky SP, Zimmer SD, Ugurbil K: P-NMR Study of the respiratory regulation in the intact myocardium. FEBS Lett 206: 257–261, 1986
42. Ugurbil K, Petein M, Maidan R, Michurski S, From A: Measurement of an individual rate constant in the presence of multiple exchanges: application to myocardial creatine kinase reaction. Biochemistry 25: 110–107, 1986
43. From A, Zimmer S, Michurski SP, Mohanakrishnan P, Ulstad VK, Thoma WJ, Ugurbil K: Regulation of oxidative phosphorylation rate in the intact heart. Biochemistry 29: 3731–3743, 1990
44. Bittl JA, Ingwall JS: Reaction rates of creatine kinase and ATP synthesis in the isolated rat heart. J Biol Chem 3512–3517, 1985
45. Bittl JA, Balski JA, Ingwall JS: Effects of norepinephrine infusion on myocardial high energy phosphate content and turnover in living rat. J Clin Invest 79: 1852–1859, 1987
46. Idele-Wenger JA, Grotyohann LW, Neely JR: Regulation of fatty acid utilization in rat heart. J Mol Cell Cardiol 14: 413–417, 1982
47. Carafoli E, Tiozzo R, Lugli G, Crovetti F, Kratzing C: The release of calcium from heart mitochondria by sodium. J Mol Cell Cardiol 6: 361–371, 1974
48. Saks VA: Creatine kinase isoenzymes and the control of cardiac contraction. In: WE Jacobus, JS Ingwall (eds.) Heart Creatine Kinase. The integration of isozymes for energy distribution. Williams and Wilkins, Baltimore-London, p. 109–126, 1980
49. Glegg JS: Properties and metabolism of the aqueous cytoplasm and its boundaries. Am J Physiol 246: R133–R151, 1984
50. Glegg JS: Cellular infrastructure and metabolic organization. In: Current Topics in Cellular Regulation 33: 3–14, 1992
51. Klingenberg M, Pfaff E: Structural and functional compartmentation in mitochondria. In: JM Tager, S Papa, E Quagriariello, EC Slater (Eds.) Regulation of Metabolic Processes in Mitochondria, Elsevier, Amsterdam-New York, p. 180–121, 1966
52. Hassinen IE, Hiltunen K: Repiratory control in isolated perfused rat heart. Role of the equilibrium relations between the mitochondrial electron carriers and the adenylate system. Biochem Biophys Acta 408: 319–330, 1975
53. Kauppinen RA, Hiltunen JK, Hassinen IE: Subcellular distribution of phosphagens in isolated perfused rat heart. FEBS Lett 112: 273–276, 1980
54. Zweier JL, Jacobus WE: Substrate-induced alterations of high energy phosphate metabolism and contractile function in the perfused heart. J Biol Chem 262: 8015–8021, 1987
55. Unitt JF, Schrader J, Brunotte F, Radda GK, Seymour AML: Determination of free creatine and phosphocreatine concentra-

tions in the isolated perfused rat heart. Biochem Biophys Acta 1133: 115–120, 1992
56. Burger R, Soboll S: Cytosolic adenylates and adenosine release in perfused working rat heart. Comparison of whole tissue with cytosolic non-aqueous fractionation analysis. Eur J Biochem 159: 203–213, 1986
57. Kushmerick M, Meyer RA, Brown TR: Regulation of oxygen consumption in fast and slow-twich muscle. Am J Physiol 263: C598–C606, 1992
58. Van B, Dounen C, Duszynsky J, Salama G, Vary TC, Lanoue KF: Effect of cardiac work on electrical potential gradient across mitochondrial membrane in perfused hearts. Am J Physiol 265: H453–H460, 1993
59. Hansford RG: Relation between mitochondrial calcium transport and control of energy metabolism. Rev Physiol Biochem Pharmacol 102: 2–72, 1985
60. Katz LA, Koretsky AP, Balaban RS: Activation of dehydrogenase activity and cardiac respiration: a P-NMR study. Am J Physiol 255: H185–H188, 1988
61. McCormack JC, England PJ: Ruthenium red inhibits the activation of pyruvate dehydrogenase caused by positive inotropic agents in the perfused rat hearts. Biochem J 214: 581–585, 1983
62. Unitt JF, McCormack JG, Reid D, MacLachlan LK, England PJ: Direct evidence for a role of mitochondrial Ca in the regulation of oxidative phosphorylation in the stimulated rat heart. Biochem J 262: 293–301, 1989
63. Hak JB, Van Beek JHGM, Eijgelshoven MHJN, Westerhof N: Mitochondrial dehydrogenase activity affects adaptation of cardiac oxygen consumption to demand. Am J Physiol 264: H448–H453, 1993
64. Ferrary R, Di Lisa R, Visioli O: The effects of ruthenium red on mitochondrial function during post-ischemic reperfusion. J Molec Cell Cardiol 14: 737–740, 1982
65. Figuenedo VM, Dresdner KP, Wolney AC, keller AM: Postischemic reperfusion injury in the isolated rat heart: effect of ruthenium red. Cardiovasc Res 25: 337–342, 1991
66. Heineman FW, Balaban RS: Effects of afterload and heart rate on NAD(P)H redox state in isolated rabbit heart. Am J Physiol 264: H433–H440, 1993
67. Harris DA, Das AM: Control of mitochondrial ATP synthesis in the heart. Biochem J 280: 561–573, 1991
68. Moravec CS, Bonds M: Effect of inotropic stimulation on mitochondrial calcium in cardiac muscle. J Biol Chem 267: 5310–5316, 1992
69. Welch GR, Glegg JS: The organisation of cell metabolism: a historical vignette. In: GR Welch, JS Glegg (eds.) The organization of cell metabolism. Plenum Press, New York-London, p. 1–5, 1986
70. Iyengar MR: Creatine kinase as an intracellular regulator. J Muscle Res Cell Motil 5: 527–534, 1984
71. Meyer RA: Linear model of muscle respiration explains monoexponential phosphocreatine change. Am J Physiol 254: C548–C553, 1988
72. Veech RL, Lawson JWR, Cornell NW, Krebs HA: Cytosolic phosphorylation potential. J Biol Chem 254: 6538–6547, 1979
73. Meyer RA: Linear dependence of muscle phosphocreatine kinetics on total creatine content. Am J Physiol 257: C1149–C1157, 1987
74. Connett RJ: Analysis of metabolic control: new insites using

scaled creatine kinase model. Am J Physiol 254: R949–R959, 1988

75. Welch GR: On the role of organized multienzyme systems in cellular metabolism: a general synthesis. Prog Biophys Mol Biol 32: 103–191, 1977

76. Sols A, Marco R: Concentrations of metabolites on binding sites. Implications in metabolic regulation. Curr Top Cell Regul 2: 227–273, 1970

77. Masters CJ: Metabolic control and the microenvironment. Curr Top Cell Regul 12: 75–107, 1977

78. Srere P: Organization of proteins within the mitochondrion. In: GR Welch (ed.) Organized multienzyme systems. Catalytic properties. Academic Press, New York-London, p. 1–163, 1985

79. Morel JE: Discussion of the state of water in the myofilament lattice and other biological systems, based on the fact that the usual concepts of colloid stability cannot explain the stability of myofilament lattice. J Theor Biol 112: 847–858, 1985

80. Srere P: Complexes of sequential metabolic enzymes. Ann Rev Biochem 56: 89–124, 1987

81. Friedrich P: Supramolecular enzyme organization: quaternary structure and beyond. Akademiai Kiado, Budapest; Pergamon Press, Oxford, 1984

82. Ottaway JH, Mowbray J: The role of compartmentation in the control of glycolysis. Curr Top Cell Regul 12: 108–208, 1977

83. Kurganov BJ, Sugrobova NP, Milman LS: Supramolecular organization of glycolytic enzymes. J Theor Biol 116: 509–526, 1985

84. Kurganov BI: The role of multienzyme complexes in integration of cellular metabolism. J Theor Biol 119: 445–455, 1986

85. Kurganov BI: The general principles of the control of functioning enzymes and multienzyme complexes. In: S Demyanovich, T Keleti, L Tron (eds.) Dynamics of biochemical systems. Akademiai Kiado, Budapest and Elsevier, Amsterdam. p.231–245, 1986

86. Kurganov BI: Principles of integration of cellular metabolism. Molec Biol (Russ.) 20: 369–377, 1986

87. Srere P, Mossbach RW (eds.) Microenvironments and metabolic compartmentation. Academic Press, New York, p. 1–310, 1978

88. Glegg JC: The physical properties and potential roles of intracellular water. In: GR Welch, JC Glegg (eds.) The organization of cell metabolism. Plenum Press. New York-London, p. 41–55, 1985

89. Welch GR, Clegg JS (eds.) The organization of cell metabolism. Plenum Press, New York-London, 1985

90. Grinwald PM: Sodium pump failure in hypoxia and reoxygenation. J Mol Cell Cardiol 24: 1393–1398, 1992

91. Morgan HE, Henderson MD, Regen DM, Park CR: Regulation of glucose uptake in muscle I. J. Biol Chem 236: 253–261, 1961

92. Gard JK, Kichura GM, Ackerman JH, Eisenberg JD, Bilodello JJ, Sobel BE, Gross RW: Quantitative P-NMR analysis of metabolite concentrations in Langendorff-perfused rabbit hearts. Biophys J 48: 803–813, 1985

93. Sommer JR, Jennings RB: Ultrastructure of cardiac muscle. In: HA Fozzard, RB Jennings, E Haber, A Katz, HE Morgan: The heart and cardiovascular system. Scientific foundations. Raven Press, New York, vol. 1, p 61–100, 1986

94. Schaper J, Meiser E, Stammler G: Ultrastructural morphometric analysis of myocardium from dogs, rats, hamsters, mice and from human hearts. Circ Res 56: 377–391, 1985

95. Barth E, Stammler G, Speiser B, Schaper J: Ultrastructural quantitation of mitochondria and myofilaments in cardiac mus-

cle from 10 different animal species including man. J Mol Cell Cardiol 24: 669–681, 1992

96. Schaper J, Hein S, Brand T, Schaper W: Contractile proteins and the cytoskeleton in isolated rat myocytes. J Appl Cardiol 1: 423–429, 1989

97. Bershadsky AD, Vaziliev JM: Cytoskeleton. New York, Plenum Press, 1988

98. Fulton AB: How crowed is cytoplasm? Cell 30: 345–347, 1982

99. Ling GN: The polarized multilayer theory of cell water and other facts of the association-induced hypothesis. In: AD Keith (ed.) The Aqueous Cytoplasm. New York, Dekkar p 23–90, 1979

100. Ling GN: In search of the physical basis of life. Plenum Press, New York, 1985

101. Franks F, Mathias S (eds.) Biophysics of the water. Wiley, New York, 1982

102. Welch GR, Somoguyi B, Damjanovich S: The role of protein fluctuations in enzyme action: a review. Prog Biophys Mol Biol 39–146, 1982

103. Friedrich P: Dynamic compartmentation in soluble multienzyme system. In: GR Welch (ed.) Organized multienzyme systems. Catalytic properties. Academic Press, New York-London, p 141–176, 1985

104. Watterson JG: A role of water in cell structure. Biochem J 248: 615–617, 1987

105. Kempner ES: In: L Nover, F Lynen, K Mothes (eds.) Cell compartmentation and metabolic chanelling. Fischer, Jena, Elsevier, Amsterdam. p 211–224, 1980

106. Schoolweth AC, LaNoue KF: The role of microcompartmentation in the regulation of glutamate metabolism by rat kidney mitochondria. J Biol Chem 225: 3403–3411, 1980

107. Vignais PV, Brandolin G, Boulay F, Block MR, Gauche I: In; A Azzi, KA Nalicz, L Wojtczack (eds.) Anion Carriers of Mitochondrial Membranes. Springer, Berlin, p 133–146, 1989

108. Klingenberg M: The ADP/ATP carrier in mitochondrial membranes. In: AN Martonoso (ed.) The Enzymes of Biological membranes, Plenum Press, New York, vol. 4, p. 511–553, 1985

109. Klingenberg M: Membrane protein oligomeric structure and transport function. Nature 290: 449–454, 1969

110. Kramer R, Klingenberg M: Reconstitution of adenine nucleotide transport from beaf heart mitochondria. Biochemistry 18: 4209–4215, 1989

111. Geisbuhler T, Altschuld RA, Ronald W, Trewyn ANN Z Ansel, Lamka K, Brierly GP: Adenine nucleotide metabolism and compartmentation in isolated adult rat heart cells. Circ Res 54: 536–546, 1984

112. Pfaller W, Guder WG, Getraunthaler G, Kotanko P, Jehart I, Purshel J: Compartmentation of ATP within renal proximal tubular cells. Biochim Biophys Acta 805: 152–157, 1984

113. Soboll S, Bunger R: Compartmentation of adenine nucleotides in the isolated working guinea pig heart stimulated by noradrenaline. Hoppe Seylers Z Physiol Che333362: 125–132, 1981

114. Humphrey SM, Buckman JE, Holliss DG: Subcellular distribution of energy metabolites in the pre-ischemic and post-ischemic perfused working rat heart. Eur J Biochem 191: 755–759, 1990

115. Infante AA, Davies RE: The effect of 2,4-dinitrofluorobenzene on the activity of striated muscle. J Biol Chem 240: 3996–4001, 1965

116. Carlson FD, Singer AJ: The mechanochemistry of muscular contraction. J Gen Physiol 44: 33–59, 1960

117. Gercken G, Schlette U: Metabolite status of the heart in acute

188

insufficiency due to 1-fluoro-2,4-dinitrobenzene. Experientia 24: 17–18, 1968

118. Gudbjarnason S, Mathes P, Ravens KG: Functional compartmentation of ATP and creatine phosphate in heart muscle. J Mol Cell Cardiol 1: 325–339, 1970

119. Neely JR, Grotyohann LW: Role of glycolytic products in damage to ischemic myocardium. Dissociation of adenosine triphosphate levels and recovery of function of reperfused ischemic hearts. Circ Res 55: 816–824, 1984

120. Hoerter JA, Lauer C, Vassort G, Gueron M: Sustained function of normoxic hearts depleted in ATP and phosphocreatine: a P-NMR study. Am J Physiol 255: C192–C201, 1988

121. Kupriyanov VV, Lakomkin VL, Kapelko VI, Steinschneider A Ya, Ruuge EK, Saks VA: Dissociation of adenosine diphosphate levels and contractile function in isovolumic hearts perfused with 2-deoxyglucose. J Mol Cell Cardiol 19: 729–740, 1987

122. Rosenkranz EL, Okamoto F, Buckberg GD, Vinten-Johansen J, Allen S, Leaf J, Bugyi H, Young H, Bernard RJ: Studies of controlled reperfusion after ischemia. II. Biochemical studies: failure of tissue adenosine triphosphate levels to predict recovery of contractile function after controlled reperfusion. J Thorac Cardiovasc Surg 92: 448–501, 1986

123. McLellan G, Weisberg A, Winegrad S: Energy transport from mitochondria to myofibril by a creatine phosphate shuttle in cardiac cells. Am J Physiol 245: C423–C427, 1983

124. Opie LH: Cardiac metabolism-emergence, decline and resurgence. Part I, II. Cardiovasc Res 26, 721–733; 817–830, 1992

125. Opie LH: Importance of glycolytically produced ATP for the integrity of the threatened myocardial cell. In: HM Piper (ed.) Pathophysiology of Severe Ischemic Myocardial Injury. Kluwer Academic Publishers, Dordrecht, The Netherlands, 41–65, 1989

126. Opie LH, Bricknell OL: Role of glycolytic flux in effect of glucose in decreasing fatty acid-induced release of lactate dehydrogenase from isolated coronary ligated rat heart. Cardiovasc Res 13: 693–702, 1979

127. Bricknell OL, Daries PS, Opie LH: A relationship between adenosine triphosphate, glycolysis and ischaemic contracture in isolated rat heart. J Mol Cell Cardiol 13: 941–945, 1981

128. Owen P, Dennis S, Opie LH: Glucose flux rate regulates onset of ischemic contracture in globally underperfused hearts. Circ Res 66: 344–354, 1985

129. Kingsley PB, Sako E, Yang MQ, Zimmer SD, Ugurbil K, Foker JE, From A HL: Ischemic contracture begins when anaerobic glycolysis stops: P-NMR study of isolated rat heart. Am J Physiol 261: H469–478, 1991

130. Jeremy RW, Ambrosio G, Pike MM, Jacobus WE, Becker LC: The functional recovery of post-ischemic myocardium requires glycolysis during early reperfusion. J Mol Cell Cardiol 25: 261–276, 1993

131. Weiss J, Hiltbrand B: Functional compartmentation of glycolytic versus oxidative metabolism in isolated rabbit heart. J Clin Invest 75: 436–447, 1985

132. Weiss JN, Lamp ST: Glycolysis preferentially inhibits ATP-sensitive K-channels in isolated guinea-pig cardiac myocytes. Science 238: 67–69, 1987

133. Koretsune Y, Marban E: Mechanism of ischemic contracture in ferret hearts: relative roles of Ca elevation and ATP depletion. Am J Physiol 258: H9–H16, 1990

134. Mercerer RW, Dunham PB: Membrane-bound ATP fuels the Na/K pump. J Gen Physiol 78: 547–568, 1981

135. Han JW, Thieleczek, Varsanyi M, Heilmeyer LMG: Compartmentalized ATP synthesis in skeletal muscle triads. Biochemistry 31: 377–384, 1992

136. Maughan D, Wegner E: The organization and diffusion of glycolytic enzymes in skeletal muscle. In: RJ Paul, G Elzinga, K Jamada (eds.) Muscle Energetics. Alan R Liss Inc New York, p 137–147, 1989

137. Levitsky DO, Levchenko TS, Saks VA, Sharov VG, Smirnov VN: The role of creatine phosphokinase in supplying energy for the calcium pump system of heart sarcoplasmic reticulum. Membr Biochem 8: 81–96, 1978

138. Rossi AM, Eppenberger HM, Volpe P, Cotfuro R, Wallimann T: Muscle-type creatine kinase is specifically bound to sarcoplasmic reticulum and can support Ca-uptake and regulate local ATP-ADP ratios. J Biol Chem 265: 5258–5266, 1990

139. Grosse R, Spitzer E, Kupriyanov VV, Saks VA, Repke KRV: Coordinate interplay between (Na,K)ATPase and creatine kinase optimizes (Na,K)antiport across the membrane of vesicles formed from plasma membrane of cardiac muscle cells. Biochem Biophys Acta 603: 142–156, 1980

140. Clark AJ, Gaddie R, Stewart CP: The anaerobic activity of the isolated frogs heart. J Physiol (London) 75: 321–331, 1932

141. Yee T, Jones DP: ATP concentration gradients in cytosol of liver cells during hypoxia. Am J Physiol 249: C385–C392, 1985

142. Jones DP: Intracellular diffusion gradients of O2 and ATP. Am J Physiol 250: C663–C675, 1986

143. Noma A: ATP regulated K-channels in cardiac muscle. Nature 305: 147–148, 1983

144. Takano M, Noma A: ATP-dependent decay and recovery of K-channels in guinea pig cardiomyocytes. Am J Physiol 258: H45–H50, 1990

145. Furukawa T, Kimura S, Furukawa N, Bassett AL, Meyeburg RJ: Role of ATP-regulated potassium channel in differential responses of endocardial and epicardial cells to ischemia. Circ Res 68: 1693–1702, 1991

146. Nichols CG, Lederer WJ: The regulation of ATP-sensitive channel activity in intact and permeabilized rat ventricular myocytes. J Physiol (London) 423: 91–110, 1990

147. Sharov VG, Saks VA, Smirnov VN, Chazov EI: An electron microscopic histochemical investigation of the localization of creatine phosphokinase in the heart. Biochim Biophys Acta 468: 495–501, 1977

148. Hodgkin AL: The conduction of nervous impulse. Liverpool University Press, Liverpool p 80–85, 1964

149. Humphrey SM, Garlick PB: NMR-visible ATP and Pi in normoxic and reperfused rat heart: a quantitative study. Am J Physiol 206: H6–H12, 1991

150. Garlick PB, Townsend RM: NMR visibility of Pi in perfused rat hearts is affected by changes in substrate and contractility. Am J Physiol 263: H497–H501, 1992

151. Takami HE, Furuya E, Tagava S, Seo Y, Murakami H, Watari H, Matsuda H, Hirose H, Yasunaru Kawashima: NMR-invisible ATP in rat heart and its change in ischemia. J Biochem 104: 35–39, 1988

152. Koretsky AP, Basus VJ, James TL, Klein MP, Weiner MV: Detection of exchange reactions involving small metabolite pools using NMR magnetization transfer techniques. Magn Res Med 2: 586–594, 1985

153. Balaban RS, Kantor HL, Ferretti JA: In vivo flux between phos-

phocreatine and ATP determined by two-dimensional phosphorus NMR. J Biol Chem 258: 12787–12789, 1983

154. Zahler R, Ingwall JS: Estimation of heart mitochondrial creatine kinase flux using magnetization transfer NMR spectroscopy. Am J Physiol 262: H1022–H1028, 1992

155. Nunnally RL, Hollis DP: ATP compartmentation in living hearts: a phosphorus nuclear magnetic resonance saturation transfer study. Biochemistry 18: 3642–3647, 1979

156. Zahler R, Bittl JA, Ingwall J: Analysis of compartmentation of ATP in skeletal and cardiac muscle using P nuclear magnetic resonance saturation transfer. Biophys J 51: 883–893, 1987

157. Horowitz SB, Miller DS: The intracellular distribution of adenosine triphosphate. In: GR Welch, JS Glegg. The Organization of Cell Metabolism. Plenum Press, New York-London, p 79–85, 1986

158. Miller DS, Horowitz SB: Intracellular compartmentation of adenosine triphosphate. J Biol Chem 261: 13911–13915, 1986

159. Ingwall JS: Phosphorus nuclear magnetic resonance spectroscopy of cardiac and skeletal muscle. Am J Physiol 242: H729–H744, 1982

160. Carmeliet E: A fuzzy subsarcolemal space for intracellular Na in cardiac cells? Cardiovasc Res 26: 433–442, 1992

161. Niggli E, Lipp P: Subcellular restricted spaces: significance for cell signalling and excitation-contraction coupling. J Muscle Res Cell Motility 14: 288–291, 1993

162. Isenberg G, Wendett-Gallitelli MF: X-ray microprobe analysis of sodium concentration reveals large transverse gradients from the sarcolemma to the centre of voltage-clamped guinea pig ventricular myocytes (abstract). J Physiol (London) 420: 86, 1990

163. Kushmerick MJ, Podolsky RJ: Ione mobility in muscle cells. Science 166: 1297–1298, 1969

164. Yoshizaki KY, Seo H, Nishikawa H, Morimoto T: Application of pulsed-gradient P-NMR on frog muscle to measure the diffusion rates of phosphorus compounds in cells. Biophys J 38: 209–211, 1982

165. Yoshizaki K, Watari H, Radda GK: Role of phosphocreatine in energy transport in skeletal muscle of bullfrog studied by P-NMR. Biochim Biophys Acta 1051: 144–150, 1990

166. Radda GK: Control, bioenergetics and adaptation in health and disease: noninvasive biochemistry from nuclear magnetic resonance. FASEB J 6: 3032–3038, 1992

167. Meyer RA, Sweeney HL, Kushmerick MJ: A simple analysis of the 'phosphocreatine shuttle'. Am J Physiol 246: C365–C377, 1984

168. Lowson JWR, Veech RL: Effects of pH and free Mg on the Keq of the creatine kinase reaction and other phosphate hydrolyses and phosphate transfer reactions. J Biol Chem 254: 6528–6437, 1979

169. Saks VA, Chernousova GB, Gukovsky DE, Smirnov VN, Chazov EI: Studies of energy transport in heart cells. Mitochondrial isozyme of creatine kinase: kinetic properties and regulatory action of Mg ions. Eur J Biochem 57: 273–290, 1975

170. Saks VA, Chernousova GB, Vetter R, Smirnov VN, Chazov EI: Kinetic properties and the functional role of particulate MM isozyme of creatine phosphokinase bound to heart muscle myofibrils. FEBS Lett 62: 293–296, 1976

171. Scholte HR: On the triple localization of creatine kinase in heart and skeletal muscle cells. Biochim Biophys Acta 305: 413–427, 1973

172. Saks VA, Chernousova GB, Voronkov YI, Smirnov VN, Chazov EI: Study of energy transport mechanism in myocardial cells. Circ Res 34–35, Suppl III 138–149, 1974

173. Saks VA, Seppet EK, Lyulina NV: Comparative investigation of the role of creatine phosphokinase isoenzymes in energy metabolism of skeletal muscle and myocardium. Biokhimiya 42: 579–588, 1977

174. Yamashita K, Yoshioka T: Profiles of creatine kinase isoenzyme composition in single muscle fibers of different types. J Muscle Res Cell Motility 12: 37–44, 1991

175. Yamashita K, Yoshioka T: Activities of creatine kinase isoenzymes in single skeletal muscle fibers of trained and untrained rats. Eur J Physiol 421: 270–273, 1992

176. Jacobs H, Heldt HW, Klingenberg M: High activity of creatine kinase in mitochondria from muscle and brain and evidence for a separate mitochondrial isoenzyme of creatine kinase. Biochem Biophys Res Commun 16: 516–521, 1964

177. Scholte HR, Weijers PJ, Wit-Peters EM: Localization of mitochondrial creatine kinase and its use for the determination of the sidedness of submitochondrial particles. Biochim Biophys Acta 291: 764–773, 1973

178. Ingwall JS: Whole-organ enzymology of the creatine kinase system in heart. Biochem Soc Transactions 19: 1006–1010, 1991

179. Apple FS, Rogers MA: Mitochondrial creatine kinase activity alterations in skeletal muscle during long-distance running. J Appl Physiol 61: 482–485, 1986

180. Yamashita K, Yoshioka T: Activities of creatine kinase isoenzymes in single skeletal muscle fibers of trained and untrained rats. Eur J Physiol 421: 270–273, 1992

181. Kapelko VI, Veksler VI, Popovich MI: Cellular mechanisms of alterations in myocardial contractile function in experimental cardiomyopathies. Biomed Sci 1: 77–83, 1990

182. Saks VA, Belikova YO, Kuznetsov AV, Khuchua ZA, Branishte TH, Semenovsky ML, Naumov VG: Phosphocreatine pathway for energy transport: ADP diffusion and cardiomyopathy. Am J Physiol Suppl 261: 30–38, 1991

183. Khuchua ZA, Ventura-Clapier R, Kuznetsov AV, Grishin MN, Saks VA: Alterations in the creatine kinase system in the myocardium of cardiomyopathic hamsters. Biochem Biophys Res Commun 165: 748–757, 1989

184. Kuznetsov AV, Saks VA: Affinity modification of creatine kinase and ADP-ATP translocase in heart mitochondria: determination of their molar stoichiometry. Biochem Biophys Res Commun 134: 359–366, 1986

185. Saks VA, Kupriyanov VV: Intracellular energy transport and control of cardiac contraction. Advances in Myocardiology. Plenum Press, New York, vol 3: 475–497, 1982

186. Baba B, Kim S, Farrell EC: Histochemistry of creatine phosphokinase. J Mol Cell Cardiol 8: 599–617, 1976

187. Schlegel J, Zurbriggen B, Wegmann G, Wyss M, Eppenberger H, Wallimann T: Native mitochondrial creatine kinase forms octameric structures. J Biol Chem 263: 16942–16953, 1988

188. Muller M, Moser R, Cheneval D, Carafoli E: Cardiolipin is the membrane receptor for mitochondrial creatine phosphokinase. J Biol Chem 260: 3839–3843, 1985

189. Cheneval D, Carafoli E: Identification and primary structure of the cardiolipin-binding domain of mitochondrial creatine kinase. Eur J Biochem 171: 1–9, 1988

190. Biermans W, Bakker A, Jacob W: Contact sites between inner and outer mitochondrial membrane: a dynamic microcompart-

190

mentation for creatine kinase activity. Biochim Biophys Acta 1018: 225–228, 1990

191. Beyer K, Klingenberg M: ADP—ATP carrier protein from beef heart mitochondria has high amounts of tightly bound cardiolipin as revealed by P-nuclear magnetic resonance. Biochemistry 24: 3821–3826, 1985

192. Bessman S, Fonyo A: The possible role of mitochondrial-bound creatine kinase in regulation of mitochondrial respiration. Biochem Biophys Res Commun 8: 361–366, 1966

193. Vial C, Godinot G, Gautheron D: Creatine kinase (EC 2.7.3.2.) in pig heart mitochondria. Properties and role in phosphate potential regulation. Biochimie (Paris) 54: 843–852, 1972

194. Jacobus WE, Lehninger AL: Creatine kinase of rat heart mitochondria. J Biol Chem 248: 4803–4810, 1973

195. Saks VA, Lipina NV, Smirnov VN, Chazov RI: Studies of energy transport in heart cells. The functional coupling between mitochondrial creatine phosphokinase and ATP-ADP translocase. Arch Biochem Biophys 173: 34–41, 1976

196. Barbour RL, Ribaudo J, Chan SHP: Effect of creatine kinase activity on mitochondrial ADP/ATP transport. J Biol Chem 259: 8246–8251, 1984

197. William CT, Yang J, Geiger PJ, Bessmann SP: Formation of creatine phosphate from creatine and 31-P- labelled ATP by isolated heart mitochondria. Biochem Biophys Res Commun 76: 882–887, 1977

198. Erickson-Viitanen S, Geiger PJ, Viitanen P, Bessmann SP: Compartmentation of mitochondrial creatine phosphokinase. J Biol Chem 257: 14405–14411, 1982

199. Gellerich FN, Saks VA: Control of heart mitochondrial oxygen consumption by creatine kinase: the importance of enzyme localization. Biochem Biophys Res Commun 105: 1473–1481, 1982

200. Gellerich FN, Khuchua ZA, Kuznetsov AV: Influence of mitochondrial outer membrane and the binding of creatine kinase. Biochim Biophys Acta 1140: 327–334, 1993

201. Jacobus WE, Moreadith RW, Vandegaer KM: Control of heart oxidative phosphorylation by creatine kinase in mitochondrial mambranes. Ann N.Y. Acad Sci 414: 73–89, 1983

202. Jacobus WE, Saks VA: Creatine kinase of heart mitochondria: changes in its kinetic properties induced by coupling to oxidative phosphorylation. Arch Biochem Biophys 219: 167–178, 1982

203. Kuznetsov AV, Khuchua ZA, Vasileva EV, Medvedeva NV, Saks VA: Heart mitochondrial creatine kinase revisited: outer mitochondrial membrane is not important for coupling of phosphocreatine production to oxidative phosphorylation. Arch Biochem Biophys 268: 176–190, 1989

204. Saks VA, Kupriyanov VV, Elizarova GV, Jacobus WE: Studies of energy transport in heart cells. The importance of creatine kinase localization for the coupling of mitochondrial phosphorylcreatine production to oxidative phosphorylation. J Biol Chem 255: 755–763, 1980

206. Saks VA, Kuznetsov AV, Kupriyanov VV, Miceli MV, Jacobus WE: Creatine kinase of rat heart mitochondria. The demonstration of functional coupling to oxidative phosphorylation in an inner membrane-matrix preparation. J Biol Chem 260: 7757–7764, 1985

207. Saks VA, Khuchua ZA, Kuznetsov AV: Specific inhibition of ATP-ADP translocase in cardiac mitoplasts by antibodies against mitochondrial creatine kinase. Biochim Biophys Acta 891: 138–144, 1987

208. Saks VA, Rosenshtraukh LV, Smirnov VN, Chazov EI: Role of creatine phosphokinase in cellular function and metabolism. Can J Physiol Pharmacol 56: 691–706, 1978

209. Kim IH, Lee HJ: Oxidative phosphorylation of creatine by respiring pig heart mitochondria in the absence of added adenine nucleotides. Biochem International 14: 103–110, 1987

210. Wallimann T, Wyss M, Brdiczka D, Nicolay K: Intracellular compartmentation, structure and function of creatine kinase isoenzymes in tissues with high and fluctuating energy demands: the 'phosphocreatine circuit' for cellular energy homeostasis. Biochem J 281: 21–40, 1992

211. Wyss M, Smeitink J, Wevers RA, Wallimann T: Mitochondrial creatine kinase: a key enzyme of aerobic energy metabolism. Biochim Biophys Acta 1102: 119–166, 1992

212. Fossel E, Hoefeler H: A synthetic functional metabolic compartment. The role of propinquity in a linked pair of immobilised enzymes. Eur J Biochem 170: 165–171, 1987

213. Aliev MK, Saks VA: Quantitative analysis of the 'phosphocreatine shuttle': I A probability approach to the description of phosphocreatine production in the coupled creatine kinase – ATP/ADP 'P' translocase – oxidative phosphorylation reactions in heart mitochondria. Biochim Biophys Acta 1143: 291–300, 1993

214. Kell DB, Westerhoff H: Catalytic facilitation and membrane bioenergetics. In: GR Welch (ed.) Organized multienzyme systems. Catalytic properties. Acad Press, New York-London, p 64–140, 1985

215. Gellerich FN, Schlame M, Saks VA: Creatine kinase of heart mitochondria: no changes in its kinetic properties after inhibition of the adenine nucleotide translocator. Biomed Biochim Acta 42; 10: 1335–1337, 1983

216. Marty I, Brandolin G, Gagnon J, Brasseur R, Vignais P: Topography of the membrane bound ADP/ATP carrier assessed by enzymatic proteolysis. Biochemistry 31: 4058–4065, 1992

217. Soboll S, Conrad A, Keller M, Hebish S: The role of mitochondrial creatine kinase system for myocardial function during ischemia and reperfusion. Biochim Biophys Acta 1100: 27–32, 1992

218. Mahler M: First-order kinetics of muscle oxygen consumption and an equivalent proportionality between QO and phosphorylcreatine level. Implication for the control of respiration. J Gen Physiol 86: 135–165, 1985

219. Flaherty JT, Weisfeldt ML, Bulkley BH, Gardner TJ, Gott VL, Jacobus WE: Mechanisms of ischemic myocardial cell damage assessed by phosphorus -31 nuclear magnetic resonance. Circulation 65: 561–571, 1982

220. Schaper J, Mulch J, Winkler ML, Schaper W: Ultrastructural, functional, and biochemical criteria for estimation of reversibility of ischemic injury. A study on the effects of global ischemia on the isolated dog heart. J Mol Cell Cardiol 11: 521–541, 1979

221. Ischiara K, Abiko Y: Rebound recovery of myocardial creatine phosphate with reperfusion after ischemia. Am Heart J 108: 1594–1597, 1984

222. Clarke K, O'Connor AJ, Willis RJ: Temporal relationship between energy metabolism and myocardial function during ischemia and reperfusion. Am J Physiol 253: H412–H421, 1987

223. Pantely GA, Malone SA, Rhen WS, Anselone CG, Arai A, Bristow JD: Regeneration of myocardial phosphocreatine in pigs despite continued moderate ischemia. Circ Res 67: 1481–1493, 1990

224. Ventura-Clapier R, Saks VA, Vassort G, Lauer C, Elizarova GV: Reversible MM-creatine kinase binding to cardiac myofibrils. Am J Physiol 253: C444–C455, 1987

225. Mommaerts WFHM: Energetics of muscular contraction. Physiol Rev 49: 427–508, 1969

226. Hsieh PS, Balaban RS: Saturation and inversion transfer studies of creatine kinase kinetics in rabbit skeletal muscle *in vivo*. Magn Res Med 7: 56–64, 1988

227. Lundberg P, Harmsen E, Clinton H, Vogel HJ: Nuclear magnetic resonance studies of cellular metabolism. Anal Biochem 181: 193–222, 1990

228. Kummel L: Ca,Mg-ATPase activity of permeabilised rat heart cells and its functional coupling to oxidative phosphorylation of the cells. Cardiovasc Res 22: 359–367, 1988

229. Saks VA, Belikova YuO, Kuznetsov AV: *In vivo* regulation of mitochondrial respiration in cardiomyocytes: specific restrictions for intracellular diffusion of ADP. Biochim Biophys Acta 1074: 302–311, 1991

230. Altschuld RC, Hohl C, Ansel A, Brierley GP: Compartmentation of K in isolated adult heart cells. Arch Biochem Biophys 209: 175–184, 1984

231. Saks A, Vasileva E, Belikova YuO, Kutnetsov AV, Lyapina S, Petrova L, Perov NA: Retarded diffusion of ADP in cardiomyocytes: possible role of mitochondrial outer membrane and creatine kinase in cellular regulation of oxidative phosphorylation. Biochim Biophys Acta 1144: 46–53, 1993

232. Veksler VI, Kuznetsov AV, Sharov VG, Kapelko VI, Saks VA: Mitochondrial respiratory parameters in cardiac tissue: a novel method of assessment using saponin-skinned fibers. Biochim Biophys Acta 892: 191–196, 1987

233. Saks VA, Kapelko VI, Kupriyanov VV, Kuznetsov AV, Lakomkin VL, Veksler VI, Sharov VG, Javadov SA, Seppet EK, Kairane C: Quantitative evaluation of relationship between cardiac energy metabolism and post-ischemic recovery of contractile function. J Mol Cell Cardiol 21: 67–78, 1989

234. Sanabe A, Tanonana K, Hanaoka Y, Katoh T, Takeo S: Regional energy metabolism of failing hearts following myocardial infarction. J Mol Cell Cardiol 25: 995–1015, 1993

235. Kunz W, Kuznetsov AV, Schultze W, Eichhorn K, Schild L, Striggow F, Grasshoff H, Neumann HW, Gellerich FN: Functional characterization of mitochondrial oxidative phosphorylation in saponin-skinned human muscle fibers. Biochim Biophys Acta 1144: 46–56, 1993

236. Kunz WS, Kuznetsov AV, Gellerich FN: Mitochondrial oxidative phosphorylation in saponin-skinned human muscle fibers stimulated by caffein. FEBS Lett 323: 188–190, 1993

237. Benz R, Kottke M, Brdiczka D: The cationically selective state of the mitochondrial outer membrane pore: a study with intact mitochondria and reconstituted mitochondrial porin. Biochim Biophys Acta 1022: 311–318, 1990

238. Colombini M: Regulation of the mitochondrial outer membrane channel, VDAC. J Bioenerg Biomembr 19: 305–358, 1987

239. Kayser H, Kratzin D, Thinnes FP, Gotz H, Schmidt WE, Eckart K, Hilschmann N: Characterization and primary structure of a 31 kD porin. Biol Chem Hoppe-Seyler 370: 1265–1278, 1988

240. Liu M, Colombini M: Voltage gating of the mitochondrial outer membrane channel VDAC is regulated by a very conservative protein. Am J Physiol 260: C371–C374, 1991

241. Sjostrand FS: The structure of mitochondrial membranes: a new concept. J Ultrastruct Res 64: 217–245, 1978

242. Nagle S: Die Bedeutung von Kreatinphosphat und Adenosintriphosphat im Hinblick auf Energiebereitstellung, -transport und verwertung im normalen und insuffizienten Herzmuskel. Klin Wschr 48: 332–339, 1970

243. Nagle S: Regelprobleme im Energiestoffwechsel des Herzmuskels. Klin Wsch 48: 1075–1089, 1970

244. Bessman SP, Geiger PJ: Transport of energy in muscle. The phosphorylcreatine shuttle. Science 211: 448–452, 1981

245. Bessman SP, Carpenter CL: The creatine-creatine phosphate energy shuttle. Annu Rev Biochem 54: 831–862, 1985

246. Wegmann G, Zanolla E, Eppenberger HM, Wallimann T: *In situ* compartmentation of creatine kinase in intact sarcomeric muscle: the acto-myosin overlap zone as a molecular sieve. J Muscle Res Cell Motil 13: 420–435, 1992

247. Bittl J, Ingwall JS: Reaction rates of creatine kinase and ATP synthesis in the isolated rat hearts. J Biol Chem 260: 3512–3517, 1985

248. Ugurbil K, Petein M, Maidan R, Michurski S, From A: Measurement of individual rate constant in the presence of multiple exchanges: application to myocardial creatine kinase reaction. Biochemistry 25: 100–107, 1986

249. Perry SB, McAuliffe JA, Balschi, Hickey PR, Ingwall JS: Velocity of creatine kinase in the neonatal rabbit heart: the role of mitochondrial creatine kinase. Biochemistry 27: 2165–2172, 1988

250. Matthews PM, Bland JL, Gadian DG, Radda GK: A P-NMR saturation transfer study of the regulation of creatine kinase in the rat heart. Biochim Biophys Acta 721: 312–320, 1982

251. Schoubridge EA, Bland JL, Radda GK: Regulation of creatine kinase during steady state isomeric twich contraction in rat skeletal muscle. Biochim Biophys Acta 805: 72–78, 1984

252. Cerdan S, Seelig J: NMR studies of metabolism. Annu Rev Biophys Biophys Chem 19: 43–67, 1990

253. Mitchell P: Compartmentation and communication in living systems. Ligand conduction: a general catalytic principle in chemical, osmotic and chemiosmotic reaction systems. Eur J Biochem 95: 1–20, 1979

254. Kupriyanov VV, Steinschneider AYa, Ruuge EK, Kapelko VI, Zueva MY, Lakomkin VL, Smirnov VN, Saks VA: Regulation of energy flux through the creatine kinase reaction *in vitro* and in perfused rat heart. Biochim Biophys Acta 805: 319–331, 1984

255. Kapelko VI, Kupriyanov VV, Novikova NA, Lakomkin VL, Steinschneider AYa, Severina MYu, Veksler VI, Saks VA: The cardiac contractile failure induced by chronic creatine and phosphocreatine deficiency. J Mol Cell Cardiol 20: 465–479, 1988

256. Zweier JI, Jacobus WE, Korecky B, Brandejs-Barry Y: Bioenergetic consequences of cardiac phosphocreatine depletion induced by creatine analogue feeding. J Biol Chem 266: 20296–20304, 1991

257. Saks VA, Kapelko VI, Ruda MYa, Semenovsky ML, Strumia E: Phosphocreatine as effective drug in clinical cardiology. In: PP de DeyN, B Marescau, V Stalon, IA Qureshi (eds.) Guanidino Compounds in Biology and Medicine, John Libbey and Company Ltd., London, p 249–252, 1992

258. Bermann MG: Non-catalytic roles of ATP in muscle metabolism and its control. J Mol Cell Cardiol 16: 191–194, 1984

259. Van Beek JGM, Westerhof N: Response time of mitochondrial oxygen consumption to heart rate steps. Am J Physiol 260: H613–H625, 1991

260. Wikman-Coffelt J, Sievers R, Coffelt R, Parmley WW: The cardic cycle: regulation and energy oscillations. Am J Physiol 245: H354–H362, 1983

261. Hochachka PV, Matheson GO: Regulating ATP turnover rates

over broad dynamic work ranges in skeletal muscles. J Appl Physiol 73: 1697–1703, 1992

262. Vignais PV: Molecular and physiological aspects of adenine nucleotide transport in mitochondria. Biochim Biophys Acta 456: 1–38, 1976

263. Brooks SPJ, Suelter CH: Association of chicken mitochondrial creatine kinase with the inner mitochondrial membrane. Arch Biochem Biophys 253: 122–132, 1987

264. Erickson-Viitanen S, Geiger PJ, Viitanen P, Bessman SP: Compartmentation of mitochondrial creatine kinase. J Biol Chem 257: 14405–14411, 1982

265. Clark JF, Kuchua ZA, Kuznetsov AV, Saks VA, Ventura-Clapier R: Compartmentation of creatine kinase isoenzymes in myometrium of gravid guinea pig. J Physiol (London) 466: 553–572, 1993

266. Levitsky DO, Levchenko TS, Saks VA, Sharov VG, Smirnov VN: The role of creatine phosphokinase in supplying energy for the calcium pump system of heart sarcoplasmic reticulum. Membrane Biochem 2: 1–8, 1978

267. Korge P, Byrd SK, Campbell KB: Functional coupling between sarcoplasmic reticulum-bound creatine kinase and Ca-ATPase. Eur J Biochem 213: 973–780, 1993

268. Krause SM, Jacobus WE: Specific enhancement of the cardiac myofibrillar ATPase by bound creatine kinase. J Biol Chem 267: 2480–2486, 1992

269. Saks VA, Lipina NV, Sharov VG, Smirnov VN, Chazov EI, Grosse R: The localization of the MM isozyme of creatine phosphokinase on the surface membrane of myocardial cells and its functional coupling to ouabain-inhibited (Na,K)-ATPase. Biochim Biophys Acta 465: 350–358, 1977

270. Fontaine E, Saks VA, Keriel C, Rigoulet M, Leverve X: Outer membrane of mitochondria controls oxidative phosphorylation in liver cells. In preparation

Note

When this article was accomplished, experiments with isolated and permeabilized hepatocytes showed that also in liver cells Km for ADP in regulation of respiration is high but can be drastically decreased by hypoosmotic treatment of cells [270]. Thus, the outer membrane of mitochondria regulates ADP movement in many types of cells.

Molecular and Cellular Biochemistry **133/134**: 193–220, 1994.
© 1994 *Kluwer Academic Publishers.*

III-2 Creatine kinase in non-muscle tissues and cells

Theo Wallimann and Wolfram Hemmer[1]

Institute for Cell Biology, Swiss Federal Institute of Technology, ETH-Hönggerberg, CH-8093 Zürich, Switzerland;
[1] Present address: University of California, San Diego, Department of Chemistry 0654, 9500 Gilman Drive, La Jolla, CA 92093-0654, USA

Abstract

Over the past years, a concept for creatine kinase function, the 'PCr-circuit' model, has evolved. Based on this concept, multiple functions for the CK/PCr-system have been proposed, such as an energy buffering function, regulatory functions, as well as an energy transport function, mostly based on studies with muscle. While the temporal energy buffering and metabolic regulatory roles of CK are widely accepted, the spatial buffering or energy transport function, that is, the shuttling of PCr and Cr between sites of energy utilization and energy demand, is still being debated. There is, however, much circumstantial evidence, that supports the latter role of CK including the distinct, isoenzyme-specific subcellular localization of CK isoenzymes, the isolation and characterization of functionally coupled *in vitro* microcompartments of CK with a variety of cellular ATPases, and the observed functional coupling of mitochondrial oxidative phosphorylation with mitochondrial CK. New insight concerning the functions of the CK/PCr-system has been gained from recent M-CK null-mutant transgenic mice and by the investigation of CK localization and function in certain highly specialized non-muscle tissues and cells, such as electrocytes, retina photoreceptor cells, brain cells, kidney, salt glands, myometrium, placenta, pancreas, thymus, thyroid, intestinal brush-border epithelial cells, endothelial cells, cartilage and bone cells, macrophages, blood platelets, tumor and cancer cells. Studies with electric organ, including *in vivo* ^{31}P-NMR, clearly reveal the buffer function of the CK/PCr-system in electrocytes and additionally corroborate a direct functional coupling of membrane-bound CK to the Na$^+$/K$^+$-ATPase. On the other hand, experiments with live sperm and recent *in vivo* ^{31}P-NMR measurements on brain provide convincing evidence for the transport function of the CK/PCr-system. We report on new findings concerning the isoenzyme-specific cellular localization and subcellular compartmentation of CK isoenzymes in photoreceptor cells, in glial and neuronal cells of the cerebellum and in spermatozoa. Finally, the regulation of CK expression by hormones is discussed, and new developments concerning a connection of CK with malignancy and cancer are illuminated. Most interesting in this respect is the observed upregulation of CK expression by adenoviral oncogenes. (Mol Cell Biochem **133/134**: 193–220,1994)

Key words: creatine kinase, functional coupling with cellular ATPases, spermatozoa, electrocytes, retina, cerebellum

Abbreviations: M-CK, B-CK and Mi-CK refer to muscle-type, brain-type and mitochondrial-type creatine kinase, respectively, with the cytosolic isoforms MM-, MB- and BB-CK forming dimers and Mi-CK forming dimers as well as octamers; BGC – Bergmann glial cell; GL – granule cell layer; ML – molecular layer; PN – Purkinje neuron; PCr – phosphoryl-creatine; Cr – creatine; cCr – cyclo-creatine; PcCr – phosphoryl-cyclo-creatine; ROS – photoreceptor rod outer segment; IS – inner segment; DNFB – 2,4-dinitro-fluoro-benzene; PEP – phosphoenolpyruvate; PK – pyruvate kinase; GPA – 3-guanidino-propionic acid

Address for offprints: T. Wallimann, Institute for Cell Biology, Swiss Federal Institute of Technology, ETH-Hönggerberg, CH-8093 Zürich, Switzerland, fax 41-1-371-28-94

Introduction

The creatine kinase/phosphoryl-creatine circuit hypothesis for energy homeostasis in muscle and other cells with intermittently high and fluctuating energy requirements

Cells require energy to survive and to carry out the multitude of tasks that characterize biological activity. Cellular energy demand and supply are generally balanced, and tightly regulated for economy and efficiency of energy use. The enzyme creatine kinase (CK; ATP: creatine N-phosphoryl-transferase, EC 2.7.3.2) omit, plays a key role in the energy metabolism of cells with intermittently high and fluctuating energy requirements, such as skeletal and cardiac muscle, neural tissues like brain and retina, or spermatozoa and electrocytes. It catalyzes the reversible transfer of the phosphoryl group from phosphorylcreatine (PCr) to ADP, to regenerate ATP. The enzyme is found in two cytosolic isoforms ('ubiquitous' B-CK and 'sarcomeric' M-CK) [1] and two mitochondrial isoforms (ubiquitous Mi_a-CK, and sarcomeric Mi_b-CK) [2–4].

Using biochemical fractionation and *in situ* immunolocalization techniques on skeletal and cardiac muscle tissue, one was able to show that in sarcomeric muscle, the CK isoenzymes, earlier considered as strictly soluble enzymes (see textbooks), are in fact not distributed evenly within the cells of these tissues [5, 6]. Instead, CK isoenzymes were found to be compartmentalized subcellularly in an isoenzyme-specific fashion [7]. The highly ordered structural organization of muscle and the prevalence in this tissue of specialized cellular ATPases, such as the acto-myosin ATPase and the SR-Ca^{2+}-ATPase, made it possible to study, by biochemical and immunohistochemical methods, the specific subcellular associations of CK with these muscular ATPases. So-called functionally coupled microcompartments could be identified and isolated *in vitro* (for recent reviews see [4] and [8] and refs. therein). For example, in sarcomeric muscle, some cytosolic M-CK is localized at the M-band [6], the sarcoplasmic reticulum (SR) and the plasma membrane. At these sites, M-CK is functionally coupled to the myofibrillar acto-myosin ATPase [9–11], the SR Ca^{2+}-ATPase [12, 13] and at the plasma membrane Na^+/K^+ ATPase [14], respectively, and utilizes PCr for *in situ* regeneration of ATP (Fig. 1). The presence of CK at these sites of high energy demand and the formation of functionally coupled microcompartments conveys ki-

netic and thermodynamic advantages to the system [8, 15]. It was demonstrated that the myofibrillar actin-activated myosin ATPase and the SR Ca^{2+}-ATPase in isolated microcompartments have priviledged access to ATP generated by bound CK, even in the presence of exogenously added ATP-supplying systems or exogenous ATP traps [9–11, 16]. Interestingly, in an *in vitro* model system, using co-immobilized CK and myosin S1, a very similar functional coupling via channelling of substrates and products between the two enzymes in an unstirred layer was observed [17].

The cytosolic as well as subcellularly associated CKs, together with the mitochondrial CK isoforms are thought to constitute an intricate cellular energy buffering and transport system interconnecting via PCr and creatine (Cr) omit, intracellular sites of high-energy phosphate production, i.e. glycolysis and oxidative phosphorylation, with sites of energy consumption, e.g. myofibrils and membrane ion pumps [8] (see Fig. 1). Such a tightly regulated communication between mitochondrial and 'cytosolic' CK isoforms has mainly been demonstrated for muscle, the tissue of choice for studying bioenergetics.

The mitochondrial CK isoenzyme [18, 19], Mi-CK, is located in the mitochondrial intermembrane space [20], where it is found along the entire inner membrane, but also at peripheral sites where inner and outer membranes are in close proximity [20, 21] (for reviews see [4, 8]). There, Mi-CK can directly transphosphorylate intramitochondrially produced ATP into PCr [22], which then is exported into the cytosol where it serves at relatively high concentration (5–40 mM, depending on the tissue) as an easily diffusible energy storage and transport metabolite (Fig. 1). Mi-CK, in contrast to the dimeric cytosolic CK isoenzymes, forms highly symmetrical, cube-like octameric structures that are characterized by a central channel running in parallel to the four-fold axis through the entire molecule [23, 24]. These Mi-CK octamers have the specific ability to peripherally bind to lipid membranes and, most importantly, to mediate contact site formation between inner and outer mitochondrial membranes *in vitro* [25]. Studies done with isolated mitochondria or permeabilized cell culture models, have shown functional coupling of mitochondrial CK with oxidative phosphorylation [22, 26–29]. This functional coupling of Mi-CK to oxidative phosphorylation occurs via the adenine nucleotide translocator (ANT) [22, 27, 29] which catalyzes the antiport of ATP versus ADP through the inner membrane. CK substrates and products also have to pass the outer mitochondrial

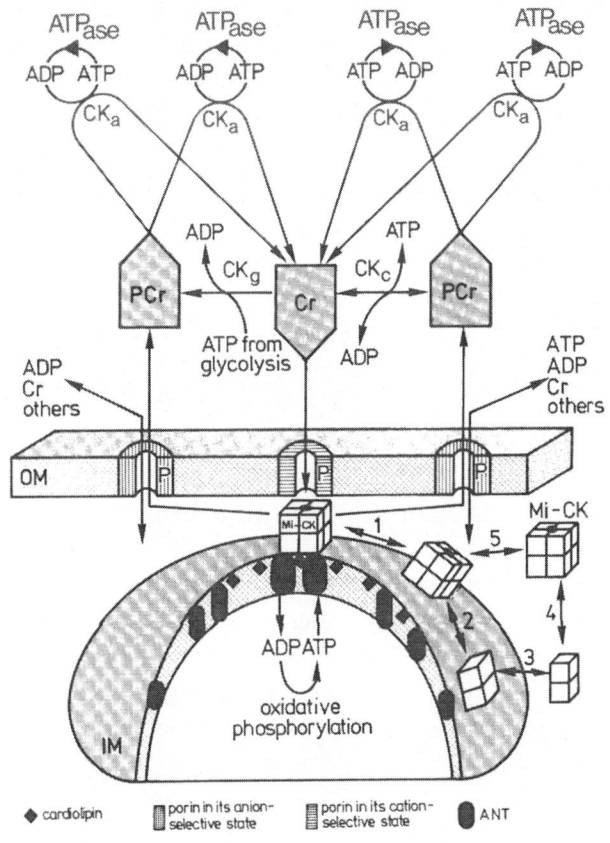

Fig. 1. The phospho-creatine circuit model for specialized cells with high and fluctuating energy metabolism. In a cell, ATP may be derived from two major synthetic pathways, that is, from oxidative phosphorylation (indicated at the bottom) and from glycogenolysis or glycolysis (GL, shown in the left middle part). Four major compartments of CK are indicated: i) strictly soluble cytosolic CK (CK_c) freely equilibrating PCr/Cr and ATP/ADP ratios in the cytosol (shown in the right middle part); ii) 'cytosolic' CK (CK_g) functionally coupled to glycolysis on the producing side of the PCr-circuit (shown in the left middle part), iii) 'cytosolic' CK (CK_a), specifically associated with subcellular structures at sites of high and fluctuating ATP requirements on the receiving end of the PCr-circuit, e.g. at the myofibrils (Refs. [9–11, 16], the sarcoplasmic reticulum (Refs. [12, 13]) and the sarcolemma (Ref. [14]), where functionally coupled CK microcompartments are formed with the myofibrillar acto-myosin ATPase, the Ca^{2+}-ATPase and the Na^+/K^+-ATPase, respectively (ATPase, at the top) [see text]; and finally, iv) mitochondrial CK (Mi-CK) being functionally coupled to oxidative phosphorylation (indicated at the bottom) (Refs. [22, 27, 30]). Note that in resting muscle, for example, the relative pool sizes of phosphoryl-creatine ([PCr] = ca. 20–40 mM) and creatine [Cr = ca. 5–15 mM] are much larger than those of the adenine nucleotides ([ATP] = ca. 3–5 mM; [ADP] = ca. 10–20 μM). Also note that PCr and Cr are smaller and less charged molecules compared to the adenine nucleotides (for Refs. see [8]). At the mitochondrial side, a cube-like Mi-CK octamer with an internal channel (Ref. [24]) is shown to interact with the inner (IM) as well as with the outer mitochondrial membrane (OM), thus stabilizing contacts between IM and OM (Ref. [25]). A Mi-CK octamer is depicted to interact with the ATP/ADP-translocator (ANT) of the IM, and with porin (P) of the OM, transiently forming a dynamic multienzyme 'channel' (Ref. [59]) at the so-called 'mitochondrial energy transfer contact sites' (Ref. [31]). The small black triangles in the IM and in association with ANT's represent cardiolipin molecules. ATP generated by oxidative phosphorylation, after transport through the IM by ANT in exchange for ADP, is transphosphorylated by Mi-CK to give PCr. Due to the functional coupling between ANT and CK (Refs. [22, 27]), PCr, as a net product of oxidative phosphorylation, leaves the mitochondrion through porin (P) of the OM in its high conductance, anion-selective state (for details see [4, 8, 31, 36]). Creatine (Cr), on the other hand, is entering at the contact sites into the intermembrane space, where the porin channel is thought to be in its cation-selective state due to capacitive coupling of the OM to the membrane potential of the IM, both being in close apposition at these sites (Ref. [31]). Possible regulatory aspects of Mi-CK in cellular energetics are depicted at the lower right, e.g. the reversible formation of contacts with ANT and P at the contact sites (arrows, number 1), the dynamic octamer/dimer equilibrium of the enzyme, while bound to the IM (arrows, number 2) or being in solution in the intermembrane space (arrows, number 4), as well as the differential pH-dependent association of the two oligomeric species of Mi-CK with the IM (3, for dimers; 5, for octamers), all observed *in vitro* (Refs. [23, 32]). According to this model, in a cellular system performing work, only small pools of adenine nucleotides (ATP and ADP) are turned over rapidly and in opposite direction at the producing (bottom and middle part of the figure) and the receiving end (top part of the figure) of the PCr-circuit. The model presented here stresses the functional coupling of ATP production with ATP utilization via CK and PCr, as well as the diffusional pathways of PCr and Cr. This pathway may be crucial for the cell at high work-load. However, parallel pathways involving a direct transport of ATP may also operate at the same time. This model, originally developed for muscle (Refs. [4, 8, 59]) is likely to be relevant also for sperm motility, photoreceptor cell function in the vertebrate retina, for smooth muscle, electrocytes, brush borders, etc., as well as for brain energetics (for reviews and references therein see Refs. [4, 8, 36, 38, 70, 71, 133, 140]).

membrane. Based on experiments showing that mitochondrial respiration can be effectively stimulated by extramitochondrial creatine [29, 30], leading to a net production of PCr by mitochondria, a functional coupling between Mi-CK and porin has also been postulated [4, 8, 31] (see Fig. 1).

The fact that Mi-CK is involved in the transphosphorylation, channelling and transport of energy-rich phosphates from mitochondria to the cytosol, together with the molecular structure of the Mi-CK octamer itself, which is very reminiscent of a 'channel protein', led to the hypothesis that Mi-CK could act as a connecting module between ANT and porin at the mitochondrial contact sites [4, 8, 31], thereby forming an efficient, tightly coupled multienzyme 'energy channel' [4, 8, 31] that combines the directed export of mitochondrial energy equivalents with the interconversion of matrix-generated ATP plus Cr into ADP plus PCr (see Fig. 1). Specific

196

features of Mi-CK, e.g. a dynamic octamer/dimer equilibrium which is influenced by physiological parameters [32], as well as the differential pH-dependent interaction of Mi-CK octamers and dimers with inner mitochondrial membranes observed *in vitro* [23, 32] (see Fig. 1) may be important parameters for the regulation of mitochondrial energetics. Based on these results, mainly obtained with muscle, the 'phospho-creatine shuttle' [16, 22, 33] or 'PCr-circuit' [8, 34] models have been formulated (see Fig. 1), and several main functions of the CK/PCr-system have been proposed (see [4, 8, 34–36].

First, the CK/PCr-system is thought to serve as a temporal energy buffer [37] keeping [ATP] and [ADP] steady and buffering [H⁺]. This temporal buffer function prevents a rapid fall of [ATP] and a large build-up of [ADP] during cellular work and at the same time avoids an intracellular acidification due to the hydrolysis of ATP during work. These functions, which have also been proposed for CK in photoreceptor cell outer segments (ROS) [38], have recently been fully confirmed by transgenic animal approaches. For example, in mice expressing significant amounts of B-CK in liver, where normally no or very little CK is present, a strong buffering effect of CK on the ATP levels and the intracellular pH has been directly demonstrated upon perfusion of the liver under anoxic conditions [39–41]. In addition, also with the transgenic liver model it was shown that [ADP] is inversely proportional to the levels of B-CK expressed in these livers and that PCr protected ATP levels from the effects of a fructose load in the same animals [40]. Furthermore, the temporal buffer function of the CK system has very recently also been confirmed in muscle with gene-targeted null-mutant transgenic mice which do not express 'cytosolic' muscle-type MM-CK [42]. These mice display a distinct phenotype with significantly reduced muscle burst activity, that is, in the first phase of muscle contraction, normal peak tension is reached, but this peak tension rapidly falls to 50–60% of the control value seen with wild-type mice [42].

Second, the PCr-circuit serves to improve the thermodynamic efficiency of ATP hydrolysis by keeping intracellular [ADP] low and by maintaining ATP/ADP ratios high at those subcellular sites where CK is functionally coupled to ATP-requiring processes, like ion pumps (see CK_a, Fig. 1 top) that are largely dependent on a high affinity of ATP hydrolysis (A) in the sarcoplasm, with $A = \Delta G°_{obs} - RT \ln ([ATP]/[ADP]*[Pi])$, where $\Delta G°_{obs}$ is the standard free energy of ATP hydrolysis, R is the gas constant and T is the absolute temperature [12, 13, 15, 43,

44]. Recent examples supporting this CK function are the observed functional and kinetic coupling of CK to the sarcoplasmic Ca^{2+}-ATPase ion pump [12, 13] and to the myofibrillar actin-activated myosin ATPase [9, 10, 29, 35, 45, 46]. The improvement by the CK system of the thermodynamic efficiency of ATP hydrolysis is important for some of these pumps, and seems crucial for the sarcoplasmic (SR) Ca^{2+}-ATPase [12, 13]. This SR pump operates with a-$\Delta G_{transport}$ of approximately 51 kJ/mol (see [44]). The free energy of ATP hydrolysis (-$\Delta G_{ATPhydrolysis}$) at physiological concentrations of ATP (5–8 mM), ADP (10–50 μM) and P_i (5–10 mM) in resting muscle may be estimated to be approximately 55 kJ/mol. This is only slightly larger than -$\Delta G_{transport}$, implying that the SR Ca^{2+} pump operates close to thermodynamic equilibrium [15, 43, 44] and depends very much on a high local ATP/ADP ratio for efficient sequestration of Ca^{2+} into the SR lumen.

Third, an important and seldom recognized consequence of the operation of the CK reaction is the net release of inorganic phosphate (P_i) from PCr. For example, during the first phases of muscle work, P_i increases proportionally to the amount of PCr hydrolysed, while [ATP] and [ADP] levels remain stable [47]. P_i exerts a regulatory effect on glycogenolysis and glycolysis, since it stimulates phosphorylase [48] and phosphofructokinase [49] and additionally relieves the inhibition of hexokinase by glucose-6-phosphate [50]. Thus, in muscle, the availability of P_i may become rate-limiting for glycogenolysis in the absence of PCr hydrolysis, a fact supported by ^{31}P-NMR work (see [51]). A functional coupling of CK with glycolysis, also postulated in our model (Fig. 1) and supported by the colocalization of CK with glycolytic enzymes in muscle [52, 53], could recently be directly demonstrated by ^{31}P-NMR methods in anoxic fish muscle [54].

Fourth, the PCr-circuit also serves as a *spatial energy buffer* or transport system. In this role, PCr is thought to function as an 'energy carrier' connecting sites of energy production, e.g. mitochondrial oxidative phosphorylation, with sites of energy utilization, whereby mitochondrial CK playing an eminent role in this process. This function of CK is supported i) by the specific subcellular compartmentation of the different CK isoenzymes in a variety of tissues, such as muscle, electrocytes, photoreceptor cells, and spermatozoa, the latter two representing highly polar cells (see below and [8]), ii) by evidence indicating subcellular compartmentation of PCr/Cr, ATP/ADP and P_i [29, 55–58], iii) by the localization, structure and functional properties of the Mi-CK octameric molecule which seems very well suited for metabolite channeling [4, 8, 59] and iv) by elegant *in vivo* [18]O labelling of the phosphoryl moieties of metabolites in intact diaphragm muscle showing that discrete adenine nucleotide pools exist in these cells and, most importantly, that the rates of appearance of [^{18}O]-PCr are consistent with a CK-catalysed phosphoryl exchange functioning in an obligatory PCr shuttle [58]. Earlier, using a similar isotope technique, the same group provided evidence for compartmentalized *in vivo* catalysis of adenylate kinase activity as well [60]. In connection with the transport function of CK, it is important to note that by speeding up 'communication' between sites of ATP production and ATP consumption, the PCr-circuit model predicts that transitions between different work states are accelerated and smoothened, that oscillations in [ADP] and [ATP] are dampened and that simultaneously the transient times for reaching a new steady-state at a given work load are reduced [4]. This is in line with the fact that in skeletal muscle and cardiomyocytes *in vivo*, the diffusion of adenine nucleotides is severely hindered compared to that of PCr and Cr [29, 61, 62]. Since upon muscle activation, [ATP] remains rather constant and [ADP] rises only little, whereas [PCr] and [Cr] can change drastically [47], the latter compounds are more likely candidates for building up gradients which would speed up trafficking of high-energy phosphates, as a PCr/Cr-shuttle system, from sites of ATP generation to sites of ATP utilization. While the transport function of this PCr-shuttle or PCr-circuit seems to be supported by ample evidence mostly from *in vitro* studies with muscle [6, 8, 18, 34, 36, 63, 64], it turned out very difficult to prove and verify this transport function of CK *in vivo*. Conflicting results have been produced by different research teams (see [35, 36, 42, 65–68]). Whereas in the glucose perfused heart, the flux through the CK reaction, as measured by ^{31}P-NMR, correlates well with the cardiac muscle performance [65, 66], no correlation, or even a negative one, has been reported for skeletal muscle [67]. By contrast, however, a clear correlation of CK-catalysed flux with brain activity has recently been demonstrated on *in vivo* brain, using NMR magnetization transfer [68] (see below). Curiously, *in vivo* ^{31}P-NMR measurements showed that MM-CK deficient transgenic mice [42] or mice expressing less than 30% of normal MM-CK (Dr. B. Wieringa, personal communication), each of which expresses normal levels of Mi-CK, still utilize PCr for muscle contraction. However, inversion transfer measurements revealed no measurable flux through the CK forward reaction [42], indicating that a

significant fraction of CK-reaction flux is inaccessible to NMR, or NMR-'invisible'. Interestingly, as soon as the MM-CK levels reached 30–40% or more, suddenly a normal 100% CK-flux was registered as in control animals, indicating that it is mainly the unbound excess of cytosolic CK which is responsible for the measured *in vivo* CK-reaction flux (see [36], Dr. B. Wieringa, personal communication). In light of this unexpected behaviour of the CK-system in NMR terms, the interpretation of earlier CK-flux measurements by ^{31}P-NMR, especially with skeletal muscle, will have to be revisited [36]. At this level of non-invasive *in vivo* measurement, the PCr-shuttle issue is certainly not yet resolved and some surprises may be ahead [36].

In recent years, CK of non-muscle cells, such as spermatozoa, electrocytes, photoreceptor and brain cells, which, like muscle cells, are also characterized by intermittently high and fluctuating energy requirements, has attracted considerable interest. Therefore, the following chapters are meant to give an update on the localization and function of various CK isoenzymes in a variety of non-muscle tissues.

Localization and function of CK isoenzymes in spermatozoa

Vertebrate sperm

Spermatozoa are highly specialized, very polar cells with a DNA-containing head, which may be round or elongated, depending on the species, and which additionally harbours the acrosomal vesicle at its very tip. Adjoining the head is a short midpiece containing mitochondria followed by a very long (50–150 µm) flagellar tail containing the axoneme, basically consisting of microtubules and dynein. The dynein motor protein uses ATP as the direct energy source for the movement of microtubules relative to each other. The flagellar wave thus generated ultimately leads to sperm movement. As shown for muscle [61], it is likely that the delivery of high energy phosphates, such as the rather bulky and negatively charged ATP molecule, from the mitochondria to the distal axoneme, may also be severely diffusion-limited in sperm [69–71]. This limitation may be overcome by the relatively large amounts of CK and total creatine (Cr) present in spermatozoa [70, 71]. In rooster and human sperm, two different types of CK isoenzymes, brain-type B-CK and mitochondrial-type Mi-CK, have been identified by native isoenzyme electrophoresis and

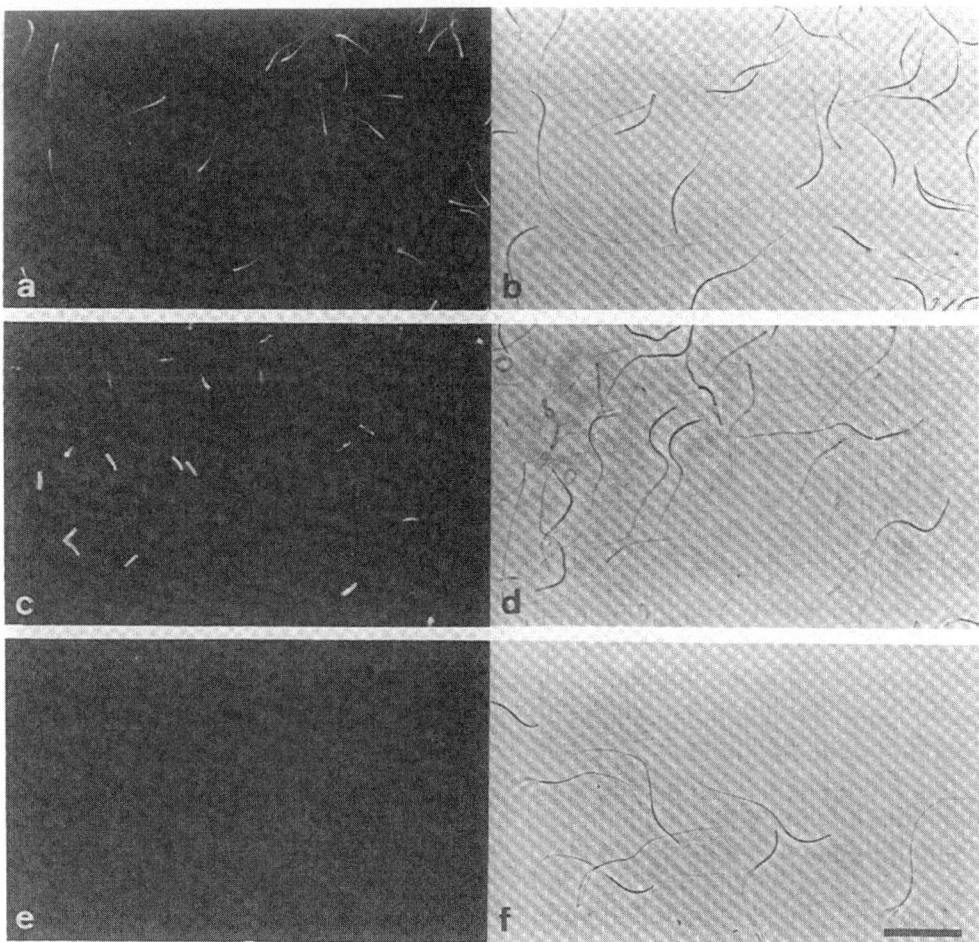

Fig. 2. Compartmentalized localization of creatine kinase isoenzymes in spermatozoa. Indirect immunofluorescence of rooster spermatozoa, stained with specific antibodies against chicken brain-type cytosolic B-CK (a, b); against chicken mitochondrial Mi-CK (c, d); or with preimmune control antibody (e, f), all followed by FITC-conjugated second antibody (Ref. [70]), P. Kaldis *et al.* in preparation). Note the specific staining by anti-B-CK antibodies of the entire sperm tail, but neither of the midpiece nor the sperm head, which in roosters are both rather elongated. The staining intensity with anti-B-CK is often tapering off somewhat towards the distal end of the sperm tail (a). Note also the specific 'box-like' staining by the anti-Mi-CK antibody of the mid-piece where the sperm mitochondria are located (c) and the negative background obtained with pre-immune antibody (e); b, d, and f are the phase contrast images corresponding to the fluorescence pictures a, c and e; bar = 20 μm.

immunoblotting [70]. In addition, by indirect immuno-fluorescence, B-CK was found along the entire length of rooster sperm tails (Fig. 2a, b), whereas Mi-CK was localized specifically to the rectangular midpiece, just behind the sperm head [70] (Kaldis *et al.* in preparation) (Fig. 2c, d), which, in rooster spermatozoa, is rather elongated (Fig. 2). Thus, an isoenzyme-specific subcellular segregation of CK isoenzymes is seen in these spermatozoa [70, 71]. Most importantly, rooster sperm motility is inhibited by CK blockers [70], indicating a dependence of sperm motility on the CK/PCr-system.

Surprisingly, in a study assessing the fertilizing potential of sperm from infertile male patients, an inverse correlation between sperm CK activity and sperm concentration in the specimens of normospermic and oligo-spermic infertile men was found [72]. Upon closer examination of sperm morphology, this puzzling fact could be explained by a higher retention of cytoplasm in spermatozoa of infertile men [72]. During normal sperm maturation, excess cytoplasm is lost to the Sertoli cells as residual bodies. Thus, the often observed higher CK content per sperm cell in infertile men can be attributed to the presence of immature or abnormal spermatozoa with irregular head size and retention of cytoplasmic CK due to incomplete sperm development [73]. While the presence in human sperm of brain-type B-CK is not disputed, the question about the identity of the second CK species is not fully settled. According to our study, the second CK species in human sperm is mitochondrial Mi-CK [70] but, according to Dr. Huszar's group [74], it was suggested to be the cytosolic muscle-type M-isoform of CK. Unexpectedly, the mitochondrial CK isoenzyme present in rooster sperm, which is a non-muscle cell, is sarcomeric muscle-type Mi_b-CK (P. Kaldis et al. in preparation). It will be intriguing to see whether, similar to lactate dehydrogenase (for refs see [70]), a possible additional, sperm-specific CK isoform [75] can be identified in human sperm in the future. Independently of the exact nature of this latter CK isoenzyme in human sperm, a low relative ratio of this 'Mi-CK' isoform to B-CK seems to be a good diagnostic sign for a low fertilizing potential of men as shown in a blinded study involving 84 infertile couples [74], indicating that CK and especially 'mitochondrial' CK is a key player also for human sperm function.

Sperm activation and energy metabolism are more complex in vertebrate spermatozoa than in sea urchin sperm [69, 71]. While sea urchin spermatozoa thrive exclusively on mitochondrial respiration by fatty acid oxidation [69, 71], vertebrate spermatozoa are motile also under low-oxygen conditions, deriving the chemical energy additionally from glycolytic pathways [76]. Furthermore, unlike sea urchin sperm, which are released into the sea water as 'dry' sperm, vertebrate spermatozoa are supplemented with relatively large volumes of seminal fluid provided by the seminal vesicles and the accessory prostate gland. The supply of substrates by seminal fluid which contains large amounts of fructose etc. [77] may be relevant for vertebrate sperm motility, since glucose, fructose and mannose were shown to support vertebrate sperm motility under either aerobic or anaerobic conditions [76]. Interestingly, seminal vesicle epithelium cells contain large amounts of PCr and Cr which are released into the seminal vesicle fluid [78]. The accumulation of these compounds in the above cells as well as in the seminal vesicle fluid is regulated by testosterone [79]. Thus, seminal vesicle epithelial cells are the first example of cells that 'secrete' PCr and Cr. An indication that external supply of high energy phosphates to vertebrate sperm via seminal vesicle fluid, representing 50–70% of the seminal plasma, may be important for human sperm motility is corroborated by the fact that upon addition of PCr to human ejaculates or isolated sperms, sperm motility as well as sperm velocity were both enhanced in vitro [80]. Blocking rooster sperm CK activity with 10–20 µM 2,4-dinitro-fluoro-benzene (DNFB) leads to significant loss of sperm motility [70]. In addition, interference with the delivery of mitochondrial ATP, by blocking the ATP/ADP translocator with 50 µM carboxyatractyloside and at the same

time inhibiting glycolysis with 20 mM deoxyglucose, leads to a complete loss of rooster sperm motility. Subsequent addition of 10 mM PCr plus 1 mM ADP, in the presence of 0.5 mM diadenosine-pentaphosphate to block adenylate kinase, led to a significant recovery of in vitro sperm motility [70], again indicating that PCr and endogenous CK can support flagellar wave movement.

It is still unclear, however, whether in vivo, external PCr in the seminal fluid can be taken up directly by intact sperms and thus be utilized as an immediate energy source, or whether it is metabolized by B-CK, which is also present both in seminal vesicle and prostate fluid [81]. In this latter case, PCr would mediate its positive effects on sperm motility [80] by an indirect mechanism. Based on these results, the controlled interference with the CK system in spermatozoa, either by supplementing CK substrates or by specifically inhibiting CK activity, may turn out to be a valuable tool for the treatment of certain types of male infertility or for male contraception, respectively, in the future. Although statistically significant differences in CK activity, which is exclusively brain-type B-CK, were found in seminal plasma of men with normal spermiograms of oligoasthenozooic sperm, the individual values show an overlap which is too wide for routine diagnostic purposes [82].

Sea urchin sperm

The most convincing evidence for a crucial role of the CK/PCr-system in sperm motility comes from studies with sea urchin sperm. These sperm can be obtained in large quantities in homogeneous form and are reversibly activatable by simply changing the ionic and pH conditions of the sea water [69]. In vivo, sea urchin sperms are spawned directly into sea water where they are activated within seconds by sodium influx via a Na^+/H^+-exchanger present in the sperm plasma membrane, leading to an increase in intracellular pH and finally of $[Ca^{2+}]$ [69]. Two different CK isoenzymes have also been observed and isolated from sea urchin spermatozoa [83], a very large unusual tail-CK with a M_r of 145'000 and a multimeric mitochondrial isoform with an apparent subunit M_r of 47'000, which turned out to be an octameric Mi-CK, similar to that of vertebrates ([4], Wyss et al., unpublished). Genetic analysis revealed that the gene for the tail-CK, named according to its location in the sperm tail, contains three contiguous but non-identical CK segments, most likely arisen by two gene duplication events, joined by non-CK-like connectors and

200

flanked by unique N- and C-termini [84] (see also [85]). The unique tail-CK is a lipophilic protein capable of interacting with sperm membrane preparations [86]. The anchoring of tail-CK to membranes and to phospholipid liposomes *in vitro* [86] is facilitated by myristoylation of the protein at the very N-terminus containing a corresponding consensus sequence [87]. In addition, some evidence suggests that tail-CK may also interact with microtubules of the axoneme [88]. Thus, the isoenzyme-specific segregation of octameric Mi-CK ([4], Wyss *et al.* unpublished) to the sea urchin head and of tail-CK to the cytosol, the plasma membrane and the axoneme, seems to represent a prerequisite for a PCr-shuttle operating also in spermatozoa.

As a matter of fact, evidence obtained by ^{31}P-NMR shows that the large pool of PCr in sperm is depleted or resynthesized, depending on whether sperm motility is induced or inhibited, respectively, by simply altering intracellular pH [89]. This indicates that CK and PCr are indeed directly involved in sperm motility. In a series of elegant experiments with live sea urchin sperm, a direct dependence of sperm motility on an intact CK system could be demonstrated [71]. By specifically blocking CK activity, but not mitochondrial respiration, with low concentrations of 2,4-dinitro-fluoro-benzene (DNFB), the tight coupling of energy utilization and energy production observed in sea urchin sperm is interrupted. Inhibition of CK activity in sperm impedes the transport of high-energy phosphates from the mitochondria to the axoneme and affects the pattern of sperm motility in the manner predicted if energy transport from the sperm head to the tail were diffusionally limited [69]. For example, by computer-assisted analysis of stroboscopic photomicrographs of DNFB-treated and control sea urchin sperm, it was shown that progressive inhibition of CK in these live sperms results in a progressive flagellar wave attenuation, whereby the distal part of the flagellum with the largest distance from the sperm head is affected first and most [90]. In addition to the attenuation of sliding velocity between flagellar microtubules in the distal region of the sperm flagellum, CK-inhibited spermatozoa generate a flagellum bending pattern with shorter wavelengths [90]. These specific alterations of sperm motility, seen after inhibition of CK activity in live sperm, and their reversal upon subsequent addition of ATP following demembranation of the same sperm preparation, provide very strong support for the CK-mediated PCr-shuttle in high energy phosphate transport in spermatozoa, with PCr and Cr being the metabolites channeled along the sperm tail [71]. In our opinion,

the whole-cell sperm model is the most convincing 'living proof', so far, for the spatial buffering or energy transport function of the CK/PCr-system proposed in the PCr-circuit model [8, 36, 69–71, 74].

Localization and function of CK in electrocytes of the electric organ of electric fish

Electric fish contain a large bi-lobed electric organ, originally derived from myogenic cells, which is composed of large disc-shaped electrocytes stacked on top of each other to form the many columns of the electric organ. These highly specialized electrocytes, unlike muscle cells, lack a contractile apparatus. Electrocytes are specifically stimulated by the release of acetylcholine from presynaptic nerve terminals of nerves originating in the electric cortex of the fish's midbrain. A large number of nicotinic acetylcholine receptors are present at the ventral, postsynaptic electrocyte membrane. Upon ligand-gated opening of the receptors, sodium ions enter the cells through the receptors, which function as ion channels. The combination of a large number of electrocytes, each approximately 100 μm in height, into many columns enables electric fish to produce a current of sufficient strength to stun or kill their prey or to defend themselves against aggressors. After electric discharge, the intracellular sodium ions are extruded from the electrocytes by the Na$^+$/K$^+$-ATPase located within the non-innervated, dorsal membrane face, which is highly invaginated by tubular infoldings, called canniculi. By this way, the surface area of the dorsal membrane system harboring the energy-requiring Na$^+$/K$^+$ ion pump is greatly increased (see [91]).

Electric organs of a variety of different electric fish contain high concentrations of CK [92, 93]. The major CK isoform in electric organ is identical to the cytosolic CK found in the skeletal muscle of the same fish, as was confirmed by protein sequencing and cDNA cloning [94–97]. The post-synaptic membrane of *Torpedo* electrocytes is specifically associated with a CK [98] showing a subunit M$_r$ of 43'000, which was also named the 'acetylcholine receptor-rich membrane-associated peripheral υ$_2$-protein' before it was identified as genuine CK [93, 98, 99]. Based on immunological cross-reactivity with heterologous anti-chicken B-CK antibodies, but not with anti-M-CK antibodies, this protein was first postulated to be brain-type B-CK [99, 100], but was subsequently shown to be genuine M-CK [97, 101]. This discrepancy can now easily be explained, since direct sequence comparison of *Torpedo* electric organ CK with the more than 20 other CK sequences known so far shows sequence motifs in the electric organ CK that are related both to M- and B-type CK's [85]. The above controversy distracted the scientific community from the facts that i) more than one CK isoenzyme is present in *Torpedo* electric organ [100], with the second isoenzyme probably being MB-CK [92, 97], and ii) that CK in electric organ, besides being a soluble cytoplasmic protein, is not only associated as peripheral υ_2-protein with the innervated, postsynaptic acetylcholine receptor-rich membrane [93, 98, 99] but, as shown by immuno-electron microscopy, is also bound to the dorsal non- innervated cannicular plasma membrane face [98] where the ouabain-sensitive Na^+/K^+-ATPase is situated [91]. The colocalization of CK at the dorsal membrane with the Na^+/K^+-pump suggests a role for CK in the regeneration of ATP, via PCr, to fuel the Na^+/K^+-ATPase during recharging of discharged electrocytes ([98], see below). This role serves to keep the local ATP/ADP ratios high in the vicinity of the ion pump, thus increasing the thermodynamic efficiency of ATP hydrolysis and of ion pumping [8, 15, 36, 43, 44]. As in muscle, an additional function of the CK system in electrocytes may be to buffer the intracellular pH, that is, while regenerating the ATP utilized for ion pumping, the CK reaction reutilizes the protons generated by the ATPase reaction [8].

The major, almost exclusive energy utilizing process in electrocytes in the Na^+/K^+-ATPase, which is fully operative after an electric discharge of the electric organ [91]. The aerobic and anaerobic metabolic synthetic pathways in this organ are rather slow [91]. Upon discharge of the electric organ, the high levels of PCr fall rapidly [102], due to the almost exclusive utilization by CK of PCr for the fueling of the fully activated Na^+/K^+-ATPase [91]. This is corroborated by experiments showing that ouabain, a blocker of this ion pump, prevents PCr depletion in discharged electric organ [103]. Most importantly, an intimate functional coupling of CK with the Na^+/K^+-ATPase has been directly demonstrated by *in vivo* saturation transfer ^{31}P-NMR measurements in the resting and stimulated electric organ, showing a highly increased flux through the CK reaction after discharge of the electric organ [103]. Thus in electrocytes, the high energy demands after electric discharge are met at the expense of PCr, and the recovery of the membrane potential during recharging is closely related to the restoration of intracellular PCr levels [102, 103].

In addition, an association between cytosolic CK and synaptic vesicles has been demonstrated in electrocytes of *Torpedo* [98], and the release of acetylcholine from *Torpedo* synaptosomes was shown to be severely affected by inhibition of CK with DNFB [104]. Since mitochondrial content of electrocytes is rather low, and so far, no Mi-CK has been detected in this tissue, a contribution of mitochondrial ATP and/or mitochondrially-derived PCr to the Na^+/K^+-ATPase seems unlikely. However, due to the fact that in electrocytes, the ATP synthetic pathways are slow, but after a discharge, large amounts of energy are needed for recharging, electric organ represents an instructive example for an almost exclusive temporal energy buffer function of the CK/PCr-system, whereas sea urchin sperm represent a similarly instructive example for the spatial buffering or energy transport function postulated in the PCr-circuit model [8, 36].

CK isoenzymes in photoreceptor cells of the retina and in the lens of the eye

Compartmentalized localization of CK isoenzymes in retina

Vertebrate photoreceptor cells of the retina, which are specialized neurons consisting of an outer segment, connected by a thin stalk to the inner segment, the nucleus and synaptic terminations, show a highly polar organization analogous to spermatozoa. The major energy requiring reactions of phototransduction (e.g. the regeneration of cGMP from ATP and GTP, hydrolysed as a consequence of photic stimulation [105]) take place in the outer segments of photoreceptor cells. The distance from the outer segment to the inner segment, where oxidative phosphorylation takes place, can range between 20–50 μm, depending on the species. The highly clustered mitochondria of photoreceptor cells are confined to the ellipsoid portion of the photoreceptor cells' inner segment [38, 106]. Since oxidative phosphorylation is crucial for photoreceptor cell function [107] the question again arises of how high-energy phosphates are transported from the inner segment through the narrow space of the connecting cilium into the outer segment, which is occupied by stacks of photo-sensitive membranes and thus imposes severe diffusion limitations even on small compounds, such as cyclic-GMP [108].

Two isoforms of CK, brain-type (B-CK) and mitochondrial CK (Mi-CK) were found in chicken and bo-

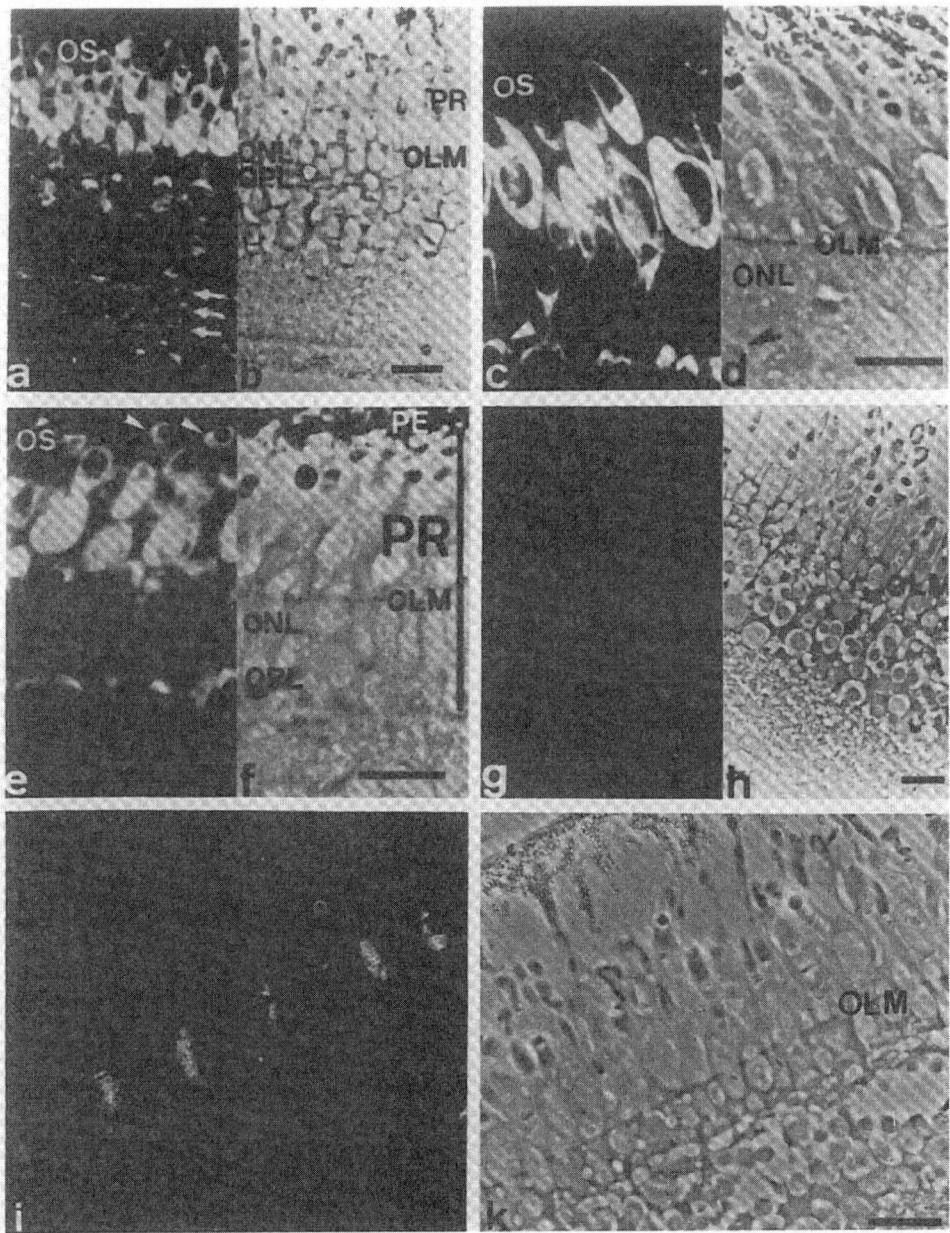

Fig. 3. Immunofluorescence localization of CK isoenzymes in chicken retina: compartmentation of cytosolic and mitochondrial CK in photoreceptor cells. Indirect immunofluorescence of thin cryosections of a retina from adult chicken prepared by the gelatine support technique (for methods see Refs. [109,110]) stained with specific antibodies against chicken brain-type cytosolic B-CK (a–h) and against mitochondrial Mi-CK (i, k). Fluorescence images (a, c, e, g, i) are displayed with their corresponding phase contrast pictures (b, d, f, h, k). Overview of a chicken retina at low magnification showing anti-B-CK staining of the different retinal layers (a, b). Note that relative to the other cell layers of the retina, the photoreceptor cell layer (PR) is stained most intensely by anti-B-CK antibody, with some stratified staining of the inner plexiform layer (triple arrows in a) and a rather strong staining also of the synaptic region with horizontal cells (arrowhead in c). Somewhat higher magnification depicts strong staining with anti-B-CK antibody of the inner and outer segments of photoreceptor cells (c, d), especially of the myoid portion above the outer limiting membrane (OLM), whereas slightly oblique sections through the photoreceptor cell layer also reveal 'ring-like' immuno-staining of outer segments (OS) (see arrowheads in e), just below the pigmented epithelium (PE). By contrast, mitochondrial Mi-CK localization is entirely restricted to the ellipsoid portion of the inner segment of the photoreceptor cells, where mitochondria are clustered in large numbers (Refs. [38, 107, 110]). Thus on the subcellular level, cytosolic B-CK, present in the inner and outer segment (Ref. [38]), and mitochondrial Mi-CK, present in the inner segment only, are spatially segregated in an isoenzyme-specific manner, thus forming the physical basis for a PCr-circuit in photoreceptor cells (Ref. [38]). PR, photoreceptor cell layer; OS, outer segment, OLM, outer limiting membrane; PE, pigmented epithelium; ONL, outer nuclear layer; OPE, outer plexiform layer; bar = 25 μm.

vine retina [38, 109]. Both of these CK isozymes are expressed at high levels but are distributed differentially in photoreceptor cell inner segments. Mi-CK is restricted to the mitochondria-rich ellipsoid portion, while B-CK is localized both to the ellipsoid and myoid portion (Fig. 3) of chicken inner segments [109].

Although some data from immunofluorescence localization studies [109, 110] (see also Fig. 3), as well as biochemical and physiological measurements [111, 112] suggested that CK is present also in outer segments, unambiguous evidence for the presence of B-CK in photoreceptor cell outer segments, as well as a quantitative analysis thereof, has only been provided recently [38]. The presence of B-CK isoenzyme in rod outer segments (ROS) of chicken retina was initially indicated by the results of immunofluorescence labelling of thin frozen sections of chicken retina (Fig. 3). However, this could only really be corroborated by immuno-gold labelling of bovine retina and has additionally been confirmed by immunoblotting and immunolabeling of isolated bovine ROS, as well as by biochemical characterization of isolated ROS [38]. Thus, while Mi-CK is restricted to the ellipsoid portion of the inner segment, B-CK is present in the inner as well as the outer segment. Therefore, in photoreceptor cells as in spermatozoa, the two CK isoenzymes are in part spatially segregated. The content of creatine kinase in isolated ROS was quantified by measuring creatine kinase activity after membrane disruption with detergent [38]. The ATP regeneration potential provided by the creatine kinase in isolated, washed bovine ROS was 1.2 ± 0.4 IU $\cdot$ mg^{-1} rhodopsin. This value was calculated to be at least an order of magnitude larger than that necessary to replenish the energy required for cGMP resynthesis in ROS, and high enough to regenerate the entire ATP pool of ROS within the time span of a photic cycle (see [38]). Since Mi-CK expression and accumulation in the chicken retina coincide with the functional maturation of the photoreceptors around the time of hatching, this enzyme may represent a good marker for terminal differentiation of these cells [110]. B-CK, on the other hand, is present from early stages of retina development and seems to be relevant for the energetics of retinal cell proliferation, migration and differentiation. The simultaneous expression of both B- and Mi-CK around the time of hatching indicates a coordinated function of the two CK isoforms as constituents of a PCr-circuit in the energetics of vision, which, in autophagous birds like the chicken, has to be operating right after hatching [110].

In photoreceptor cells, which depend on mitochon-

drial oxidative phosphorylation, ATP produced in inner segment mitochondria may be transphosphorylated by Mi-CK to provide PCr for those fractions of cytosolic CK which are associated specifically with sites of energy consumption, e.g. the Na$^+$/K$^+$-ATPase in the plasma membrane of the inner segment or various ATP-requiring processes in the outer segment, where this bound CK regenerates the ATP and maintains high local ATP/ADP ratios. Outer segment CK, which in part may also be associated with the plasma membrane, would play an important role in phototransduction by providing energy for the visual cycle, maintaining high local ATP/ADP ratios and consuming protons produced by ATPases located in the outer segment and thus preventing an acidification of the outer segment compartment (for details see [38]).

The remarkable ATP-regeneration potential provided by B-CK present in ROS indicates that CK-dependent ATP generation may play a major role in many aspects of ATP function in ROS. For example, ATP is not only utilized as an energy source to regenerate cGMP, which is hydrolysed during photic stimulation, but also seems to be a regulatory factor influencing cGMP levels by rapidly quenching light-induced phosphodiesterase activity [113]. This dual role of ATP, as a direct energy source and as a regulatory molecule, has been confirmed by independent groups [114, 115].

Thus, due to its ability to regulate ATP levels in photoreceptor cells, CK in ROS may be involved directly in ATP-mediated regulatory control of phototransduction. Such multiple potential roles of CK in phototransduction, a process that is tightly regulated at many levels, implies that B-CK activity, its intracellular distribution, or even both properties are also likely to be subject to control. A possible regulatory mechanism affecting either of the above properties may be protein phosphorylation. B-CK of rat [116], mouse [117] and chicken [118] are indeed phosphoproteins. The heterogeneity of bovine ROS B-CK observed on 2D gels [38] was similar to that reported for B-CK from other species, suggesting that some of the B-CK in bovine ROS is also phosphorylated. Phosphorylation was shown to alter the kinetic properties of B-CK [117, 118]. This post-translational modification might also regulate the distribution of B-CK between membranes and cytosol, as has been demonstrated for other proteins [119, 120].

CK in the cuboidal epithelium of the eye's lens

Interestingly enough, neonatal rat and human lens were reported to express only a cathodic variant of CK, while near the time of sexual maturation, a dramatic increase in the expression of B-CK and to some extent also of M-CK was noted [121]. It is possible that in lens, as in other tissues such as brain, uterus, placenta, amnion, decidua and mammary gland, B-CK expression is also stimulated by hormones (see [122]). This is supported by the fact that differentiation of lens epithelial cells has been demonstrated in response to vitreous liquid and serum, which both contain hormones [123]. B-CK was localized to the cuboidal epithelial cells of the adult rat lens [121]. These cuboidal epithelial cells, which cover the anterior surface of the lens, have important ion-transport transport functions. Therefore, it is possible that B-CK localized in these cells is functionally associated with areas of high transport ATPase activity.

Localization and functions of CK isoenzymes in brain

CK isoenzymes in brain

Although the relative distribution of CK in different areas or cell types of the brain has been investigated in numerous studies, there is still no consistent and complete overview of the localization of CK isoenzymes in the brain. Regional variations in CK activity with comparably high levels in the cerebellum were reported in studies using native isoenzyme electrophoresis [124] or enzymatic CK activity measurements of either tissue extracts [125] or cultured brain cells [126]. In particular, the molecular layer of the cerebellar cortex contains high levels of CK activity [125, 127], consistent with the recent ^{31}P-NMR findings which indicate that grey matter shows a higher flux through the CK reaction and higher PCr concentrations as compared to white matter [128]. In contrast, high levels of either CK activity or corresponding mRNA were shown in cultured oligodendrocytes [126, 129], typical glial cells of the white matter. B-CK, the major 'cytosolic' CK isoenzyme present in brain [1], has been characterized extensively [118, 130–133]. For chicken B-CK, considerable heterogeneity was found; there are two major B-type subunits and additional subspecies arising from alternative ribosomal initiation [134] and post-translational modifications [118, 132, 135]. Already in the sixties, it was reported that CK activity is

associated with brain mitochondria [18, 136]. This enzyme activity was characterized as a genuine brain mitochondrial CK (Mi-CK) [137–139] and later identified as the so-called 'ubiquitous' Mi_a-CK isoform [2, 3, 139]. Mi_a-CK, which was also characterized extensively [4, 139], is localized preferentially as octamers in contact sites of brain mitochondria [20, 31]. In several early studies, the presence of muscle-type M-CK in brain was also postulated (for review see [133, 140]). M-CK has indeed been demonstrated recently in postmortem human brain extracts by biochemical isolation and protein sequencing [141] and in chicken cerebellum by immunoprecipitation, immunoblotting and immunofluorescence analysis [133, 140].

Having a carefully characterized set of highly specific antibodies against chicken CK isoenzymes at hand, the cellular distribution and localization of all chicken CK isoenzymes within the chicken cerebellum was investigated [133]. In addition, the localization, accumulation and developmental appearance of CK isoenzymes during maturation of the rat brain was studied, and these data were correlated with *in vivo* ^{31}P-NMR CK reaction flux measurements [140, 166].

Brain-type creatine kinase isoenzyme in Bergmann glial cells of the cerebellum

The localization of CK isoenzymes is most advanced in cerebellum due to the relatively 'simple', stratified structure of this part of the brain composed of well characterized cell types. Anti-B-CK staining was found in all layers of the cerebellar cortex as well as in the deeper nuclei of the cerebellum, indicating that a high proportion of the cerebellar cell types contain B-CK. The labeling was most intense in Bergmann glial cells (BGC) (Fig. 4a, b, small arrow-heads in b point to BGC cell bodies). The processes of these cells, lying in the vicinity of Purkinje neurons (large arrow), span radially through the entire molecular layer and finally form, with their endfeet, the membrana limitans, which is also stained heavily (Fig. 4a, arrowheads). Thus, the morphology of BGCs is perfectly matched by the intense anti-B-CK staining pattern. Besides BGC, some other cell types in the molecular layer, such as basket cells and neurons in the deeper nuclei, contain B-CK (for details see [133, 140]). Additionally, structures in the granular cell layer, likely to be glomeruli [142] and astrocytes contain significant anti-B-CK immunoreactivity (Fig. 4a, b), whereas cerebellar white matter appears to contain rather low levels of B-

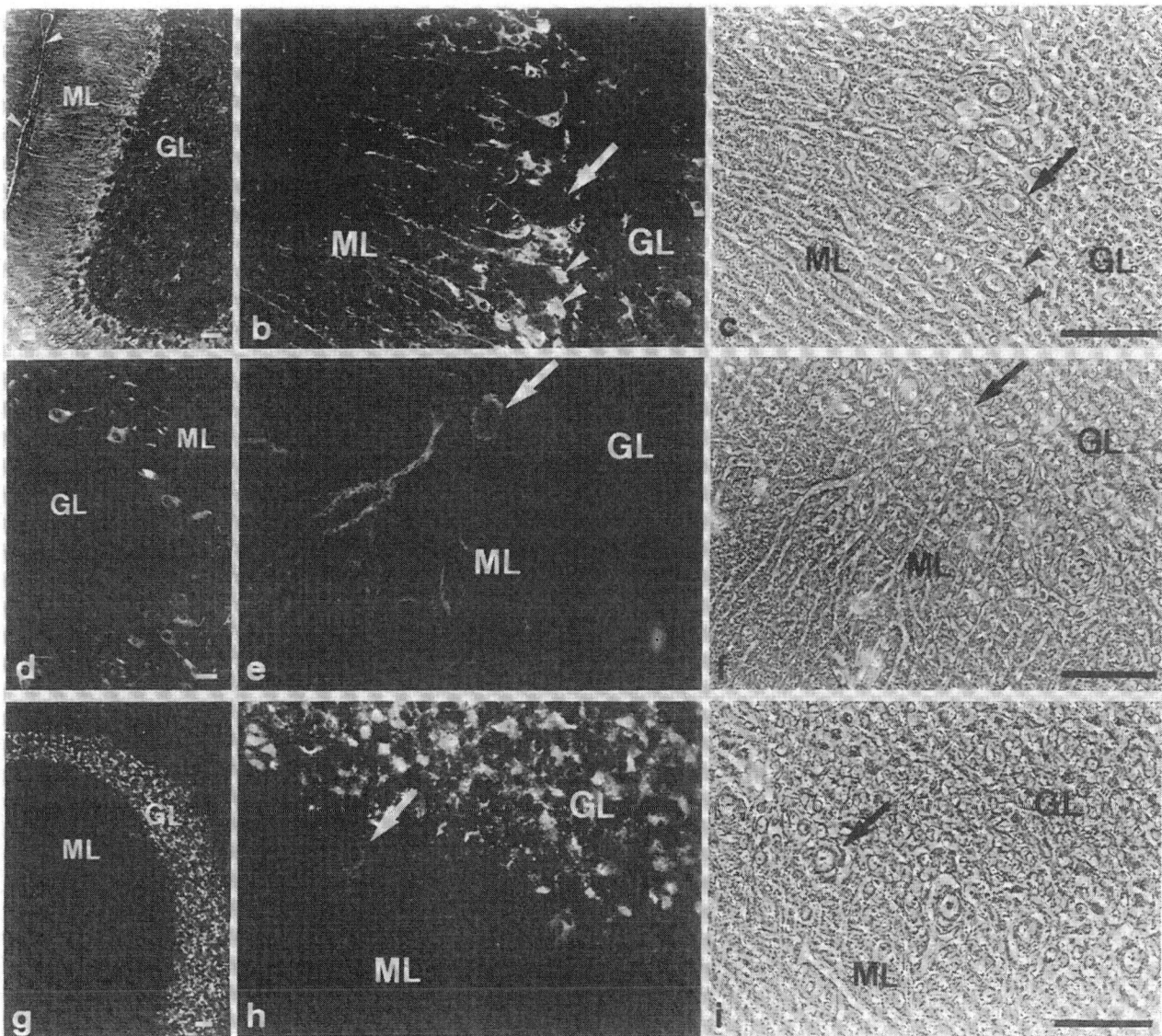

Fig. 4. Localization of brain-type B-CK, muscle-type M-CK and mitochondrial CK in chicken cerebellum. Cerebelli from adult chicken were fixed and embedded in paraffin by standard techniques and labeled by indirect immunofluorescence staining, using specific antibodies against chicken brain-type B-CK, muscle-type M-CK and against ubiquitous mitochondrial Mi$_a$-CK (for details see [133, 140]). Low magnification immunofluorescence overviews of cerebellar regions (a, d, g), higher magnifications of immunostained pictures (b, e, h), and the corresponding phase contrast pictures in (c, f, i), shown after staining for brain-type B-CK (panels a–c), for muscle type M-CK (panels d–f), and for mitochondrial Mi$_a$-CK (panels g–i), all followed by rhodamine-conjugated second antibody. Control sections, incubated only with preimmune sera, displayed no significant staining (not shown here, see [133, 140]). Small arrow-heads indicate the cell bodies of Bergmann glial cells (BGCs, in b and c) and the membrana limitans gliae (in a) which are both strongly stained by anti-B-CK antibodies. Large arrows indicate the Purkinje neurons (PNs) which are strongly stained by anti-M-CK antibodies (e and f), weakly stained by anti-Mi$_a$-CK antibodies (h and i), but remain unstained by anti-B-CK antibodies (b and c). Note that the anti-M-CK staining in the proximal processes of the Purkinje neuron is not uniform, but vesicular (e). This would be consistent with a staining of the endoplasmic reticulum network that is highly enriched in this region of Purkinje neuron. Note also the relatively strong staining by anti-Mi$_a$-CK of the glomeruli in the granular layer (GL) of the cerebellum. ML and GL refer to molecular and granular cell layers of the cerebellum, respectively; bar = 50 μm. An extensive study concerning the immunohistochemical localization of CK in chicken brain, including more details on the exact localization of the different CK isoenzymes, as well as on the characterization and specificity of the antibodies used was published elsewhere [133].

206

CK. The latter finding is consistent with previous histochemical and ^{31}P-NMR data [127, 128].

The BGC is a specialyzed type of astroglial cell. It provides the migratory pathway for granule cell migration from the external to the internal granule cell layer during cerebellar development [143, 144]. Another main function of these cells is the proposed ATP-dependent spatial buffering of potassium ions [145, 146], released during the electrical activity of neurons. This function is also reflected by the morphology of BGC, which envelop the synaptic sites of Purkinje neuron dendrites with the exception of the precise sites at which Purkinje spines make contact with parallel or climbing fibers [143]. Since BGC processes directly face the cerebrospinal fuid at the membrana limitans, these cells were suggested to be responsible for releasing the K$^+$ ions, taken up via ATP-driven Na$^+$/K$^+$-ion pumps from the extracellular space around the highly active Purkinje neurons, into the subdural space, which acts as a K$^+$ sink [145]. It is therefore reasonable to assume that the high B-CK content of BGC (Fig. 4a, b) reflects their high energy demands in relation to spatial K$^+$ buffering [147]. In this respect, it is interesting to note that Müller cells, representing a functionally and morphologically specialized astrocyte cell type found in the vertebrate retina, were also proposed to be involved in spatial K$^+$ buffering [145]. Like BGC, the Müller cells also contain significant amounts of B-CK [109]. The presence of B-CK in astrocytes (for details see [126,133,140]) may be related to the energy requirements of these cells for metabolic interactions with neurons, e.g. tricarboxylic acid (TCA) cycle metabolite and neurotransmitter trafficking [147].

Muscle-type creatine kinase isoenzymes in
Purkinje neurons of the cerebellum

Purkinje neurons play a very important role in brain function. Receiving excitatory input from parallel fibers and climbing fibers, they represent the sole neuronal output structures of the cerebellar cortex. A remarkable feature of Purkinje neurons (PN) is that a single PN makes hundreds of synaptic contacts to a single climbing fiber. Climbing fiber impulses evoke complex Ca^{2+}-spikes and prolongued Ca^{2+}- mediated depolarizations in Purkinje cell dendrites [148] which in turn are thought to play a central role in the mechanism of cerebellar motoric learning [149]. The presence of the 'unusual' muscle-type M-CK (see Fig. 4d,e and [133, 140]) and most likely also of the muscle-type Mi$_b$-CK (P. Kaldis, personal communication) in Purkinje neurons of the chicken brain may reflect an adaptation of Purkinje neurons to their very special energy requirements. It is known that Purkinje neurons specifically express a whole variety of enzymes involved in Ca^{2+}-homeostasis (for references see [133]). Interestingly, several of these proteins are also muscle-type isoforms, e.g. the skeletal muscle-type ryanodine receptor [150] and calsequestrin, a typical protein of the sarcoplasmic reticulum [151, 152]. In addition, Purkinje neurons contain the highest concentration of sarcoplasmic/endoplasmic reticulum Ca^{2+}-ATPase (SERCA) found in any non-muscle cell type [153] and also preferentially express a muscle-specific isoform of this enzyme, that is SERCA2b [154]. Recent *in vivo* ^{31}P-NMR saturation transfer experiments, showing that dihydropyridine calcium antagonists reduce the consumption of high-energy phosphates and concomitantly decrease the CK reaction flux in rat brain [155], strongly support the above conclusions that CK is directly or indirectly coupled to energetic processes needed for Ca^{2+} homeostasis or to cellular processes triggered by this second messenger. Thus, the presence of muscle-type M-CK in Purkinje neurons fits well into the general picture that PNs display some 'muscle-like' characteristics and may also reflect the better suitability of M-CK, compared to other cytosolic CK isoenzymes, to associate with certain subcellular structures, e.g. with the endoplasmic reticulum membrane system, in these cells. The vesicular immunofluorescence staining pattern seen in the proximal processes of PNs (Fig. 4e) would be consistent with staining of endoplasmic reticulum vesicles which are prominent in this region of the cell. M-CK was shown in skeletal muscle to be associated with the ATP-dependent Ca^{2+}-pump [12, 13]. Thus, the role M-CK plays in muscle, that is, i) preferentially supplying the Ca^{2+}-pump of the sarcoplasmic reticulum with ATP [12, 13] and ii) keeping local ATP/ADP ratios high in the vicinity of the Ca^{2+}-pump, thereby increasing the thermodynamic efficiency of this ion pump [8,15,44], may apply to M-CK bound to the endoplasmic reticulum in Purkinje neurons as well.

Creatine kinase isoenzymes in the glomerular structures
of the cerebellum

The granule layer of the cerebellum, especially the glomerular structures, contains high levels of Mi$_a$-CK as well as B-CK, as judged from the intensities of anti-CK antibody staining (Fig. 4a, b and g, h). These structures, forming intimate synaptic as well as glial-neuron interactions also called 'neuropils', are known to be rich in mitochondria and to display a very high energy metabolism. Large amounts of energy are needed in these structures for restoration of potassium ion gradients partially broken down during neuronal excitation, as well as for metabolite and neurotransmitter trafficking between glial cells and neurons (for review see [147]). Thus, the localization of both B- and Mi_a-CK isoforms within these structures may be an indication that part of the energy consumed in these giant complexes of mossy fiber, Golgi cell and granule cell synapses (for more details concerning the localization of CK, see [133]), might be provided by a 'PCr-circuit', as it has been proposed for other excitable cells [8, 38, 70, 71].

Creatine kinase in neurons

In brain, B-CK has also been found in association with synaptic vesicles [156], as well as with the plasma membranes [157]. Since a similar association between B-CK and synaptic vesicles as well as the plasma membrane has been demonstrated in electrocytes of *Torpedo* [98], the electrocyte system serves as a good analogy for the function of this portion of B-CK in brain. Convincing data concerning a direct functional coupling of CK with the Na^+/K^+-ATPase have been obtained by *in vivo* ^{31}P-NMR studies on electric fish, which showed that CK and the membrane-bound Na^+/K^+-ATPase are tightly coupled in the resting as well as in the stimulated electric organ [103]. Additionally, CK bound to synaptic vesicles in electrocytes is involved in neurotransmitter release [104]. Thus, the fractions of CK that are bound to synaptic vesicles and to the plasma membrane in neurons may also be involved in neurotransmitter release, as well as in the maintenance of membrane potentials and the restoration of ion gradients before and after electrical discharge, both in conjunction with the Na^+/K^+-ATPase [68, 103]. This is consistent with the fact that high energy turnover and, concomitantly, high CK concentrations have been found in those regions of the brain that are rich in synaptic connections, e.g. in the molecular layer of the cerebellum, in the glomerular structures of the granule layer and also in the hippocampus (Hemmer, unpublished observation). In neurons, the Na^+-extrusion activity facilitated via the neuron-specific Na^+/K^+-ATPase [158] is especially high in the synaptic region (see [159]). This is also true for Ca^{2+}-extrusion activity, mediated either via the plasma membrane Ca^{2+}-pump responsible for net extrusion of calcium out of neurons, or via the Na^+/Ca^{2+}-exchanger. The Na^+/Ca^{2+}-exchange is driven by the Na^+ gradient which in turn is maintained by ATP indirectly through the operation of the Na^+/K^+-ATPase (for review see [160]). The observation that a rise in CK levels, observed in a fraction of brain containing nerve endings and synapses, parallels the neonatal increase in Na^+/K^+ ATPase is also suggestive that higher levels of PCr and CK are characteristic of regions in which energy expenditure for processes such as ion pumping are large [159]. In addition, protein phosphorylation which plays an important role in brain function is also thought to consume a sizable fraction of the total energy available to these cells [159].

Finally, CK, together with nerve-specific enolase, belongs to a group of proteins known as slow component b (SCb). These proteins are synthesized in neuronal cell bodies and are directed by axonal transport to the axonal extremities [161, 162]. The question of whether CK participates in the actual energetics of axonal transport remains to be answered. However, the association of a fraction of 'soluble' CK with SCb proteins shows an intracellular compartmentation of the enzyme also in neurons. In addition, during preparation of neuron-specific enolase, brain CK co-chromatographes with the latter glycolytic enzyme [163], indicating a functional coupling of brain CK with glycolysis as was demonstrated in muscle [8, 52–54].

Interestingly, in differentiating primary cell cultures of neuronal cells, some B-CK was localized to the nuclei [126], whereas in adult chicken and rat brain, after *in situ* immunofluorescence staining with our anti-CK antibodies [133, 140], the nuclei of most neuronal cells, which are, however, fully differentiated, remained unstained.

Postnatal accumulation of mitochondrial CK in rat brain and in vivo *function of CK*

In the altricial neonates (mouse, rat, rabbit, pig and human), marked quantitative and qualitative changes in the physiology of ATP metabolism occur postnatally (see [164]). Similarly, rather dramatic postnatal increases in total CK activity and PCr content were noted. For example, in the narrow time-window between days 12–15 of postnatal development of mouse and rat, i) the *in vivo* rate of CK-catalysed ATP synthesis increases 4-fold, as measured by saturation transfer ^{31}P-NMR [165, 166], ii) the brain develops the capacity to increase ATP synthesis by oxidative phosphorylation in response to sudden changes in energy demand [164] and iii) a population of cerebral brain mitochondria appears with tight contacts between inner and outer membranes [165]. Since Mi-CK has been identified in isolated contact site membrane fractions of brain mitochondria [20, 31] and since octameric Mi-CK was shown to be able to induce contact formation between isolated mitochondrial membranes [25], the appearance of the population of mitochondria described above may be related to the expression and accumulation of Mi-CK in these mitochondria. In rat brain, an increase of Mi-CK activity by 4–6 fold, which is higher than the increase of total CK or B-CK activity over this time period, takes place between days 12–20 of postnatal development, concomitant with a corresponding 4 fold increase in the *in vivo* rate of CK-catalysed reaction flux [166]. These observations, showing that the developmental appearance of Mi-CK parallels the maturational changes in brain energy metabolism, suggest that Mi-CK, and CK in general, are critical in the control of cellular ATP metabolism in the adult brain [166]. The fact that in the developing brain, B-CK expression is acutely stimulated by vitamin D metabolites, but not by estrogen [167, 168] is certainly interesting in the context of brain development.

The interpretation that CK plays a key role in the energetics of the adult brain is supported by very recent *in vivo* ^{31}P-NMR magnetization transfer measurements showing that the pseudo first-order rate constant of the CK reaction (in the direction of ATP synthesis) as well as the CK flux (J_f) correlate with brain activity, which was measured by EEG as well as by the amount of deoxy-glucose phosphate formed in the brain after administration of deoxy-glucose [68]. These data show that *in vivo* the CK/PCr-system serves not merely as a temporal energy buffer [37], but has also a spatial energy buffer or transport function [8] with Mi-CK functioning as a key player in the intricate energy distribution system [4, 8] also in brain [166].

Localization and function of CK in uterus and placenta

Two major CK isoenzymes, B-CK [169] and ubiquitous Mi-CK, have been identified both in uterus and placenta [170]. In addition, a recent analysis of the myometrium of gravid guinea pigs revealed, besides the prevalence of BB-CK and the relatively low concentration of Mi-CK, the presence of MB- and MM-CK in this tissue [171]. During pregnancy, the total CK activity in myometrium increases almost by a factor of three, but the distribution of isoenzymes does not significantly change with gestation, except that the contribution of Mi-CK increases from trace activity in the non-gravid uterus to 5%! in the gravid uterus [171], which is indicative for a role of Mi-CK in the gravid myometrium. In detergent skinned fibres from myometrium, significant amounts of BB-CK, but not MM- or MB-CK, remain specifically bound [171], indicating an isoenzyme-specific compartmentation of CK in myometrium, with a fraction of BB-CK located at the myofilaments. Contraction experiments with chemically skinned fibres showed that in the presence of 10 mM PCr plus 250 µM ADP, in the absence of ATP, non-gravid and gravid uterine fibres are able to support 43% and 65% of maximal tension, respectively, via the endogenous CK system [171].

These results indicate that the presence of B-CK at the myofilaments and of Mi-CK in mitochondria, as well as their elevated specific activity during pregnancy, will lead to better interrelation and coupling between oxidative metabolism and force of contraction during increased demand for parturition [171]. This notion is also supported by the fact that *in vivo*, the uterine PCr levels increase significantly before labour, when large metabolic demands are made on the contracting myometrium. [PCr] remains high during parturition and returns to the much lower prepartum levels within a week after birth [172].

The structural and functional reorganization of the gestating uterus is under hormonal control. As a matter of fact, B-CK is *the* major estrogen-induced protein in uterus [173], and by CAT-assays, a functional estrogen enhancer was demonstrated to be contained within a 2.9 kb fragment of the 5′-upstream flanking region of the B-CK gene [174]. Very recently, it has been reported that the expression of both B-CK and Mi-CK in uterus as well as placenta are in fact hormone-regulated [170] and that both CK isoforms can be induced very rapidly by estrogen *in vivo* and *in vitro* [168, 170, 173, 175]. *In vivo*, B-CK and Mi-CK expression are highly coordinated both with respect to time during pregnancy and after delivery [170]. Interestingly, the hormone-induced increase in specific CK activity in experimentally estrogen-stimulated rat uterus is paralleled by changes in uterine PCr concentration [176]. Most importantly, this increase, partly due to an upregulation of Mi-CK as well, is paralleled by an increase in CK reaction flux from PCr to γ-ATP [177]. Taken together, these results suggest an important role of the PCr/CK system in uterus and placenta for the maintenance and termination of pregnancy [171].

Localization and function of CK in intestinal brush border epithelial cells and in endothelial cells

CK-supported contraction of intestinal brush-borders

In intestinal epithelial cells, B-CK and Mi-CK were also found to be compartmentalized subcellularly. While B-CK is distributed diffusely throughout the cytoplasm of these cells, it is concentrated distinctly in the brush border terminal web region, where the contractile-ring myosin is located. The mitochondrial Mi-CK isoenzyme, on the other hand, is specifically confined to the mitochondria just subjacent to the terminal web region [178].

Glycerol-permeabilized brush borders, with their cytoskeleton intact retain significant amounts of B-CK, which is indicative for a relatively strong subcellular association of a fraction of the enzyme with internal structures of brush borders [178]. Such brush border preparations do contract *in vitro* by virtue of their circumferential ring myosin, and the extent of contraction can be measured by the curvature of the brush borders at the level of the circumferential ring. Maximal contraction can be supported equally well either by an externally added ATP-regeneration system [phosphoenol-pyruvate (PEP)/pyruvate kinase] or by simply adding PCr in the presence of micromolar concentrations of ADP [178]. This indicates that endogenous B-CK is sufficient for delivery of ATP to the contractile-ring myosin and for ATP regeneration. Most importantly, whereas PEP-pyruvate-kinase-supported contraction is efficiently blocked by external addition of an ATP trap (hexokinase and glucose), the PCr-ADP-supported contraction is not at all inhibited [179], thus demonstrating that ATP is preferentially supplied, via PCr, to the circumferential-ring myosin by endogenous CK bound at the terminal web. This preferential supply of ATP imparts to this myosin a selective energetic advantage over other cellular ATPases. Thus, similar to muscle, CK-coupled contractile-ring myosin appears to be one end of an energy circuit that supplies the energy for brush border contraction [179].

CK in epithelial and hair cells of the inner ear

The organ of Corti and the stria of the inner ear are known to contain high concentrations of CK [180], possibly involved in the energetics of ion transport and auditory sensation. By immuno-histochemical analysis, marginal cells of the cochlear stria vascularis, as well as dark cells and transitional cells of the vestibular system, all contain an abundance of CK [181]. These cell types concentrate K^+ in the endolymph of the inner ear against a large gradient and depend, for such ion transport, on a ouabain-sensitive basolateral Na^+/K^+-ATPase. Thus, in analogy to the well-known cases in other cells, it is reasonable to speculate that CK in strial marginal cells and dark cells is also coupled to this ion pump and maintains a thermodynamically favourable ATP/ADP ratio in the vicinity of the pump. High levels of CK have also been demonstrated in the cochlea's inner and outer phalangeal (Deiter's) cells and, although at lower levels, in the sensory hair cells [181]. CK in Deiter's cells may have a function in the re-uptake of K^+-effluxing from the sensory hair cells analogous to the role played by CK in Bergmann glia cells and Müller cells of brain and retina, respectively (see previous chapters).

CK in endothelial cells

Endothelial cells lining the inside of blood vessels, when maintained under normoxic conditions, express various CK isoenzymes (BB-, MM- and Mi-CK) and possess significant stores of PCr [182]. If these cells are exposed for prolonged periods of time to hypoxic conditions, they exhibit a significant reduction in their PCr stores and dramatically upregulate their glucose transport activity [182]. Thus, even though very little is known about the function of CK in these cells, endothelial cells obviously seem to require PCr to meet their metabolic demands, especially under metabolic stress. It will be interesting in the future to localize the CK isoenzymes at the subcellular level in endothelial cells and to establish functional assays to probe for CK function.

CK in kidney and rectal salt gland

Kidney

PCr and CK have received almost no attention in kidney, probably because the levels in whole kidney are ve-

210

ry low as compared to muscle and brain. For example, CK activity in kidney compared to muscle is lower by at least a factor of 100 [183]. However, by microdissection of different regions of the kidney, it was shown that [PCr] and CK activity vary by as much as 5–10-fold and 100-fold, respectively, depending on the segments along the nephron. The distal convoluted tubules containing highest [PCr] and specific CK activity, with the latter reaching some 5% and 20% of muscle and brain CK activity, respectively [183]. Thus, it emerged that PCr and CK must be confined to a few cell types of the whole kidney. The first results obtained with immuno-peroxidase staining for CK in kidney were in agreement with this notion in that CK was mainly confined to the epithelial cells of the thick ascending limb of the Henle's loop and the collecting tubules [184]. In a subsequent study, using isoenzyme-specific anti-CK peptide antibodies, two CK isoenzymes, brain-type B-CK and ubiquitous Mi-CK were identified and characterized [185]. By immunohistochemical staining, it was demonstrated that both CK isoforms are co-localized on a cellular level in the inner stripe of the outer medulla of rat kidney, mainly confined to the distal tubules of nephrons [185]. Interestingly, this distribution of CK in kidney corresponds in general to the region of greatest ATP utilization, oxygen consumption and sodium transport by the Na^+/K^+-ATPase [183]. Thus, besides providing an energy buffer during periods of low oxygen tension, characteristic of the renal medulla, the cellular co-localization of the two CK isoforms in a region of high energy turnover related to ion transport indicates a more active role for the CK/PCr-system in the distal nephron [185], e.g. it may serve as an energy shuttle system to provide energy for sodium transport and may in addition fulfil other CK functions postulated in the PCr-circuit model [8]. This is supported by the fact that upon hypoxia of the kidney, similar to muscle, electric organ, brain and macrophages (see this article), PCr in the distal nephron is also depleted more rapidly and to a greater extent than ATP [183].

Rectal salt gland

The highly specialized rectal gland for sodium excretion, found in elasmobranchs such as dogfish and sharks, was shown to contain very high levels of PCr, and high expression levels of brain-type B-CK as well as of ubiquitous Mi-CK [186]. This CK isoform composition is the same as in mammalian kidney [186]. As in kidney, these two isoforms are colocalized on a cellular level and are

found to be concentrated at the basal region of the tubule cells where the sodium transport ATPase is also located [186]. Thus, the two CK isoforms in these sodium-secreting tubule cells may provide the components of a shuttle system for the regeneration of energy required for sodium transport and may keep the thermodynamic efficiency of this process high. Such a role is corroborated by the fact that stimulation of rectal gland sodium-secretion by cyclic AMP causes a rapid decrease in PCr levels with little or no change in [ATP] [187], suggesting that sodium-excretion in this salt gland is tightly coupled to PCr-hydrolysis in order to maintain high local ATP/ADP ratios in the vicinity of the ion pump.

CK in adipose tissue

Brown and white adipose tissue both contain PCr and CK activity, with the specific activity of CK in brown fat being approximately 50 times higher than in white fat tissue [188]. In brown adipose tissue, which is responsible for heat generation through a process called 'non-shivering thermogenesis', the CK activity is in the same order of magnitude as that found in cardiac or nerve tissue. So far, the CK isoenzymes in this tissue have not been identified, but the function of CK may be directly related to thermogenesis. As a matter of fact, based on *in vivo* Cr-uptake experiments, it was shown that the labelling of the total Cr pool with radioactive Cr proceeded much faster in adipose tissue than in skeletal muscle [188]. It was concluded that fatty acids and free Cr may be synergistic in promoting mitochondrial respiration for thermogenesis [188].

CK in pancreas

The possible importance of CK in endocrine tissues may have been overlooked in the past. The fact that PCr [189] and CK activity [190] were identified in the islets of Langerhans of pancreas, with B-CK as the major CK isoenzyme [190], indicates that PCr and CK may play a role in the secretion of insulin and/or glucagon. In pancreas, the energy required for insulin exocytosis was assumed to be supplied by ATP synthesized in acinar cells. However, ^{31}P-NMR measurements of cerulein-stimulated rat pancreas indicate i) that ATP for insulin exocytosis was derived from phosphorylation of ADP via PCr by pancreatic CK and ii) that large amounts of PCr are synthesized during the first minutes after cerulein-stimulation [191]. During this process, the PCr levels reach a maximum at 10 min, then fall between 10–20 min after stimulation and finally return to control levels [191].

CK in thymus, thyroid and liver

In addition to expressing variable amounts of B-CK, the thymus was shown, quite surprisingly, to contain also muscle-specific M-CK [192], as well as another muscle-specific protein, the 165'000 M_r M-band protein. Nodules, embedded in the thymic reticulum, stain in situ positively for both of these proteins. If thymus tissue was dissociated into single cells and cultured at high density, myotubes which were morphologically similar to those from muscle-derived cultures appeared and stained positively for the two muscle proteins mentioned above [192]. Obviously, in the thymus, there are some myoid cells present which are capable of forming myotubes. This explains the occurrence of muscle-specific proteins in this tissue.

Already very early, the presence of CK in thyroid has been reported, and the isoenzymes found in this tissue have been tentatively identified as MM- (MB-) and BB-CK [193]. The invariable presence of significant amounts of CK in the thyroid of several species including man suggests that this enzyme may have a role in thyroid tissue metabolism or hormone biosynthesis, however, no follow-up study on the localization and function of this enzyme in this tissue has been published.

Surprisingly, the presence in human liver of mitochondrial CK, showing electrophoretic properties similar to those of human cardiac Mi-CK, has also been reported, and Mi-CK was actually purified from the mitochondrial fraction of human liver [194]. By contrast, all our efforts by native CK isoenzyme analysis, affinity labelling of CK with radioactive N-bromoacethyl-3,3',5-triiodo-L-thyronine, immunoblotting of extracts from whole liver tissue or from highly enriched mitochondrial fractions from chicken, mouse and rat livers [195], as well as Northern-blot analysis of chicken liver with Mi-CK-specific cDNA probes [M. Stolz, IZB, ETH, personal communication], did not reveal Mi-CK in liver. If anything, only minute amounts of B-CK, which easily could have been derived from blood vessels (smooth muscle, and endothelial cells) or blood cells (macrophages and blood platelets) have occasionally been found in this tissue. This raises the interesting question of whether human liver is an exception with respect to CK content or whether the expression of Mi-CK in human livers, as described by the above authors, may be linked to a specific unrecognized disease affecting the "healthy" patient whose liver was taken post-mortem.

CK in cartilage and bone

Cartilage cells

The presence of PCr in chondrocytes was demonstrated by ^{31}P-NMR measurements of superfused resting zone cartilage from the growth plates of bones from young animals [196]. Extraction and chemical analysis revealed the highest amounts of PCr in the proliferative region of cartilage, but no PCr is present in calcified cartilage [197]. In sharp contrast to developing skeletal muscle, where a transition from BB-CK in embryonic to MM-CK in adult muscle takes place, exclusively MM-CK was found in resting and proliferating cartilage, while MB and BB-CK are clearly the predominant CK's in hypertrophic cartilage [197]. The levels of CK activity are directly related to chondrocyte maturation, with CK activity increasing with the progression of chondrocyte hypertrophy [197]. As in uterus, the expression of B-CK in cartilage cell cultures is regulated by estrogen [198] and also by calciotrophic agents, such as certain vitamin D metabolites [199]. Most importantly, lowering of the energy status of developing cartilage and bone in vivo, either by feeding of rats with the creatine analogue 3-guanidino-propionic acid (GPA), or by addition of GPA to cultured chondrocytes in vitro, leads to a significant inhibition of normal cartilage development and differentiation, respectively [200].

Bone cells

B-CK activity in bone cells in culture is also stimulated, like in cartilage cells, by some, but not all vitamin D metabolites [201] and, in addition, by parathyroid hormone and prostaglandin E_2 [202]. Most intriguing, however, is the direct and sex-specific stimulation of B-CK expression by sex steroids in rat bone [203]. Whereas in bones of female rats, 17β-estradiol (E_2), but not testosterone, stimulates the appearance of B-CK activity, the inverse situation is observed in bones of male rats, that is, testosterone stimulates B-CK synthesis, while E_2 is ineffective. This indicates that gonadal steroids may contribute to stimulating bone growth and to maintaining a balanced bone-turnover, with CK being directly involved in the energetics of these processes [200].

Very recently Ch'ng and Ibrahim (1994) [225] have shown that when a rat osteoblastic cell line is induced to differentiate with 1,25-dihydroxyvitamin D_3, the expression of B-CK is upregulated by a two-fold increase in the transcription rate and by translational modulation involving increased B-CK mRNA stability as well as binding of a cytosolic factor to the highly conserved 3'-untranslated region of the B-CK mRNA [225]. The trans-acting factor shown to prevent completion of translation of the mRNA [226] may prevent diffuse and indiscriminant expression of B-CK in the cell and target the protein-RNA-complex to the appropriate subcellular compartment for local expression of B-CK [225]. Most importantly, inhibition of the up-regulation of B-CK in osteoblasts with anti-sense RNA targeted to the B-CK mRNA results in failure to attain the mature osteoblastic/osteocytic phenotype (Ujihara and Ch'ng, manuscript in preparation).

The importance of CK in eneragy metabolism is generally reflected by the stringent regulation of its expression, both developmentally and spatially. The molecular melchanisms, both at the transcriptional and translational level, of B-CK expression in osteoblasts obviously play significant roles in osteoblast-osteocyte physiology which is characterized by changes in cytoskeleton and increased activity of membrane pumps in differentiated osteoblasts-osteocytes [225].

CK in macrophages blood platelets

CK was first identified in phagocytic white blood cells, that is, in rabbit alveolar macrophages, by DeChatelet *et al.* in 1973 [204]. Later, it was found that mouse macrophages contain B-CK as the major CK isoenzyme, whereas the same *human* cells seem to express B-CK and possibly also Mi-CK [205]. The expression of CK in mouse and human mononuclear phagocytes is a developmentally regulated process occurring during *in vivo* and *in vitro* differentiation of monocytes into macrophages [205]. The latter cells display relatively high specific CK activity and accumulate PCr in 3–5-fold molar excess over ATP [205]. The induction of CK during differentiation of monocytes into macrophages in cell culture occurs independently of the concentration of Cr in the medium. The size of the intracellular PCr pool in macrophages, however, is directly proportional to the [Cr] of the culture medium [206]. Macrophage differen-

tiation is also associated with a marked increase in the Cr transporting capacity [205]. This is evident that macrophages, like muscle and some neurons, do not synthesize their own Cr, but accumulate the compound via a transmembrane Cr-transporter that has recently been cloned from rabbit brain and functionally expressed in COS-7 cells [207]. Most important, however, is the finding that during phagocytosis in both resident and thioglycollate-elicited mouse peritoneal macrophages, ATP levels remain fairly stable, while PCr concentration decreases by 40–50% [208]. These results demonstrate that the ATP consumed by macrophages during phagocytosis is replenished by CK via PCr.

B-CK [209] as well as PCr [210] were identified also in isolated, washed blood platelets which can therefore be a possible source of CK in blood plasma and serum. The fact that aerobically generated ATP preferentially disappears from blood platelets during thrombin-induced aggregation and clotting [211] would indicate that mitochondrial CK, although not identified in these cells so far, may also be contained in blood platelets and that CK may have a function in the above processes.

CK in malignant tumors and cancer cells

As shown above, brain-type B-CK is a major enzyme of cellular energy metabolism in non-muscle cells. In this context, it is interesting that B-CK is overexpressed in a wide range of solid tumors and tumor cell lines. For example, highly elevated levels of this enzyme have been detected in tumor samples obtained from patients with small-cell lung carcinoma, colon and rectal adenocarcinoma, breast and prostate carcinoma, as well as neuroblastoma [212–216]. As a result of its elevated level, B-CK can be used as a diagnostic marker for small-cell lung carcinoma [213].

In many cancer patients with one of the carcinomas mentioned above, B-CK, and in certain cases also Mi-CK, are released into the serum, whereby elevated levels of CK in tumors and/or in the serum [217] are associated with untreated, progressive or metastatic disease. Thus, highly elevated levels of CK in the serum of such patients seem to be an adverse prognostic indicator (for refs. see [4, 218, 219]). Most likely, a rather bulky tumor or an advanced stage of malignancy is required for the continuous release of routinely detectable CK into the serum [215]. The presence of Mi-CK in the serum of cancer patients with a variety of adenocarcinomas [220] and with certain melanomas [221] was suggested to be a reliable tumor marker, with Mi-CK-positive individuals showing significantly higher mortality rates compared to Mi-CK-negative ones [4, 222].

As mentioned above, B-CK is an estrogen-regulated protein in uterus, placenta and other specialized cells. This holds true also for human breast cancer explants and cultured breast cancer cells that are estrogen-induced. In such tumors and cells, a positive relationship between B-CK and estrogen receptor content has been observed [216]. This indicates that a quantitative evaluation of B-CK activity in human breast cancer may be of potential value for the detection of hormone-responsive tumors, as well as for the strategy of treatment.

It seems that increased amounts of PCr are used by a variety of malignant cells as an energy source. This is corroborated by the fact that malignant cells often contain highly elevated [PCr] and that a broad spectrum of cancer cells derived from different solid tumors is growth-inhibited by the creatine analogue, cyclo-creatine (cCr) [218, 219]. It was shown that in these cells, cCr accumulates as phosphoryl-cCr (PcCr), with thermodynamic and kinetic properties different from those of PCr. In tumor cells that express high levels of CK, the tumor-growth inhibition is seen at extracellular concentrations of cCr ranging between 2–8 mM, which led to an intracellular accumulation of the compound reaching 30–50 mM [219], indicating that the active form may be PcCr. By contrast, tumor cell lines expressing low levels of CK accumulate much less PcCr and, consequently, are growth-inhibited only at 1-fold higher concentrations of this Cr analogue [219]. However, the fact that Cr itself, although only when fed to cancer-bearing rats at higher concentrations than cCr (up to 5% of the total diet), shows a similar inhibitory effect on tumor growth *in vivo* [218], argues against the proposition that cCr inhibits tumor growth by disturbing the energy metabolism. Therefore, the question remains of whether the observed inhibition of tumor growth rates by cCr and especially by Cr is due a direct disturbance of energy metabolism by the respective phosphagens, PcCr and PCr, in tumor suppression, or to effects of the phosphagens on other cellular processes.

Most interesting with respect to a possible CK/cancer connection is the observation that B-CK gene expression is induced by adenovirus type 5 [223]. It was shown that the 5′-upstream promotor of the human B-CK gene has a strong similarity to the adenovirus promotor for the E2E gene, the latter encoding a 72 kDa single-stranded-DNA-binding protein which is crucial for virus replication. The expression of the E2E gene is highly dependent on the adenovirus products of the E1a region coding for two major oncogenic nuclear phosphoproteins (for refs. see [223]). As it turned out, both B-CK activity and B-CK mRNA levels are highly induced by the oncogenic products of the adenovirus E1a gene region [223]. These very exciting results suggest that the control of B-CK expression lies along the path that is closely linked to cell growth control. The induction by an oncogene of a cellular gene for energy metabolism, such as B-CK, may be of significance for the metabolic events that take place after oncogenic activation [223] and thus may turn out to be relevant for the etiology of the transformed phenotype of cancer cells which are known to be able switch from one energy-producing pathway to another. For additional information concerning the identification of CK isoenzymes in normal and diseased non-muscle tissues and cells of man, the reader is referred to references [214, 224].

In conclusion, the 'ubiquitous' presence of B-CK in many non-muscle tissues, with B-CK in most cases being co-expressed with ubiquituous Mi_a-CK, seems to indicate a multitude of functions for the CK system in these tissues and cells. In some of the very specialized tissues or cells, like sperm, electric organ and photoreceptor cells, the identification of the isoenzymes, their subcellular localization, as well as some of the functions of the CK/PCr-system have been elaborated quite extensively. In these well studied tissues and cells, the respective functions of CK are more obvious than in other, less specialized tissues and cells. It will be extremely intriguing to see the phenotypes of transgenic mice with the B-CK and/or Mi_a-CK genes knocked-out, if such null mutants should turn out viable (for the phenotype of M-CK null mutant transgenic mice see [42]). If our hypotheses of the CK/PCr-systems are correct, one would expect to see some severe phenotypes in B-CK knock-out mice. For example, sperm motility (fertility), vision, brain development and brain functions would all be expected to be severely impaired, unless compensatory measures are taken by the affected tissues and cells, as has often be seen in the recent past with gene-targeted knock-out mice. However, for this sort of work, well-trained organ and cell physiologists, being able to test subtle physiological differences in transgenic versus normal mice, will be in great demand again in the near future. Thus, sophisticated molecular biology and organ physiology shall embrace each other more closely than ever before and, in a combined effort, will help to solve some of the puzzles of life.

214

Acknowledgements

Prof. Hans Eppenberger is acknowledged for his support and continuous interest in this work, Mrs. Else Zanolla and Mr. Philipp Kaldis for excellent technical assistance and for communicating unpublished results, respectively. We are also grateful to Dr. Markus Wyss and Dr. Elizabeth Furter-Graves for carefully reading the manuscript and for discussion. This work was supported by Swiss National Science Foundation grant No. 31-33907.92 to T.W., by an ETH-graduate training grant for W.H. and by financial support from the Swiss Society for Muscle Diseases and from the Helmut-Horten Foundation.

References

1. Eppenberger HM, Dawson DM, Kaplan NO: The comparative enzymology of CKs. I) Isolation and characterization from chicken and rabbit tissues. J Biol Chem 242: 204–209, 1967
2. Hossle JP, Schlegel J, Wegmann G, Wyss M, Böhlen P, Eppenberger JM, Wallimann T, Perriard JC: Distinct tissue-specific mitochondrial creatine kinases from chicken brain and striated muscle with a conserved CK framework. Biochem Biophys Res Commun 151: 408–416, 1988
3. Haas RC, Strauss AW: Separate nuclear genes encode sarcomer-specific and ubiquitous human mitochondrial creatine kinase isoenzymes. J Biol Chem 265: 6921–6927, 1990
4. Wyss M, Smeitink J, Wevers RA, Wallimann T: Mitochondrial creatine kinase: a key enzyme of aerobic metabolism. Biochim Biophys Acta 1102: 119–166, 1992
5. Turner DC, Wallimann T, Eppenberger HM: A protein that binds specifically to the M-line of skeletal muscle is identified as the muscle form of creatine kinase. Proc Natl Acad Sci USA 70: 702–705, 1973
6. Wallimann T, Turner DC, Eppenberger HM: Localization of creatine kinase isoenzymes in myofibrils. I Chicken skeletal muscle. J Cell Biol 75: 297–317, 1977
7. Wallimann T, Moser H, Eppenberger HM: Isoenzyme-specific localization of M-line-bound CK in myogenic cells. J Muscle Res Cell Motil 4: 429–441, 1983
8. Wallimann T, Wyss M, Brdiczka D, Nicolay K, Eppenberger HM: Intracellular compartmentation, structure and function of creatine kinase isoenzymes in tissues with high and fluctuating energy demands: the 'phospho-creatine circuit' for cellular energy homeostasis. Biochem J 281: 21–40, 1992
9. Wallimann T, Schlösser T, Eppenberger HM: Function of M-line-bound CK as intramyofibrillar ATP regenerator at the receiving end of the phosphocreatine shuttle in muscle. J Biol Chem 259: 5238–5246, 1984
10. Krause SM, Jacobus WE: Specific enhancement of the cardiac myofibrillar ATPase by bound creatine kinase. J Biol Chem 267: 2480–2486, 1991
11. Ventura-Clapier R, Mekhfi H, Vassort G: Role of creatine kinase in force development in chemically skinned rat cardiac muscle. J Gen Phys 89: 815–837, 1987
12. Rossi AM, Eppenberger HM, Volpe P, Cotrufo R, Wallimann T: Muscle-type MM-creatine kinase is specifically bound to sarcoplasmic reticulum and can support Ca^{2+}-uptake and regulate local ATP/ADP ratios. J Biol Chem 265: 5258–5266, 1990
13. Korge P, Byrd SK, Campbell KB: Functional coupling between sarcoplasmic-reticulum-bound creatine kinase and Ca^{2+}-ATPase. Eur J Biochem 213: 973–980, 1993
14. Grosse R, Spitzer E, Kupriyanov VV, Saks VA, Repke KRH: Coordinate interplay between (Na^+/K^+)-ATPase and CK optimizes (Na^+/K^+)-antiport across the membrane of vesicles formed from the plasma membrane of cardiac muscle. Biophys Acta 603: 142–156, 1980
15. Kammermeier H: Why do cells need phosphocreatine and a phosphocreatine shuttle? J Mol Cardiol 19: 115–118, 1987
16. Wallimann T, Eppenberger HM: Localization and function of M-line-bound creatine kinase: M-band model and creatine phosphate shuttle. In: JW Shay (ed.) Cell and Muscle Motility Vol. 6. Plenum Publ. Co., New York, 1985, pp 239–285
17. Arrio-Dupont M, Bechet JJ, d'Albis A: A model system of coupled activity of co-immobilized creatine kinase and myosin. Eur J Biochem 207: 951–957, 1992
18. Jacobus WE, Lehninger AL: Creatine kinase of rat heart mitochondria. Coupling of creatine phosphorylation to electron transport. J Biol Chem 248: 4803–4810, 1973
19. Jacobs M, Heldt HW, Klingenberg M: High activity of CK in mitochondria from muscle and brain. Evidence for a separate mitochondrial isoenzyme of CK. Biochem Biophys Res Commun 16: 516–521, 1964
20. Kottke M, Adams V, Wallimann T, Kumar-Nalam V, Brdiczka D: Location and regulation of octameric mitochondrial creatine kinase in the contact sites. Biochim Biophys Acta 1061: 215–225, 1991
21. Jacob WJ, Biermans W, Bakker A: Mitochondrial contact sites: a dynamic compartment for creatine kinase activity. In: PP De Deyn, B Marescau, V Stalon, IA Qureshi (eds) Guanidino Compounds in Biology and Medicine. J. Libbey, London, 1992, pp 165–174
22. Jacobus WE; Respiratory control and the integration of heart high-energy phosphate metabolism by mitochondrial creatine kinase. Annu Rev Physiol 47: 707–725, 1985
23. Schlegel J, Wyss M, Eppenberger HM, Wallimann T: Functional studies with the octameric and dimeric form of mitochondrial creatine kinase: differential pH-dependent association of the two oligomeric forms with the inner mitochondrial membrane. J Biol Chem 265: 9221–9227, 1990
24. Schnyder T, Gross H, Winkler HP, Eppenberger HM, Wallimann T: Structure of the mitochondrial creatine kinase octamer: high resolution shadowing and image averaging of single molecules and formation of linear filaments under specific staining conditions. J Cell Biol 112: 95–101, 1991
25. Rojo M, Hovius R, Demel RA, Nicolay K, Wallimann T: Mitochondrial creatine kinase mediates contact formation between mitochondrial membranes. J Biol Chem 266: 20290–20295, 1991
26. Erickson-Viitanen S, Geiger PJ, Viitanen P, Bessman SP: Compartmentation of mitochondrial creatine phosphokinase. II The importance of the outer mitochondrial membrane for mitochondrial compartmentation. J Biol Chem 257: 14405–14411, 1982
27. Saks VA, Kuznetsov AV, Kupriyanov VV, Miceli MV, Jacobus

WE: Creatine kinase of rat heart mitochondria: the demonstration of functional coupling to oxidative phosphorylation in an inner membrane preparation. J Biol Chem 260: 7757–7764, 1985

28. Gellerich FN, Schlame M, Bohnensack R, Kunz W: Dynamic compartmentation of adenine nucleotides in the mitochondrial intermembrane space of rat-heart mitochondria. Biochim Biophys Acta 890: 117–126, 1987

29. Saks VA, Belikova YO, Kuznetsov AV, Kuchua ZA, Branishte TH, Semenovsky ML, Naumov VG: Phosphocreatine pathway for energy transport. ADP diffusion and cardiomyopathy. Am J Physiol Suppl (Oct) 261: 30–38, 1991

30. Bessman SP, Fonyo A: The possible role of the mitochondrial bound creatine kinase in regulation of mitochondrial respiration. Biochem Biophys Res Commun 22: 597–602, 1966

31. Brdiczka D: Contact sites between mitochondrial envelope membranes. Structure and function in energy- and protein-transport. Biochim Biophys Acta 1071: 291–321, 1991

32. Gross M, Wallimann T: Kinetics of assembly and dissociation of mitochondrial creatine kinase octamers. A fluorescence study. Biochem 32: 13933–13940, 1993

33. Bessman SP, Carpenter CL: The creatine-creatine phosphate energy shuttle. Annu Rev Biochem 54: 831–862, 1985

34. Wallimann T, Schnyder T, Schlegel J, Wyss M, Wegmann G, Rossi AM, Hemmer W, Eppenberger HM, Quest AFG: Subcellular compartmentation of creatine kinase isoenzymes, regulation of CK and octameric structure of mitochondrial CK: important aspects of the phosphoryl-creatine circuit. In: RJ Paul, G Elzinga, K Yamada (eds) Progress in Clinical and Biological Research, Vol. 315: 'Muscle Energetics'. A.R. Liss Inc., New York, 1989, pp 159–176

35. Saks VA, Ventura-Clapier R: Biochemical organization of energy metabolism in muscle. J Biochem Organization 1: 9–29, 1992

36. Wallimann T: Dissecting the role of creatine kinase. The phenotype of gene knockout mice deficient in a creatine kinase isoform sheds new light on the physiological role of the phosphocreatine circuit. Curr Biol 4: 42-46, 1994

37. Meyer RA, Sweeney HL, Kushmerick MJ: A simple analysis of the 'phosphocreatine shuttle'. Am J Physiol 246: C365–C377, 1984

38. Hemmer W, Riesinger I, Wallimann T, Eppenberger HM, Quest AFG: Brain-type creatine kinase in photoreceptor cell outer segments: role of a phosphocreatine circuit in outer segment energy metabolism and phototransduction. J Cell Sci 106: 671–684, 1993

39. Koretsky AP, Brosnan JM, Chen L, Chen J, VanDyke T: NMR detection of creatine kinase expressed in liver of transgenic mice: determination of free ADP levels. Proc Natl Acad Sci USA 87: 3112–3116, 1990

40. Brosnan JM, Chen L, Wheeler CE, VanDyke T, Koretsky AP: Phosphocreatine protects from a fructose load in transgenic mouse liver expressing creatine kinase. Am J Physiol 260: C1191–C1200, 1991

41. Miller KR, Halow JM, Koretsky AP: Phosphocreatine protects transgenic mouse liver expressing creatine kinase from hypoxia and issemia. Am J Physiol 265 (6 Pt 1): C1544–1551, 1993

42. Van Deursen J, Heerschap A, Oerlemans F, Ruitenbeek W, Jap P, terLaak H, Wieringa B: Skeletal muscles of mice deficient in M-creatine kinase lack burst activity. Cell 74: 621–631, 1993

43. Hasselbach W, Oetliker H: Energetics and electrogenicity of the sarcoplasmic reticulum pump. Annu Rev Physiol 45: 325–339, 1983

44. Läuger P: Ca^{2+}-pump sarcoplasmic reticulum. In: Electrogenic ion pumps, Vol. 5. Sinnauer Assoc. Inc. Publishers, Sunderland, Mass., USA, 1991, pp 226–251

45. Ventura-Clapier R, Veksler VK, Elizarova GV, Mekhfi H, Levitskaya EL, Saks VA: Contractile properties and creatine kinase activity of myofilaments following ischemia and reperfusion of the rat heart. Biochem Med Metabolic Biol 38: 300–310, 1987

46. Ventura-Clapier R, Saks VA, Vassort G, Lauer C, Elizarova GV: Reversible MM-creatine kinase binding to cardiac myofibrils. Am J Physiol 253: C444–C455, 1987

47. Gudbjarnason S, Mathes P, Ravens KG: Functional compartmentation of ATP and creatine phosphate in heart muscle. J Mol Cell Cardiol 1: 325–339, 1970

48. Morgan HE, Parmeggiani A: Regulation of glycogenolysis in muscle. III Control of muscle glycogen phosphorylase activity. J Biol Chem 239: 2440–2445, 1964

49. Passoneau JV, Lowry OH: Phosphofructokinase and the Pasteur effect. Biochem Biophys Res Commun 7: 10–15, 1992

50. Rose IA, Warms JVB, O'Conell EL: Role of inorganic phosphate in stimulating the glucose utilization of human red blood cells. Biochem Biophys Res Commun 15: 33–37, 1964

51. Meyer RA, Brown TR, Krilowicz BL, Kushmerick MJ: Phosphagen and intracellular pH changes during contraction of creatine-depleted rat muscle. Am J Physiol 250: C264–C274, 1986

52. Dillon PF, Clark JF: The theory of diazymes and functional coupling of pyruvate kinase and creatine kinase. J Theor Biol 143: 275–284, 1990

53. Wegmann C, Zanolla E, Eppenberger HM, Wallimann T: In situ compartmentation of creatine kinase in intact sarcomeric muscle: the acto-myosin overlap zone as a molecular sieve. J Muscle Res Cell Motil 13: 420–435, 1992

54. VanWaarde A, Van den Thillart G, Erkelens C, Addink A, Lugtenburg J: Functional coupling of glycolysis and phosphocreatine utilization in anoxic fish muscle. J Biol Chem 265: 914–923, 1990

55. Saks VA, Ventura-Clapier R, Huchua ZA, Preobrazhensky AN, Emelin IV: Creatine kinase in regulation of heart function and metabolism. I. Further evidence for compartmentation of adenine nucleotides in cardiac myofibrillar and sarcolemmal coupled ATPase-creatine kinase systems. Biochim Biophys Acta 803: 254–264, 1984

56. Savabi F: Free creatine available to the creatine phosphate energy shuttle in isolated rat atria. Proc Natl Acad Sci USA 85: 7476–7480, 1988

57. Gellerich FN, Schlame M, Bohnensack R, Kunz W: Dynamic compartmentation of adenine nucleotides in the mitochondrial intermembrane space of rat-heart mitochondria. Biochim Biophys Acta 890: 117–126, 1987

58. Zeleznikar RJ, Goldberg ND: Kinetics and compartmentation of energy metabolism in intact skeletal muscle determined from ^{18}O-labelling of metabolite phosphoryls. J Biol Chem 266: 15110–15119, 1991

59. Wyss M, Wallimann T: Metabolite channelling in aerobic metabolism. J Theor Biol 158: 129–132, 1992

60. Zeleznikar RJ, Heyman RA, Graeff RM, Walseth TF, Dawis SM, Butz EA, Goldberg ND: Evidence for compartmentalized adenylate kinase catalysis serving a high energy phosphoryl

216

transfer function in rat skeletal muscle. J Biol Chem 265: 300–311, 1990

61. Yoshizaki K, Watari H, Radda G: Role of phosphocreatine in energy transport in skeletal muscle of bullfrog studied by ^{31}P-NMR. Biochim Biophys Acta 1051: 144–150, 1990

62. Saks VA, Belikova YO, Kuznetsov AV: In vivo regulation of mitochondrial respiration in cardiomyocytes: specific restrictions for intracellular diffusion of ADP. Biochim Biophys Acta 1074: 302–311, 1991

63. Saks VA, Rosenstraukh LV, Smirnov VN, Chazov EI: Role of creatine phosphokinase in cellular function and metabolism. Can J Physiol Pharmacol 56: 691–706, 1978

64. Bessman SP, Geiger PJ: Transport of energy in muscle. The phosphorylcreatine shuttle. Science 211: 448–452, 1981

65. Bittl JA, Ingwall JS: Reaction rates of creatine kinase and ATP synthesis in the isolated rat heart. J Biol Chem 260: 3512–3517, 1985

66. Perry SB, McAuliffe J, Balschi JA, Hickey PR, Ingwall JS: Velocity of the CK reaction in the neonatal rabbit heart: role of mitochondrial creatine kinase. Biochem 27: 2165–2172, 1988

67. Brindle KM, Blackledge MJ, Challiss JRA, Radda GK: ^{31}P NMR magnetization-transfer measurements of ATP turnover during steady-state isometric contraction in the rat hind limb in vivo. Biochem 28: 4887–4893, 1989

68. Sauter A, Rudin M: Determination of creatine kinase kinetic parameters in rat brain by NMR-magnetization transfer: correlation with brain function. J Biol Chem 268: 13166–13171, 1993

69. Shapiro B: The existential decision of a sperm. Cell 49: 293–294, 1987

70. Wallimann T, Moser H, Zurbriggen B, Wegmann G, Eppenberger HM: Creatine kinase isoenzymes in spermatozoa. J Muscle Res Cell Motil 7: 25–34, 1986

71. Tombes RM, Shapiro BM: Metabolite channeling: a phosphocreatine shuttle to mediate high energy phosphate transport between sperm mitochondria and tail. Cell 41: 325–334, 1985

72. Huszar G, Corrales M, Vigue L: Correlation between sperm creatine phosphokinase activity and sperm concentration in normospermic and oligospermic men. Gamete Res 19: 67–75, 1988

73. Huszar G, Vigue L: Incomplete development of human spermatozoa is associated with increased creatine phosphokinase concentration and abnormal head morphology. Mol Reprod Dev 34: 292–298, 1993

74. Huszar G, Vigue L, Morshedi M: Sperm creatine phosphokinase M-isoform ratios and fertilizing potential of men: a blinded study of 84 couples treated with in vitro fertilization. Fertil Steril 57: 882–888, 1992

75. Garber AT, Winkfein RJ, Dixon GH: A novel creatine kinase cDNA whose transcript shows enhanced testicular expression. Biochim Biophys Acta 1087: 256–258, 1990

76. Hammerstedt RH, Lardy HA: The effect of substrate cycling on the ATP yield of sperm glycolysis. J Biol Chem 258: 8759–8768, 1983

77. Leonardi D, Colpi GM, Campana A, Balerna M: Protein characterization of multi-fraction split-ejaculates. Some physicochemical properties of prostatic and vesicular proteins. Acta Eur Fertil 14: 181–189, 1983

78. Lee HJ, Fillers WS, Iyengar MR: Phosphocreatine, an intracellular high-energy compound, is found in the extracellular fluid of the seminal visicles in mice and rats. Proc Natl Acad Sci USA 85: 7265–7269, 1988

79. Lee H, Gong C, Wu S, Iyengar MR: Accumulation of phosphocreatine and creatine in the cells and fluid of mouse seminal vesicles is regulated by testosteron. Biol Reprod 44: 540–545, 1991

80. Fakih H, MacLusky N, DeCherney A, Wallimann T, Huszar G: Enhancement of human sperm motility and velocity in vitro: effects of calcium and creatine phosphate. Fertil Steril 46: 938–944, 1986

81. Kavanagh JP, Darby C: Creatine kinase and ATPase in human seminal fluid and prostatic fluid. J Reprod Fertil 68: 51–56, 1983

82. Asseo PP, Panidis DK, Papadimas JS, Ikkos DG: Creatine kinase in seminal plasma of infertile men: activity and isoenzymes. Int J Androl 4: 431–439, 1981

83. Tombes RM, Shapiro BM: Enzyme termini of a phospho-creatine shuttle: purification and characterization of two creatine kinase isoenzymes from sea urchin sperm. J Biol Chem 262: 16011–16019, 1987

84. Wothe DD, Charbonneau H, Shapiro BM: The phosphocreatine shuttle of sea urchin sperm: flagellar creatine kinase resulted from a gene triplication. Proc Natl Acad Sci USA 87: 5203–5207, 1990

85. Mühlebach SM, Gross M, Wirz T, Wallimann T, Perriard JC, Wyss M: Sequence homology and structure predictions of the creatine kinase isoenzymes. Mol Cell Biol 133/134: 245–262, 1994

86. Quest AFG, Shapiro B: Membrane-associated sperm flagellar creatine kinase cytosolic isoforms in a phosphocreatine shuttle. J Biol Chem 266: 19803–19811, 1991

87. Quest AFG, Chadwick JK, Wothe DD, McIlhinney RAJ, Shapiro BM: Myristoylation of flagellar creatine kinase in the sperm phosphocreatine shuttle is linked to its membrane association properties. J Biol Chem 267: 15080–15085, 1992

88. Tombes RM, Farr A, Shapiro BM: Sea urchin sperm creatine kinase: the flagellar isoenzyme is a microtubule-associated protein. Exp Cell Res 178: 307–317, 1988

89. Christen R, Schackmann RW, Dahlquist FW, Shapiro BM: ^{31}P-NMR analysis of sea urchin sperm activation: reversible formation of high energy phosphate compounds by changes in intracellular pH. Exp Cell Res 149: 289–294, 1983

90. Tombes RM, Brokaw CJ, Shapiro BM: CK-dependent energy transport in sea urchin spermatozoa. Flagellar wave attenuation and theoretical analysis of high energy phosphate diffusion. Biophys J 52: 75–86, 1987

91. Blum H, Nioka S, Johnson RG: Activation of the Na$^+$,K$^+$-ATPase in Narcine brasiliensis. Proc Natl Acad Sci USA 87: 1247–1251, 1990

92. Carneiro LH, Hasson-Voloch A: Creatine kinase from the electric organ of Electrophorus electricus: Isoenzyme analysis. Int J Biochem 15: 111–114, 1983

93. Barrantes FJ, Mieskes G, Wallimann T: A membrane-associated CK identified as an acidic species of the non-receptor, peripheral v-proteins in Torpedo acetylcholine receptor membranes. FEBS Lett 152: 270–275, 1983

94. West BL, Babbitt PC, Mendez B, Baxter JD: Creatine kinase protein sequence encoded by a cDNA made from Torpedo californica electric organ mRNAs. Proc Natl Acad Sci USA 81: 7007–7012, 1984

95. Giraudat J, Devillers-Thiery A, Perriard JC, Changeux JP: Complete nucleotide sequence of Torpedo marmorata mRNA coding for the 43 kDa v$_2$-protein: muscle-specific CK. Proc Natl Acad Sci USA 81: 7313–7317, 1984

96. Witzemann K: Creatine phosphokinase: isoenzymes in *Torpedo marmorata.* Europ J Biochem 150: 201–210, 1985

97. Perryman MB, Knell JD, Ifegwu J, Roberts R: Identification of the 43-kDa polypeptide associated with acetylcholine receptor-enriched membranes and MM-creatine kinase. J Biol Chem 260: 9399–9404, 1985

98. Wallimann T, Walzthöny D, Wegmann G, Moser H, Eppenberger HM, Barrantes FJ: Subcellular localization of creatine kinase in *Torpedo* electrocytes: association with acetylcholine receptor-rich membranes. J Cell Biol 100: 1063–1072, 1985

99. Barrantes FJ, Mieskes G, Wallimann T: Creatine kinase activity in the *Torpedo* electrocyte and in the non-receptor, peripheral v-proteins from acetylcholine receptor-rich membranes. Proc Natl Acad Sci USA 80: 5440–5444, 1983

100. Barrantes FJ, Braceras A, Caldironi HA, Mieskes G, Moser H, Toren CE, Roque ME, Wallimann T, Zechel A: Isolation and characterization of acetylcholine receptor membrane-associated and soluble electrocyte creatine kinase. J Biol Chem 260: 3024–3034, 1985

101. Gysin R, Yost B, Flanagan SD: Creatine kinase isoenzymes in *Torpedo californica*: absence of the major brain isoenzyme from nicotinic acetylcholine receptor membrane. Biochem 25: 1271–1278, 1986

102. Borroni E: Role of creatine phosphate in the discharge of the electric organ of *Torpedo marmorata.* J Neurochem 43: 795–798, 1984

103. Blum H, Balschi JA, Johnson RG: Coupled *in vivo* activity of creatine phosphokinase and the membrane-bound (Na$^+$/K$^+$)-ATPase in the resting and stimulated electric organ of the electric fish *Narcine brasiliensis.* J Biol Chem 266: 10254–10259, 1991

104. Dunant Y, Loctin F, Marsal J, Müller D, Parducz A, Rabasseda X: Energy metabolism and quantal acetylcholine release: effects of Botulinum toxin, 1-fluoro-2,4-dinitrobenzene, and diamide in the *Torpedo* electric organ. J Neurochem 50: 431–439, 1988

105. Ames A, Walseth TF, Heyman RA, Barad M, Graeff RM, Goldberg ND: Light-induced increases in cGMP metabolic flux correspond with electrical responses of photoreceptor cells. J Biol Chem 261: 13034–13042, 1986

106. Hughes JT, Jerome D, Krebs HA: Ultrastructure of the avian retina: an anatomical study of the retina of the domestic pigeon (*Columba livia*) with particular reference to the distribution of mitochondria. Exp Eye Res 14: 189–205, 1972

107. Buono RJ, Sheffield JB: Changes in distribution of mitochondria in the developing chick retina. Exp Eye Res 53: 187–198, 1991

108. Olson A, Pugh EN: Diffusion coefficient of cyclic GMP in Salamander rod outer segments estimated with two fluorescent probes. Biophys J 65: 1335–1352, 1993

109. Wallimann T, Wegmann G, Moser H, Huber R, Eppenberger HM: High content of creatine kinase in chicken retina: compartmentalized localization of creatine kinase isoenzymes in photoreceptor cells. Proc Natl Acad Sci USA 83: 3816–3819, 1986

110. Wegmann G, Huber R, Zanolla E, Eppenberger HM, Wallimann T: Differential expression and localization of brain-type and mitochondrial creatine kinase isoenzymes during development of the chicken retina: Mi-CK as a marker for differentiation of photoreceptor cells. Differentiation 46: 77–87, 1991

111. Dontsov AE, Zak PP, Ostrovskii MA: Regeneration of ATP in outer segments of frog photoreceptors. Biochem (USSR) 43: 471–474, 1978

112. Schnetkamp PPM, Daemen FJM: Transfer of high-energy phos-phates in bovine rod outer segments. Biochim Biophys Acta 672: 307–312, 1981

113. Sitaramayya A, Liebman PA: Mechanism of ATP quench of phosphodiesterase activation in rod disk membranes. J Biol Chem 258: 1205–1209, 1983

114. Sather WA, Detwiler PB: Intracellular biochemical manipulation of phototransduction in detached rod outer segments. Proc Natl Acad Sci USA 84: 9290–9294, 1987

115. Fesenko EE, Krapivinsky GB: Cyclic GMP binding sites and light control of free cGMP concentration in vertebrate rod photoreceptors. Photobiochem Photobiophys 13: 345–358, 1986

116. Mahadevan LC, Whatley SA, Leung TKC, Lim L: The brain form of a key ATP-regulating enzyme, creatine kinase, is a phosphoprotein. Biochem J 222: 139–144, 1984

117. Chida K, Tsunenaga M, Kasahara K, Kohno Y, Kuroki T: Regulation of creatine phosphokinase B activity by protein kinase C. Biochem Biophys Res Commun 173: 346–350, 1990

118. Quest AFG, Soldati T, Hemmer W, Perriard JC, Eppenberger HM, Wallimann T: Phosphorylation of chicken brain-type creatine kinase affects a physiologically important kinetic parameter and gives rise to protein microheterogeneity *in vivo*. FEBS Lett 269: 457–464, 1990

119. Kühn H, Hall SW, Wilden U: Light-induced binding of 48-kDa protein to photoreceptor membranes is highly enhanced by phosphorylation of rhodopsin. FEBS Lett 176: 473–478, 1984

120. Mascarelli F, Raulais D, Courtois Y: Fibroblast growth factor phosphorylation and receptors in rod outer segments. EMBO J 8: 2265–2273, 1989

121. Friedman DL, Hejtmancik JF, Hope JN, Perryman B: Developmental expression of creatine kinase isoenzymes in mammalian lens. Exp Eye Res 49: 445–457, 1989

122. Golander A, Binderman I, Kaye AM, Nimrod A, Sömjen D: Stimulation of creatine kinase activity in rat organs by human growth hormone *in vivo* and *in vitro*. Endocrinol 118: 1966–1970, 1986

123. Piatigorsky J: Lens differentiation in vertebrates. Differentiation 19: 134–153, 1981

124. Chandler WL, Fine JS, Emery M, Weaver D, Reichenbach D, Clayson KJ: Regional creatine kinase, adenylate kinase and lactate dehydrogenase in normal canine brain. Stroke 19: 251–255, 1988

125. Maker HS, Lehrer GM, Silides DJ, Weiss C: Regional changes in cerebellar creatine phosphate metabolism during late maturation. Exp Neurol 38: 295–300, 1973

126. Manos P, Bryan GK, Edmond J: Creatine kinase activity in postnatal rat brain development and in cultured neurons, astrocytes and oligodendrocytes. J Neurochem 56: 2101–2107, 1991

127. Kahn MA: Effect of calcium on creatine kinase activity of cerebellum. Histochem 48: 29–32, 1976

128. Cadoux-Hudson TA, Blackledge MJ, Radda GK: Imaging of human brain creatine kinase activity *in vivo*. FASEB J 3: 2660–2666, 1989

129. Molloy GR, Wilson CD, Benfield P, deVellis J, Kumar S: Rat brain creatine kinase messenger RNA levels are high in primary cultures of brain astrocytes and oligodendrocytes and low in neurons. J Neurochem 59: 1925–1932, 1992

130. Quest AFG, Eppenberger HM, Wallimann T: Purification of brain-type creatine kinase (B-CK) from several tissues of the chicken: B-CK subspecies. Enzyme 41: 33–42, 1989

131. Quest AFG, Eppenberger HM, Wallimann T: Two different B-

218

type creatine kinase subunits dimerize in a tissue-specific manner. FEBS Lett 262: 299–304, 1990

132. Hemmer W, Glaser SJ, Hartmann GR, Eppenberger HM, Wallimann T: Covalent modification of creatine kinase by ATP: evidence for autophosphorylation. In: LMG Heilmeyer (ed.) Cellular Regulation by Protein Phosphorylation. NATO ASI Series Vol. H56. Springer, Berlin, 1991, pp 143–147

133. Hemmer W, Zanolla E, Furter-Graves EM, Eppenberger HM, Wallimann T: Creatine kinase isoenzymes in chicken cerebellum: specific localization of brain-type CK in Bergmann glial cells and muscle-type CK in Purkinje neurons. Eur J Neurosci 6: 538–549, 1994

134. Soldati T, Schäfer BW, Perriard JC: Alternative ribosomal initiation gives rise to chicken brain-type CK isoproteins with heterogeneous amino termini. J Biol Chem 265: 4498–4506, 1990

135. Hemmer W, Skarli M, Perriard JC, Wallimann T: Effect of okadaic acid on protein phosphorylation patterns of chicken myogenic cells with special reference to creatine kinase. FEBS Lett 327: 35–40, 1993

136. Swanson Ph: The particulate adenosine triphosphate creatine phosphotransferase from brain: its distribution in subcellular fractions and its properties. J Neurochem 14: 343–356, 1967

137. Booth RFG, Clark JB: Studies on the mitochondrially bound form of rat brain creatine kinase. Biochem J 170: 145–151, 1978

138. Wevers RA, Reutlingsperger CPM, Dam B, Soons JBJ: Mitochondrial creatine kinase in the brain. Clin Chim Acta 119: 209–223, 1982

139. Wyss M, Schlegel J, James P, Eppenberger HM, Wallimann T: Mitochondrial creatine kinase from chicken brain. Purification, biophysical characterization and generation of heterodimeric and heterooctameric molecules with subunits of other creatine kinase isoenzymes. J Biol Chem 265: 15900–15908, 1990

140. Hemmer W, Wallimann T: Functional aspects of creatine kinase in brain. Dev Neurosci 15 (3–5), 1993

141. Hamburg RJ, Friedman DL, Olson EN, Ma TS, Cortez MD, Goodman C, Puleo PR, Perryman MB: Muscle creatine kinase isoenzyme expression in adult human brain. J Biol Chem 265: 6403–6409, 1990

142. Palay S, Chan-Palay V: Cerebellar cortex, cytology and organization. Springer Verlag, New York, 1974

143. Rakic P: Neuron-glia relationship during granule cell migration in developing cerebellar cortex. A Golgi and electronmicroscopic study in Maccacus rhesus. J Comp Neurol 141: 283–312, 1971

144. Hatten ME: Riding the glial monorail: a common mechanism for glial-guided neuronal migration in different regions of the developing mammalian brain. Trends Neurosci 13: 179–184, 1990

145. Newman EA: Regulation of potassium levels by glia cells in the retina. Trends Neurosci 8: 156–159, 1985

146. Reichenbach A: Glial K$^+$-permeability and CNS K$^+$-clearance by diffusion and spatial buffering. In: NJ Abott (ed.) Glial-Neuronal Interaction. Acad. Sci., New York, 1991, pp 272–286

147. Hertz L, Peng L: Energy metabolism at the cellular level of the CNS. Can J Physiol Pharmacol 70: S145–S157, 1991

148. Knöpfel T, Vranesic I, Staub C, Gähwiler BH: Climbing fibre responses in olivo-cerebellar slice cultures. II Dynamics of cytosolic calcium in Purkinje cells. Eur J Neurosci 3: 343–348, 1991

149. Ito M: The cellular basis of cerebellar plasticity. Curr Opinion Neurobiol 1: 616–620, 1991

150. Kuwajima G, Futatsugi A, Niinobe M, Nakanishi S, Mikoshiba K: Two types of ryanodyne receptors in mouse brain: skeletal

muscle type exclusively in Purkinje cells and cardiac muscle type in various neurons. Neuron 9: 1133–1142, 1992

151. Villa A, Podini P, Clegg DO, Pozzan T, Meldolesi J: Intracellular Ca^{2+}-stores in chicken Pukinje neurons: differential distribution of low affinity-high capacity Ca^{2+}-binding protein, calsequestrin, of Ca^{2+}-ATPase and of ER lumenal protein, Bip. J Cell Biol 113: 779–791, 1991

152. Takei K, Stukenbrok H, Metcalf A, Mignery JA, Sudhof TC, Volpe P, DeCamilli P: Ca^{2+}-stores in Purkinje neurons: endoplasmic reticulum subcompartments demonstrated by the heterogeneous distribution of the InsP$_3$-receptor, Ca^{2+}-ATPase, and calsequestrin. J Neurosci 12: 489–505, 1992

153. Michelangeli F, DiVirgilio F, Villa A, Podini P, Meldolesi J, Pozzan T: Identification, kinetic properties and intracellular localization of the (Ca^{2+}-Mg^{2+})-ATPase from the intracellular stores of chicken cerebellum. Biochem J 275: 555–561, 1991

154. Campbell AM, Wuytack F, Fambrough DM: Differential distribution of the alternative forms of the sarcoplasmic/endoplasmic reticulum (Ca^{2+}-Mg^{2+})-ATPase, SERCA 2b and SERCA 2a, in the avian brain. Brain Res 605: 67–76, 1993

155. Rudin M, Sauter A: Dihydropyridine calcium antagonists reduce the consumption of high-energy phosphates in the rat brain. A study using combined ^{31}P/^{1}H magnetic resonance spectroscopy and ^{31}P saturation transfer. J Pharmacol Exper Therapeutics 251: 700–706, 1989

156. Friedhoff AJ, Lerner MH: CK isoenzyme associated with synaptosomal membrane and synaptic vesicles. Life Sci 20: 867–872, 1977

157. Lim L, Hall C, Leung T, Mahadevan L, Whatley S: Neuron-specific enolase and CK are protein components of rat brain synaptic plasma membranes. J Neurochem 41: 1177–1182, 1983

158. Sweadner KJ: Two molecular forms of Na$^+$/K$^+$-stimulated ATPase in brain. J Biol Chem 254: 6060–6067, 1979

159. Erecinska M, Silver IA: ATP and brain function. J Cerebr Blood Flow and Metabolism 9: 2–19, 1989

160. Carafoli E: Intracellular calcium homeostasis. Annu Rev Biochem 56: 395–433, 1987

161. Brady ST, Lasek RJ: Nerve-specific enolase and creatine kinase in axonal transport: 'Soluble proteins' and axoplasmic matrix. Cell 23: 515–523, 1981

162. Oblinger MM, Brady ST, McQuarrie IG, Lasek RJ: Cytoplasmic differences in the protein composition of the axonally transported cytoskeleton in mammalian neurons. J Neurol 7: 433–462, 1987

163. Gerbitz KD, Deufel T, Summer J, Thallemer J, Wieland OH: Brain-specific proteins: creatine kinase BB isoenzyme is cochromatographed during preparation of neuron-specific enolase from human brain. Clin Chim Acta 133: 233–239, 1983

164. Holtzman D, McFarland EW, Jacobs D, Offutt MC, Neuringer LJ: Maturational increase in mouse brain creatine kinase reaction rates shown by phosphorus magnetic resonance. Dev Brain Res 58: 181–188, 1991

165. Holtzman D, Herman MM, Desantel M, Lewiston N: Effects of altered osmolality on respiration and morphology of mitochondria from the developing brain. J Neurochem 33: 453–460, 1979

166. Holtzman D, Tsuji M, Wallimann T, Hemmer W: Functional maturation of creatine kinase in rat brain: a hypothesis. Dev Neurosci 15 (3–5), 1993

167. Binderman I, Harel S, Earon Y, Tomer A, Weisman Y, Kaye AM, Sömjen D: Acute stimulation of creatine kinase activity by vita-

min D metabolites in the developing cerebellum. Biochim Biophys Acta 972: 9–16, 1988

168. Bergen HT, Pentecost BT, Dickerman HW, Pfaff DW: *In situ* hybridization for creatine kinase-B messenger RNA in rat uterus and brain. Mol Cell Endocrinol 92: 111–119, 1993

169. Iyengar MR, Fluellen CE, Iyengar CW: Creatine kinase from the bovine myometrium: purification and characterization. J Muscle Res Cell Motility 3: 231–246, 1982

170. Payne RM, Friedman DL, Grant JW, Perryman BM, Strauss AW: Creatine kinase isoenzymes are highly regulated during pregnancy in rat uterus and placenta. Am J Physiol 265: E624–E635, 1993

171. Clark JF, Khuchua Z, Kuznetsov A, Saks VA, Ventura-Clapier R: Compartmentation of creatine kinase isoenzymes in myometrium of gravid guinea pigs. J Physiol (Lond) 466: 553–572, 1993

172. Dawson MJ, Wray S: Changes in phosphorus metabolism in the rat uterus following parturition. J Physiol (Lond) 336: 19–20, 1983

173. Reiss NA, Kaye AM: Identification of the major component of the estrogen-induced protein of rat uterus as the BB-isoenzyme of CK. J Biol Chem 256: 5741–5749, 1981

174. Spatz M, Waisman A, Kaye AM: Responsiveness of the 5′-flanking region of the brain type isozyme of creatine kinase to estrogens and anti-estrogens. J Steroid Biochem Mol Biol 41: 711–714, 1992

175. Iyengar MR, Fluellen CE, Iyengar CW: Increased creatine kinase in the hormone-stimulated smooth muscle of the bovine uterus. Biochem Biophys Res Commun 94: 948–954, 1980

176. Degani HT, Shaer A, Victor TA, Kaye AM: Estrogen-induced changes in high-energy phosphate metabolism in rat uterus: ^{31}P-NMR studies. Biochem 23: 2572–2577, 1984

177. Degani HT, Victor TA, Kaye AM: Effects of 17β-estradiol on high energy phosphate concentration and the flux catalysed by creatine kinase in immature rat uteri: ^{31}P nuclear magnetic resonance studies. Endocrinol 122: 1631–1638, 1988

178. Keller TCS, Gordon PV: Discrete subcellular localization of a cytoplasmic and a mitochondrial isoenzyme of creatine kinase in intestinal epithelial cells. Cell Motil Cytoskeleton 19: 169–179, 1991

179. Gordon PV, Keller TCS: Functional coupling to brush border creatine kinase imparts a selective energetic advantage to contractile ring myosin in intestinal epithelial cells. Cell Motil Cytoskel 21: 38–45, 1992

180. Thalmann R, Miyoshi T, Thalmann I: The influence of ischemia upon energy reserves of inner ear tissue. Laryngoscope 82: 2249–2272, 1972

181. Spicer SS, Schulte BA: Creatine kinase in epithelium of the inner ear. J Histochem Cytochem 40: 185–192, 1992

182. Loike JD, Cao L, Brett J, Ogawa S, Silverstein SC, Stern D: Hypoxia induces glucose transporter expression in endothelial cells. Am J Physiol 263: C326–C333, 1992

183. Bastin J, Cambon N, Thomson M, Lowry OH, Burch HB: Change in energy reserves in different segments of the nephron during brief ischemia. Kidney Internatl 31: 1239–1247, 1987

184. Ikeda K: Localization of brain-type creatine kinase in kidney epithelial cell subpopulations in rat. Experientia 44: 734–735, 1988

185. Friedman DL, Parryman MB: Compartmentation of multiple forms of creatine kinase in the distal nephron of the rat kidney. J Biol Chem 266: 22404–22410, 1991

186. Friedman DL, Roberts R: Purification and localization of brain-type CK in sodium chloride transporting epithelia of the spiny dogfish, *Squalus acanthias*. J Biol Chem 267: 4270–4276, 1992

187. Epstein FH, Stoff JS, Silvia P: Mechanism and control of hyperosmotic NaCl-rich secretion by the rectal gland of *Squalus acanthias*. J Exp Biol 106: 25–41, 1983

188. Berlet HH, Bonsmann I, Birringer H: Occurrence of free creatine, phosphocreatine and creatine phosphokinase in adipose tissue. Biochim Biophys Acta 437: 166–174, 1976

189. Gosh A, Ronner P, Cheong E, Khalid P, Matschinsky FM: The role of ATP and free ADP in metabolic coupling during fuel-stimulated insulin release from islet β-cells in the isolated perfused rat pancreas. J Biol Chem 266: 22887–22892, 1991

190. White KC, Babbitt PC, Buechter DD, Kenyon GL: The principle islet of the Coho Salmon (*Oncorhyncus kisutch*) contains the BB isoenzyme of creatine kinase. J Prot Chem 11: 489–494, 1992

191. Aired S, Creach Y, Palevody C, Esclassan J, Hollande E: Creatine phosphate as energy source in the cerulein-stimulated rat pancreas: study by ^{31}P nuclear magnetic resonance. Int J Pancreatol 10: 81–95, 1991

192. Perriard JC, Rosenberg UB, Wallimann T, Eppenberger HM, Caravatti M: The switching of creatine kinase gene expression during myogenesis. In: ML Pearson, HF Epstein (eds) Monograph on Muscle Development: Molecular and Cellular Control. Cold Spring Harbor, 1982, pp 237–245

193. Graig FA, Smith JC: Creatine phosphokinase in thyroid: isoenzyme composition compared with other tissues. Science 156: 254–255, 1967

194. Kanemitsu F, Kawanishi I, Mizushima J: Characteristics of mitochondrial creatine kinases from normal human heart and liver tissues. Clin Chim Acta 119: 307–317, 1982

195. Wyss M, Wallimann T, Köhrle J: Selective labelling and inactivation of creatine kinase isoenzymes by the thyroid hormone analogue N-bromoacetyl-3,3′,5-triiodo-L-thyronine. Biochem J 291: 463–472, 1993

196. Kvam BJ, Pollesello P, Vittur F, Paoletti S: ^{31}P-NMR studies of resting zone cartilage from growth plate. Magn Res Med 25: 355–361, 1992

197. Shapiro IM, Debolt K, Funanage VL, Smith S, Tuan RS: Developmental regulation of creatine kinase activity in cells of the epiphyseal growth cartilage. J Bone Mineral Res 7: 493–500, 1992

198. Sömjen D, Weisman Y, Mor Z, Harell A, Kaye AM: Regulation of proliferation of rat cartilage and bone by sex steroid hormones. J Steroid Biochem Mol Biol 40: 717–723, 1991

199. Sömjen D, Kaye AM, Binderman I: 24R,25-dihydroxyvitamine D stimulates creatine kinase BB activity in chick cartilage cells in culture. FEBS Lett 167: 281–284, 1984

200. Funanage VL, Carango P, Shapiro IM, Tokuoka T, Tuan RS: Creatine kinase activity is required for mineral deposition and matrix synthesis in endochondral growth cartilage. Bone and Mineral 17: 228–236, 1992

201. Sömjen D, Weisman Y, Binderman I, Kaye AM: Stimulation of creatine kinase BB activity by 1α,25-dihydroxycholecalciferol and 24R,25-dihydroxycholecalciferol in rat tissues. Biochem J 219: 1037–1041, 1984

202. Sömjen D, Kaye AM, Binderman I: Stimulation of creatine kinase BB activity by parathyroid hormone and by prostaglandin E$_2$ in cultured bone cells. Biochem J 225: 591–596, 1985

203. Sömjen D, Weisman Y, Harell A, Berger E, Kaye AM: Direct and sex-specific stimulation by sex steroids of creatine kinase ac-

220

tivity and DNA synthesis in rat bone. Proc Natl Acad Sci USA 86: 3361–3365, 1989

204. DeChatelet LR, McCall CE, Shirley PS: Creatine phosphokinase activity in rabbit alveolar macrophages. Infection and Immunity 7: 29–34, 1973

205. Loike JD, Kozler VF, Silverstein SC: Creatine kinase expression and creatine phosphate accumulation are developmentally regulated during differentiation of mouse and human monocytes. J Exp Med 159: 746–757, 1984

206. Loike JD, Somes M, Silverstein SC: Creatine uptake, metabolism, and efflux in human monocytes and macrophages. Am J Physiol 251: C128–C135, 1986

207. Guimbal C, Kilimann MW: A Na$^+$-dependent creatine transporter in rabbit brain, muscle, heart and kidney. cDNA cloning and functional expression. J Biol Chem 268: 8418–8421, 1993

208. Loike JD, Kozler VF, Silverstein SC: Increased ATP and creatine phosphate turnover in phagocytosing mouse peritoneal macrophages. J Biol Chem 254: 9558–9564, 1979

209. Shibata S, Kobayashi B: Blood platelets as a possible source of creatine kinase in rat plasma and serum. Thromb Haemost 39: 701–706, 1978

210. Shibata S: Creatine phosphate in rat blood platelets. Thromb Haemost 39: 707–711, 1978

211. Matsui S, Watanabe Y, Kobayashi B: Preferential disappearance of aerobically generated ATP from platelets during thrombin-induced aggregation. Thromb Diath Haemorrh 32: 441–456, 1974

212. Feld RD, Witte DL: Presence of creatine kinase BB isoenzyme in some patients with prostatic carcinoma. Clin Chem 23: 1930–1932, 1977

213. Gazdar AF, Sweig MH, Carney DN, VanSteirten AC, Baylin SB, Minna JD: Levels of creatine kinase and its BB isoenzyme in lung cancer specimens and cultures. Cancer Res 41: 2773–2777, 1981

214. World LE, Li C-Y, Homburger HA: Localization of the B and M polypeptide subunits of creatine kinase in normal and neoplastic human tissues by an immuno peroxidase technique. Am J Clin Pathol 75: 327–332, 1981

215. Tsung SH: Creatine kinase activity and isoenzyme pattern in various normal tissues and neoplasms. Clin Chem 29: 2040–2043, 1983

216. Scambia G, Santeusanio G, Panici PB, Iacobelli S, Mancuso S: Immunohistochemical localization of creatine kinase BB in primary cancer: correlation with estrogen receptor content. J Cancer Res Clin Oncol 114: 101–104, 1988

217. Silverman LM, Dermer GB, Zweig MH, VanSteirteghem AC, Tökes ZA: Creatine kinase BB: a new tumor-associated marker. Clin Chem 25: 1432–1435, 1979

218. Miller E, Evans AE, Cohn M: Inhibition of rate of tumor growth by creatine and cyclocreatine. Proc Natl Acad Sci USA 90: 3304–3308, 1993

219. Lillie JW, O'Keefe M, Valinski H, Hamlin A, Varban ML, Kaddurah-Daouk R: Cyclocreatine inhibits growth of a broad spectrum of cancer cells derived from solid tumors. Cancer Res 53: 3172–3178, 1993

220. Pratt R, Vallis LM, Lim CW, Chisnall WN: Mitochondrial creatine kinase in cancer patients. Pathol 19: 162–165, 1987

221. DeLuca M, Hall N, Rice R, Kaplan NO: Creatine kinase isoenzymes in human tumors. Biochem Biophys Res Commun 99: 189–195, 1981

222. Kanemitsu F, Kawanishi I, Mizushima J, Okigaki T: Mitochondrial creatine kinase as a tumor associated marker. Clin Chim Acta 138: 175–183, 1984

223. Kaddurah-Daouk R, Lillie JW, Daouk GH, Green MR, Kingston R, Schimmel P: Induction of a cellular enzyme for energy metabolism by transforming domains of adenovirus E1a. Mol Cell Biol 10: 1476–1483, 1990

224. Lang H: Creatine kinasae isoenzymes: Pathophysiology and clinical application. In: H Lang (ed.). Springer Verlag, New York, 1981

225. Ch'ng JLC, Ibrahim B: Transcriptional and posttranscriptional mechanisms modulate creatine kinase expression during differentiation of osteoblast cells. J Biol Chem 269: 2336–2341, 1994

226. Ch'ng JLC, Shoemaker DL, Schimmel P, Holmes EW: Reversal of CK translational repression by 3′-untranslaed sequences. Science 248: 1003–1006, 1992

Molecular and Cellular Biochemistry **133/134**: 221–232, 1994.
© 1994 *Kluwer Academic Publishers.*

III-3 The creatine kinase system in smooth muscle

Joseph F. Clark
Department of Biochemistry, University of Oxford, South Parks Road, Oxford OX1 3QU, England

Abstract

Despite the energetic flux being much lower in smooth muscle compared to striated muscles (such as the heart and skeletal muscle) creatine kinase (CK) has been found present and active in all smooth muscles studied to date. A complete CK circuit has been identified, with CK found in the mitochondria, contractile elements, membrane pumps and the cytoplasm. CK isoenzymes are coupled to many cellular energetic processes and appears to be involved in energy production and consumption by acting as an energy transducer. The CK system responds to pathological insults and development (e.g. hypertrophy and gestation respectively) by changes in sub-cellular distribution localization, isoenzymes, and specific activity. The conclusion from these observations is that creatine kinase is intimately involved in the energetic system of smooth muscle. (Mol Cell Biochem **133/134**: 221–232, 1994)

Key words: uterus, taenia coli, artery, ADP, ATP, respiration, phosphocreatine, skinned smooth muscle

Abbreviations: CK – creatine kinase; Mi-CK – mitochondrial creatine kinase; Cr – creatine; PCr – phosphocreatine; NMR – nuclear magnetic resonance; SHR – spontaneously hypertensive rat; β-GPA – β-guanidinopropionic acid

Introduction

Creatine kinase (CK) is found in 4 different isoforms: three are cytosolic and one mitochondrial. Though sub-forms of these have been characterized I will stick to the convention of M and B monomers and one mitochondrial CK (Mi-CK) in this review. The M and B monomers combine to form three dimers the MM and BB homodimers and the MB heterodimer. Mi-CK is found on the outside of the inner-mitochondrial membrane, while the other three forms are frequently localized in cellular organelles and the cytoplasm.

The specific activity of CK in smooth muscle ranges from 1–300 IU/g ww with BB-CK frequently being the predominant isoenzyme. The diversity of the isoenzyme distribution and activities of CK makes its role in smooth muscle metabolism a subject worthy of study.

Presented here is evidence that CK is specifically localized at the sites of energy utilization and production, functioning as an integral component of the energetics and contractility of smooth muscle.

To address the coupling of the creatine kinase enzymes in smooth muscle one would be required to closely examine their role and specialization in the microcompartments within the smooth muscle cell. This was the ethos behind a series of experiments performed on the creatine kinase system in smooth muscle by this author [1–12] carried out on the taenia coli, carotid artery, aorta, gestating and quiescent uterus of the guinea-pig, pig and rat. These and other studies are discussed below.

Address for offprints: J.F. Clark, Department of Biochemistry, University of Oxford, South Parks Rd, Oxford OX1 3QU, England

Smooth muscle creatine kinase metabolism

Energetics and contractility

Smooth muscle is an interesting tissue, from an energetic view point, because of its diverse metabolism and contractility. Its energetic turnover is generally considered to be feeble compared to striated muscle. Paul and collaborators [13–15] have studied vascular contractility and its coupling to energetics and determined the ATP costs per contraction by measuring ATP flux and oxygen consumption. The ATP:O ratio was approximately 3 when they measured the time course of energy utilization during steady-state isometric contraction. The results showed that oxygen consumption (ATP synthase) did not correlate with force maintenance and force generation. Rather O_2 consumption paralleled the velocity of tension generation. This was referred to as an increased energy utilization in the pre-steady-state conditions resulting in tension maintenance with low energy costs. The economy of the maintained tension was estimated to be approximately twice that of tension development. It may be that the different contractile costs of vascular smooth muscle is related to the differences in the smooth muscle contractile protein interactions compared to skeletal muscle. Differences in the contractile types of smooth muscle have been described by Paul as tonic and phasic smooth muscles [13]. The taenia coli is a phasic smooth muscle because of its rhythmic contractions where the carotid artery which maintains tension for prolonged periods is an example of a tonic smooth muscle.

Hai and Murphy [16, 17] determined the relationship of myosin light chain phosphorylation and energy fluxes. This is related to the ability of the smooth muscle to have attached but dephosphorylated cross-bridges. These slowly cycling cross-bridges are able to maintain tension with high economy of energy consumption. This state of the cross-bridges was termed latch [18] and has a 5 fold slower rate of detachment than phosphorylated cycling cross-bridges. Thus tension is maintained at low Ca^{++} with low ATP consumption. The smooth muscle however, has a high cost for generating tension, as discussed above. Thus it appears that the energetic system of vascular smooth muscle is economical for maintaining tension while there is a high cost during tension development.

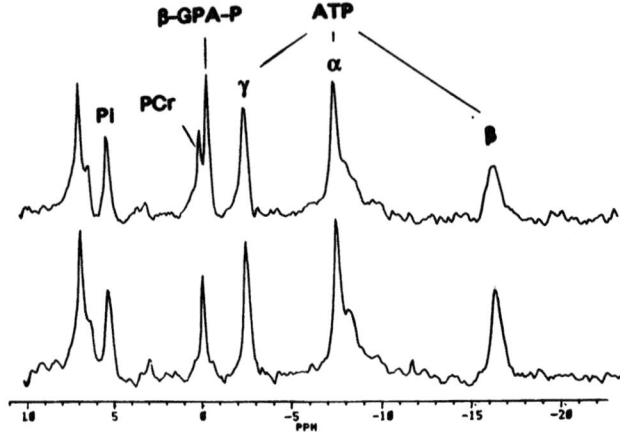

Fig. 1. A stack plot of porcine carotid arteries perfused with a Krebs buffer containing 50 mM β-GPA and 0.1 mM Pi. The bottom spectrum is the control spectrum and the top is after 12 hrs in the presence of β-GPA. β-GPA-P is the phosphorylated form of β-GPA seen due to phosphorylation by CK.

ATP and ADP in smooth muscle

Vermue and Nicolay [19] used 31-P NMR to show that maximally contracted taenia caeci from the guinea pig and spontaneously active, had similar energetics. They demonstrated that PCr decreased rapidly and preceded ATP during stimulation with a concomitant increase in Pi. They estimated that free [ADP] increased to 0.7 mM under these conditions. Fisher and Dillon [20] used NMR to directly determine free ADP in hypoxic porcine carotid artery. They reported a calculated resting ADP of 0.028 mM if fully oxygenated tissue, to a measurable level of 0.12 mM ADP in the hypoxic vascular smooth muscle. Their NMR observable ADP allowed them to directly determine the concentration of ADP and the *in vitro* value of CK equilibrium constant: $K_{CK} = 7.6 \times 10^8 \text{ M}^{-1}$. The precision of this determination required that the CK reaction was at equilibrium during the measurements with precise metabolite concentrations. If the CK reaction is rapid compared to the rate of PCr change, then the reaction may still be considered to be at or near equilibrium. Therefore despite a slowly changing [PCr], equilibrium may still be assumed and the above K_{CK} regarded as valid. The critical unknown in most determinations of free ADP using the CK reaction is the use of the appropriate K_{CK} value. Equilibrium consists of the sum of the creatine kinases in the mitochondria and cytoplasmic. Therefore there may be a steady state flux through the CK reaction of energy production and utilization.

Figure 1 is a stack plot of 2 NMR spectra taken from porcine carotid arteries of control (bottom) and after 12

hours of perfusion in the presence of the creatine analogue β-guanidinopropionic acid (β-GPA). β-GPA enters the cells and is phosphorylated by CK producing
β-GPA-P which is observed upfield of PCr. From these
experiments it was found that when CK phosphorylates
β-GPA there is a decrease in lactate, possibly because of
increased oxidative phosphorylation. Oxidative phosphorylation may be stimulated because of a decrease in
ATP and PCr (nonetheless maintaining the ATP/PCr
ratio). These experiments show that β-GPA may be acting as a phosphate sink, and that the CK system facil-
itates a decrease in both PCr and ATP while maintaining
their ratio. The smooth muscle must therefore be controlling this ratio in an effort to protect or maintain normal cellular function.

The role of creatine kinase in smooth muscle is fascinating because the levels of PCr are very low compared
to striated and other smooth muscles, ranging from 0.5
to 4.4 mM [5, 7, 19, 20]. Because of these low levels of
PCr, the concept of PCr as an energy pool (to buffer
[ATP]) may be less meaningful than in skeletal muscle
[4, 20]. What then is CK doing in smooth muscle and why
is it there? This question is, of course, related to the size
of the PCr pool with respect to its rate of utilization.
Therefore, if the main role of the creatine kinase system
was to buffer and maintain [ATP], then one might expect a larger reservoir of PCr. However [PCr] in the porcine carotid artery was 60% that of [ATP] as seen in
NMR experiments [1, 5, 20]. Clark *et al.* [10, 11] proposed
that, if the role of CK in smooth muscle was more intimately involved in contraction and relaxation, then one
might predict that CK would be specifically bound to the
contractile filaments and associated with constituents of
the cross-bridge cycle thereby obviating the need for a
high [PCr].

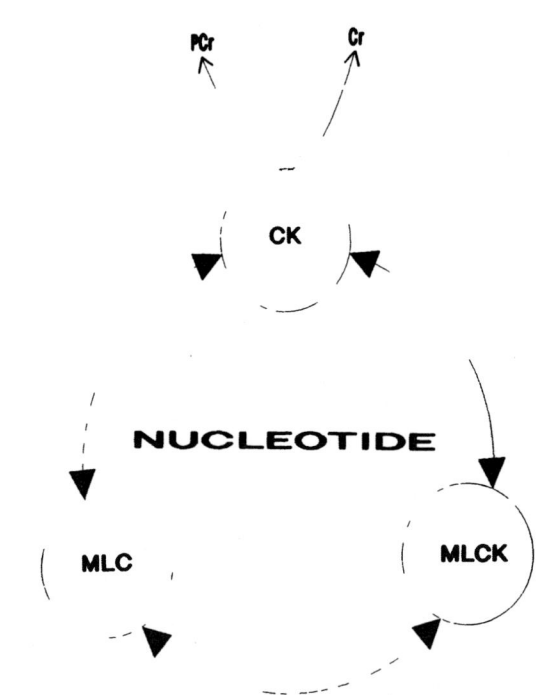

Fig. 2. A schematic diagram representing how creatine kinase might
act as an energy transducer at the contractile proteins in the presence
of nucleotide. In this energetic microcompartment, creatine kinase
could be supplying ATP to myosin light chain kinase (MLCK) to phosphorylate myosin light chains (MLC) by phosphorylating ADP produced by the contractile proteins. A bound and energetically available
pool of nucleotide would be beneficial in such a compartment because
the ADP produced might be rephosphorylated and dephosphorylated
at a rate rapid enough to maintain a functional microcompartment.

ported the concept of differing cross-bridge cycling
rates. This may, in part account for the variable costs for
force output and maintenance. Figure 2 demonstrates
how a bound and energetically active nucleotide could
be involved in smooth muscle contractility.

Bound nucleotide in smooth muscle

Butler *et al.* [21] used permeabilized rabbit portal vein to
show that resting smooth muscle ADP is bound to myosin in the form of an M.ADP.Pi complex. Myosin bound
ADP was released however, and the release rate increased, by phosphorylation of myosin coupled with
myosin actin interaction. They estimated 60–70 μM ra-
dio labeled ADP could be bound to myosin [22]. The
release of ADP was biexponential with the first exponential component containing one third of the total
ADP, at a rate of 5–10 times faster than that of the second
release component [21]. They suggested that this sup-

Localization of creatine kinase

Table 1 shows some literature values of CK isoenzymes
found in various smooth muscles. One should note that
the tendency is for the tonic vascular smooth muscle to
have lower specific activities compared to phasic
smooth muscles. There are however, intriguing differences seen with the low specific activities in the stomach
and non-gravid uterus. The non-gravid uterus is exposed
to much lower energy demands and thus may have less
of a need for the CK system. The results for the stomach
may be due to the sampling technique used in obtaining
human tissue samples or due to metabolic reasons as yet

224

undefined. Nonetheless, the metabolic demands of a tissue seem to correlate to the CK specific activity in smooth muscles.

BB-CK is the main isoform in most smooth muscles although significant amounts of MB and MM-CK are present. In cardiac muscle, the significance of the different isoforms of CK has been related to their intracellular localization rather than to kinetic differences [23]. Kinetic differences nonetheless could have some significance to subcellular isoenzyme localization in smooth muscle (see, BB-CK Kinetics). It seems logical to hypothesize that the multiple isoforms of CK may be localized within the cell to perform specific functions related to other localized proteins and metabolites. For example, perhaps Mi-CK is in the mitochondria to produce PCr, BB-CK at the contractile proteins to act as a transducer of PCr to ATP, MM-CK with membranes and glycolytic enzymes and soluble MB-CK in the cytoplasm (see NMR Studies of Smooth Muscle CK). On the other hand, the heart is characterized by having different CK isoenzymes which are found specifically localized at sarcoplasmic reticulum plasma membrane, myofilaments, mitochondria and glycolytic complexes [8, 23–27]. Despite the evidence for subcellular localization of CK, Meyer et al. [28] proposed that the mechanism of facilitated diffusion may explain spatial as well as temporal buffering of [ATP] in fast skeletal muscle.

In fast skeletal muscle there is a high [PCr] which can act as a buffer or reservoir for high energy phosphates and the buffering capacity of the system is aided by the high CK activity found in the cytoplasm at or near equilibrium. Fast skeletal muscle contains predominantly MM-CK with a specific activity of 2000 IU/g ww in contrast to smooth muscle which contains 1–300 IU/g ww and frequently all four of the isoenzymes usually with BB-CK dominating. This is an example of how the energetics and CK system of a tissue are differing concomitantly with the demands placed upon them.

Skinned smooth muscle

Clark et al. [10] used the taenia coli to examine the creatine kinase system associated with the contractile proteins. Using Triton X-100 skinned fibers they showed that BB-CK was selectively bound to the contractile proteins. Figure 3 shows the specific activity of BB-CK remaining in Triton skinned rat carotid artery. After skinning there is no detectable MM or MB-CK and only BB-CK remaining. In Triton skinned carotid, uterus and taenia coli BB-CK was the isoenzyme specifically bound to the contractile fibers. After skinning, 22% of the CK activity remained bound to the taenia coli fibers. Intact taenia coli was found to be dominated by BB-CK fol-

Table 1. Smooth muscle creatine kinase activity table

Tissue	Activity*	% MM-CKΦ	% MB-CK	% BB-CK	Ref
Aorta	26	Tr	Tr	++++	29
Abdominal aorta	21	9	35	56	53
Thoracic aorta	8	20	36	43	53
Aorta	1	39	7	54	58
Aorta	2	25	9	66	53
Carotid artery	1	24	40	36	7
Carotid artery	2	56	2	42	57
Carotid artery	47	10	33	57	12
Bladder	35	2–7	3–5	89–93	58
Stomach	23	3	6	91	58
GI Tract	140		3	97	57
Colon	125	4		96	56
Ileum	161	3	1	96	56
Taenia coli	163	Tr	+	+++	10
Taenia caeci	153	Tr	++	+++	29
Non-gravid uterus	74	20	10	70	11
Gravid uterus	196	22	8	70	11
Vas deferens	329	Tr	++	+++	29

* Specific activity is IU/g ww.
Φ Mitochondrial creatine kinase frequently was not reported.
Tr Trace activity observed.

225

TRITON SKINNED RAT CAROTID

Isolated Mitochondria from Gravid Guinea Pig

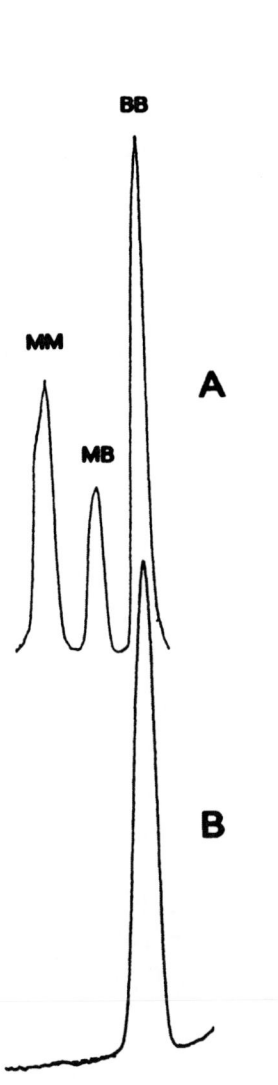

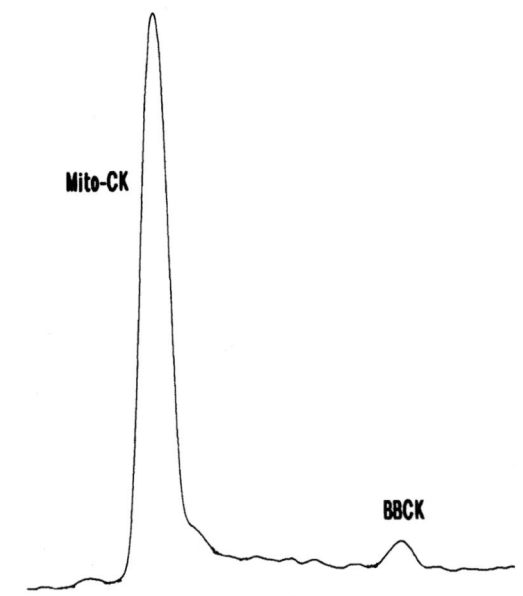

Fig. 4. Densitometer trace taken from isolated uterine mitochondria. Mi-CK is the dominant isoenzyme of CK remaining after isolation with trace amounts of BB-CK observed.

Fig. 3. Densitometer traces taken from rat carotid arteries before (A) and after exposure to 1% Triton X-100 for 1 hr. In trace A (the whole arteries) three isoenzymes of CK can be distinguished. After skinning only BB-CK is resolved in the trace (B). These experiments are consistent with the results reported by Clark *et al.* [12] for the guinea pig carotid artery.

lowed by MB and MM (Table 1). Mitochondrial CK was not observed in this study though Ishida *et al.* [29] found Mi-CK in enriched mitochondrial fractions. Similar results can be seen in Fig. 4 from uterine mitochondria. Bound CK in skinned fibers was capable of producing sufficient ATP to enable the fibers to produce 59% of maximal tension. Tension was generated in a stimulating solution with pCa 4.5 in the presence of 12 mM PCr 250 μM ADP without exogenous ATP. When the skinned fibers were put in a high tension rigor state the bound creatine kinase was able to produce 52% relaxa-

tion compared to control relaxation. They went on to show that the creatine kinase system was capable of generating 41% of maximal tension without exogenous ADP. This, they concluded, was due to a bound form of nucleotide associated with the contractile machinery being phosphorylated and hydrolyzed in a microcompartment containing MLCK, CK and MLC (Fig. 2). Their hypothesis was that such an enzyme cluster may enable efficacious functioning of the contractile system in spite of the low concentrations of ATP and PCr.

The CK system is also capable of relaxing skinned smooth muscles from a high tension rigor state. High tension rigor is induced in smooth muscle by removing high energy phosphates and calcium from pre-contracted fibers. ATP is required to relax these fibers and endogenous CK can produce sufficient ATP to produce a significant relaxation.

Interestingly, it was found that the Triton skinned taenia coli was able to have complete relaxation after the high tension rigor conditions where the carotid was capable of only 60% relaxation. It has been speculated that the tonic vascular smooth muscle is not capable of relaxing completely because of a greater sensitivity to rigor conditions. It was reported by Nishiye *et al.* [30] that the rate of relaxation in permeabilized smooth muscle is effected by ADP and that this effect was greater in the tonic smooth muscle compared to phasic smooth muscle. It may therefore be that the carotid is unable to

226

completely relax from the high tension rigor conditions because of its greater sensitivity to ADP during relaxation.

In the experiments by Nishiye *et al.* [30] if the fibers were unable to relax in the presence of ADP (with a K_d of 1–2 μM) then relaxation under physiological conditions would be hindered by ADP. Under these conditions, the role of CK could be very important in keeping the ADP concentration low enough for the fibers to relax when necessary. CK bound to the contractile proteins would have the role of maintaining the action of the contractile apparatus by supplying ATP and preventing the accumulation of ADP as well. Indeed Clark *et al.* [55] concluded that the creatine kinase system was functioning as an energetic defense mechanism in the heart and smooth muscle by its ability to keep the [ATP]/[ADP] ratio high.

The ability of the skinned fibers to produce tension and relaxation via an endogenous form of CK is strong evidence that there is a microcompartment of CK associated with the contractile machinery. There is convincing evidence as well that there is nucleotide associated within this functional microcompartment, so an energetic arrangement like the one represented in Fig. 2 is indeed quite plausible.

Clark *et al.* [10–12] have studied the creatine kinase system using Triton X-100 skinned smooth muscle fibers from the guinea pig carotid, taenia coli, and uterus. In every case studied there was significant and specific binding of BB-CK to the contractile proteins. As discussed above the percentage of bound CK after Triton treatment was consistently 22–23% of the total specific activity in the unskinned fibers. However, in the non-gravid uterus only 9% of the total specific activity of CK remained after Triton treatment while in the gestating guinea pig uterus the percentage increased to 22%.

Again, this was specifically bound BB-CK with the loss of all soluble enzymes such as MM, MB and Mi-CK.

From Table 2 a distinct difference between the role and function of creatine kinase in the other skinned smooth muscles compared to the skinned non-gestating guinea pig uterus can be clearly seen. The specific activity of creatine kinase bound after Triton skinning is less than half that of the other tissues. The much lower quantity of bound creatine kinase could then lead to the lower percent tension generation and decrease the extent of relaxation from the high tension rigor conditions. It becomes evident that the degree of relaxation and tension generation is dependent upon the activity of bound creatine kinase. The next logical question to ask is: what role does the bound creatine kinase have in the tissue *in vivo*?

One can begin to address the question of the role of bound creatine kinase *in vivo* by comparing the data in Table 2. The non-gravid uterus has relatively low energetic and contractile demands with concomitantly low percentage of bound creatine kinase. The result is the decreased ability of the non-gravid skinned uterine fibers to generate tension or relax the tissue from rigor via endogenous creatine kinase. The apparent relocation of soluble creatine kinase to the contractile proteins during gestation indicates that the energetic demand of the tissue is affecting the creatine kinase system. Creatine kinase localized to the contractile apparatus may act to maintain the ATP/ADP ratio.

Vascular smooth muscle

Scott and Coburn [31] stated that allosteric actions of Cr or PCr did not mediate 'oxidative metabolism-contraction coupling' in the rabbit aorta. They claimed that an invariant relationship between PCr and Cr to the force

Table 2. Smooth muscle contractility and creatine kinase

Tissue	Total CK*	Skinned CK*	Percent CK left	Tension generation#	Percent relaxation+
Carotid artery	47	11	23	50	65
Taenia coli	163	36	22	59	51
Gravid uterus	196	44	22	65	61
Non-gravid uterus	74	6.7	9	43	47

* Specific activity of creatine kinase reported as IU/g ww. Skinned CK is the specific activity of creatine kinase after one hour of Triton X-100.
# Tension generation is percent maximal tension generation in a solution containing 250 μM ADP, pCa 4.5, and 12 mM PCr. Thus ATP produced only comes from endogenous creatine kinase.
+ Percent relaxation is relaxation from high tension rigor conditions in a solution containing 250 μM ADP, pCa 9 and 12 mM PCr. ATP produced can only come from endogenous creatine kinase compared to complete relaxation which was produced by addition of exogenous ATP. For complete details of experiments see references 10–12.

generated is a result of a decreased oxidative metabolism and energy production. Furthermore they found that Cr or PCr as an energy store could not be controlling the coupling between metabolism and contraction [32]. Ishida and Paul [33] found that in taenia caeci, and to a lesser extent in the porcine coronary arteries [34], hypoxia induced a significant fall in tension which preceded the fall in [ATP] but closely paralleled the fall in [PCr]. They did not however, speculate on the possible role of CK at the contractile filaments for tension generation or relaxation. It is interesting to note that the tonic vascular smooth muscle, with a lower specific activity of CK, exhibited less hypoxia induced relaxation compared to the taenia. It is unlikely that this difference is due to a difference in the localization of CK at the contractile filaments because the percentage of CK associated with Triton skinned fibers of the taenia coli and carotid artery was 22 and 23%, respectively [10, 12]. Paul's reported differences may however be due to the lower tension costs of vascular smooth muscle compared to the taenia [33, 34].

BB-CK kinetics

The above results of Clark et al. [10] were supported by experiments performed by Clark [7] where BB-CK was purified from porcine carotid arteries. It was found that BB-CK from the carotid arteries had a high K_m for ATP. A low affinity for ATP indicates that the BB-CK would run in the direction of ATP production. Knowing that BB-CK is localized at the contractile filaments, and in light of the relatively low concentration of ATP and PCr, it therefore seems quite plausible that BB-CK would kinetically favor ATP production. This tendency towards ATP production would make CK's role in energy transduction at the contractile proteins more efficacious.

Membrane bound CK

Hardin et al. [27] used plasma membrane vesicles from porcine stomach smooth muscle to show that the (Ca^{++}/Mg^{++}) ATPase is functionally linked to glycolytic activity. They found that the calcium pump preferentially utilizes nascent ATP synthesized from membrane bound energetic enzymes. Within this enzymic localization at the plasma membrane they found creatine kinase activity as well. Evidence from Barrantes et al. [35] also showed that CK can be associated with membranes in an

enzymic complex. It however remains to be determined which isoenzyme is associated in the membranes of smooth muscle but it appears that CK may function as an energy transducer for plasma membrane pumps (see Ishida et al., Ch. I-3, this volume).

Creatine kinase and the gestating uterus

The creatine kinase system in the gestating uterus undergoes biochemical and physiological changes and there is much evidence that the creatine kinase system is an integral component in that developmental process [11]. The evidences from the literature regarding the creatine kinase system in the uterus are discussed below emphasizing the potential benefits to the tissue energetics during gestation.

Iyengar et al. [36] found that estrogen stimulation of the bovine myometrium is accompanied by a doubling of the creatine kinase activity. They also found that the microsomal fraction of creatine kinase increased by four fold. Riess and Kaye [37] showed that the induced protein (a protein seen to greatly increase during gestation) of the rat uterus was indeed BB-CK. Isolated BB-CK from the bovine myometrium has a K_m for MgADP and PCr of 0.12 and 0.7 mM, respectively [38]. The relatively high affinity for PCr was taken to indicate a possible role of creatine kinase in harnessing the energy reserves of the uterine smooth muscle. The capacity to do this may thus be increased during gestation due to the increased BB-CK, aiding in the transducer role of the CK system at the contractile filaments.

Figure 5 represents the CK isoenzyme profiles taken from a gravid uterine horn in the absence of fetus, and contralateral gravid uterine horn in the presence of the fetus. Gestation is accompanied by an increase in specific activity and increased Mi-CK and BB-CK. Interestingly, the presence of a fetus in the uterine horn produced an increase in Mi-CK, specific activity and a further shift towards BB-CK. More work needs to be done to determine if the presence of the fetus may be producing a physical stimulus in the form of stretch to produce these changes or whether there is local humeral control from the placenta to the myometrium. Nonetheless, the creatine kinase system is responding to the hormonal stimulation seen during gestation but there is a further response seen in the presence of the fetus.

In the non-gestating uterus the specific activity of Mi-CK was 17.5 µg/g ww in the mitochondria while during gestation it was increased 8-fold to 140 µg/g ww [11]. The

228

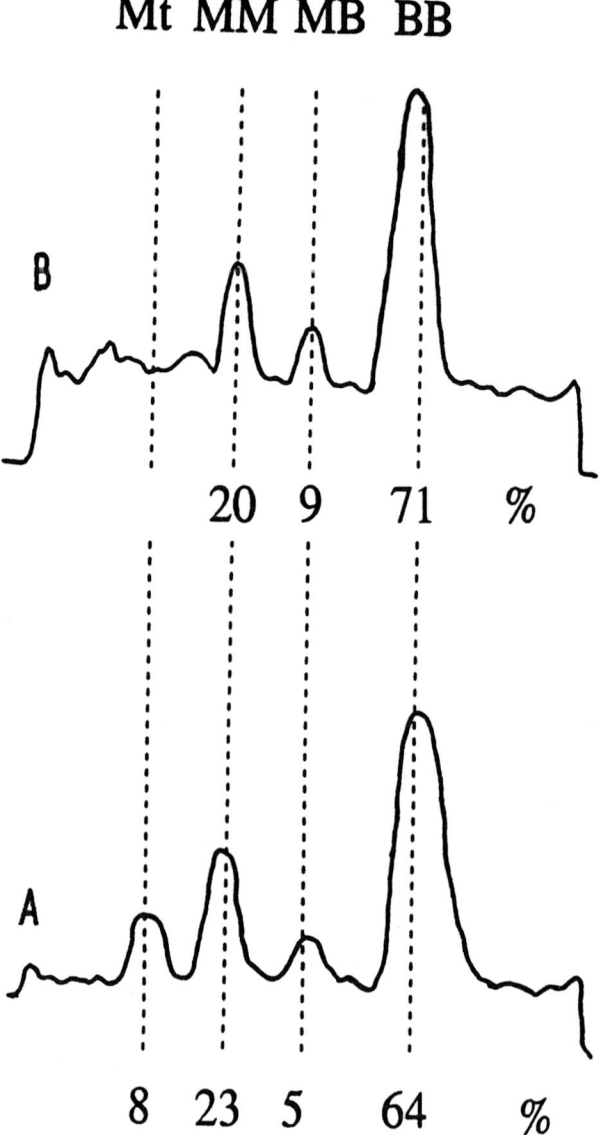

Mt MM MB BB

B

20   9   71   %

A

8   23   5   64   %

Fig. 5. Densitometer traces taken from uterine fibers. Fibers were obtained from gravid uteri. The two traces were taken from the contralateral uterine horns of the same guinea pig which had developing fetuses in only one horn. In the gravid uterus, the isoenzymes of CK are distinctly different in the presence (A) and absence (B) of the fetus. There is an increase in Mt-CK and an increase in total activity of CK of about 2 fold due to the presence of the fetus.

maximal rate of oxygen consumption increased by 3 fold while in the mouse uterus, and cytochrome oxidase increased by only 50% [39]. Thus, Mi-CK increases disproportionately greater than other mitochondrial enzymes indicating that Mi-CK is increasing per mitochondrion. The role of the Mi-CK appears to become increasingly important in the uterus during gestation in parallel with the cytosolic CK system. A parallel development of the Mi-CK and myofibrillar CK systems has been reported by Ventura-Clapier *et al.* [40] in the neonatal heart. They concluded that the CK system was

more completely developed in the guinea pig at birth than the rat because the guinea pig is more fully developed and active at birth compared to the rat. Similar reasons may be behind the changes seen in the uterus during gestation to render the mitochondrial CK system better equipped to respond to and deal with the energetic demands required during delivery. This could include maintaining the ATP/ADP ratio and supporting the contractile apparatus.

Clark *et al.* [11] found an absence of creatine stimulated respiration in guinea pig uterine fibers. Creatine stimulated respiration occurs in mitochondria when Mi-CK is functionally coupled to oxidative phosphorylation (for more details see [41–43]). This lack of creatine stimulated respiration may be due to the mitochondria having unrestricted diffusion of ADP. Thus ADP produced by Mi-CK is able to diffuse out of the mitochondria instead of being transported back into the mitochondrial matrix. Such a situation is strikingly different than that reported for the heart [41, 43].

NMR and uterine energetics

NMR has been used in several studies of uterine energetics [39, 44–46]. Using the NMR technique of magnetization transfer to study the rate of exchange of ATP and PCr through creatine kinase, Degani *et al.* [46] found that the creatine kinase reaction is at or near equilibrium. Magnetization exchange experiments can measure chemical exchange as well as exchange of magnetization due to NOE, but it does not distinguish whether the exchange is from mitochondrial or other creatine kinase exchanges. It is plausible that the apparent equilibrium condition is due to the energy supply and demand being equivalent in the whole cell. The NMR observations therefore do not preclude a tendency towards a unidirectional flux through BB-CK, at the contractile proteins, towards ATP production or through Mi-CK towards PCr. More discussion on CK equilibrium will follow below.

NMR has also been used in the rat uterus [44, 45] to show that PCr increased by 40% in the gestating uterus while ATP increased by 30% without a change in Pi. The increased [PCr] concomitant with the increased specific activity of creatine kinase supports the hypothesis of Iyengar [47] where creatine kinase is a major determinant of the content of PCr by virtue of its ability to act as an intracellular binding protein for creatine from the extracellular fluid as well as sequestering ADP. There are

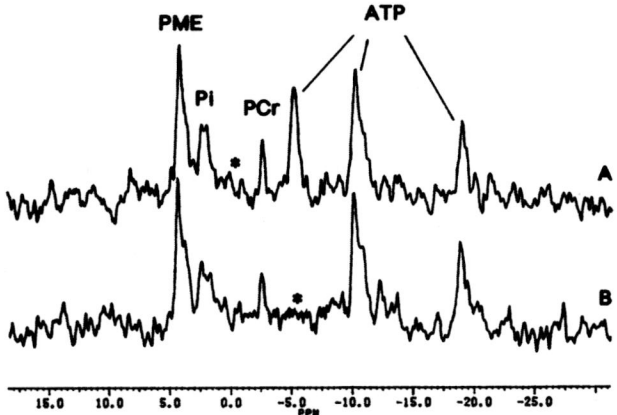

Fig. 6. NMR spectra of porcine carotid arteries. * indicates the position of selective saturation of the resonance. When the gamma ATP peak was saturated (B) that peak was decreased due to a randomization of magnetization. This randomization of magnetization is caused by the phosphorus nuclei of the gamma ATP effecting the net magnetization of PCr during exchange and is seen as a decrease in the peak intensity of PCr compared to spectrum A.

however recent experiments in skeletal muscle which refute this hypothesis because Brosnan *et al.* [26] increased cytosolic creatine kinase in skeletal muscle without increasing [PCr]. During gestation though, the specific activity of creatine kinase increases, possibly leading to an increase in [PCr], while the mitochondrial capacity increases along with an increase in the ATP concentration. The above situation makes the CK system well suited for maintaining cellular ATP while keeping ADP low [25, 48].

NMR studies of smooth muscle creatine kinase

There have been several reports utilizing NMR to study the creatine kinase system in smooth muscle [7, 19, 46, 49]. Vermue and Nicolay [19] suggested that there is a relatively slow flux through the creatine kinase system in intestinal smooth muscle. Yoshizaki [49] used saturation transfer measurements to further investigate this observation by determining the flux through creatine kinase in the bullfrog stomach. They found that the pseudo first order rate constant from PCr to ATP was 0.16 s^{-1} and the reverse was 0.13 s^{-1}. These are very similar to the values reported by Clark *et al.* in the porcine carotid artery of 0.17 and 0.13^{-1}, respectively [3, 6, 7]. In Fig. 6 are two representative spectra from an NMR saturation transfer experiment where the gamma ATP was saturated and the magnetization exchange observed with the decrease in the PCr peak. The conclusion drawn from both studies was that the creatine kinase reaction is at

equilibrium in the tissue at rest. Recently Brosnan *et al.* [26] showed an increase in flux through the creatine kinase reaction in transgenic mouse skeletal muscle with increased BB-CK. The increased flux could not be explained exclusively by the increase in specific activity of CK though the authors did not address the possibility of the NMR experiments measuring a more 'NMR visible' CK pool. Clark and Dillon [6] examined the flux during unloaded stimulation in the porcine carotid artery. They again found that the creatine kinase reaction was at equilibrium assuming a two site exchange (even under stimulated conditions). Thus the ATPase reaction was significantly less than the creatine kinase reaction. They estimated that the flux through ATPase in these experiments could not exceed 0.04 µmoles/sec/g ww while Ishida *et al.* [33] determined the approximate flux due to ATPase to be 0.012 µmoles/sec/g ww. Therefore, even under stimulated conditions the creatine kinase system appears to be at equilibrium in unloaded vascular smooth muscle. In a tissue where the sites of energy production and energy consumption are spatially separated, this conclusion may mean that energy production and energy utilization are closely matched or that there is an other fraction of the PCr pool rapidly exchanging at or near equilibrium. A pool of rapidly interconverting PCr and ATP within the cell could function to enhance the cellular responsiveness to energetic demands through the creatine kinase system [50, 51].

Creatine kinase isoenzymes and pathology

In cardiac and skeletal muscle during pathology it has been found that the isoenzymes of creatine kinase shift towards the B isoform. This isoenzyme shift has been termed the fetal shift because the B isoenzyme is dominant during fetal life. In the heart the fetal shift is seen with hypertrophy and hypertension. However, with this pathology, there is also vascular hypertrophy [52]. It was found by Boehm *et al.* [53] that vascular smooth muscle energetics are sensitive to these changes and responds with a characteristic fetal shift. In the aorta taken from the spontaneously hypertensive rat (SHR), the BB-CK isoenzyme increased from 65.8% to 81.0%. MB-CK changed significantly from 9.1% to 2.5% and MM-CK decreased from 24.6% to 16.2% (but this change was not significant). In the carotid artery of the same SHR rats there was a significant shift towards BB (54.0% to 80.6%) and away from MM (34.0% to 15.9%). Again there was a significant decrease in MB-CK from 12% to

230

0.5%. This decrease in MB-CK is interesting because it is indicative of a selectivity in the vascular smooth muscle towards the BB dimer and not only towards the B monomer.

In parallel experiments, it was demonstrated that in vascular smooth muscle from rats whose hypertrophy was due to chronic anaemic hypoxia, there was again a fetal shift [53–55]. Chronic anemic hypoxia is caused by feeding rats an iron deficient diet, producing cardiac hypertrophy with a characteristic fetal shift. In the aortas of these animals there was a significant shift towards the BB isoenzyme (57.7% to 64%) and away from MM (28.6% to 21.9%). There was however, no significant difference in the percent MB-CK and no significant differences in the specific activities. As can be seen from the above discussion, the BB isoenzyme is the dominant isoenzyme in vascular smooth muscle. In spite of its dominance in the control tissue, there is a further shift towards more BB-CK in the artery during hypertrophy and hypertension. It is also evident that the creatine kinase system in vascular smooth muscle is responsive to pathology with slightly different responses in the carotids compared to the aorta. One question therefore still remains and that is regarding the polymorphic nature of the creatine kinase system in the artery along the vascular tree.

Polymorphism of creatine kinase

The polymorphism of the creatine kinase system along the vascular tree was examined using the abdominal and thoracic aorta of the guinea pig (Table 1). It was found that the abdominal aorta had significantly more BB-CK (56%) compared to the thoracic aorta (42.7%). There was also significantly less MM-CK in the abdominal aorta compared to thoracic (20.0% and 9.3%, respectively). There was however, no significant difference in the percent activity of MB-CK between the two tissues (35.8% and 34.6%, respectively). It appears that there is greater BB-CK distally along the vascular tree because of the greater values of BB-CK found in the carotid and abdominal aorta compared to the thoracic aorta. The reason for these differences in the creatine kinase system along the vascular tree is not known nor is it known why there is a fetal shift during the pathological process. However, the plasticity of the creatine kinase system is intriguing. Possibly, the creatine kinase system and indeed smooth muscle energetics may be responding and adapting to the metabolic demands placed upon it. This

is consistent with the changes seen with increased blood pressure in the SHR and increased flow/sheer forces in the anemic vessels. It is also consistent with the energetic changes seen in the hypertrophic heart. We have seen a shift away from the BB-CK isoenzyme in the abdominal aorta below the band of an aortic banded guinea pig (unpublished observations). But, one may ask, what are the responses of the creatine kinase system in other smooth muscle under conditions of increased demand? This question was addressed by Clark *et al.* [11] using the gestating guinea pig uterus and was discussed in section creatine kinase and gestating uterus (Creatine kinase and the gestating uterus, see p. 227, above) and is represented in Fig. 5.

Conclusions

Though care must be taken when one makes generalizations regarding smooth muscle, it seems reasonable to conclude that the creatine kinase system is an integral component of smooth muscle energetics. There is specific localization of CK isoenzymes at the mitochondria, contractile machinery and membrane pumps thus completing the components of the creatine kinase circuit. BB-CK is the dominant isoenzyme in smooth muscle and its dominance increases during pathology and gestation. Mitochondrial creatine kinase is coupled to oxidative phosphorylation and BB-CK is coupled to the contractile proteins. In phasic smooth muscle such as the taenia coli and gestating uterus, BB-CK is bound to energetically available nucleotide in the Triton skinned fybers, thus suggesting an energetic microcompartment. Though the specific activity of CK, and the concentration of PCr, in the porcine carotid artery is extremely low, the flux through the CK reaction is at equilibrium as indicated by NMR experiments. The CK system appears to be acting as an energy transducer between the sites of energy production and energy consumption.

Acknowledgements

This chapter was made possible only through the help and support of all my friends and collaborators. Thank you all. Thanks go to Prof. G.K. Radda and the members of the MRC Biochemical and Clinical Magnetic Resonance Unit Oxford, for supporting me during the preparation of this manuscript; M.L. Field for his stimulating discussions and for previewing early versions of this manuscript and E. Boehm for his assistance with the densitometer measurements. Thanks also to D. Harris, K. Clarke, P.F. Dillon and G. Kemp for previewing this manuscript. Lastly I would like to thank the editors for giving me the opportunity to contribute this chapter and for all their help and support along the way.

References

1. Clark JF, Dillon PF: Mechanical and metabolic toxicity of 3-(trimethylsilyl) propanesulfonic acid to porcine carotid arteries. Biochim Biophys Acta 1014: 235–238, 1989

2. Clark JF, Dillon PF: Purification of creatine kinase from porcine carotid arteries. FASEB J 31: A1292, 1989

3. Clark JF, Dillon PF: Porcine carotid artery creatine kinase kinetics using NMR saturation transfer. Biophys J 57: 156a, 1990

4. Clark JF, Khuchua Z, Kuznetsov AV, Ventura-Clapier R: Creatine kinase in the gestating Guinea-pig uterus. J Physiol 438: 102P, 1991

5. Clark JF, Harris GI, Dillon PF: Multisite saturation transfer using DANTE and continuous wave. Magnetic Resonance in Medicine 17: 274–278, 1991

6. Clark JF, Dillon PF: Creatine kinase kinetics of porcine carotid arteries using 31 P NMR saturation transfer. Under review

7. Clark JF: Kinetics of porcine carotid artery brain isoform creatine kinase *in situ* and *in vitro*. Thesis, Michigan State University, 1990

8. Dillon PF, Clark JF: The theory of diazymes and functional coupling of pyruvate kinase and creatine kinase. J Theor Biol 143: 257–284, 1990

9. Clark JF, Khuchua Z, Kuznetsov AV, Ventura-Clapier R, Saks VA: Creatine kinase function in mitochondria isolated from gravid and non-gravid guinea-pig uteri. Under review

10. Clark JF, Khuchua Z, Ventura-Clapier R: Creatine kinase binding and possible role in chemically skinned guinea-pig taenia coli. Biochim Biophys Acta 1100: 137–145, 1992

11. Clark JF, Khuchua Z, Kuznetsov AV, Saks VA, Ventura-Clapier R: Compartmentation of creatine kinase isoenzymes in myometrium of gravid guinea-pig. J Physiol Cambridge 466: 553–573, 1993

12. Clark JF, Khuchua Z, Boehm E, Ventura-Clapier R: Creatine kinase activity associated with the contractile proteins of the guinea pig carotid artery. In press. J. Mus. Res. Cell Motil.

13. Paul RJ: Chemical energetics of vascular smooth muscle. Handbook of physiology Section 2: The cardiovascular system, Vascular smooth muscle 201–234, 1980

14. Paul RJ: Functional compartmentalization of oxidative and glycolytic metabolism in vascular smooth muscle. Am J Physiol 244: C399–C409, 1983

15. Paul RJ, Krisanda JM, Hellstrand P: Relations among oxygen consumption, phosphagen and contractility in vascular smooth muscle. Smooth muscle contractility. Marcell-Dekker Inc 245–257, 1984

16. Hai CM, Murphy RA: Cross-bridge phosphorylation and regulation of latch state in smooth muscle. Am J Physiol 254: C99–C106, 1988

17. Hai CM, Murphy RA: Regulation of shortening velocity by cross-bridge phosphorylation in smooth muscle. Am J Physiol 255: C86–C94, 1988

18. Dillon PF, Askoy M, Driska SP, Murphy RA: Myosin phosphorylation and the cross-bridge cycle in arterial smooth muscle. Science 211: 495–497, 1981

19. Vermue NA, Nicolay K: Energetics of smooth muscle taenia caecum of guinea-pig a 31 P-NMR study. FEBS Lett 156: 293–297, 1983

20. Fisher MJ, Dillon PF: Direct determination of ADP in hypoxic porcine carotid artery using 31P NMR. NMR in Biomed 1: 121–126, 1988

21. Butler TM, Pacifico DS, Siegman MJ: ADP release from myosin in permeabilized smooth muscle. Am J Physiol 256: C59–C66, 1989

22. Butler TM, Siegman MJ, Moors SU, Narayan SR: Myosin product complex in the resting state and during relaxation of smooth muscle. Am J Physiol 258: C1092–1099, 1990

23. Saks VA, Rosenshtraukh LV, Smirnov VN, Chazov EI: Role of creatine phosphokinase in cellular function and metabolism. Can J Physiol Pharmacol 56: 691–706, 1978

24. Ventura-Clapier R, Saks VA, Vassort G, Lauer C, Elizarova G: Reversible MM-creatine kinase binding to cardiac myofibrils. Am J Physiol 253: C444–C455, 1987

25. Wallimann T, Wyss M, Brdiczka D, Nicolay K, Eppenberger HM: Intracellular compartmentation, structure and function of creatine kinase isoenzymes in tissues with high and fluctuation energy demands: the phosphocreatine circuit for cellular energy homeostasis. Biochem J 281: 21–40, 1992

26. Brosnan MJ, Raman SP, Chen L, Koretsky AP: Altering creatine kinase isoenzymes in transgenic mouse muscle by overexpression of the B subunit. Am J Physiol 264: C151–C160, 1993

27. Hardin CD, Raeymaekers L, Paul RJ: Comparison of endogenous and exogenous sources of ATP in fueling Ca^{++} uptake in smooth muscle plasma membrane vesicles. J Gen Physiol 99: 21–40, 1992

28. Meyer RA, Sweeney HL, Kushmerick MH: A simple analysis of the 'phosphocreatine shuttle'. Am J Physiol 246: C365–C377, 1984

29. Ishida Y, Wyss M, Hemmer W, Wallimann T: Identification of creatine kinase isoenzymes in the guinea-pig: presence of mitochondrial creatine kinase in smooth muscle. FEBS Lett 283: 37–43, 1991

30. Nishiye E, Somlyo AV, Torok K, Somlyo AP: The effects of MgATP on cross-bridge kinetics: a laser flash photolysis study of guinea pig smooth muscle. J Physiol (Lond) 460: 247–271, 1993

31. Scott DP, Coburn RF: Phosphocreatine and oxidative metabolism-contraction coupling in rabbit aorta. Am J Physiol 257: H597–H602, 1989

32. Scott DP, Davidheiser S, Coburn RF: Effects of phosphocreatine on force and metabolism in rabbit aorta. Am J Physiol 253: H461–H465, 1987

33. Ishida Y, Paul RJ: Effects of hypoxia on high-energy phosphagen content, energy metabolism and isometric force in guinea-pig taenia caeci. J Physiol 42: 41–56, 1990

34. Ishida Y, Hashimoto M, Paul RJ: Does a limitation of energy supply to the contractile apparatus underlie the relaxation induced by hypoxia and smooth muscle? Regulation and contraction of smooth muscle. Alan R Liss, 1987, pp 463–464

35. Barrantes FJ, Braceras A, Caldiron HA, Mieskes G, Moser H, Toren EC, Roque ME, Wallimann T, Zechel A: Isolation and characterization of acetylcholine receptor membrane-associated (nonreceptor v_2-protein) and soluble electrocyte creatine kinases. J Biol Chem 260: 3024–3034, 1985

36. Iyengar MR, Fluellen CE, Jyengar CWL: Increases creatine ki-

nase in the hormone-stimulated smooth muscle of the bovine uterus. Biochem Biophys Res Commun 94: 948–954, 1980

37. Reiss NA, Kaye AM: Identification of the major component of the estrogen-induced protein of rat uterus as the BB isoenzyme of creatine kinase. J Biol Chem 256: 5741–5749, 1981

38. Iyengar MR, Fluellen CE, Iyengar CL: Creatine kinase from the bovine myometrium: purification and characterization. J Musc Res Cell Motil 2: 231–246, 1982

39. Dawson MJ, Raman J: Uterine metabolism and energetics. In: ME Carsten, JD Miller (eds) Uterine Function. Plenum Publishing Corp, 1990, pp 35–69

40. Ventura-Clapier R, Hoerter JA, Kuznetsov AV, Khuchua Z, Clark JF: Perinatal development of the creatine kinase system in mammalian heart. In: PP DeDeyn, B Marescau, B Stalon, IA Qureshi (eds) Guanidino Compounds in Biology and Medicine. John Libbey and Company, 1992, pp 195–204

41. Veksler VI, Kuznetsov AV, Sharov VG, Kapelko VI, Saks VA: Mitochondrial respiratory parameters in cardiac tissues: a novel method of assessment by using saponin skinned fibers. Biochim Biophys Acta 892: 191–196, 1987

42. Veksler VI, Murat I, Ventura-Clapier R: Ventricular skinned fibers study of mechanical and mitochondrial function in hereditary and diabetic cardiomyopathies. Can J Physiol Pharmacol 69: 852–858, 1991

43. Saks VA, Belikova YO, Kuznetsov AV: *In vivo* regulation of mitochondrial respiration in cardiomyocytes: specific restrictions for intracellular diffusion of ADP. Biochim Biophys Acta 1074: 302–311, 1991

44. Dawson MJ, Wray S: The effects of pregnancy and parturition on phosphorus metabolites in rat uterus studied by 31P NMR. J Physiol 368: 19–31, 1985

45. Dawson MJ, Wray S: 31 P NMR studies of isolated rat uterus. J Physiol (Lond) 336: 19P–20P, 1983

46. Degani H, Shaer A, Victor RA, Kaye AM: Estrogen-induced changed in high-energy phosphate metabolism in rat uterus: 31P NMR studies. Biochem 23: 2572–2577, 1984

47. Iyengar MR: Creatine kinase as an intracellular regulator. J Musc Res Cell Motil 5: 527–534, 1984

48. Wyss M, Smeitink J, Wevers RA, Wallimann T: Mitochondrial creatine kinase: a key enzyme of aerobic energy metabolism. Biochim Biophys Acta 1102: 119–166, 1992

49. Yoshizaki K, Radda GK, Inubushi T, Chance B: 1H and 31 P-NMR studies on smooth muscle of bullfrog stomach. Biochim Biophys Acta 928: 26–44, 1987

50. Crabtree B, Newsholme A: Sensitivity of a near equilibrium reaction in a metabolic pathway to changes in substrate concentration. Eur J Biochem 89: 19–32, 1978

51. Crabtree B, Newsholme EA: A quantitative approach to metabolic control. Curr Top Cell Regul 25: 21–76, 1985

52. Camilleri JP, Berry CL, Fiessinger JN, Bariety J: Diseases of the arterial wall. Springer Verlag, Berlin-Heidelberg, 1989

53. Boehm EA, Clark JF, Radda GK: Creatine kinase isoenzyme changes in the vascular smooth muscle during hypertension. J Mol Cell Cardiol 24: S108, 1992

54. Field ML, Clark JF, Henderson C, Seymour A-M, Radda GK: Alterations in the myocardial creatine kinase system during chronic anaemic hypoxia. Cardiovasc Res, in press

55. Clark JF, Field ML, Boehm EA, Radda GK: The cardiovascular creatine kinase system in chronic anaemic hypoxia. In: Gnaiger, Schneeweiss, Lenz (eds) Adaptive Mechanisms in Hypoxia. Springer-Verlag, Vienna Austria, in press

56. Tsung SH: Creatine kinase isoenzyme patterns in human tissue obtained at surgery. Clin Chem 22: 173–175, 1976

57. Lang H: Creatine Kinase Isoenzymes. Springer-Verlag, 1981

58. Jockers-Wretou E, Pfleiderer G: Quantitation of creatine kinase isoenzymes in human tissues and sera by an immunological method. Clin Chim Acta 58: 223–232, 1975

PART IV

MOLECULAR BIOLOGY OF
CREATINE KINASES

Molecular and Cellular Biochemistry **133/134**: 235–243, 1994.

IV-1 Expression of the mitochondrial creatine kinase genes

R. Mark Payne[1] and Arnold W. Strauss[1,2]
Departments of[1] Pediatrics; and[2] Molecular Biology and Pharmacology, Washington University School of Medicine, St. Louis, MO, USA

Abstract

Mitochondrial Creatine Kinase (MtCK) is responsible for the transfer of high energy phosphate from mitochondria to the cytosolic carrier, creatine, and exists in mammals as two isoenzymes encoded by separate genes. In rats and humans, sarcomere-specific MtCK (sMtCK) is expressed only in skeletal and heart muscle, and has 87% nucleotide identity across the 1257 bp coding region. The ubiquitous isoenzyme of MtCK (uMtCK) is expressed in many tissues with highest levels in brain, gut, and kidney, and has 92% nucleotide identity between the 1254 bp coding regions of rat and human. Both genes are highly regulated developmentally in a tissue-specific manner. There is virtually no expression of sMtCK mRNA prior to birth. Unlike cytosolic muscle CK (MCK) and brain CK (BCK), there is no developmental isoenzyme switch between the MtCKs. Cell culture models representing the tissue-specific expression of either sMtCK or uMtCK are available, but there are no adequate developmental models to examine their regulation. Several animal models are available to examine the coordinate regulation of the CK gene family and include 1) Cardiac Stress by coarctation (sMtCK, BCK, and MCK), 2) Uterus and placenta during pregnancy (uMtCK and BCK), and 3) Diabetes and mitochondrial myopathy (sMtCK, BCK, and MCK). We report the details of these findings, and discuss the coordinate regulation of the genes necessary for high-energy transduction. (Mol Cell Biochem **133/134**: 235–243, 1994)

Key words: creatine kinase, mitochondria, metabolism, creatine phosphate shuttle, gene expression, muscle

Introduction

Mitochondria provide the major energy generating system in mammalian cells, both through glycolysis and the citric acid cycle, and the β-oxidation of fatty acids [1]. Different tissues have widely varying energy requirements which is reflected in the differing density of mitochondria among tissues, the differing substrates employed to generate energy, and the differing requirements for oxygen and susceptibility to hypoxic or ischemic insult. Mitochondria from different tissues contain a variable subset and/or ratio of enzymes which allow rapid adaptation to stress, injury, and development. Thus, a major goal of our work has been to understand the biogenesis of mitochondria, particularly the *coordinate regulation* of the nuclear genes encoding mitochondrial proteins. Expression of these genes changes during development and stress, and differs among tissues with various energy requirements. We believe, therefore, that specific transcription factors provide the coordinated expression of the gene products responsible for energy transduction in these tissues.

To understand the factors and mechanisms whereby energy transduction is regulated, we have sought to determine the tissue distribution and regulation of those key enzymes responsible for energy generation. We

Address for offprints: A.W. Strauss, Department of Pediatrics, Washington University School of Medicine, Box 8116, St. Louis, MO, 63110, USA

have detailed the expression of the two mitochondrial creatine kinase (MtCK) isoenzymes in both rats and humans, and have found the distribution of one isoenzyme to be limited to the sarcomeric tissues of heart and skeletal muscle (sMtCK). The other isoenzyme of MtCK is expressed in many disparate tissues throughout the body, and has been termed 'ubiquitous' to reflect this distribution (uMtCK). Both isoenzyme mRNA's are highly regulated during development and with changing metabolic needs, in a tissue-specific manner. Finally, we have established cell-culture systems to examine the regulation of these two genes.

Tissue-specific distribution in humans and rats

The first MtCK gene was isolated using a cDNA probe from dog MCK to screen a human genomic library under reduced stringency and was later determined to be the gene for the uMtCK isoenzyme [2]. The initial RNA hybridization studies were done using a 312-bp KpnI-BanII genomic fragment which encodes all of the second exon (210-bp of coding region) and part of the second intron. These studies demonstrated abundant MtCK mRNA in human heart (ventricle), skeletal muscle, and ileum. However, further RNA blot analysis using a 140-bp SalI-BamHI human placenta-derived MtCK cDNA fragment encoding only 5'-untranslated region (UTR) showed abundant mRNA in placenta and jejunum only, and no signal in heart or skeletal muscle. Thus, clearly a second MtCK gene was expressed in sarcomeric tissues which shared coding region identity with the cDNA for MtCK isolated from placenta.

Screening of a human heart cDNA library with the 312-bp KpnI-BanII genomic fragment (above) isolated the cDNA for the second MtCK isoenzyme [3]. This cDNA had high coding region identity with the cDNA for the placental derived MtCK, but no conservation of the 3'- or 5'-UTR between the two isoenzymes, nor were the transit peptide sequences conserved. RNA blot analysis with this heart-derived MtCK cDNA showed signal only in heart and skeletal muscle. Probing of the same RNA blot with the 5'UTR region of the uMtCK cDNA showed signal in kidney, placenta, jejunum, uterus, and brain (faintly). RNA blot analysis with the entire uMtCK cDNA did not reveal additional signals in new tissues strongly suggesting that only two different genes encoded the MtCK isoenzymes. Thus, in human tissues, mRNA for the uMtCK gene is expressed in intestine (ileum, jejunum), brain, kidney, placenta, uterus, skeletal muscle, and heart. The mRNA for the sMtCK gene is expressed *only* in heart and skeletal muscle and is, therefore, a marker of sarcomeric tissues.

The surprising finding of two genes encoding MtCK in the human prompted us to isolate the equivalent cDNAs for the rat for analysis of tissue distribution, tissue-specific mitochondrial uptake, and response to stress and development [4]. As with the human mRNAs for MtCK, coding region identity is high between the rat mRNAs but there is no conservation of the 5'- and 3'-UTRs. Conservation of nucleotide sequence, however, is extremely high between respective rat and human MtCK isoforms, including the transit peptide region.

RNA blot analysis of an adult rat tissue panel with the 3'-UTR of the rat sMtCK cDNA demonstrated expression of sMtCK mRNA only in heart and skeletal muscle. Thus, as in the human, mRNA for the sMtCK gene is expressed *only* in sarcomeric tissues. Probing of this same RNA blot with the 3'-UTR for the rat uMtCK mRNA demonstrates expression only in small intestine and brain, with lesser expression in uterus, kidney, and aorta. No expression of uMtCK mRNA is found in adult heart or skeletal muscle. A coding region probe derived from the rat sMtCK cDNA demonstrates the combined expression patterns of both MtCK mRNAs making it unlikely that a third isoenzyme of MtCK exists.

To examine tissues which had shown little or no MtCK signal using total RNA, we performed RNA blot analysis of poly(A) selected RNA from rat liver, uterus, and testes. As a control, total RNA from gut epithelium was included on the blot. Abundant signal is present for the uMtCK-specific probe in gut epithelium, but there is no signal from liver, uterus, or testes. A coding region probe detects faint MtCK message expression in liver, uterus, and testes, as well as a strong signal from the gut epithelium. Predictably, there is no sMtCK mRNA in these tissues. Because no significant cytosolic CK has been detected in liver [5], this suggests that the faint MtCK signal from liver may represent blood elements (e.g. macrophages) or fibroblasts from connective tissue in the liver. Several other possibilities exists for the low expression of uMtCK in this experiment: a) either alternative splicing of the 3'-UTR for uMtCK is present, or a third form of MtCK is expressed in these tissues. Neither of these possibilities has been supported by current data. b) The mRNA for uMtCK may undergo processing and not possess a poly(A) tail. Thus, it was not recovered by oligo dT affinity chromatography. This possibility is intriguing and suggests that the mRNA for

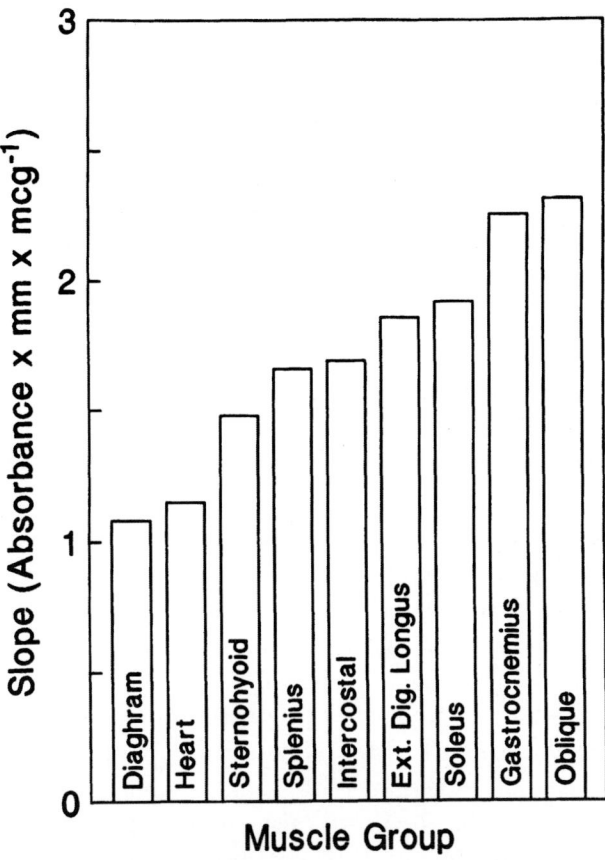

Fig. 1. Muscle-specific sMtCK mRNA expression. Total RNA prepared from nine individual muscles was blotted onto nitrocellulose using a dot-blot apparatus, and probed with rat sMtCK specific cDNA. The autoradiogram was scanned with an LKB Utrascan laser densitometer. The signal intensity was divided by the micrograms (mcg) of total RNA per sample to generate the slope of the line. Thus, a steeper slope represents greater sMtCK mRNA expression per mcg of total RNA. Diaghram, diaghramatic muscle; Heart, whole heart; Sternohyoid, Splenius, Soleus, and Gastrocnemius muscles are as marked; Intercostal muscle taken from right thoracic wall between the 5th and 6th ribs; Ext. Dig. Longus, external digitorum longus; Oblique, internal and external abdominal oblique muscles.

uMtCK, as with certain other mRNAs such as vasopressin in brain, may be regulated based on its 3′UTR, e.g., translation may be more active in the absence of the poly (A) tail [6]. In summary, tissue distribution of MtCK in rat is very similar to human and is tissue-specific: sMtCK is expressed only in heart and skeletal muscle and is, therefore, a marker of sarcomeric tissues, uMtCK mRNA is expressed in many tissues such as intestine, uterus, kidney, testes, brain, and aorta. No significant expression of either MtCK mRNA is found in lung, liver, or spleen.

To compare and quantify the expression of sMtCK mRNA in specific sarcomeric tissues, total RNA from a rat skeletal muscle survey was probed with sMtCK cRNA by dot blot analysis. Variation of sMtCK mRNA content among muscle groups is less than 2.5 fold (Fig. 1). Gastrocnemius and abdominal internal and external oblique muscles have the highest slopes, demonstrating greater expression of sMtCK mRNA in these tissues than in heart or diaphragm. Predominately fast twitch (extensor digitorum longus) or slow twitch (soleus) muscles have similar sMtCK mRNA content. Thus, among sarcomeric tissues, variation in content of sMtCK mRNA is relatively small. These results are congruent with MCK expression, as MCK activity varies less than 6 fold among sarcomeric tissues, and cardiac tissue typically has about one-third the MCK activity found in skeletal muscle [7].

The tissue-specific expression of sMtCK mRNA closely parallels the expression of MCK mRNA. MCK mRNA is detected in abundance in adult heart and skeletal muscle, with low levels in lung. It is developmentally regulated in both heart and skeletal muscle [5]. Similarly, the tissue-specific expression of uMtCK mRNA in the adult rat is virtually identical to expression of BCK mRNA. BCK mRNA is abundant in adult brain, gut and heart, with lesser levels in lung, skeletal muscle, and kidney. Neither MCK nor BCK mRNAs have been detected in liver. We have shown coordinate expression in rat heart of cytosolic CK mRNAs with MtCK mRNAs under conditions of pressure overload [6]. Based on these comparisons with the expression of cytosolic CKs in both rats and humans, it is likely that ubiquitous and sarcomeric MtCK mRNAs will be coordinately regulated with BCK and MCK, respectively, with changes in energy requirements in many tissues.

Developmental expression of MtCK

Our finding of tissue-specific expression of two MtCK genes with presumably identical function prompted us to examine the developmental expression of these two genes, and compare them with the cytosolic isoenzymes, BCK and MCK. Expression of the cytosolic CK's is highly regulated during development: BCK predominates in embryonic tissues and drops in the postnatal period thereby creating an isoenzyme switch with MCK [5]. MCK is markedly upregulated in sarcomeric tissues after birth. We used dot blot RNA hybridization to quantitatively examine mRNA expression of the two MtCK genes in rats for five tissues from late gestation into adulthood (Fig. 2).

Our results demonstrated that mRNA's for the cytosolic CK's, MCK and BCK, are expressed in many tis-

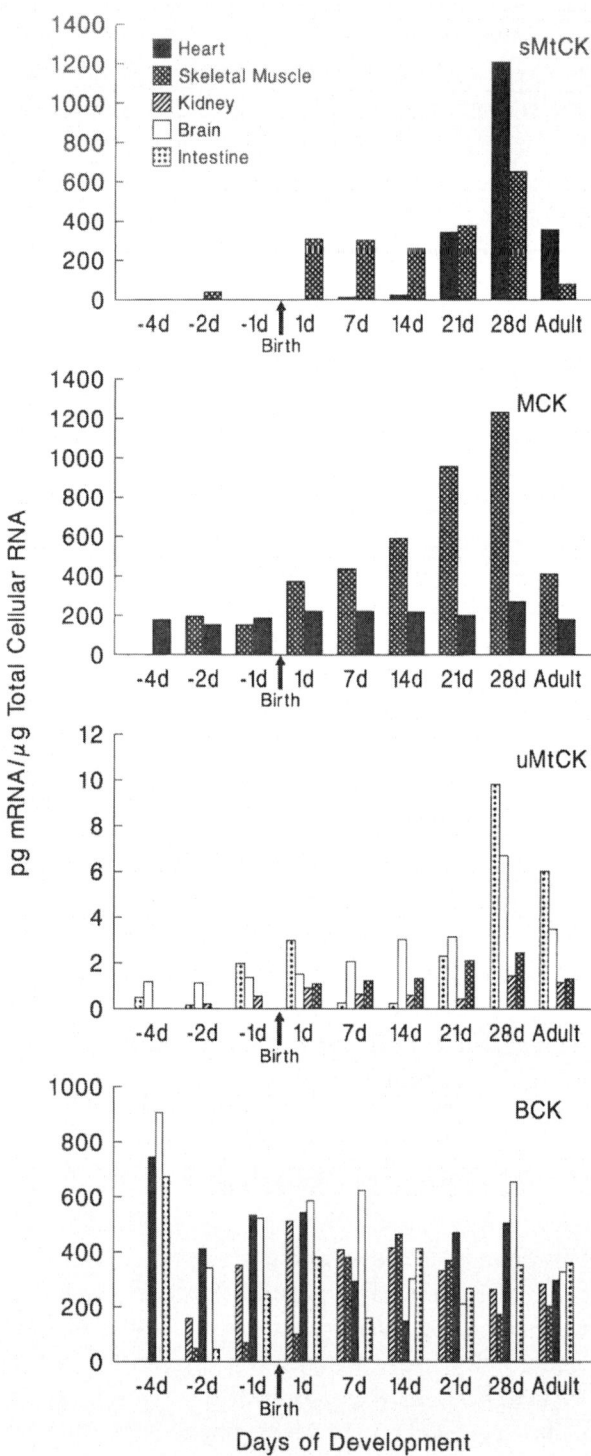

Fig. 2. Developmental expression of creatine kinase mRNA's. mRNA expression of all four CK genes in rat tissues is shown for late gestation to adult ages. Time points are shown as days prior to birth: − 4 d, − 2 d, − 1 d, and days after birth: 1 d, 7 d, 14 d, 21 d, 28 d, and adult (3 months old). Birth is denoted by the arrow, and mRNA is expressed as pg mRNA/µg total cellular RNA. Note that there was no skeletal muscle for the − 4 d time point. Brain, kidney, and heart represent whole organs. Skeletal muscle was taken from the proximal hindquarters, and intestinal samples taken from the ileum.

sues throughout development. This has been described by other investigators for both mRNA and protein expression [5, 8]. The MtCK mRNA's, however, are detectable in quantity only after birth although significant expression of uMtCK mRNA is found in prenatal intestine and brain. Because brain receives a disproportionate share of oxygenated blood *in utero*, it may well have regions of highly active oxidative respiration. Expression of uMtCK mRNA in prenatal gut is low and may reflect the numerous potential hormone response elements present within the gene [9].

Developmental expression of MtCK mRNA in rat heart is remarkable. There is virtually *no* MtCK in heart until 7 d after birth, and significant quantities of sMtCK mRNA do not appear until three weeks after birth. This developmental pattern correlates with the appearance of MtCK *activity* in mouse and rat heart [8, 10] and strongly suggests that transcriptional events regulate sMtCK expression in sarcomeric tissues. Both MCK and BCK mRNAs are heavily expressed in pre- and postpartum rat heart. However, significant evidence supports the hypothesis of a coordinated system for energy transduction in sarcomeric tissues during development. For example, Hoerter *et al.* examined the development of the CK system in perinatal rabbit heart, and observed that while MCK and sMtCK are expressed at different stages, their *functional* activities appear coordinately between mitochondria and myofibrils [11]. Thus early cardiac development is characterized by progressive compartmentalization of the CK isoenzymes: MtCK targeted to mitochondria, and cytosolic MCK becomes bound to myofibrils. Further evidence for coordinated maturation of a high-energy system in heart is found in the developmental expression of the adenine nucleotide translocases (ANT). Of the three isoforms that are expressed, ANT2 and ANT3 are found in many tissues at early (fetal) stages, whereas ANT1 is specific for mature sarcomeric tissues [12]. ANT and sMtCK are expressed in a 1:1 molar ratio in cardiac mitochondria [13], and are functionally coupled [14]. Thus, our data strongly suggest that the late appearance of sMtCK in neonatal rat heart signals the shift from glycolytic (fetal) to oxidative (mature) respiration. However, sMtCK, and thus, a highly active CK shuttle, is *not* required during the embryologic development of sarcomeric tissues.

The time course for the appearance of sMtCK mRNA in developing rat heart is important because terminal differentiation of cardiac myocytes in the rat occurs between two or three weeks postpartum [15]. In rat heart, for example, DNA replication ceases by 17 d postpartum, and thus, cardiac growth by myocyte hyperplasia stops as well [16]. Furthermore, the contractile properties of the rat heart reach functional maturity by 16 d postpartum [17], and correlate with maturation of the M-band in cardiac myofibrils [18]. These changes correspond with the progressive innervation of the heart, which reaches an adult pattern by the third week of postnatal development [19]. Claycomb et al. has suggested that the terminal differentiation and proliferation of myocytes are controlled by adrenergic innervation, and that norepinephrine and cAMP serve as chemical mediators of this control [20]. In the human sMtCK gene, there are numerous cAMP responsive elements [21] and stimulation of C_2C_{12} cells with mediators which increase cAMP causes a marked increase in sMtCK mRNA expression.[1] Thus, the dramatic upregulation of sMtCK in developing rat heart may be under adrenergic neural control, and correspond with terminal differentiation of cardiac myocytes.

The expression of sMtCK mRNA in skeletal muscle is markedly upregulated within four hours of birth. This suggests a rapid development of a mature system for energy transduction in these tissues. It also suggests that terminal differentiation occurs in skeletal muscle much earlier than in cardiac muscle. Prenatal expression of sMtCK in skeletal muscle has not been found in altricious animals, with the exception of fetal human quadriceps muscle. However, it is expressed prenatally in myocardium and skeletal muscle of precocious animals such as guinea pigs and sheep, thus indicating rapid maturation of energy systems in these animals [22]. The progressive postnatal increase in MCK and sMtCK mRNAs may reflect increasing binding sites on mitochondria and myofibrils as these organelles proliferate [11]. There is no isoenzyme switch between sMtCK and uMtCK in skeletal muscle as there is for the cytosolic enzymes, MCK and BCK [5].

Expression of uMtCK mRNA is 10- to 100-fold less than the other three CK mRNAs in all tissues we have examined [9]. This may reflect a lower energy requirement in these tissues, or their ability to rely on glycolysis. The highest level of uMtCK mRNA is found in intestine. Characteristics of this tissue include both a significant cell turnover (epithelium) and heavy energy requirements for metabolite transport. uMtCK mRNA and protein are strongly expressed in the epithelium of in-

testine but not in the smooth muscle or lymph tissue of the wall.[2] Similar patterns of highly localized expression for uMtCK mRNA are present in brain and kidney [23]. This localized expression of uMtCK mRNA indicates that energy flux within these cells is very high compared to other components of the whole organ. Further examination of tissues expressing uMtCK mRNA using techniques that demonstrate cell-specific expression, e.g., in situ hybridization, will likely show highly regionalized and developmental expression of uMtCK mRNA in all organs expressing this isoenzyme.

Ubiquitous MtCK mRNA is clearly present in prenatal brain. This is in contrast to published reports that there is no MtCK activity in fetal brain, however, both of these studies used small amounts of tissue for the measurement of uMtCK [10, 24]. Because this mRNA expression represents whole brain tissue, it is possible that uMtCK is expressed regionally. Alternatively, uMtCK expression in brain may be under translational control, as we have demonstrated for rat uterus and placenta [9]. Other investigators have noted that rat brain shows a rapid increase in the establishment of synapses and neuronal growth in the neonatal period. Thus, as in the heart, CK gene expression in brain may be under neuronal control during a period of rapid growth and increasing energy requirements.

Thus, during development, the tissue-specific expression of the two MtCK mRNA's is maintained. There is no isoenzyme switch between uMtCK and sMtCK as there is for the cytosolic CK's, BCK and MCK. Both MtCK mRNA's are expressed primarily after birth, and are coordinate with their respective cytosolic CK mRNA's for the tissue of expression.

Cell models of MtCK mRNA expression

To further define regulatory mechanisms of MtCK gene expression, we have investigated cell culture systems which express either uMtCK or sMtCK mRNA and protein. BC$_3$H1 cells are derived from a chemically induced mouse brain neoplasm, and have muscle like characteristics. They can be induced to differentiate with serum withdrawal which also stimulates marked upregulation of MCK and BCK mRNA [25]. MtCK mRNA expres-

[1] Klein SC, Payne RM, Kelly DP, Dong Y, Strauss AW, data in submission.

[2] Payne RM, and Friedman DL, work in progress.

240

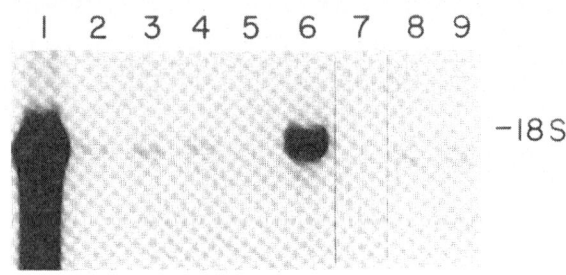

Fig. 3. sMtCK mRNA expression in AT-1 and C_2C_{12} cells. 5 µg per lane of total RNA were separated on a 1% agarose formaldehyde gel and blotted onto nitrocellulose. The blot was hybridized to the rat sMtCK specific cDNA under stringent conditions. Lanes: 1) rat heart. 2) AT-1 cells, control. 3) AT-1 cells, confluent and beating. 4) AT-1 cells, confluent and exposed to isoproterenol. 5) AT-1 cells, confluent and exposed to forskolin. 6) Neonatal rat heart cultured fibroblasts. 7) C_2C_{12} cells, control. 8) C_2C_{12} cells after 20 hr exposure to forskolin (15 µM). 9) C_2C_{12} cells after 20 hr exposure to 8-bromo-cAMP (0.33 µM).

sion in these cells follows a similar course of upregulation reaching greatest abundance 48 h after serum withdrawal [2]. Because the uMtCK cDNA probe utilized in this experiment shared highly conserved sequences with the sMtCK cDNA, we are unable to identify induction of specific MtCK isoforms in this system. However, these findings are consistent with coordinate regulation of MtCK and cytosolic MCK expression.

We have further defined expression of sMtCK mRNA in two sarcomeric cell lines having either skeletal muscle, or cardiac specific characteristics. C_2C_{12} cells (ATCC CRL 1772) are derived from normal adult mouse leg muscle, and can be induced to differentiate rapidly from myoblasts to myotubes with expression of characteristic muscle proteins. They express CK isoforms [26], however, the myoblasts do not express detectable levels of sMtCK mRNA. With differentiation to myotubes, low level expression of sMtCK mRNA is present, but uMtCK mRNA expression is not [21]. Furthermore, when confluent C_2C_{12} myoblasts are deprived of serum to induce differentiation and stimulated with agents that increase cAMP, either 8-bromo-cAMP or forskolin, a dramatic increase in sMtCK mRNA expression of the endogenous sMtCK gene occurs (Fig. 3). Clearly, the sMtCK gene is responsive to cAMP stimulation, although the cAMP responsive elements and Ap-2 homologs located in the 5′-flanking region of the human sMtCK gene are unaffected by cAMP stimulation.[1]

We have the examined the transgenic mouse atrial tumor cells, AT-1, for sMtCK mRNA expression [27]. These cells express adult cardiac specific contractile ele-

ments, as well as connexin43, the major gap junction protein of cardiac myocytes. As shown in Fig. 3, AT-1 cells express sMtCK mRNA at relatively low levels compared to adult rat heart total RNA. Furthermore, growth to confluency, and stimulation with cAMP stimulating agents such as isoproterenol and forskolin, does not increase sMtCK mRNA expression. Comparing these results with expression in C_2C_{12} cells, we conclude that control of sMtCK gene expression is different between these two sarcomere-derived cell lines, thus offering tissue-specific cell culture models for both skeletal and heart muscle.

Additional cell lines in which we have found expression of sMtCK mRNA include neonatal rat heart fibroblasts, and A7r5 cells. Neonatal rat heart fibroblasts are prepared by trypsinizing minced whole hearts and using a process of differential adherance to separate fibroblasts from primary neonatal cardiocytes. The fibroblasts grow well in standard media, passage easily, and express sMtCK mRNA strongly. A7r5 cells are derived from embryonic rat thoracic aorta (smooth muscle), and express CK after cessation of cell division. They also express abundant sMtCK mRNA. Thus, at least four cell lines with tissue-specific characteristics express sMtCK mRNA. We are using C_2C_{12} cells and AT-1 cells as sarcomeric models for investigation of sMtCK gene regulation.

We have also investigated two cell lines for expression of the uMtCK gene. AN3-CA cells (ATCC, HTB 111) are derived from a human endometrial adenocarcinoma and are sensitive to estrogen stimulation. They express the uMtCK gene mRNA and when stimulated with β-estradiol, will upregulate uMtCK mRNA expression within 100 minutes of exposure reaching a peak expression after 48 hours of stimulation by estrogen. Because the uMtCK gene contains several putative estrogen response elements, and is clearly regulated in a manner consistent with hormonal regulation in vivo [9], we are using this cell line to examine both promoter and hormone response elements in the human uMtCK gene. A second cell line, Hep G2 (ATCC, HB 8065) is derived from a human hepatocellular carcinoma and expresses uMtCK mRNA. Preliminary data demonstrate that the promoter for the human uMtCK gene is highly regulated in this cell line.

Animal models of MtCK expression and regulation

Earlier data had shown that the two MtCK genes were highly regulated in a tissue-specific and developmental manner. To examine sMtCK expression in response to changing energy needs, we used the animal model of cardiac pressure overload induced by banding the suprarenal aorta [6]. Four days after banding, the left ventricular myocardium from the hypertensive rats was compared with normotensive, sham-operated rats for isoenzyme activities by chromatography; M and B creatine kinase subunit protein by Western blot; and MCK, BCK, and sMtCK mRNA by Northern blot analysis. Although total creatine kinase *activity* increased in the hypertensive animal, the relative proportions of the three CKs did not change. Furthermore, the mRNA for MCK, BCK, and sMtCK were increased compared to the normotensive animal, but their expression relative to each other did not change. Thus, the increased energy requirements in acute pressure overload in heart are met by a generalized induction of CK mRNA and subunit protein, and not by an isoenzyme switch. We have extended the length of aortic banding to 28 days, and observed no significant shift in isoenzyme contribution to CK mass or activity.[3] Total CK activity decreases by 28 days in the banded rats, probably reflecting myocardial scarring from chronic pressure overload.

Earlier results had suggested that uMtCK may be expressed in rat uterus in a manner consistent with hormonal regulation [4]. Thus, we examined expression of CK isoenzymes in rat uterus and placenta to evaluate their contribution to energy metabolism during pregnancy and delivery [9]. uMtCK mRNA expression is high in prepartum uterus, but rapidly falls after delivery to a nadir at 7 days postpartum. Prepartum BCK mRNA expression is coordinate with uMtCK, but has a 15-fold greater expression than uMtCK. Both CK mRNAs rise by 17 days postpartum. Both BCK and uMtCK mRNA expressions are strongly induced in placenta at 20 days of gestation with a rapid fall immediately prior to delivery. Protein expression of BCK and uMtCK in uterus and placenta is also coordinate, however, total CK *activity* in both tissues reflects BCK expression. Analysis of mRNA and protein expression indicates that significant post-transcriptional regulation of both CKs occurs in a tissue-specific fashion. By immunohistochemistry, BCK and uMtCK protein expression are highly localized in the placenta and endometrium of prepartum uterus. This expression shifts entirely to the uterine smooth muscle by 17 days postpartum. Based on these data, it is clear that CK in nonsarcomeric tissues can be rapidly regulated to accomodate high energy demands. The highly regulated expression of these energy related proteins in uterus and placenta suggests an important role for CK in the maintenance and termination of pregnancy. We are currently analyzing human placenta for expression of CK during normal and abnormal conditions of pregnancy.

Finally, we have examined the effects on CK gene expression, of two disease states which affect myocyte function. Diabetes alters sarcomeric contractility in both cardiac muscle and skeletal muscle [28]. In streptozotocin induced diabetes in rats, sMtCK mRNA expression was reduced substantially (> 70%) in both slow-twitch (soleus) and fast-twitch (extensor digotorum longus) muscle, but was unaffected in heart muscle. Total CK activity in diabetic myocardium fell by 33%. The widespread adverse effect of diabetes on sarcomeric tissues was reversible with insulin. Thus, reduced contractile performance seen in diabetes may be partially explained by alterations in energy transduction. The molecular mechanisms underlying this downregulation of CK gene expression in diabetes are unknown, but could involve a change in substrate utilization for energy production, or loss of gene regulation by cell-surface receptors for insulin. A second disease state affecting myocyte function is azidothymidine (AZT) induced cardiotoxicity in rats. This drug may also have a cumulative effect with the human immunodeficiency virus to produce the dilated cardiomyopathy associated with AIDS. In rats, AZT disrupts cardiac mitochondrial ultrastructure and expression of mitochondrial cytochrome b mRNA in a time- and dose-dependent manner, but does not affect expression of nuclear encoded mitochondrially targeted genes such as sMtCK, or contractile genes [29]. The mechanism of AZT cardiotoxicity, therefore, may relate to inhibition of mitochondrial DNA replication while the nuclear enzymes associated with high-energy transduction remain unaffected. Thus, for the two animal models studied, loss of efficient, high-energy generating mechanisms in sarcomeric tissues may result in myopathies which are not fatal. It is quite possible that further investigation of skeletal and cardiac myopathies in animals and humans will reveal defects of CK gene ex-

[3] Payne RM, Abendschein DR, and Strauss AW, unpublished data.

pression that are either primarily responsible for, or related to, the myopathic disease.

Conclusion

We have analyzed CK isoenzyme mRNA expression in both rat and human tissues, as well as during periods of stress. The cytosolic components of the CK shuttle, MCK and BCK, are expressed *coordinately* in their tissue distribution with the mitochondrial CKs, sMtCK and uMtCK, respectively. sMtCK mRNA is found in sarcomeric tissues only. uMtCK mRNA is expressed in many tissues similar to BCK. We have not found a marked shift in isoenzyme contribution for the two disease models studied (heart and skeletal muscle). The non-sarcomeric CKs are highly regulated in uterus and placenta suggesting an important role for high-energy production in pregnancy and delivery. All four CK isoenzymes are highly regulated during development, with the MtCK mRNAs being expressed primarily after birth. Expression of sMtCK mRNA in developing rat heart may signal the terminal differentiation of cardiac myocytes, and the shift from glycolytic to oxidative respiration. Cell models for analysis of gene regulation are detailed for sMtCK, including heart specific, and skeletal muscle specific cell models. Cell models appropriate for uMtCK gene regulation studies include characteristics of this gene, such as responsiveness to hormonal stimulation, and constitutive expression.

References

1. Tzagoloff A, Meyers AM: Genetics of mitochondrial biogenesis. Annu Rev Biochem 55: 249–285, 1968
2. Haas RC, Korenfeld C, Zhang Z, Perryman B, Roman D, Strauss AW: Isolation and characterization of the gene and cDNA encoding human mitochondrial creatine kinase. J Biol Chem 264: 2890–2897, 1989
3. Haas RC, Strauss AW: Separate nuclear genes encode sarcomere-specific and ubiquitous human mitochondrial creatine kinase isoenzymes. J Biol Chem 265: 6921–6927, 1990
4. Payne RM, Haas RC, Strauss AW: Structural characterization and tissue-specific expression of the mRNAs encoding isoenzymes from two rat mitochondrial creatine kinase genes. Biochim Biophys Acta 1089: 352–361, 1991
5. Trask RV, Billadello JJ: Tissue-specific distribution and developmental regulation of M and B creatine kinase mRNAs. Biochim Biophys Acta 1049: 182–188, 1990
6. Maciejewski-Lenoir D, Jirikowski GF, Sanna PP, Bloom FE: Reduction of exogenous vasopressin RNA poly(A) tail length increases its effectiveness in transiently correcting diabetes insipid-

us in the Brattleboro rat. Proc Natl Acad Sci USA 90: 1435–1439, 1993
7. Smith AF: Separation of tissue and serum creatine kinase isoenzymes on polyacrylamide gel slabs. Clin Chem Acta 39: 351, 1972
8. Hall N, DeLuca M: Developmental changes in creatine phosphokinase isoenzymes in neonatal mouse hearts. Biochem Biophys Res Commun 66: 988–994, 1975
9. Payne RM, Friedman DL, Grant JW, Perryman MB, Strauss AW: Creatine kinase isoenzymes are highly regulated during pregnancy in rat uterus and placenta. Amer J Physiol 265 (Endocrinol Metab 28): E624–E635, 1993
10. Norwood WI, Ingwall JS, Norwood CR, Fossel ET: Developmental changes of creatine kinase metabolism in rat brain. Am J Physiol 244: C205–C210, 1983
11. Hoerter JA, Kuznetsov A, Ventura-Clapier R: Functional development of the creatine kinase system in perinatal rabbit heart. Circ Res 69: 665–676, 1991
12. Lunardi J, Hurko O, Engel WK, Attardi G: The multiple ADP/ATP translocase genes are differentially expressed during human muscle development. J Biol Chem 267: 15267–15270, 1992
13. Kuznetsov AV, Saks VA: Affinity modification of creatine kinase and ATP-ADP translocase in heart mitochondria: determination of their molar stoichiometry. Biochem Biophys Res Commun 134: 359–366, 1986
14. Brooks SP, Suelter CH: Association of chicken mitochondrial creatine kiase with the inner mitochondrial membrane. Arch Biochem Biophys 253: 122–132, 1987
15. Claycomb WC: Cardiac-muscle hypertrophy. Differentiation and growth of the heart cell during development. Biochem J 168: 599–601, 1977
16. Claycomb WC: Biochemical aspects of cardiac muscle differentiation. J Biol Chem 250: 3229–3235, 1975
17. Hopkins SF Jr, McCutcheon EP, Wekstein DR: Postnatal changes in rat ventricular function. Circ Res 32: 685–691, 1973
18. Anversa P, Olivetti G, Bracchi P-G, Loud AV: Postnatal development of the M-band in rat cardiac myofibrils. Circ Res 48: 561–568, 1981
19. Inversen LL, DeChamplain J, Glowinski J, Axelrod J: Uptake, storage and metabolism of norepinephrine in tissues of the developing rat. J Pharmacol Exp Ther 157: 509–516, 1967
20. Claycomb WC: Biochemical aspects of cardiac muscle differentiation. J Biol Chem 251: 6082–6089, 1976
21. Klein SC, Haas RC, Perryman MB, Billadello JJ, Strauss AW: Regulatory analysis and structural characterization of the human sarcomeric mitochondrial creatine kinase gene. J Biol Chem 266: 18058–18065, 1991
22. Smeitink J, Ruitenbeek W, van Lith T, Sengers R, Trijbels F, Wevers R, Sperl W, de Graaf R: Maturation of mitochondrial and other isoenzymes of creatine kinase in skeletal muscle of preterm born infants. Ann Clin Biochem 29: 302–306, 1992
23. Friedman DL, Perryman MB: Compartmentation of multiple forms of creatine kinase in the distal nephron of the rat kidney. J Biol Chem 266: 22404–22410, 1991
24. Webster KA, Gunning P, Hardeman E, Wallace DC, Kedes L: Coordinate reciprocal trends in glycolytic and mitochondrial transcript accumulations during the *in vitro* differentiation of human myoblasts. J Cell Physiol 142: 566–573, 1990
25. Spizz G, Roman R, Strauss A, Olson EN: Serum and fibroblast growth factor inhibit myogenic differentiation through a mecha-

nism dependent on protein synthesis and independent of cell proliferation. J Biol Chem 261: 9483–9488, 1986

26. Trask RV, Strauss AW, Billadello JJ: Developmental regulation and tissue-specific expression of the human muscle creatine kinase gene. J Biol Chem 263: 17142–17149, 1988

27. Lanson NA Jr, Glembotski CC, Steinhelper ME, Field LJ, Claycomb WC: Gene expression and atrial natriuretic factor processing and secretion in cultured AT-1 cardiac myocytes. Circulation 85: 1835–1841, 1992

28. Su C-Y, Payne M, Strauss AW, Dillmann WH: Selective reduction of creatine kinase subunit mRNAs in striated muscle of diabetic rats. Am J Physiol 263 (Endocrinol Metab 26): E310–316, 1992

29. Lewis W, Papoian T, Gonzalez B, Louie H, Kelly DP, Payne RM, Grody WW: Mitochondrial ultrastructural and molecular changes induced by Zidovudine in rat hearts. Laboratory Investigation 65: 228–236, 1991

Molecular and Cellular Biochemistry **133/134**: 245–262, 1994.
© 1994 *Kluwer Academic Publishers*.

IV-2 Sequence homology and structure predictions of the creatine kinase isoenzymes

S.M. Mühlebach, M. Gross, T. Wirz, T. Wallimann, J.-C. Perriard and M. Wyss[1]
Institute for Cell Biology, ETH Hönggerberg, CH-8093 Zürich, Switzerland; [1] Department of Transplant Surgery, Research Division, University Hospital, Anichstr. 35, A-6020 Innsbruck, Austria

Abstract

Comparisons of the protein sequences and gene structures of the known creatine kinase isoenzymes and other guanidino kinases revealed high homology and were used to determine the evolutionary relationships of the various guanidino kinases. A 'CK framework' is defined, consisting of the most conserved sequence blocks, and 'diagnostic boxes' are identified which are characteristic for anyone creatine kinase isoenzyme (e.g. for vertebrate B-CK) and which may serve to distinguish this isoenzyme from all others (e.g. from M-CKs and Mi-CKs). Comparison of the guanidino kinases by near-UV and far-UV circular dichroism further indicates pronounced conservation of secondary structure as well as of aromatic amino acids that are involved in catalysis. (Mol Cell Biochem **133/134**: 245–262, 1994).

Key words: creatine kinase, arginine kinase, protein sequence comparison, evolution, CK framework, 'diagnostic boxes', secondary structure prediction

Abbreviations: GuaK – guanidino kinase; CK – creatine kinase; B- and M-CK – brain and muscle cytosolic CK isoenzyme; Mi-CK – mitochondrial CK isoenzyme; ArgK – arginine kinase; Cr – creatine; PCr – phosphorylcreatine; PArg – phosphorylarginine

Introduction

Guanidino kinases (GuaKs)[1] in general and creatine kinases (CKs) in particular are found throughout the animal kingdom [1–4]. Vertebrate tissues contain exclusively CK isoenzymes, while tissues of invertebrates mostly display arginine kinase (ArgK) activity. Spermatozoa of many invertebrates, however, comprise CK as the main or even sole guanidino kinase. Annelida are also an exception to the general picture, in as far as a variety of different guanidino compounds and guanidino kinases were found in this phylum, including arginine (Arg), creatine (Cr), glycocyamine (= guanidinoacetate), taurocyamine, hypotaurocyamine and lombricine, always

together with the corresponding kinases. Of this whole class of guanidino kinases, the (mostly) complete primary structures of 26 different CK isoenzymes, of lobster tail muscle ArgK, and of a *Schistosoma mansoni* (trematode) guanidino kinase with unknown substrate specificity have been reported until the end of October 1993, together with amino acid sequences of some short protein fragments. The goals of the present article are (i) to align and compare the known guanidino kinase sequences, (ii) to define a 'CK framework', based on the evolutionarily most conserved parts of the molecules, (iii) to identify isoenzyme-specific residues or sequence

Address for offprints: S.M. Mühlebach, Institute for Cell Biology, ETH Hönggerberg, CH-8093 Zürich, Switzerland

blocks, and (iv) to correlate particular residues or sequence stretches with physiological functions of the guanidino kinases. These interpretations are complemented by secondary structure predictions and by the identification by CD spectroscopy of evolutionarily conserved Trp and Tyr residues which are important for catalysis and for the structural integrity of the molecules.

Characterization of the guanidino kinases

In birds and mammals, four different nuclearly encoded CK isoforms are found, all displaying a protomer M_r of approximately 40'000. These isoforms are expressed tissue-specifically and differ in subcellular localization as well. The two cytosolic isoforms B- (for brain) and M-CK (for muscle) form dimeric molecules (MM-, MB- and BB-CK; [5]). In contrast, the two mitochondrial isoforms Mi_a- and Mi_b-CK ('a' for the more acidic and 'b' for the more basic Mi-CK isoform; also called ubiquitous and sarcomeric Mi-CK, respectively) form dimeric and octameric molecules which, depending on the experimental conditions, are readily interconvertible (see [6]). Although heterodimeric and heterooctameric molecules between Mi_a- and Mi_b-CK can be generated *in vit-*

$\longrightarrow$

Fig. 1. Alignment and comparison of the known CK protein sequences. The primary structures of 26 CK isoenzymes are compared, with each repeat of the triplicated sea urchin tail CK sequence being analyzed separately. The sequences used in this study were from:

Humbck	human B-CK [47, 49, 81]
Dogbck	dog B-CK [82]
Rabbck	rabbit B-CK [83]
Ratbck	rat B-CK [84]
Moubck	mouse B-CK [52]
Chibbck	chicken B_b-CK [85, 86]
Chiback	chicken B_a-CK [8]
Hummck	human M-CK [51, 63]
Dogmck	dog M-CK [87]
Rabmck	rabbit M-CK [88]
Ratmck	rat M-CK [89]
Moumck	mouse M-CK [65]
Chimck	chicken M-CK [90, 91]
Troutck	CK isoenzyme from rainbow trout displaying enhanced testicular expression [21]
Tormarck	CK from the electrocytes of *Torpedo marmorata* (marbled electric ray; [23])
Torcalck	CK from the electrocytes of *Torpedo californica* (pacific electric ray; [92])
Xenlaeck3	*Xenopus laevis* CK-III [38]
Xenlaeck4	*Xenopus laevis* CK-IV [37]
Humubimick	human ubiquitous Mi-CK [53]
Ratubimick	rat ubiquitous Mi-CK [93]
Mouubimick	mouse ubiquitous Mi-CK [55]
Chimiack	chicken Mi_a-CK (a = more acidic pI; ubiquitous; [39, 40])
Humsarmick	human sarcomeric Mi-CK [94]
Ratsarmick	rat sarcomeric Mi-CK [93]
Chimibck	chicken Mi_b-CK (b = more basic pI; sarcomeric; [62])
Spfckd1,2,3	flagellar CK from the sea urchin *Strongylocentrotus purpuratus*, with part 1 including amino acids 60–433, part 2 including residues 434–807, and part 3 encompassing residues 808–1174 [31]

Dots in the listed sequences stand for inserted gaps, introduced in order to allow an optimal alignment of the different CKs, or for unknown parts of the primary structures (N-termini of *X. laevis* CK-III and CK-IV). Hyphens stand for amino acids which are identical in 24 (residues 1–81) or 25 (residues 82–395) out of the 28 sequences, with the corresponding amino acid being listed in the consensus sequence at the bottom. If at a particular position only two different amino acids are found in all sequences, both of them are listed in the consensus sequence, one above the other. If there is no consensus, or if an amino acid in a given sequence differs from that in the consensus sequence, it is listed in the respective sequence itself. The N-terminal methionine residue in the cytosolic CK sequences is cleaved off after synthesis.

The filled bars (numbered 1 to 6) below the consensus sequence mark the regions with the most pronounced sequence conservation. The open diamonds mark the so-called 'reactive cysteine' Cys-283, the two residues (Cys-74 and Lys-196) which in cross-linking experiments of rabbit MM-CK were shown to be structurally close to Cys-283, as well as Asp-340 which, as suggested by affinity labelling experiments with an alkylating ATP analogue, may be involved in Mg^{2+} binding. Boxed amino acid residues either are isoform-specific or allow a clear-cut distinction between mitochondrial and cytosolic CK isoenzymes. Extended stretches (≥ 5 residues) with isoenzyme-specific sequence patterns are termed 'diagnostic boxes' (see the text) and are designated by A, B, C, . . . I. If a box comprises two (or more) different isoenzymes, and if, within this particular box, residues are conserved in an isoenzyme-specific pattern, they are represented by bold letters. Finally, filled diamonds mark putative phosphorylation sites.

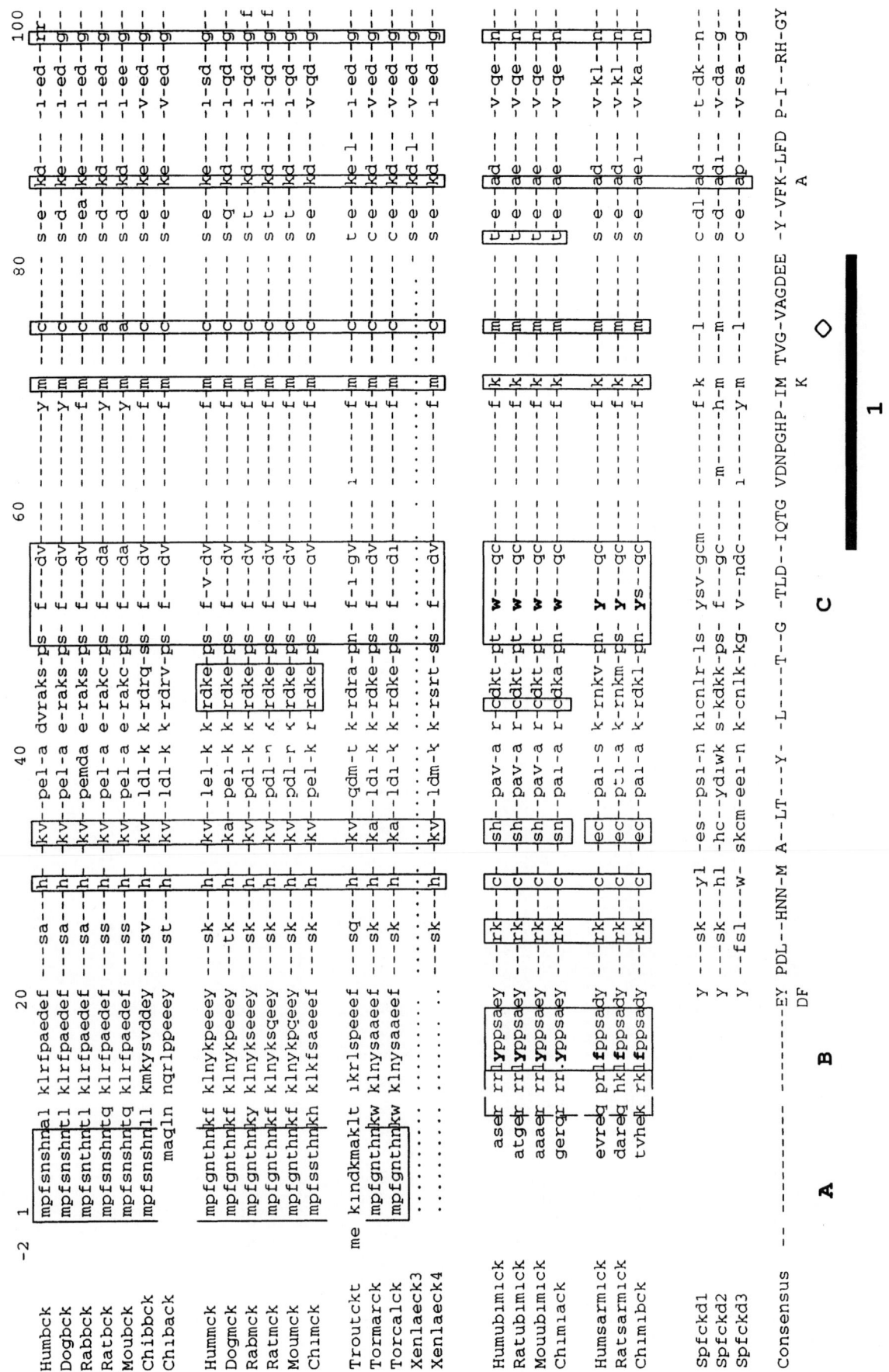

Fig 1

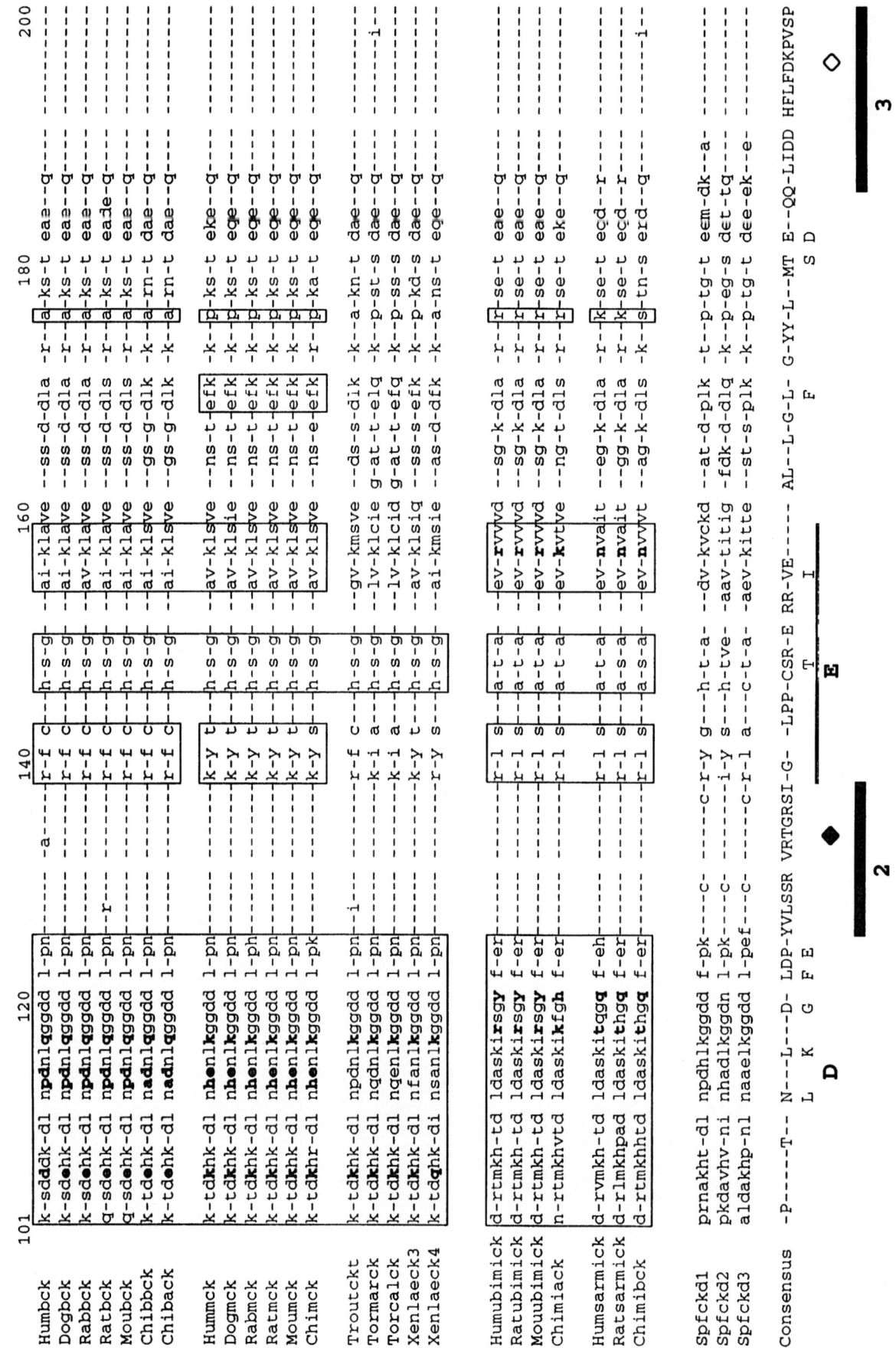

Fig. 1

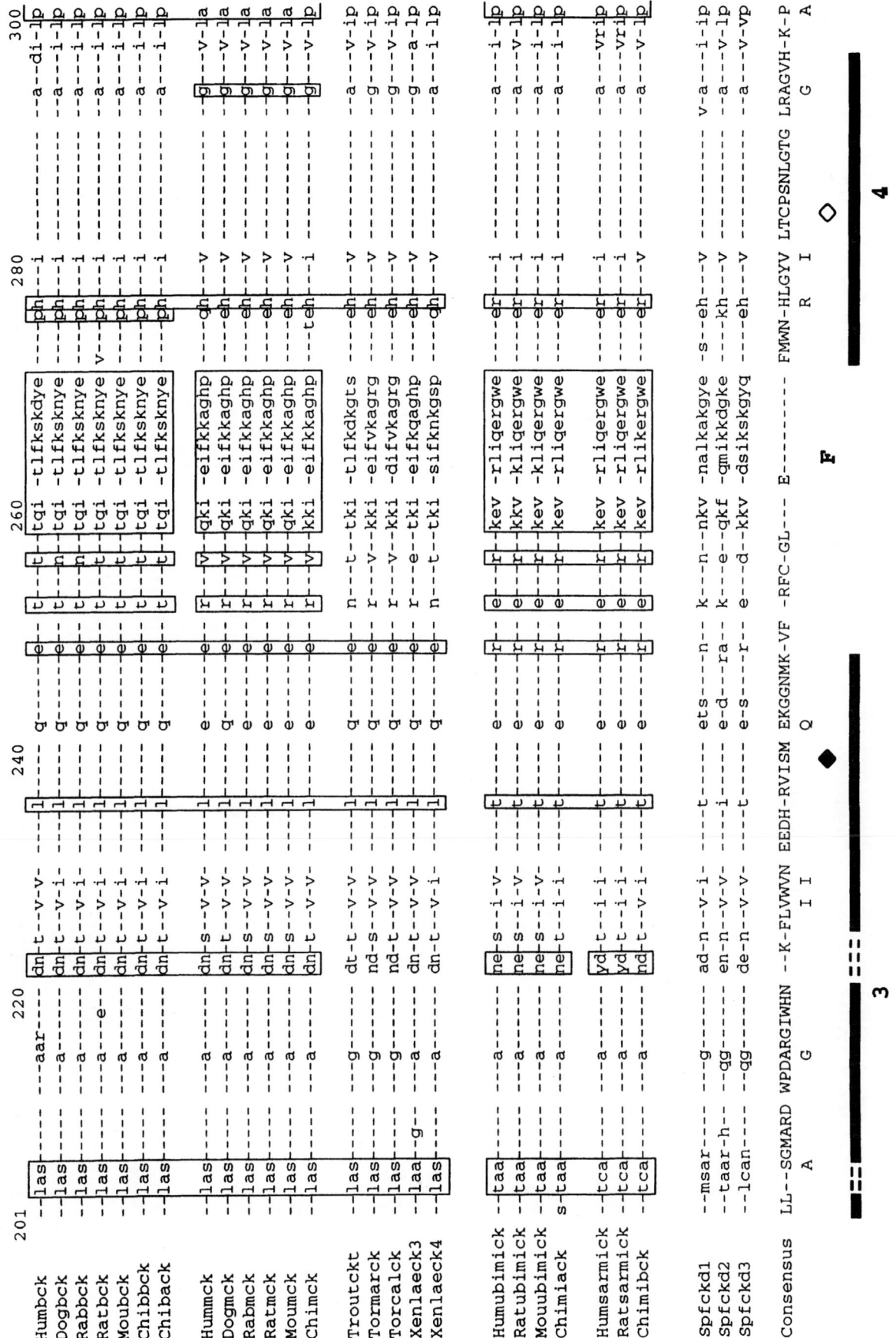

Fig. 1

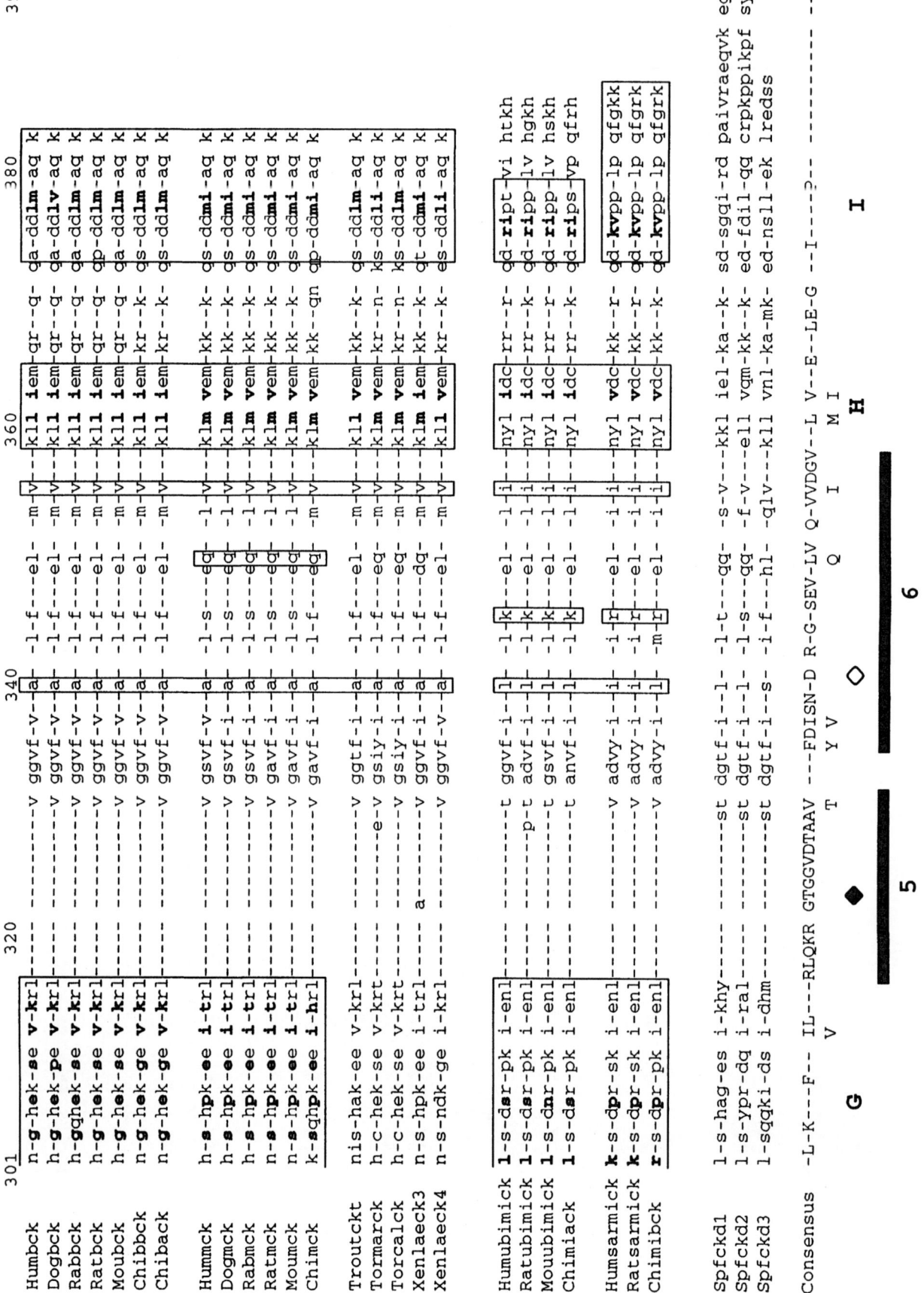

Fig. 1

ro [7], they have not been detected so far *in vivo*. Likewise, heterodimer formation between a mitochondrial and a cytosolic CK isoenzyme is also excluded *in vivo*. Among the cytosolic CK isoenzymes of mammals and birds, chicken B-CK seems to be unique, in as far as alternative splicing of the single B-CK gene produces two isoforms, B_a- and B_b-CK (see Fig. 1), differing in approximately 20 of the first 50 N-terminal amino acids [8]. Further heterogeneity of B-CK, as evidenced, for example, by 2D-electrophoresis experiments [9], was shown to be due to alternative ribosomal initiation at internal transcriptional start sites (at Met-12, Met-30, Leu-36 and Met-70 of chicken B_b-CK; [10]) or to post-translational phosphorylation (as demonstrated for chicken, mouse and rat B-CK; [10–14]).

Lower vertebrates like fish and frogs also contain several CK isoenzyme loci, all giving rise to dimeric molecules with an M_r of approximately 80'000. While some frogs were suggested to display a CK isoenzyme system similar to mammals and birds [15–19], the CK isoenzymes of *Xenopus* and other pipid frogs do not fit into the M-/B-/Mi_a-/Mi_b-CK classification. The five CK isoforms observed in pipid frogs are therefore termed CK-I to CK-V and give rise to up to nine different bands on zymograms [20]. CK-I and possibly also CK-V (see [6]) are located within the mitochondria, while CK-II, CK-III and CK-IV are clearly cytosolic.

The four CK isoforms of teleost fish are termed CK-A to CK-D and are all of cytoplasmic origin. CK-A, CK-C and CK-D are expressed predominantly in striated muscle, stomach, and testis, respectively, while CK-B is expressed ubiquitously or is confined to neural tissues [16]. The trout CK listed in Fig. 1, showing enhanced testicular expression [21], thus most probably represents a CK-D isoform. In primitive fish species, only two CK isoforms are found which obviously correspond to CK-A and CK-C of teleost fish. While CK-A is again restricted to striated muscle, CK-C in primitive fish is expressed ubiquitously [16]. Since *Torpedo* electrocytes were shown by isoenzyme and 2D-electrophoresis to contain the same major CK isoenzyme as muscle [22–25], the two *Torpedo* CKs listed in Fig. 1 very likely represent CK-A isoforms.

Based on a comparison of the tissue-specificity of expression of the various isoenzymes, it has been hypothesized that CK-II of frogs and CK-A of fish correspond to M-CK of mammals and birds, while CK-IV and CK-C would correspond to B-CK [20]. The other CK isoenzymes of frogs and fish seem to be the result of (additional) gene duplications. Clearly, more biochemical and molecular genetic work is needed to unequivocally prove the phylogenetic relationships of the various CK isoenzymes.

Sea urchin spermatozoa contain two different CK isoenzymes. The mitochondrial CK is confined to the midpiece of the sperm and is, like the vertebrate Mi-CKs, an octameric molecule composed of subunits with an M_r of 44'000–50'000 [26–29]. The tail CK isoenzyme, however, is located along the sperm tail and is a monomeric protein with an M_r of 140'000–155'000 [26–28, 30] which most likely originates from triplication of an ancestral CK gene ([31]; see Fig. 1). These findings suggest that the separation into mitochondrial and cytosolic CK isoenzymes occurred before the start of divergence between echinoderms and vertebrates.

Relatively little is known about the invertebrate guanidino kinases. While for the lobster tail muscle ArgK shown in Fig. 1 [32], substrate specificity and biochemical properties have been investigated intensively (see [33]), the physiological substrate of the duplicated guanidino kinase from the trematode *Schistosoma mansoni* is not known [34]. Analysis of homogenates simply indicated that this guanidino kinase displays low CK, but no ArgK activity.

Table 1. Amino acid sequence comparisons of the guanidino kinases known so far

B-CKs among each other	88–98
M-CKs among each other	89–99
B-CKs versus M-CKs	77–82
Ubiquitous Mi-CKs among each other	91–98
Sarcomeric Mi-CKs among each other	89–96
Sarcomeric versus ubiquitous Mi-CKs	82–84
Mi-CKs versus B- and M-CKs	60–65
X. laevis CK-III versus B-CKs	79–83
X. laevis CK-III versus M-CKs	87–91
Torpedo CKs versus B-CKs	76–80
Torpedo CKs versus M-CKs	83–86
X. laevis CK-IV versus B- and M-CKs	84–89
Trout CK versus B- and M-CKs	78–87
Trout/*Torpedo*/*X. laevis* CKs versus vertebrate Mi-CKs	61–67
Sea urchin CK repeats among each other	66–70
Sea urchin CK repeats versus vertebrate CKs	60–69
Lobster ArgK versus all CKs	38–44
Lobster ArgK versus Flukeckp1	45
Lobster ArgK versus Flukeckp2	40
Flukeckp1 versus all CKs	33–38
Flukeckp2 versus all CKs	32–35

The degree of identity (in %) in primary structure between any two guanidino kinases was determined using the program 'Distances' of the GCG software package, with the threshold for a match set at 1.5. For discussion of the results see the text.

LobArgkin
Flukeckp1
Flukeckp2
Cefrag14a3
Cefrag00136

GuaK-Consensus

CK-Consensus

```
              1                  20                   40                   60                   80                  100
LobArgkin          mad aatiak-eeg  fkkleaatdc  ksllk---sk  d-fds[l]ka--  -s-----ldv  g****vgl[yl]  pd[ae]-sl[a]  p[l]--i-e--
Flukeckp1          mqves-qnl       qakirndern  hsltk---[ta]  d-vkkyqat-  -s-g--aqc  l****lpr.s  v[n]hay-[pga]  [cd]n-et-r  df-av-a-
Flukeckp2     qeynkgapeg  vmpvep-tyl  akllegasie  kcytr---[tp]  e-ikkydg-r  -th---ahm  q****yvhel  [gf]n-irt-i  dyl--l-c--
Cefrag14a3    ......... ........ ........ ........ -d[l]kd--  -k--n-ldv  g****x-lds  pxa[e]-tl-[k]  p[l]--l-q--
Cefrag00136   ......... ........ ........ ........ ...... *****

GuaK-Consensus  --------  ------L---  ------I----  ----KYL-  I---V-N--  -****-  --AY--F-  --FDP-I-DY
                -2                 16                   36                   56                   76                   96
CK-Consensus   --EYPDL--H  NN-MA--LT-  --Y--L---  T--G-TLD--  P-IMTVG-VA  GCEE-Y-VFK  -LFDP-I--R
                               DF                                          K                          A
```

```
              101                120                  140                  160                  180                 200
LobArgkin     -[kg]fkqtd--  -akd---.v  -e-t-[vi]----  -c---m--yp  -n-c-t[ea]qy  k-m-e-v-st  -sg-e.--[lk]  k-v[gg]k-[ld]--
Flukeckp1     -kvpdgkiq-  -ksn---.l  y-nl[v]v----  -l--tv--fg  -g-t-tket[h]  i-l-n-i-ta  -hn-s.--ye  -t-[y]----  cq  rgq[ng]tskrh
Flukeckp2     -gvkdsaf-  -apt---.l  [k]lp-g----  -v---v--fl  i[pt]imsktd[t]  ikl-qvi-ga  -kg-t.--ha  -t-[y]----  e[]-drkq-ve-
Cefrag14a3    -rg[fa]p[da]-q  -n[cd]l-xkt  -a..lv----  -c---lq-yp  -n-c-seany  l-mgiqgqgy  ..-  .........
Cefrag00136   ...*  ........  -a.  -e-k-in--i  ..m-s-vkai  fdnitdp-l[a]  -k-f--d-[t]  k-i[dg]q-[lk]

GuaK-Consensus  H------KH  P---FGD--  S---F-DLDP  -G-F--STRV  R-GRS-EG-  -E-E-K-S--  L--L-GE--  -E----L--D
                 97                113                  131                  151                  170                 190
CK-Consensus   H-GY-P----  -*T--N-**-  -L---D-LDP  **-YVLSSRV  RTGRSI-G--  LPP-CSR-ER  R-VE-----A  L--L-G-L-  E--QQ-LIDD
                               L         K      G  F      E                              T    I         F      SD
```

```
              201                220                  240                  260                  280                 300
LobArgkin     ---*-eg-r  f-qa-nac-y  --a---y--  dn------c  --l-------  m-d-gq--  r--vsavnd-  [e]*****-rvp  -shh-[rl]-fl  -f--t----
Flukeckp1     ---*rnd-n  v-rd-[ggyid]  --t-----i  kq-k---i  --------[v]  [k]-rd-ia--  k--adaiqel  s*****-slk  -af[n]-[rl]-[fl]  -f-[s]--t--
Flukeckp2     ---*nd-[E]  v-rd-[ggy]d  --v------  ns-----vc  ---m---  q-[l]-aa--  k--ie[g]ina-  g*****-smk  -ahs-Ky-[yi]  -c--[s]--s--
Cefrag14a3    .....*  ........  ........  ........  ........  -* ******
Cefrag00136   ---*-eg-r  f-qa-nac-y  --k------  nq------  e-[i]-l---  e--[n]vgq-l  e--ikgvkt-  x******-qap

GuaK-Consensus  HFLF*K--D-  -L--A---R-  WP-GRGIFHN  --KTFLVW-N  EEDH-RIISM  Q-GG-L--VY  -RL------I  -*****K---  F---D--G--  T-CP-NLGT-
                191                210                  230                  250                  270                 290
CK-Consensus   HFLFDKPVSP  LL--SGMARD  WPDARGIWHN  --K-FLVWVN  EEDH-RVISM  EKGGNMK-VF  -RFC-GL---  E------  FMWN-HLGYV  LTCPSNLGTG
                               A         G               I  I                 Q                              R     I
```

Fig. 2

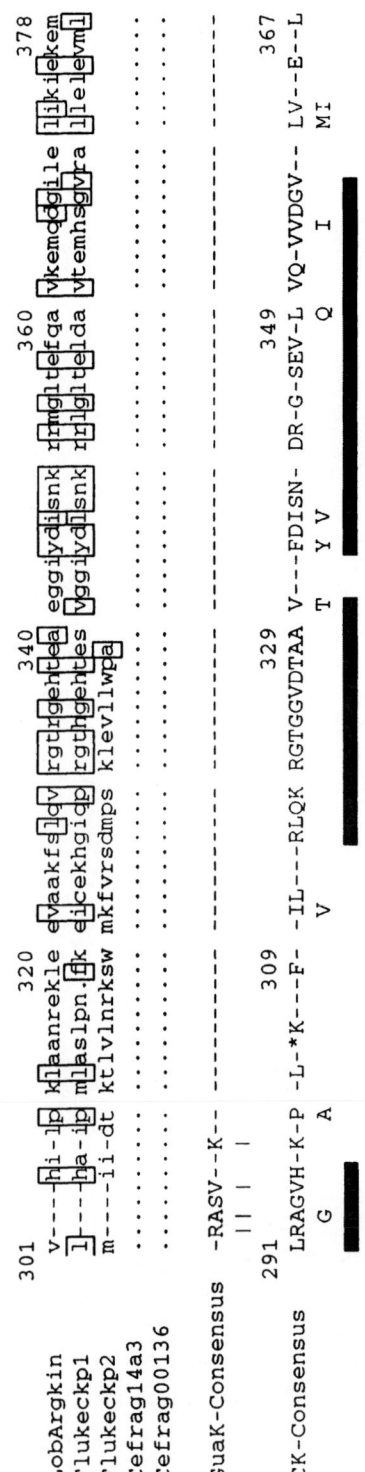

Fig. 2. Comparison of the invertebrate guanidino kinases with the CK consensus sequence. The primary structures of lobster ArgK [32], of a guanidino kinase from the trematode *Schistosoma mansoni* (Flukeckp1 represents residues 1–352 and Flukeckp2 represents residues 353–675 of this 'duplicated' guanidino kinase; [34]), and of two guanidino kinase sequence fragments from *Caenorhabditis elegans* (which were derived from single reads of expressed sequence tags; [95, 96]) were aligned and compared to the CK consensus sequence from Fig. 1.

The dots stand either for gaps that were introduced in order to allow an optimal sequence alignment or for parts of the primary structure that have not been sequenced so far (N- and C-termini of the *C. elegans* fragments). The hyphens in the invertebrate guanidino kinase sequences stand for amino acids that are identical in three out of the five sequences, with the corresponding amino acid being written in the guanidino kinase (GuaK) consensus sequence below. If no consensus is observed, or if an amino acid differs from the consensus sequence, it is written in the respective sequence itself.

Stars represent gaps that were introduced for an optimal alignment of the invertebrate guanidino kinases to the CK consensus sequence. Identical amino acids between the guanidino kinase and CK consensus sequences are indicated by vertical lines, while boxed residues represent additional amino acids in the individual guanidino kinase sequences that are identical to the CK consensus sequence.

vis [37, 38] and the chicken Mi$_a$-CK sequences [39, 40]. The alignment was done using the programs 'Bestfit', 'Pileup', 'Lineup' and 'Pretty' available in the GCG software package. The degree of identity between any two protein sequences (Table 1) was calculated using the program 'Distances', with the threshold for a match being set at 1.5 based on the default amino acid comparison table of the GCG software package [41, 42]. The phylogenetic tree of the guanidino kinases (Fig. 3) was constructed using the T.NJM26 program of the MacT (Macintosh's Trees) program package [43]. Tree construction is based on a distance matrix method commonly known as neighbor joining method [44]. Sequences were aligned in the manner shown in Fig. 2, i.e. gaps were allowed, but were not taken into account for the distance calculations. The distance between any two sequences was calculated by summing the *minimal* number of base substitutions required to convert one amino acid in another at each site compared, and by dividing this sum by the total number of amino acids compared ('Fitch' option; see [45]). Therefore, the distance represents the *average* number of mutations per site needed to get from one sequence to the other.

Programs and databases

All protein sequences compared in this study are derived from nucleic acid sequences. They were extracted from the GenBank and EMBL nucleotide sequence libraries using the programs of the GCG software package [35, 36], with the exception of the two *Xenopus lae-*

254

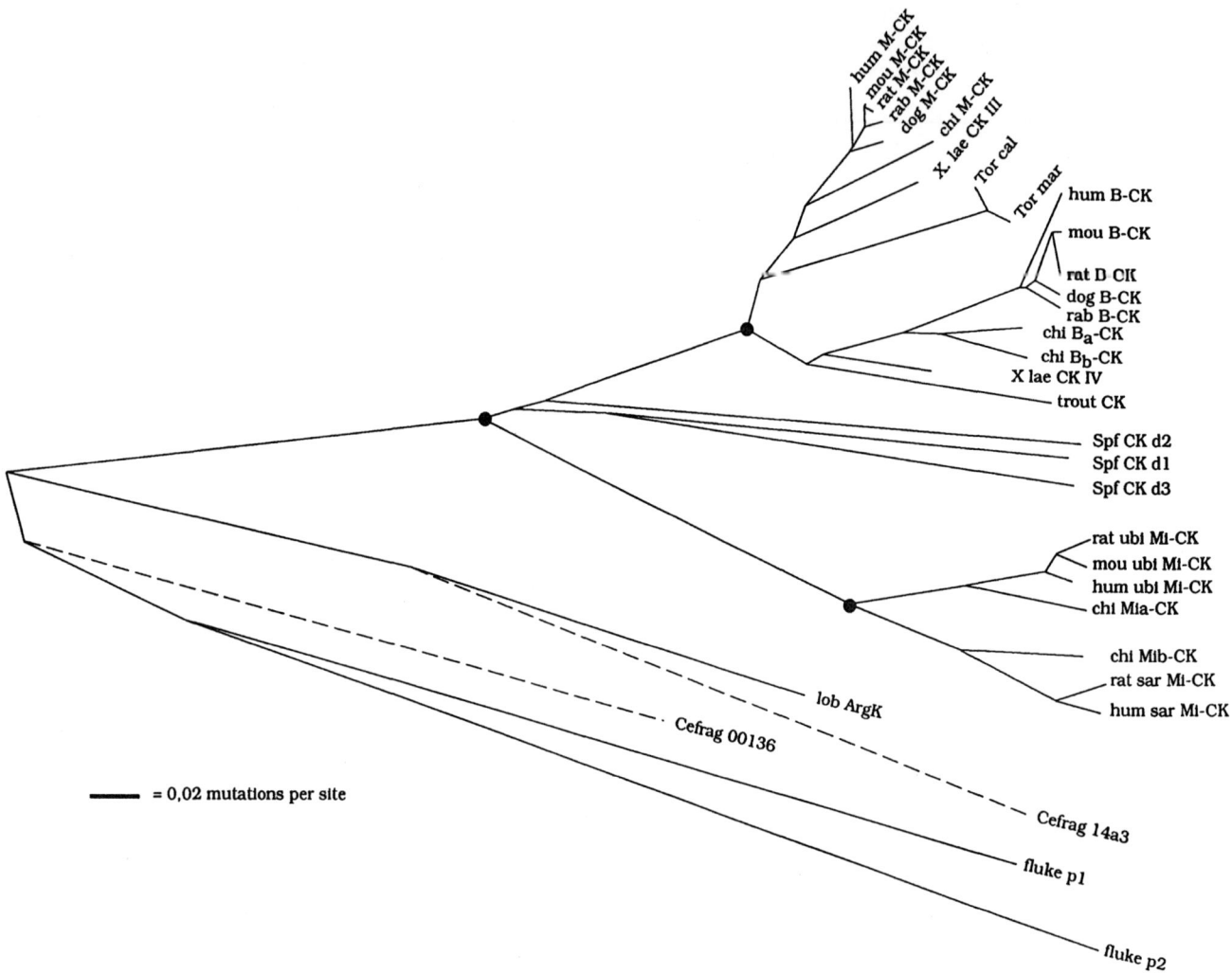

Fig. 3. Evolutionary tree of the guanidino kinases. The phylogenetic tree of the guanidino kinase sequences shown in Figs. 1 and 2 was constructed as described under 'Programs and Databases'. The distance between any two sequences approximates the *average* number of (base) mutations per site needed to convert one sequence into the other. The bar at the bottom represents an 'evolutionary distance' of 0.02 mutations per site. The three dots mark the three principal gene duplication events discussed in the text.

Comparison of the known guanidino kinase protein sequences – evolutionary relationships

Figure 1 shows an alignment of the 28 CK protein sequences or sequence repeats known so far, and Fig. 2 shows an alignment of the invertebrate guanidino kinases to the CK consensus sequence. The three repeats of the triplicated sea urchin tail CK (Spfckd1–3; Fig. 1) and the two repeats of the duplicated guanidino kinase from *Schistosoma mansoni* (Flukeckp1,2; Fig. 2) are analysed separately. Some 20–80 amino acids at the N-termini of the *Xenopus* CKs (Fig. 1) escaped sequencing so far, and for the guanidino kinase(s) of *Caenorhabditis elegans* (Fig. 2), only the listed fragments can be aligned reasonably. The amino acid identity scores between different guanidino kinases are summarized in Table 1, and a tentative evolutionary tree for the guanidino kinases is shown in Fig. 3.

In mammals and birds, the known CK sequences can easily be grouped into the four isoenzyme classes (B-, M-, Mi_a- and Mi_b-CK) that have been postulated earlier on the basis of tissue distribution, electrophoretic behaviour and biochemical characterization. The amino acid identities within each single class range from 88 to 99%. Evidently, a higher degree of homology is observed between the B- and the M-CKs (77–82%) on one hand and between the Mi_a- and the Mi_b-CKs (82–84%) on the other hand than between the cytosolic and the mitochondrial CK isoforms (60–65%), indicating that during evolution, a first gene duplication event resulted in a primordial cytosolic and a primordial mitochondrial CK isoenzyme (Fig. 3). Gene duplications giving rise to the

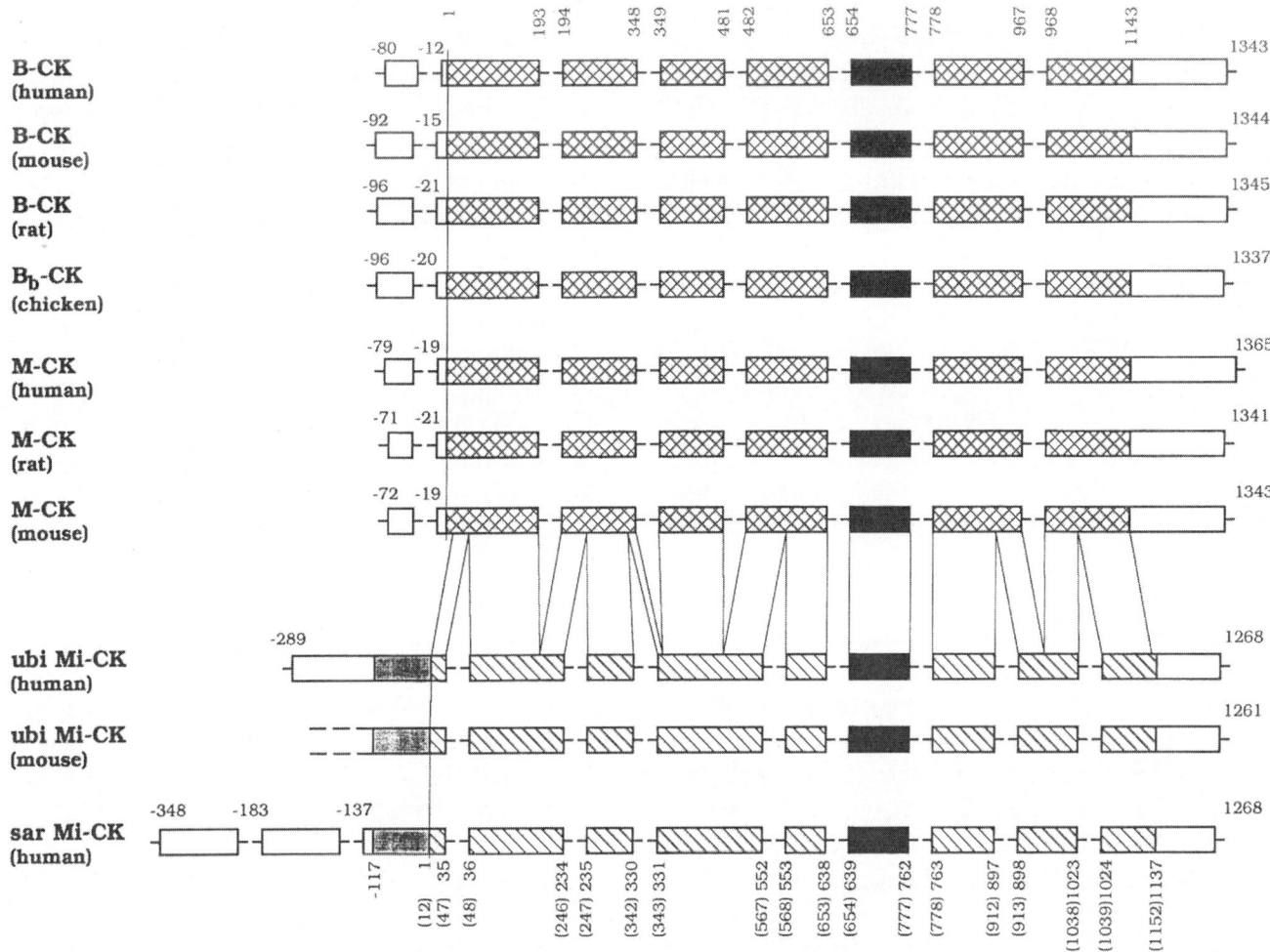

Fig. 4. Schematic representation of the gene structures of the vertebrate CK isoenzymes investigated so far. The lengths of the exons are drawn in scale, while those of the introns are not. The protein coding regions within the exons of cytosolic and mitochondrial CK isoenzymes are cross-hatched and hatched, respectively, while the single exon which is conserved in all mammalian and avian CK isoenzymes (exon 8 of sarcomeric Mi-CK, exon 6 of the other CKs) is represented by a filled box. The coding region for the Mi-CK transit peptide (which is responsible for the import into the mitochondrial intermembrane space) is shaded. Numbers above and below the respective exons refer to the corresponding nucleotide positions within the protein coding (cDNA) sequences. In the case of the cytosolic CK isoenzymes, the first nucleotide of the start codon is given number 1, while for the Mi-CKs, 1 represents the first nucleotide of the codon for the very N-terminal amino acid of the *mature* protein (i.e. lacking the transit peptide). Below the numbers for the Mi-CK isoenzymes, the respective positions within the cDNA sequences of the cytosolic CKs are listed in brackets.

The respective gene structures are from: human B-CK [47–49]; mouse B-CK [52]; rat B-CK [50]; chicken B_b-CK [8]; human M-CK [51]; rat M-CK [50]; mouse M-CK [46]; human ubiquitous Mi-CK [53]; mouse ubiquitous Mi-CK [55]; and human sarcomeric Mi-CK [54]. For simplicity, the additional exon coding for the amino terminus of chicken B_a-CK [8] has not been considered in this diagram.

multiple cytosolic and mitochondrial CK isoenzymes observed in vertebrates must have occurred at a later stage during evolution. This interpretation is favoured by the likely presence of a single cytosolic and a single mitochondrial CK isoenzyme in echinoderm species [27] and by the gene structures of the mammalian and avian CK isoenzymes investigated so far (Fig. 4). The lengths of the exons as well as the location of the splice sites within the coding region are identical for all M- and B-CKs on one hand [8, 46–52] and for all Mi-CKs on the other hand [53–55]. If, however, the gene structure of the Mi-CKs is compared to that of the cytosolic CKs, pronounced differences are observed, with only the location of a single exon (exon 8 of human Mi_b-CK, exon 6 of the Mi_a- and the cytosolic CKs; coding for amino acids 219–259 in Fig. 1) being conserved.

The *X. laevis* isoenzymes CK-III and CK-IV, the two *Torpedo* CKs as well as the trout CK can be assigned to the branch of the cytosolic vertebrate CK isoenzymes (Table 1; Fig. 3), since the amino acid identities to the B-

256

and M-CKs of mammals and birds (76–91%) are much higher than to the Mi-CKs (60–67%). *X. laevis* CK-III and the *Torpedo* CKs clearly correspond to mammalian and avian M-CKs, while *X. laevis* CK-IV and the trout CK seem to be related somewhat more closely to B-CK than to M-CK (Table 1; Fig. 3). The trout CK is peculiar in as far as the first 17 amino acids at the N-terminus (see Fig. 1) differ completely from those of the other vertebrate cytosolic CK isoenzymes, perhaps with the exception of chicken B$_a$-CK.

The homologies of the repeats of the triplicated sea urchin CK with each other (66–70% identity) are only slightly higher than those of the sea urchin CK repeats to the vertebrate CKs (60–69%) and of the vertebrate cytosolic CKs to the Mi-CKs (60–67%), implicating that branching into a mitochondrial and a cytosolic CK isoenzyme as well as the triplication of the cytosolic sea urchin CK occurred approximately at the time when echinoderms and vertebrates started to diverge. The two repeats of the guanidino kinase from the parasitic trematode *Schistosoma mansoni* are more homologous to lobster tail muscle ArgK (40–45% identity) than to all CKs known (32–38%). Since extracts of *Schistosoma* cercaria, however, display low CK, but no ArgK activity [34], a different guanidino substrate specificity has to be postulated for this guanidino kinase [32]. The two sequence fragments from *Caenorhabditis elegans* display more pronounced homology to lobster ArgK (63% for the Ce14a3 fragment, 68% for the Ce00136 fragment) than to CK (32–42%, 44-50%) or to the *Schistosoma* guanidino kinase (26 and 33%, 40 and 51%), suggesting that they represent ArgK isoenzymes. Finally, amino acid sequencing of tryptic fragments of a guanidinoacetate (glycocyamine) kinase from the polychaete *Neanthes diversicolor* revealed higher homology to the corresponding parts of the CKs than of the other guanidino kinases [56]. As a matter of fact, guanidinoacetate, among the natural guanidino compounds, also displays the most pronounced structural homology to Cr.

In summary, the evolutionary tree of the guanidino kinases shown in Fig. 3 'visualizes' the homologies observed between the sequences of the various guanidino kinases and, in addition, agrees with most biochemical findings. Nevertheless, it should be regarded as tentative. Clearly, a much larger number of guanidino kinases, especially in the invertebrate phyla, have to be cloned and/or sequenced before a correct tracking of the evolutionary relationships as well as a reliable estimation of the time points in evolution when gene duplication events occurred will be possible. However, one

clear-cut postulate of the analyses presented here is that all vertebrates, including fish, amphibia, and reptiles, contain mitochondrial CK isoenzymes.

It has been hypothesized earlier that phosphorylarginine (PArg) and ArgK are phylogenetically older than the other phosphagens and guanidino kinases (see [2, 3]). The facts that (i) ArgK is by far the most widespread guanidino kinase in invertebrate phyla, that (ii) Arg is a key component of basic metabolism, while the other guanidino compounds necessitated the evolution of additional enzymes for their biosynthesis, and that (iii) most ArgKs are monomeric proteins in fact support the notion that a primordial monomeric ArgK represents the common ancestor of all guanidino kinases [2]. However, the postulate that phosphorylcreatine (PCr) represents a functional improvement over PArg, this explaining the *apparent* switch from ArgK to CK at the transition from invertebrates to vertebrates [3], is untenable since PCr and CK are found in a large variety of invertebrate spermatozoa [2]. In these invertebrates, PCr is often the sole phosphagen in the sperm cells, while other tissues contain PArg or other phosphagens. Due to the different thermodynamic properties of the various phosphagens, it is more reasonable to assume that PCr is better suited for *some* species and cell types, while PArg and the other phosphagens are advantagous for others [4, 6].

Recently, it has been postulated, on the basis of proteolysis and small-angle X-ray scattering experiments as well as of sequence comparisons [57–61] that guanidino kinases are structurally similar to 3-phosphoglycerate kinase. Accordingly, each guanidino kinase protomer would consist of two domains with M$_r$'s of 20'000–25'000 which are separated by a deep cleft. The two substrates are bound on either side of the cleft, and binding of the second substrate is likely to result in a closure of the cleft, thus allowing a direct in-line transfer of a phosphate group during catalysis, in an environment excluding water. Whether and how closely the guanidino kinases and 3-phosphoglycerate kinase are also related evolutionarily remains to be established.

CK framework

The amino acid sequences of the CK isoenzymes display six blocks with extensive homologies, flanked by seven regions that are more variable (Fig. 1). Among the latter are the N-terminus and the C-terminus. It has been suggested previously that the highly conserved parts form the 'framework' of the molecule, being involved in basic functions like substrate binding and catalysis, while the variable segments may be responsible for isoenzyme- or species-specific functions like oligomer formation or interaction of the CK isoenzymes with subcellular structures (either with membranes or with other proteins; [62]). The six highly conserved blocks are indicated by bars below the CK consensus sequence in Fig. 1 and are numbered 1 to 6. Compared to blocks 1–5, block 6 is somewhat less well defined, but still 19 out of 24 amino acids are conserved.

As can be seen in Fig. 1, several residues in blocks 1–6 are only imperfectly conserved. While some of the 'deviations' may be real, others are certainly due to sequencing artifacts. This has, for instance, been demonstrated already [51] for six amino acids (5 of which are in the conserved blocks 2, 3 and 5) of the human M-CK sequence published by Perryman *et al.* [63].

Block 4 contains the highly reactive and absolutely conserved Cys-283, alkylation of which is always paralleled by a very pronounced or even complete loss of enzymatic activity (for references see [64]). In earlier publications, it has therefore often been suggested that this residue is 'essential' for and possibly directly involved in catalysis. Recently, however, site-directed mutagenesis of Cys-283 of chicken Mi$_b$-CK demonstrated that this residue is not involved in catalysis itself, but that it is necessary for synergism in substrate binding and that it may also provide a negative charge for maximum enzymatic activity [64]. Block 4 also comprises a putative adenine nucleotide binding motif (glycine-rich loop) LGXGXXGXV [65, 66], but it is not yet clear whether it is functional or not. The importance of block 4 for CK and guanidino kinase function in general is further supported by the fact that short peptide sequences around Cys-283 of taurocyamine kinase, lombricine kinase and glycocyamine kinase also show high sequence conservation [56, 67, 68].

Blocks 1 and 3 contain the two residues Cys-74 and Lys-196 which, on the basis of cross-linking experiments with rabbit MM-CK, have been implicated to be structurally close to the highly reactive Cys-283 [69, 70]. While Lys-196 is absolutely conserved in all CK sequences (Fig. 1), Cys-74 is replaced by an Ala in rat and mouse B-CK, and by a Met or Leu in all Mi-CK and sea urchin CK sequences. This suggests that at least the latter residue is not essential for CK function.

Asp-340 in block 6 of chicken Mi$_b$-CK has been labelled by an alkylating ATP analogue, and it has been suggested that this residue is involved in the binding of

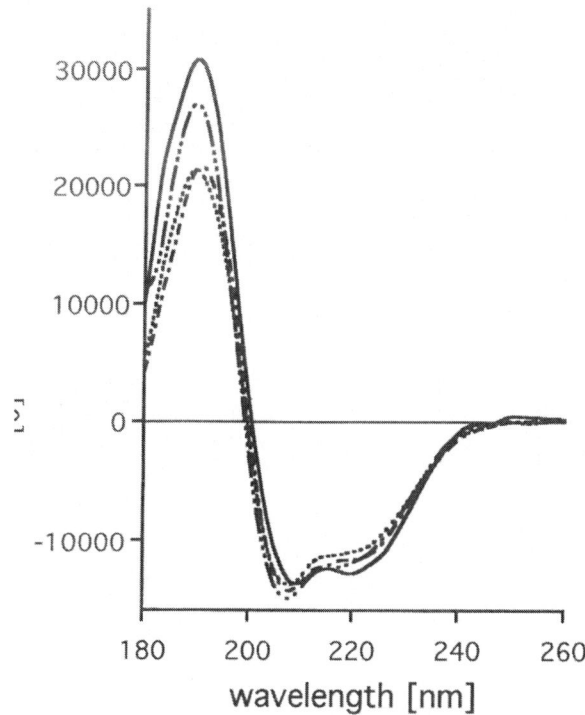

Fig. 5. Far-UV CD spectra of guanidino kinases. Spectra of octameric chicken heart Mi$_b$-CK (——), chicken brain BB-CK (-··-), rabbit muscle MM-CK (···) and lobster tail muscle ArgK (-··) were recorded on a Jasco J-710 dichrograph at a protein concentration of 0.5 mg/ml in 50 mM potassium phosphate buffer, pH 7.0. Cylindrical quartz cells with 0.02 cm path length were used. Residual molar ellipticities ([Θ]) are given in deg cm^2 dmole^{-1}.

the Mg^{2+} ion [71]. Again, this residue is absolutely conserved in all CK sequences. Finally, no functions have yet been assigned to blocks 2 and 5, except that they contain, together with block 3, three putative phosphorylation sites (Thr-133, Ser-239 and Thr-322) which are absolutely conserved among all CKs and are also found in some of the invertebrate guanidino kinase sequences.

A comparison of the CK framework with the other guanidino kinase sequences reveals that block 1 is 'missing' in the invertebrate guanidino kinases, while in blocks 2–6, pronounced homology is observed (Fig. 2). It is therefore tempting to speculate that block 1 determines the guanidino substrate specificity of the guanidino kinases and, in particular, the Cr specificity of the CKs. In accordance with this notion, it has recently been postulated, on the basis of biochemical evidence, that the N-terminal and C-terminal halves (= domains) of the CK molecule are involved in Cr and MgATP binding, respectively [61]. In the invertebrate guanidino kinases known so far, Cys-283 and Lys-196 (only in Flukeckp1, Lys-196 is replaced by Arg) are also highly conserved, while Cys-74 and Asp-340 are not.

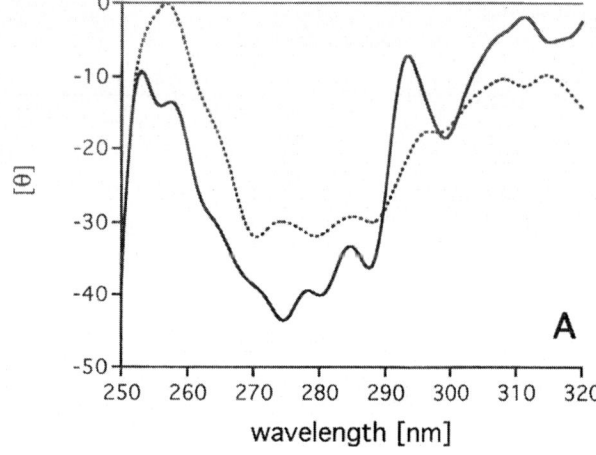 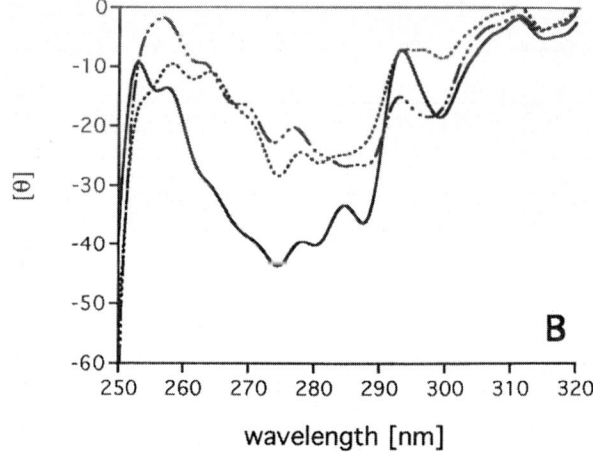

Fig. 6. Near-UV CD spectra of guanidino kinases. (A) Creatine (Mi$_b$-CK, ——) and arginine kinase (lobster, – – –). (B) Mi$_b$-CK (——), MM-CK (···), and BB-CK ----) [same proteins as in Fig. 5]. The spectra were recorded on a Jasco J-710 dichrograph at a protein concentration of 1 mg/ml in 50 mM potassium phosphate buffer, pH 7.0. Cylindrical quartz cells with 1 cm path length were used. Residual molar ellipticities ([(Θ)] are given in degcm²dmole⁻¹.

Isoenzyme-specific boxes

While the six highly conserved blocks of the 'CK framework' often do not allow a distinction between isoenzymes, the more variable segments in between frequently contain single residues or even extended peptide stretches that either allow a clear-cut distinction between cytosolic and mitochondrial CKs or are specific for any one isoenzyme (M-, B-, Mi$_a$- or Mi$_b$-CK). The most instructive of these residues or peptide stretches are boxed in Fig. 1, with purely isoenzyme-specific residues within larger boxes being written in bold. Extended stretches (≥ 5 residues) with isoenzyme-specific sequence patterns are termed 'diagnostic boxes' (A, B, C, ... I), since they are likely to allow the correct assignment of a new vertebrate CK sequence to one of the four isoenzyme classes (boxes A, B, D, G, H, I) or at least to either the mitochondrial or cytosolic CK isoenzymes (boxes C, E, F). For instance, screening of the Gen-EMBL sequence data libraries with blocks A and I yielded exclusively the expected CK sequences with 100% identity, showing that they are absolutely specific for the respective CK isoenzymes.

It is very likely that the 'diagnostic' residues or boxes are responsible for isoenzyme-, cytosolic CK-, or Mi-CK-specific properties, e.g. for octamer formation, membrane binding, interaction with the myofibrillar M-band or other subcellular sites. Accordingly, limited proteolysis and site-directed mutagenesis experiments have shown that the very N-terminus of chicken Mi$_b$-CK, particularly the positively charged residues 5–7, is important for octam-er formation and stability [72]. Further mutagenesis experiments on chicken Mi$_b$-CK have shown that Trp-269 in block F of Fig. 1 is also important for octamer formation and stability [73], indicating that blocks B and F may form the complementary surface areas on neighboring dimers that interact with each other within the Mi-CK octamer. None of the other diagnostic boxes has been linked with a function of the molecule so far.

Structural homologies in the guanidino kinase family – conservation of secondary structure and of aromatic side chains

The high sequence homology among the members of the guanidino kinase family is paralleled by a number of structural similarities which are revealed by spectroscopic investigations. Circular dichroism (CD) measurements are particularly well suited for a structural comparison of the guanidino kinases.

Secondary structure compositions of proteins can be evaluated and compared by measuring CD spectra in the far-UV wavelength range (λ ≤ 240 nm). Using a library of reference spectra of proteins with known three-dimensional structures, the secondary structure composition of the protein of interest can be computed by a variety of algorithms [74]. The far-UV CD spectra of the guanidino kinases reveal a high α-helical content, the spectra closely resembling those of lysozyme or myoglobin. All guanidino kinases, including the mammalian and avian octameric Mi-CK and dimeric cytosolic CK isoenzymes, as well as the invertebrate (mostly monomeric) arginine, lombricine and taurocyamine kinases, display nearly superimposable far-UV CD spectra (Fig. 5 and [75]), indicating an almost perfect conservation of secondary structure elements. This leads to the conclusion that all the distinctive properties of the individual members of the family (guanidino substrate specificity, ability to form oligomeric molecules, specific localization within the cell and association with subcellular structures) do not require major adaptations of secondary structure, but can be achieved by relatively small differences in primary sequence (e.g. the isotype-specific sequence boxes of the CKs described above).

Since the far-UV CD spectra of the guanidino kinases are nearly indistinguishable, a representative secondary structure calculation was performed for chicken Mi_b-CK, using the variable selection algorithm introduced by Manavalan & Johnson [76], with 22 proteins in the original reference database. The procedure resulted in a prediction of 37% α-helix, 30% antiparallel β-sheet, no parallel β-sheet, 15% turns, and 18% other structures.

From CD spectra in the UV wavelength range above 240 nm (near-UV), information about the aromatic amino acid residues, in particular Tyr and Trp, can be obtained. Although the near-UV CD spectra of the various guanidino kinases (Fig. 6; see also [75]) show distinctive characteristics, they also exhibit several common, conserved features. The most obvious one is a conserved negative Cotton band at 295–300 nm, which, from its position, can clearly be assigned to (a) Trp residue(s). Since magnetic resonance, fluorescence [77], CD [78], and chemical modification data [79] have shown that in CK, a Trp residue positioned close to the adenine nucleotide substrate is essential for enzyme activity, it might be assumed that the CD band originates from this structurally and functionally conserved Trp. The only two Trp residues common to ArgKs and CKs are amino acids 211 and 228 (numbering according to Fig. 1). They are the only Trp residues present in lobster ArgK, whereas CK has two (cytosolic CKs) or three (Mi-CKs) additional indole side chains. Consequently, either residue 211 or 228 should be the Trp essential for catalytic activity. Site-directed mutagenesis studies on Mi_b-CK have indeed confirmed that Trp-228 is essential for catalysis, with even the conservative replacement by Phe leading to ≥ 99% inactivation. However, these experiments further showed that Trp-211 is also important for the structural integrity of the active site and that in fact this latter

residue is the origin of the Cotton band at 295–300 nm [73].

In the region below 290 nm, all CKs and ArgKs show a characteristic pattern of three negative Cotton peaks at 288, 280 and 274 nm, superimposed on a broad negative band extending from 255 to 290 nm. This region mainly reflects the environment of Tyr residues. For CK, an involvement of Tyr in enzyme activity has also been demonstrated [80], suggesting that the common CD pattern reflects the structural and functional conservation of tyrosyl side chains among the guanidino kinases.

Acknowledgement

Prof. H.M. Eppenberger is gratefully acknowledged for continuous support. We are also indebted to Dr. M. Messerli, Prof. R. Rieger (Innsbruck) and D. Gilligan for valuable discussion and comments on the manuscript. This work was supported by research grants from the ETH Zürich (No. 0-20-064-90 to J.-C.P.), from the Swiss National Science Foundation (No. 31-27756.89 to J.-C.P., No. 31-33907.92 to T.W., postdoctoral fellowship No. 823A-037106 to M.W.), from the Swiss Society for Muscle Diseases and from the Helmut Horten Foundation (to T.W.).

References

1. Robin Y: Biological distribution of guanidines and phosphagens in marine annelida and related phyla from California, with a note on pluriphosphagens. Comp Biochem Physiol 12: 347–367, 1964
2. Watts DC: Evolution of phosphagen kinases. In: E Schoffeniels (ed.) Biochemical Evolution and the Origin of Life. North-Holland Publishing Company, Amsterdam and London, 1971, pp 150–173
3. Watts DC: Evolution of phosphagen kinases in the chordate line. Symp Zool Soc Lond 36: 107–127, 1975
4. Ellington WR: Phosphocreatine represents a thermodynamic and functional improvement over other muscle phosphagens. J Exp Biol 143: 177–194, 1989
5. Eppenberger HM, Dawson DM, Kaplan NO: The comparative enzymology of creatine kinases. J Biol Chem 242: 204–209, 1967
6. Wyss M, Smeitink J, Wevers RA, Wallimann T: Mitochondrial creatine kinase: A key enzyme of aerobic energy metabolism. Biochim Biophys Acta 1102: 119–166, 1992
7. Wyss M, Schlegel J, James P, Eppenberger HM, Wallimann T: Mitochondrial creatine kinase from chicken brain. J Biol Chem 265: 15900–15908, 1990
8. Wirz T, Brändle U, Soldati T, Hossle JP, Perriard J-C: A unique chicken B-creatine kinase gene gives rise to two B-creatine kinase isoproteins with distinct N termini by alternative splicing. J Biol Chem 265: 11656–11666, 1990

260

9. Rosenberg UB, Eppenberger HM, Perriard J-C: Occurrence of heterogenous forms of the subunits of creatine kinase in various muscle and nonmuscle tissues and their behaviour during myogenesis. Eur J Biochem 116: 87–92, 1981

10. Soldati T, Schäfer BW, Perriard J-C: Alternative ribosomal initiation gives rise to chicken brain-type creatine kinase isoproteins with heterogeneous amino termini. J Biol Chem 265: 4498–4506, 1990

11. Mahadevan LC, Whatley SA, Leung TKC, Lim L: The brain isoform of a key ATP-regulating enzyme, creatine kinase, is a phosphoprotein. Biochem J 222: 139–144, 1984

12. Quest AFG, Soldati T, Hemmer W, Perriard J-C, Eppenberger HM, Wallimann T: Phosphorylation of chicken brain-type creatine kinase affects a physiologically important kinetic parameter and gives rise to protein microheterogeneity in vivo. FEBS Lett 269: 457–464, 1990

13. Chida K, Kasahara K, Tsunenaga M, Kohno Y, Yamada S, Ohmi S, Kuroki T: Purification and identification of creatine phosphokinase B as a substrate of protein kinase C in mouse skin in vivo. Biochem Biophys Res Commun 173: 351–357, 1990

14. Hemmer W, Skarli M, Perriard J-C, Wallimann T: Effect of okadaic acid on protein phosphorylation patterns of chicken myogenic cells with special reference to creatine kinase. FEBS Lett 327: 35–40, 1993

15. Eppenberger ME, Eppenberger HM, Kaplan NO: Evolution of creatine kinase. Nature 214: 239–241, 1967

16. Fisher SE, Whitt GS: Evolution of isozyme loci and their differential tissue expression. J Mol Evol 12: 25–55, 1978

17. Fisher SE, Shaklee JB, Ferris SD, Whitt GS: Evolution of five multilocus isozyme systems in the chordates. Genetica 52/53: 73–85, 1980

18. Klemann SW, Pfohl RJ: Creatine phosphokinase in Rana pipiens: Expression in embryos, early larvae and adult tissue. Comp Biochem Physiol 73B: 907–914, 1982

19. Legssyer A, Arrio-Dupont M: Mitochondrial isoenzyme of creatine kinase in frog heart. Comp Biochem Physiol 89B: 251–255, 1988

20. Wolff J, Kobel HR: Creatine kinase isozymes in pipid frogs: Their genetic bases, gene expressional differences, and evolutionary implications. J Exp Zool 234: 471–480, 1985

21. Garber AT, Winkfein RJ, Dixon GH: A novel creatine kinase cDNA whose transcript shows enhanced testicular expression. Biochim Biophys Acta 1087: 256–258, 1990

22. Barrantes FJ, Mieskes G, Wallimann T: Creatine kinase activity in the Torpedo electrocyte and in the non-receptor, peripheral ν-proteins from acetylcholine receptor-rich membranes. Proc Natl Acad Sci USA 80: 5440–5444, 1983

23. Giraudat J, Devillers-Thiery A, Perriard J-C, Changeux JP: Complete nucleotide sequence of Torpedo marmorata mRNA coding for the 43,000-dalton ν₂ protein: Muscle-specific creatine kinase. Proc Natl Acad Sci USA 81: 7313–7317, 1984

24. Witzemann V: Creatine phosphokinase: Isoenzymes in Torpedo marmorata. Eur J Biochem 150: 201–210, 1985

25. Gysin R, Yost B, Flanagan SD: Creatine kinase isoenzymes in Torpedo californica: Absence of the major brain isoenzyme from nicotinic acetylcholine receptor membranes. Biochemistry 25: 1271–1278, 1986

26. Tombes RM, Shapiro BM: Metabolite channelling: A phosphorylcreatine shuttle to mediate high energy phosphate transport between sperm mitochondrion and tail. Cell 41: 325–334, 1985

27. Tombes RM, Shapiro BM: Enzyme termini of a phosphocreatine shuttle: Purification and characterization of two creatine kinase isozymes from sea urchin sperm. J Biol Chem 262: 16011–16019, 1987

28. Tombes RM, Shapiro BM: Energy transport and cell polarity: Relationship of phosphagen kinase activity to sperm function. J Exp Zool 251: 82–90, 1989

29. Wyss M: Biochemical and physiological aspects of mitochondrial creatine kinase. Diss. No. 9777, ETH Zürich, Switzerland, 1992

30. Ratto A, Shapiro BN, Christen R: Phosphagen kinase evolution. Eur J Biochem 186: 195–203, 1989

31. Wothe DD, Charbonneau H, Shapiro BM: The phosphocreatine shuttle of sea urchin sperm: Flagellar creatine kinase resulted from a gene triplication. Proc Natl Acad Sci USA 87: 5203–5207, 1990

32. Dumas C, Camonis J: Cloning and sequence analysis of the cDNA for arginine kinase of lobster muscle. J Biol Chem 268: 21599–21605, 1993

33. Morrison JF: Arginine kinase and other invertebrate guanidino kinases. In: PD Boyer (ed.) The Enzymes. Academic Press, New York, 1973, pp 457–486

34. Stein LD, Harn DA, David JR: A cloned ATP:guanidino kinase in the trematode Schistosoma mansoni has a novel duplicated structure. J Biol Chem 265: 6582–6588, 1990

35. Devereux J, Haeberli P, Smithies O: A comprehensive set of sequence analysis programs for the VAX. Nucl Acids Res 12: 387–395, 1984

36. Genetics Computer Group, GCG: Program Manual for the GCG Package. Version 7, April 1991. 575 Science Drive, Madison, Wisconsin, 53711, USA

37. Robert J, Wolff J, Jijakli H, Graf J-D, Karch F, Kobel HR: Developmental expression of the creatine kinase isozyme system of Xenopus: maternally derived CK-IV isoform persists far beyond the degradation of its maternal mRNA and into the zygotic expression period. Development 108: 507–514, 1990

38. Robert J, Barandun B, Kobel HR: A Xenopus laevis creatine kinase isozyme (CK-III/III) expressed preferentially in larval striated muscle: cDNA sequence, developmental expression and subcellular immunolocalization. Genet Res Camb 58: 35–40, 1991

39. Wirz T: Genetische Grundlagen der Heterogenität von zytosolischen und mitochondrialen Kreatin Kinasen. Diss. No. 9409, ETH Zürich, Switzerland, 1991

40. Mühlebach SM, Brändle U, Wirz T, Egli A, Perriard J-C: in preparation, 1994

41. Schwartz RM, Dayhoff MO: Matrices for detecting distant relationships. In: MO Dayhoff (ed.) Atlas of Protein Sequence and Structure. National Biomedical Research Foundation, Washington, DC, 1978, pp 353–358

42. Gribskov M, Burgess RR: Sigma factors from E. coli, B. subtilis, phage SPO1, and phage T4 are homologous proteins. Nucl Acids Res 14: 6745–6763, 1986

43. Lüttke A, Fuchs R: MacT: Apple Macintosh programs for constructing phylogenetic trees. Comput Applic Biosci 8: 591–594, 1992

44. Saitou N, Nei M: The neighbor-joining method: A new method for reconstructing phylogenetic trees. Mol Biol Evol 4: 406–425, 1987

45. Fitch WM, Margoliash E: Construction of phylogenetic trees. Science 155: 279–284, 1967

46. Jaynes JB, Chamberlain JS, Buskin JN, Johnson JE, Hauschka SD: Transcriptional regulation of the muscle creatine kinase gene and

261

regulated expression in transfected mouse myoblasts. Mol Cell Biol 6: 2855–2864, 1986

47. Mariman ECM, Broers CAM, Claesen CAA, Tesser GI, Wieringa B: Structure and expression of the human creatine kinase B gene. Genomics 1: 126–137, 1987

48. Mariman ECM, Schepens JTG, Wieringa B: Complete nucleotide sequence of the human creatine kinase B gene. Nucleic Acids Res 17: 6385, 1989

49. Daouk GH, Kaddurah-Daouk R, Putney S, Kingston R, Schimmel P: Isolation of a functional human gene for brain creatine kinase. J Biol Chem 263: 2442–2446, 1988

50. Benfield PA, Graf D, Korolkoff PN, Hobson G, Pearson ML: Isolation of four rat creatine kinase genes and identification of multiple potential promoter sequences within the rat brain creatine kinase promoter region. Gene 63: 227–243, 1988

51. Trask RV, Strauss AW, Billadello JJ: Developmental regulation and tissue-specific expression of the human muscle creatine kinase gene. J Biol Chem 263: 17142–17149, 1988

52. Van Deursen J, Schepens J, Peters W, Meijer D, Grosveld G, Hendriks W, Wieringa B: Genetic variability of the murine creatine kinase B gene locus and related pseudogenes in different inbred strains of mice. Genomics 12: 340–349, 1992

53. Haas RC, Korenfeld C, Zhang Z, Perryman B, Roman D, Strauss AW: Isolation and characterization of the gene and cDNA encoding human mitochondrial creatine kinase. J Biol Chem 264: 2890–2897, 1989

54. Klein SC, Haas RC, Perryman MB, Billadello JJ, Strauss AW: Regulatory element analysis and structural characterization of the human sarcomeric mitochondrial creatine kinase gene. J Biol Chem 266: 18058–18065, 1991

55. Steeghs K, Peters W, Wieringa B: Structure of the murine mitochondrial creatine kinase ubiquitous gene. Unpublished/extracted from GenEMBL sequence data libraries under accession number Z13968, 1992

56. Furukohri T, Fujimoto K, Susuki T: Glycocyamine kinase from the polychaete, *Neanthes diversicolor*. Isolation, purification and tryptic digestion of glycocyamine kinase. Mem Fac Sci Kochi Univ Ser D (Biol) 8: 85–94, 1987

57. Pickover CA, McKay DB, Engelman DM, Steitz TA: Substrate binding closes the cleft between the domains of yeast phosphoglycerate kinase. J Biol Chem 254: 11323–11329, 1979

58. Dumas C, Janin J: Conformational changes in arginine kinase upon ligand binding seen by small-angle X-ray scattering. FEBS Lett 153: 128–130, 1983

59. Morris GE, Cartwright AJ: Monoclonal antibody studies suggest a catalytic site at the interface between domains in creatine kinase. Biochim Biophys Acta 1039: 318–322, 1990

60. Morris GE, Jackson PJ: Identification by protein microsequencing of a proteinase-V8-cleavage site in a folding intermediate of chick muscle creatine kinase. Biochem J 280: 809–811, 1991

61. Wyss M, James P, Schlegel J, Wallimann T: Limited proteolysis of creatine kinase. Biochemistry 32: 10727–10735, 1993

62. Hossle JP, Schlegel J, Wegmann G, Wyss M, Böhlen P, Eppenberger HM, Wallimann T, Perriard J-C: Distinct tissue specific mitochondrial creatine kinases from chicken brain and striated muscle with a conserved CK framework. Biochem Biophys Res Commun 151: 408–416, 1988

63. Perryman MB, Kerner SA, Bohlmeyer TJ, Roberts R: Isolation and sequence analysis of a full-length cDNA for human M creatine kinase. Biochem Biophys Res Commun 140: 981–989, 1986

64. Furter R, Furter-Graves EM, Wallimann T: Creatine kinase: The reactive cysteine is required for synergism but is nonessential for catalysis. Biochemistry 32: 7022–7029, 1993

65. Buskin JN, Jaynes JB, Chamberlain JS, Hauschka SD: The mouse muscle creatine kinase cDNA and deduced amino acid sequences: comparison to evolutionarily related enzymes. J Mol Evol 22: 334–341, 1985

66. Taylor SS, Buechler JA, Yonemoto W: cAMP-dependent protein kinase: framework for a diverse family of regulatory enzymes. Annu Rev Biochem 59: 971–1005, 1990

67. DerTerrossian E, Desvages G, Pradel L-A, Kassab R, van Thoai N: Comparative structural studies of the active site of ATP:guanidine phosphotransferases. The essential cysteine tryptic peptide of lombricine kinase from *Lumbricus terrestris* muscle. Eur J Biochem 22: 585–592, 1971

68. Brevet A, Zeitoun Y, Pradel L-A: Comparative structural studies of the active site of ATP:guanidine phosphotransferases. The essential cysteine tryptic peptide of taurocyamine kinase from *Arenicola marina*. Biochim Biophys Acta 393: 1–9, 1975

69. Mahowald TA: Identification of an epsilon amino group of lysine and a sulfhydryl group of cysteine near the reactive cysteine residue in rabbit muscle creatine kinase. Fed Proc 28: 601, 1969

70. Babbitt PC, Kenyon GL, Kuntz ID, Cohen FE, Baxter JD, Benfield PA, Buskin JD, Gilbert WA, Hauschka SD, Hossle JP, Ordahl CP, Pearson ML, Perriard J-C, Pickering LA, Putney SD, West BL, Zivin RA: Comparisons of creatine kinase primary structures. J Protein Chem 5: 1–14, 1986

71. James P, Wyss M, Lutsenko S, Wallimann T, Carafoli E: ATP binding site of mitochondrial creatine kinase. FEBS Lett 273: 139–143, 1990

72. Kaldis P, Furter R, Wallimann T: The N-terminal heptapeptide of mitochondrial creatine kinase is important for octamerization. Biochemistry 33, 952–959, 1994

73. Gross M, Furter-Graves EM, Eppenberger HM, Wallimann T, Furter R: The tryptophan residues of mitochondrial creatine kinase: Trp-223, Trp-206, and Trp-264 in active site and quarternary structure formation, Prot Sci 3, in press, 1994

74. Yang JT, Chuen-Shang CW, Martinez HM: Calculation of protein conformation from circular dichroism. Meth Enzymol 130: 208–256, 1986

75. Oriol C, Landon M-F: Le dichroisme circulaire de diverses phosphagène phosphotransférases. Biochim Biophys Acta 214: 455–462, 1970

76. Manavalan P, Johnson WC: Variable selection method improves the prediction of protein secondary structure from circular dichroism spectra. Anal Biochem 167: 76–85, 1987

77. Vasak M, Nagayama K, Wüthrich K, Mertens M, Kägi JHR: Creatine kinase. Nuclear magnetic resonance and fluorescence evidence for interaction of ADP with aromatic residue(s). Biochemistry 18: 5050–5055, 1979

78. Kägi JHR, Li T-K, Vallee BL: Extrinsic Cotton effects in complexes of creatine kinase with adenine coenzymes. Biochemistry 10: 1007–1015, 1971

79. Zhou H-M, Tsou C-L: An essential tryptophan residue for rabbit muscle creatine kinase. Biochim Biophys Acta 830: 59–63, 1985

80. Fattoum A, Kassab R, Pradel L-A: The tyrosyl residues in creatine kinase. Modification by iodine. Biochim Biophys Acta 405: 324–339, 1975

81. Villarreal-Levy G, Ma TS, Kerner SA, Roberts R, Perryman MB: Human creatine kinase: Isolation and sequence analysis of cDNA

262

clones for the B subunit, development of subunit specific probes and determination of gene copy number. Biochem Biophys Res Commun 144: 1116–1127, 1987

82. Billadello JJ, Kelly DP, Roman DG, Strauss AW: The complete nucleotide sequence of canine brain B creatine kinase mRNA: Homology in the coding and 3′ noncoding regions among species. Biochem Biophys Res Commun 138: 392–398, 1986

83. Pickering L, Pang H, Biemann K, Munro H, Schimmel P: Two tissue-specific isozymes of creatine kinase have closely matched amino acid sequences. Proc Natl Acad Sci USA 82: 2310–2314, 1985

84. Benfield PA, Henderson L, Pearson ML: Expression of a rat brain creatine kinase-beta-galactosidase fusion protein in *Escherichia coli* and derivation of the complete amino acid sequence of rat brain creatine kinase. Gene 39: 263–267, 1985

85. Hossle JP, Rosenberg UB, Schaefer B, Eppenberger HM, Wallimann T, Perriard JC: The primary structure of chicken B-creatine kinase and evidence for heterogeneity of its mRNA. Nucl Acids Res 14: 1449–1463, 1986

86. Kwiatkowski RW, Ehrismann R, Schweinfest CW, Dottin RP: Accumulation of creatine kinase mRNA during myogenesis: Molecular cloning of a B-creatine kinase cDNA. Dev Biol 112: 84–88, 1985

87. Roman D, Billadello J, Gordon J, Grace A, Sobel B, Strauss A: Complete nucleotide sequence of dog heart creatine kinase mRNA: Conservation of amino acid sequence within and among species. Proc Natl Acad Sci USA 82: 8394–8398, 1985

88. Putney S, Herlihy W, Royal N, Pang H, Aposhian HV, Pickering L, Belagaje R, Biemann K, Page D, Kuby S, Schimmel P: Rabbit muscle creatine phosphokinase: cDNA cloning, primary structure, and detection of human homologues. J Biol Chem 259: 14317–14320, 1984

89. Benfield PA, Zivin RA, Miller LS, Sowder R, Smythers GW, Henderson L, Oroszlan S, Pearson ML: Isolation and sequence analysis of cDNA clones coding for rat skeletal muscle creatine kinase. J Biol Chem 259: 14979–14984, 1984

90. Kwiatkowski RW, Schweinfest CW, Dottin RP: Molecular cloning and the complete nucleotide sequence of the creatine kinase-M cDNA from chicken. Nucl Acids Res 12: 6925–6934, 1984

91. Ordahl CP, Evans GL, Cooper TA, Kunz G, Perriard J-C: Complete cDNA-derived amino acid sequence of chick muscle creatine kinase. J Biol Chem 259: 15224–15227, 1984

92. West BL, Babbitt PC, Mendez B, Baxter JD: Creatine kinase protein sequence encoded by a cDNA made from *Torpedo californica* electric organ mRNA. Proc Natl Acad Sci USA 81: 7007–7011, 1984

93. Payne RM, Haas RC, Strauss AW: Structural characterization and tissue-specific expression of the mRNAs encoding isoenzymes from two rat mitochondrial creatine kinase genes. Biochim Biophys Acta 1089: 352–361, 1991

94. Haas RC, Strauss AW: Separate nuclear genes encode sarcomere-specific and ubiquitous human mitochondrial creatine kinase isoenzymes. J Biol Chem 265: 6921–6927, 1990

95. McCombie WR, Adams MD, Kelley JM, FitzGerald MG, Utterback TR, Khan M, Dubnick M, Kerlavage AG, Venter J, Fields C: *Caenohabditis elegans* expressed sequence tags reveal gene families and potential disease gene homologues. Unpublished/extracted from GenEMBL data library under accession number M79599, 1992

96. Waterston R, Martin C, Craxton M, Huynh C, Coulson A, Hillier L, Durbin R, Green P, Shownkeen R, Halloran N, Metzstein M, Hawkins T, Wilson R, Berks M, Du Z, Thomas K, Thierry-Mieg J, Sulston J: A survey of expressed genes in *Caenorhabditis elegans*. Nature Genet 1: 114–123, 1992

Molecular and Cellular Biochemistry **133/134**: 263–274, 1994.
© 1994 *Kluwer Academic Publishers.*

IV-3 Approaching the multifaceted nature of energy metabolism: inactivation of the cytosolic creatine kinases via homologous recombination in mouse embryonic stem cells

J. van Deursen[1] and B. Wieringa
Department of Cell Biology and Histology, University Nijmegen, The Netherlands; [1] Present address: Department of Cell Biology and Genetics, Erasmus University, Rotterdam

Abstract

To study the physiological role of the creatine kinase/phosphocreatine (CK/PCr) system in cells and tissues with a high and fluctuating energy demand we have concentrated on the site-directed inactivation of the B- and M-CK genes encoding the cytosolic CK protein subunits. In our approach we used homologous recombination in mouse embryonic stem (ES) cells from strain 129/Sv. Using targeting constructs based on strain 129/Sv isogenic DNA we managed to ablate the essential exons of the B-CK and M-CK genes at reasonably high frequencies. ES clones with fully disrupted B-CK and two types of M-CK gene mutations, a null (M-CK⁻) and leaky (M-CK^l) mutation, were used to generate chimaeric mutant mice via injection in strain C57BL/6 derived blastocysts. Chimaeras with the B-CK null mutation have no overt abnormalities but failed to transmit the mutation to their offspring. For the M-CK⁻ and M-CK^l mutations successful transmission was achieved and heterozygous and homozygous mutant mice were bred. Animals deficient in MM-CK are phenotypically normal but lack muscular burst activity. Fluxes through the CK reaction in skeletal muscle are highly impaired and fast fibres show adaptation in cellular architecture and storage of glycogen. Mice homozygous for the leaky M-CK allele, which have 3-fold reduced MM-CK activity, show normal fast fibres but CK fluxes and burst activity are still not restored to wildtype levels. (Mol Cell Biochem **133/134**: 263–274, 1994)

Key words: energy metabolism, creatine kinase B (B-CK), creatine kinase M (M-CK), gene targeting, embryonic stem cells, muscle performance, ^{31}P-NMR

Introduction

Many mammalian cells contain phosphocreatine at a level several fold higher than that of ATP. This serves as an additional reservoir of chemical energy, and the pools of PCr and ATP in a cell can exchange high-energy phosphates via the reaction catalysed by CK isoenzymes: PCr + ADP + (H⁺) ⇌ Cr + ATP. Substrates and enzymes involved in the CK reaction form the so-called CK/PCr system [1–4].

According to the generally held belief, the CK/PCr system (i) serves to buffer changes in ATP and ADP levels at periods of high cellular activity (temporal energy buffering) and (ii) acts in the transport of energy (spatial

Address for offprints: B. Wieringa, Department of Cell Biology and Histology, Faculty of Medical Sciences, University Nijmegen, P.O. Box 9101, 6500 HB Nijmegen, The Netherlands

energy buffering). Other proposed functions of the CK/PCr system are direct consequences of these activities and may include (iii) proton buffering at the onset of ATP hydrolysis, (iv) optimization of the thermodynamic efficiency of ATP synthesis and hydrolysis and (v) constituting a reservoir of anorganic phosphate [reviewed in refs 1–4 and see refs in this issue].

Most, if not all of these functions require delicate regulation of the intracellular distribution of the metabolites and the CK enzymes, especially the cytosolic CK isoforms. CKs, therefore, may be classified as ambiquitous enzymes [5] as they can generally be found at two or more locations in cells with a high and fluctuating energy demand. In turn, the intracellular distribution may represent a regulatory mechanism which results in alteration of kinetic parameters of the PCr-ATP conversion reaction to suit changing needs in energy metabolism.

Five different CK isoenzymes are currently known to exist in mammalian cells and tissues. Three of them are located in the cytoplasm or subcellular compartments and two are found specifically in the mitochondria. The cytosolic CK isoenzymes, BB-, BM- and MM-CK, are dimeric molecules which are build up of two types of CK protein subunits, the M- (for Muscle) subunit and the B- (for Brain) subunit. These two subunits are encoded by individual genes, called the B- and M-CK genes. In man, they span approximately 3.2 kilobase pairs (kbp) and 17.5 kbp, and have been assigned to chromosomes 14 and 19, respectively. The expression of these cytosolic CK genes is tissue-specific and developmentally controlled resulting in typical tissue- and stage-specific isoenzyme patterns [6–19]. High levels of BB-CK homodimers are found for example in very early embryos (morula, blastocysts), in neurons and other cells in the CNS, retina, heart, the thick ascending limb of the loop of Henle in kidney, osteoclasts, uterus and in spermatozoa [ref. 4 and refs therein]. High amounts of BB-CK are also present in smooth muscles, however, here it is coexpressed with BM- and MM-CK isoforms. Moderate BB-CK activity is observed in several tumors, adipose tissue, and white blood cells whereas minimal BB-CK has been reported in many other tissues [4]. It may be not surprising therefore, that the regulation of BB-CK activity is extremely complex. Not only are transcriptional and translational principles involved in its expression, also the modulation of enzymatic activity by protein modification (phosphorylation) and non-covalent protein and membrane associations do play a role. Several reports emphasize the role of B-CK in generating ATP for sodium transport and it has been suggested that B-CK is closely associated to Na$^+$/K$^+$-ATPases in various tissues [20, 21]. B-CK studies may therefore form a paradigm for the study of other enzymes with a role in the ATP regulatory networks.

The distribution of MM-CK activity yields a more simple picture. High activity is exclusively found in mammalian skeletal and cardiac muscle. Recently, low MM-CK expression has also been observed in certain areas of the brain and in epithelium cells of the inner ear. In developing muscle there is a transition from BB-CK and MM-CK expression, characterised by a transitional period where there is a pronounced expression of BM-CK heterodimers [22]. MM-CK dimers predominate in adult heart, but interestingly, small amounts of BM-CK and traces of BB-CK remain expressed during the adult stages. The relative amounts of these CK isoforms are known to undergo changes at different stages of mammalian heart development [4].

Within muscle fibres, the majority of MM-CK dimers are located in the cytoplasm as soluble molecules, another significant fraction is bound to the sarcomeric M line, where it is functionally linked to myosin ATPase. MM-CK is also found in sarcomeric I bands where it is colocalized with the enzymes involved in glycolysis. These glycolytic enzymes function in complexes loosely associated with actin filaments, especially in fast fibers. Moreover, MM-CK is also associated with the sarcoplasmatic reticulum to provide maximal energy for the resorption of Ca^{2+} ions via ATP dependent calcium pumps. Finally, MM-CK is also localized in the sarcolemma, where it functions – like BB-CK – in the regeneration of ATP for sodium-potassium ATPase pumps [reviewed in 4, 23–25].

The two mitochondrial CK isoforms, called ubiquitous and sarcomeric Mi-CK, are located within the mitochondrial intermembrane space. *In vitro*, both Mi-CK isoforms appear in two distinct interconvertible oligomeric forms, dimers and octamers. Both oligomeric molecules are known to catalyse the CK reaction *in vitro*, but evidence has accumulated over the past few years that octamers are the only Mi-CK molecules with catalysing activity in mitochondria of living cells [see 16 and refs therein]. The protein subunits of the two Mi-CK isoenzymes are encoded by two independent nuclear genes, the ubiquitous and the sarcomeric Mi-CK gene, which are located on human chromosomes 15 and 5, span about 5.5 and 37 kbp, and are build up of 9 and 11 exons, respectively [9, 10, 12, 15, 18]. The expression patterns of the ubiquitous and sarcomeric Mi-CK genes generally parallel the expression of the B- and M-CK genes but have not been studied in great detail until now [but see refs Strauss and Payne, this issue].

The observation of a strict spatio-temporal regulation of CK activity during development, the adaptation of intracellular localization in accordance to energy demand, as well as the mere presence of CK isoenzymes in quite diverged species of the animal kingdom suggest an important and crucial biological role for the CK/PCr system. For many of the functional postulates about its significance, however, critical *in vivo* tests have yet to be performed.

A powerful new approach for studying the significance of gene products in their natural context *in vivo*, is to generate CK knock-out mice via targeted recombination in murine ES cells [26–29]. ES cells offer the possibility of genetic modifications by which both gain- and loss-of-function experiments can be performed. In addition, they provide the option for studying the effects of these modifications in the whole animal as well as in *in vitro* model systems for development [30], and at the cellular level. We have decided to use this system to functionally ablate each of the individual CK isoenzymes of mouse, while leaving the other components of the CK/PCr system in this animal intact. The best way to paralyse both the energy buffer and the energy transport function of CK/PCr system by gene targeting is to disrupt the M-CK or the B-CK gene. In theory, this will completely block the CK catalysed transfer of high-energy phosphates from PCr to ADP. Inactivation of mitochondrial CK will prevent production of PCr via the mitochondrial route, but PCr synthesis via CK associated with compartments of glycolytic activity or solubilized in the cytoplasm can still ensue. Therefore, we decided to concentrate primarily on the genes for cytosolic CKs, as the most strategic targets for complete inactivation of the CK/PCr system and report here on their mutagenic inactivation via homologous DNA recombination. Similar experiments involving the inactivation of the Mi-CK genes in ES cells will be reported elsewhere (K. Steeghs *et al.* in preparation). Parameters that may act upon the quite different frequencies of mutagenic inactivation for B-CK and M-CK genes are mentioned briefly. For MM-CK, mice with a complete enzyme deficiency were obtained. Unanticipated findings regarding the phenotypic appearance of mice deficient in M-CK and other results are presented in condensed format as they have described in detail elsewhere [31–34 and van Deursen *et al.* submitted].

Materials and methods

Construction of the B-CK targeting vectors

As starting material recombinant phage DNAs containing the conplete B-CK gene of two different mouse strains were used. Isolation of the B-CK genomic DNAs proved difficult and several libraries, including a Balb/c-derived lambda phage library were screened unsuccessfully. Ultimately, screening of a genomic library of mouse hybrid [CBA × C57BL/6] in lambda FIX II that was plated on the PLK17 (mrcA-) *E. coli* host strain yielded one type of hybridizing phage containing the entire functional B-CK gene. Based on typing of diagnostic strain specific [TG]-repeat length variation we concluded that the cloned B-CK gene originated from inbred strain CBA [33]. Furthermore, by using an intron 5 specific B-CK probe of the CBA-derived gene, another B-CK genomic DNA was isolated from a mouse strain 129/Sv derived genomic library (again in low yield).

The replacement vector 129-pRV7.0 was derived from a 7.4 kbp genomic ApaI-HincII fragment originating from the 129/Sv genomic DNA phage [33]. An allelic 8.6 kbp ApaI/ClaI segment from the CBA-derived phage served as basis for the construction of vector CBA-pRV8.5. These segments contain all exon sequences of the functional B-CK gene. To construct 129-pRV7.0, a 401-base pair (bp) SmaI fragment, spanning exon 2 and part of intron 1 and 2 [33], was replaced by a 1.2 kbp SmaI-HincII fragment carrying the neo® resistance cassette [27]. Moreover, a 2.0 kbp herpes simplex virus thymidine kinase (hsv-tk) gene cassette was attached to the 3′end of the B-CK gene as a negatively selectable marker. In this cassette the polyoma enhancer PyF441 is driving the expression of the tk gene which carries the polyadenylation signal of pMCneo-polyA [27]. To construct CBA-pRV8.5 a 161 bp SacII fragment spanning part of exon 3 and intron 3 was replaced by a 1.2 XhoI/BamHI fragment. The hsv-tk gene cassette was attached to the 5′end of the 8.6 kbp ApaI-ClaI fragment. The resulting vectors are detailed in Fig. 1. For all cloning steps standard methodology was used [35].

Construction of M-CK targeting vectors

The replacement vector 129-pRV8.3 was derived from a 9.0 kbp genomic BamHI fragment spanning exons 1–5 of the M-CK gene, that originated from a mouse inbred

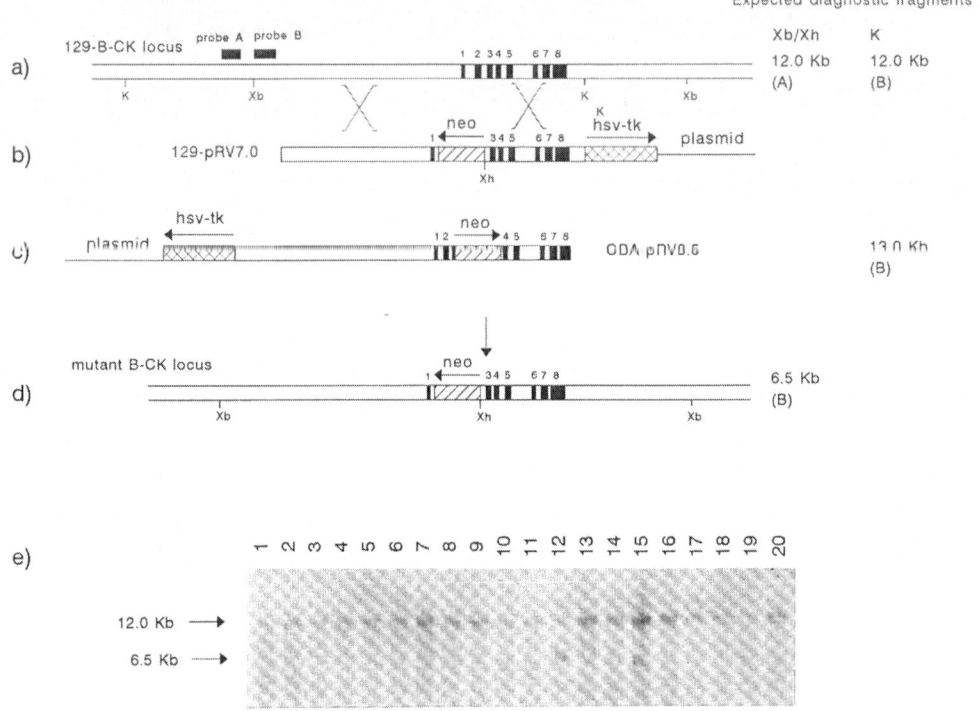

Fig. 1. Schematic diagram showing (a) the structure of the B-CK locus in the 129/Sv mouse inbred strain, (b) and (c) the mouse strains 129/Sv and CBA derived targeting vectors 129-pRV7.0 and CBA-pRV8.5, respectively, and (d) the predicted structure of the B-CK locus after targeted mutagenesis. Only the mutant B-CK after homologous recombination with the 129-pRV7.0 vector is shown, as the CBA-pRV8.5 vector failed to yield homologous recombinants. (e) Southern blot analysis of XbaI/XhoI doubly cleaved genomic DNAs from twenty randomly chosen ES cell clones which were transfected with the 129-pRV7.0 vector DNA. One single clone resulted from a homologous recombination event (diagnostic 6.5 kbp fragment in lane 12). To the right, the anticipated sizes of XbaI/XhoI or KpnI restriction fragments are given. More detailed sequence information on the strain specific differences in B-CK genomic DNAs can be found in ref. 33.

strain 129/Sv genomic DNA library in phage EMBL3 [34]. A 0.9 kbp EcoRI-EcoRV fragment, encompassing exon 2 and part of introns 1 and 2, was removed and replaced by a 2.0 kbp hygroB® cassette [34]. Furthermore, a 2.0 kbp hsv-tk cassette was attached to the 3′ end of the M-CK gene. BALB/c-pRV8.3 was constructed similarly, except that the 9.0 kbp BamHI fragment originated from a mouse strain Balb/c genomic DNA library. To enable the typing of targeting events by Southern blot analyses, the unique XhoI site in the 5′ polylinker sequence of the hygroB® cassette in Balb/c-pRV8.3 was replaced by an EcoRV site. All further details regarding these two constructs have been described before [34].

Electroporation and tissue culture

129-RV7.0, CBA-RV8.5, 129-RV8.3 and Balb/c-pRV8.3 vector DNAs were prepared for transfection, purified and linearized with restriction enzyme HincII, ClaI, BamHI and BamHI, respectively, phenol-chloroform

extracted again, ethanol precipitated and suspended in H_2O at a concentration of 0.5–1.0 µg/µl. Targeting vector DNA was introduced into AB-1 ES cells (kindly provided by Dr. A. Bradley, Baylor College of Medicine, Houston, USA) by electroporation with a Ta750 transfection apparatus (Krüss GmbH Hamburg, Germany). Batches of about 4–8.10^6 AB1 cells were suspended in 0.4 ml electroporation buffer [10 mM potassium phosphate (pH 7.1)/0.28 M sucrose/1 mM $MgCl_2$/200 µg/ml BSA (Boehringer Mannheim GmbH)] and mixed with 10 µg/ml vector DNA. The cells were given 24 pulses [17 µs each, 1.5 kV/cm], incubated for 1 min at room temperature, and were then plated onto a 9 cm tissue culture dish carrying SNLH9 feeder cells in 10 ml of medium (Dulbecco's modified Eagle's medium with 15% fetal calf serum, 2 mM glutamine, 1 mM sodium pyruvate and 0.1 mM β-mercaptoethanol). SNLH9 feeder cells were clonally derived from cell line SNL76/7 after transfection with vector pG3-hygroB DNA [32, 34] containing the hygroB® gene [34]. 24 hr after electroporation ES cells were refed with medium containing 140 µg/ml hygromycin B (dry powder, Sigma) and 0.2 µM FIAU (1-[2-deoxy, 2-fluoro-β-D-arabinofuranosyl, a kind gift of Bristol Myers). Eight to ten days after transfection individual colonies were picked and expanded on 60% (v/v) BRL-conditioned medium (to remove SNLH9 feeder cells) for DNA preparation and storage. To determine the negative selection enrichment factor, cells were first grown for ten days on medium containing 250 µg/ml G418 or 140 µg/ml hygromycin B only and then colonies were scored. Subsequently, individual distinct colonies were picked and grown on G418 of hygromycin B plus FIAU medium for 1 week and then scored again by Southern blot analysis.

Southern blot analyses of targeted clones

For identification of B-CK mutants, genomic DNAs (approximately 2 µg each) of individual clones were digested with diagnostic restriction enzymes (XhoI/XbaI for 129-pRV7.0 and KpnI for CBA-pRV8.5) and DNA fragments were resolved on a 0.7% agarose gel, transferred to Hybond and hybridized with ^{32}P-labeled 5′ external probes (probe A, a 300-bp RsaI fragment for 129-pRV7.0 and probe B, a 500 bp SacII-HincII fragment, for CBA-pRV8.5, see Fig. 1). For identification of M-CK mutants, genomic DNAs were subjected to restriction enzyme digestion with KpnI, EcoRV and XbaI/XhoI. Each digest was fractionated on a 0.7% agarose gel, transferred to Hybond (Amersham), and hybridized with ^{32}P-labeled probes as previously described [31–34]. After hybridization, blots were washed at a final stringency of 0.05 × SSC, 0.1% SDS at 65° C and exposed to X-O mat film for 3–5 days at −70° C.

Generation of chimaeric mice

8–12 ES cells of the targeted clones were injected into the blastocoele cavity of C57BL/6 embryos at 3.5 days post coitum (dpc). Injected embryos were allowed to recover 30–120 min (37° C) before transfer into the uteri horns of pseudopregnant hybrids (CBA × C57BL/6) at 2.5 dpc [28].

Results

Targeting of the B-CK gene in murine ES cells
The chromosomal structure of the mouse B-CK gene is

illustrated in Fig. 1A. To disrupt the B-CK gene, we initially constructed the replacement vector CBA-pRV8.5 (Fig. 1C). In this vector part of exon 3 and intron 3 were replaced by a 1.2 kbp fragment containing the neo® selection marker. Targeting with this vector was completely unsuccessful, as no recombinants out of 1328 colonies screened were found. Therefore, another vector, 129-pRV7.0 (Fig. 1B), based on isogenic strain 129/Sv DNA, was constructed. This latter vector includes the neo® cassette flanked by 4.5 kb of 5′ homology and 2.5 kb of 3′ homology to the endogenous B-CK gene. To allow selection against random integration, a hsv-tk gene driven by the MC1 promoter was added to the 3′ end of this construct [27]. Appropriate targeting would result in inactivation of the B-CK gene, as part of intron 1 and 2 and exon 2 – containing the ATG translation start codon – are deleted from the vector. 5′ linearized vector DNA was electroporated into AB-1 cells followed by double selection in the presence of the drugs G418 and FIAU [27, 34]. A small part of the transfected cells was selected exclusively with G418 to assess the enrichment factor. We estimate that negative selection with FIAU yielded an approximately 3-fold enrichment for targeting events. DNAs of 304 doubly resistant transformants were digested with two restriction enzymes, XhoI and XbaI, and analyzed on Southern blots to identify clones carrying a targeted replacement of the B-CK locus. Hybridization was performed with a probe in the 5– end of the B-CK gene, a chromosomal region that was not included in the targeting vector. With this probe, homologous recombinants were expected to show both the wild-type 12.0 kbp XbaI fragment as well as a mutant 6.5 kbp XbaI/XhoI fragment (see Fig. 1E). The analysis revealed that 11 (frequency about 3–4%) cell clones contained the replacement of one wild-type allele by the mutant incoming DNA. A typical example of the Southern blot screening procedure is depicted in Fig. 1E and shows one of the targeted clones (AB-1-295; lane 12) amongst 19 nontargeted ES cell lines.

Targeting of the M-CK gene in ES cells
The two types of – mouse strain specific – M-CK targeting vectors depicted in Fig. 2B and C have been described in detail elsewhere [34]. BamHI linearized targeting constructs were introduced into AB-1 ES cells and simultaneous selection in hygromycin B and the pyrimidine derivative FIAU yielded colonies which were clonally expanded, and used for isolation of genomic DNA to identify homologous recombinants by Southern blot analyses. Correctly targeted clones presented

268

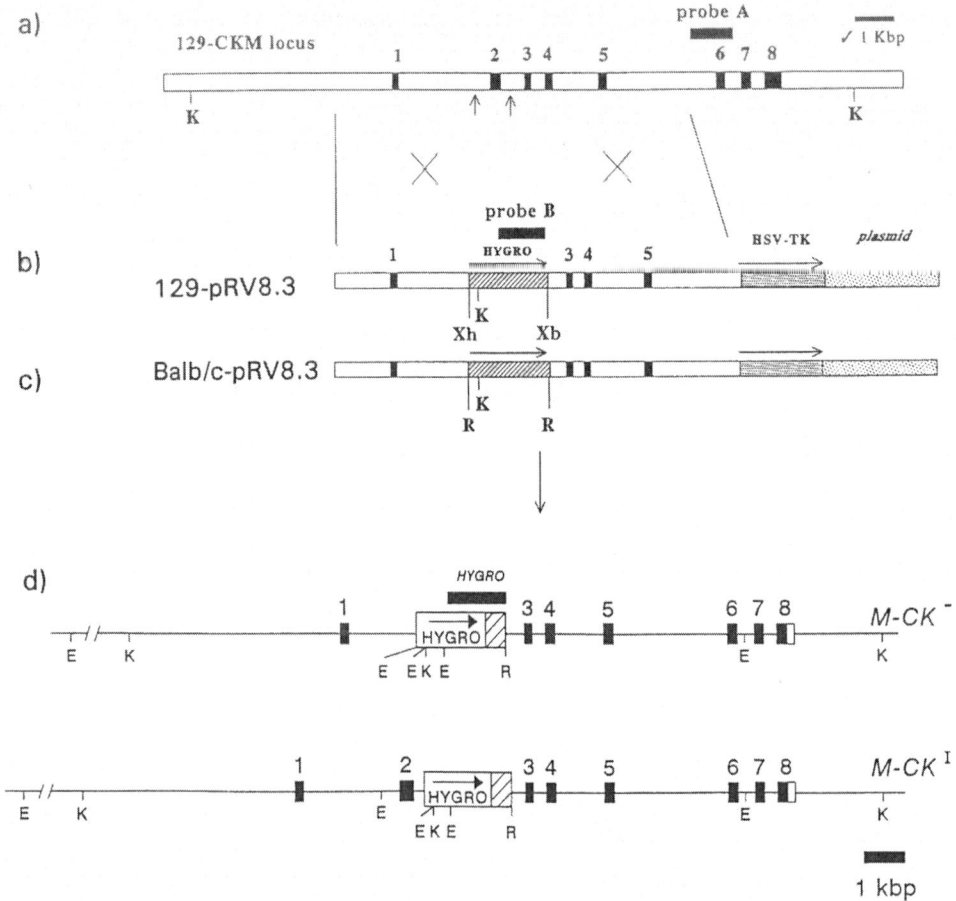

Fig. 2. Schematic diagram showing (a) the structure of the M-CK locus in the 129/Sv mouse inbred strain, (b) and (c) the mouse strains 129/Sv and Balb/c derived targeting vectors 129-pRV8.3 and Balb/c-pRV8.3, respectively. In (d) the anticipated structures of the M-CK loci as present in the M-CK⁻ null mutant (top) and the M-CK^I partial mutant (bottom) are shown. Note that in the M-CK^I mutant the hygro® cassette does not replace the second exon but is located entirely within the gene's second intron.

both a diagnostic 10.7 kbp KpnI mutant fragment as well as the wildtype 18.0 kbp fragment [34]. In addition one clone with a somewhat different mutated M-CK locus, called M-CK^I, that originated from an unexpected insertion instead of a replacement event, was identified (Fig. 2D). In two experiments with 129-pRV8.3, we scored a total of 25 independent homologous integration events out of 209 colonies screened (frequency 12%). In contrast, in three individual experiments with construct Balb/c-pRV8.3, amongst 357 hygroB® and FIAU® colonies analyzed in total, no homologous recombinants were observed.

Generation of B-CK mutant chimaeras and their phenotypic characterisation

Cells of all 11 clones were injected into 3.5 day-old C57BL/6 blastocysts (24–40 embryos per clone). Injected embryos were transferred to pseudopregnant foster mothers. The number of offspring per clone was above

8. Two clones, AB-1-114 and AB-1-176 gave rise to chimaeras with an ES cell contribution, as estimated by the amount of agouti pigment in the coat, that ranged from 5 to 85%. The chimaeras obtained from the other 9 clones injected had less than 50% agouti coat pigment and were not analysed further. To screen for germline transmission, two male chimaeras of clone AB-1-114 (~ 65% and ~ 80% agouti), and 4 male chimaeras of AB-1-176 (~ 50%, ~ 65%, ~ 75% and ~ 85% agouti) were mated to C57BL/6 females until approximately 50 pups were born of each male. Unfortunately, none of the litters contained progeny with the ES cell-derived agouti pigmentation. Hence, we cannot draw any conclusion about the possible detrimental effects of heteryzogous or homyzogous B-CK mutations. However, as overtly normal chimaeras with a very high percentage of mutant cells were identified, we expect the ablation of one B-CK gene to have – if any – one minor phenotypic effects on growth and development. Furthermore, we conclude

that pluripotent stem cells are not inhibited in their developmental program if one of their B-CK genes is missing.

Generation of M-CK deficient mice and their phenotypic characterization

Altogether thirteen of the M-CK mutant ES cell lines were injected into C57BL/6 host blastocysts. One clone in which the pertinent exon 2 region of the M-CK gene was fully deleted, generated offspring with extensive coat colour chimerism that transmitted the coat colour marker through the germline [see for details ref. 31]. Southern blot analysis of DNA isolated from tail tips revealed that, as expected, approximately 50% of the offspring carried one disrupted M-CK and one wildtype allele. These heterozygous mutants were bred to generate homozygous M-CK deficient mice. From the inspection of litter sizes and the general appearance of the animals we conclude that homozygous M-CK mutant mice were indistinguishable from their heterozygous and wildtype littermates and displayed no overt skeletal muscular or behavioural abnormalities (see photograph in Fig. 3). Homozygous mice, when interbred, are fertile and give rise to litters of normal size indicating that there is no loss in utero or death directly after birth. Similar results were obtained with crosses on a 129/Sv inbred genetic background. Northern blot analysis and electrophoretic zymogram analysis on cellulose-acetate strips revealed the complete lack of M-CK mRNA and the complete ablation of M-CK enzymatic activity in our mice [given in ref. 31]. A more detailed screening of skeletal muscle and heart for alterations in levels of other CK mRNAs and proteins showed that the lack of M-CK did not result in compensatory expression of any of the other, i.e. brain specific 'cytosolic' B-CK or Mi-CK, isoforms.

Also from the M-CK[I] mutation, that resulted from targeted insertion of the hygroB® cassette in the second M-CK intron while leaving the endogenous exon 2 segment intact (Fig. 2D, bottom line), germline chimaeras were obtained. Homozygous M-CK[I] mice typically express about 1/3rd of the MM-CK activity of a normal wild type mouse in muscle and heart tissues. A more detailed account on the characterisation and use of these animals will be given elsewhere (van Deursen *et al.* submitted).

Energy metabolites and energy flow in resting muscle of M-CK deficient mice

The bioenergetic status of resting skeletal muscles of mice with lowered or absent MM-CK activity was deter-

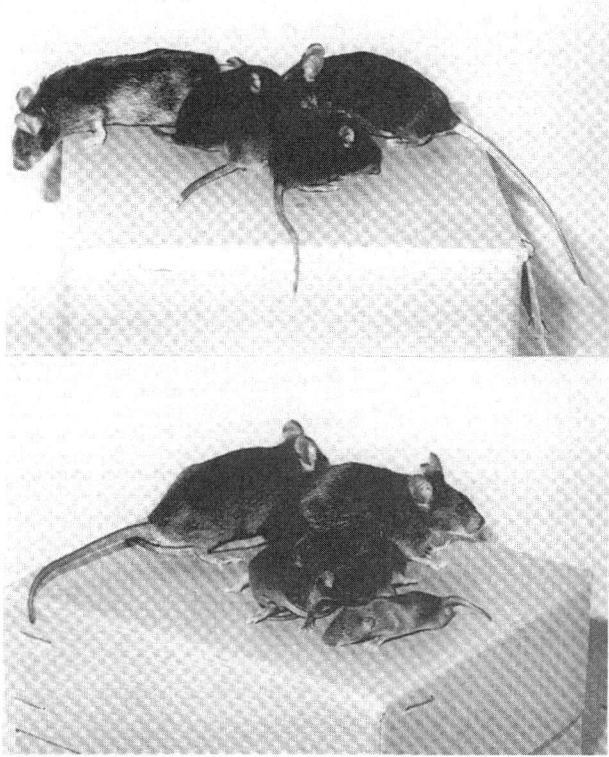

Fig. 3. One chimaeric male carrying the M-CK⁻ mutation, one normal C57BL/6 female and their progeny (two C57BL/6 and one heterozygote M-CK⁺/⁻male). Bottom Phenotypic normal heterozygote and homozygote 129/Sv M-CK⁻ mutant mice and their phenotypic normal offspring.

mined by chemical methods. The absolute levels of ATP in gastrocnemius-plantaris-soleus (GPS) muscle groups in these mutants were not significantly different compared to controls (Table 1). High-energy phosphates in resting skeletal muscle of anesthetized mice were also studied by means of ^{31}P-NMR spectroscopy [36–42]. Typical fully relaxed ^{31}P-NMR spectra (8 s interpulse delay) of wildtype, M-CK[I/I] (not shown) and M-CK⁻/⁻ upper hind limb muscles (both representative for eight experiments performed in each group) revealed that the relative peak areas of the PCr, ATP and P$_i$ resonances of mutant and normal mice were indistinguishable (see Fig. 4). If skeletal muscle ATP levels measured chemically and by *in vivo* ^{31}P-NMR are assumed to be similar it follows that the concentrations of free ATP, PCr and P$_i$ appear to be normal in all animals tested.

We measured the reverse rate of the CK reaction in intact skeletal muscles at rest by means of ^{31}P-NMR inversion transfer (PCr inversion) technique [31]. In these experiments we found that the area of the ATP γ-phos-

270

*Table 1.*Phenotypic comparison of muscles of M-CK deficient mice

Mode of typing	Control	M-CK$^{-/-}$	M-CK$^{l/l}$
Biological:			
– M-CK expression (protein/mRNA)	100%	0%	± 35%
Physical (NMR):			
– P$_i$	5.8 ± 1.7	4.6 ± 1.3	5.1 ± 0.9
– PCr	44.9 ± 2.9	46.7 ± 0.6	46.6 ± 2.2
– α/β/γ-ATP	16.4 ± 1.6	16.2 ± 2.0	16.1 ± 1.4
– P-transfer	100%	< 5%	< 5%
Chemical:			
– ATP	100%	90–100%	90–100%
– Glycogen	100%	160%	100%
Histological:			
– Type 1, 2A/2B fibre distribution	normal	normal	normal
– Mitochondrial content/volume	100%	150–180%	100%
– Intermyofibrillar mitochondria in type 2 fibres	±	+++	±
Physiological:			
– Initial twitch force	normal	normal	normal
– Burst activity	normal	lowered	intermediate
– Endurance performance (steady state level)	normal	increased	not tested

All phenotypings were performed on hind limb muscles. Values are given as % of normal values (100% in wildtype) or as subjective scores in wording (normal, increased, lowered) or symbols (± or +++). NMR data are given as percentages and relate the relative peak areas of P$_i$, PCr, and ATP to the sum of peak areas in resonance spectra as shown in Fig. 4. (ATP values are means of the α, β and γ-ATP areas; P-transfer = rate of high-energy phosphate exchange through the CK reaction). For chemically determined ATP and glycogen 100% levels were 7.88 ± 0.50 µmol/g wet weight (n = 7) and 26.1 ± 5.8 µmol glucose/g wet weight (n = 6), respectively. Further quantitative details can be found in ref. 31 and van Deursen *et al.* submitted.

phate resonance remained unaffected over a longer period of time (Fig. 5) in MM-CK deficient animals. As anticipated, in the converse γ-ATP inversion transfer experiment (γ-ATP inversion), also no flux of labeled high-energy phosphates from γ-ATP to PCr was seen in M-CK$^-$ mice. Surprisingly, in M-CK$^{l/l}$ mice, high-energy phosphate exchange between PCr and ATP was also completely undetectable (at least 20-fold reduced compared to wild type; data not shown). These magnetization transfer experiments show that the fast exchange of potential energy between PCr and ATP as observed in wild type muscles, normally comes entirely from the MM-CK catalyzed reaction. The role of Mi-CK isoforms in this process can therefore be neglected. Moreover, low levels of M-CK apparently do not suffice to restore the CK fluxes to normal.

Increased mitochondrial and glycogen content in M-CK lacking muscles

Data regarding the further (ultra)structural-histological and histo-chemical analysis of the GPS group and the heart ventricle in M-CK deficient mice have been described in detail elsewhere [31]. Briefly, the size (mean diameter) and the distribution of the three different fibre type populations was identical in gastrocnemius, plantaris and soleus muscles of M-CK deficient and wildtype mice. Also M-CK$^{l/l}$ mice had wildtype appearance. However, we noticed that type 2A and 2B fibers of mice completely lacking M-CK had a clearly increased intermyofibrillar mitochondrial activity reflected in the identification of conspicuous dense rows of mitochondria, located between individual myofibrils (not shown). Such rows, which normally are characteristic for small sized slow-twitch fibres, are exceptional in fast-twitch fibres of wildtype animals, where mostly isolated mitochondria are noticed between myofibrils. In contrast, the mitochondrial distribution in type 1 fibres of M-CK lacking and wildtype mice was similar. We found glycogen content and mitochondrial reference enzymes that act in the Krebs cycle and the respiratory chain to be 60–100% higher in a homogenate of M-CK deficient upper hind limb muscles as compared to control muscles. The respiratory rates paralleled this trend, indicating that in general the aerobic ATP producing capacity of the entire M-CK deficient upper hind limb musculature was about two-fold enlarged. Histochemical staining (Schiffs-base, PAS) showed that the increase in glycogen was found in type 2A and 2B muscle fibres only. No impairment of anaerobic generation of energy in these fibres was noticed in any of the mice with complete or

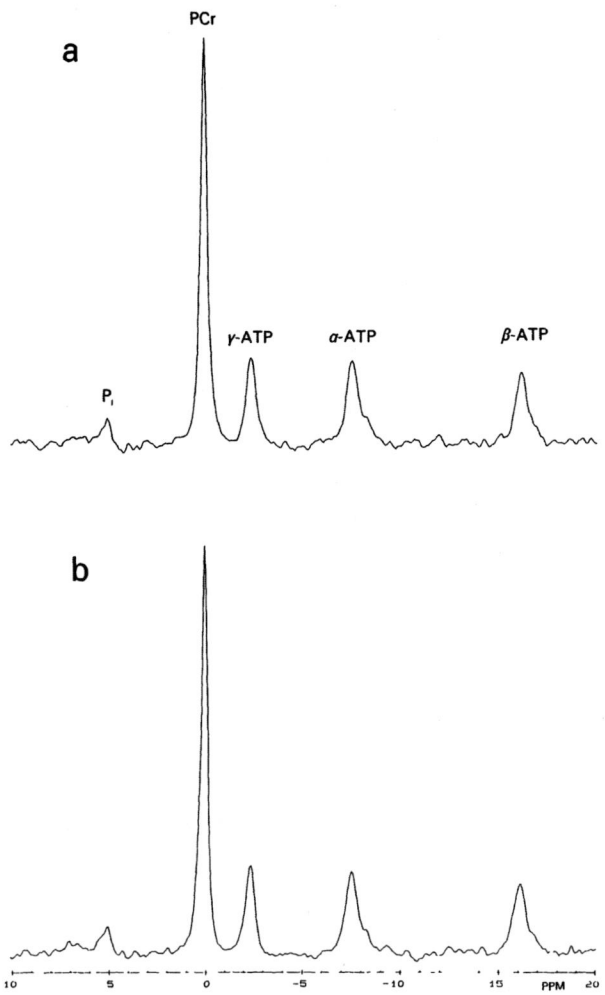

Fig. 4. Examples of ³¹P NMR spectra of hind leg muscles of wildtype (a) and M-CK deficient mice (b) at rest. Each spectrum is the average of 128 free induction decays excited with 70° magnetization pulse angles at 15 s intervals. Resonance assignments are indicated in the figure.

partial lack of M-CK [31]. Conspicuously, in the M-CK^{I/I} mutants with partial M-CK deficiency, levels of mitochondrial indicator enzymes (cytochrome C oxidase; citrate synthase) were not significantly altered compared to wildtype.

M-CK deficient muscles have adaptations for endurance exercise

One of the most significant differences between M-CK mutant and wiltype muscles was in the muscle performance, in the contractile characteristics of the GPS muscle group. Though the initial isometric twitch force of the very first contraction was not significantly different between mutant and control muscles there was a very quick decline in twitch force immediately after the onset of stimulation (1 or 5 Hz) in mutant muscle [31]. The loss of twitch force was most steep at high-intensity stimulation and occurred always within the first 4–8 contrac-

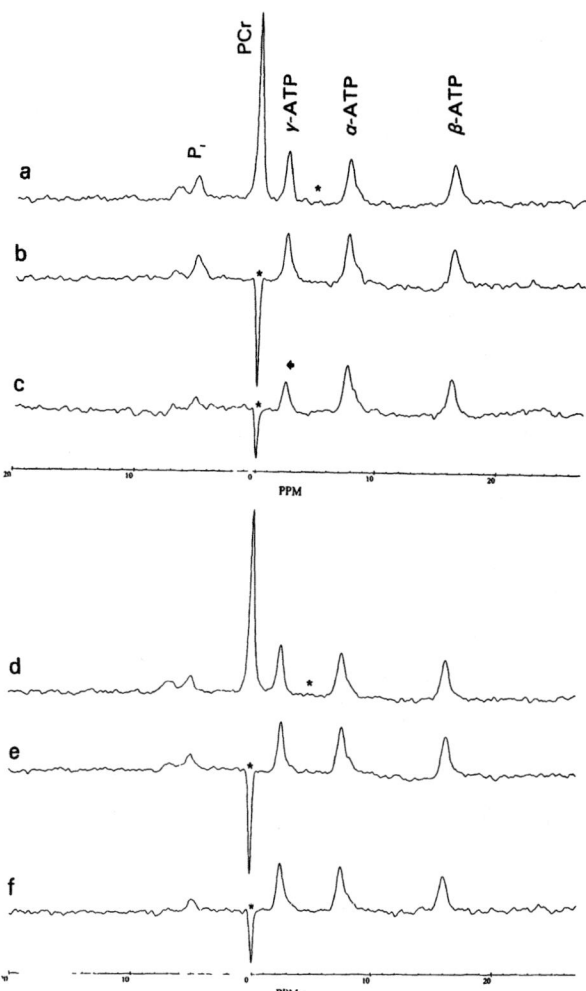

Fig. 5. ³¹P NMR PCr inversion experiments on mouse upper hind leg muscles. Representative examples of inversion transfer experiments (average of 96 scans) on wildtype (a–c) and M-CK deficient muscles (d–f). * indicates the centra frequency of the inversion. (a–d) Inversion controls for PCr (300 ms delay time). (b and e) Inversion of PCr sampled 0.5 ms after the inversion pulse. (c and f) Inversion of PCr sampled 300 ms after the inversion. Note the high reduction of the γ-ATP peak area 300 ms after inversion of PCr in wildtype muscles, due to transfer of inverted P nuclei from PCr to ATP (the arrow marks the γ-ATP level before inversion of PCr). In contrast, the γ-ATP peak did not reduce in case of M-CK deficient muscles.

tions (depending on the intensity of stimulation). When muscle force in mutants recovered and remained at steady-state levels under continuous stimulation, the performance of mutant muscle was significantly better. Ultimately, steady state twitch force levels in M-CK muscles were 10–15% higher than in controls upon continuous stimulation (both at 1 or 5 Hz). M-CK^I mutant mice showed an intermediate phenotype with less rapid initial decline in twitch force and intermediate steady-state force levels (Table 1). To us these considerable reductions of initial muscle force after every single twitch suggest that the thermodynamic efficiency of ATP hydroly-

272

sis might be readily affected by inappropriate local ATP/ADP ratios occurring early during work performance [see 31 and van Deursen *et al.* submitted for more detail).

Discussion

Gene activation by homologous recombination is the technique of choice to study gene function in the context of the living cells and whole organisms. Two major problems in gene targeting and ES cell technology are often encountered, namely (i) a low efficiency of homologous recombination and (ii) inability of ES cells to colonize the germline following a gene targeting procedure. As reported now by several groups including ours [34] the first problem can be minimalized if constructs are used which are derived from isogenic DNA, i.e. DNA from the same inbred mouse strain as from which the ES cells are derived. Also the use of vectors carrying sufficient homology to the endogenous locus and the implementation of positive/negative selection procedures [27, 29] does help to alleviate the problems.

Yet, though the overall characteristics of our B-CK and M-CK targeting vectors were not highly different, we observed a conspicuous difference in targeting frequency between these two loci. Despite the fact that the B-CK is highly expressed in ES cells [van Deursen, unpublished and see also 43] whereas the M-CK is fully repressed at this stage, targeting frequencies on average were 3–5 fold better for the M-CK gene. We conclude that there is no preference for recombination at actively transcribed loci, at least not for the members of the CK family of genes, and other factors like the chromosomal context could play a role.

The second problem mentioned, that of failure in germline transmission was only observed with our B-CK chimaeras. Certainly, the choice and maintenance of an ES cell line with good germline potential is very important, but as we had no problems in obtaining transmitting M-CK chimaeras we do not know how our handlings during growth and selection of the AB-1 ES cells may have caused the specific problems with B-CK. As an additional – and not very likely – possibility we have to consider that ablation of one B-CK allele does not influence somatic differentiation but specifically affects the formation and differentiation of germ cells in our chimaeras. Human patients which carry cytogenetically visible (i.e. large) chromosome 14q deletions spanning the B-CK gene and surrounding regions have been described (D. Cox, personal communication). To our knowledge, indeed no severe consequences of the loss of one entire B-CK gene have been reported for these patients. For the time being, therefore, the most trivial explanation is that our 11 ES cell clones simply lost the ability for germline contribution and more clones should be tested before definite conclusions can be drawn.

More definite statements can be given on the biological relevance of the M-CK gene and its product. In contrast to our expectation, M-CK deficient mice were completely normal and only lacked the ability to sustain short periods of maximal muscle activity. Our oldest animals are now well over one year old and therefore also longevity seems to be unchanged in those mutants that have not been sacrificed for phenotyping. Strikingly, any adverse effects of the MM-CK absence were counterbalanced in muscle by adaptations in the intermyofibrillar mitochondrial content and the glycogen level. These adaptations were most significant in type 2A and 2B, fast fibres, showing the high plasticity of muscle architecture to adapt to changes in metabolic pathways.

We have shown that the combined approach via reverse genetics and ES cells can provide us with new and unanticipated insight in the complex parameters of the CK/PCr system. Gradual decrease in MM-CK levels gives a gradual loss in the ability to sustain maximal muscle output during short periods of high muscle work. Similar approaches as reported here are currently used to inactivate the Mi-CK genes of mouse in our laboratory and have been reported by others [44, and see 29 and refs therein]. Ultimately, this should enable us to generate mice with all possible defects - or combinations thereof – in the four-membered CK gene family. In turn, these animals then can serve as models for studies as detailed elsewhere in this issue.

Acknowledgements

The authors thank David Iles, Arend Heerschap, Jan Schepens, Frank Oerlemans, Wim Ruitenbeek and Paul Jap for help and encouragement and for assistance in the studies and preparation of this manuscript. This work was supported by a program grant from the Dutch Organization for Scientific Research (Medical Sciences).

References

1. Bessman SP, Geiger PJ: Transport of energy in muscle: the phosphorylcreatine shuttle. Science 211: 448–452, 1982

2. Bessman SP, Carpenter LC: The creatine-creatine phosphate energy shuttle. Ann Rev Biochem 54: 831–862, 1985

3. Saks VA, Rosenshtraukh LV, Smirnov VA, Chazov IE: Role of creatine phosphokinase in cellular function and metabolism. Can J Physiol Pharmacol 56: 691–706, 1978

4. Wallimann T, Wyss M, Brdiczka D, Nicolay K, Eppenberger HM: Intracellular compartmentation, structure and function of creatine kinase isoenzymes in tissues with high and fluctuating energy demands: the 'phosphocreatine circuit' for cellular energy homeostasis. Biochem J 281: 21–40, 1992

5. Wilson JE: Ambiquitous enzymes: Variation in intracellular distribution as a regulatory mechanism. TIBS 3: 124–125, 1978

6. Benfield PA, Graf D, Korolkoff PN, Hobson G, Pearson ML: Isolation of four rat creatine kinase genes and identification of multiple potential promoter sequences within the rat brain creatine kinase promoter region. Gene 63: 227–243, 1988

7. Buskin JN, Jaynes JB, Chamberlain JS, Hauschka SD: The mouse muscle creatine kinase cDNA and deduced amino acid sequence: comparison to evolutionarily related enzymes. J Mol Evol 22: 334–341, 1985

8. Schlegel J, Zurbriggen B, Wegmann G, Wyss M, Eppenberger HM, Wallimann T: Native mitochondrial creatine kinase forms octameric structures. Isolation of two interconvertible mitochondrial creatine kinase forms. Dimeric and octameric mitochondrial creatine kinase: Characterization, localization, and structure function relationships. J Biol Chem 32: 16942–16953, 1988

9. Haas RC, Korenfeld DC, Zhang Z, Perryman MB, Roman D, Strauss AW: Isolation and characterisation of the gene and cDNA encoding human mitochondrial creatine kinase. J Biol Chem 265: 6921–6927, 1989

10. Haas RC, Strauss AW: Separate nuclear genes encode sarcomere-specific and ubiquitous human mitochondrial creatine kinases. J Biol Chem 265: 6921–6927, 1990

11. Quest AFG, Eppenberger HM, Wallimann T: Purification of brain-type creatine kinase (B-CK) from several tissues of the chicken: B-CK subspecies. Enzyme 41: 33–42, 1989

12. Klein SC, Haas RS, Perryman MB, Billadello JJ, Strauss AW: Regulatory element analysis and structural characterization of the human sarcomeric mitochondrial creatine kinase gen. J Biol Chem 266: 18058–18065, 1991

13. Mariman ECM, Broers CAM, Claesen CAA, Tesser GI, Wieringa B: Structure and expression of the human creatine kinase B gene. Genomics 1: 126–137, 1987

14. Perryman MB, Kerner SA, Bolhmeyer TJ, Roberts R: Isolation and sequence analysis of a full-length cDNA for human M creatine kinase. Biochem Biophys Res Commun 140: 981–989, 1986

15. Payne RM, Haas RC, Strauss AW: Structural characterization and tissue-specific expression of the mRNAs encoding isoenzymes from two rat mitochondrial creatine kinase genes. Biochim Biophys Acta 1089: 352–361, 1991

16. Wyss M, Smeitink J, Wevers RA, Wallimann T: Mitochondrial creatine kinase: a key enzyme of aerobic energy metabolism Biochim Biophys Acta 1102: 119–166, 1992

17. Povey S, Inwood M, Tanyar A, Bobrow M: The expression of human creatine kinase isoenzymes in human cultured cells. Ann Hum Genet 43: 15–26, 1979

18. Stallings RL, Olson EAWS, Thompson LH, Bachinski L, Siciliano MJ: Human creatine kinase genes on chromosomes 15 and 19 and proximity of the gene for the muscle form to the genes for apolipoprotein C2 and excision repair. Am J Hum Genet 43: 144–151, 1988

19. Smeets H, Bachinski L, Coerwinkel M, Schepens J, Hoeijmakers J, van Duin M, Grezschik K-H, Weber CA, de Jong P, Siciliano MJ, Wieringa B: A long-range restriction map of the human chromosome 19q13 region: Close physical linkage between CKMM and the ERCC1 and ERCC2 genes. Am J Hum Genet, 1990

20. Korge P, Byrd SK, Campbell KB: Functional coupling between sarcoplasmic-reticulum bound creatine kinase and Ca-ATPase. Eur J Biochem 213: 973–980, 1993

21. Friedman FL, Roberts R: Purification and localization of brain-type creatine kinase in sodium chloride transporting epithelia of the spiny dogfish squalus acanthias. J Biol Chem 267: 4270–4276, 1992

22. Perriard JC, Rosenberg UB, Wallimann T, Eppenberger HM, Caravatti M: The switching of creatine kinase gene expression during myogenesis. In: ML Pearson, HF Epstein (eds) Muscle Development Molecular and Cellular Control. Cold Spring Harbor, New York, 1982, pp 237–245

23. Bessman SP, Yang WCT, Gieger PJ, Erikson-Viitanen S: Intimate coupling of creatine phosphokinase and myofibrillar ATPase. Biochem Biophys Res Commun 96: 1414–1420, 1980

24. Ventura-Clapier R, Saks VA, Vassort G, Lauer C, Elizarova G: Reversible MM-creatine kinase binding to cardiac myofibrils. Am J Physiol 253: C444–C455, 1987

25. Saks VA, Lipina NV, Sharov VG, Smirnov VN, Chazov E, Grosse R: The localization of the MM isozyme of creatine phospho-kinase on the surface membrane of myocardial cells and its functional coupling to ouabain-inhibited (Na/K) ATPase. Biochem Biophys Acta 465: 550–558, 1977

26. Capecchi MR: The new mouse transgenetics: altering the genome by gene targeting. Trends in Genet 5: 70–76, 1989

27. Thomas KR, Capecchi MR: Site-directed mutagenesis by gene targeting in mouse embryo-derived stem cells. Cell 51: 503–512, 1987

28. Bradley A: Production and analysis of chimeric mice. In: EJ Robertson (ed.) Teratocarcinomas and Embryonic Stem Cells: A Practical Approach. IRL Press, Oxford, UK, 1987, pp 113–151

29. Pascoe WS, Kemler R, Wood SA: Genes and functions: trapping and targeting in embryonic stem cells. Biochim Biophys Acta 1114: 209–221, 1992

30. Robbins J, Doetschman T, Jones WK, Sanchez A: Embryonic stem cells as a model for cardiogenesis. Trends Cardiovasc Med 2: 44–50, 1992

31. Van Deursen J, Heerschap A, Oerlemans F, Ruitenbeek W, Jap P, ter Laak H, Wieringa B: Skeletal muscles of M-creatine kinase deficient mice lack burst activity. Cell 74: 621–631, 1993

32. Van Deursen J, Lovell-Badge R, Oerlemans F, Schepens J, Wieringa B: Modulation of gene activity by consecutive gene targeting of one creatine kinase M allele in mouse embryonic stem cells. Nucl Acids Res 19: 2637–2643, 1991

33. Van Deursen J, Schepens J, Peters W, Meijer D, Grosveld G, Hendriks W, Wieringa B: Genetic variability of the murine creatine kinase B gene locus and related pseudogenes in different inbred strains of mice. Genomics 12: 340–349, 1992a

34. Van Deursen J, Wieringa B: Targeting of the creatine kinase M gene in embryonic stem cells using isogenic and nonisogenic vectors. Nucl Acids Res 20: 3815–3820, 1992b

35. Sambrook J, Fritsch EF, Maniatis T: Molecular Cloning: A Lab-

oratory Manual. Cold Spring Harbor Laboratory, Cold Spring Harbor, New York, 1989

36. Ackerman JJH, Grove TH, Wong GG, Godown DG, Radda GK: Mapping of metabolites in whole animals by ^{31}P NMR using surface coils. Nature 283: 167–170, 1980

37. Heerschap A, Bergman AH, Vaals JJ, Wirtz P, Loermans HMT, Veerkamp JH: Alterations in relative phosphocreatine concentrations in preclinical mouse muscular dystrophy revealed by *in vivo* NMR. NMR Biomed 1: 27–31, 1988

38. Hsieh PS, Balaban RS: Saturation and inversion transfer studies of creatine kinase kinetics in rabbit skeletal muscle *in vivo*. Magn Reson Med 7: 56–64, 1988

39. Gadian DG, Radda GK, Brown TR, Chance EM, Dawson MJ, Wilkie DR: The activity of creatine kinase in frog skeletal muscle studied by saturation-transfer nuclear magnetic resonance. Biochem J 194: 215–228, 1981

40. Meyer RA, Brown TR, Krilowicz BL, Kushmerick MJ: Phosphagen and intracellular pH changes during contraction of creatine-depleted rat muscle. Am J Physiol 250: C264–C274, 1986

41. Koretsky AP, Basus VJ, James TL, Klein MP, Weiner MW: Detection of exchange reactions involving small metabolite pools using NMR magnetization transfer techniques: relevance to subcellular compartmentation of creatine kinase. Magn Res Med 2: 586–594, 1985

42. Meyer RA, Brown TR, Kushmerick MJ: Phosphorus nuclear magnetic resonance of fast- and slow-twitch muscles. Am J Physiol 248: C279–C287, 1985

43. Iyengar MR, Iyengar CW, Chen HY, Brinster RL, Bornslaeger E, Schultz RM: Expression of creatine kinase isoenzyme during oogenesis and embryogenesis in the mouse. Develop Biol 96: 263–268, 1983

44. Brosnan MJ, Sasikala P, Raman, Lihong Chen, Koretsky A: Alteration of the creatine kinase isoenzyme distribution in transgenic mouse muscle by overexpression of the B subunit. Am J Physiol 264: C151–C160, 1993

PART V

DEVELOPMENT AND PATHOLOGICAL ALTERACTIONS OF CREATINE KINASES

Molecular and Cellular Biochemistry **133/134**: 277–286, 1994.
© 1994 *Kluwer Academic Publishers.*

V-1 Compartmentation of creatine kinases during perinatal development of mammalian heart

Jacqueline A. Hoerter, Renée Ventura-Clapier and Andrey Kuznetsov[1]
Cardiologie Cellulaire et Moléculaire, CJF INSERM 92-11, Université Paris-Sud, Faculté de Pharmacie, F-92296 Chatenay Malabry, France; [1] Laboratory of Bioenergetics, Cardiology Research Center, 3, Cherepkovskaya Str 15A, Moscow 121552, Russia

Abstract

Maturation of the cardiac cell is characterized by increasing diversity of isozymic expression of creatine kinases. Expression of the M-CK isozyme always precedes that of mitochondrial isozyme (mi-CK), however the expression of an isoform does not inform about its localization or cellular function. The functional role of isozymes binding to sites of energy utilization and production characteristic of the adult myocardium can be evidenced by the functional coupling of M-CK to myofibrillar ATPase and mito-CK to translocase in Triton X-100 and saponin skinned fibers. Functional activity of M-CK and mito-CK were investigated during perinatal development. Both functional activities appear during late fetal life in species mature at birth like guinea pig, and in the first postnatal weeks in immature species like rat or rabbit. Thus, the functional activity of bound CK isozymes is not associated with birth *per se* but with the general process of cell maturation. Localization of CK in the cytosol appears optimal for the transfer of glycolytic production of ATP to sites of utilization in an immature heart. During cell maturation, the increasing contribution of oxidative phosphorylation to ATP production, the apparition and binding of mi-CK to mitochondria, the binding of M-CK to myofibrils, turn the cell in a compartmentalized system of energy production. This provides the cellular basis for energy transfer by the PCr-Cr-CK system between sites of ATP production and utilization. Compartmentation of both Ca handling and energy turnover leads to a highly structured cell organization and could be essential for the efficiency of heart function. (Mol Cell Biochem **133/134**: 277–286, 1994)

Key words: fetal and neonatal development, creatine kinase isozymes, skinned cardiac fibers, mitochondrial respiration, mechanics, intracellular compartments, myofibrils

Introduction

Perinatal development offers a unique opportunity to try to further understand the role of the localization of the specific CK isozymes in cardiac cell function. This chapter deals with the functional evidence of a perinatal change in CK isozymes distribution in association with maturation of the mammalian cardiac cell. The devel-

Address for offprints: J. Hoerter, Cardiologie Cellulaire et Moléculaire, CJF INSERM 92-11, Université Paris-Sud, Faculté de Pharmacie, F-92296 Chatenay Malabry, France

opmental pattern of expression of the various mRNAs is described by Payne and Strauss (Chapter IV-1).

There are considerable species variations in the time course of development of the CK system in muscle. However, in all species, the embryonic muscle contains only BB-CK isozyme. Differentiation of myoblasts is associated with the expression of MM-CK isozyme (for a review see [1]) as well as muscular isoforms of adenine nucleotide translocator ANT1 and mitochondrial ATP-synthase β subunit [2]. In heart, MM-CK first appears in the proximal part of the outflow tract and in the right but not the left trabeculae. Such an heterogeneity of MM-CK distribution between the two ventricles will faint in the last third of gestation where the MM isozyme can be evidenced in the whole myocardium at the exception of pulmonary veins or coronary sinus [3]. Fetal development of the skeletal muscle and heart cell is then associated with a sharp increase in total CK activity and the progressive increase in the specific muscular isozyme MM-CK at the expense of the BB-CK [1, 4]. This shift in the expression of B- towards M-mRNA occurs earlier in heart than in skeletal muscle [5]. In both skeletal and cardiac muscles and in all species, the synthesis of MM-CK [4, 6] and the expression of M-mRNA always precede those of mitochondrial CK (mi-CK); in sarcomeric tissues (heart and skeletal muscle) a sarcomeric mi-CK is encoded by a specific gene in a coordinated manner with M-CK (see Chapter IV-1, Fig. 2). Thus, although coordinately expressed in striated muscle, the two main isoforms linked to the sites of production and utilization of energy appear at different time during development. However, the presence of an isoform does not tell about its localization or function.

In the adult heart, CK isoenzymes are located on specific fixation sites: MM-CK close to the ATPases of sarcolemma, sarcoplasmic reticulum and myofibrils, and mi-CK close to the ATP-ADP translocase. This compartmentation of CK isozymes and the functional coupling of CK with sites of ATP production and utilization are the basis of the role of CK in energy transfer and participates in the permanent adequation of ATP production and utilization in the adult myocardial cell [7–10]. Such organization of CK isozymes with high amounts of CK bound to myofibrils and mitochondria is characteristic of oxidative slow muscles with sustained activity, whereas abundance of cytosolic CK isozyme is found in glycolytic skeletal muscles with rapid and short activity. Fetal heart mainly depends on high glycolytic activity for its energetic requirements, whereas oxidative phosphorylation capacities increase during perina-

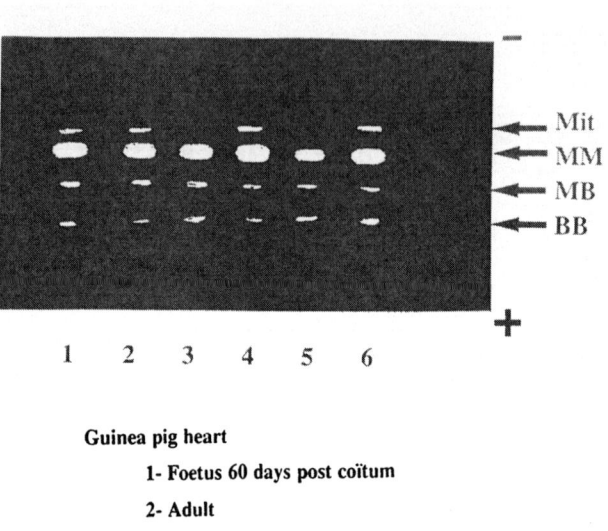

Guinea pig heart
  1- Foetus 60 days post coïtum
  2- Adult

Rat heart
  3- Newborn 7 day post partum
  4- Adult

Rabbit heart
  5- Newborn 1 day post partum
  6- Adult

Fig. 1. Species differences in the developmental pattern of CK isozymic profiles in guinea pig, rabbit and rat heart. Electrophoresis on agarose gel. Buffer in mM: Tris/HCl 60, diethyl barbital 10, sodium barbital 50, EGTA 1, DTT 1, 0.1% Triton X-100 (pH 9.0); 200 V, 1 hour at 4° C.

tal development. Thus in relation to this maturation process, it is interesting to evidence the compartmentation of CK isozymes and their functional activity by studying the coupling of MM-CK to myofibrillar ATPase and of mi-CK to respiration during perinatal development [11].

To understand if the diversity of CK isozymes is related to birth *per se* or to a general process of cell maturation, we will compare species which are fully mature at birth like sheep or guinea pig and develop high autonomic level of physical activity (weanling ca 1 week) and immature species like rabbit, rat or mice which are immobile and born dumb and without fur (weanling ca 3 to 4 weeks).

Total creatine kinase activity and isozymic profile

Marked increase in total CK activity occurs during embryonic and foetal development. In some species, CK activity still increases after birth. For example in mouse, CK activity per wet weight doubles from birth to adult

[12] and in rat it doubles in the first 3 postnatal weeks [13] and doubles again from 3 weeks to adult in the absence of change in other mitochondrial enzymes [14]. In other species like rabbit [11], guinea pig [15] or sheep [4, 6], no major changes occur in total CK activity from 90% of gestation to adult.

Moreover major species differences occur in the isozymic profile. Figure 1 shows that, in mature species like guinea pig, the isozymic pattern is close to the adult one: the four creatine kinase isoenzymes, MM, MB, BB and mi-CK are already present in the foetus at 60 days post coïtum (dpc), i.e. at 90% gestation. Similar isozymic pattern is detected at the end of gestation in the fetal sheep, another mature species [6]. By contrast immature species express only MM, BB and MB: mitochondrial CK is not detected either on the day of birth in rabbit heart (1 day post partum, 1 dpp) or after one week in the newborn rat (7 dpp). Thus in all animals in late gestation or in first postnatal weeks, MM-CK is present whether or not mi-CK is expressed.

Functional properties of myofibrillar bound creatine kinase

Since MM-CK is expressed before birth in all mammalian species, it is interesting to know if the MM isozyme is, as in the adult heart, associated with the contractile proteins and thus able to efficiently rephosphorylate MgADP produced by the contractile activity. In the adult heart and skeletal muscle, MM-CK is bound to myofibrils, at the level of M-line where it deserves both structural and enzymatic roles [16]. In rat heart, at birth no evidence of M-line [17] nor of MM-CK bound to myofibrils [18] is observed by immunocytochemistry; the proteins of the M-line progressively appear only during the first postnatal weeks [17]. One way to evidence the functional consequence of CK binding to the myofibrils is to study the relaxation of rigor tension by CK substrate in fibers skinned with Triton X-100, a preparation allowing to probe the contractile function of the myofibrillar compartment [19, 20]. A stepwise decrease in MgATP in the absence of calcium and PCr, leads to the appearance of rigor tension. In the adult, the presence of PCr in the medium shifts the dependency of the rigor tension on MgATP concentration towards much lower MgATP concentrations [21]. This shift is due to local ADP rephosphorylation by bound CK. It can be taken as an index of CK efficacy reflecting the functional state of bound CK (see Chapter II-5). Figure 2 presents

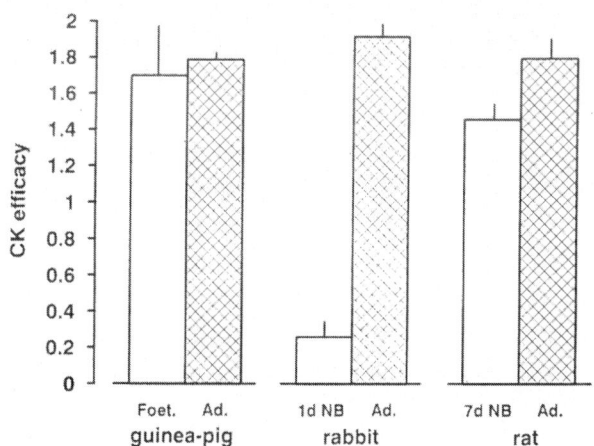

Fig. 2. Increase in myofibrillar creatine kinase efficacy during perinatal development of guinea pig, rabbit and rat heart. Myofibrillar CK efficacy in Triton X-100 skinned fibers of left papillary muscle (same ages as Fig. 1). Rigor tension is induced by decreasing MgATP concentration. The sensitivity of myofibrils to lack of ATP is characterized by the concentration of MgATP inducing half maximal rigor force: $pMgATP_{50}$ in log unit (see Fig. 3 of Chapter II-5 for curves in adult heart). CK efficacy is expressed as the shift in $pMgATP_{50}$ upon addition of 12 mM PCr for the various groups of animals. Bound myofibrillar CK is completely functional in foetal guinea pig, whereas low activity is observed at birth in rabbit; in the rat, a week after birth, CK efficacy is lower than in the adult.

the functional data related to isozymic composition of the three species described in Fig. 1. Similar CK efficacy are found in the adult rabbit, rat and guinea pig heart, however considerable species differences exist in the developmental pattern of bound myofibrillar CK activity. In foetal guinea-pig heart, at the end of gestation, bound MM-CK is already fully active: myofibrillar CK efficacy was 1.70 ± 0.03 (n = 3) in foetus and was not significantly different from that of adult guinea-pig (1.79 ± 0.05, n = 5) [15]. By contrast myofibrillar CK efficacy was very low at birth in rabbits (0.26 ± 0.08, n = 4), it progressively reaches adult levels only 17 days after birth (1.45 ± 0.06, n = 7) [11]. Similarly, in rat heart 7 days after birth, myofibrillar CK efficacy (1.46 ± 0.05, n = 8) was lower than in adult (1.80 ± 0.09, n = 11, p < 0.01). This poor efficacy of the newborn rat myofibrillar CK is in agreement with the late development of M-line between 1 and 11 dpp [17]; in this very immature species full myofibrillar CK activity is not even achieved after 3 postnatal weeks [22]. Since the kinetic constants of CK, including its apparent Km for ADP are similar in newborns and adults [23], the increase in functional myofibrillar CK activity directly translates into an increased capacity of ADP rephos-

280

phorylation and pH buffering close to myofibrillar AT-Pase during perinatal development.

Factors involved in MM-CK binding are presently unknown. Perinatal increase in myocardial contractility [24–25] is associated with quantitative accumulation of myofibrils and increase in their structural organization into aligned characteristic myofilaments as well as accumulation of sarcoplasmic vesicles [26–29]. Moreover qualitative changes involve genetic expression of mRNAs coding for numerous isozymes of the intrinsic and regulatory contractile proteins as well as proteins involved in excitation-contraction coupling (for a review see [30]). Intrinsic contractile properties studied in Triton skinned fibers allow to probe the changes which are specific for myofibrillar development independently from those due to the excitation process. Maximal force and active stiffness per muscle cross area as well as rate of cross bridge cycling increase during perinatal development [31]. At the end of gestation both fetal ventricles have similar mass, cardiac output and perform similar work. Birth and early postnatal life induce major circulatory changes leading to opposite change in resistance and cardiac output of both ventricles [32]. Since the time course of CK binding to myofibrils is similar in both ventricles, the changes in blood pressure and load associated with birth *per se* do not appear to be major determinants of CK binding to myofibrils. Clear confirmation that cardiac load changes induced by birth are not responsible for MM-CK binding to myofibrils is given by full myofibrillar CK efficacy already observed in the foetus of a mature species like the guinea pig. Right now, there seems to be only one MM-CK gene. However, two isoforms of MM appear to be translated *in vitro*, only one of them decreasing in cardiac hypertrophy [33]. Thus transcription or translation control of CK appears much more complex than previously believed. Further work is needed to understand if binding is due to the expression of a slightly different MM isozyme, to a change of the contractile protein or to the appearance of a protein anchoring cytosolic MM-CK on the myofilaments.

Functional activity of mitochondrial creatine kinase

High level of expression of sarcomeric specific mitochondrial CK is a characteristic of slow oxidative muscles with high energy demand like myocardium and slow twitch skeletal muscle (see Chapter IV-1 and [10]). Fast

glycolytic muscle express low levels of mi-CK. A shift toward slow oxidative type of fibers and a several fold increase in mi-CK is the usual adaptation of fast twitch muscle to chronic stimulation [34] or endurance training, for example in marathon runners [35]. Conversely, in several species including man, muscular dystrophy increases cytosolic MB and BB activity together with a marked decrease in mi-CK activity [36].

As described earlier, the expression of mi-CK is a late event in the process of heart development. Mi-CK appears at the end of gestation in mature species like guinea pig or lamb and in the first postnatal weeks in species born immature like mouse, rat or rabbit [4, 6, 11–15]. Since we showed that the functional activity of MM-CK in myofibrils appears much later than the expression of the isozyme, the same question applies to the expression and functional activity of mi-CK. To study the appearance of the functional activity of CK in mitochondria and the perinatal development of respiration, mitochondrial function can be assessed in saponin skinned fibers where mitochondria are kept in their environment [37]. This technique avoids the process of mitochondrial isolation and excludes damages of the outer membrane, a structure which is fundamental for mitochondrial properties (see Chapter II-1, II-2 and III-1). The respiration rate (Vmax, state 3), measured in the presence of both 1 mM ADP and 20 mM Cr results from the amount of mitochondria and from the maximal activity of oxidative phosphorylations. Maximal respiration differs in adult rat, rabbit and guinea pig (Fig. 3a) reflecting species differences in both the enzymatic activities of oxidative phosphorylation and in the mitochondrial content [38, 39].

The developmental pattern of mitochondrial properties appears markedly different. In guinea pig, foetal respiration is about half of the adult. At the end of gestation, the foetal guinea pig cell morphology is similar to the adult, thus lower respiration rates are due to poor activity of oxidative enzymes [40, 41]. In the rabbit, one day after birth, the maximal capacity of respiration in state III is equivalent to the adult. However in rat heart, the respiratory capacities have not reached the adult level by the first postnatal week (Fig. 3a). The ADP/O ratio was constant along perinatal development showing no change in coupling ratio in foetal, newborn and adult mitochondria [42–44]. Transiently high maximal respiratory capacities (state 3), which can even surpass the adult respiration have been described in newborn or late foetal mitochondria of rabbit [11, 25, 44–46], lamb [47], guinea pig [41] and newborn dogs [47] for all types of

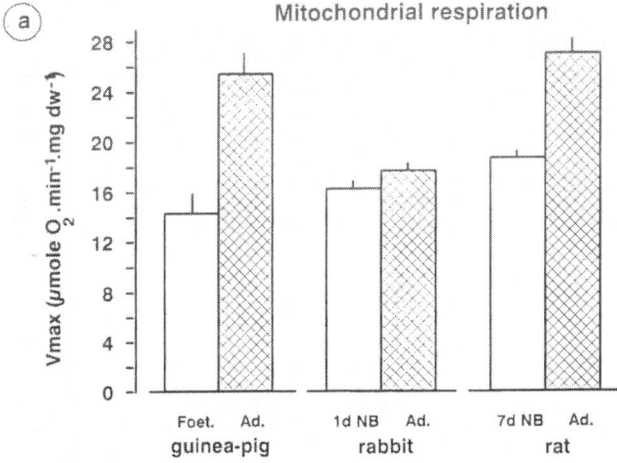

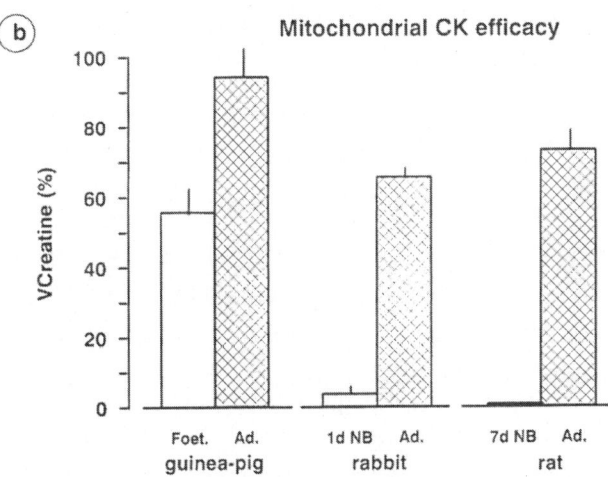

Fig. 3. Mitochondrial maximal rate of respiration rate and mitochondrial CK efficacy during perinatal development of guinea pig, rabbit and rat heart. Rates of respiration and stimulation by creatine in saponin skinned cardiac fibers (same ages as Fig. 1). (a) Maximal respiration rates (V_{max}) were measured in the presence of both 20 mM creatine and 1 mM ADP and expressed as a function of fiber dry weight. (b) Mitochondrial CK efficacy: VCreatine (%) represents the stimulation of mitochondrial respiration induced by addition of 20 mM creatine at submaximal ADP concentration (0.1 mM) as a function of age in the three species. It is expressed as the percentage of stimulation by Cr of the submaximal respiration observed without Cr. The medium assay contained in mM: glutamate 5, malate 2, Pi 3, free Mg 3, albumin 2 mg/ml, imidazole 20 (pH 7.0), EDTA 10 (pCa 7), ionic strength 160 mosm, temperature 22° C.

substrate supported respiration. Another characteristic of the newborn mitochondria is a high activity of Ca uptake suggesting that in presence of a sparse sarcoplasmic reticulum, the immature mitochondria could *in vivo* participate in Ca homeostasis of the foetal and newborn cells [44, 45].

The functional activity of mi-CK can be also evi-denced on saponin fibers. Subsequent addition of 20 mM creatine, to a submaximal respiration rate (achieved by 0.1 mM ADP) activates mitochondrial CK and increases the ADP concentration in the vicinity of the inner mitochondrial membrane thus enhancing respiration rate in the adult heart (see Chapter III-1). The increase in submaximal respiration after addition of Cr: Vcreatine (expressed in % of respiration without Cr) can be taken as an index of mitochondrial CK efficacy. Figure 3b shows the functional activity of mitochondrial CK. In foetal guinea pig heart, the percentage of increase in respiration following creatine addition amounted already to $55.6 \pm 7.1\%$, that is more than half of the adult ($94.2 \pm 8.2\%$). Thus, mi-CK is already functional in foetal guinea pig whereas it is hardly observable in rabbit one day after birth, and not even detected in the 7 day newborn rat heart (Fig. 3b). In the latter at 3 weeks, mi-CK activity of isolated mitochondria is at most a third of the adult activity [22].

The development of several glycolytic or mitochondrial respiratory enzymes is regulated at the pretranslational level and modulated by oxygen tension in myoblasts. Coordinate and reciprocal expression of several respiratory and glycolytic mRNAs is observed during myogenesis. Hypoxic environment induces an over expression of glycolytic transcripts and a decline in mitochondrial transcripts [48]. A marked increase in mitochondrial relative volume [26–29], and in enzymes of the oxidative pathway occurs at birth [40–43]. This sudden increase in synthesis of respiratory chain proteins observed in the first hours of extra-uterine life is abolished in foetus delivered in hypoxic environment [49]. On the opposite, birth-induced increase in blood pO_2 does not appear to be a major determinant of CK binding to mitochondria since mi-CK activity appears well before birth in mature species like lamb or guinea pig and in the first postnatal weeks in immature species [4, 6, 11, 12, 22] and Fig. 3b. In adult heart, CK is bound to cardiolipin close to the translocase [50]. It is interesting to notice that although marked change in phospholipid composition of mitochondrial membrane occurs in the neonatal rabbit, the cardiolipin content remains constant and similar to the adult [44]. Thus the binding of mi-CK to the external face of the inner membrane is not due to increased binding sites for CK. Careful parallel examination of mi-CK expression and functional activity will be needed to ascertain the exact time course of expression of mi-CK mRNA, synthesis of the isozyme and its binding to inner mitochondrial membrane. It is already obvious from our present knowledge, that whether this happens during pre- or postnatal period, these events occurs in a very short duration (probably hours) as opposed to a delay of weeks between the expression of MM-CK and its binding to myofibrils.

We observed in rabbit heart the parallel apparition and development of functional coupling of MM-CK to myofibrillar ATPase and mi-CK to respiration even though the isoforms are not expressed at the same time. This led us to hypothesize that both specific sites of myocardial energy transfer system characteristic of the adult heart developed together probably under some common hormonal induction related to the general process of cell maturation. However, more recent data obtained in the rat heart showed that myofibrillar CK was already bound to myofilaments 7 days after birth, whereas nor mi-CK isozyme neither its functional coupling could be evidenced at this stage (Fig. 2, 3b). Careful re-examination of the timing of functional development of CK binding to myofibrils or mitochondria in rabbit also shows that full binding of myofibrillar CK precedes by 1 to 3 days full activity of the mitochondrial CK (see Fig. 7 of [11]). Thus the process of induction of mi-CK and binding of MM-CK to myofibrils do not appear to be under common induction. Further work is needed to understand the responsibility of trophic factors like innervation or hormonal status in the CK isozymic developmental pattern and activities.

Creatine kinase binding as part of the maturation of the cardiac cell function (Fig. 4)

In a variety of adult tissues, the concentration of PCr has been shown to be correlated with CK activity whatever isozyme is expressed [51]. It is of interest to notice that, during perinatal development, total PCr and creatine myocardial content markedly increases together with CK activity in all species [4, 12]. This massive increase in creatine moieties occurs with minor changes in adenine pool. Accumulation of Cr and PCr occurs in two steps. For example in mouse heart, a first increase is observed during the last third of gestation at a time of rapid increase in CK activity (and accumulation of MM). A second fourfold rise in creatine pool occurs in the first month, concomitant with the apparition of mi-CK [4]. In a mature species like lamb, the creatine pool has reached its adult value one week after birth, in agreement with full maturation of CK activity and isozymic distribution [4, 6]. Such an accumulation of Cr and PCr needs the conjugated presence of creatine kinase and of sarcolemmal mechanism to import Cr since no synthesis occurs in heart cell. Addition of creatine to cultures of cardiac and skeletal muscle increase their PCr content [52] (this observation has even been among the earlier suggestions for a role of the PCr-Cr-CK system in muscle cell). However, surprisingly few data exist on the developmental aspect of creatine transport in myocardium. Whether both the recently cloned Na creatine transporter [53] and CK isozymes expression are coordinately regulated and how both mechanisms interfere with accumulation of creatine pool remain unknown.

The only way to assess CK activity in the whole heart it to measure its kinetics by ^{31}Phosphorus NMR spectroscopy. The use of magnetization transfer techniques has been widely applied in the adult heart to evaluate the modification of CK in pathological or adaptive situations (see Chapter I.1 and for a recent review [54]). This experimental approach has been very actively applied by the group of Ingwall in the rabbit neonatal cardiac development. In isovolumic perfused heart, CK velocity increases three fold between birth and 3 weeks postnatal. Strikingly the increase in CK velocity occurs at constant total CK activity but appears clearly related to the amount of mitochondrial CK activity and increased creatine pool [55]. If in vitro measurement of CK velocity by 31 P NMR is straightforward, the interpretation of in vivo data requires knowledge of the isozymic CK distribution, the total CK activity and the concentration of the substrates (not mentioning the problem of NMR visibility of substrates under physiological

$\longrightarrow$

Fig. 4. Perinatal development of the myocardial cell. Maturation of the cardiac cell is characterized by complexification of structures involved both in Ca handling and energy turnover. Contraction and relaxation depend on the exchange of Ca with the extracellular compartment in the immature heart. Increase in the amount of myofibrils and complexification of Ca handling system by the development of sarcoplasmic reticulum (SR) and the apparition T-tubules enhance the efficacy of the excitation-contraction coupling by a compartmentation of Ca movements. Increase in the amount of mitochondria and a shift of ATP production from a predominantly cytosolic glycolysis to oxidative phosphorylation of fatty acids enhance susceptibility of the adult heart to hypoxia. Cytosolic CK isozymes of the immature heart buffers glycolytic ATP production. Compartmentation of CK isozymes during cell maturation includes the apparition and binding to mitochondria of mi-CK, the binding of M-CK to myofibrils and provides the cellular basis for energy transfer by the PCr-Cr-CK system between sites of ATP production and utilization.

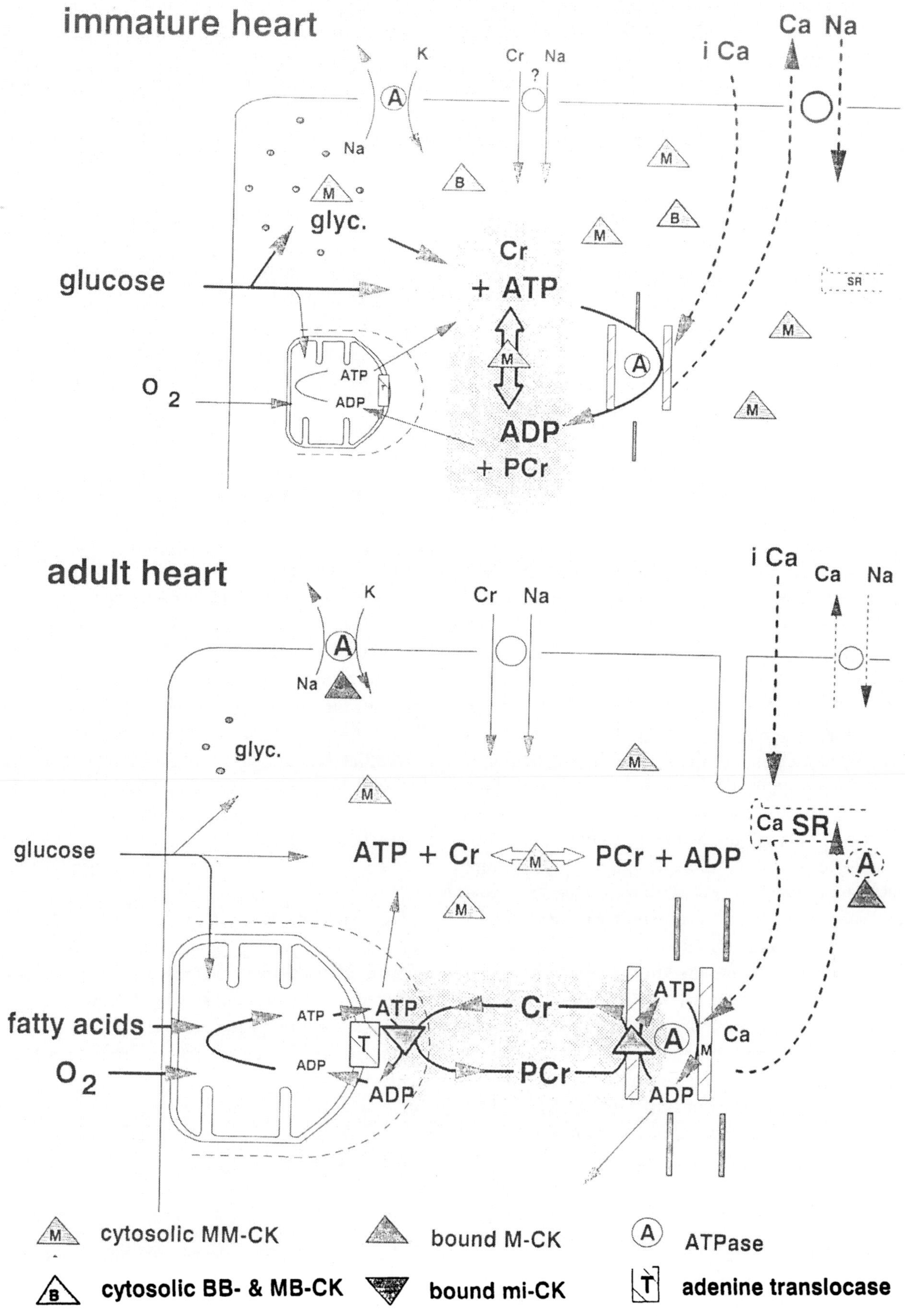

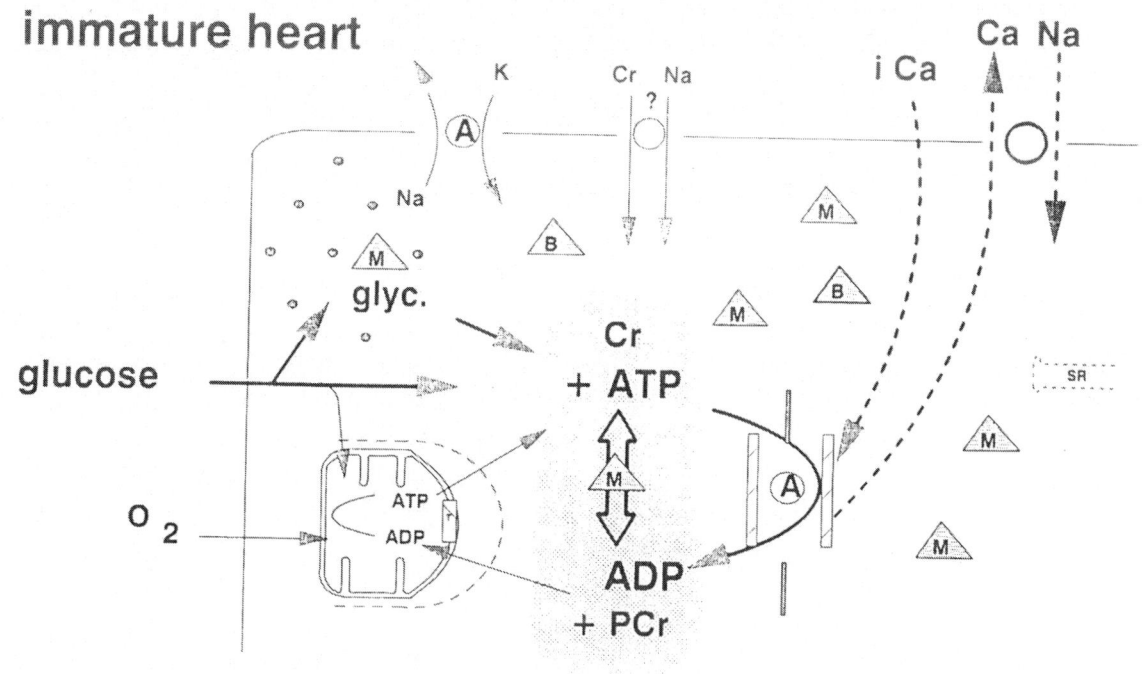

immature heart
K
Cr Na
?
i Ca
Ca Na
Na
glyc.
glucose
Cr
+ ATP
ATP
ADP
O 2
ADP
+ PCr
SR
adult heart
K
Cr Na
i Ca
Ca Na
Na
glyc.
glucose
ATP + Cr
PCr + ADP
Ca SR
fatty acids
O2
ATP
ATP
ADP
ADP
T
Cr
PCr
ATP
ADP
Ca
cytosolic MM-CK
bound M-CK
ATPase
cytosolic BB- & MB-CK
bound mi-CK
adenine translocase

or pathological conditions which is beyond the topic of this paper). This led to the elaboration of mathematical model and method of saturation transfer allowing to analyze the contribution of mi-CK flux to energetic pathway in various working conditions [56, 57]. During the first neonatal weeks, when mi-CK appears, its flux increases in response to increased muscle work and ATP synthesis while total CK flux remains constant. In summary, the appearance during postnatal rabbit development of functional coupling of MM-CK to myofibrillar ATPase and mi-CK to translocase that we evidenced on intracellular compartments in skinned fibers is contemporary of the increase in CK velocity measured by NMR spectroscopy, which probes the function of CK in myocardial energetics.

Embryonic and foetal immature hearts express significant amounts of BB-CK isozyme. Foetal and newborn immature hearts depend mainly for their ATP production on high activity of glycolytic and glycogenolytic enzymes which confers to the developing heart its well known resistance to hypoxia [24, 58, 59]. Probably more important than the presence of BB isozyme is the fact that, at this time, CK expression is restricted to BB, MB and MM which are localized in cytosol i.e. close to sites of ATP production. Indeed, in other glycolytic tissues like white skeletal muscle or smooth muscle, CK is associated with clusters of glycolytic enzymes either in the cytosol [60] or loosely bound to the I-band of myofibrils [61]. Such associations might exists in immature hearts and could participate in PCr synthesis in the absence of mi-CK. Thus both the CK cytosolic localization and the expression of isozymes appears optimal for the transfer of glycolytic ATP production in an immature cell with low creatine pool. CK, in the developing myocardium would have principally, as in white skeletal muscle, a role of efficient temporal and spatial buffering of glycolytic ATP production and utilization.

In all species perinatal growth is characterized by a marked decrease in the heart weight to body weight ratio. For example in rabbit this ratio is 3 times higher in the fetus at term than in the adult [11]. This suggests that the heart cell develops strategies to increase its efficiency during perinatal development. Although marked increase in the mitochondrial relative volume and the activity of oxidative enzyme pathway occurs at birth, the newborn heart can not support its contractile function solely on using fatty acids [62]. The ability to efficiently oxidize other substrates than glucose as fuels for ATP synthesis postnatally develops both in immature and mature species [40, 62]. This high versatility of substrate

oxidation, characteristic of the adult heart, could be fundamental for its efficiency. Moreover, the foetal cell is small, that is of same size as an amphibian heart cell and similarly depends mainly on external calcium for its activation [29, 63]. Postnatal growth is early achieved by hyperplasia: for example in rabbit, it is only after a week that cells starts to hypertrophy together with incrasing complexity of structures involved in excitation-contraction coupling. Sarcoplasmic reticulum Ca release [64] increases as well as the number of ryanodine and dihydropyridine receptors [65]. Ca-ATPase activity and functional activity of the SR appear [29, 63, 66], together with increase in the expression of mRNA encoding for the specific isoform of Ca ATPase [67], finally the T-tubules appear. The increase in cell volume and the development of SR structure occur before birth in precocious animals and after birth in immature species: this turns the cell in a compartmented Ca handling system with an efficient amplification of Ca signal.

In the same way the binding of CK to myofibrils and mitochondria will turn the cell into a compartmented system of energy turnover earlier in development in the most mature animals. Both compartmentation of Ca handling and energy turnover participate in cellular maturation which leads to spatial organization that is highly structured and integrated. Multienzyme complexes constitute a high degree of compartmentation which increases the efficacy of reactions [68–69]. The development of such integrated multienzyme systems could be essential for the rapid neonatal adaptation to increased metabolic demand and for the continuous adequation of energy production and utilization characteristic of the adult mammalian cell.

Acknowledgements

This work was supported by grants from 'Fondation de France', 'Fondation pour la Recherche Médicale' and 'Association Française contre les myopathies'. R. Ventura-Clapier is supported by 'Centre National de la Recherche Scientifique'.

References

1. Eppenberger HM, Perriard J-C, Wallimann T: Analysis of creatine kinase during muscle differentiation. In: M Rattazi, JC Scandalios, GS Whitt (eds) Isozymes: Current Topics in Biological and

Medical Research. Vol. 7: Molecular structure and regulation. Alan Liss, New York, 1983, pp 19–38

2. Stepien G, Torroni A, Chung AB, Hodge JA, Wallace DC: Differential expression of adenine nucleotide translocator isoforms in mammalian tissues and during muscle cell differentiation. J Biol Chem 267: 14592–14597, 1992

3. Hasselbaink HDJ, Labruyère WT, Moorman AFM, Lamers WH: Creatine kinase isozyme expression in prenatal rat heart. Anat Embryol 182: 195–203, 1990

4. Ingwall JS, Kramer MF, Friedman WF: Developmental changes in heart creatine kinase. In: WE Jacobus, JS Ingwall (eds) Heart Creatine Kinase. Williams & Wilkins Co., Baltimore London, 1980, pp 9–16

5. Trask RV, Billadello JJ: Tissue-specific distribution and developmental regulation of M-creatine and B-creatine kinase messenger RNAs. Biochim Biophys Acta 1049: 182–188, 1990

6. Ingwall JS, Kramer MF, Woodman D, Friedman WF: Maturation of energy metabolism in the lamb: Changes in myosin ATPase and creatine kinase activities. Pediat Res 15: 1128–1133, 1981

7. Saks VA, Chernousova GB, Voronkov YI, Smirnov VN, Chazov EI: Study of energy transport mechanism in myocardial cells. Circ Res 34–35, supIII: 138–149, 1974

8. Bessman SP, Carpenter CL: The creatine-creatine phosphate energy shuttle. Ann Rev Biochem 54: 831–862, 1985

9. Jacobus WE: Respiratory control and integration of heart high energy phosphate metabolism by mitochondrial creatine kinase. Ann Rev Physiol 47: 707–725, 1985

10. Walliman T, Wyss M, Brdiczka D, Nicolay K: Intracellular compartmentation, structure and function of creatine kinase isoenzymes in tissues with high and fluctuating energy demands: the 'phosphocreatine circuit' for cellular energy homeostasis. J Biochem 281: 21–40, 1992

11. Hoerter JA, Kuznetsov A, Ventura-Clapier R: Functional development of the creatine kinase system in perinatal rabbit heart. Circ Res 69: 665–676, 1991

12. Hall N, DeLuca M: Developmental changes in creatine phosphokinase in neonatal mouse hearts. Biochem Biophys Res Commun 66: 988–994, 1975

13. Baldwin KM, Cooke DA, Cheadle WG: Enzyme alterations in neonatal heart during development. J Mol Cell Cardiol 9: 651–660, 1977

14. Dowell RT: Mitochondrial component of the phosphoryl-creatine shuttle is enhanced during rat heart perinatal development. Biochem Biophys Res Commun 141: 319–325, 1986

15. Ventura-Clapier R, Hoerter JA, Kuznetsov A, Khuchua Z, Clark JF: Perinatal development of the creatine kinase system in mammalian heart. In: PP De Heyn, B Marescau, V Stalon, IA Qureeshi (eds) Guanidino Compounds in Biology and Medicine. John Libbey & Comp. Ltd., 1992, pp 195–204

16. Walliman T, Eppenberger HM: Localization and function of M-line bound creatine kinase: M-band model and creatine phosphate shuttle. J Muscle Res Cell Motil 6: 239–285, 1985

17. Anversa P, Olivetti G, Bracchi PG, Loud AV: Postnatal development of the M-band in rat cardiac myofibrils. Circ Res 48: 561–568, 1981

18. Carlsson E, Kjorell U, Thornell LE: Differentiation of the myofibrils and the intermediate filament system during postnatal development of the rat heart. Eur J Cell Biol 27: 62–73, 1982

19. Ventura-Clapier R, Vassort G: Role of myofibrillar creatine kinase in the relaxation of rigor tension in skinned cardiac muscle. Pflügers Arch 404: 157–161, 1985

20. Ventura-Clapier R, Mekhfi H, Vassort G: Role of creatine-kinase in force development in chemically skinned rat cardiac muscle. J Gen Physiol 89: 815–837, 1987

21. Veksler VI, Kapelko VI: Creatine kinase in regulation of heart function and metabolism. II. The effects of phosphocreatine on the rigor tension of EGTA treated rat myocardial fibers. Biochim Biophys Acta 803: 265–270, 1984

22. Dowell RT: Phosphorylcreatine shuttle enzymes during perinatal heart development. Bioch Med Metabol Biol 37: 374–384, 1987

23. Dowell RT, Fu MC: Perinatal enhancement of cardiac myofibrillar creatine kinase activity without change in enzyme Km. Am J Physiol 265 (Cell Physiol 34): C375–C378, 1993

24. Friedman WF: The intrinsic physiologial properties of the developing heart. Progr Cardiovasc Res 15: 87–111, 1972

25. Nakanishi T, Jarmakani JM: Developmental changes in myocardial mechanical function and subcellular organelles. Am J Physiol 246: H615–H625, 1984

26. Legato MJ; Ultrastructural changes during normal growth in the dog and rat ventricular myofiber. In: M Lieberman, T Sano (eds) Developmental and physiological correlates of cardiac muscle. Raven Press, New York, 1975, pp 249–273

27. Smith HE, Page E: Ultrastructural changes in rabbit heart mitochondria during the perinatal period: neonatal transition to aerobic metabolism. Dev Biol 57: 109–117, 1977

28. Olivetti G, Anversa P, Loud AV: Morphometric study of early postnatal development in the left and right ventricular myocardium of the rat. II Tissue composition, capillary growth and sarcoplasmic alterations. Circ Res 46: 503–512, 1980

29. Hoerter J, Mazet F, Vassort G: Perinatal growth of the rabbit cardiac cell: possible implications for the mechanism of relaxation. J Mol Cell Cardiol 13: 725–740, 1981

30. Lompre AM, Mercadier J-J, Schwartz K: Changes in gene expression during cardiac growth. Intern Rev Cytol 124: 137–186, 1991

31. Murat I, Hoerter J, Ventura-Clapier R: Developmental changes in effects of halothane and isoflurane on contractile properties of rabbit cardiac skinned fibers. Anesthesiology 73: 137–145, 1990

32. Rudolph AM, Heymann MA: Fetal and neonatal circulation and respiration. Ann Rev Physiol 36: 187–207, 1974

33. Boheler KR, Dillmann WH: Cardiac response to pressure overload in the rat: The selective alteration of in vitro directed RNA translation products. Circ Res 63: 448–456, 1988

34. Schmitt T, Pette D: Increased mitochondrial creatine kinase in chronically stimulated fast twitch rabbit muscle. FEBS Lett 188: 341–344, 1985

35. Apple FS, Rogers MA: Mitochondrial creatine kinase activity alterations in skeletal muscle during long distance running. J Appl Physiol 61: 482–485, 1986

36. Mahler M: Progressive loss of mitochondrial creatine phosphokinase in muscular dystrophy. Biochem Biophys Res Commun 88: 895–906, 1979

37. Veksler VI, Kuznetsov AV, Sharov VG, Kapelko VI, Saks VA: Mitochondrial respiratory parameters in cardiac tissues: a novel method of assessment by using saponin skinned fibers. Biochim Biophys Acta 892: 191–196, 1987

38. Bass A, Stejskalova M, Ostadal B, Samanek M: Differences between atrial and ventricular energy supplying enzymes in five mammalian species. Physiol Rev 42: 1–6, 1993

39. Bart E, Stämmler G, Speiser B, Shaper J: Ultrastructural quantifi-

286

cation of mitochondria and myofilaments in cardiac muscle from 10 different animal species including man. J Mol Cell Cardiol 24: 669–681, 1992

40. Rolph TP, Jones CT, Parry D: Ultrastructural and enzymatic development of fetal guinea pig heart. Am J Physiol 243 (Heart Circ Physiol 12): H87–H93, 1982

41. Barrie SE, Harris P: Myocardial enzyme activities in guinea pigs during development. Am J Physiol 233 (Heart Circ Physiol 2): H707–H710, 1977

42. Warshaw JB: Cellular energy metabolism during fetal development. J Cell Biol 41: 651–657, 1969

43. Wells RJ, Friedman WF, Sobel BE: Increased oxidative metabolism in the fetal and newborn lamb heart. Am J Physiol 222: 1488–1493, 1972

44. Wolf WJ, Rex KA, Geshi E, Sordahl LA: Postnatal changes in heart mitochondrial calcium and energy metabolism. Am J Physiol 261 (Heart Circ Physiol 30): H1–H8, 1991

45. Sordahl LA, Crow CA, Kraft GH, Schwartz A: Some ultrastructural and biochemical aspects of heart mitochondria associated with development: fetal and cardiomyopathic tissue. J Mol Cell Cardiol 4: 1–10, 1972

46. Werner JC, Whitman V, Musselman J, Schuler HG: Perinatal changes in mitochondrial respiration in the rabbit heart. Biol Neonate 42: 208–216, 1982

47. Mela L, Goodwin CW, Miller LD: Correlation of mitochondrial cytochrome c concentration and activity of oxygen availability in the newborn. Biochem Biophys Res Commun 64: 384–389, 1975

48. Webster KA, Gunning P, Hardeman E, Wallace DC, Kedes L: Coordinate reciprocal trends in glycolytic and mitochondrial transcript accumulations during the in vitro differentiation of human myoblasts. J Cell Physiol 142: 566–573, 1990

49. Hallman M; Changes in mitochondrial respiratory chain proteins during perinatal development. Evidence of the importance of environmental oxygen tension. Biochim Biophys Acta 253: 360–372, 1971

50. Muller M, Moser R, Cheneval D, Carafoli E: Cardiolipin is the membrane receptor for mitochondrial creatine phosphokinase. J Biol Chem 260: 3839–3843, 1985

51. Iyengar MR: Creatine kinase as an intracellular regulator. J Muscle Res Cell Motil 5: 527–534, 1984

52. Seraydarian MW, Abbott BC: Creatine and the control of energy metabolism in cardiac and skeletal muscle cells in culture. J Mol Cell Cardiol 6: 405–413, 1974

53. Guimbal C, Kilimann MW: A Na-dependent creatine transporter in rabbit brain, muscle, heart and kidney. cDNA cloning and functional expression. J Biol Chem 268: 8418–8421, 1993

54. Ingwall JS: Is cardiac failure a consequence of decreased energy reserve? Circulation 87 sup VII: VII.58–VII.62, 1993

55. Perry SB, McAuliffe J, Balschi JA, Hickey PR, Ingwall JS: Velocity of the creatine kinase reaction in the neonatal rabbit heart: role of mitochondrial creatine kinase. Biochemistry 27: 2165–2172, 1988

56. McAuliffe JJ, Perry SB, Brooks EE, Ingwall JS: Kinetics of the creatine kinase reaction in the neonatal rabbit heart: an empirical analysis of the rate equation. Biochemistry 30: 2585–2593, 1991

57. Zahler R, Ingwall JS: Estimation of heart mitochondrial creatine kinase flux using magnetization transfer NMR spectroscopy. Am J Physiol 262 (Heart Circ Physiol 31): H1022–H1028, 1992

58. Hoerter JA, Opie LH: Perinatal changes in glycolytic function in response to hypoxia in the incubated or perfused rat heart. Biol Neonate 33: 144–161, 1978

59. Jarmakani JM, Nagatomo T, Nakazawa M, Langer GA: Effect of hypoxia on myocardial high-energy phosphates in the neonatal mammalian heart. Am J Physiol 235: H475–H481, 1978

60. Kupriyanov VV, Seppet EK, Emelin IV, Saks VA: Phosphocreatine production coupled to glycolytic reactions in the myocardial cell cytosol. Biochim Biophys Acta 592: 197–210, 1980

61. Maughan D, Wegner E: On the organization and diffusion of glycolytic enzymes in skeletal muscle. In: RJ Paul, G Elzinga, K Jamada (eds) Muscle energetics. Alan R Liss Inc., 1989, pp 137–147

62. Lopashuk GD, Spafford ME: Energy substrate utilization by isolated working hearts from newborn rabbits. Am J Physiol 258 (Heart Circ Physiol 27): H1274–H1280, 1990

63. Chin TK, Friedmann WF, Klitzner TS: Developmental changes in cardiac myocyte calcium regulation. Circ Res 67: 574–579, 1990

64. Fabiato A, Fabiato F: Calcium induced release of calcium from the sarcoplasmic reticulum of skinned cells from adult human, dog, cat, rabbit, rat and frog hearts and from fetal and newborn rat ventricles. Ann NY Acad Sci 307: 491–522, 1978

65. Wibo M, Bravo G, Godfraind T: Postnatal maturation of excitation contraction coupling in rat ventricle in relation to the subcellular localization and surface density of 1,4 dihydropyridine and ryanodine receptors. Circ Res 68: 662–673, 1991

66. Mahony L, Jones LR: Developmental changes in cardiac sarcoplasmic reticulum in sheep. J Biol Chem 261: 15257–15265, 1986

67. Lompre A-M, Lambert F, Lakatta EG, Schwartz K: Expression of Ca^{2+} ATPase and calsequestrin genes in rat heart during ontogenic development and aging. Circ Res 69: 1380–1388, 1991

68. Hervagault JF, Thomas D: Theoretical and experimental studies on the behavior of immobilized multienzyme systems. In: GR Welch (ed.) Organized multienzyme systems: Catalytic properties. London Academic Press, 1985, pp 381–418

69. Srere P: Complexes of sequential metabolic enzymes. Ann Res Biochem 56: 89–124, 1987

Molecular and Cellular Biochemistry **133/134**: 287–298, 1994.
© 1994 *Kluwer Academic Publishers.*

V-2 *In situ* study of myofibrils, mitochondria and bound creatine kinases in experimental cardiomyopathies

Vladimir Veksler[1] and Renée Ventura-Clapier[2]
[1] *Laboratory of Experimental Cardiac Pathology, Cardiology Research Center, Moscow, Russia;* [2] *Laboratoire de Cardiologie Cellulaire et Moléculaire, INSERM CJF 92-11, Faculté de Pharmacie, Université de Paris-Sud, Châtenay-Malabry, France*

Abstract

Human cardiomyopathy has been extensively studied in the last decade, and knowledge of the functional and structural alterations of the heart has grown. However, understanding of the pathogenesis has come mostly from experimental studies. A number of work have been designed to elucidate if alterations of the contractile apparatus of cardiac cells contribute to the impairment of heart mechanics in cardiomyopathies. As well, an important question is to be solved: whether energy supply of the contraction-relaxation cycle is sufficient in the myopathic heart. Use of cardiac fibers skinned by different techniques allows to evaluate functional ability of myofibrils, mitochondria and bound creatine kinase which plays an important role in cardiomyocyte energy metabolism. The data presented in this chapter show that experimental cardiomyopathies of various types have some common features. These are an increase in calcium sensitivity of myofibrils and a depression of functional activity of mitochondrial creatine kinase. Possible mechanisms and physiological significance of these changes are discussed. (Mol Cell Biochem **133/134**: 287–298, 1994)

Key words: skinned fibers, mitochondrial respiration, Ca sensitivity, cardiac chronic pathologies, contractility, creatine kinase functional coupling

Introduction

In spite of extensive studies of the pathogenesis of human cardiomyopathy, the mechanisms of development of this disease is still not clear. However, experimental studies are contributing significantly to the knowledge of the functional and structural alterations in the myopathic heart. A number of models have been developed to reproduce this class of diseases in animals. Several following examples show the variety of the approaches to investigate experimental cardiomyopathies.

Hereditary forms of cardiomyopathies in hamsters and turkeys are considered to be very useful models of genetically-determined myocardial diseases. Some strains of Syrian hamsters at the age of 50–250 days in 100% of cases demonstrate signs of cardiac insufficiency associated with ventricular dilation or hypertrophy. Dilated form of spontaneous cardiomyopathy could be also screened in young turkeys. Diabetic cardiomyopathy has been reproduced in many studies by injections of streptozotocin. A diabetic-like type of cardiomyopathy has been reproduced in food-restricted rats [1]. A very interesting model of creatine-deficient cardiomyopathy could been obtained by chronic feeding of animals with

Address for offprints: Valimir Veksler, INSERN CJF 92-11 INSERN, Université Paris-Sud, 92 296 Châtenay-Malabry, France

a special creatine-free diet containing high quantity of β-guanidinopropionate. As a result, muscles of these animals have very low content of creatine and phosphocreatine so that intracellular creatine kinase (CK) system is depressed. A number of works were carried out in animals to elucidate the pathogenesis of adriamycin (doxorubicin)- induced cardiomyopathy. Adriamycin is an anthracycline antibiotic which is effective for various types of malignant neoplasms, however its clinical use is severely restricted by its cardiotoxicity.

Contractile apparatus in the myopathic heart

Regardless of their nature, all these types of diseases show impairments in cardiac pump function. Studies on isolated perfused hearts or isolated cardiac muscles demonstrated that cardiomyopathic myocardium has lower systolic force, diminished velocities of contraction and relaxation, decreased compliance and increased diastolic tone [for review see 2–4]. To check if the inability to generate normal force is a result of abnormalities in the contractile apparatus, function of myofibrils was studied *in situ*, using chemically skinned cardiac fibers.

Skinned fibers, i.e. strips of tissue treated with various detergents to destroy the sarcolemma, allow to investigate functional properties of cellular organelles by changing composition of the intracellular medium. To study the myofibrillar function, treatment with Triton X-100 is more often used. This detergent solubilizes all the cellular membrane structures, while myofibrils are still intact. Measurements of mechanical activity of such preparations under standard conditions of activation permit to evaluate functional state of the contractile system.

We have tested functional properties of Triton-skinned ventricular fibers obtained from animals with different kinds of experimental cardiomyopathies. Isometric tension developed by these fibers was measured at various free calcium concentrations at 22° C. All solutions contained (in mM): EGTA 10, free Mg^{2+} 3, MgATP 3–5, phosphocreatine 12–15, dithiothreitol 0.3, imidazole 30 (pH 7.0–7.1), ionic strength 160, adjusted with potassium acetate or propionate.

Measurements of the absolute force developed by skinned fibers obtained from hamsters with hereditary (CHF-146, and UM-X7.1 strains) cardiomyopathy and diabetic rats [5, 6] have shown that at equal sarcomere

length (2.2 μ) in the absence of calcium the values of resting tension are not different from respective controls. This means that passive properties of diseased myocardium are not changed and there is not any significant contribution of connective tissue proliferation.

Activation of the fibers with high Ca^{2+} (pCa 4.5) concentration induced maximal isometric tension development. We found [5, 6] that only in one type of cardiomyopathies (strain UM-X7.1, dilated form of hereditary cardiomyopathy [7]) the maximal Ca-activated force was significantly decreased (by 36%). In two other types of the disease there were no differences in maximal force, as compared to respective controls. Probably, some forms of experimental cardiomyopathies are associated with the loss of contractile material, since Herzig *et al.* [8] also showed a decrease in maximal calcium-activated force in one strain of hereditary myopathic hamsters.

Measurements of isometric force generated by skinned fibers for a stepwise increase in free calcium concentration [3, 5, 6] give the possibility to evaluate the calcium sensitivity of myofibrils. Figure 1 demonstrates pCa/tension relationships in 4 various types of cardiomyopathy (adriamycin-induced; hereditary, strain CHF-146; creatine-deficient; diabetic). As can be seen, in all cases myofibrillar calcium sensitivity was increased, that is at equal [Ca^{2+}], diseased myofibrils generated higher force as compared to respective controls. Only in one type of cardiomyopathies (hereditary dilated form, hamster strain UM-X7.1) calcium sensitivity of myofibrils did not differ from control. We have also found (in collaboration with I.V. Orekhova) that rat cardiac fibers in hypothyroidism, induced by administration of propylthiouracil, also are characterized by an increase in calcium sensitivity. Hypothyroid fibers skinned with saponin developed half-maximal force at [Ca^{2+}] which was by 21% lower than that it needed for half-maximal force in normal fibers. Similar significant shift in the force/pCa relation to lower [Ca^{2+}] in skinned fibers isolated from hypothyroid rat heart was observed by Gibson *et al.* [9]. Very recently Khandoudi *et al.* [10] also reported small but significant increase in calcium sensitivity of myofibrils in experimental diabetic cardiomyopathy.

One group of authors did not find any increase in calcium sensitivity of myofibrils in skinned fibers isolated from ventricles of cardiomyopathic humans [11] and turkeys [12]. However, in these works, the calcium sensitivity was rather high even in respective controls. For example, half-maximal force developed by human skinned

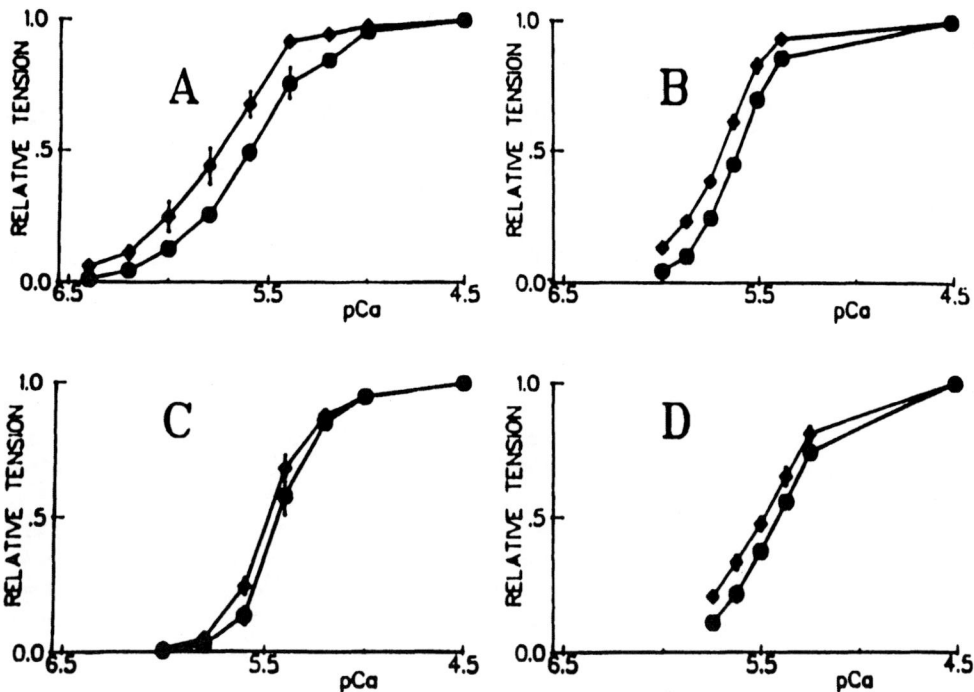

Fig. 1. Effect of calcium on relative isometric force developed by skinned ventricular fibers isolated from animals with adriamycin-induced (A), hypertrophic hereditary (B), creatine-deficient (C), and diabetic (D) cardiomyopathies (♦) and from respective control animals (●). Each value is mean ± s.e.m. Where standard error bars are not shown, they did not extend beyond symbol. (Reprinted with permission from [3]).

muscles at 1.06 μM Ca^{2+}. This was, possibly, due to the fact that the force/[Ca^{2+}] curves were obtained at a high sarcomere length, corresponding to a muscle length at which maximal active twitch force was achieved. The same authors reported lower calcium sensitivity of human skinned ventricular muscle, when force/[Ca^{2+}] relation was determined at the slack length (half-maximal tension was recorded at 1.45 μM [Ca^{2+}]) [13]. However, normal and failing myocardium were shown to have different responses to stretching [14]. Increase in sarcomere length indices a significant rise in calcium sensitivity in normal muscle, but not in failing myocardium. This means that differences in calcium sensitivity between normal and diseased myofibrils are more pronounced at a short sarcomere length and could be attenuated when the muscle is stretched.

It is not known, what is the cause of changes in myofibrillar calcium sensitivity. Nevertheless, several hypotheses were put forward. First, the increase in calcium sensitivity could be related to changes in troponin T isoform expression. Multiple troponin T subunits have been described [15–17] in several species. Their expression could be different in various stages of development and in different regions of the heart. A positive correlation has been found between pCa for half-maximal tension developed by newborn rabbit skinned cardiac fibers and relative amount of troponin T_2 [18]. Increase in

the myocardial expression of the T_2 isoform has been reported for human heart failure [19], where higher calcium sensitivity of myofibrils has been observed as well [14]. Anderson *et al.* [19] suggest that the increase in the relative amount of troponin T_2 in the failing human heart may be a recapitulation of ontogenic changes in troponin T expression.

Other proposed mechanisms of myofibrillar calcium sensitivity modulation under pathological conditions are related to the function of thick filament. One hypothesis considers that redistribution of myosin light chain isoforms could influence the activation of myofilaments by calcium. In the adult heart, ventricles and atria contain their characteristic light chain isoforms (VLC$_1$ and VLC$_2$ for ventricles; ALC$_1$ and ALC$_2$ for atria). Atrial light chains are normally expressed only in atrium, but ALC$_1$'s transiently appear in fetal ventricle [20, 21]. Development of cardiomyopathy is associated with the expression of atrial-like light chains in ventricles [22]. Very recently a correlation has been shown between the calcium sensitivity of myofibrils and expression of the atrial myosin light chain 1 in human diseased ventricles [23]. However, a cause-effect relationship between these events has not been yet established.

The next hypothesis [24, 25] relates the changes in myofibrillar calcium sensitivity with redistributions in myosin heavy chain isoforms. According to this hypoth-

Table 1. Indices of myofibrillar creatine kinase functional activity in skinned ventricular fibers from control and myopathic hearts

	$pMgATP_{50}$ in the absence of PCr	$pMgATP_{50}$ in the presence of PCr	CKeff	Time constant (ms) of tension recovery after quick stretch in the presence of	
				PCr + ATP	PCr + ADP
Control hamsters	3.59 ± 0.04	5.26 ± 0.03	1.69 ± 0.03	12.6 ± 1.0	13.4 ± 1.4
Cardiomyopathic CHF-146 hamsters	3.64 ± 0.05	$5.14 \pm 0.03**$	$1.49 \pm 0.05**$	13.8 ± 1.0	$18.4 \pm 1.1**$
Controls hamsters	3.44 ± 0.05	5.24 ± 0.12	1.80 ± 0.08	15.4 ± 0.8	16.7 ± 0.8
Cardiomyopathic UM-X7.1 hamsters	3.49 ± 0.07	5.05 ± 0.03	1.56 ± 0.12	$20.5 \pm 1.4*$	$20.2 \pm 1.1*$
Control rats	3.53 ± 0.04	5.13 ± 0.07	1.60 ± 0.10	20.1 ± 1.0	22.0 ± 1.3
Diabetic rats	3.72 ± 0.08	5.29 ± 0.05	1.56 ± 0.06	$24.4 \pm 1.0**$	$25.7 \pm 1.3*$

$pMgATP_{50}$ values ($pMgATP$ at which 50% rigor tension developed) were determined in the absence and in the presence of 12 mM phosphocreatine (PCr). CKeff, index of myofibrillar creatine kinase activity taken as the difference between the $pMgATP_{50}$ values in the presence and in the absence of PCr. Time constant of tension recovery was calculated as time for an *e*-fold change in tension assuming the process of tension recovery to be exponential. Time constant of tension recovery after quick stretch was determined at pCa 4.5 and 12 mM PCr + 3.16 mM MgATP or 12 mM PCr + 0.25 mM ADP. Values are means ± s.e.m. for 7–20 preparations. Statistical analysis was made between diseased animals and age-matched controls. CHF-146 myopathic hamsters and respective controls were obtained from Canadian Hybrid Farms (Halls Harbour, N.S., Canada); UM-X7.1 myopathic hamsters and respective controls were obtained from INSERM U-2 (Limeil-Brevannes, France). * = p < 0.05; ** = p < 0.01.

esis these changes are mediated by modulations of cross-bridges kinetics. In fact, human atria expressing the specific form of myosin heavy chains and having higher cross-bridges cycling rate than ventricles, show lower calcium sensitivity of myofilaments. Furthermore, several kinds of experimental cardiomyopathies (hereditary, diabetic) in rodents are associated with a redistribution of ventricular myosin heavy chain isoforms. Normally expressed V_1 isoenzyme is partially replaced by V_3 form [26, 27] possessing lower ATPase activity. This is considered to decrease the cross-bridge cycling rate and, in such way, could increase the sensitivity of myofibrils to calcium.

However, other data do not give a possibility to relate directly myosin heavy chain isoenzyme redistribution to the changes in myofibrillar calcium sensitivity. Heart failure in humans due to cardiomyopathy is associated with an increase in myofibrillar calcium sensitivity [14] and, probably, with a decrease in cross-bridge cycling rate, because the number of cross-bridges interactions during the isometric twitch is severely reduced in human non-skinned failing myocardium [28]. At the same time, there is no myosin heavy chain isoforms shift [29–31].

Earlier, we evaluated cross-bridge cycling rate in skinned ventricular fibers taken from animals with hereditary and diabetic cardiomyopathies [5, 6]. Quick stretches were applied to skinned fibers activated with a high calcium concentration. In dilated form of hamster hereditary cardiomyopathy and in diabetes in rats a decrease in the rate of tension recovery after stretch was observed (Table 1). This indicated a slowing of cross-

bridges cycling rate, apparently due to changes in myosin heavy chains isoenzyme pattern. Similar slowing of cross-bridges turnover rate well correlated with pronounced myosin isoenzymic shift to V_3 was observed in pressure overload-indiced hypertrophy in rats, but this was not accompanied by changes in myofibrillar calcium sensitivity [32]. Therefore, myosin heavy chain redistribution appears not to be the cause for calcium sensitivity modulation.

Altogether, these data do not allow to make any clear conclusion as to the cause of myofibrillar calcium sensitivity modulation. Increase in troponin T_2 isoform content, appearance of atrial-like myosin light chains in the ventricle, changes in myosin heavy chain isoenzymes probably represent general adaptive rearrangements in cardiac cells towards the fetal status. Elevated calcium sensitivity of myofilaments is also a feature of fetal myocardium [33, 34]. Under conditions of chronic pathology, when an imbalance exists between mechanical workload and energy supply, expression of fetal form of proteins appears to adjust functional properties of cardiomyocytes. The increase of calcium sensitivity of myofibrils would facilitate the tension development of cardiac muscle, and, in such way, partially compensate for the decreased contractility of the cardiomyopathic heart.

Functional state of myofibrillar creatine kinase

Important role of CK system in intracellular energy metabolism served as a reason to extensively study the enzyme function under conditions of myopathy. Total CK activity was shown to decrease in cardiomyopathic hearts. The enzyme activity in cardiac tissue of hereditary myopathic hamsters of CHF-146 strain [34], diabetic [36–38] and food-restricted rats [39] was decreased by 30–40%. A prominent decline in total CK activity has been found in cardiac biopsy samples taken from patients with dilated cardiomyopathy [40]. Analysis of the isoenzyme distribution has shown that percent representation of MM form does not change in diabetic [36, 37] and hereditary cardiomyopathic [35] hearts of animals as well as in human cardiomyopathic tissue [40]. However, hereditary and diabetic cardiomyopathies manifest different responses in B subunit content. Diabetes induces a significant decrease in percent representation of BB and MB forms [36, 37, 41], whereas hereditary cardiomyopathy leads to a relative increase in activity of these isoenzymes [35, 41]. Relative increase in MB creatine kinase isoenzyme has been found by Ingwall et al. [42] in ventricles of patients with coronary artery disease (chronic ischemia) and left ventricular hypertrophy due to aortic stenosis. These authors supposed that accumulation of MB creatine kinase represents a favorable adaptation to chronic energy deficiency, since the affinity of the MB isoenzyme for phosphocreatine is higher than the affinity of MM isoenzyme [43, 44]. They also assumed that hypoxia is a stimulus for activated synthesis of B subunit. In line with this assumption, Awaji et al. [41] considered that increased B subunit content in hamster hereditary cardiomyopathy is a reflection of myocardial hypoperfusion, whereas the decrease in B subunit content in diabetes represents another type of adaptation, which is not related to ischemic or hypoxic events.

Functional state of myofibrillar bound CK in situ in experimental cardiomyopathies was studied using Triton-skinned ventricular fibers. A stepwise decrease in MgATP concentration in myofibrillar environment without calcium gives rise to actomyosin crossbridges formation and to so-called rigor force generation (see chapter by Ventura- Clapier and Veksler in this issue). This isometric force developed is a characteristic of MgATP deficiency in the myofibrillar compartment. The presence of phosphocreatine in the medium shifts very sharply the dependency of the rigor tension on MgATP concentration towards a much lower MgATP concentration [45–47]. Since this shift is thought to be due to ADP rephosphorylation as a consequence of myofibrillar CK activity, one can take the value of the

shift as an index reflecting the functional state of CK [47].

Table 1 shows the values of $pMgATP_{50}$ ($pMgATP$ at which 50% rigor tension develops) in the absence and in the presence of phosphocreatine in various cardiomyopathies. It can be seen that the differences between these values, CKeff's (indices of myofibrillar CK functional activity), in cardiomyopathies and in respective age-matched controls are close. Only in hereditary myopathic strain CHF-146, the index of myofibrillar CK activity was significantly lower than in control.

The state of myofibrillar CK was also evaluated in the experiments with quick stretch of skinned fibers. Triton-treated muscles were activated with calcium (pCa 4.5) in the presence of 12 mM phosphocreatine and 0.25 mM MgADP, so that actomyosin ATPase was supplied only with MgATP generated by the CK reaction. Since the rate of tension recovery after a quick length change is dependent on the supply of actomyosin ATPase with MgATP, under these conditions the time constant of tension recovery can be considered as a parameter reflecting CK activity [47]. Data presented in Table 1 show that the time constant of tension recovery after quick stretch of control fibers in the presence of phosphocreatine and MgADP is close to that in the presence of relatively high concentrations of MgATP and phosphocreatine. This indicates a high efficacy of CK under control conditions. Myopathic fibers demonstrated a lower rate of tension recovery, as it was mentioned above. In two models of cardiomyopathies, replacement of a high ATP concentration by 0.25 mM MgADP in the presence of phosphocreatine did not markedly change the rate of tension recovery (Table 1). Only fibers isolated from CHF-146 myopathic hamsters responded to the change of ATP for ADP by an increase in time constant of tension recovery.

These data show that development of hereditary cardiomyopathy in hamsters of CHF-146 strain is followed by some depression in the functional activity of myofibrillar CK. Another model of hereditary myopathy and diabetes are not associated with a change in functional state of the bound enzyme, in spite of the fact that, at least in diabetes [36, 37] and in diabetic-like cardiomyopathy induced by food restriction [39], specific activity of the cardiac MM isoform decreases. It appears that myofibrillar CK is a rather conservative and stable part of the CK system in myocytes. Severe ischemia followed by reperfusion, which is known to induce considerable disorders in cardiac CK system, does not change properties of the enzyme bound to myofibrils [48]. Myofibrils isolated from skeletal muscle of chickens with muscular dystrophy and from muscle of humans with Duchenne dystrophy have also demonstrated normal CK activity [49].

Mitochondrial function in the myopathic heart

Experimental cardiomyopathies are characterized by an altered energy metabolism. In myopathic hamsters, tissue levels of ATP and phosphocreatine has been shown to decrease considerably in the isolated perfused hearts [50–52] as well as in hearts *in situ* [52], whereas inorganic phosphate is increased and intracellular pH is decreased [50, 53], thus indicating a metabolic deficiency. Myocardial high-energy phosphate content was shown to be depleted in other models of cardiomyopathies (diabetic, adriamycin-, norepinephrine-induced) [3, 54–56]. A number of studies have been undertaken to determine whether or not a defect in the oxidative phosphorylation exists in cardiomyopathic hearts. Mitochondria isolated from hearts of rats with streptozotocin-induced diabetes were shown to have lower State 3 respiration rate and decreased oxidative phosphorylation capacity (rate of ADP phosphorylation) [57, 58]. The State 3 respiration rate and respiratory control index of mitochondria isolated from hearts of genetically diabetic mice and littermate controls were dependent on substrates used as well as on technique of mitochondria preparation [59]. In some cases, diabetic mitochondria demonstrated lower ADP-stimulated oxygen consumption (for example, in the presence of pyruvate + malate as substrates).

The experiments with isolated mitochondria from hearts of hereditary cardiomyopathic hamsters gave disparate results. Two groups of workers [60–62] observed decreased respiratory control and lower oxygen consumption rate in animals without signs of cardiac failure; hamsters with manifestation of congestive failure exhibited decreases in both of these parameters. Panagia *et al.* [63] showed that mitochondrial respiratory activity in cardiomyopathy was depressed only during severe stages of heart failure. Wrogemann and colleagues [64, 65] consistently failed to detect any abnormality of oxidative phosphorylation in cardiomyopathic mitochondria.

These discrepancies were suggested to be a consequence of the presence of two metabolically distinct populations of mitochondria (subsarcolemmal and interfibrillar) having different properties. Indeed, State 3 respiration is markedly decreased in interfibrillar mito-

chondria obtained from hereditary myopathic hamsters, whereas subsarcolemmal mitochondria have normal respiration rate [66]. These authors, however, have not found any differences in respiratory control indices of mitochondrial subpopulations isolated from normal and diseased hearts.

Moreover, as has been mentioned above, method of isolation of mitochondria could influence the properties of mitochondria even in normal tissue. In addition, in case of pathology, changes in diseased myocardium could affect processes of mitochondria isolation so that characteristics of mitochondrial fractions obtained could be rather different from the properties of these organelles *in situ*.

A few years ago, investigation of mitochondrial functioning in experimental cardiomyopathies was undertaken using a method of saponin-skinned fibers [67] for the study of respiration of total cardiac mitochondrial population without its isolation [5, 6]. Bundles of ventricular fibers were treated with saponin to selectively destroy the integrity of the sarcolemma under conditions which preserve intactness of mitochondria. Respiratory activity of the mitochondria *in situ* was measured at 22° C in a medium, which had the composition close to that of diastolic intracellular milieu (0.1 µM free Ca^{2+}, 3 mM free Mg^{2+}, 20 mM taurine, 3 mM phosphate, 2–5 mg/ml fatty acid free bovine serum albumin, 20 mM imidazole, pH 7.0). Physiological ionic strength as high as 160 mM was adjusted by addition of potassium-2-(N-morpholino) ethanesulphonate, respiratory substrates were 5 mM glutamate and 2 mM malate.

Figure 2 shows an oxygraphic trace of respiration of mitochondria in saponin-skinned fibers. The fibers were transferred into the medium without phosphate acceptor. The basal respiration rate was accelerated by the addition of 100 µM ADP. Oxygen consumption rate under these conditions was far from the maximal one since apparent K_m for ADP estimated in skinned fibers and cardiomyocytes is quite high (around 300 µM) [40, 68, see also the chapter by Saks *et al.* in this issue]. There was no transition to State 4, since skinned fibers possessed a high ATPase activity. Further addition of 20 mM creatine, which activates CK reaction catalyzed by the mitochondrial isoenzyme, enhanced the respiration rate further. The relative extent of stimulation of respiration by creatine in the presence of low ADP concentration could be taken as an index of functional activity of mitochondrial creatine kinase (see below). The maximal rate of oxygen consumption was observed after addition of 1 mM ADP. Carboxyatractyloside, an inhibitor of

293

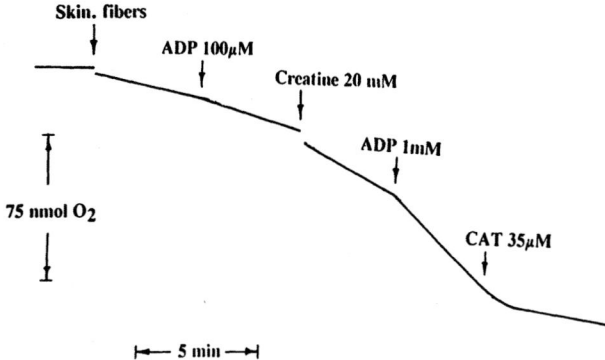

Fig. 2. An oxygraph trace of respiratory activity of mitochondria in saponin-skinned rat ventricular fibers. The arrows indicate time of additions of fibers, ADP, creatine, and carboxyatractyloside (CAT) into the medium (volume 3 ml).

adenine nucleotide translocase, depressed the respiration rate considerably, indicating a good integrity of the inner mitochondrial membrane.

Studies [5, 6, 69] on maximal ADP-stimulated oxygen consumption by total mitochondrial population *in situ* in three different groups of cardiomyopathic animals revealed a variety in respiratory activity in different forms of diseases (Fig. 3A). Saponin-skinned fibers isolated from ventricles of hamsters (175–200 days old) with transitory hypertrophic cardiomyopathy (strain CHF-146) and with dilated form of cardiomyopathy (strain UM-X7.1) demonstrated normal maximal ADP-stimulated oxygen consumption rate. It is interesting to compare these data with those obtained by Khuchua *et*

al. [35] in mitochondria isolated from the hearts of myopathic hamsters of CHF-146 strain. These authors found that cytochrome aa₃ content per mg of mitochondrial protein was not decreased, and State 3 respiration rate did not significantly differ from control. This indicated normal status of the mitochondrial respiratory chain in myopathic cardiac cells. At the same time, cytochrome aa₃ content per gram of cardiac tissue decreased by 37%. Consequently, the myopathic hearts had lower content of respiratory units. This could be explained by the fact that the myopathic CHF-146 hearts had extensive foci of fibrosis and necrosis [70]. Apparently, the myopathic heart homogenate used for cytochrome determination contained lower percentage of myocyte constituents. On the other hand, thin bundles of cardiac cells prepared for respiration measurements, probably, did not include significant quantities of non-muscle tissue. This is also confirmed by the observation that myopathic skinned fibers developed maximal calcium-activated force similar to control value [5].

In contrast, State 3 respiration rate of mitochondria in skinned fibers isolated from diabetic rat hearts was significantly depressed. Similar decrease (by 30–60%) in maximal ADP-stimulated respiration was recently reported [71] for digitonin-treated cardiomyocytes from diabetic rats; the degrees of inhibition of respiration were greater in the presence of NAD-linked substrates. Insulin treatment of rats corrected the abnormalities in cell respiration caused by diabetes. Some depression in

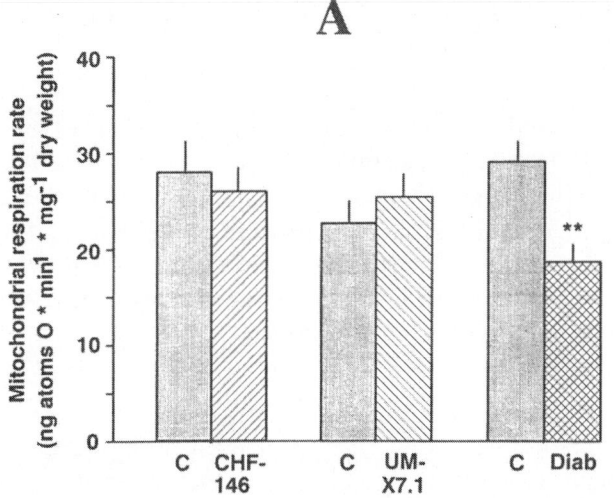
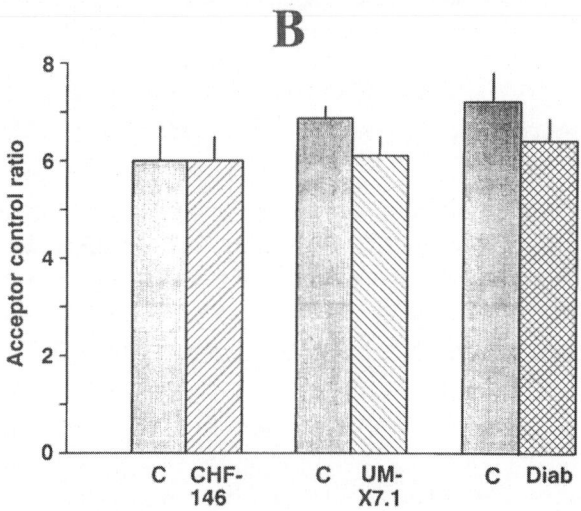

Fig. 3. Maximal rate of oxygen consumption (A) and acceptor control ratio (B) of mitochondria in saponin-skinned ventricular fibers isolated from cardiomyopathic hamsters of CHF-146 (175–200 days) and UM-X7.1) (175–200 days) strains, diabetic (diab) rats and respective controls. Maximal respiration rate was determined in the presence of 1 mM ADP and 20 mM creatine. Acceptor control ratio was taken as the ratio of maximal respiration rate to basal respiration rate in the absence of added ADP. CHF-146 myopathic hamsters and respective controls were obtained from Canadian Hybrid Farms (Halls Harbour, N.S., Canada); UM-X7.1 myopathic hamsters and respective controls were obtained from INSERM U-2 (Limeil-Brevannes, France). ** = p < 0.01.

294

mitochondrial State 3 respiration rate was found in diabetic-like cardiomyopathy induced by a food restriction [39].

Interestingly, acceptor control ratio (the ratio of maximal ADP-stimulated respiration rate to the respiration rate in the absence of ADP) did not change in experimental cardiomyopathies (Fig. 3B). Therefore, one can conclude that oxidation-phosphorylation coupling is not impaired in cardiac mitochondria in cardiomyopathy, at least in the experimental conditions used – low [Ca^{2+}], neutral pH, etc.

Thus, the studies on total mitochondrial population *in situ* show that in different forms of cardiomyopathies there are different patterns of mitochondrial function. Skinned fibers from two strains of myopathic hamsters demonstrated rather normal respiratory activities. Severe depression of maximal ADP-stimulated respiratory capacity of skinned fibers and cardiomyocytes in diabetes indicates considerable derangements in oxidative metabolism and deserves further detailed investigations.

In situ studies of mitochondrial function in skinned fibers have been also carried out in human cardiomyopathies [40]. Biopsy samples were obtained during angiographic examination from patients with dilated cardiomyopathy of various degrees of congestive heart failure. The data obtained by these authors show that the maximal ADP-stimulated respiration rate of fibers under standard conditions decreased only in patients with very pronounced cardiac insufficiency.

Functional state of mitochondrial creatine kinase

Several studies have shown a decrease in specific activity of mitochondrial creatine kinase (mi-CK) in cardiac cells in cardiomyopathy. Activity of the mitochondrial isoenzyme in cardiac homogenates was found to be lowered in hereditary myopathy (hamster strain CHF-146) [35], diabetes [36, 37], catecholamine-induced cardiomyopathy [41], hypothyroidism [72]. Only in hearts of myopathic Bio 14.6 hamsters, mi-CK activity was observed to be normal [41]. Specific activity of the enzyme in cardiac mitochondria isolated from myopathic CHF-146 hamsters [35] and diabetic rats [58] was lowered by 32–35%.

Functional state of mi-CK can be well estimated by use of saponin-treated cardiac fibers or cells. Functional coupling between CK and ATP-ADP translocase results

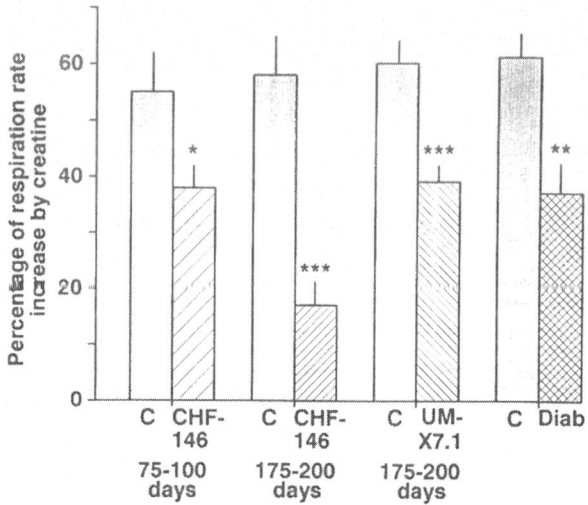

Fig. 4. Functional activity of mitochondrial creatine kinase in skinned fibers in various models of experimental cardiomyopathies. Percentages of the respiration rate increase was determined in the presence of 100 µM ADP after an addition of 20 mM creatine. Values are means ± s.e.m. Statistical analysis was made between diseased animals and age-matched controls. * = p < 0.05; ** = p < 0.01; *** = p < 0.001.

in decrease of apparent K_m of oxidative phosphorylation for ADP in the presence of creatine. Therefore, provided that this coupling exists, mitochondrial respiration rate at submaximal ADP concentrations should be increased after addition of creatine. It is important to mention, that in skinned fibers coupled CK reaction is capable to accelerate respiration rate even in the presence of a powerful ADP-regenerating system (cellular ATPases). Functional activity of CK could be assessed by relative enhancement of oxygen consumption rate by creatine.

Studies on skinned fibers isolated from ventricles in cardiomyopathy [4–6, 40, 72] consistently revealed an impairment of functional status of mi-CK in experimental animals and humans. In the presence of 100 µM ADP, addition of 20 mM creatine to respiring skinned fibers from normal hamsters and rats led to an increase in oxygen consumption rate by about 55–60% (Fig. 4). This value (an index of functional mi-CK activity) was significantly diminished in various models of experimental cardiomyopathies. Studies on two ages of CHF-146 myopathic hamsters showed that CK function impaired progressively, the creatine-stimulated respiration being more depressed in older animals.

Using the same technique, Seppet *et al.* [72] found a depression in mi-CK state in hypothyroidism. In their experiments, creatine addition decreased the K_m for ADP in skinned fibers from euthyroid rats 2.23 times, whereas in fibers from hypothyroid rats the K_m value was lowered by creatine only 1.58 times.

Investigation of CK function was also carried out using biopsy samples taken from patients with different classes of heart failure due to dilated cardiomyopathy [4, 40]. A clear linear correlation was found between stimulation of respiration by creatine and 1) left ventricular end-diastolic pressure (negative correlation), and 2) left ventricular ejection fraction (positive correlation). Thus, more severe heart failure was associated with more pronounced depression of mi-CK functional state.

It is worthwhile to emphasize a difference between the functional activity of mi-CK and its specific activity. The former is a consequence of the functional coupling between this enzyme and adenine nucleotide translocase; under these conditions, in the presence of active cellular ATPases in skinned cells (there is no respiratory State 4), creatine is able to increase the local ADP concentration in the vicinity of translocase. In contrast, specific activity gives an information only about the presence of chemically active enzyme in the mitochondrial compartment. Of course, this is a prerequisite for the enzyme to be coupled, but not a sufficient one. Therefore, one could not always expect to have a direct correlation between functional and specific activities.

Such a discordance can be found in a paper of Seppet et al. [72]. Hyperthyroidism in rats slightly diminished specific activity of mi-CK. At the same time, analysis of skinned fiber respiration showed that in hyperthyroidism creatine decreased K_m for ADP significantly more than in control (3.25 times vs 2.23 times), thus indicating an elevated mi-CK functional activity. An inverse situation was observed by Clark et al. [73]. They found an increase in mi-CK specific activity in guinea-pig myometrium during gestation, but creatine had no effect on respiration of saponin-skinned fibers.

Compared to specific activity, the functional activity of the enzyme is more sensitive to changes in cellular metabolism. Acute reversible ischemia of perfused rat heart induces a rapid depression of the mi-CK functional activity [67, 74], but reperfusion is able to restore it [74]. This suggests that, despite an impairment of functional coupling, CK probably does not leave mitochondrial compartment during ischemia. Very recently [75], we investigated effects of ischemic factors (elevated concentration of inorganic phosphate and lowered pH) on functional activity of the enzyme. High [Pi] (20 mM) exerted a very strong depressive action on the functional activity of mi-CK in skinned rat cardiac fibers. This result is consistent with data showing that high [Pi] causes

a solubilization of mi-CK [76] in isolated mitochondria. However, if skinned fibers, which demonstrated very weak mi-CK functional activity in high [Pi] solution, were transferred into a solution with normal (3 mM) [Pi], the functional ability of mi-CK was completely restored. These data clearly show that mi-CK solubilized by high [Pi] does not leave the mitochondrial compartment, apparently being present in the intermembrane space. However, the redistribution between bound and solubilized forms of the enzyme substantially depressed its functional ability. This is in agreement with data of Kuznetsov et al. [77], where a dissociation of mi-CK from the inner membrane of isolated mitochondria into the intermembrane space by isotonic KCl has led to a reversible uncoupling between mi-CK and oxidative phosphorylation.

The second ischemic metabolic factor, lowered pH, had an opposite effect on the function of mi-CK in skinned fibers. At low [Pi] in the acid (pH 6.6) medium, creatine stimulated respiration even more than at normal (7.1) pH. This effect could not be attributed to an activation of mi-CK by protons, as it has been shown that the optimum pH for this enzyme in rat heart is 8.0 in the direction of phosphocreatine formation [78]. Most likely, the influence of acidosis is mediated by changes in affinity of the enzyme to the inner membrane structures [76]. This suggestion was confirmed by the experiments where high [Pi] did not depress the functional activity of mi-CK in the acid medium [75]. Lowering of pH seems to affect the same process as elevated [Pi] does, but in the opposite direction, that is to prevent the solubilization of mi-CK.

Thus, the functional activity of mi-CK in cardiac muscle could be controlled by state of the enzyme in the intermembrane space and, of course, by the specific activity of mi-CK as a consequence of expression of the enzyme. Cardiomyopathies of various genesis with chronically lowered cardiac mechanical work have been shown to associate with a depression in mi-CK functional activity. It is not clear what is the cause for this depression. Nevertheless, one can mark a relationship between the activity of mi-CK and long-term functional status of ventricular muscle. Moreover, this relationship exists not only for cardiac muscle, but for skeletal one as well. Mi-CK specific activity per unit of mitochondrial protein decreases in skeletal muscle of dystrophic chickens with disorders of mechanical function [79, 80]. Accordingly, this disease is accompanied by a loss of regulation by creatine on mitochondrial respiration [80].

In turn, long-term increase in muscular mechanical

function induces a rise in mi-CK activity. Hyperthyroidism, known to elevate cardiac work, led to an increase in functional activity of mi-CK in rat ventricle [72]. In respiring skinned fibers, creatine decreased K_m for ADP 3.25 times (in fibers isolated from euthyroid rats only 2.23 times). Chronic stimulation of a rabbit fast-twitch skeletal muscle induced 4-fold increase in mi-CK specific activity [81]. Endurance training of rats resulted in a significant rise of mi-CK activity in fast-twitch muscles [82]. In human gastrocnemius muscle obtained from long-distance runners, mi-CK activities were significantly greater than in nonrunning control skeletal muscle and significantly increased during training for and after a marathon race [83].

All these data allow to suggest that mi-CK activity, specific, and, apparently, more precisely, functional one, is a reflection of long-term level of intracellular energy fluxes. This activity rises under conditions of elevated workload to meet the increased energy demands. Diminished activities of cellular ATPases (first of all, myofibrillar ATPase) in cardiomyopathy and, therefore, lowered energy turnover are accompanied with an attenuation of mi-CK function, a crucial point of energy flux regulation.

References

1. Savabi F, Kirsch A: Diabetic type of cardiomyopathy in food-restricted rats. Can J Physiol Pharmacol 70: 1040–1047, 1992
2. Kapelko VI, Veksler VI, Popovich MI: Cellular mechanisms of alterations in myocardial contractile function in experimental cardiomyopathies. Biomedical Science 1: 77–83, 1990
3. Kapelko VI, Veksler VI, Popovich MI, Ventura-Clapier R: Energy-linked functional alterations in experimental cardiomyopathies. Am J Physiol Suppl (Oct) 261: 39–44, 1991
4. Kapelko VI, Popovich MI, Veksler VI, Ventura-Clapier R, Khuchua ZA, Saks VA: Subcellular basis for increased diastolic stiffness in experimental cardiomyopathies. In: M Nagano, N Takeda, NS Dhalla (eds). The Cardiomyopathic Heart. Raven Press, Ltd, New York, 1994, pp 185–195
5. Veksler VI, Ventura-Clapier R, Lechene P, Vassort G: Functional state of myofibrils, mitochondria and bound creatine kinase in skinned ventricular fibers of cardiomyopathic hamsters. J Mol Cell Cardiol 20: 329–342, 1988
6. Veksler VI, Murat I, Ventura-Clapier R: Creatine kinase and mechanical and mitochondrial functions in hereditary and diabetic cardiomyopathies. Can J Physiol Pharmacol 69: 852–858, 1991
7. Kapelko VI, Parmley WW, Wu S, Stone RD, Jasmin G, Wikman-Coffelt J: Increased left ventricular diastolic stiffness in the early phase of hereditary cardiomyopathy. Am Heart J 116: 765–770, 1988
8. Herzig JW, Gerber W, Salzmann R: Heart failure and Ca^{++} activation of the cardiac contractile system: hereditary cardiomyopathy in hamsters (BIO 14.6), isoprenaline overload and the effect of APP 201–533. Basic Res Cardiol 82: 326–340, 1987
9. Gibson LM, Wendt IR, Stephenson DG: Contractile activation properties of ventricular myocardium from hypothyroid, euthyroid and juvenile rats. Pflügers Arch 422: 16–23, 1992
10. Khandoudi N, Guo AC, Chesnais M, Feuvray D: Skinned cardiac fibres of diabetic rats: contractile activation and effects of 2,3-butanedione monoxime (BDM) and caffeine. Cardiovasc Res 27: 447–452, 1993
11. Hajjar RJ, Gwathmey JK: Cross-bridge dynamics in human ventricular myocardium. Regulation of contractility in the failing heart. Circulation 86: 1819–1826, 1992
12. Gwathmey JK, Hajjar RJ: Calcium-activated force in a turkey model of spontaneous dilated cardiomyopathy: adaptive changes in thin myofilament Ca^{2+} regulation with resultant implications on contractile performance. J Mol Cell Cardiol 24: 1459–1470, 1992
13. Gwathmey JK, Hajjar RJ: Effect of protein kinase C activation on sarcoplasmic reticulum function and apparent myofibrillar Ca^{2+} sensitivity in intact and skinned muscles from normal and diseased human myocardium. Circ Res 67: 744–752, 1990
14. Schwinger RHG, Böhm M, Koch A, Uhlmann R, Schmidt U, Morano I, Rüegg JC, Erdmann E: Attenuated force-tension relationship in the failing human myocardium. J Mol Cell Cardiol 25 (Suppl I): S20, 1993
15. Cooper TA, Ordahl CP: A single cardiac troponin T gene generates embryonic and adult isoforms via developmentally regulated alternate splicing. J Biol Chem 260: 11140–11148, 1985
16. Anderson PAW, Moore GE, Nassar R: Developmental changes in the expression of rabbit left ventricular troponin T. Circ Res 63: 742–747, 1988
17. Anderson PAW, Oakeley AE: Immunological identification of five troponin T isoforms reveals an elaborate maturational troponin T profile in rabbit myocardium. Circ Res 65: 1087–1093, 1988
18. Nassar R, Malouf NN, Kelly MB, Oakeley AE, Anderson PAW: Force-pCa relation and troponin T isoforms of rabbit myocardium. Circ Res 69: 1470–1475, 1991
19. Anderson PAW, Malouf NN, Oakeley AE, Pagani ED, Allen PD: Troponin T isoform expression in humans. A comparison among normal and failing adult heart, fetal heart, and adult and fetal skeletal muscle. Circ Res 69: 1226–1233, 1991
20. Cummins P, Price KM, Littler WA: Foetal myosin light chain in human ventricle. J Muscle Res Cell Motil 1: 357–366, 1980
21. Price KM, Littler WA, Cummins P: Human atrial and ventricular myosin light-chain subunits in the adult and during development. Biochem J 191: 571–580, 1980
22. Hirzel OH, Tuchschmid CR, Schneider J, Krayenbuehl HP, Schaub MC: Relationship between myosin isoenzyme composition, hemodynamics, and myocardial structure in various forms of human cardiac hypertrophy. Circ Res 57: 729–740, 1985
23. Morano I, Hädicke K, Böhm M, Erdmann E: Expression of the atrial myosin light chain 1 in the human ventricle increased Ca^{2+}-sensitivity. J Mol Cell Cardiol 25 (Suppl): S131, 1993
24. Morano I, Bletz C, Wojciechowski R, Rüegg JC: Modulation of crossbridge kinetics by myosin isoenzymes in skinned human heart fibers. Circ Res 68: 614–618, 1991
25. Morano I: Myosin light chain phosphorylation and myosin isoenzyme expression regulate cardiac calcium sensitivity by modulation of cross-bridge cycling kinetics. In: JA Lee, DG Allen (eds). Modulation of Cardiac Calcium Sensitivity. A New Approach to

Increasing the Strength of the Heart. Oxford University Press Inc., New York, 1993, pp 178–196

26. Malhotra A, Karell M, Scheuer J: Multiple cardiac contractile protein abnormalities in myopathic Syrian hamsters (BIO 53:58). J Mol Cell Cardiol 17: 95–107, 1985

27. Pollack PS, Malhotra A, Fein FS, Scheuer J: Effects of diabetes on cardiac contractile proteins in rabbits and reversal with insulin. Am J Physiol 251: H448–H454, 1986

28. Hasenfuss G, Mulieri LA, Leavitt BJ, Allen PD, Haeberle JR, Alpert NR: Alteration of contractile function and excitation-contraction coupling in dilated cardiomyopathy. Circ Res 70: 1225–1232, 1992

29. Mercadier JJ, Bouveret P, Gorza L, Schiaffino S, Clark WA, Zak R, Swynghedauw B, Schwartz K: Myosin isoenzymes in normal and hypertrophied human ventricular myocardium. Circ Res 53: 52–62, 1983

30. Feldman AM, Ray PE, Silan CM, Meercer JA, Minobe W, Bristow MR: Selective gene expression in failing human heart. Circulation 83: 1866–1872, 1991

31. Solaro RJ, Powers FM, Gao L, Gwathmey JK: Control of myofilament activation in heart failure. Circulation 87 (Suppl VII): 38–43, 1993

32. Ventura-Clapier R, Mekhfi H, Oliviero P, Swynghedauw B: Pressure overload changes cardiac skinned-fiber mechanics in rat, not in guinea pigs. Am J Physiol 254: H517–H524, 1988

33. McAuliffe JJ, Gao L, Solaro RJ: Changes in myofibrillar activation and troponin C Ca^{2+} binding associated with troponin T isoform switching in developing rabbit heart. Circ Res 66: 1204–1216, 1990

34. Reiser PJ, Westfall M, Solaro RJ: Developmental transition in myocardial troponin-T (TnT) isoforms correlates with a change in Ca^{2+} sensitivity. Biophys J 57: 549a, 1990

35. Khuchua ZA, Ventura-Clapier R, Kuznetsov AV, Grishin MN, Saks VA: Alterations in the creatine kinase system in the myocardium of cardiomyopathic hamsters. Biochem Biophys Res Commun 165: 748–757, 1989

36. Popovich BK, Boheler KR, Dillmann WH: Diabetes decreases creatine kinase enzyme activity and mRNA level in the rat heart. Am J Physiol 257: E573–E577, 1989

37. Savabi F, Kirsch A: Alteration of the phosphocreatine energy shuttle components in diabetic rat heart. J Mol Cell Cardiol 23: 1323–1333, 1991

38. Su C-Y, Payne M, Strauss AW, Dillmann WH: Selective reduction of creatine kinase subunit mRNAs in striated muscle of diabetic rats. Am J Physiol 263: E310–E316, 1992

39. Kirsch A, Savabi F: Effect of food restriction on the phosphocreatine energy shuttle components in rat heart. J Mol Cell Cardiol 24: 821–830, 1992

40. Saks VA, Belikova YO, Kuznetsov AV, Khuchua ZA, Branishte TH, Semenovsky ML, Naumov VG: Phosphocreatine pathway for energy transport: ADP diffusion and cardiomyopathy. Am J Physiol Suppl (Oct) 261: 30–38, 1991

41. Awaji Y, Hashimoto H, Matsui Y, Kawaguchi K, Okumura K, Ito T, Satake T: Isoenzyme profiles of creatine kinase, lactate dehydrogenase, and aspartate aminotransferase in the diabetic heart: comparison with hereditary and catecholamine cardiomyopathies. Cardiovasc Res 24: 547–554, 1990

42. Ingwall JS, Kramer MF, Fifer MA, Lorell BH, Shemin R, Grossman W, Allen PD: The creatine kinase system in normal and diseased human myocardium. New Engl J Med 313: 1050–1054, 1985

43. Wong PCP, Smith AF: Biochemical differences between the MB and MM isoenzymes of creatine kinase. Clin Chim Acta 68: 147–158, 1976

44. Szasz G, Gruber W: Creatine kinase in serum. 4. Differences in substrate affinity among the isoenzymes. Clin Chem 24: 245–249, 1978

45. Veksler VI, Kapelko VI: Creatine kinase in regulation of heart function and metabolism. II. The effect of phosphocreatine on the rigor tension of EGTA-treated rat myocardial fibers. Biochim Biophys Acta 803: 265–270, 1984

46. Ventura-Clapier R, Vassort G: Role of myofibrillar creatine kinase in the relaxation of rigor tension in skinned cardiac muscle. Pflügers Arch 404: 157–161, 1985

47. Ventura-Clapier R, Mekhfi H, Vassort G: Role of creatine kinase on force development in chemically skinned rat cardiac muscle. J Gen Physiol 89: 815–837, 1987

48. Ventura-Clapier R, Veksler VI, Elizarova GV, Mekhfi H, Levitskaya EL, Saks VA: Contractile properties and creatine kinase activity of myofilaments following ischemia and reperfusion of the rat heart. Biochem Med Metab Biol 38: 300–310, 1987

49. Feit H, Fuseler J, Cook JD: Myofibrillar creatine kinase in Duchenne and avian muscular dystrophy. Biochem Med 29: 355–359, 1983

50. Sievers R, Parmley WW, James T, Wikman-Coffelt J: Energy levels at systole vs. diastole in normal hamster hearts vs. myopathic hamsters hearts. Circ Res 53: 759–766, 1983

51. Whitmer JT: Energy metabolism and mechanical function in perfused hearts of Syrian hamsters with dilated or hypertrophic cardiomyopathy. J Mol Cell Cardiol 18: 307–317, 1986

52. Wikman-Coffelt J, Sievers R, Parmley WW, Jasmin G: Verapamil preserves adenine nucleotide pool in cardiomyopathic hamster. Am J Physiol 250: H22–H28, 1986

53. Markiewicz W, Wu SS, Parmley WW, Higgins CB, Sievers R, James TL, Wikman-Coffelt J, Jasmin G: Evaluation of the hereditary Syrian hamster cardiomyopathy by ^{31}P nuclear magnetic resonance spectroscopy: improvement after acute verapamil therapy. Circ Res 59: 597–604, 1986

54. Allison TB, Bruttig SP, Crass MF III, Eliot RS, Shipp JC: Reduced high-energy phosphate levels in rat hearts. I. Effects of alloxan diabetes. Am J Physiol 230: 1744–1750, 1976

55. Miller TB Jr: Cardiac performance of isolated perfused hearts from alloxan diabetic rats. Am J Physiol 236: H808–H812, 1979

56. Kapelko VI, Popovich MI, Sharov VG, Kostin SI, Schulzhenko VS, Golikov MA, Saks VA: The ultrastructural, metabolic and functional alterations of the heart at prolonged adriamycin treatment. J Appl Cardiol 4: 79–89, 1989

57. Pierce GN, Dhalla NS: Heart mitochondrial function in chronic experimental diabetes in rats. Can J Cardiol 1: 48–54, 1985

58. Savabi F: Mitochondrial creatine phosphokinase deficiency in diabetic rat heart. Biochem Biophys Res Commun 154: 469–475, 1988

59. Kuo TH, Giacomelli F, Wiener J: Oxidative metabolism of Polytron versus Nagarse mitochondria in hearts of genetically diabetic mice. Biochim Biophys Acta 806: 9–15, 1985

60. Lochner A, Opie LH, Brink AJ, Bosman AR: Defective oxidative phosphorylation in hereditary myocardiopathy in the Syrian hamster. Cardiovasc Res 3: 297–307, 1968

61. Jasmin G, Eu HY: Cardiomyopathy of hamster dystrophy. Ann N Y Acad Sci 317: 46–58, 1979

62. Proschek L, Jasmin G: Hereditary polymyopathy and cardiomyo-

298

pathy in the Syrian hamster: 2. Development of heart necrotic changes in relation to defective mitochondrial function. Muscle Nerve 5: 26–32, 1982

63. Panagia V, Lee SL, Singh A, Pierce GN, Jasmin G, Dhalla NS: Impairment of mitochondrial and sarcoplasmic reticular functions during the development of heart failure in cardiomyopathic (UM-X7.1) hamsters. Can J Cardiol 2: 236–247, 1986

64. Wrogemann K, Blanchaer MC, Jacobson BE: Oxidative phosphorylation at various stages of the genetically determined cardiomyopathy in the Syrian hamster. Recent Adv Studies Cardiac Struct Metabol 3: 467–478, 1973

65. Wrogemann K, Blanchaer MC, Thakar JH, Mezon BJ: On the role of mitochondria in the hereditary cardiomyopathy of the Syrian hamster. Recent Adv Studies Cardiac Struct Metabol 6: 231–241, 1975

66. Hoppel CL, Tandler B, Parland W, Turkaly JS, Albers LD: Hamster cardiomyopathy. A defect in oxidative phosphorylation in the cardiac interfibrillar mitochondria. J Biol Chem 257: 1540–1548, 1982

67. Veksler VI, Kuznetsov AV, Sharov VG, Kapelko VI, Saks VA: Mitochondrial respiratory parameters in cardiac tissue: a novel method of assessment by using saponin-skinned fibers. Biochim Biophys Acta 892: 191–196, 1987

68. Saks VA, Belikova YO, Kuznetsov AV: In vivo regulation of mitochondrial respiration in cardiomyocytes: specific restrictions for intracellular diffusion of ADP. Biochim Biophys Acta 1074: 302–311, 1991

69. Khuchua ZA, Kuznetsov AV, Vasilyeva EV, Ventura-Clapier R, Clark J, Steinschneider AY, Korchazhkina OV, Lakomkin VL, Branishte T, Ruuge EK, Kapelko VI, Saks VA: The creatine kinase system and cardiomyopathy. Am J Cardiovasc Pathology 4: 223–234, 1992

70. Hunter EG, Hughes V, White J: Cardiomyopathic hamsters, CHF 146 and CHF 147: a preliminary study. Can J Physiol Pharmacol 62: 1423–1428, 1984

71. Tanaka Y, Konno N, Kako KJ: Mitochondrial dysfunction observed in situ in cardiomyocytes of rats in experimental diabetes. Cardiovasc Res 26: 409–414, 1992

72. Seppet EK, Kairane CB, Khuchua ZA, Kadaya LY, Kallikorm AP, Saks VA: Hormone regulation of cardiac energy metabolism. III. Effect of thyroid state on distribution of creatine kinase isoenzymes and creatine-controlled respiration in cardiac muscle. J Appl Cardiol 6: 301–311, 1991

73. Clark JF, Khuchua Z, Kuznetsov A, Saks VA, Ventura-Clapier R: Compartmentation of creatine kinase isoenzymes in myometrium of gravid guinea-pig. J Physiol 466: 553–572, 1993

74. Saks VA, Kapelko VI, Kupriyanov VV, Kuznetsov AV, Lakomkin VL, Veksler VI, Sharov VG, Javadov SA, Seppet EK: Quantitative evaluation of relationship between cardiac energy metabolism and post-ischemic recovery of contractile function. J Mol Cell Cardiol 21 (Suppl): 67–78, 1989

75. Veksler VI, Ventura-Clapier R: Ischemic metabolic factors – high inorganic phosphate and acidosis – modulate mitochondrial creatine kinase functional activity in skinned cardiac fibers. J Mol Cell Cardiol 26: 335–339, 1994

76. Vial C, Font B, Goldschmidt D, Gautheron DC: Dissociation and reassociation of creatine kinase with heart mitochondria; pH and phosphate dependence. Biochem Biophys Res Commun 88: 1352–1359, 1979

77. Kuznetsov AV, Khuchua ZA, Vassil'eva EV, Medved'eva NV, Saks VA: Heart mitochondrial creatine kinase revisited: the outer mitochondrial membrane is not important for coupling of phosphocreatine production to oxidative phosphorylation. Arch Biochem Biophys 268: 176–190, 1989

78. Jacobus WE, Lehninger AL: Creatine kinase of rat heart mitochondria. Coupling of creatine phosphorylation to electron transport. J Biol Chem 248: 4803–4810, 1973

79. Mahler M: Progressive loss of mitochondrial creatine phosphokinase activity in muscular dystrophy. Biochem Biophys Res Commun 88: 895–906, 1979

80. Bennett VD, Hall N, DeLuca M, Suelter CH: Decreased mitochondrial creatine kinase activity in dystrophic chicken breast muscle alters creatine-linked respiratory coupling. Arch Biochem Biophys 240: 380–391, 1985

81. Schmitt T, Pette D: Increased mitochondrial creatine kinase in chronically stimulated fast-twitch rabbit muscle. FEBS Lett 188: 341–344, 1985

82. Yamashita K, Yoshioka T: Activities of creatine kinase isoenzymes in single skeletal muscle fibres of trained and untrained rats. Pflügers Arch 421: 270–273, 1992

83. Apple F, Rogers MA: Mitochondrial creatine kinase activity alterations in skeletal muscle during long-distance running. J Appl Physiol 61: 482–485, 1986

Molecular and Cellular Biochemistry **133/134**: 299–309, 1994.
© 1994 *Kluwer Academic Publishers.*

V-3 Thyroid hormones and the creatine kinase system in cardiac cells

Enn K. Seppet and Valdur A. Saks[1]

Department of Pathophysiology, Medical Faculty, University of Tartu, Tartu; [1] Laboratory of Bioenergetics, Institute of Chemical and Biological Physics, Tallinn, Estonia

Abstract

The paper reviews the current evidence on the role of thyroid hormones in regulating the creatine kinase energy transfer system at multiple structures in cardiac cells. 1) Thyroid hormones modulate the overall synthesis of phosphocreatine (PCr) by increasing the rate of mitochondrial oxidative phosphorylation. 2) Thyroid hormones regulate the total activity of creatine kinase and its isoenzyme distribution. In comparison with normal thyroid state (euthyroidism), hypothyroidism is characterized by decreased total creatine kinase activity owing to diminished fraction of creatine kinase. On the other hand, hyperthyroidism, while causing no change in total creatine kinase activity, leads to increased fractions of neonatal isoforms of creatine kinase, and, in case of prolonged hyperthyroidism, to decreased fraction of mitochondrial creatine kinase. The latter change is associated with partial uncoupling between mitochondrial creatine kinase and adenine nucleotide translocase reflected by decreased PCr/O ratio. 3) Hyperthyroidism leads to increased passive sarcolemmal permeability due to which the leakage of creatine along its concentration gradient occurs. As a result of (i) increased sarcolemmal permeability for creatine, (ii) uncoupling of mitochondrial PCr synthesis, and (iii) increased energy utilization rate the steady state intracellular PCr content decreases under hyperthyroidism which, in turn, increases the myocardial susceptibility to hypoxic damage. Thyroid state also modulates the protective effects of exogenous PCr on energetically depleted myocardium. (Mol Cell Biochem **133/134**: 299–309, 1994)

Key words: hypothyroidism, hyperthyroidism, oxidative phosphorylation, phosphocreatine synthesis, creatine kinase, isoenzymes, creatine transport, intracellular energy transfer, myocardium

Introduction

Changes in the thyroid state are accompanied by marked alterations in energy metabolism of the myocardium. In comparison with normal thyroid state (euthyroid), hyperthyroidism is characterized by increased rate of energy utilization [1]. This is due to the higher rates of ATP hydrolysis in the ATPase reactions of actomyosin, Na^+-K^+-pump and sarcoplasmic reticulum Ca^{2+} pump [2–8]. In contrast, these reactions occur much slower in hypothyroid heart, leading to decreased energy utilization rates [2–8]. Thyroid hormone-dependent changes in energy utilization lead to corresponding alterations in glycogenolysis, glycolysis and oxidative phosphorylation: their rates are higher in hyperthyroid and lower in hypothyroid myocardium [9, 10–14]. To date, the mechanism by which a close relationship between thyroid-dependent changes in utilization and production of energy is achieved is not defined. However, an increasing number of experimental evidence suggests

Address for offprints: Enn K. Seppet, M.D., Ph.D., Department of Pathophysiology, Medical Faculty, University of Tartu, 18 Ülikooli St., EE2400 Tartu, Estonia

that specific action of thyroid hormones on phospho-creatine (PCr) energy transfer mechanism may play a role. It has been demonstrated that thyroid state modulates this mechanism a) by altering the intracellular contents of creatine and PCr, b) by changing the distribution of creatine kinase isoenzymes, and c) by modulating the coupling between mitochondrial creatine kinase and adenine nucleotide translocase (ANT).

Creatine transport in hyperthyroid myocardium

Total creatine concentration in cardiomyocytes which is in a range of 12–16 mM, as calculated from its tissue content (8-10 µmole/g wet weight) and 0.6 ml of intracellular water per gram wet weight, exceeds the creatine concentration in blood serum (0.05–0.2 mM) by a factor about 100. This concentration gradient is maintained by a mechanism of active transport which has been characterized for skeletal [15, 16] and cardiac [17] muscles. Seppet et al. [17] used a Langendorff perfusion technique to study the rates of incorporation of [1-^{14}C]creatine separately into intracellular creatine and PCr in rat myocardium. In one set of experiments, the [1-^{14}C]creatine uptake was registered in the presence of 0–0.6 mM creatine in the perfusion medium (active transport). In another set, the perfusion medium contained 50 mM of creatine, i.e. exceeding its concentration in the intracellular space, to test the rate of passive creatine transport. The results showed that similarily to skeletal muscle [15, 16] and erythrocytes [18] an energy-dependent active transport mechanism obeying Michaelis-Menten kinetics is responsible for extracellular creatine uptake in cardiac muscle. In euthyroid hearts, the following parameters of active creatine uptake were found: K_m = 0.05 mM which is lower than that in skeletal muscle (0.2–0.5 mM, [15, 16]); V_{max} = 20 nmole/min/g dry tissue weight, which is close to that parameter in skeletal muscle [15, 16]. The rate of passive creatine uptake was negligible in euthyroid heart [17]. In comparison with euthyroid heart, hyperthyroid myocardium is characterized by markedly decreased creatine and PCr contents [13, 17, 19–22], whereas ATP and ADP contents remain unchanged [13, 17]. That hyperthyroidism leads to decreased tissue creatine content has also been demonstrated for rat skeletal muscles [22]. It has been shown [17] that, in rat myocardium, the level of creatine and PCr depletion depends on a dose of excess L-thyroxine and a duration of hormone treatment. The maximal dec-

rement (50% of euthyroid level) in a sum of tissue creatine and PCr contents was registered after 1 week of L-thyroxine administration in a daily dose of 1 µg/g B.W. Prolongation of hormone administration did not significantly attenuate the depletion of creatine. Cessation of hormone administration led to a restoration of both creatine and PCr pools in the myocardium [17]. It has been suggested that the inhibition of creatine transport system accounts for decreased tissue creatine and PCr in hyperthyroid heart [21, 22]. However, the results of Seppet et al. [17] argue against this hypothesis. They showed that both the maximal rate of active creatine uptake and its passive movement along its concentration gradient were enhanced under hypothyroidism. The rate of passive transport was increased to 0.4 µmol/min/g dry weight. In parallel, the penetration of colloidal lanthanum into the cells of hyperthyroid heart was observed which shows that the increased passive permeability for creatine is due to the sarcolemmal damage. The fact that cessation of L-thyroxine administration was followed by the restoration of tissue creatine pools reflects the reversibility of structural changes in the sarcolemma under excess thyroid hormones. Perfusion of hyperthyroid rat hearts with 50 mM creatine significantly restored the creatine content in the cells. It was concluded that hyperthyroidism increases the membrane permeability due to which the leakage of creatine from the cells along its concentration gradient occurs. Apparently, this process will be equilibrated by an enhanced active uptake of extracellular creatine to set up a new but lower steady-state creatine content in the cells.

Effects of thyroid hormones on distribution of creatine kinase isoenzymes

There are only few data describing the effects of thyroid hormone on creatine kinase activity and isoenzyme distribution in cardiac muscle. According to Kessler-Icekson [23], addition of L-thyroxine to neonatal rat cardiac culture exerted no effect on creatine kinase activity and isoenzyme distribution. However, Brik et al. [24] have shown that exposure of neonatal rat cardiac culture to 1 µM L-thyroxine or 0.1 µM L-triiodothyronine induced a 20% decrease in MM creatine kinase isoenzyme together with a 40% increase in BB isoenzyme without changes in total creatine kinase activity. Schyler and Yarbrough [25, 26] have compared the effects of in vivo changes in thyroid state on expression of creatine kinase and myosin. On the basis of mRNA analysis they concluded that thyroid hormones appear to play a critical role in regulating expression of the isoforms of myosin but not of creatine kinase. In these studies, however, the influence of thyroid state on mitochondrial creatine kinase was not addressed. Recently, this aspect has been studied in detail by Seppet *et al.* [27]. In their studies, the total creatine kinase activity in euthyroid rat ventricle averaged 512 μmoles/min/g wet weight, of which 29.6, 53.1, 11.5 and 4.6% corresponded to mitochondrial, MM, MB and BB isoenzymes respectively, which conforms to earlier analyses [28–30]. Hypothyroidism in rat was induced by adding PTU (0.05%) for 6 weeks into drinking water. Compared to euthyroid group, the hearts of hypothyroid rats showed decreased total creatine kinase activity (by 22%) owing exclusively to diminished portion of mitochondrial creatine kinase. Correspondingly, the tissue cytochrome levels were decreased by 15–20%. Creatine kinase activity was also measured in skinned myocardial fibers treated with saponin. This agent removes SL from the cardiac cells but keeps the interfibrillar mitochondria intact [31]. In these preparations, hypothyroidism led to 42% decrease in total creatine kinase activity. Thus, it is likely that the lower content of mitochondrial membranes was a reason for restrained activity of the mitochondrial creatine kinase in hypothyroid myocardium. That hypothyroidism decreases total creatine kinase activity has also been shown in neonatal rat skeletal muscle [32].

Hyperthyroidism caused by daily s/c injections of L-thyroxine in a dose of 1 μg/g B.W. to rats during either 1 or 3 weeks did not affect the total creatine kinase activity, as compared to euthyroid state [27]. However, after 1 weeks of hyperthyroidism, the relative activity of MB isoenzyme (15% of total) became markedly higher than in euthyroid group with insignificant alterations in other isoenzymes. Thus, while supporting the data of Brik *et al.* [24] on increased MB activity, our results did not confirm the hypothesis, that MB activity increases at the expense of MM creatine kinase [24]. Prolongation of hyperthyroidism up to 3 weeks was accompanied by a further increase in MB isoenzyme activity (17% of total). In addition, the percentage of BB isoenzyme increased almost twice, and of mitochondrial creatine kinase decreased by 15% compared to euthyroid group. The myocardial contents of cytochromes in 1 or 3 week of hyperthyroid groups were similar to those in euthyroid group. Hence, the decrease in the activity of mitochondrial creatine kinase after 3 weeks of hyperthyroidism occurred

independently of the tissue content of mitochondrial membranes, which suggests that a partial dissociation of this enzyme's molecules from inner mitochondrial membrane and/or its inactivation could occur under long lasting hyperthyroidism.

Effects of thyroid hormones on mitochondrial oxidative phosphorylation and phosphocreatine synthesis

Oxidative phosphorylation

Compared to mitochondria from the liver, the influence of thyroid hormones on these organelles in the heart is far less studied. In both tissues, hypo- and hyperthyroidism result in decreased and increased levels of state 3 respiration, respectively, as compared to euthyroid state [9, 11, 12, 33–37]. Hyperthyroidism has been shown not to affect the coupling of oxidation to phosphorylation as suggested by unchanged ADP/O ratio when compared with euthyroid mitochondria. In hypothyroid mitochondria of liver and heart, however, both unchanged [38, 39] and decreased ADP/O ratios have been observed [12, 40–42]. The reason for depressed ADP/O ratio in hypothyroid mitochondria is not clear. It is suggested that the concentrations of mitochondrial protein [43] and Ca^{2+} ions in the polarographic medium, as well as the level of ADP-ribosylation of inner membrane components [40, 44–46] may play a role in controlling the oxidation to phosphorylation coupling in hypothyroid mitochondria. The respiration rate of normal mitochondria in state 3 is controlled by the cytochrome c oxidase, dicarboxylate carrier and ANT [47] whereas the state 4 respiration is limited by slow transmembrane proton leakage [48]. Alterations in thyroid state probably affect all these phases, since they lead to the changes in transmembrane protonic electrochemical potential difference [35, 48], ANT [42, 49, 50], ATP synthase [51], matrix enzymes [33, 52], cytochrome reduction levels [36] and contents of cytochromes [11, 36, 53–55] in both liver and cardiac mitochondria. These changes could possibly be caused by the multiple mechanisms. 1) One of them could be realized via thyroid hormone action on the enzymes controlling the mitochondrial lipid composition. Compared to euthyroid state, hyperthyroidism is characterized by increased phospholipid and decreased cholesterol contents together with the enhanced unsaturation index of fatty acids in the liver and heart mitochondria [56, 57]. The concentrations of cardiolipin and phosphatidylserine have increased by more than 50% [58, 59]. These changes lead to increased membrane fluidity due to which the mobility and activity of several mitochondrial enzyme proteins (ANT [60], pyruvate and citrate carriers [58, 59]) increase. By contrast, hypothyroidism is characterized by increased cholesterol and depressed phospholipid contents, as well as decreased unsaturation index of fatty acids. These changes are accompanied by the decreased activities of ANT and oxidative phosphorylation [49, 60–62]. 2) Thyroid hormones appear to modulate oxidative phosphorylation via direct binding to ANT [61]. As a result the overall rate of the oxidative phosphorylation increases. 3) The receptor-mediated regulation of biosynthesis of mitochondrial proteins seems also an important mechanism of thyroid hormone action on the processes of oxidative phosphorylation [52, 63, 64]. Most likely this mechanism is involved in thyroid hormone effects on cytochrome content in the mitochondria. Interestingly, a clear positive correlation exists between the mitochondrial content of cytochrome aa_3 and a respiration rate: They are higher in hyperthyroid and lower in hypothyroid mitochondria [11, 12, 27, 36, 65]. This fact shows that alterations in aa_3 are of critical importance for the enhanced rates of electron transport in hyperthyroid mitochondria. Possibly, the elevated content of cytochrome aa_3 as a component of cytochrome c oxidase, favours oxidation of cytochrome c in hyperthyroid mitochondria [36].

Phosphocreatine synthesis

The effects of thyroid state on mitochondrial PCr synthesis have recently been estimated [12]. In these experiments, mitochondria were incubated in polarographic medium containing 25 mM creatine and different concentrations of MgATP. Addition of ATP to creatine-containing medium led to the concentration-dependent activation of respiration and PCr synthesis, both occurring linearly with time in each group of mitochondria studied. In contrast, MgATP concentration remained constant, at essentially its initial value. The maximal rates (V_{max}) of creatine-activated oxygen utilization and PCr synthesis in euthyroid mitochondria were 0.50 µg-atoms O/mg/min and 1.29 µmoles/mg/min, respectively. Euthyroid mitochondria exhibited identical values of PCr/O and ADP/O (2.70). Also, the apparent K_m for MgATP were equal in the reactions of oxygen consump-

tion and PCr synthesis (64.9 µM) with values several times lower than what has been found for solubilized mitochondrial creatine kinase [66]. These findings are indicative of the tight coupling between mitochondrial creatine kinase and ADP-ATP exchange due to which all of the ATP synthesized in the reactions of oxidative phosphorylation was transformed into PCr [66, 67]. It could be argued that the functional parameters of the isolated mitochondria may differ from those *in situ*. Therefore, the ADP-stimulated respiration was measured in the absence and presence of exogenous creatine in the myocardial fibers which were made permeable for substrates by skinning with saponin [27]. The results show that creatine markedly stimulated the myocardial respiration and increased the affinity of mitochondria to exogenous ADP. Thus, the effective coupling between mitochondrial creatine kinase and ANT is characteristic of mitochondria in the intracellular milieu as well.

In normal heart mitochondria, a constant molar ratio exists between cytochrome aa_3, ANT, and mitochondrial creatine kinase, which is an important pre-requisite for establishing functional coupling between oxidative phosphorylation and creatine kinase [68, 69]. Considering that thyroid hormones stimulate the rate limiting reactions of oxidative phosphorylation at the level of cytochrome aa_3 and ANT [42, 49, 50], thereby increasing the amount of ATP available for mitochondrial creatine kinase, it may be believed that thyroid control over PCr formation could be established by hormonal effects on oxidative phosphorylation only. However, the results of Seppet *et al.* [12, 27] suggest that coupling between mitochondrial creatine kinase might represent an additional specific target for thyroid hormones.

The V_{max} of creatine-activated oxygen utilization was decreased by 36% in hypothyroid mitochondria, while the V_{max} of PCr synthesis was significantly less depressed (22%) [12]. Consequently, the PCr/O ratio reached values characteristic or normal mitochondria and was significantly higher than ADP/O ratio. Hence, the impaired coupling between oxidation and phosphorylation in hypothyroid mitochondria that was revealed in the absence of creatine, was being normalized in conditions of PCr synthesis. The fact that, during PCr synthesis, the hypothyroid mitochondria also showed decreased values of K_m for MgATP relative to that in euthyroid mitochondria, explains the mechanism for normal PCr/O ratio: Interaction of mitochondrial creatine kinase with ANT allowed to increase the effective concentration of MgATP in the active center of creatine kinase pushing the reaction in the direction of PCr synthesis. As a result, less ATP would be lost in the futile ATPase reactions and more PCr would be synthesized i.e. PCr/O ratio becomes higher than ADP/O ratio. That hypothyroidism does not impair the coupling between mitochondrial creatine kinase and ANT was demonstrated also in skinned fibers, where maximal rate for creatine-stimulated respiration 33.22 ng- atom O/mg dry wt/min was close to that in euthyroid fibers [27]. In hyperthyroidism, the mitochondrial function appeared to depend on the duration of hyperthyroid state. After 1 week of L-thyroxine administration, the increment in the maximal rates of respiration and PCr production of isolated mitochondria were roughly equal, and the ADP/O and PCr/O ratios were identical and equal to the euthyroid ratios. The respiration rates in the presence and absence of creatine in skinned hyperthyroid fibers were similar to the corresponding rates in euthyroid fibers [27]. These results show normal coupling between the processes of oxidative phosphorylation and PCr synthesis. Thus, the increased rate of PCr synthesis was exclusively due to enhanced rate of oxidative phosphorylation after 1 week of hyperthyroidism. However, when hormone administration was prolonged for 2 weeks, the imbalance between PCr synthesis and respiration became evident as indicated by a decreased PCr/O ratio compared to the ADP/O ratio. Also, a decreased affinity of creatine-dependent respiration for MgATP was detected after 2 weeks of L-thyroxine treatment. Thus, mitochondrial creatine kinase failed to keep pace with the increased rate of ATP production and limited overall PCr synthesis. Probably, this defect is based on decreased mitochondrial creatine kinase activity in mitochondria occurring during longer than 2 week of hyperthyroid period [27].

When the values of PCr/O ratio were plotted versus creatine-stimulated respiration for each individual assay, the strong reciprocally linear relationship between these parameters was revealed. The shift from higher to lower values of PCr/O was obtained in response to change from hypothyroid to hyperthyroid states. It was concluded that thyroid state controls the efficiency of coupling of mitochondrial creatine kinase with ANT [27].

The pathogenic role of altered energy transport in hypothyroid and hyperthyroid heart

Figure 1 summarizes the effect of thyroid state on PCr energy transfer system in cardiac cells. In normal cardiac cells (Fig. 1A), energy produced in reactions of oxidative phosphorylation in form of ATP molecules, is transformed into PCr due to functional coupling between ANT and mitochondrial creatine kinase in the space between inner and outer mitochondrial membrane. PCr, diffusing in the extra- mitochondrial medium, is used for rephosphorylation of ADP near ATPases of energy- utilizing structures (e.g. myofibrils) due to coupling these ATPases to MM-creatine kinase isoenzyme. Creatine, liberated in these reactions diffuses back to mitochondria where it initiates the new cycle of PCr synthesis and, via ADP, liberated in creatine kinase reaction, stimulates the processes of oxidative phosphorylation. Thus, PCr/Cr cycle effectively links the places of energy utilization and reactions of energy production to each other. The faster is the rate of ATP hydrolysis, the faster is ATP production in the mitochondria, and vice versa. The Fig. 1A also shows the role of active transport of creatine in cardiac cells: to establish high content of i/c creatine and 2) to compensate a small transsarcolemmal loss of this compound.

Under hypothyroidism, the rate of PCr utilization is low as compared to euthyroid state [2–8]. Correspondingly, the oxidative phosphorylation occurs slower [11–14]. This change is balanced by the slower rates of oxidative phosphorylation, mostly due to decreased synthesis of the components of the respiratory assembly. At the same time, however, hypothyroid myocardium is capable to maintain high levels of PCr due to 1) normal tissue content of creatine (Seppet *et al.*, unpublished), and 2) effective mitochondrial synthesis of PCr.

Figure 1B depicts the following main alterations in hyperthyroid myocardium: 1) the massive leak of creatine from the cell due to increased passive permeability of sarcolemma; 2) increased activity of transsarcolemmal creatine pump, unable, however, to keep pace with increased outward flow of creatine. On the other hand, dysproportionate alterations in cytochrome aa$_3$ and mitochondrial creatine kinase activity in conditions of long lasting hyperthyroidism suggests the partial dissociation of mitochondrial creatine kinase from the ANT. Consequently, a part of ATP generated in the reactions of oxidative phosphorylation cannot be transformed into

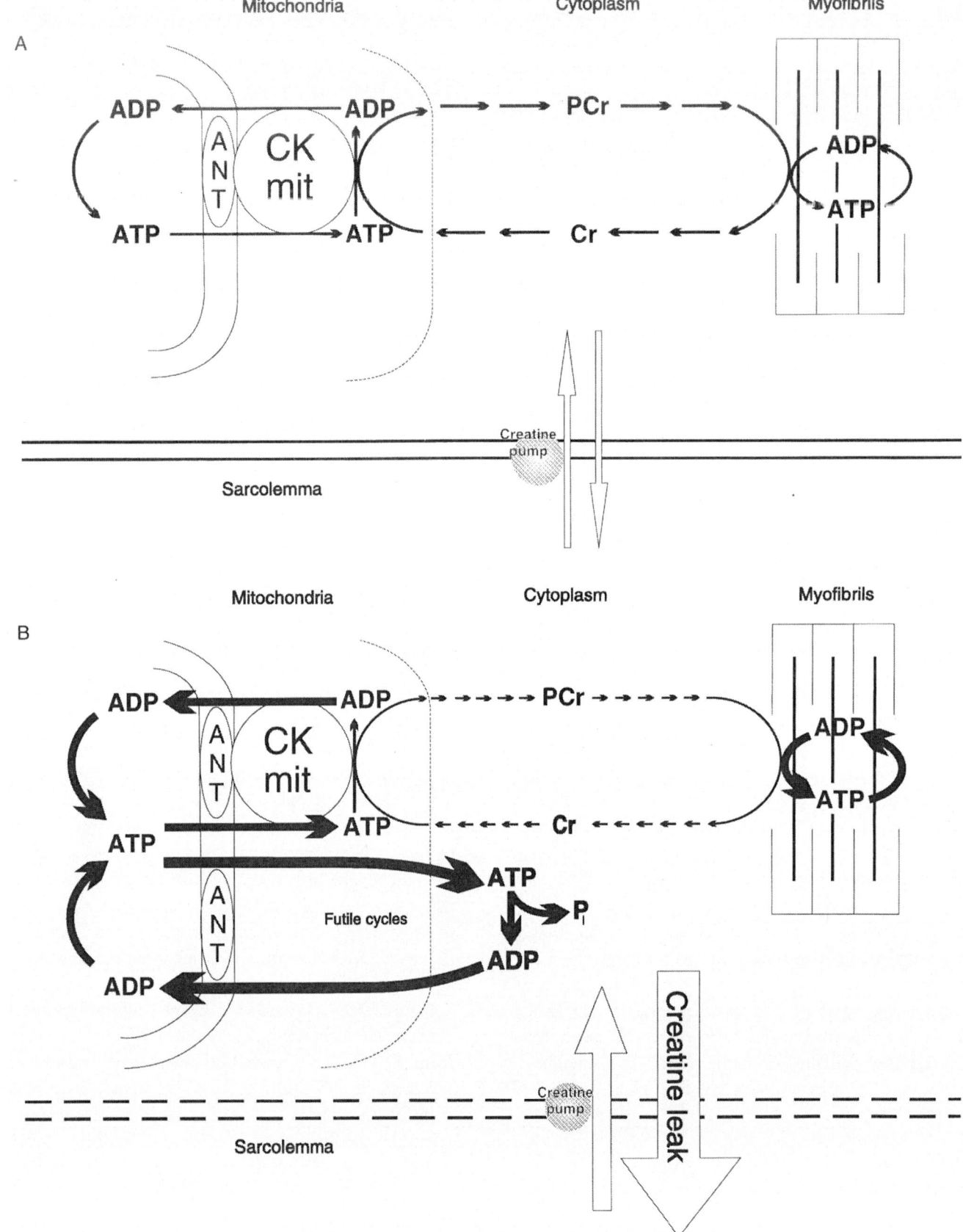

PCr, and, therefore, is easily hydrolyzed in the futile cycles both in the intramembrane space and outside the mitochondria. These changes together result in decreased i/c PCr and creatine pools. The figure also demonstrates the assumption, that the PCr/Cr cycling must occur faster in order to ensure increased rate of oxidative phosphorylation to balance the increased ATP utilization.

In conditions of hypoxia or ischemia, however, the decreased intracellular PCr content together with impaired mitochondrial function may play a crucial pathogenetic role in hyperthyroid myocardium. This is evident from the facts that in these conditions, compared to euthyroid myocardium, hyperthyroid myocardium is characterized by faster decrease in contractile function and development of contracture [17, 70, 80], and that ischemic contracture is related to depletion of ATP in the myofibrillar compartment leading to rigor complex formation [71, 72].

It is important to notice that changes occurring in hyperthyroid myocardium resemble those described for cardiac hypertrophy and cardiomyopathy in general. In fact, a significant decrease in the total creatine content occurs during a development of cardiac hypertrophy and cardiomyopathy [67, 73, 74]. On the other hand, similarly to hyperthyroidism, a shift towards augmented MB and BB isoenzymes of creatine kinase in rat heart occurs in response to chronic pressure overload [29], spontaneous hypertension [75] and experimentally induced renal hypertension [30]. During chronic pressure overload [29], these changes were accompanied by a decrease in MM isoenzyme activity. The hearts of rats with spontaneous hypertension [75] and renal hypertension [30] have shown the decreased content of mitochondrial creatine kinase as well. This impairment is considered to be indicative of the decompensated state of myocardial failure [75] and seems to be correlated with hypertrophy rather than blood pressure [30]. Therefore, one could assume that changes in creatine kinase isoenzyme composition, registered in this study, reflect an unspecific adaptation of hyperthyroid heart to increased workload and some reversal to fetal expression of creatine kinase. We can also make a tentative suggestion that the decrease in mitochondrial creatine kinase activity may determine the boundaries beween compensated and uncompensated stages of myocardial failure during hyperthyroidism.

Effect of exogenous phosphocreatine on heart muscle contractility modulated by hyperthyroidism and extracellular calcium concentration

Exogenous PCr has shown to exert a positive inotropic effect in hypodynamic heart [76] and to protect cardiac muscle against ischemic damage [77–79]. The increased permeability of cardiomyocyte membrane for creatine in hyperthyroidism suggests that the response of myocardium to exogenous PCr could also be modulated. Seppet et al. [80] have studied the effect of 10 mM PCr added to superfusate, on contractile function of euthyroid and hyperthyroid papillary muscles in conditions of normoxia as well as hypoxia with glucose deprivation at different extracellular Ca^{2+} concentrations. In fresh aerobic eu- and hyperthyroid papillary muscles, developing a maximal DT, PCr did not influence their mechanical performance. However, it significantly augmented the DT of exhausted by long-term work papillary muscles (by 160–180% of control), with faster response in hyperthyroid group. Probably, this effect of exogenous PCr may be realized through the activation of slow Ca^{2+} current, as it occurs in hypodynamic frog ventricular muscle [76].

The effect of exogenous PCr on hypoxic papillary muscles was dependent both on thyroid state and extracellular Ca^{2+} concentration. At normal Ca^{2+} concentration (2.5 mM), exogenous PCr exerted beneficial influence both on eu- and hyperthyroid papillary muscles

Fig. 1. Effect of thyroid state on the function of PCr energy transport system in the rat myocardium. A. Euthyroid myocardium. ATP, synthesized in the matrix of mitochondria is transformed into PCr which diffuses to myofibrils to be used for rephosphorylation of ADP liberated in the actomyosin ATPase reaction. Creatine (Cr) diffuses back to the mitochondria where it initiates a new cycle of PCr synthesis. Empty arrows indicate the movement of creatine inward via active transport mechanism (creatine pump) and outward due to leak. ANT – adenine nucleotide translocase; CK mit – mitochondrial creatine kinase. It is assumed that under hypothyroid conditions, the overall rate of PCr shuttle cycling is lower than in euthyroid myocardium (not shown). B. Hyperthyroid myocardium. In comparison with the euthyroid myocardium, the thicker arrows indicate increased ATP-ADP turnover in mitochondria and myofibrils. Note the partial uncoupling between ANT and CKmit due to which ATP is hydrolyzed in futile cycles before reaching the myofibrils. In comparison with the euthyroid myocardium, the arrows symbolizing PCr/Cr cycling are thinner, in accordance with lower tissue content of these compounds. The meaning of other symbols is the same as in A. For further explanations, see the text.

306

which was manifested: a) in a slower decrease in DT, b) in a later development of hypoxic contracture, and c) in a faster restoration of contractile force during posthypoxic aerobic resuperfusion. Compared to the euthyroid one the hyperthyroid myocardium was more sensitive to a positive inotropic action of PCr within the first 15 min of hypoxic superfusion, when PCr potentiated the DT up to the values of euthyroid heart muscle, but less sensitive during posthypoxic period of aerobic resuperfusion in which the recovery of those parameters was delayed. Probably, PCr potentiated the contractility of both muscle groups by stabilizing the structure of sarcolemma [81], decreasing the breakdown of membrane phospholipids [82, 83] and suppressing the catabolism of adenine nucleotides [79]. Since PCr can diffuse into hypoxic cardiomyocytes [84, 85], it may also support the transport function of SL by restoring subsarcolemmal ATP pools through the membrane bound MM-creatine kinase. In the presence of supranormally high extracellular Ca^{2+} concentration (5 mM), PCr paradoxically stimulated the development of hypoxic contracture both in euthyroid and hyperthyroid papillary muscles. In contrast to the euthyroid myocardium, exhibiting a complete recovery of mechanical parameters during posthypoxic aerobic resuperfusion in the presence of PCr, the contractile function of hyperthyroid myocardium was not restored during that phase of experiment. The mechanism of such a negative action of PCr on hypoxic heart could be explained in terms of those disturbances which are known to occur during energy-dependent Ca^{2+}-overload of cardiomyocytes [86]. Thus, impairment of the SL structure during hypoxia may favour penetration of PCr together with excess Ca^{2+} into the cell. This process may be accelerated by hyperthyroidism which increases the passive permeability of sarcolemma to La^{3+} and creatine [17]. As a result, the intracellular Ca^{2+} concentration would increase up to the levels sufficient to stimulate the accumulation of Ca^{2+} and phosphate into the mitochondria at the expense of energy [87] derived from PCr after its diffusion into cells. These processes lead to an irreversible impairment of reactions of oxidative phosphorylation, acidification of intracellular milieu, excretion of cytoplasmic enzymes and, finally, to the development of contracture [88]. Thus, it could be concluded that the level of Ca^{2+} overload of cardiomyocytes may be of essential importance for determining the effectiveness of the cardioprotective action of exogenous PCr, particularly in hyperthyroid heart.

Acknowledgements

The technical assistance of Dr. Ave Minajeva is gratefully acknowledged.

References

1. Skelton CL, Pool PE, Seagren SC, Braunwald E: Mechanochemistry of cardiac muscle. V. Influence of thyroid state on energy utilization. J Clin Invest 50: 463–473, 1971
2. Morkin E, Flink IL, Goldman S: Biochemical and physiological effects of thyroid hormone on cardiac performance. Progress Cardiovasc Dis 25: 435–463, 1983
3. Philipson KD, Edelman JS: Thyroid hormone control of Na^+-K^+-ATPase and K^+ dependent phosphatase in rat heart. Am J Physiol 232: C196–C201, 1977
4. Horowitz B, Hensley CB, Quintero M, Azuma KK, Putnam D, McDonough AA: Differential regulation of Na,K-ATPase 1, 2, and subunit mRNA and protein levels by thyroid hormone. J Biol Chem 265: 14308–14314, 1990
5. Daly MJ, Seppet EK, Vetter R, Dhalla NS: Membrane abnormalities and changes in cardiac cations due to alterations in thyroid status. In: B Korecky, NS Dhalla (eds) Subcellular Basis of Contractile Failure. Boston, Kluwer Academic Publishers, pp 173–191, 1990
6. Suko J: The calcium pump of cardiac sarcoplasmic reticulum. Functional alterations of different levels of thyroid state in rabbits. J Physiol 228: 563–581, 1973
7. Limas CJ: Calcium transport ATPase of cardiac sarcoplasmic reticulum in experimental hyperthyroidism. Am J Physiol 235: H745–H751, 1978
8. Dillmann WH: Biochemical basis of thyroid hormone action in the heart. Am J Med 88: 626–630, 1990
9. Piatnek-Leunissen DA, Leunissen RLA: Heart mitochondrial function in acute and chronic hyperthyroidism in rats. Circ Res 25: 171–181, 1969
10. Read LC, Wallace PG, Berry MN: Effects of thyroid state on respiration of perfused rat and guinea pig hearts. Am J Physiol 253: H519–H523, 1987
11. Nishiki K, Erecinska M, Wilson DF, Cooper S: Evaluation of oxidative phosphorylation in hearts from euthyroid, hypothyroid and hyperthyroid rats. Am J Physiol 235: C212–C219, 1988
12. Seppet EK, Kadaya LY, Kallikorm AP, Saks VA: Hormone regulation of cardiac energy metabolism. II. The effect of thyroid state on the coupling between the reactions of oxidative phosphorylation and phosphocreatine synthesis in rat heart mitochondria. J Appl Cardiol 5: 367–381, 1990
13. Seymour A-M L, Eldar H, Radda GK: Hyperthyroidism results in increased glycolytic capacity in the rat heart. A. ^{31}P-NMR study. Biochim Biophys Acta 1055: 107–116, 1990
14. Seppet EK, Kadaya LY, Hata T, Kallikorm AP, Saks VA, Vetter R, Dhalla NS: Thyroid control over membrane processes in rat heart. Am J Physiol, Suppl (Oct) 26: 66–71, 1991
15. Fitch CD, Shields RP: Creatine metabolism in skeletal muscle. I. Creatine movement across muscle membranes. J Biol Chem 241: 3611–3614, 1966

16. Fitch CD, Shields RP, Payne WF, Dacus JM: Creatine metabolism in skeletal muscle. III. Specificity of the creatine entry process. J Biol Chem 243: 2024–2027, 1968

17. Seppet EK, Adoyaan AJ, Kallikorm AP, Chernousova GB, Lyulina NV, Sharov VG, Severin VV, Popovich MI, Saks VA: Hormone regulation of cardiac energy metabolism. I. Creatine transport across cell membranes of euthyroid and hyperthyroid rat heart. Biochem Med 34: 267–279, 1985

18. Syllm-Rapoport I: Creatine transport into human red blood cells. Acta Biol Med Ger 36: 3–4, 1977

19. Bodansky MJ: Effect of thyroid and thyroxine on concentration of creatine in heart, muscle, liver and testes of albino rat. J Biol Chem 109: 615–622, 1935

20. Buccino RA, Spann JF Jr, Pool PE, Sonnenblick EH, Braunwald E: Influence of the thyroid state on the intrinsic contractile properties and energy stores of the myocardium. J Clin Invest 46: 1669–1682, 1967

21. Kurahashi M: Tissue specificity of inhibitory action of excess thyroid hormone on creatine transport in the rat. Japan J Physiol 28: 603–610, 1978

22. Fitch CD, Coker R, Dinning JS: Metabolism of creatine-1-C_{14} by vitamin E-deficient and hyperthyroid rats. Am J Physiol 198: 1232–1234, 1960

23. Kessler-Icekson G: Effect of triiodothyronine on cultured neonatal rat heart cells: beating rate, myosin subunits and CK-isozymes. J Mol Cell Cardiol 20: 649–655, 1988

24. Brik H, Alkaslassi L, Harell D, Sperling O, Shainberg A: Thyroxine-induced redistribution of creatine kinase isoenzymes in rat cardiomyocytes cultures. Experientia 45: 592–594, 1989

25. Schyler GT, Yarbrough LR: Comparison of myosin and creatine kinase isoforms in left ventricles of young and senescent Fischer 344 rats after treatment with triiodothyronine. Mech Ageing Dev 56: 39–48, 1990

26. Schyler GT, Yarbrough LR: Changes in myosin and creatine kinase mRNA levels with cardiac hypertrophy and hypothyroidism. Basic Res Cardiol 85: 481–494, 1990

27. Seppet EK, Kairane CB, Khuchua ZA, Kadaya LY, Kallikorm AP, Saks VA: Hormone regulation of cardiac energy metabolism. III. Effect of thyroid state on distribution of creatine kinase isoenzymes and creatine-controlled respiration in cardiac muscle. J Appl Cardiol 6: 301–311, 1991

28. Saks VA, Rosenstraukh LV, Smirnov VN, Chazov EI: Role of creatine phosphokinase in cellular function and metabolism. Can J Physiol Pharmacol 56: 691–706, 1978

29. Younes A, Schneider JM, Bercovici J, Swynghedauw B: Redistribution of creatine kinase isoenzymes in chronically overloaded myocardium. Cardiovasc Res 19: 15–19, 1984

30. Smith SH, Kramer MF, Reis I, Bishop SP, Ingwall JS: Regional changes in creatine kinase and myocyte size in hypertensive and nonhypertensive cardiac hypertrophy. Circ Res 67: 1334–1344, 1990

31. Veksler VV, Kuznetsov AV, Sharov VG, Kapelko VG, Saks VA: Mitochondrial respiratory parameters in cardiac tissue: a novel method for assessment by using saponin-skinned fibers. Biochim Biophys Acta 892: 191–196, 1987

32. Simonides WS, van Hardeveld C: The postnatal development of sarcoplasmic reticulum Ca^{2+} activity in skeletal muscle of the rat is critically dependent on thyroid hormone. Endocrinology 124: 1145–1153, 1989

33. Soboll S: Thyroid hormone action on mitochondrial energy transfer. Biochim Biophys Acta 1144: 1–16, 1993

34. Bronk JR: Thyroid hormones: control of terminal oxidation. Science 141: 816–818, 1963

35. Shears SB, Bronk JR: The influence of thyroxine administered *in vivo* on the transmembrane protonic electrochemical potential difference in rat liver mitochondria. Biochem J 178: 505–507, 1979

36. Horrum MA, Tobin RB, Ecklund RE: Thyroxine-induced changes in rat liver mitochondrial cytochromes. Mol Cell Endocrinol 41: 163–169, 1985

37. Verhoven AJ, Kamer P, Groen AK, Tager JM: Effects of thyroid hormone on mitochondrial oxidative phosphorylation. Biochem J 226: 183–192, 1985

38. Hoch FL: Early action of injected L-thyroxine on mitochondrial oxidative phosphorylation. Proc Natl Acad Sci USA 58: 506–512, 1967

39. Sterling K, Brenner MA, Sakurada T: Rapid effect of triiodothyronine on the mitochondrial pathway in rat liver *in vivo*. Science 210: 340–342, 1980

40. Palacios-Romero R, Mowbray J: Evidence for the rapid direct control both *in vivo* and *in vitro* of the efficiency of oxidative phosphorylation by 3,5,3'-tri-iodo-thyronine in rats. Biochem J 184: 527–538, 1979

41. Keogh JM, Matthews PM, Seymour AM, Radda GK: A phosphorus-31-nuclear magnetic resonance study of effects of altered thyroid state on cardiac bioenergetics. Adv Myocardiol 6: 299–309, 1985

42. Crespo-Armas A, Mowbray J: The rapid alteration by tri-iodo L-thyronine *in vivo* of both the ADP/O ratio and H^+/O ratio in hypothyroid-rat liver mitochondria. Biochem J 241: 657–661, 1987

43. Hafner RP, Brand MD: Hypothyroidism in rats does not lower mitochondrial ADP/O and H^+/O ratios. Biochem J 250: 477–484, 1988

44. Corrigall J, Tselentis BS, Mowbray J: The efficiency of oxidative phosphorylation and the rapid control by thyroid hormone of nicotinamide nucleotide reduction and transhydrogenation in intact rat liver mitochondria. Eur J Biochem 141: 435–440, 1984

45. Thomas WE, Crespo-Armas A, Mowbray J: The influence of nanomolar calcium ions and physiological levels of thyroid hormone on oxidative phosphorylation in rat liver mitochondria. A possible signal amplification control mechanism. Biochem J 247: 315–320, 1987

46. Thomas WE, Mowbray J: Evidence for ADP-ribosylation in the mechanism of rapid thyroid hormone control of mitochondria. FEBS Lett 223: 279–283, 1987

47. Groen AK, Wanders RJA, Westerhoff HV, Van der Meer R, Tager JM: Quantification of the contribution of various steps to the control of mitochondrial respiration. J Biol Chem 257: 2754–2757, 1982

48. Shears SB, Bronk JR: Ion transport in liver mitochondria from normal and thyroxine-treated rats. J Bioenerg Biomembr 12: 379–393, 1980

49. Babior BM, Creagan S, Ingbar S, Kipnes RS: Stimulation of mitochondrial adenosine diphosphate uptake by thyroid hormones. Proc Nat Acad Sci USA 70: 98–102, 1973

50. Mowbray J, Corrigall J: Short-term control of mitochondrial adenine nucleotide translocator by thyroid hormone. Eur J Biochem 139: 95–99, 1984

51. Das AM, Harris DA: Control of mitochondrial ATP synthase in

rat cardiomyocytes: effects of thyroid hormone. Biochim Biophys Acta 1096: 294–290, 1991

52. Tanaka T, Morita H, Koide H, Kawamura K, Takatsu T: Biochemical and morphological study of cardiac hypertrophy. Effects of thyroxine on enzyme activities in the rat myocardium. Bas Res Cardiol 80: 165–174, 1985

53. Bronk JR: Thyroid hormone: effects on electron transport. Science 153: 638–639, 1966

54. Volfin P, Kaplay SS, Sanadi DR: Early effect of thyroxine *in vivo* on rapidly labeled mitochondrial protein fractions and respiratory control. J Biol Chem 244: 5631–5635, 1969

55. Katyare SS, Joshi MV, Fatterpaker P, Sreenivasan A: Effect of thyroid deficiency on oxidative phosphorylation in rat liver, kidney, and brain mitochondria. Arch Biochem Biophys 182: 155–163, 1977

56. Hoch FL: Thyroid control over biomembranes. VII. Heart muscle mitochondria from L-triiodothyronine-injected rats. J Mol Cell Cardiol 14: 81–90, 1982

57. Muller MJ, Seitz HJ: Thyroid hormone action on intermediary metabolism. Part II: Lipid metabolism in hypo- and hyperthyroidism. Clin Wochenschr 62: 49–55, 1984

58. Paradies G, Ruggiero FM: Effect of hyperthyroidism on the transport of pyruvate in rat-heart mitochondria. Biochim Biophys Acta 935: 79–86, 1988

59. Paradies G, Ruggiero FM: Enhanced activity of the tricarboxylate carrier and modification of lipids in hepatic mitochondria from hyperthyroid rats. Arch Biochem Biophys 278: 425–430, 1990

60. Hoch FL: Adenine nucleotide translocation in liver mitochondria of hypothyroid rats. Arch Biochem Biophys 178: 535–545, 1977

61. Sterling K: The molecular mechanism of thyroid hormone action at the cellular level. In: LC van Middlesworth (ed) The Thyroid Gland: Practical Clinical Treatise. Year Book, Chicago, pp 203–229, 1986

62. Sterling K, Milch PO, Brenner MA, Lazarus JH: Thyroid hormone action: The mitochondrial pathway. Science 197: 996–999, 1977

63. Mutvei A, Husman B, Andersson G, Nelson BD: Thyroid hormone and not growth hormone is the principle regulator of mammalian mitochondrial biogenesis. Acta Endocrinol (Copenh) 121: 223–228, 1989

64. Leung ACF, McKee EE: Mitochondrial protein synthesis during thyroxine-induced cardiac hypertrophy. Am J Physiol 258: E511–E518, 1990

65. Winder WW, Holloszy JO: Response of mitochondria of different types of skeletal muscle to thyrotoxicosis. Am J Physiol 232: C180–C184, 1977

66. Saks VA, Kupriyanov VV, Elizarova G, Jacobus WE: Studies of energy transport in heart cells. The importance of creatine kinase localization for the coupling of mitochondrial phosphorylcreatine production to oxidative phosphorylation. J Biol Chem 255: 755–763, 1980

67. Wyss M, Smeitink J, Wevers RA, Wallimann T: Mitochondrial creatine kinase: a key enzyme to aerobic energy metabolism. Biochim Biophys Acta 1102: 119–166, 1992

68. Saks VA, Kuznetsov AV, Huchua ZA, Kupriyanov VV: Compartmentation of adenine nucleotides and phosphocreatine shuttle in cardiac cells: some new evidence. Adv Exp Med Biol 194: 103–116, 1986

69. Kupriyanov VV, Elizarova GV, Saks VA: Determination of the molar content of creatine kinase in heart mitochondria using SH-reagents. Biokhimia 46: 930–941, 1981 (In Russian)

70. Palacios J, Kiran S, Powell WJ: Effect of hypoxia on mechanical properties of hyperthyroid cat papillary muscle. Am J Physiol 237: H293–H298, 1979

71. Hearse DJ, Garlick PB, Humphrey SM: Ischemic contracture of the myocardium: mechanisms and prevention. Am J Cardiol 39: 986–993, 1977

72. Ventura-Clapier R, Vassort G: Electrical and mechanical activities of frog heart during energetic deficiency. J Muscle Res Cell Motil 1: 429–444, 1980

73. Reilly PJ, Cooksey JD: Cardiac energy stores and creatine in experimental cardiac hypertrophy. Proc Soc Exp Biol Med 161: 193–198, 1979

74. Khuchua ZA, Ventura-Clapier R, Kuznetsov AV, Grishin MN, Saks VA: Alterations in the creatine kinase system in the myocardium of cardiomyopathic hamsters. Biochem Biophys Res Commun 165: 748–757, 1989

75. Ingwall JS, Fossel ET: Changes in the creatine kinase system in the hypertrophied myocardium of the dog and rat. In: NR Alpert (ed) Myocardial Hypertrophy and Failure. Persp Cardiovasc Res, Vol 7, pp 601–617. Raven Press, New York, 1983

76. Rosenstraukh LV, Saks VA, Undrovinas AI, Chazov EI, Smirnov VN, Sharov VG: Studies of energy transport in heart cells. The effect of creatine phosphate on the frog ventricular contractile force and action potential duration. Biochem Med 19: 148–164, 1978

77. Hearse DJ, Stewart DA, Braimbridge MV: Cellular protection during myocardial ischemia. The development and characterization of a procedure for the induction of reversible ischemic arrest. Circulation 54: 193–202, 1976

78. Fagbemy O, Kane KA, Parratt JR: Creatine phosphate suppresses ventricular arrhythmias resulting from coronary artery ligation. J Cardiovasc Pharmacol 4: 53–58, 1982

79. Saks VA, Javadov SA, Pozin E, Preobrazhensky AN: Biochemical basis of the protective action of phosphocreatine on the ischemic myocardium. In: VA Saks, YG Bobkov, E Strumia (eds) Creatine Phosphate, Biochemistry, Pharmacology and Clinical Efficiency. Edizioni Minerva Medica, Torino, pp 95–111, 1987

80. Seppet EK, Eimre MA, Kallikorm AP, Saks VA: Effect of exogenous phosphocreatine on heart muscle contractility modulated by hyperthyroidism and extracellular calcium concentration. J Appl Cardiol 3: 369–380, 1988

81. Sharov VG, Javadov SA, Beskrovnova NN, Tolokolnikov AV, Kryzhanovsky SA, Kaverina NV, Bobkov YG: Ultrastructural aspects of protective effect of exogenous phosphocreatine on ischemic myocardium. In: VA Saks, YG Bobkov, E Strumia (eds) Creatine Phosphate, Biochemistry, Pharmacology and Clinical Efficiency. Edizioni Minerva Medica, Torino, pp 112–122, 1987

82. Anukhovsky EP, Javadov SA, Preobrazhensky AN, Beloshapko GG, Rosenstraukh LV, Saks VA: Effect of phosphocreatine and related compounds on the phospholipid metabolism of ischemic heart. Biochem Med Metab Biol 35: 327–334, 1986

83. Javadov SA, Preobrazhensky AN, Saks VA: Action of phosphocreatine on lysophosphoglyceride levels under total ischemia in rat myocardium. Biokhimia 51: 668–674, 1986

84. Breccia A, Fini A, Girotti S: Intracellular distribution of double-labelled creatine phosphate in the rabbit myocardium. Curr Ther Res 37: 1205–1215, 1985

85. Preobrazhensky AN, Javadov SA, Saks VA: Study on the mecha-

nisms of the protective action of phosphocreatine on the ischemic myocardium. Biokhimia 51: 675–683, 1986

86. Ruigrok TJC, Boink ARTJ, Spies F, Blok J, Maas AHJ, Zimmer-mann ANE: Energy dependency of the calcium paradox. J Mol Cell Cardiol 10: 991–1002, 1978

87. Brierley GP, Murer E, Bachmann E: Studies on ion transport. III.

The accumulation of calcium and inorganic phosphate by heart mitochondria. Arch Biochem Biophys 105: 89–102, 1964

88. Ganote CE, Nayler WG: Contracture and calcium paradox. J Mol Cell Cardiol 17: 733–745, 1985

PART VI

METABOLIC REGULATION: THEORETICAL BASIS

Molecular and Cellular Biochemistry **133/134**: 313–331, 1994.
© 1994 *Kluwer Academic Publishers.*

VI-1 Control theory of metabolic channelling

Boris N. Kholodenko,[1] Marta Cascante[2] and
Hans V. Westerhoff[3,4]

[1] *A.N. Belozersky Institute of Physico-Chemical Biology, Moscow State University, 119899 Moscow, Russia;*
[2] *Department de Bioquimica i Fisiologia, Universitat de Barcelona, Marti i Franques 1, Barcelona, Spain;*
[3] *E.C. Slater Institute, University of Amsterdam, Plantage Muidergracht 12;* [4] *Department of Molecular Biology, H5, The Netherlands Cancer Institute, Plesmanlaan 121, NL-1066 CX, The Netherlands*

Abstract

Various factors appear to control muscle energetics, often in conjunction. This calls for a quantitative approach of the type provided by Metabolic Control Analysis for intermediary metabolism and mitochondrial oxidative phosphorylation. To the extent that direct transfer of high energy phosphates and spatial organization plays a role in muscle energetics however, the standard Metabolic Control Theory does not apply, neither do its theorems regarding control.

This chapter develops the Control Theory that does apply to the muscle system. It shows that direct transfer of high energy phosphates bestows a system with enhanced control: the sum of the control exerted by the participating enzymes on the flux of free energy form the mitochondrial matrix to the actinomyosin may well exceed the 100% mandatory for ideal metabolic pathways. It is also shown how sequestration of high energy phosphates may allow for negative control on pathway flux. The new control theory gives methods functionally to diagnose the extent to which channelling and metabolite sequestration occur. (Mol Cell Biochem **133/134**: 313–331, 1994)

Key words: metabolic channeling, control analysis, muscle energy metabolism, metabolite / enzyme sequestration

Introduction

Quantitative approaches have led to appreciable advances in understanding the control of cellular metabolism. In the framework of these approches the intuitive concept of rate-limiting step has been substituted by the more subtle definition of control exerted by *any* enzyme on the flux. The quantitative formulation was introduced by Higgins [1] and, in more elaborated form, by Kacser and Burns [2, 3], Savageau [4] and Heinrich and Rapoport [5, 6]. It was renamed to flux control coefficient of an enzyme by Burns *et al.* [7] in the context of metabolic control analysis.

Treating cellular metabolism as a network of enzyme reactions control analysis used to be based on the following important assumptions: (1) the rate of any reaction is proportional to the concentration of enzyme (a property often termed additivity [8]), (2) different enzymes are independent catalysts, (3) the concentrations of enzyme-bound metabolites can be neglected as compared to their free concentrations (substitution of the latter by the total metabolite concentrations greatly simplifies control analysis [9, 10]).

These assumptions, which are true for simple or 'ideal' [11] metabolic pathways result in the equivalence of different ways of defining and measuring the control coefficients of enzymes [12–15]. Moreover, they allow one to describe metabolic processes at the 'macroscopic' level of 'block' (i.e., the enzyme) reactions without going to the details of reaction mechanisms. With these assumptions great advances were achieved in derivation and interpretation of the control coefficients of enzymes, and, moreover, the ('global') pathway control properties have been expressed in terms of the ('local') regulatory properties of the individual enzyme reactions and of the structure of a pathway [14–18].

However, cellular metabolism can not be always reduced to a number of independent enzyme reactions occurring in a well-stirred continuous tank reactor (for which the assumptions of the classical theory hold true). In highly organized cellular metabolic pathways direct enzyme-enzyme interactions and enzyme associations take place (for reviews see [19–22]). They lead to a restricted diffusion or / and even to a direct transfer of metabolites between interacting enzymes [19, 23, 24]. In some pathways such direct transfer is even standard, for example, in parts of the electron transfer chains of free energy transducing membranes and in the bacterial phosphotransferase systems for sugar uptake [25].

Membrane linked free energy transduction involving proton pumping [26] has also been proposed to involve direct transfer of protons or associated electric fields between adjacent proton pumps (see [27] as a review). Also in this case such transfer would normally occur in parallel to the generation of a proton electrochemical potential difference between the two aqueous bulk phases bordering the energy coupling membrane. Consequently, conclusive evidence for such proton channelling is still rare.

More recently, it has been shown that a somewhat neglected structural feature of mitochondria, i.e., the outer mitochondrial membrane, warrants greater interest. Molecules such as adenine nucleotides permeate this membrane through a pore to which hexokinase may bind. Inner and outer mitochondrial membranes are in close contact at certain sites. In the context of channelling of high-Gibbs-energy phosphates the issue has been raised of the relative location of adenine nucleotide translocase, creatine kinase, adenylate kinase, hexokinase, contact site and porin [28–32].

A number of authors have extended the control analysis approach to special cases of the regulation of systems with enzyme-enzyme interaction and metabolic channelling (see, e.g. [8, 33–36]). However, no control theory dealing with the general case of partial channelling, dynamic and / or static, has been developed. Following the work [11] we here develop the essentials of such a control theory for a pathway with metabolic channelling at steady state.

Quantitative definitions of control

The loose question whether the particular enzyme i does or does not control the flux can be made both operational and quantitative by rephrasing it as follows: If we change the activity of the enzyme i by, say, 1% what will be the percentage change in that flux at the steady state? If there is no variation in the flux, clearly this enzyme does not control the flux at all. In case the flux also changes by 1% the enzyme i may be called rate-limiting: the flux completely follows the activity of that enzyme. A strict mathematical definition extrapolates the percent variation to an infinitesimal one and relates a fractional change (dJ / J) in the steady-state flux to the fractional modulation (de_i / e_i) of the enzyme concentration [3]:

$$C_{e_i}^J = \left(\frac{dJ / J}{de_i / e_i} \right)_{sys} = \left(\frac{d \ln|J|}{d \ln e_i} \right)_{sys} \qquad (1)$$

The subscript sys signifies that differentiation conditions require the steady state of the system and allow the concentrations of metabolites to adjust accordingly.

The dimensionless coefficient $C_{e_i}^J$ is called the flux control coefficient of the enzyme i [7]. As was mentioned, if the control coefficient equals zero, the enyme (i) is non-rate-limiting and in case it equals unity the enzyme is rate-limiting. Thus, these two categories of the older qualitative analysis are retained in the new terminology. More importantly, however, in any intermediate case the control coefficient can take any value and it can be considered as a quantitative indicator of the relative degree to which the enzyme controls (or limits) the overall flux.

In ordinary (ideal) metabolic pathways there is a one-to-one correspondence between the enzymes and reactions, and an other definition for the control coefficient [5, 6] compares a relative change of the local rate $(\delta v_i / v_i)$ through a reaction i with a variation $(\delta J / J)$ of the system's (steady-state) flux caused by a change in the rate v_i. The control coefficient defined in this manner is designated by $C_{v_i}^J$ [8], and is referred to as the control coefficient with respect to the activity or rate of an enzyme to emphasize its relation to a 'local' rate, v_i. A short-hand notation [5, 6, 8] for this definition reads:

$$C_{v_i}^J = \frac{(dJ / J)_{sys}}{(\partial v_i / v_i)_{enz}} \qquad (2)$$

The subscript enz signifies that a perturbation in the local rate $(\partial v_i / v_i)$ is considered as if the enzyme i reaction were in isolation from the pathway (therefore, only the *i*-th reaction is required to be at steady state). The subscript *sys* signifies that after an initial perturbation of the local rate the pathway as a whole attains a new steady state. The mathematically more explicit expression considers changes in an enzyme rate as caused by a variation in any parameter p_i which affect only the rate v_i, $(\partial v_i / \partial p_i \neq 0, \partial v_j / \partial p_i \, 0$ for any $j \neq i)$:

$$C_{v_i}^J = \frac{(d \ln |J| / dp_i)_{sys}}{(\partial \ln|v_i| / \partial p_i)_{enz}} = \frac{v}{J} \cdot \frac{(dJ / dp_i)_{sys}}{(\partial v_i / \partial p_i)_{enz}}, \quad (3)$$

(When taking the derivative $\partial v_i / \partial p_i$ all the concentrations of metabolites should be kept at the values of the initial steady state of the pathway).

In an analogous manner (see Eqs 1–3) the concentration control coefficients can be defined. They measure the response of the concentration of a particular metabolite to changes in a specific parameter (e.g., e_i or p_i:

$$C_{e_i}^{x_k} = \left(\frac{dx_k / x_k}{de_i / e_i} \right)_{sys} = \left(\frac{d \ln x_k}{d \ln e_i} \right)_{sys} \quad (4)$$

$$C_{v_i}^x = \frac{(d \ln |x_k| / dp_i)_{sys}}{(\partial \ln|v_i| / \partial p_i)_{enz}} = \frac{v}{J} \cdot \frac{(dx_k / dp_i)_{sys}}{(\partial v_i / \partial p_i)_{enz}}, \quad (5)$$

where x_k is the concentration of a metabolite. For lack of space we shall mainly deal with the control of the fluxes through a system.

In ideal pathways the definitions of the control coefficients with respect to the enzyme rate do not depend on the choice of a parameter p_i [12, 14, 15, 37]. Considering in Eq. 3 the enzyme concentration (e_i) as a parameter (p_i) one can see, that in ideal pathways (where a reaction rate is proportional to the enzyme concentration, $v_i \propto e_i$, see Introduction) the control coefficient with respect to the rate ($C_{v_i}^J$) is completely equivalent to that one with respect to the enzyme concentration ($C_{e_i}^J$). However, in non-ideal pathways, for example, in metabolic pathways with high enzyme concentrations and moiety conservations, the coefficients $C_{v_i}^J$ and $C_{e_i}^J$ may differ significantly ([9, 10], see also Section 3). Moreover, the control coefficient in a non-ideal pathway, as defined according to Eq. 3 (or Eq. 5), may depend on the choice of parameter p_i ([10, 38], see an example in Section 5). This dependence appears due to direct or indirect (i.e., via the moiety conservation involving different enzymes bound to the same substrate moiety) interactions between enzymes. Indeed, below we shall see that the difference in the control coefficients can be used to diagnose channelling (see also [34, 36, 38]).

One of the best known and intuitively most understandable properties of the flux control coefficients is that when summing over all enzymes of a system they add up to unity:

$$\sum_i C_{e_i}^J = 1 \quad (6)$$

In ideal pathways this property, called the summation theorem, is valid independently of the structure of a pathway and the local kinetic properties of the enzymes. Indeed, in such pathways any steady-state flux is a homogeneous first-order function of the enzyme concentrations (i.e., if the enzyme concentrations simultaneously change by a factor (α), all fluxes in the system will change by the same factor α), and the summation theorem (Eq. 6) is a simple consequence of Euler's theorem (cf. [15, 39 40]). However, if a metabolic pathway is non-ideal, the sum in Eq. 6 may be not equal to unity. The behavior of the sum of the flux control coefficients in different systems will be analyzed below. We shall see that measuring this sum can give a deeper insight into regulatory properties of the real cellular pathways.

Static and dynamic channelling: partly channelled pathways

The traditional view enzymes in an aqueous solution obtain their substrates from a well-stirred bulk phase and return their products back to that phase. As was mentioned in the Introduction the existence of alternative forms of metabolic organization including enzyme-enzyme interactions and enzyme associations requires to extend this naive concept. The term 'channelling' has been used by a number of authors to indicate a sequence of chemical conversions taking place within enzyme complexes without releasing the intermediate metabolites into the bulk phase. The reaction sequence in such a 'channelled' phase may be catalyzed by various types of enzyme associations. In the case of so called 'static' channelling the enzymes in the complex are tightly bound, i.e. such complexes exist longer than the mean passage time of metabolites through that part of the pathway. In an extreme case different monomers can be linked covalently as, for instance, in the pyruvate dehydrogenase and tryptophan synthase enzyme complexes [41]. Obviously, in the latter case the concentrations of different 'domains' of the complex can not be varied independently, so that the question how variations of the

316

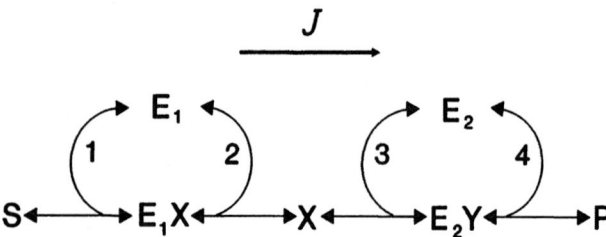

Fig. 1. 'Ideal' pathway of two enzymes. The concentrations of the initial substrate, S, and the end product, P, are constants. X is the intermediate in the bulk phase. Steps 1, 2 and 3, 4 are E_1- and E_2-dependent elemental processes, respectively. The positive direction of the flux J (from the substrate S to the product P) is indicated.

concentrations of different enzymes (domains) affect the flux becomes inoperational. Much of the current debate in the literature is more concerned with weak enzyme associations, often requiring the presence of the common intermediate [42–45]. In this case the amount of complex enzymes depends dynamically on the metabolite concentrations and the flux through that part of the pathway. If such an association of enzymes leads to a direct transfer of an intermediary metabolite, it is referred to as 'dynamic channel'. In the following we develop a general control theoretical treatment applicable to an arbitrary physico-chemical mechanism of channelling (static or dynamic, different schemes that are possible in general are described in detail in [45, 46]).

For a moment we return to a simple metabolic pathway in which channelling is absent (Fig. 1). Traditionally this pathway is treated in terms of two consecutive, enzyme catalyzed reactions without going to the detailed enzyme mechanisms. Each of the reactions, e.g., reaction 1, has a control coefficient over the pathway flux defined by considering a small change in the total concentration of enzyme (e_1), see Eq. 1. Here we note, that in the framework of a 'microdescription' (see Fig. 1) such an increase of e_1 corresponds to a simultaneous proportional increase in all the forward and reverse (pseudo-) first order rate constants of the enzyme 1. To make this statement exact, we need to define the control coefficients of the *elemental processes (steps)* of the enzyme 1 with respect to the overall steady-state flux through the pathway (C_i^J):

$$C_i^J = (\text{d} \ln|J| / \text{d} \ln k_i^+)_{sys}, \ k_i^- / k_i^+ = \text{const}, \ i = 1,2 \quad (7)$$

where the differentiation conditions are such that the forward (k_i^+) and the reverse (k_i^-) rate constants of the elemental process i are changed by the same factor, all other parameters being kept constant. It should be noted that this definition [cf., 47, 48] does not affect mi-

croscopic reversibility. An equivalent, but more general definition of the *elemental* control coefficient ($C_i^{J_i}$) can be given in a manner similar to Eq. 3. In this case, however, the rate (v_i) becomes the rate of the ith elemental process rather than the rate of the entire enzyme i reaction. Most importantly, when addressing the *elemental processes* Eq. 3 defines the control coefficient, C_i^J, as *general* quantity, independent of a special choice of a parameter p_i, for both ideal and *non-ideal cellular* pathways [Kholodenko *et al.*, submitted]:

$$C_i^J = \frac{(\text{d} \ln |J| / \text{d} p_i)_{sys}}{(\partial \ln|v_i| / \partial p_i)_{proc}} \quad (8)$$

The analogous definition for the elemental control coefficient over any concentration x_k reads:

$$C_i^{x_k} = \frac{(\text{d} \ln x_k \text{d} p_i)_{sys}}{(\partial \ln|v_i| / \partial p_i)_{proc}}, \quad (9)$$

Now, for the pathway of Fig. 1, the control coefficients relative to both the concentration ($C_{e_1}^J$) or the activity ($C_{ve_1}^J$) of the enzyme 1 can be written as:

$$C_{e_1}^J = C_{ve_1}^J = C_1^J + C_2^J, \quad (10)$$

Here the symbol ve_1 is used to emphasize that the control coefficient is related to the entire enzyme e_1 rate rather than to the rate of the elemental step.

For the pathway of Fig. 1 the summation theorem (Eq. 6) can be written as:

$$C_{e_1}^J + C_{e_2}^J = C_1^J + C_2^J + C_3^J + C_4^J = = 1 \quad (11)$$

This reformulation of the summation theorem for simple pathways may be useful for the cases where one is attempting to understand what the implication is of a regulation of one of the elemental steps in an enzyme catalyzed reaction on the flux through the metabolic pathway. However, here we merely use it as a prelude to the control theoretical treatment for the channelled systems in Fig. 2. Figure 2a represents the example of a static channel which was analyzed by Sauro and Kacser [33]. It was assumed that the enzymes form the complex ($Q = E_1E_2$) independently of their interactions with metabolite molecules. Complex Q catalyzes the direct conversion of S to P, denoted as the step 5 in Fig. 2a. In this such a static channel, the number of enzyme molecules involved in the channel does not vary with the metabolic flux. Figure 2b denotes the more general case of dynamic channelling, where the extent of channeling depends on the relative rates of collision of E_1X and E_2 compared

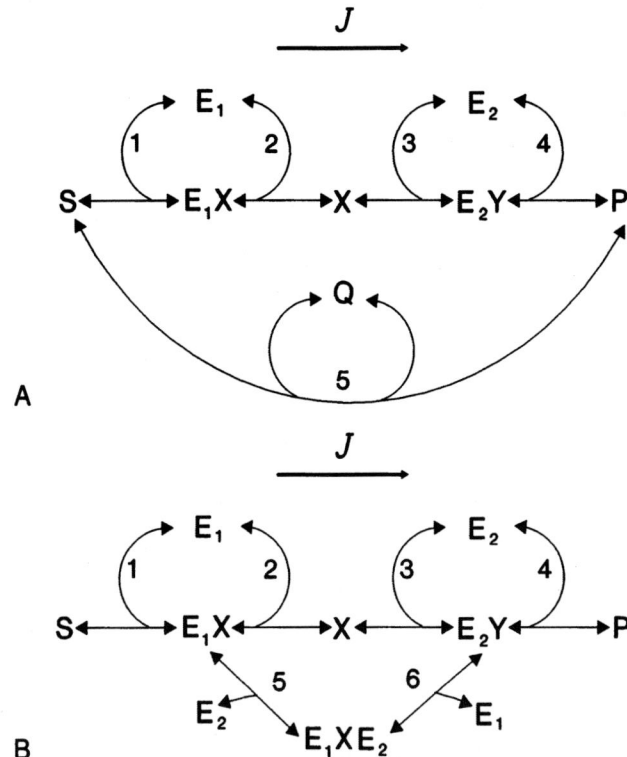

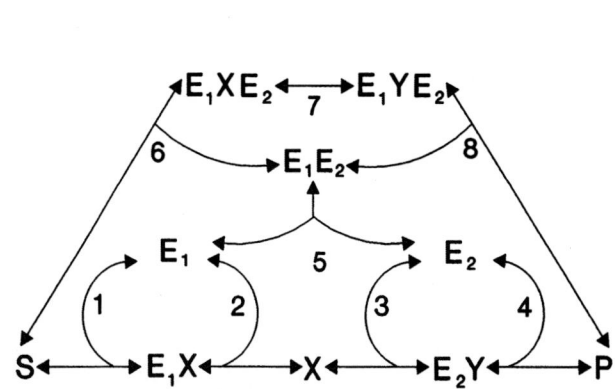

Fig. 3. 'Static' channel where only the free ('empty') forms of the enzymes 1 and 2, E_1 and E_2, can associate into the complex.

Fig. 2. 'Static' (a) and 'dynamic' (b) channels. The dynamic complex E_1XE_2 is formed only after binding X to E_1, while the static complex Q is formed independently of the presence of a common intermediate. In either system the upper route represents the usual reaction pathways through the bulk phase intermediate X, catalyzed by free enzymes, and the lower routes represent the 'channelling'. (a): steps 1, 2, 5 and 3, 4, 5 are E_1- and E_2-dependent processes, respectively. (b): steps 1, 2, 5, 6 and 3, 4, 5, 6 are E_1- and E_2-dependent processes, respectively.

to the rate of dissociation of E_1X into E_1 plus X and also on the other rate constants. The reaction sequence in the 'channelled phase' corresponds to steps 5 and 6 and is represented by the lower route in Fig. 2b.

Figure 2 may illustrate the problem of defining control coefficients of the participating enzymes in the case of channelling. For instance, enzyme 1 participates in two rather than one reaction (the channelled reaction from S to P as well as the reaction from S to X). Indeed, this is where classical control theory got stuck. Sauro and Kacser [33] were the first who for the case of static channelling (Fig. 2a) have indicated a solution. Assuming thermodynamic equilibrium between the separate enzymes and the complex they calculated the elasticities of partial reactions (i.e., proceeding in the bulk phase and in the channelled phase) with respect to the total enzyme concentrations. However, only under special conditions these elasticities which are referred to as protein (π) elasticities [8, 33] depend only on the equilibrium constant and the total amounts of the enzymes. This approximation fails for the more general case of a

dynamic channel of Fig. 2b. Moreover, one should take care applying this method to a static channel if association / dissociation of enzyme monomers can no longer be assumed *significantly slower* than the catalytic reactions.

Figure 3 shows a pathway which possesses features of a static channel. At any steady state of the system the step 5 of the complex formation is at thermodynamic equilibrium. However, if this step is not *significantly slower* than other steps and if the concentrations of the enzyme forms with bound ligands are comparable to the concentrations of the corresponding free enzyme forms, the π elasticities of partial reactions should depend on the flux and the concentrations of intermediates and can not be calculated only in terms of the equilibrium constant and the total amount of the enzymes. For, to fulfill the conditions necessary for the calculation of the π elasticities via equilibrium parameters, one would have to assume that the affinity of E_1 for E_2 does not depend upon whether a ligand (X or Y) has bound. This is not consistent with a scheme of Fig. 3, already because it requires the existence of a complex E_1XE_2Y. It can be shown, that the correct determination of the π elasticities requires an analysis at the level of the elemental processes of a system [Kholodenko et al., in preparation].

Our consideration of Fig. 1 above now suggests a solution to this dilemma: one should recognize that Figs 2a, 2b and 3 are still networks of chemical conversions and that they may be treated in terms of control coefficients with respect to the elemental processes (Eq. 8). These processes correspond to transitions between different states of enzymes, or to sequences of such transitions that are not interrupted by branches. Consequently, there are five, six and eight elemental flux control coefficients for the systems in Figs 2a, 2b and 3, respectively. For example, for Fig. 2b:

$$C_6^J = \partial \ln |J| \, / \, \partial \ln k_6^+, \text{ where } k_6^- / k_6^+ = \text{const},\qquad(12)$$

Because the flux through the system is a homogeneous function of all the elemental rate constants the following summation theorem holds for this channelled system [cf., 39]:

$$C_1^J + C_2^J + C_3^J + C_4^J + C_5^J + C_6^J = 1,\qquad(13)$$

i.e., the sum of the flux control coefficients continues to equal 1 provided the sum is over all the elemental processes.

Now from the level of the elemental processes of a pathway (the 'microdescription' level) we return to the level of enzyme reactions. The concept of elemental control coefficients allows us to assign an analogue of the control coefficient with respect to the enzyme activity ($C_{v_i}^J$) to the *individual* enzymes of a channel. Suppose, we simultaneously change the elemental rate constants of all processes in which the enzyme i is involved, by the same factor. Considering the corresponding change in the steady state flux J we define the '*impact*' control coefficient, $^{\text{imp}}C_{e_i}^J$, as the sum of the elemental control coefficients over the processes in which *any of the forms of the enzyme i* is involved (E_i-dependent processes):

$$^{\text{imp}}C_{e_i}^J = \sum_{\substack{\text{all } E_i\text{-dependent} \\ \text{processes } k}} C_k^J \qquad(14)$$

This coefficient evaluates the total impact enzyme i has on the flux J (the term 'E_i-dependent' refers to any form of enzyme i, monomeric or complexed, that contains E_i-moiety). In Sections 4 and 5 we explain this terminology more thoroughly by considering the responses of channelled pathways to signal molecules and the experimental methods of measuring the control coefficients (see also [38, 47]). Here we note that 'channelled' elemental steps are the E_i- and E_j-dependent processes simultaneously when enzymes i and j form a complex involved in catalytic transformations. In fact, for the scheme of Fig. 2a:

$$^{\text{imp}}C_{e_1}^J = C_1^J + C_2^J + C_5^J, \; ^{\text{imp}}C_{e_2}^J = C_3^J + C_4^J + C_5^J + \qquad(15)$$

Consequently, the channelled step 5 contributes to the

control exerted by either enzyme. For the scheme of Fig. 2b:

$$^{\text{imp}}C_{e_1}^J = C_1^J + C_2^J + C_5^J + C_6^J,$$

$$^{\text{imp}}C_{e_2}^J = C_3^J + C_4^J + C_5^J + C_6^J + \qquad(15)$$

We note that, the definition of the impact control coefficient $^{\text{imp}}C_{e_1}^J$ (by modulation of activities of all E_1-dependent processes) does not correspond to just a change in the total concentration of the enzyme 1 at constant distribution of this enzyme over all its forms and at a constant concentration of the enzyme 2. Indeed, the concomitant change in the form E_1XE_2 violates the conservation constraint imposed on the total concentration of the enzyme 2. So, in cases of enzyme-enzyme interactions as well as in other non-ideal pathways there is a difference between the control coefficients defined in terms of modulations of activity and those defined in terms of modulations of the enzyme concentration (cf. Eqs 1, 2 and 11). Both definitions are important since they refer to different experimental methods of determining the control coefficients (see Section 4). Here, for the channelled pathways of Fig. 2 we relate these two types of the control coefficients (see Section 6 for the corresponding equations and their derivation for the general case):

$$C_{e_1}^J + C_{e_2}^J \cdot e_{12}^{\text{comp}} / e_2 = {}^{\text{imp}}C_{e_1}^J$$

$$C_{e_1}^J \cdot e_{12}^{\text{comp}} / e_2 + C_{e_2}^J = {}^{\text{imp}}C_{e_2}^J \qquad(17)$$

Here $e_{12}^{\text{comp}} = [E_1E_2]$ or $[E_1XE_2]$ for the static (Fig. 2a) or the dynamic (Fig. 2a) channel, respectively, e_1 and e_2 are the total concentrations of the enzymes. In view of Eqs 15 or 16 one can use Eq. 17 to express the control coefficients with respect to enzyme concentrations into the control coefficients of the elemental processes. This then allows one also to evaluate the expected magnitude of the sum of the enzyme control coefficients. For the dynamic channel the result reads:

$$C_{e_1}^J + C_{e_2}^J = (1 + C_5^J + C_6^J)/(1 + e_{12}^{\text{comp}}/e) \qquad(18)$$

For simplicity we here considered the case where the total concentrations of the two enzymes are equal (e). Considering J as the overall flux through the pathway the relative contribution of the channelled steps to flux control can be expressed as following (see the branch theorems in [15, 16]):

$$(C_5^J + C_6^J)/J_{chan} = (C_2^J + C_3^J)/(J - J_{chan})$$

where J_{chan} denotes the flux through the channel. Using Eq. 13 one arrives at:

$$C_{e_1}^J + C_{e_2}^J = \{1 + (J_{chan}/J) \cdot (1 - (C_1^J + C_4^J))\}/(1 + e_{12}^{comp}/e) \quad (19)$$

In the case of static channelling (Fig. 2a) C_1^J and C_4^J disappear from this expression. Eq. 19 shows that the magnitude of the sum of the flux control coefficients can vary from less than unity to two depending on the ratio of the channelled and bulk-phase fluxes and the kinetic properties of the enzymes involved.

Responses of channelled pathways to regulatory signals

External effectors often play a role as signal molecules causing the system to modify its behavior in order to meet altered environmental requirements. In order to understand the cells regulatory structure it is important to realize how the response of the whole system is related to the 'local' response of the reactions directly affected by the signallers. By analyzing the responses to signals we shall arrive at the methods of experimental diagnosis of channelling (see Sections 5 and 6). We now show that the response to signals of 'channelled' pathways can differ drastically from the response of the corresponding non-channelled pathways.

The response of the steady-state flux (J toward an external signal (effector) is quantified by the response coefficient (R_σ^J), defined as log – log derivative of the flux with respect to the concentration of 'signal' molecules (σ) (cf. Eq. 1) [3]:

$$R_\sigma^J = \left(\frac{dJ/J}{d\sigma/\sigma} \right)_{sys} = \left(\frac{d \ln|J|}{d \ln \sigma} \right)_{sys} \quad (20)$$

In ideal pathways a metabolic response to a signal is determined by the flux control coefficient and by the elasticity coefficient ($\varepsilon_{\sigma_1}^{e_i}$) of the rate of the receptor ('target') enzyme (i) with respect to this signal [3]:

$$R_{\sigma_i}^J = \left(\frac{d \ln|J|}{d \ln e_i} \right)_{sys} \cdot \left. \frac{\partial \ln|v_i|}{\partial \ln \sigma_i} \right)_{enz} = C_{e_i}^J \cdot \varepsilon_{\sigma_i}^{e_i} \quad (21)$$

This relationship requires that signal molecules σ_i affect only one enzyme i in the pathway, as denoted by the subscript i.

In metabolic pathways with enzyme-enzyme interactions Eq. 21 is no longer valid. One of the reasons is lack of a one-to-one correspondence between enzymes and independent reactions. Now, one can guess that to derive a general expression for the responses of non-ideal pathways one may venture to the level of the elemental processes, i.e., use the 'microdescription' of a pathway. The elasticity coefficient of the elemental process k with respect to the signal molecules present at concentration σ_i will be designated as

$$\varepsilon_{\sigma_i}^k = (\partial \ln|v_k|/\partial \ln \sigma_i)_{proc}$$

Then, one can find the response of the flux to a signal specific to its receptor enzyme (i) via the control coefficients (C_k^J) of the E_i-dependent elemental processes (v_k) and their elasticities ($\varepsilon_{\sigma_i}^{e_i}$), (the response theorem [49]):

$$R_{\sigma_i}^J = \sum_{\substack{\text{all } E_i\text{-dependent} \\ \text{processes } k}} C_k^J \varepsilon_{\sigma_i}^k \quad (22)$$

As was mentioned above, some of the elemental processes may depend on two enzymes simultaneously, if these enzymes form a complex involved in catalytic transformations.

Applying Eq. 22 to the schemes of Figs 4 and 5 one can see that additional terms appear in the response of channelled pathways to signallers. For example, the flux responses to a signal (σ_1) affecting the enzyme 1 in the simple (non-channelled) pathway (Fig. 4) and in the 'static' channel (Fig. 5a) are, respectively:

$$R_{\sigma_1}^J = C_1^J \cdot \varepsilon_{\sigma_1}^1 + C_2^J \cdot \partial_{\sigma_1}^2 = C_{e_1}^J \cdot \partial_{\sigma_1}^{e_1} \quad (23)$$

(the superscript e_1 denotes the elasticity of the entire enzyme rate with respect to signal molecules), and

$$R_{\sigma_1}^J = C_1^J \cdot \varepsilon_{\sigma_1}^1 + C_2^J \cdot \partial_{\sigma_1}^2 + C_5^J \cdot \partial_{\sigma_1}^5 \quad (24)$$

Obviously, the term $C_5^J \cdot \partial_{\sigma_1}^5$ is absent in the response of the pathway of two unchannelled enzyme reactions, Eq. [23]. In the response ($R_{\sigma_2}^J$) of the channelled flux to a signal σ_2 affecting the enzyme 2 an analogous additional term ($C_5^J \cdot \partial_{\sigma_2}^5$) appears.

The flux responses to signaller for the 'dynamic' channel (Fig. 5b) are:

$$R_{\sigma_1}^J = C_1^J \cdot \varepsilon_{\sigma_1}^1 + C_2^J \cdot \partial_{\sigma_1}^2 + C_5^J \cdot \partial_{\sigma_1}^5 + C_6^J \cdot \partial_{\sigma_1}^6$$

$$R_{\sigma_2}^J = C_1^J \cdot \varepsilon_{\sigma_2}^1 + C_4^J \cdot \partial_{\sigma_2}^4 + C_5^J \cdot \partial_{\sigma_2}^5 + C_6^J \cdot \partial_{\sigma_2}^6 \quad (25)$$

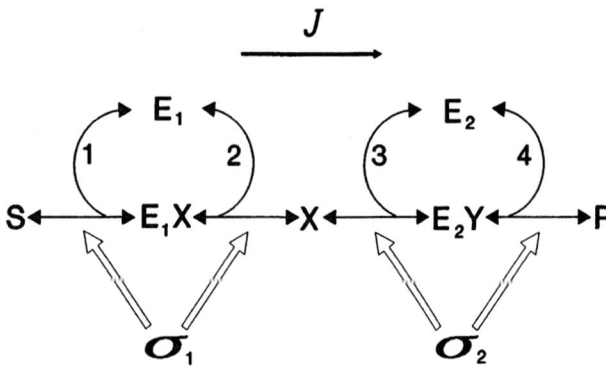

Fig. 4. How signal molecules affect an ideal pathway, σ_1 and σ_2 are the signal molecules, specific to the receptor enzymes 1 and 2, respectively. For other details see the legend to Fig. 1.

Eqs 23–25 already suggest that the response of channelled pathways to signals differs from that of non-channelled ones. To compare the responses it is convenient to normalize them to the response of the receptor enzyme by itself. In the light of definitions of the control theory (see Eqs 3 and 20), this normalized response will be called the 'signal transduction' coefficient of the enzyme i, $^{R}C_{e_i}^{J}(\sigma_i)$. This signal transduction coefficient is equal to the ratio of the response of the whole pathway to the response of the 'isolated' enzyme:

$$^{R}C_{e_i}^{J}(\sigma_i) = R_{\sigma_i}^{J_i} = \sum_{\substack{\text{all } E_i\text{-dependent} \\ \text{processes } k}} C_k^{J} \cdot \varepsilon_{\sigma_i}^{k} / \partial_{\sigma_i}^{e_i} \qquad (26)$$

here the superscript e_i denotes the elasticity of the rate of the enzyme i considered as if in isolation from the pathway. In ideal systems the signal transduction coefficient coincides with the classical control coefficient of the enzyme (see Eq. 23). In non-ideal pathways the value of the former may depend not only on the pathway properties but also on the peculiarities of signal molecules, as emphasized by the argument σ_i of the signal transduction coefficient.

Here we will consider the simple case when the elasticities $(\varepsilon_{\sigma_i}^{k})$ of all E_i-dependent elemental processes to the signal are equal to each other and to the elasticity of the reaction catalyzed by the enzyme i in 'isolation' $(\partial_{\sigma_i}^{e_i})$:

$$\varepsilon_{\sigma_i}^{k} = \partial_{\sigma_i}^{e_i}, \text{ if } k \text{ is an } E_i\text{-dependent process} \qquad (27)$$

It follows from Eqs 26 and 27 that in this case the expression for the signal transduction coefficient will be following:

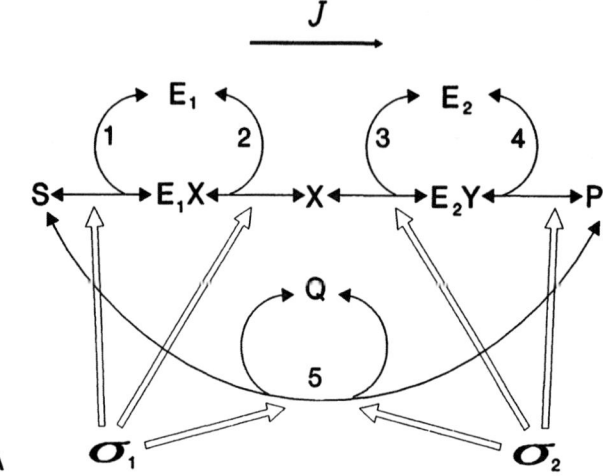

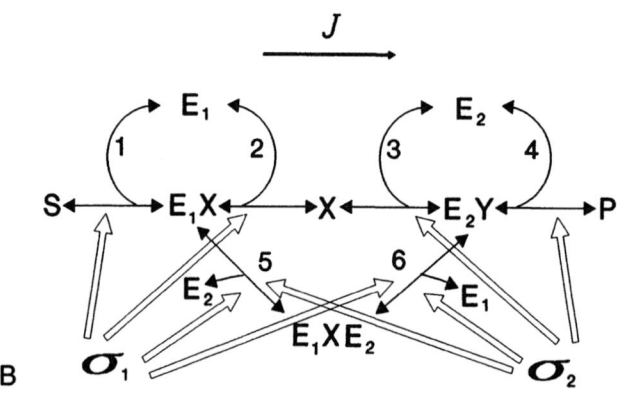

Fig. 5. How signal molecules affect 'channelled' pathways. (a): 'static' and (b) 'dynamic' channels. σ_1 and σ_2 are the signal molecules, specific to the receptor enzymes 1 and 2, respectively. Other details see in the legend to Fig. 2.

$$^{R}C_{e_i}^{J}(\sigma_i) = \sum_{\substack{\text{all } E_i\text{-dependent} \\ \text{processes } k}} C_k^{J} = {}^{\text{imp}}C_{e_i}^{J} \qquad (28)$$

Thus, for signals satisfying Eq. (27), the signal transduction coefficient coincides with the impact control coefficient (Section 3). Accordingly, the enzyme then transduces such signals via all the processes in which any of its forms is involved, and the contribution of every process is determined only by its flux control coefficient.

It is worth mentioning that treating in this manner some inhibitors as signals, one may indicate a way of measuring the control coefficients (see Section 5). Here, for signals satisfying Eq. (27) we compare the signal transduction coefficients of enzymes in ideal and channelled pathways, see Figs 4 and 5, respectively. It follows from the equations above (see Eqs 23–28) that for either enzyme the same additional term is present in the channelled pathway equating

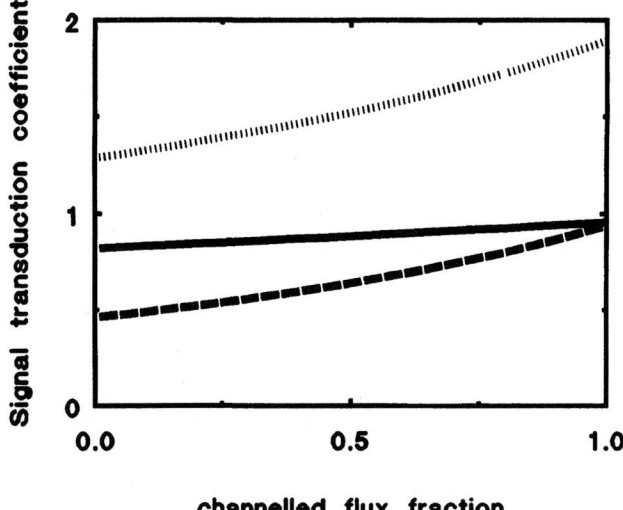

Fig. 6. Channelling enhances response to a signal. Signal transduction coefficients for enzyme 1 (solid line) and 2 (dashed line) of Fig. 5b and their sum (dotted line) were calculated as functions of the fraction of the total flux running through the channel. Leakage of bulk phase (X) and of channelled (E_1XE_2) intermediates was added to the scheme shown in Fig. 5b. Total concentrations of enzymes 1 and 2 were set to 1. Elemental rate constants were equal to: $k_1^+ = 10$, $k_1^- = 0.1$, $k_2^+ = 10\alpha$, $k_2^- = \alpha$, $k_3^+ = \alpha$, $k_3^- = \alpha$, $k_4^+ = 10$, $k_4^- = 0.01$, $k_5^+ = 100\beta$, $k_5^- = \beta$, $k_6^+ = 0.0045\beta$, $k_6^- = 0.045\beta$, and for 'channelled' (subscript lc) and 'unchannelled' (subscript l) leaks considered as irreversible steps, $k_{lc} = 2.316 \, 10^{-3}\beta$, $k_l = 0.05\alpha$. α and β were chosen such that the total output flux (J_{out}, to product P) and the total leak flux remained constant (at 0.3869 and $0.323J_{out}$, respectively). The channelled flux fraction was defined as the flux from E_1XE_2 to E_2Y divided by J_{out} (in the case considered it also equals to the channelled flux before the leak reaction divided by J_{in}). The signal transduction and control coefficients were calculated by increasing elemental rate constants (k_1^+, k_1^-, k_2^+, k_2^-, k_5^+, k_5^-, k_6^+, k_6^-, k_{lc}) for enzyme E_1 and (k_3^+, k_3^-, k_4^+, k_4^-, k_5^+, k_5^-, k_6^+, k_6^-, k_{lc}) for enzyme E_2 by 0.01%.

$$C_5^J = J_{chan}/J,$$

for the static channel (Fig. 5a) or,

$$C_5^J + C_6^J = (J_{chan}/J) \cdot (1 - (C_1^J + C_4^J)),$$

for the dynamic channel (Fig. 5b). In case a significant fraction J_{chan}/J of the flux flows through the channel, the signal transduction coefficient of each of the enzymes 1 and 2 can be close to unity, whereas only one of them can have such a value in the pathway of two unchannelled enzyme reactions.

In realistic cases the intermediates may be subject to leakage. In such a case the input flux (J_{in}, the substrate consumption) differs from the output flux (J_{out}, synthesis of the product), the difference equals the sum of the leaks of bulk phase (X) and channelled (E_1XE_2) intermediates. Figure 6 shows how the signals can control the output flux at various degrees of channelling in the case

when 'channelled' and 'unchannelled' leaks are proportional to the corresponding 'channelled' and 'unchannelled' fluxes. We can see that the signal transduction coefficient of either enzyme ($^RC_{e_i}^{Jout}$) increases in parallel with the fraction of the flux that runs through the channel.

In a simple case only the bulk phase intermediate is subject to leakage. Then, the signal transduction coefficient ($^RC_{e_i}^{Jout}$) of either enzyme includes an additional term which in the case of the static channel is equal to:

$$C_5^{Jout} = J_{chan}/J_{out},$$

for the dynamic channel,

$$C_5^{Jout} + C_6^{Jout} = (J_{chan}/J_{out}) \cdot (1 - C_4^J) - (J_{chan}/J_{in}) \cdot C_1^J$$

Under the condition that most of the bulk-phase flux ends up in the leak and the binding of S and P to E_1 and E_2, respectively, are near equilibrium, the signal transduction coefficient of each of the enzymes E_1 and E_2 should be close to unity even if the channelled flux is much smaller than total bulk phase flux but comparable to the output flux. We conclude that channelling tends to increase the response of a pathway to signals.

Experimental methods for determining the control coefficients

Classical definition of the enzyme's control coefficient, Eq. 1, has the operational implication of modulating the concentration of that enzyme. However, direct titration of the amount of an enzyme is possible only in a reconstituted system [50]. In native systems a change in the concentration of the enzyme of interest can be achieved by genetic means, for instance by comparing heterocaryons of *Neurospora crassa* [51], by causing a gene to be expressed from a plasmid [52], or by modulating the expression of the chromosomal gene [53]. This approach has the complication that the modulation of a single gene may result (although non-directly) in changes of the concentrations of several enzymes. On the other hand, Eq. 3 allows one to measure the control coefficients both in intact and reconstituted systems using inhibitors or other modificators specific to single enzymes [54–56].

In ideal pathways, the result of such a measurement does not depend on the type of inhibitor used, provided that the inhibitor (I_i) only affects the target enzyme (i).

However, this is not true for non-ideal systems, for instance with direct enzyme-enzyme interactions. To emphasize this we called the coefficients, determined according to classical theory, the 'apparent' control coefficients and denoted them by a left upper index 'app' (cf. Eq. 26) [38, 47]:

$$^{app}C_{e_i}^J(I_i) = \frac{(d \ln |J|/dI_i)_{sys}}{(\partial \ln |v_i|/\partial I_i)_{enz}} = \frac{V_i}{J} \cdot \frac{(dJ/dI_i)_{sys}}{(\partial v_i/\partial I_i)_{enz}} \quad (29)$$

Here the elasticity of the rate (v_i) of the 'target' enzyme to the inhibitor is considered as if the enzyme i were in isolation from the pathway at the same metabolite concentrations as in the steady state of the system,

$$\varepsilon_{\sigma_i}^{e_i} = (\partial \ln|v_i|/\partial I_i)_{enz} \quad (30)$$

Since usually the inhibitor, I_i, is absent in the intact system, all the derivatives in Eq. 29 should be calculated at $I_i = 0$. Therefore, we used the derivatives with respect to I_i rather than to $\ln I_i$, in order to avoid indefiniteness connected with the logarithmic derivatives at zero inhibitor concentration (the usual definition of the elasticity coefficients in the control analysis uses logarithmic derivatives, cf. Eq. 21). For instance, if I_i is an irreversible inhibitor binding to the enzyme i in the stoichiometry ratio 1 : 1, Eq. 29 takes the form [54]:

$$^{app}C_{e_i}^J(I_i) = -\frac{(I_{end})}{J} \cdot \frac{\partial J}{\partial I_i}\Big|_{I_i=0} \quad (31)$$

Here I_{end} denotes the inhibitor concentration necessary to completely suppress the enzyme (in the titration curve I_{end} is determined as a concentration extrapolating to where flux J no longer depends on I_i), $\partial J/\partial I_i|_{I=0}$ is determined by the initial slope of the titration curve.

We start with a simple numerical example showing that the result of an experiment on titrating a channelled pathway with inhibitor does depend on the type of that inhibitor even if the latter affects only a single enzyme [47, Kholedenko *et al.*, submitted]. We consider a simple dynamic channel (Fig. 2b), and suppose that in order to measure the control coefficient of enzyme 1 on the total flux through the pathway (i.e. the net production rate of P) one uses different specific inhibitors of the enzyme: (i) a purely non-competitive inhibitor (I_1^{nc}) that binds to all the forms of the enzyme 1 independently of their interactions with ligand or enzyme 2 molecules and (ii) a purely competitive inhibitor (I_1^c) that binds only to the free form, E_1, which is not complexed with any ligand or enzyme 2. The corresponding formulas for the calculation of the control coefficients using purely non-compet-

itive and purely competitive inhibitors (I_1^{nc} and I_1^c) read (see, e.g., [54]):

$$^{app}C_{e_1}^J(I_1^{nc}) = -\frac{(K_1^{nc})}{J} \cdot \frac{\partial J}{\partial I_1^{nc}}\Big|_{I_1^{nc}=0} \quad (32)$$

$$^{app}C_{e_1}^J(I_1^c) = -\frac{(K_1^c)}{J} \cdot (1 + \frac{S}{K_M^S} + \frac{X}{K_M^X}) \cdot \frac{\partial J}{\partial I_1^c}\Big|_{I_1^c=0} \quad (33)$$

Here K_1^{nc} and K_1^c are the inhibition constants of the inhibitors I_1^{nc} and I_1^c, respectively. (To ensure the validity of Eqs 32 and 33 the values of the inhibition constants should exceed significantly the total concentration of enzymes, or I should be interpreted as the free inhibitor concentration [47].) K_M^S and K_M^X are Michaelis constants for the substrate (S) and product (X) of the enzyme 1, respectively.

The titration curve (the dependence of steady-state flux (J) on the inhibitor concentration) was simulated by solving the equations determining the steady state of the pathway of Fig. 2b for the following set of parameters (in dimensionless units): all the elemental rate constants equaled 1, except for $k_1^- = k_4^- = 0.5$; the total concentrations of enzymes were taken $e_1 = e_2 = 1$; $K_1^{nc} = 10$, $K_1^c = 10$, $S = 10$, p = 1. Then, Eqs 32 and 33 were applied. The apparent control coefficient of the enzyme 1 on the flux converged to 0.51 when determined using the purely non-competitive inhibitor (I_1^{nc}) and to 0.40 when determined by use of the purely competitive inhibitor (I_1^c). Clearly, contrary to the classical results, the flux control coefficient in this channelled system depends on how it is determined.

A way how to determine the apparent control coefficients in the general case of channelling was outlined in Section 4. We can consider molecules of an inhibitor (I_i) as signal molecules affecting only enzyme i. Then, in view of Eq. 26, the formula for the apparent control coefficient, Eq. 29, takes the form:

$$^{app}C_{e_i}^J(I_i) = \sum_{\substack{\text{all } E_i\text{-dependent} \\ \text{processes } k}} C_k^J \cdot \varepsilon_{I_i}^k / \varepsilon_{I_i}^{e_i} \quad (34)$$

Here the elasticity coefficients ($\varepsilon_{I_i}^k$) of the elemental processes with respect to the inhibitor are defined as:

$$\varepsilon_{I_i}^{V_k} = (\partial \text{ LN}|V_k|/\partial I_i)_{PROC} \quad (35)$$

(similarly as above the derivatives with respect to I_i rather than to $\ln I_i$ are used here in order to avoid indefiniteness at zero inhibitor concentration).

There are various types of inhibitors and various mechanisms of inhibition (see, e.g., [57, 58]). For pathways with enzyme-enzyme interactions this variety may even be greater since the response of the pathway flux to the inhibitor titration strongly depends on how that inhibitor affects the enzyme-enzyme complexes. For instance, in the traditional case a non-competitive inhibitor is thought of as an inhibitor that does not replace metabolites at their binding sites in the enzyme molecule. A *purely* non-competitive inhibitor is usually considered to affect only the parameter V_{max} of the steady-state rate equation, which in the framework of 'microdescription' corresponds to a simultaneous proportional decrease in all the forward and reverse (pseudo-) first order rate constants of the affected enzyme (cf., [57]). However, the addition of such an inhibitor specific to the enzyme i to a channelled pathway results in an equal relative inhibition of the activities of all E_i-dependent elemental processes only if the following conditions are fulfilled: (i) this inhibitor binds to all the enzyme i forms with the same binding constant, (ii) its binding to any enzyme i form transforms the latter into a less active (or inactive) state, (iii) its binding does not change the ability of the enzyme i to form complexes with the other enzymes of the pathway. In such cases the elasticities of all the E_i-dependent elemental processes are equal to each other and to the elasticity of the reaction catalyzed by the enzyme i in 'isolation' (cf. Eq. 28):

$$\varepsilon_{I_i^{nc}}^{k} = \varepsilon_{I_i^{nc}}^{f} \text{ if } k \text{ is an } E_i\text{-dependent process} \qquad (36)$$

where the symbol I_i^{nc} denotes a purely non-competitive inhibitor (cf. the numerical example above). It follows from Eqs 34 and 36 that titrating the system with such an inhibitor we measure the impact control coefficient of the enzyme i:

$$^{app}C_{e_i}^{J}(I_i^{nc}) = {}^{imp}C_{e_i}^{J}, \qquad (37)$$

This control coefficient can not be measured by a direct variation in the enzyme concentration, e.g., by modulating the gene encoding the enzyme. Importantly, the impact control coefficient can differ drastically from the control coefficient with respect to the enzyme concentration (see Fig. 6, where Eq. 28 is assumed to be fulfilled).

A competitive inhibitor competes with a substrate (product) for the binding sites at the enzyme molecule. In terms of a 'mechanistic' model the molecules of such an inhibitor displace the substrate molecules from the binding site. A *purely* competitive inhibitor is thought to affect only the parameters K_m-s of the steady-state rate equation (cf., [57]). Generalizing this definition one can consider such a purely competitive inhibitor (I_i^c) of the enzyme i that binds only to the form, E_i, which is not complexed with any other metabolite or enzyme of a pathway. Such an inhibitor will suppress the rates only those elemental processes in which the 'empty' form, E_i, is a 'substrate' or 'product'. Hence, in a channelled system Eq. 36 is not valid for a purely competitive inhibitor (I_i^c) of the enzyme i, and the control coefficient of the enzyme i measured with I_i^c will differ of that of determined by use of Eq. 37 and a purely non-competitive inhibitor I_i^{nc}. Indeed, we have witnessed a difference between these apparent control coefficients in the numerical example above. Such a difference becomes especially notable in the cases of very tight-binding and irreversible (covalently bound) inhibitors to which Eq. 31 can be applied. It can be shown, that titrating a channelled pathway with an irreversible purely competitive inhibitor one determines (by virtue of Eq. 31) the classical control coefficient ($C_{e_i}^{J}$) with respect to the enzyme concentration, whereas titrating that pathway with irreversible purely non-competitive inhibitor (and using Eq. 31) one measures the *impact* control coefficient ($^{imp}C_{e_i}^{J}$) of the enzyme i [Kholodenko and Westerhoff, submitted].

Thus, one may conclude that the finding the different flux control coefficients for the same enzyme using different inhibitors may be an indication that this enzyme is involved in a partially channelled pathway. An other such indication can be any difference between the sum of enzyme control coefficients and unity. A more thorough analysis of the problem can be found in the next Section.

Implementation of control theory to the experimental diagnosis of channelling

We saw already (see Section 3, Eq. 19) that when a direct transfer of the intermediate takes place, the sum of flux control coefficients ($C_{e_i}^{J}$) over all enzymes generally differs from unity. The contrary is not true, however [33]. In fact, suppose that enzyme association merely leads to a change in kinetics of the constituent enzymes, and channelling of the intermediate does not occur, Fig. 7. In this case there is only a bulk phase pool of the intermediate (X), and all reaction pathways include diffusional steps

324

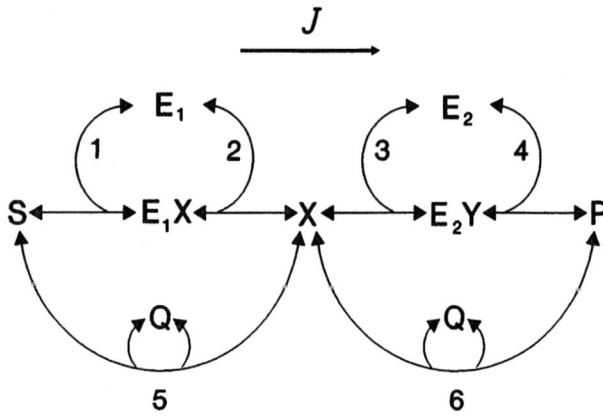

Fig. 7. Pathway of two enzymes capable to form a complex (Q), where channelling is absent. In this system there is only a bulk phase pool of the intermediate (X), and all reaction pathways include diffusional steps to and from X.

to (from) X. However, since the ratios between concentrations of the free enzymes and the complex changes with a change in the total enzyme concentration, also in this case, the sum of the flux control coefficients will differ from unity for this system (as well as for the systems with channelling shown in Fig. 2).

Therefore, the sum of the classical control coefficients $(C_{e_i}^J)$ of the enzymes may not be used as the sole criterion for discriminating between direct transfer and a free diffusion mechanism. In the particular case of equal concentrations of both enzymes, Sauro & Kacser [33] proposed to distinguish between channelling and free diffusion by following the bulk phase intermediate X. When the total enzyme concentration, $e_1 + e_2$, is varied the concentration of X does change in the system shown in Fig. 7 (in which the direct transfer is absent), whereas it does not change in the case of a static channel (Fig. 2a).

However, this criterion fails in systems with dynamic complexes. As was mentioned, in this case an enzyme complex is formed only if one of the enzymes has already bound the common intermediate, Fig. 2b. It has been shown, that even if the total concentrations of both enzymes are equal, the concentration of the bulk phase intermediate X will change upon a change in the total enzyme concentration $e_1 + e_2$ [47].

Our consideration of the responses of the flux to signal (Fig. 6) or inhibitor molecules (Section 5) already suggests that the sum of the signal transduction control coefficients (apparent control coefficients in the case of inhibitor titration) seemed to be useful for discrimination between channelling and a free diffusion mechanism. The question arises: what is this sum equal to for the system shown in Fig. 7?

We shall now consider titrations of pathways with en-

zyme-enzyme complexes and a direct (Fig. 2) or a free diffusion mechanism (Fig. 7) of metabolite transfer with purely non-competitive inhibitors. For channelled pathways, upon binding with the complex E_1E_2, an inhibitor specific to enzyme 1 may inhibit channelling. Similarly a second inhibitor, binding specifically to 2 may do the same (see Fig. 5 with the signal molecules σ_1 and σ_2 considered now as inhibitors specific to the enzyme 1 and 2, respectively). We saw that such an inhibition of channelling by either inhibitor made the apparent control coefficient of both enzyme 1 and enzyme 2 include the elemental control coefficients of the shared 'channelled' steps (when Eq. 36 is fulfilled). On the other hand, in the case of a free diffusion mechanism of metabolite transfer (Fig. 7), binding of an inhibitor specific for enzyme 1, to the complex E_1E_2, may affect the catalytic steps in the reaction catalyzed by the enzyme 1 (which is bound to E_2) without affecting the reaction catalyzed by the enzyme 2 (which is bound to E_1). In other words, for this enzyme complex that catalyzes a free diffusion mechanism, Eq. 36 is not fulfilled. Then, the sum of the apparent control coefficients will exceed unity for the channelled enzymes (Fig. 2) and will be equal to unity for the complexed enzymes with free diffusion (Fig. 7). We conclude that depending on inhibitor type the sum of the apparent control coefficients determined by inhibitor titration can be used for experimental discrimination between channelling and a free diffusion mechanism [47].

The availability of inhibitors with the properties described above is rare. As an alternative to inhibitor titration an irreversible modification of the enzyme catalytic site could be performed in a separate sample (by chemical or genetic methods). Such a modification should suppress the enzyme catalytic activity only, leaving unchanged the ability of the enzyme to associate with other enzymes. Different amounts of the modified enzyme $(e_i^{mod}$, which has lost the catalytic activity) should then be added to the system under study at the constant total concentration (e_i^t) of the active and inactive enzyme:

$$e_i^{mod} + e_i = e_i^t$$

The dependence of the flux (J) on the amount of the modified enzyme (e_i^{mod}) is an analog of the titration curve with specific irreversible inhibitor of the enzyme i (the concentration of e_i^{mod} plays the role of the inhibitor concentration). Transforming Eq. 31 for the irreversible inhibitor to this case, one finds:

$$^{app}C_{e_i}^J = - \frac{e_i^t}{J} \left(\frac{\partial J}{\partial e_i^{mod}} \right) \Bigg|_{e_i^{mod}=0} \qquad (38)$$

The sum of the apparent control coefficients measured using the modification enzyme technique proves to *exceed unity* for the channelled pathways and to *equal unity* for the pathways with enzyme complexes but with a free diffusion mechanism of metabolite transfer.

Control of partly or completely channelled systems in general

The conceptual shift underlying the approach developed here is to view any metabolic pathway as a number of 'elemental' processes with (well-defined) forward and reverse rate constants. Notably, this implies that we no longer agglomerate all processes catalyzed by the enzyme into 'separate' reactions catalyzed in 'isolation' from the system. The general case of 'non-ideal' cellular pathway which includes r enzyme reactions, is then treated as the network of chemical conversions where the 'metabolites' (system variables) are both free metabolites and different enzyme forms (states) that include enzyme-bound metabolites and enzyme-enzyme complexes (see Section 3 and [11]). These conversions are called the elemental processes of the system.

Let n be the number of the elemental processes in a pathway (typically $n > r$), and m be the total number of free metabolites and different enzyme forms (system variables). Obviously, there are some constraints on the variation of the concentrations ($x_1, x_2, \ldots, x_m$) of these substances, and not all concentrations (x_k) are independent. At least, there are r constraints, corresponding to the moiety conservations of every enzyme (i), which may have the following form:

$$e_i = E_i + E_iS + \ldots E_iE_j + E_iSE_j + \ldots, i = 1, 2, \ldots, r \qquad (39)$$

Here instead of x_k the chemical symbols are used for the concentrations: E_i designates free enzyme, E_iS designates the enzyme-substrate form, E_iE_j and E_iSE_j designate enzyme-enzyme and enzyme-substrate-enzyme complexes, respectively. Such participants of the enzyme i moiety-conserved cycle are called the enzyme i forms (or states, cf. [59]). Obviously, complexes involving two enzymes i and j (e.g., E_iE_j and E_iSE_j) enter both enzyme i and enzyme j moiety-conserved cycles, as they are enzyme i and enzyme j forms simultaneously.

In what follows it is convenient to present the total enzyme concentration e_i as the sum of the concentrations of the monomeric enzyme forms (e_i^{mon}) (which may be complexed only with metabolites), and complexes (e_i^{comp}) of the enzyme i with any other enzyme (j),

$$e_i = e_i^{mon} + \sum_{j \neq i} e_{ij}^{comp} = e_i^{mon} + e_i^{comp} \qquad (40)$$

Here e_i^{comp} is the part of the enzyme i concentration complexed with all the other enzymes. For the sake of simplicity, we only consider heterodimer complexes (in particular, here we do not consider homodimers e_ie_i). The generalization for the case of several enzyme molecules is straight-forward.

A common feature of metabolic pathways is the presence of substrate moiety-conserved cycles [60], the interconversion of NAD$^+$ and NADH may serve as an example. In an arbitrary metabolic pathway s such substrate moiety-conserved cycles may be present.

$$T_i = \sum_{k=1}^{m} \gamma_{ik}x_k, \ i^v 1, 2, \ldots, s \qquad (41)$$

Here T_i designates the total concentration of i-th conserved substrate (not enzyme) moiety, and γ_{ik} can in most cases be interpreted as the number of moieties of type i in the metabolite x_k. Note that x_k entering Eq. 41 correspond to the concentrations of both free and (for a different value of k) enzyme-bound metabolites. For a mathematically inclined reader we also note that the rank of the m rows by n columns stoichiometric matrix $\mathbf{N}$ of the metabolic network is equal to m-r-s. Eqs 39 and 41 describe $r + s$ linear dependencies between the rows of $\mathbf{N}$ (see, e.g., [14]).

Any elemental process (step) in the network under consideration is a mono- or bimolecular reaction with respect to the enzyme forms. Moreover, if any of the enzyme j forms enter the left-hand side of chemical equation of the elemental process, then some other form of the same enzyme j will enter the right-hand side of that chemical equation. Hence, the rate (v_i) of every i-th elemental process is a homogeneous function of zero or first order with respect to the concentrations of the forms of any enzyme j (the order equals 0 or 1 depending on whether enzyme j forms participate in i^{th} elemental process). All the elemental processes with rates depending on the enzyme j forms were called E_j-dependent processes in Section 3.

Now we define n parameters ξ_i, $i = 1, 2, \ldots, n$ each modulating the activity only of the i-th process,

$$v_i(x, \xi_i) = \xi_iv_i(x, 1) \qquad (42)$$

According to Eq. 8 of Section 3 the elemental flux control coefficient of any process can be defined as following (cf. Eq. 7):

$$C_i^J = \mathrm{dln}|J|/\mathrm{dln}\,\xi_i\Big|_{\xi_i=1},\ i = 1, 2, \ldots, n \qquad (43)$$

Note, that for the classical 'macrodescription' of a pathway the parameter ξ_i plays the role of the activity or the concentration of the enzyme and the control coefficient defined by Eq. 43 coincides with the 'classical' control coefficient (see Eq. 1).

Relating control by enzymes to control by the elemental processes

In section 3 for simple examples of channelled pathways we related the elemental to the classical control coefficients (see Eqs 14–17). To elucidate the particular features of the control in the general case of pathways which may involve channelling and moiety sequestrating, we apply a method of perturbation of the steady-state [15, 61]. Let in the initial steady state the concentrations and parameters be perturbed as the following: (i) every concentration involved in the enzyme i moiety-conserved cycle (see Eq. 39) is increased by a factor λ_i:

$$E_i(\lambda_i) = \lambda_i E_i,\ E_i S(\lambda_i) = \lambda_i E_i S,\ E_i E_j(\lambda_i) = \lambda_i E_i E_j,$$
$$E_i SE_j(\lambda_i) = \lambda_i E_i SE_j,\ldots,\ i = 1, 2, \ldots, r \qquad (44)$$

(ii) parameters, ξ_k, which correspond to the rates of E_i-dependent elemental processes (in which any of the forms of the enzyme i participates) are decreased by the same factor λ_i:

$$\xi_k(\lambda_i) = \xi_k/\lambda_i,$$
$$\text{if } \xi_k \text{ corresponds to } E_i\text{-dependent process} \qquad (45)$$

Since the rates v_k of E_i-dependent processes are homogeneous first-order functions of the concentrations of the enzyme i forms (see above), all the rates in the new steady state will be equal to the initial non-perturbed rates. However, the parameters $\mathbf{e}, \mathbf{T}, \xi$ do differ between the old and the new steady state (see Eqs 44 and 45). The new values of e_j and T_l are:

$$e_i(\lambda_i) = \lambda_i \cdot e_i;$$

$$\text{for } j \neq i, e_j(\lambda_i) = e_j^{\mathrm{mon}} + \sum_{k \neq j, i} e_{ik}^{\mathrm{comp}} + \lambda_i \cdot e_{ik}^{\mathrm{comp}} = e_j + (\lambda_i - 1) \cdot e_{ik}^{\mathrm{comp}};$$

$$T_l(\lambda_i) = T_l + (\lambda_i - 1) \cdot T_l^{\mathrm{e_i}}, l = 1, 2, \ldots, s \qquad (46)$$

here $T_l^{\mathrm{e_i}}$ designates the part of l-th conserved moiety bound to all forms of the enzyme i. Since the steady-state fluxes (J) are functions of $(\mathbf{e}, \mathbf{T}, \xi)$ one can write:

$$0 = \mathrm{dln}|J|/\mathrm{dln}\,\lambda_i = \sum_{j=1}^{r} C_{e_j}^J \cdot \mathrm{dln}e_j/\mathrm{dln}\,\lambda_i + \sum_{l=1}^{s} R_{T_l}^J \cdot \mathrm{dln}T_l/$$
$$\mathrm{dln}\lambda_i + \sum_k C_k^J \cdot \mathrm{dln}\xi_k/\mathrm{dln}\,\lambda_i; \qquad (47)$$

here $R_{T_l}^J$ is the flux response coefficient to a change in l-th substrate total T_l,

$$R_{T_l}^J = \mathrm{dln}|J|/\mathrm{dln}T_l,$$

and the third sum in Eq. 47 includes only E_i-dependent elemental processes (in which the enzyme i forms participate). Substituting into Eq. 47 the derivatives with respect to $\ln \lambda_i$ calculated at $\lambda_i = 1$ (using Eqs 44–46), we obtain:

$$C_{e_i}^J + \sum_{j \neq i} C_{e_j}^J \cdot e_{ji}^{\mathrm{comp}}/e_j + \sum_{l=1}^{s} R_{T_l}^J \cdot T_l^{\mathrm{e_i}}/T_l =$$
$$\sum_{\substack{\text{all } E_i\text{-dependent} \\ \text{processes } k}} C_k^J = {}^{\mathrm{imp}}C_{e_i}^J,\ i = 1, 2, \ldots, r \qquad (48)$$

The right-hand side of this equation represents the *impact* control coefficient of the enzyme i which has been introduced in Section 3, see Eq. 14. The left-hand side contains the classical enzyme's i control coefficient, the control coefficients of those enzymes that are complexed by enzyme i, and the response coefficients to a modulation of conserved metabolites complexed by enzyme i. Note, that applying Eq. 48 to channelled pathways in Fig. 2 we arrive at Eq. 17 relating classical and elemental control coefficients for these simple channels.

Concentration control

To relate the classical concentration control coefficients ($C_{e_i}^{x_k}$, Eq. 4) to the corresponding control coefficients of the elemental processes ($C_e^{x_k}$, Eq. 9) one should take into account that after perturbation of the parameters according to Eqs 45 and 46, in the new steady state all the concentrations except the enzyme i forms (see Eq. 44) will have the same values as in the initial steady state. Formally:

$$d \ln x_k = \delta(x_k, E_i) d \ln \lambda$$

Here $\delta(x_k, E_i) = 1$, if x_k is any of the forms of the enzyme i, otherwise, $\delta x_k, E_i) = 0$ (the argument E_i signifies that any of the enzyme i forms, monomeric or complexed, contains E_i-moiety).

Using the same approach as before, that is considering the steady-state concentrations as functions of the parameters $\mathbf{e}$, $\mathbf{T}$, ξ one obtains:

$$C_{e_i}^{x_k} + \sum_{j \neq i} C_{e_j}^{x_k} e_{ji}^{comp}/e_j + \sum_{l=1}^{s} R_{T_l}^{x_k} T_l^{\xi}/T_l =$$

$$\delta(x_k, E_i) + \sum_{\substack{\text{all } E_i\text{-dependent} \\ \text{processes}_j}} C_j^{x_k} =$$

$$\delta(x_k, E_i) + {}^{imp}C_{e_i}^{x_k} \quad i=1, 2, \ldots, r \quad (49)$$

where $R_{T_l}^{x_k}$ is the response coefficient of the concentration x_k to a change in l-th substrate total T_l,

$$R_{T_l}^{x_k} = d \ln x_k/d \ln T_l$$

General summation theorems for the flux and concentration control coefficients

Summing Eq. 48 over all r enzymes and rearranging one obtains:

$$\sum_{i=1}^{r} C_{e_i}^{J} \cdot (1+\frac{e_i^{comp}}{e_i}) = 1+\sum c_i^J - \sum_{l=1}^{s} R_{T_l}^{J}(\frac{T_l^{mon} 2 T_l^{comp}}{T_l}) \quad (50)$$

here the sum of $T_{l_i}^{\xi}$ over all enzymes (i) is subdivided into two parts: T_l^{mon} bound to the monomeric enzyme forms and T_l^{comp} bound to the enzyme-enzyme complexes (note that T_l^{comp} enter twice the sum of T_l^{ξ} over all enzymes). The additional to unity sum in the right-hand side of Eq. 50 is taken over all 'protein interaction' steps, i.e., all the elemental processes in which two different enzyme moieties participate (either as monomeric or complexed enzymes).

The summation theorem for the concentration control coefficients is obtained by summing Eq. 14 over all r enzymes of the pathway:

$$\sum_{i=1}^{r} C_{e_i}^{x_k}(1+\frac{e_i^{comp}}{e_i}) = \delta(x_k, E) + \sum_{\substack{\text{channelled} \\ \text{elemental} \\ \text{step}_l}} C_l^{x_k} -$$

$$\sum_{l=1}^{s} R_{T_l}^{x_k} \cdot (\frac{T_l^{mon} + 2T_l^{comp}}{T_l}) \quad (51)$$

here $\delta(x_k, E) = 1$, if x_k is any of the forms of *any* of the pathway enzymes, otherwise, $\delta(x_k, E) = 0$.

The equations derived above predict dramatic changes in the control properties of the enzymes in real cellular pathways when compared with the 'ideal' pathways considered by the classical metabolic control theory. In ideal pathways both sums on the left-hand sides of Eq. 48 and Eq. 49, respectively vanish. In that case these formulas state that the control coefficient of an enzyme over the pathway flux or the metabolite concentration can be expressed as the sum of the elemental control coefficients of all steps in the reaction cycle of this enzyme (cf. Eq. 10).

In pathways with direct enzyme-enzyme interactions the first sum on the left-hand side of Eq. 48 or Eq. 49 reflect the contribution to the *impact* control coefficient of the enzyme i of all those enzymes with which enzyme i is complexed. In pathways with high enzyme concentrations and moiety conservations but without direct enzyme-enzyme interactions the first sum on the left-hand side of Eq. 48 or Eq. 49 equals zero, and the right-hand side of Eq. 48 or Eq. 49, respectively, represent the control coefficient with respect to the enzyme rate or activity [9, 10, 33]. In this case Eqs 48 and 49 show that the control coefficient with respect to the enzyme concentration can be significantly less than the control coefficient with respect to the enzyme activity due to sequestration of metabolites. Indeed, in such cases the flux control coefficient with respect to the enzyme activity can even take negative values when its 'classical' analog is positive [10].

Eqs 50 and 51 present the summation theorems for the flux and concentration control coefficients valid now for an arbitrary pathway. In ideal pathways the sum of the flux control coefficients of the enzymes equals 1 (see Eq. 6, [Kacser]), and the sum of the concentration control coefficients equals zero or unity depending on whether the control is considered over the concentration of any free metabolite (substrate) or any enzyme form. Eqs 50 and 51 show that sequestrating of moiety-conserved metabolites by binding them to the enzymes present in high concentrations can significantly decrease these sums. On the other hand, channelling of metabolites can affect the sum of the flux control coefficients in two different modes depending on the average complexed fraction of enzymes and on the control exerted by 'channelled' steps. The latter may reach as much as 1, so in a pathway without substrate moiety-conservations and small life time of enzyme-enzyme complexes the sum of the flux control coefficients of the enzymes can reach 2 [48]. We shall illustrate this phenomenon considering the group transfer pathway which is an important aspect of cellular bioenergetics system.

Example. Perfect dynamic channels: group-transfer pathways

We address here highly organized cellular pathways involving the transfer of a chemical group through a series of proteins, i.e. group-transfer pathways (or relay pathways). Examples include the bacterial phosphotransferase system transferring a phosphate group from phosphoenolpyruvate to a sugar molecule whilst transporting the latter across the membrane [25], and the electron transport chain in mitochondrial and bacterial membranes transferring an electron. Such a group transfer pathway can be considered as a perfect dynamic channel in which a transferred group is not released into the bulk aqueous phase until it reaches the end of the reaction sequence.

Figure 8 shows the group-transfer pathway where a group P is transferred between r enzymes from the donor SP to the ultimate acceptor W. Each enzyme (i) interacts with two adjacent enzymes (i-1) and (i+1), with which enzyme i may form complexes, designated by $Q_{i-1} = E_{i-1}PE_i$ and $Q_i = E_iPE_{i+1}$. The complexes of enzymes 1 and r with the 'boundary' substrates are designated as $Q_0 = SPE_1$ and $Q_r = E_rPW$, respectively. Different forms of every enzyme (i) participate in the four elemental processes numbered as $2i$-1, $2i$, $2i$+1, $2i$+2 (see Fig. 8). Applying Eq. 48 to a group transfer pathway one arrives at:

$$C_{e_{i-1}}^J \cdot \frac{Q_{i-1}}{e_{i-1}} + C_{e_i}^J + C_{e_{i+1}}^J \cdot \frac{Q_i}{e_{i+1}} = C_{2i-1}^J + C_{2i}^J + C_{2i+1}^J + C_{2i+2}^J =$$

$$^{\text{imp}}C_{e_i}^J, \quad i = 2, 3, \ldots, r-1 \qquad (52)$$

For the initial ($i = 1$) and ultimate ($i = r$) enzymes one obtains instead of Eq. 52:

$$C_1^J + C_2^J \cdot \frac{Q_1}{e_2} = {}^{\text{imp}}C_{e_1}^J; \quad C_{e_{r-1}}^J \cdot \frac{Q_{r-1}}{e_{r-1}} + C_{e_r}^J = {}^{\text{imp}}C_{e_r}^J \qquad (53)$$

As for the general case, the equations obtained allow one to express the control coefficients of the enzymes of a group transfer pathway in terms of the control coefficients of the elemental processes and the fractions of the

enzymes involved in enzyme-enzyme complexes. Here we shall use Eqs 52 and 53 merely to derive the summation theorem for group transfer pathways (it can be derived from the general Eq. 50 as well). Summing these equations and adding the control exerted by the initial (S and SP) and ultimate (W, WP) substrates designated by,

$$C_{e_0}^J = \frac{d \ln |J|}{d \ln [S]} + \frac{d \ln |J|}{d \ln [SP]};$$

$$C_{e_{r+q}}^J = \frac{d \ln |J|}{d \ln [W]} + \frac{d \ln |J|}{d \ln [W]}$$

one obtains [Kholodenko and Westerhoff, submitted]:

$$C_{e_0}^J + \sum_{i=1}^r C_{e_i}^J \cdot (1 + \frac{Q_{i-1} + Q_i}{e_i}) + C_{e_{r+1}}^J = 2 \qquad (54)$$

Hence, in a group transfer pathway (perfect dynamic channel) the control exerted by the enzymes (including 'boundaries') on the pathway flux *always* exceeds unity. We saw, however, that in the general case of channelling it can be less than unity (see Section 3, Eq. 19).

Discussion

Metabolic control analysis and Biochemical Systems Theory have been great assets for the understanding of the control of intermediary metabolism, mitochondrial oxidative phosphorylation [reviews 39, 62]. It has founded the notion that a total flux control of 100% tends to be distributed among the enzymes participating in the pathway. It has also provided the rationale for any observed distribution of control in terms of enzyme kinetic properties. Consequently, it would seem time to apply these successful approaches experimentally to muscle and more specifically to the phosphoryl transfer from mitochondria through creatine kinase to the actinomyosin ATPase.

The bad news of this chapter, i.e., that standard MCA may not be applicable to muscle energy metabolism, may have reduced the enthusiasm somewhat. Because of the possible superorganization of the high free-energy phosphate metabolism in muscle, the standard Metabolic Control Analysis may not be directly applicable. For, it assumes metabolic pathways to be ideal in the sense of consisting of chemical conversions of well-mixed metabolites by independent enzyme activities. Instead, metabolite channelling (i.e., direct transfer not involving metabolite pools) by creatine kinase isoen-

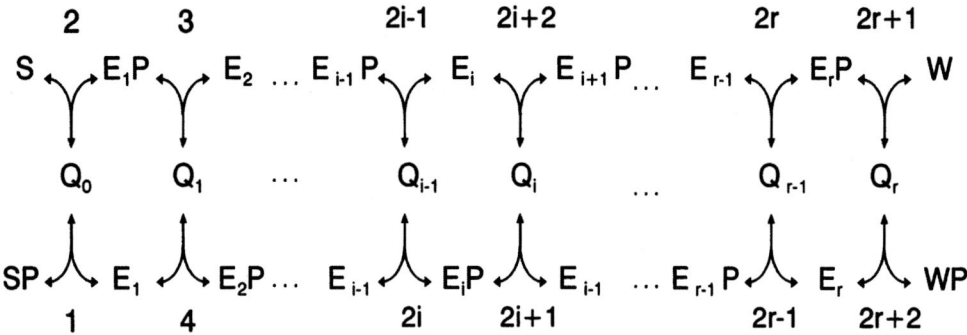

Fig. 8. Group-transfer (-relay) pathway. A group P is transferred between r enzymes from the donor SP to the ultimate acceptor W. Enzyme-enzyme complexes are referred to by $Q_i = E_i PE_{i+1}$. The complexes of enzymes 1 and r with the 'boundary' substrates are designated as $Q_0 = SPE_1$ and $Q_r = E_r PW$, respectively. For every enzyme (i) E_2-dependent elemental processes are numbered as $2i–1$, $2i$, $2i+1$, $2i+2$.

zymes may be of prime importance for aerobic energy metabolism [32]. And, the relative localization of contact sites between mitochondrial membranes, creatine kinase, hexokinase and the adenine nucleotide translocase may have strong implications for control.

By the terms 'superorganized' metabolism we here refer to such a level of organization that reaches beyond that of biochemical pathways inclusive of gene-expression regulation and covalent modification. Superorganization involves phenomena such as direct transfer of metabolites between enzymes, complex formation between enzymes [63], sequestration of metabolites by enzymes and restricted diffusion.

The good news of this chapter is that the analogue of MCA that is able to deal with such superorganized metabolism has now been developed. And, it has turned out that such metabolism has quite distinct control properties.

For instance, the sum of the flux control coefficients by enzymes is the golden standard of 100% in ideal pathways. In superorganized pathways it tends to exceed 1 because of direct transfer of metabolites and it tends to be lower than 1 because of metabolite sequestration. We showed how measuring the responses of a channelled pathway to external effectors (e.g., inhibitors) can enhance the insight into regulatory properties of the pathway.

As a consequence, control studies may be used to determine the functional extent to which metabolism is organized. Early on, such a use was pioneered by Baum and colleagues [64] and Kell and colleagues in studies of possible superorganization in chemiosmotic coupling [34, 36].

The analysis presented here may remove an important limitation of metabolic control theory. Moreover, it provides new definitions that should facilitate the quantitative characterization of metabolite channelling, without taking the system apart. The theory developed here also allows one to analyze implications of channelling for the regulation of cellular metabolism.

More in particular, determining for group transfer pathways the sum of the flux control coefficients of enzymes and in particular its deviation from 1 and 2 may serve as a measure of the extent of complexation of the proteins in the pathway. This may be one of the few methods by which such complexation can be demonstrated in cases where cell disruption leads to dissociation of such complexes [22].

The present study also suggests that in case of a dynamic organization of the components of the electron transfer chain in membranes [65], the sum of the control coefficients on the flux of reducing equivalents should equal two [cf. 48], whereas in the case of a static organization [cf. 66] it should equal unity. Existing data lack sufficient resolution to decide.

Mitochondrial creatine kinase is an attractive alternative candidate to prove that metabolite channelling by dynamic enzyme complexes occurs. For, various studies suggest that the adenine nucleotide translocase, mitochondrial creatine kinase of the intermembrane space and porin of the outer mitochondrial membrane channel high-energy phosphates from the matrix to the cytosol [23, 32].

A second asset of the new control theory is that it suggests that the muscle system might exhibit extra regulation precisely because of its superorganization properties. In 'simple' metabolic pathways the response to an external signal is readily described in terms of the effect of the signal on its receptor enzyme and the control exerted by that enzyme. We have shown that in the response of 'channelled' pathways to such a signal, additional terms appear that reflect the direct enzyme-enzyme interactions [cf. 67]. They tend to enhance the responsiveness of the pathway. The normalized value of

330

the response was called the signal transduction coefficient. We showed that in channelled pathways these coefficients are usually larger than in corresponding non-channelled (simple) pathways.

We conclude that the time is ripe for a quantitative understanding of control and regulation of muscle physiology [see for instance 68]. The control theory developed here may serve to suggest an approach. It will certainly help in understanding why control in such systems is aberrant from that in ideal pathways.

Acknowledgements

This work was supported by the Russian University Grant, by the Spanish Advice Commission for the Scientific and Technical Research (CICYT, PB 92-0852) and by the Netherlands Organization for Scientific Research.

References

1. Higgins JJ: Dynamics and control in cellular reactions. In: B Chance, RW Estabrook, JR Williamson (eds) Control of Energy Metabolism. Academic Press, New York, 1965, pp 13–46
2. Kacser H, Burns JA: The control of the flux. In: DD Davies (ed.) Rate Control of Biological Processes. Cambridge Univ. Press, London, 1973, pp 65–104
3. Kacser H, Burns JA: Molecular democracy: who shares the controls? Biochem Soc Trans 7: 1149–1160, 1979
4. Savageau MA: The behavior of intact biochemical control systems. Current Topics in Cellular Regulation 6: 63–130, 1972
5. Heinrich R, Rapoport TA: A linear steady-state treatment of enzymatic chains. General properties, control and effector strength. Europ J Biochem 42: 89–95, 1974
6. Heinrich R, Rapoport TA: Mathematical analysis of multi-enzyme systems: II. Steady state and transient control. Bio Systems 7: 130–136, 1975
7. Burns JA, Cornish-Bowden A, Grown AK, Heinrich R, Kacser H, Porteous JW, Rapoport SM, Rapoport T, Stucki JW, Tager JM, Wanders RJA, Westerhoff HV: Control analysis of metabolic systems, Trends Biochem Sci 10: 16, 1985
8. Kacser H, Sauro HM, Acerenza L: Enzyme-enzyme interactions and control analysis. The case of non-additivity: monomer-oligomer associations. Eur J Biochem 187: 481–491, 1990
9. Fell DA, Sauro HM: Metabolic control analysis. The effects of high enzyme concentrations. Eur J Biochem 192: 183–187, 1990
10. Kholodenko BN, Lyubarev AE, Kurganov BI: Control of metabolic flux in a system with high enzyme concentrations and moiety-conserved cycles. The sum of the flux control coefficients can drop significantly below unity. Eur J Biochem 210: 147–153, 1992
11. Kholodenko BN, Westerhoff HV: Metabolic channelling and control of the flux. FEBS Lett 320: 71–74, 1993
12. Heinrich R, Rapoport SM, Rapoport TA: Metabolic regulation and mathematical models. Prog Biophys Mol Biol 32: 1–83, 1977
13. Groen AK, Wanders RJA, Westerhoff HV, Van der Meer R, Tager JM: Quantification of the contribution of various steps to the control of mitochondrial respiration. J Biol Chem 257: 2754–2757, 1982
14. Reder C: Metabolic control theory: a structural approach. J Theor Biol 135: 175–201, 1988
15. Kholodenko BN: Control of molecular transformations in multienzyme systems: quantitative theory of metabolic regulation. Mol Biol (USSR) 22: 1238–1256, 1988 (English transl, 22: 990–1005)
16. Fell DA, Sauro HM: Metabolic control and its analysis: additional relationships between elasticities and control coefficients. Eur J Biochem 148: 555–561, 1985
17. Cascante M, Franco R, Canela EI: Use of implicit methods from general sensitivity theory to develop a systematic approach to metabolic control. II. Complex systems. Mathem Biosci 94: 289–309, 1989
18. Westerhoff HV, Hofmeyr J-H S, Kholedenko BN: Getting to the inside of cells using metabolic control analysis. Biophys Chem, 1993 (in press)
19. Srere PA: Complexes of sequential metabolic enzymes. Ann Rev Bioch 56: 89–124, 1987
20. Srere PA, Ovadi J: Enzyme-enzyme interactions and their metabolic role. FEBS Lett 268: 360–364, 1989
21. Welch GR, Clegg JS, eds: The organization of cell metabolism. New York, Plenum Press, 1986
22. Clegg JS: Cellular infrastructure and metabolic organization. In: Current Topics in Cellular Regulation 33: 3–14, 1992
23. Wallimann T, Wyss M, Brdiczka D, Nicolay K, Eppenberger HM: Intracellular compartmentation, structure and function of creatine kinase isoenzymes in tissues with high and fluctuating energy demands: the 'phosphocreatine circuit' for cellular energy homeostasis. Biochemical J 281: 21–40, 1992
24. Srivastava DK, Bernard SA: Biophysical chemistry of metabolic reaction sequences in concentrated enzyme solution and in the cell. Ann Rev of Biophys and Biophys Chem 16: 175–204, 1987
25. Postma PW, Lengeler JW: Phosphoenolpyruvate:carbohydrate phosphotransferase system of bacteria. Microbiol Rev 49: 232–269
26. Skulachev VP: Membrane bioenergetics. Springer, Berlin, 1988
27. Kell DB, Westerhoff HV: Catalytic facilitation and membrane bioenergetics. In: GR Welch (ed.) Organized Multienzyme Systems. Academic Press, New York, 1985, pp 63–138
28. Gellerich FN, Khuchua ZA, Kuznetsov AV: Influence of the mitochondrial outer membrane and the binding of creatine kinase to the mitochondrial inner membrane on the compartmentation of adenine nucleotides in the intermembrane space of rat liver mitochondria. Biochim Biophys Acta 1140: 327–334, 1993
29. Saks VA, Rosenstraukh LV, Smirnov VN, Chazov EI: Role of creatine phosphokinase in cellular function and metabolism. Can J Physiol Pharmacol 56: 691–706, 1978
30. Saks VA, Vasileva E, Belicova Yu O, Kuznetsov AV, Lyapina S, Petrova L, Perov NA: Retarded diffusion of ADP in cardiomyocites: possible role of mitochondrial outer membrane and the creatine kinase in cellular regulation of oxidative phosphorylation. Biochim Biophys Acta 1144: 134–148, 1993
31. Gellerich FN: The role of adenylate kinase in dynamic compartmentation of adenine nucleotides in the intermembrane space of rat heart mitochondria. FEBS Lett 297: 55–58, 1992
32. Wyss M, Wallimann T: Metabolite channelling in aerobic energy metabolism. J Theor Biol 158: 129–132, 1992

33. Sauro HM, Kacser H: Enzyme-enzyme interactions and control analysis. The case of non-independence: heterologous associations. Eur J Biochem 187: 493–500, 1990

34. Westerhoff HV, Kell DB: A control theoretical analysis of inhibitor titration assays of metabolic channelling. Comm Mol Cell Biophys 5: 57–107, 1988

35. Easterby JS: Temporal analysis of the transition between steady states. In: A Cornish-Bowden, ML Cardenas (eds) Control of Metabolic Processes, pp 291–290, Plenum Press, New York, 1990

36. Kell DB, Westerhoff HV: In: PA Srere, ME Jones, CK Mathews (eds) Structural and Organizational Aspects of Metabolic Regulation, pp 273–289, Wiley-Liss, New York, 1990

37. Schuster S, Heinrich R: The definitions of metabolic control analysis revisited. BioSystems 27: 1–15, 1992

38. Westerhoff HV, Van Dam K: Thermodynamics and Control of Biological Free-Energy Transduction. Elsevier, Amsterdam, 1987

39. Giersch C: Control analysis of metabolic networks. 1. Homogeneous functions and the summation theorems for control coefficients. Eur J Biochem 174: 509–513, 1988

40. Kholodenko BN, Westerhoff HV: 1992. In: Mazat JP, Schuster S, Rigoulet M (eds) Modern Trends in BioThermoKinetics. Proc 5th BTK Meeting, Bordeaux. Plenum, New York and London, in press

41. Yanofsky C: Tryptophan synthetase of *E. coli*: a multifunctional, multicomponent enzyme. Biochim Biophys Acta 1000: 133–137, 1989

42. Gutfreund H, Chock PB: Substrate channelling among glycolytic enzymes – fact or fiction. J Theor Biol 152: 117–121, 1991

43. Wu X, Gutfreund H, Lakatos S, Chock PB: Substrate channelling in glycolysis: a phantom phenomenon. Proc Natl Acad Sci USA 88: 497–501, 1991

44. Brooks SPJ, Storey KB: Re-evaluation of the glycerol-3-phosphate dehydrogenase/L-lactate dehydrogenase enzyme system. Evidence against the direct transfer of NADH between active sites. Biochem J 278: 875–881, 1991

45. Smolen P, Keizer J: Kinetics and thermodynamics of metabolite transfer between enzymes. Biophys Chem 38: 241–263, 1990

46. Ovady J: Physiological significance of metabolic channelling. J Theor Biol 152: 1–22, 1991

47. Kholodenko BN: Control theory of 'non-classical' enzyme systems and methods for the study of metabolic channelling. Biochemistry (USSR) 58: 512–528, 1993

48. Van Dam K, Van der Vlag J, Kholodenko BN, Westerhoff HV: The sum of the control coefficients of all enzymes on the flux through a group-transfer pathway can be as high as two. Eur J Biochem 212: 791–799, 1993

49. Kholodenko BN: How do external parameters control fluxes and concentrations of metabolites? An additional relationship in the theory of metabolic control. FEBS Lett 232: 383–386, 1988

50. Torres NV, Mateo F, Melendez-Hevia E, Kacser H: Kinetics of metabolic pathways. A system *in vitro* to study the control of flux. Biochem J 234: 169–174, 1986

51. Flint HJ, Tateson RW, Barthelmess IB, Porteous DJ, Donachie WD, Kacser H: Control of the flux in the arginine pathway of Neurospora crassa. Modulations of enzyme activity and concentrations. Biochem J 200: 231–246, 1981

52. Walsh K, Koshland DE Jr: Proc Natl Acad Sci USA 82: 3577–3581, 1985

53. Jensen PR, Westerhoff HV, Michelsen O: Excess capacity of H^+-ATPase and inverse respiratory control in E. coli. EMBO J 12: 1277–1282, 1993

54. Groen AK, Wanders RJA, Westerhoff HV, Van der Meer R, Tager JM: Quantification of the contribution of various steps to the control of mitochondrial respiration. J Biol Chem 257: 2754–2757, 1982

55. Kholodenko B, Zilinskiene V, Borutaite V, Ivanovene L, Toleikis A, Praskevicius A: The role of adenine nucleotide translocator in regulation of oxidative phosphorylation in heart mitochondria. FEBS Lett 223: 247–250, 1987

56. Rigoulet M, Averet N, Mazat J-P, Gurin B, Cohadon F: Redistribution of the flux-control coefficients in mitochondrial oxidative phosphorylation in the course of brain edema. Biochim Biophys Acta 932: 116–123, 1988

57. Kelety T: Basic enzyme kinetics. Academiai Kiado, Budapest, 1986

58. Cornish-Bowden A: Principles of enzyme kinetics. Butterworth, London, 1976

59. Hill TL: Free energy transduction in biology. Academic Press, New York, 1977

60. Hofmeyr J-H S, Kacser H, Van der Merwe KJ: Metabolic control analysis of moiety conserved cycles. Eur J Biochem 155: 631–641, 1986

61. Kholodenko BN: Metabolic control theory. New relationships for determining control coefficients of enzymes and response coefficients of system variables. J Nonlinear Biol 1: 107–126, 1991

62. Fell DA: Metabolic control analysis: a survey of its theoretical and experimental development. Biochem J 286: 313–330, 1992

63. Kurganov BI: Allosteric enzymes: kinetic behaviour. John Wiley & Son, 1978

64. Baum H: In: S Fleischer, Y Hatefi, DM McLennan, A Tzagoloff (eds) Molecular Biology of Membranes. Plenum, New York, pp 243–262, 1978

65. Hackenbrock CR, Gupte SS: 1988. In: JJ Lemasters, CR Hackenbrock, RG Thurman, HV Westerhoff (eds) Integration of Mitochondrial Function. Plenum Press, New York, pp 15–22

66. Ferguson-Miller S, Rajarathnam K, Kochman J, Schindler M: 1988. In: JJ Lemasters, CR Hackenbrock, RG Thurman, HV Westerhoff (eds) Integration of Mitochondrial Function. Plenum Press, New York, pp 23–32

67. Kholodenko BN, Demin O, Westerhoff HV: 'Channelled' pathways can be more sensitive to specific regulatory signals. FEBS Lett 320: 75–78, 1993

68. Hak JB, Van Beek JHGM, van Wijhe MH, Eijgelshoven MHJ, Westerhof N: Reduced cardiac ATP-synthetic capacity slows metabolic regulation and reduces contractility. Circ Res, in press, 1994

Molecular and Cellular Biochemistry **133/134**: 333–346, 1994.
© 1994 *Kluwer Academic Publishers.*

VI-2 Mathematical modeling of intracellular transport processes and the creatine kinase systems: a probability approach

M.K. Aliev[1] and V.A. Saks[2]

[1] *Laboratory of Experimental Cardiac Pathology, Cardiology Research Center, Moscow, Russia;* [2] *Laboratory of Bioenergetics, Institute of Chemical and Biological Physics, Ravala 10, EEOOO1 Tallinn, Estonia*

Abstract

A probability approach was used to describe mitochondrial respiration in the presence of substrates, ATP, ADP, Cr and PCr. Respiring mitochondria were considered as a three-component system, including: 1) oxidative phosphorylation reactions which provide stable ATP and ADP concentrations in the mitochondrial matrix; 2) adenine nucleotide translocase provides exchange transfer of matrix adenine nucleotides for those from outside, supplied from medium and by creatine kinase; 3) creatine kinase, starting these reactions when activated by the substrates from medium. The specific feature of this system is close proximity of creatine kinase and translocase molecules. This results in high probability of direct activation of translocase by creatine kinase-derived ADP or ATP without their leak into the medium. In turn, the activated translocase with the same high probability directly provides creatine kinase with matrix-derived ATP or ADP. The catalytic complexes of creatine kinase formed with ATP from matrix together with those formed from medium ATP provide activation of the forward creatine kinase reaction coupled to translocase activation. Simultaneously the catalytic complexes of creatine kinase formed with ADP from matrix together with those formed from medium ADP provide activation of the reverse creatine kinase reaction coupled to translocase activation. The considered probabilities were arranged into a mathematical model. The model satisfactorily simulates the available experimental data by several groups of investigators. The results allow to consider the observed kinetic and thermodynamic irregularities in behavior of structurally bound creatine kinase as a direct consequence of its tight coupling to translocase. (Mol Cell Biochem **133/134**: 333–346, 1994)

Key words: phosphocreatine, creatine kinase, ATP/ADP translocase, mitochondrion, oxidative phosphorylation, mathematical modeling, heart

Introduction

Mathematical modeling of the multienzyme systems forming metabolic networks is a task which needs development of new approaches to overcome its complexity. One of the systems which await description is the creatine kinase system. The kinetics of the soluble creatine kinase is well known for a long time [1–3]. However, in cardiomyocytes, skeletal muscle cells and also in brain and other specific cell types creatine kinase is specifically compartmentalized [4–6]. The behavior of the structurally bound enzymes is controlled and highly dependent on their microenvironment, mostly by the surrounding enzymes. In mitochondria, creatine kinase is con-

Address for offprints: Dr. M.K. Aliev, 121552, Russia, Moscow, Cardiology Research Center, 3rd Cherepkovskaya str. 15A

334

trolled by adenine nucleotide translocase and in other structures it is controlled by ATPases [4, 5]. It is known since 1975 [3] that not only the behavior of the creatine kinase in mitochondria under conditions of oxidative phosphorylation is not governed by substrate concentration in the medium and soluble enzyme kinetics, but even the direction of the creatine kinase reaction may be different depending on the oxidative phosphorylation which very significantly increases the rate of phosphocreatine production and decreases the rate of the reverse reaction of ATP formation (see the chapters by Soboll *et al.* and Saks *et al.* in this volume).

Kinetically, the control of oxidative phosphorylation over mitochondrial creatine kinase reaction is manifested as change in dissociation constant for MgATP from the ternary enzyme-substrate complexes [7, 8], and, thermodynamically, it is manifested as a maintenance of the reaction in the direction opposite to that predicted from mass action ratio in the medium and equilibrium constant value [3, 9]. All that evidence was taken to show the functional coupling between mitochondrial creatine kinase and adenine nucleotide translocase: translocase directs ATP molecules to the active site on creatine kinase and simultaneously removes the reaction product – ADP [2–14]. The functioning of this complex has not been described quantitatively in sufficient details: till 1993 only short and rather general model has been published in 1976 [15]. In 1993 we were able to develop a mathematical model quantitatively describing the forward reaction of aerobic phosphocreatine synthesis coupled to oxidative phosphorylation in cardiac mitochondria [16]. The present work is a continuation of the previous one to develop a general mathematical model quantitatively describing the process of aerobic phosphocreatine synthesis (forward and reverse reactions of mitochondrial creatine kinase) coupled to oxidative phosphorylation in cardiac mitochondria.

Mathematical model of creatine kinase reaction coupled to adenine nucleotide translocation and oxidative phosphorylation

The model is based on a probability approach to the description of the enzyme-enzyme interaction. First part of model is a kinetic equation of the creatine kinase reaction itself, which is then incorporated into more general model.

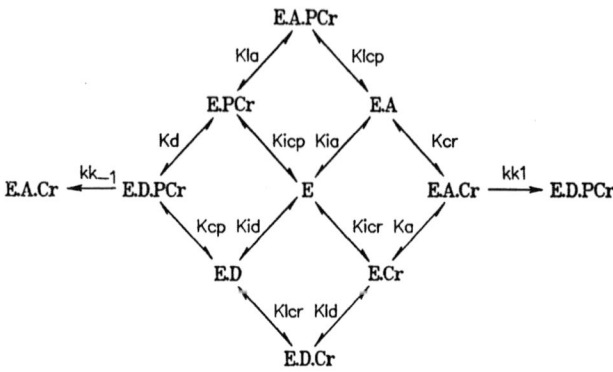

Fig. 1. Kinetic scheme of the creatine kinase reaction mechanism. Details in text.

The mitochondrial creatine kinase reaction

The creatine kinase reaction mechanism is of rapid-equilibrium random binding BiBi type, according to Cleland's classification [2, 17, 18]. This mechanism is schematically illustrated in Fig. 1. In this figure the dissociation constants, K, of enzyme-substrate complexes are given for primary complexes with an index 'i' (*ini*tial), and for ternary complexes only with the symbol for the substrate: 'a' (A) for *A*TP, 'd' (D) for *A*DP, 'cr' (Cr) for *c*reatine and 'cp' (PCr) for *p*hospho*c*reatine. The central complex for the forward reaction is E.A.Cr which is converted into that for the reverse reaction E.D.PCr with the rate constant kk_1. Besides these enzymatically active complexes, there exist dead-end complexes E.D.Cr and with low probability E.A.PCr. Dissociation constant of substrates of these complexes are given with the symbol 'I' (their formation is *I*nhibitory for the reaction).

According to the rapid equilibrium random binding BiBi type mechanism the binding and dissociation of substrates and products is very rapid and the reaction rate is determined by interconversion of the ternary complexes [2, 17, 18]. Interconversion of the central ternary complex E.A.Cr into enzyme-product complex E.D.PCr occurs with the rate constant kk_1, and the equations for this reactions are given in several earlier works [2, 3, 15]. In the equilibrium or steady state the distribution of the enzyme between enzyme-substrate complexes and free state can be expressed in the terms of probabilities for the purpose of modeling of coupled reactions. For example, the probability, (P), of the existence of the enzyme in the free state, (E), is given by

$$P(E) = [E]/[Etot] = 1/\{1 + [Cr]/Kicr + [A]/Kia + [A] \times [Cr]/(Kia \times Kcr) + [PCr]/Kicp + [PCr] \times [A]/(Kicp \times$$

$$KIa) + [D]/Kid + [D] \times [PCr]/(Kid \times Kcp) + [D] \times [Cr]/(Kid \times KIcr)\} = 1/Den \quad (1)$$

where Etot designates the *total* concentration of *enzyme* and Den designates the *den*ominator.

The probabilities of the existence of the enzyme in different enzyme-substrate complexes are given by the following equations:

$$P(E.A) = [E.A]/[Etot] = [A]/(Kia*Den) \quad (2)$$

$$P(E.D) = [E.D]/[Etot] = [D]/(Kid*Den) \quad (3)$$

$$P(E.Cr) = [E.Cr]/[Etot] = [Cr]/(Kicr*Den) \quad (4)$$

$$P(E.PCr) = [E.PCr]/[Etot] = [PCr]/(Kicp*Den) \quad (5)$$

$$P(E.A.Cr) = [E.A.Cr]/[Etot] =$$
$$[A]*[Cr]/(Kia*Kcr*Den) \quad (6)$$

$$P(E.PCr.A) = [E.PCr.A]/[Etot] =$$
$$[PCr]*[A]/(Kicp*KIa*Den) \quad (7)$$

$$P(E.PCr.D) = [E.PCr.D]/[Etot] =$$
$$[PCr]*[D]/(Kcp*Kid*Den) \quad (8)$$

$$P(E.D.Cr) = [E.D.Cr]/[Etot] =$$
$$[D]*[Cr]/(Kid*KIcr*Den) \quad (9)$$

The effective concentration of each complex is expressed as a product of its probability and the total concentration of the enzyme, Etot. The rate of reaction product formation, ADP and PCr, in the *forward* reaction is given by the following equation:

$$V_{fw} = Etot*P(E.A.Cr)*kk_1 \quad (10)$$

The rate of ATP and Cr formation in the *reverse* reaction is

$$V_r = Etot*P(E.D.PCr)*kk_{-1} \quad (11)$$

The *net* rate of ADP and PCr production by CK is the difference between the rates in the forward and reverse directions

$$V_{net} = V_{fw} - V_r \quad (12)$$

This routine consideration is valid for the soluble creatine kinase [2]. In the coupled system the creatine ki-

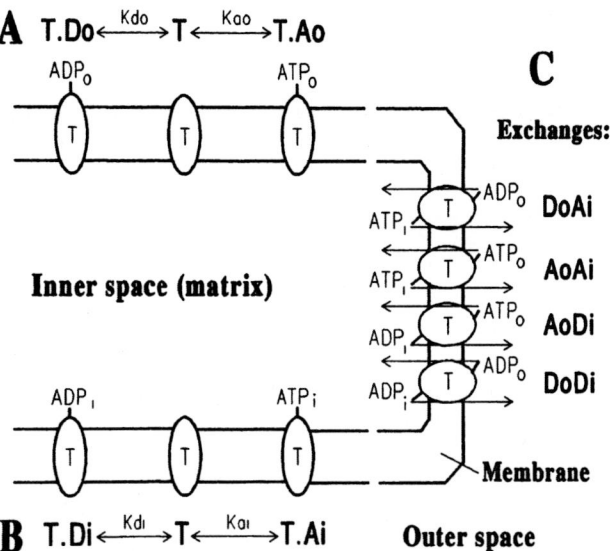

Fig. 2. Schematic presentation of the mechanism of mitochondrial adenine nucleotide translocase. The ellipses show the translocase molecules penetrating the inner mitochondrial membrane. The binding of ATP or ADP only at the outer (o) or inner (i) sides of the membrane in a competetive manner to each other does not result in adenine nucleotide translocation (parts A and B of the figure). The exchange translocation occurs only after their simultaneous binding to translocase at both sides of the membrane (part C). Four possible exchanges are shown.

nase reaction is governed by interaction with other mitochondrial systems – adenine nucleotide translocase and oxidative phosphorylation [3, 7–16].

The mitochondrial adenine nucleotide translocase reaction

Mitochondrial adenine nucleotide translocase provides exchange transfer of matrix ATP or ADP for outside ATP or ADP when adenine nucleotides are simultaneously bound to both sides of translocase [19]. This mechanism is schematically illustrated in Fig. 2, where translocase (T) molecular complexes are depicted as ellipses piercing the mitochondrial inner membrane. Translocase binds ATP and ADP both on outer space side (Fig. 2A) and inner matrix side (Fig. 2B). The dissociation constants for *A*TP (Ka) and *A*DP (Kd) are supplied by special symbol to indicate the orientation of the adenine nucleotide binding sites: *i* for the *i*nner side and *o* for *o*uter side. The substrates for T also are supplied by these symbols to discriminate between their locations. ATPo and ADPo bind on the outer side of T (Fig. 2A) with probabilities (P(T.A)o and P(T.D)o, respectively) estimated from the simple equations 13 and 14 of concurrent binding [20]:

336

$$P(T.D)o = Do/Kdo / (1 + Ao/Kao + Do/Kdo) \quad (13)$$

$$P(T.A)o = Ao/Kao / (1 + Ao/Kao + Do/Kdo) \quad (14)$$

The equations for the probabilities of ATP and ADP binding on the inner side of T (P(T.A)in and P(T.D)in, respectively) are analogous to Eqs 13–14:

$$P(T.A)in = Ai/Kai / (1 + Ai/Kai + Di/Kdi) \quad (15)$$

$$P(T.D)in = Di/Kdi / (1 + Ai/Kai + Di/Kdi) \quad (16)$$

When translocase molecular complexes are occupied simultaneously by adenine nucleotides on both sides of T, translocase provides exchange transfer of adenine nucleotides. The kind of exchange is determined and signed by the pair of bound adenine nucleotides, DoAi, AoDi, AoAi and DoDi (Fig. 2C). In this nomenclature the importing substrate is listing before the exporting one. The probabilities of the formation of these effective exchange complexes of T can be estimated simply as a product of two independent partial probabilities, the probabilities of T occupation by substrates from medium, P(T.A)o and P(T.D)o, and by substrates from matrix, P(T.A)in and P(T.D)in. So the probability of DoAi exchange, $P(T)_{DoAi}$ can be evaluated [16] as

$$P(T)_{DoAi} = P(T.D)o * P(T.A)in \quad (17)$$

In a similar way we can estimate the probabilities for AoDi exchange ($P(T)_{AoDi}$, Eq. 18), DoDi exchange ($P(T)_{DoDi}$, Eq. 19) and AoAi one ($P(T)_{AoAi}$, Eq. 20):

$$P(T)_{AoDi} = P(T.A)o * P(T.D)in \quad (18)$$

$$P(T)_{DoDi} = P(T.D)o * P(T.D)in \quad (19)$$

$$P(T)_{AoAi} = P(T.A)o * P(T.A)in \quad (20)$$

The rate constants for four main kinds of T exchanges may be settled as kt1 (constant of translocation) for DoAi exchange, kt2 for AoDi exchange; kt3 and kt4 for electroneutral DoDi and AoAi exchange, respectively. These rate constants, while equal in deenergized mitochondria, differ manifold in energized state, especially the constants for electrogenic DoAi and AoDi exchanges [21].

The effective concentration of each complex of translocase is expressed as a product of its probability and the total concentration of T, N_T. The rates (v) of adenine nu-

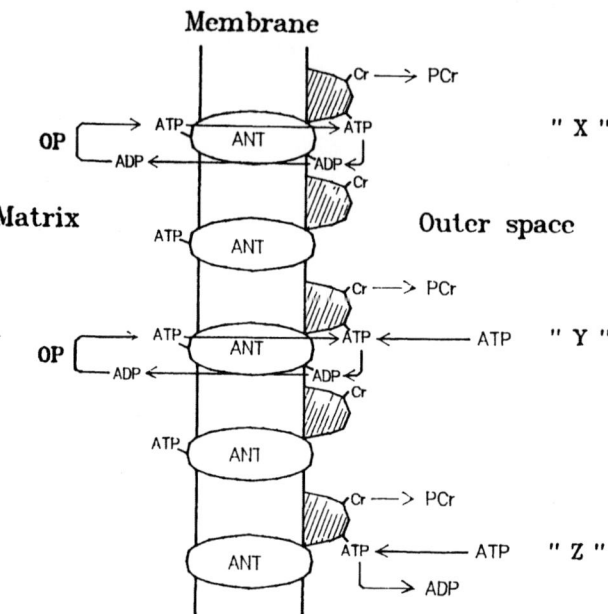

Fig. 3. Schematic representation of three possible combinations of ATP sources for mitochondrial phosphocreatine production in the forward reaction [16].

In the energized state (pathway X) significant population of adenine nucleotide translocase (ANT) molecules bind ATP from mitochondrial matrix and in the presence of catalytic amounts of ADP from mitochondrial creatine kinase (Mi-CK (shaded areas) directly transports ATP across the membrane directly to the active site of mi-CK. In this pathway 'X' this mitochondrial ATP is a single source of ATP for the creatine kinase reaction.

Besides, ATP can also bind from the surrounding medium, which is the second possible source for ATP for mi-CK. Phosphocreatine production from these sources is illustrated as pathway 'Y'. Both in 'X' and 'Y' pathways ADP, produced simultaneously with phosphocreatine, PCr, is preferentially directed back to translocase because of close spatial proximity of mi-CK to ANT.

Finally, in the absence of oxidative phosphorylation (OP), phosphocreatine is produced solely from ATP from the surrounding medium as shown in pathway 'Z'.

cleotide translocation in different exchanges of translocase can be given by following four Eqs 21–24 for four kinds of exchange:

$$v_{TDoAi} = N_T * P(T)_{DoAi} * kt1 \quad (21)$$

$$v_{TAoDi} = N_T * P(T)_{AoDi} * kt2 \quad (22)$$

$$v_{TDoDi} = N_T * P(T)_{DoDi} * kt3 \quad (23)$$

$$v_{TAoAi} = N_T * P(T)_{AoAi} * kt4 \quad (24)$$

This simple consideration is valid for the translocase, activated by substrates from medium. In the coupled system the translocase reaction, as the creatine kinase one, is governed by interaction with other mitochondrial systems – creatine kinase and oxidative phosphorylation [16].

The creatine kinase reaction coupled to adenine nucleotide translocase and oxidative phosphorylation

Before considering a complete system it is necessary to briefly account for the model from our previous publication [16], because the algorithm found in that paper is basic for present study. Figure 3, taken from that paper, illustrates the basic features of current model for the simplest case, the mitochondria, respiring in medium with two substrates of CK forward reaction, ATP and Cr. In model, the respiring mitochondria are considered as a three-component system, including: 1) oxidative phosphorylation (OP) reactions which provides stable ATP and ADP concentrations in the mitochondrial matrix [16]; 2) adenine nucleotide translocase (ANT) which provides exchange of matrix ATP for outside CK-supplied ADP when both substrates are simultaneously bound to translocase [19]; and 3) CK (shaded areas in Fig. 3), starting these reactions when activated by the substrates from medium. The specific feature of this system is a close proximity of CK and translocase molecules [13,14]. This results in a high probability of direct activation of translocase by CK-derived ADP without its leak into medium (concept of direct channeling, pathways X and Y in Fig. 3). In turn, the activated translocase with the same high probability directly provides creatine kinase with matrix-derived ATP. This 'local' ATP (A_{loc}), when forming effective E.A_{loc}.Cr complexes with CK (pathway X in Fig. 3), provides sustained PCr production from mitochondrial ATP. When CK can not accept or retain locally supplied ATP molecules, they diffuse out into solution (pathway Y in Fig. 3). In the pathway Z PCr is exclusively produced from ATP in the medium and the ADP formed is released into medium, so this pathway is solely governed by substrate concentrations in medium and well defined soluble CK kinetics [2, 3, 15, 17, 18] (see Eq. 6). The enzyme functioning in states X and Y, in the regimen of T activation exclusively by CK-supplied 'local' ADP, has been accounted [16] by two main equations.

The probability of activation of DoAi exchange of T by 'local' ADP (Dl) from CK, $P(T)_{DIA}$, was expressed [16] as a product of probability of CK activation, P(CK)ef, the probability of CK-derived 'local' ADP to meet with T on its outer side, $P(T_{Dloc})$out, and the probability

of a T occupation by ATP on its inner (matrix) side, P(T.A)in:

$$P(T)_{DIa} = P(CK)ef * P(T_{Dloc})out * P(T.A)in \quad (25)$$

The total probability of CK activation (P(CK)ef) by ATP from medium (P(E.A.Cr)) and by T-supplied 'local' ATP (Al) can be expressed by a simple equation:

$$P(CK)ef = P(E.A.Cr) + P(T)_{DIA} * P(CK_{Alact}) \quad (26)$$

where multiplier $P(CK_{Alact})$ designate the integral probability of E.Al.Cr complex formation (Aloc activation of CK) from the complexes E.Cr, E and E.PCr. This multiplier can be calculated from a separate set of equations:

$$P(CK_{Alact}) = P(CK_{Aloc}) * [P(E.Cr) + P(E)*Pc1 + P(E.PCr)*Pc2*Pc1] \quad (27)$$

$$Pc1 = [Cr]*K_{Cr+}/([Cr]*K_{Cr+} + k_{-1}) \quad (28)$$

$$Pc2 = k_{-4}/(k_{-4} + k_{-3}) \quad (29)$$

where $P(CK_{Aloc})$ designates the probability for T-supplied ATP to meet with CK; Pc1 (Partition coefficient [22]) is the probability of CK.A_{loc} complex transformation to E.A_{loc}.Cr complex; Pc2 is the probability of CK.PCr.A_{loc} complex transformation to CK.A_{loc} complex; K_{Cr+} is the diffusion-limited association rate constant for Cr; k_{-1} is the rate constant for ATP dissociation from CK.A_{loc} complex; k_{-3} and k_{-4} are the rate constants for ATP and PCr dissociation, respectively, from E.A_{loc}.PCr complex. The detailed description of these equations can be find in our previous paper [16].

Taken together, the expressions 25 and 26 represent a system of two equations with two unknown arguments. They can be easily solved to obtain single analytical expression with only one unknown variable, P(CK)ef or $P(T)_{DIA}$ [16]. The AoAi exchange on T in our previous publication [16] was ignored for simplicity.

In the present study we consider all exchanges on T in the system with a complete set of substrates, ATP, ADP, PCr, Cr. The above considered equations 21–24 for these exchanges are valid only for direct activation of T by ATP and ADP from medium. In the coupled system T can also be activated by CK-supplied 'local' ATP and ADP. So we must distinguish between 'local' ATP and ADP (Al and Dl) and 'medium' ATP and ADP (Am and Dm, respectively) activations of T: DoAi exchange (Eqs

338

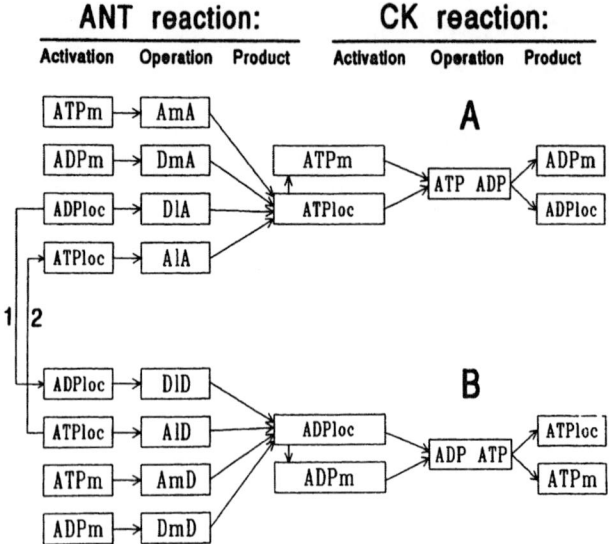

ANT reaction:

| Activation | Operation | Product |

CK reaction:

| Activation | Operation | Product |

Fig. 4. Schematic presentation of the interaction of adenine nucleotide translocase and creatine kinase in the complete system.

A. Interaction between translocase and creatine kinase under conditions of the forward creatine kinase reaction which is initiated by both ATP from the medium, ATPm, and from translocase, ATPloc. In turn, the product of the forward creatine kinase reaction, ADPloc activates the exchange DoAi which is designated as DlA exchange to make it different from the exchange initiated by ADP in the medium, DmA exchange. The second source for the local activation of the translocase is the reverse creatine kinase reaction – this path is shown by the arrow 2.

B. Interaction between translocase and creatine kinase under conditions of the reverse creatine kinase reaction of creatine production which is initiated by ADP from the medium, ADPm, and by ADP from different exchange processes, ADPloc. In turn, the creatine kinase reaction by producing ATP activates locally AlD exchange by translocase. Another source for the local activation of the translocase is the forward creatine kinase reaction (arrow 1). Translocase is also activated by substrates from the medium. It is important to note that different ways of activation of translocase result in local activation of creatine kinase.

17, 21, Fig. 2C) must be a sum of DlA and DmA exchanges; AoDi exchange (Eqs 18, 22, Fig. 2C) must be a sum of AlD and AmD exchanges; DoDi exchange (Eqs 19, 23, Fig. 2C) must be a sum of DlD and DmD exchanges; AoAi exchange (Eqs 20, 24, Fig. 2C) must be a sum of AlA and AmA exchanges. The main relations between these exchanges and CK activation in the coupled system are schematically depicted in Fig. 4.

In Fig. 4, part A depicts the forward direction of CK reaction, the synthesis of PCr and ADP from ATP and Cr. Translocase in this part of Scheme exports ATP from matrix in 4 species (AmA, AlA, DmA, DlA) of two main kinds (AoAi, DoAi) of exchange. We designate this main *in vivo* direction of T activity as forward. In the previous paper [16] only one (DlA) of 4 forward directed species of translocase exchanges has been consid-

ered. In the complete system 'local' ATP (ATPloc) for direct CK activation can be derived not only from DlA exchange on T, but also from DmA, AmA and AlA exchanges (pathway like X on Fig. 3). When CK can not accept or retain locally supplied ATP molecules, they can diffuse out into solution (pathway like Y in Fig. 3, ATPloc → ATPm path in Fig. 4A). The unique feature of a complete system is participation of a reverse CK activity in the stimulation of a forward CK reaction through AlA exchange: a CK-derived 'local' ATP from reverse reaction (arrow 2 in Fig. 4) may directly activate AoAi (AlA) exchange of T (concept of direct channeling), activated translocase in turn with the same high probability will provide CK with ATP for the forward reaction (pathway like X in Fig. 3, AlA → ATPloc path in Fig. 4A). Such interference, although complicating the calculations, makes possible the simulation of drastic changes in an apparent equilibrium constant of CK reaction in the presence of oxidative phosphorylation (see the last part of the Results).

Part B of Fig. 4 depicts the reverse direction of CK reaction, the synthesis of ATP and Cr from ADP and PCr. Translocase here exports ADP from matrix in 4 species (DmD, DlD, AmD, AlD) of another two main kinds (DoDi, AoDi) of exchange. This direction of T functioning is taken as reverse. Part B of scheme is a mirror image of part A of Fig. 4, the system is completely symmetrical. Forward CK reaction participates in reverse CK reaction activation through the DlD exchange on translocase (arrow 1 on Fig. 4), the sources of activation of reverse CK reaction are also multiple.

The mathematical realization in our previous paper [16] was very simple, we solved a system of two linear equations, one of them being for T (Eq. 25) and the other for CK (Eq. 26). This simplicity resulted mainly from accepted time accordance in system: the rate constants of DoAi exchange of T and of CK forward reaction had been taken equal as well as amounts of functioning complexes of CK and T [16]. So all disturbances in the distribution of CK complexes (the changes in relative amounts of E, E.Cr and E.PCr complexes of CK occurring on its local activation, see Eq. 27) could be realized and accomplished during single catalytic cycle of CK and closely associated co-running cycle of T.

The situation is quite different in the complete system, where the rate constants of T and CK differ manifold. For example, the rate constant for CK reverse reaction exceeds that for forward reaction about 4 times [3, 23], but the latter, while coinciding with rate constant for DoAi exchange of T [16], exceeds the rate constant for

DoDi or AoAi exchange of T about several times [21]. It means that at the time of the next steady state DoAi-cycle activation a certain amount of T will be still occupied by concomitant AoAi exchange complexes of T, which are more lasting in time. On the other hand, the next going steady state brief cycle of CK reverse activation will come at the moment, when disturbances in the distribution of CK complexes, arising from preceding local activation of CK forward reaction, will be still persistent. Such disturbances must be taken into account.

The shortest way to formulate the steady state equations for complete system is to mark the disturbed steady state parameters by special symbol 's' (steady state Probability) to distinguish them from undisturbed initial ones. Accordingly $P(T.A)$in in Eq. 25 may be named as $sP(T.A)$in; $P(E.A.Cr)$ and $P(CK_{Alact})$ in Eq. 26 may be named as $sP(E.A.Cr)$ and $sP(CK_{Alact})$, respectively; $P(E)$, $P(E.Cr)$ and $P(E.PCr)$ in Eq. 27 may be named as $sP(E)$, $sP(E.Cr)$ and $sP(E.PCr)$, respectively. Now we are able to formulate a set of equations for the forward activation of T and CK in the complete coupled system (Fig. 4, part A). These equations are numbered separately with preceding symbol 'f' (forward) for their separation from the rest of preceding ones.

$$P(T)_{DmA} = P(T.D)o * sP(T.A)in \tag{f1}$$

$$P(T)_{AmA} = P(T.A)o * sP(T.A)in \tag{f2}$$

$$P(T)_{DIA} = P(CK_{ef})fw * P(T_{Dloc})out * sP(T.A)in \tag{f3}$$

$$P(T)_{AIA} = P(CK_{ef})r * P(T_{Aloc})out * sP(T.A)in \tag{f4}$$

$$P(T_{ef})fw = P(T)_{DmA} + P(T)_{AmA} + P(T)_{DIA} + P(T)_{AIA} \tag{f5}$$

$$P(CK_{ef})fw = sP(E.A.Cr) + P(T_{ef})fw * sP(CK_{Alact}) \tag{f6}$$

$$sP(CK_{Alact}) = P(CK_{Aloc}) * [sP(E.Cr) + sP(E)*Pc1 + sP(E.PCr)*Pc2*Pc1] \tag{f7}$$

Equations f1–f4, describing the probabilities of activation of all forward directed 4 species of T exchange (Fig. 4A), are written on the basis of equations 17 (f1), 20 (f2) and 25 (f3, f4). Eq. f5 designate the sum of these probabilities, the total probability of T activation in the forward direction, $P(T_{ef})fw$. Eq. f6 describes the total probability of activation of CK forward reaction in the complete system, $P(CK_{ef})fw$, it is written on the basis of Eq. 26. Eq. f7 is rewritten on the basis of Eq. 27.

Based on the own symmetry in CK (Fig. 1) and T (Fig.

2) reaction schemes, the forward and reverse reaction schemes in the complete system are also perfectly symmetric (Fig. 4). Based on this important event, we can formulate a set of equations for reverse activation of T and CK in the complete coupled system (Fig. 4, part B) as a mirror image of the corresponding equations f1–f7 for forward direction. To stress this symmetry, the corresponding equations have been numbered by the same numbers, but preceded with symbol 'r' (reverse).

$$P(T)_{AmD} = P(T.A)o * sP(T.D)in \tag{r1}$$

$$P(T)_{DmD} = P(T.D)o * sP(T.D)in \tag{r2}$$

$$P(T)_{AID} = P(CK_{ef})r * P(T_{Aloc})out * sP(T.D)in \tag{r3}$$

$$P(T)_{DID} = P(CK_{ef})fw * P(T_{Dloc})out * sP(T.D)in \tag{r4}$$

$$P(T_{ef})r = P(T)_{AmD} + P(T)_{DmD} + P(T)_{AID} + P(T)_{DID} \tag{r5}$$

$$P(CK_{ef})r = sP(E.D.PCr) + P(T_{ef})r * sP(CK_{Dlact}) \tag{r6}$$

$$sP(CK_{Dlact}) = P(CK_{Dloc}) * [sP(E.PCr) + sP(E)*Pc3 + sP(E.Cr)*Pc4*Pc3] \tag{r7}$$

In this set, Equations r1–r4 describe the probabilities of activation of all reversely directed 4 species of T exchange (Fig. 4B). $P(T_{Aloc})$out in Eqs r3–f4 designates the general probability for CK-supplied 'local' ATP (Aloc) to meet with T on its outer side. Eq. r5 designates the sum of these probabilities, the total probability of T activation in the reverse direction, $P(T_{ef})r$. Eq. r6 describes the total probability of activation of CK reverse reaction in the complete system, $P(CK_{ef})r$. $P(CK_{Dloc})$ in Eq. r7 designates the general probability for T-supplied local ADP in the reverse reaction to meet with CK. $sP(CK_{Dlact})$ in Eqs r6–r7 designates the steady state integral probability of effective reverse E.Dl.PCr complex formation by T-derived 'local' ADP (Dloc activation of CK). This parameter is a sum of probabilities of E.Dloc.PCr complex formation from steady state E.PCr complex of CK (the expression $sP(E.PCr) * P(CK_{Dloc})$ in Eq. r7), free CK (the expression $sP(E) * Pc3 * P(CK_{Dloc})$ in Eq. r7) and from E.Cr complex of CK (the expression $sP(E.Cr) * Pc4 * Pc3 * P(CK_{Dloc})$ in Eq. r7). The equation r7 is analogous to Eq. 27 for forward coupled reaction of CK, the detailed description of the latter can be found in our previous paper [16]. The probabilities Pc3 and Pc4 (Eq. r7) have never been described elsewhere. To do

340

this, we must consider Schemes 1 and 2, which are the fragments of general scheme of CK reaction in Fig. 1.

$$E \xrightarrow[\underset{k_{dif}}{\overset{k_{+5} = [D]*K_{d+}}{\rule{3cm}{0pt}}}]{} E.Dloc \xrightarrow[\underset{k_{-6}}{\overset{k_{+6} = [PCr]*K_{cp+}}{\rule{3cm}{0pt}}}]{} E.D.PCr \xrightarrow{kk_{-1}}$$

Scheme 1. Elementary steps in the conversion of E.Dloc, the complex of mitochondrial ADP with free CK_{mit}.

When 'local' ADP from T binds to free CK (E), E.Dloc complex is formed. This complex may dissociate with the rate constant k_{-5} with the subsequent leakage of ADP from intermembrane space by diffusion process, characterized by diffusion constant k_{dif}. On the other hand, the complex E.Dloc may react with PCr; if the second order rate constant for this reaction is K_{cp+}, the pseudo first order rate constant $k_{+6} = [PCr]*K_{cp+}$ may be used to describe the conversion of E.Dloc into the effective ternary complex E.ADP.PCr for reverse CK reaction. A partition coefficient [22] Pc3 describes the part of E.Dloc which is converted into this complex:

$$Pc3 = [PCr]*K_{cp+}/([PCr]*K_{cp+}+k_{-5}) \qquad (b8)$$

$$E.Cr \xrightarrow[\underset{k_{dif}}{\overset{k_{+7} = [D]*K_{d+}}{\rule{3cm}{0pt}}}]{} E.Cr.Dloc \xrightarrow[\underset{k_{+8} = [Cr]*K_{cr+}}{}]{k_{-8}} E.Dloc$$

Scheme 2. Elementary steps in the conversions of E.Cr.Dloc, the complex of mitochondrial ADP with E.Cr.

When 'local' ADP from T binds to E.Cr complex of CK, the formed E.Cr.ADPloc dead-end complex may dissociate to E.Cr one with the rate constant k_{-7}, with subsequent diffusion of ADP from the intermembrane space (Scheme 2). Alternatively, these dead-end complexes may release Cr with the rate constant k_{-8} and be converted into E.Dloc. The probability of production of E.Dloc from the dead-end complex E.Cr.Dloc is described by a partition coefficient Pc4:

$$Pc4 = k_{-8} / (k_{-8} + k_{-7}) \qquad (b9)$$

The formed E.Dloc complex will be subsequently transformed to effective E.Dloc.PCr with known probability Pc3.

Traditionally, kinetic equation of the reaction give the distribution of the enzyme between different enzyme-substrate complexes in the steady state at any moment of the time. In the microscopic time scale creatine kinase reaction is considered as a continuous subsequence of

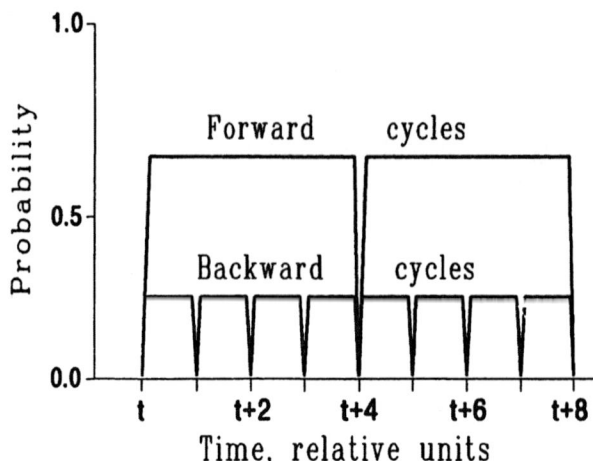

Fig. 5. The proposed timescale of the creatine kinase reaction. The arbitrary time unit is the duration of the reverse creatine kinase reaction. The vertical lines represent formation and dissociation of the catalytically active complexes of the forward and reverse creatine kinase reactions, the horizontal lines show the lifetime of the complexes and their relative levels. For details see text.

catalytic cycles the rate of which is determined by interconversion of ternary enzyme-substrate complexes E.ATP.Cr ↔ E.ADP.PCr (Fig. 5). In Fig. 5 we give the detailed graphic illustration of the basic parameter of the enzymology – molecular enzyme activity, determined as a number of substrate molecules converted into products within a time unit. In Fig. 5 the time required for one cycle of the reverse creatine kinase reaction was taken arbitrary as a unit of relative time. Since the reaction constant for the forward creatine kinase reaction is four times less than that of the reverse reaction, the duration of the reaction cycles in forward direction is four relative time units. To simplify calculation, we have accepted the ratio of the reaction cycle duration in the forward and reverse directions to be equal to 4 : 1 (the actual ratios have been measured as 4.2 ± 0.2 : 1 [23]) and the reaction cycles have been assumed to be synchronized in the time.

These assumptions are based on the fact that in contrast to chemical catalysis, the enzymatic catalysis involves conformational changes in the enzyme molecules (formation of the 'working' or transition-state conformation of the ternary enzyme-substrate complexes, see [18, 24]). Assumption of synchronization is arbitrary and needs to be proven. General consideration of this question has shown that the degree of synchronization may vary from zero to unity [25]. In the coupled system the necessary synchronization may be achieved by the continuous chain of passes of substrate at the supertiny time intervals from CK to T and consequently from T to CK and so on.

The enzyme molecules which perform catalysis do not freely exchange the substrates with the surrounding medium [24]. Consequently, if CK_{mit} at a time moment $t + 1$ (Fig. 5) receives ATP or ADP from translocase, the enzyme molecules already involved in catalysis cycle will be excluded from the process of activation by the substrate. Thus, in the process of activation by the ATP or ADP supplied by translocase only available complexes E, E.Cr and E.PCr could be involved. Their share, however, will decrease with formation of a new amount of catalytically active ternary complexes E.ATP.Cr and the distribution of the CK_{mit} between enzyme-substrate complexes will be very different from that given by the equations 1–9 of CK activation by substrates from medium.

All calculations were carried out in IBM-compatible PC/AT 286/287 computer. The numerical solution of the system of equations was found by a method which together with the listing of used Pascal program is too large to be shown in this paper but is available from M.K. Aliev on request.

Choice of the parameters for modeling

The following parameters were used. For the mitochondrial forward creatine kinase reaction all the data were taken from Jacobus and Saks [7] paper: Kia = 0.75 mM; Kicr = 26 mM; Kicp = 1.6 mM; Ka = 0.15 mM; Kcr = 5.2 mM; KIcp = 24 mM; KIa = 11.25 mM was calculated from the thermodynamic equation: Kicp * KIa = Kia * KIcp. $Vfw_{max} = N_{CK} * kk_1 = 1.0$ μmol/(min * mg of protein). For the reverse CK reaction we have accepted KIcr = Kcr [3]; Kcp = 1.0 mM; Kid = 0.208 mM; Kd = 0.13 mM. The published values of Kcp are ranged from 0.5 mM [3, 23] to 1.35 mM [26]. Kid and Kd values have been settled to attain the experimental value of apparent equilibrium constant of CK reaction 0.0133 at pH 7.4 [9]. The experimental data on Kd are considerably lower, 0.051 mM [3] and 0.076 mM [26]. KId = 0.042 mM was calculated from thermodynamic equation: KId * Kicr = Kid * KIcr. $Vr_{max} = N_{CK} * kk_{-1} = Vfw_{max} * 4.0$ (see preceding section).

For calculation of k_{-1} the value of diffusion-limited association rate constant for ATP, $K_{a+} = 2 * 10^7$ $M^{-1}S^{-1}$ [27] was used. From that value we found $k_{-1} = 2 * 10^7 M^{-1}S^{-1} * 7.5 * 10^{-4}$ M $= 15 * 10^3 S^{-1}$. For calculation of k_{+2} the value of the constant was taken to be twice the K_{a+} value, $K_{Cr+} = 4 * 10^7$ $M^{-1}S^{-1}$, as the PCr molecules diffuse about 2-fold

faster than larger ATP ones [28]. k_{-3} was calculated as $K_{a+} * KIa = 2 * 10^7$ $M^{-1}S^{-1} * 11.25 * 10^{-3}$ M $= 225 * 10^3 S^{-1}$. k_{-4} was calculated as $K_{cp+} * KIcp = 960 * 10^3 S^{-1}$. k_{-5} was calculated as $K_{cp+} * Kicp = 4 * 10^7 M^{-1}S^{-1} * 1.6 * 10^{-3}$ M $= 64 * 10^3 S^{-1}$. k_{-7} was calculated as $K_{d+} * KId = 840$ S^{-1}. k_{-8} was calculated as $K_{cr+} * KIcr = 1040 * 10^3 S^{-1}$.

$P(T.A)_{in} = 0.9$ was found by a method of best approximation to the experimental data [16]. The value of this parameter was taken to be constant in each particular experiment, as the matrix ATP/ADP ratio has been revealed constant, about 4, on stimulation of mitochondrial respiration from 0 to 75% of maximum [29]. $P(T.D)_{in}$ was taken to be $1.0 - 0.9 = 0.1$.

The probabilities of creatine kinase derived local ADP or ATP to meet with translocase, $P(T_{Dloc})_{out}$ and $P(T_{Aloc})_{out}$, respectively, and of translocase derived total ATP or ADP to meet with creatine kinase, $P(CK_{Aloc})$ and $P(CK_{Dloc})$, respectively, were taken to be equal to 1.0 (concept of direct channeling).

Kdo and Kao for translocase has been taken to be 0.02 mM [10] and 20 mM [30], respectively. kt1, the rate constant for DoAi exchange of translocase, has been taken equal to kk1, the rate constant for CK forward reaction [16]. Based on modeling of Kramer and Klinkenberg [21] data on kinetics of reconstituted adenine nucleotide translocase (our unpublished results) we have arbitrarily chosen the following rate constants for the remaining exchanges of translocase: kt2 = kt1/2; kt3 = kt1/3; kt4 = kt1/3. The differing rate constants for translocase exchanges has been previously considered by Bohnensack in his kinetic model of mitochondrial adenine nucleotide translocase [31]. The number of functioning molecular complexes of CK_{mit} and T has been taken equal [16, 32] (see the chapters by Saks *et al.* in this volume).

The concentrations of ATP, ADP, Cr and PCr were as experimentally used. When indicated, the concentration of ADP was taken to be zero.

Results and discussion

As already reported [16], we have simulated all experimental data of Jacobus and Saks [7], and in the Table 1 the final results of simulation (right column of the Table) are compared with experimental ones (left column). The most important feature of data is the theoretical confirmation that in the case of mitochondrial CK reaction coupled with oxidative phosphorylation the value

Table 1. Experimental [7] and simulated kinetic constants for rat heart mitochondrial creatine kinase as a function of oxydative phosphorylation (OP)

Partial CK reactions	Constants	Without OP		With OP	
		Experimental	Used in model	Experimental	Simulated
E.MgATP ↔ E + MgATP	Kia, mM	0.75 ± 0.06	0.75	0.29 ± 0.04	0.15
E.Cr.MgATP ↔ E.Cr + MgATP	Ka, mM	0.15 ± 0.01	0.15	0.014 ± 0.005	0.015
E.Cr ↔ E + Cr	Kicr, mM	28.8 ± 8.45	26.0	29.4 ± 12.0	--
E.MgATP.Cr ↔ E.MgATP + Cr	Kcr, mM	5.20 ± 0.30	5.20	5.20 ⊥ 2.30	5.20
E.PCr ↔ E + PCr	Kicp, mM	1.60 ± 0.20	1.60	1.40 ± 0.20	0.91
E.MgATP.Pcr ↔ E.MgATP + PCr	KIcp, mM	20–50	24.0	20–50	24.0
$V1_{max}$, μmol ATP*min^{-1}*mg^{-1}		1.06 ± 0.20	1.00	0.99 ± 0.07	0.90

An oxygraph measurements were conducted at 30° C in medium, containing 0.25 M sucrose, 10 mM HEPES, pH 7.4, 0.2 mM EDTA, 5 mM K$^+$ phosphate, 5 mM K$^+$ glutamate, 2 mM K$^+$ malate, 3.3 mM Mg(OAc)$_2$, 0.3 mM dithiothreitol, 1.0 mg/ml bovine serum albumin, 0.5–0.1 mg/ml of the rat heart mitochondrial protein. In respiring mitochondria the ATP/ADP translocation rates were calculated from oxygen consumption rates. In mitochondria, which respiration was completely inhibited by pretreatment with 10 μM rotenone and 5 μg/mg oligomycin, the creatine kinase reaction rates were determined spectrophotometrically, using a phosphoenolpyruvate (2 mM) – pyruvate kinase (2 IU/ml) system to regenerate exogenous ATP. Mean values and SD are given for five experiments [7]. Simulation parameters are indicated in the 'Choice of parameters for modeling' section.

of Km for MgATP (Ka) is decreased about ten times, while the apparent affinity of the E.MgATP complex to creatine (1/Kcr) is not changed. That shows an apparent specific elevation of the affinity of system to MgATP. At the same time in the presence of oxidative phosphorylation the calculated value of the apparent Kia, the dissociation constant for MgATP from E.MgATP complex, is lower (0.15 mM) than the corresponding Kia from experiment (0.29 mM). The model describes the tendency of lowering of this constant by oxidative phosphorylation [7] and confirms that an alteration of this constant by oxidative phosphorylation is much less than decrease in Ka under these conditions. Thus, the data of Table 1 illustrate the functional coupling between oxidative phosphorylation and the creatine kinase reaction in heart muscle mitochondria. Because of this functional coupling the creatine kinase reaction exerts effective acceptor control over respiration, and in this tightly coupled system even the apparent Km for ADP may become meaningless, if the ATP and creatine are present and the creatine kinase reaction in the intermembrane is a source of ADP for oxidative phosphorylation. This is illustrated by data given in the Fig. 6A, which shows the calculated rates of oxidative phosphorylation and ATP translocation across the inner mitochondrial membrane (thick solid line) and PCr production in the coupled reaction (thin solid line) as a function of added ADP concentration in the medium which already contains creatine and ATP to activate the creatine kinase reaction. The dotted line is a control, it shows the dependence of the respiration rate on the ADP concentration in the absence of ATP and creatine. In accordance with the ex-

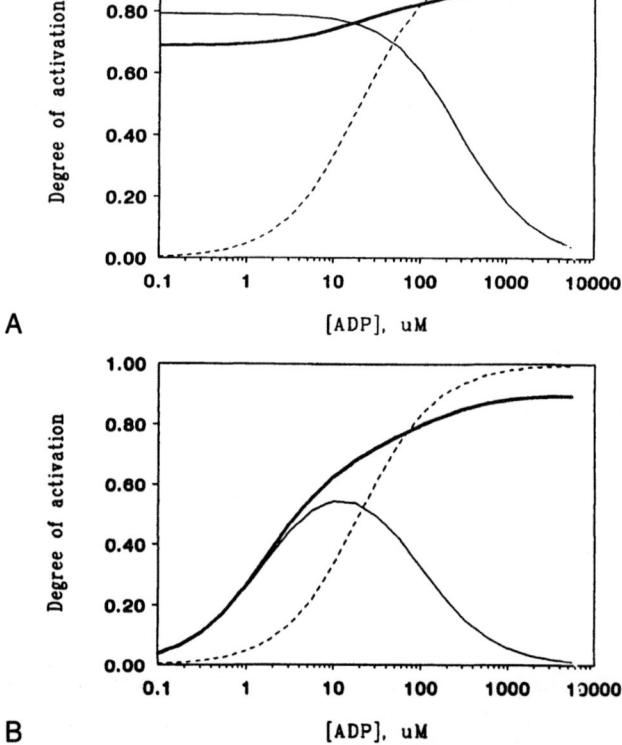

A

B

Fig. 6. A. Calculated dependences of the rate of mitochondrial respiration and ATP export (thick solid line), and aerobic phosphocreatine synthesis (thin solid line) on the external ADP concentrations under conditions of the forward creatine kinase reaction in the presence of 0.75 mM ATP and 25 mM creatine. The dotted line gives the calculated dependence of respiration on external ADP in the absence of ATP and creatine, Km for translocase was taken to be 20 μM.

B. Calculated dependences of the heart mitochondrial respiration rates on external ADP concentrations in absence (dotted line) and presence of 25 mM creatine (solid line). Thick solid line – respiration and ATP export, thin solid line – phosphocreatine production.

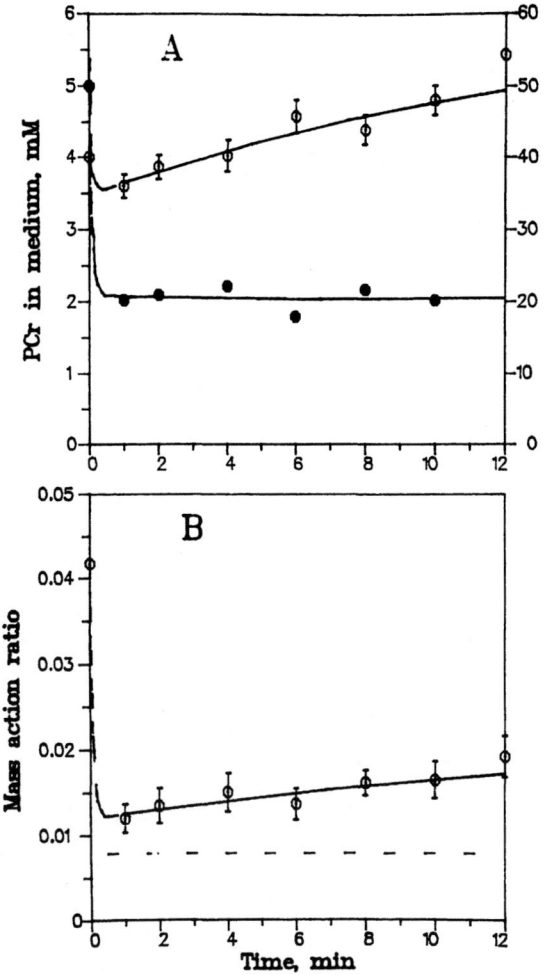

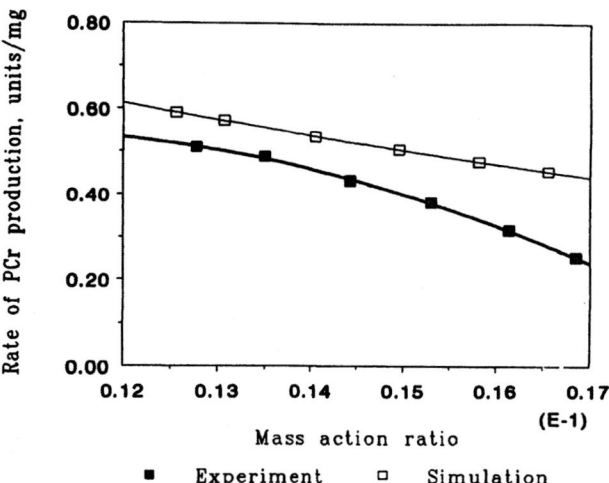

Fig. 7. Analysis of the experimental data from ref. 9 on respiration-induced shift of the creatine kinase reaction out of equilibrium. A and B: reproduction of data from ref. 9 with permission. A: Changes in PCr concentration (open circles) and ADP concentrations (closed circles). B: Change in the mass action ratio (PCr) × (MgADP)/(Cr) (MgATP) of the creatine kinase reaction in the medium which initially contained 0.12 mM ATP, 0.05 mM ADP, 40 mM creatine and 4 mM PCr and substrates for respiration. Dotted line in B: apparent equilibrium constant equal to 0.0079 [9]. C: Experimental (closed squares) and calculated (open squares) rates of phosphocreatine production coupled to the oxidative phosphorylation at different mass action ratios of the creatine kinase reaction. Experimental values were obtained from the Figs 7A and B for different time points. Note the good fitting of the experimental and calculated data.

perimental results, the simulated data show that in the presence of 0.75 mM ATP and 25 mM creatine the oxidative phosphorylation reaction (steady state ATP production and translocation) is almost maximally activated in the absence of ADP. Data in Fig. 6A show that the effective acceptor control of respiration by mitochondrial creatine kinase is mathematically well described by the model developed in this work. At low ADP concentrations starting from zero the creatine kinase reaction controls oxidative phosphorylation (in our model both ATP production and its translocation out of mitochondria), but at higher ADP concentrations the control is shifted to the external ADP due to the reversal of the creatine kinase reaction. If ATP is absent and ADP is taken to be added in the presence of creatine or in its absence (Fig. 6B), the calculations according to the model showed shift in apparent Km value for ADP from 20 μM (no creatine, dotted line) to 3.5 μM (with 25 mM creatine, thick solid line) due to increased adenine nucleotide turnover in the coupled reactions, that is well consistent with the data in the literature [10].

Finally, Fig. 7 shows the results of simulation of respi-

ration-induced shift of mitochondrial creatine reaction off the equilibrium experimentally described several years ago [9] and discussed also in this issue in a chapter II-3 by Soboll et al. Figures 7A and 7B reproduce the data from ref. 9: dotted line in Fig. 7B shows the experimentally observed value of apparent equilibrium constant, K eq = (PCr) × (MgADP)/(Cr) × (MgATP) which was equal to 0.0079 [9]. In the presence of oxidative phosphorylation the mass action ratio, instead of approaching apparent Keq value, deviated in direction of higher values (Fig. 7B) due to the aerobic PCr production (Fig. 7A). Figure 7C shows the experimentally determined (closed squares) and simulated (open squares) rates of aerobic PCr production in heart mitochondria at different mass action ratios obtained for different time-points shown in Figs 7A and B. Both the experimental and simulated values of PCr production are close to each other and positive instead of being negative as it should be according to the requirements of thermodynamics of the soluble system, since the mass action ratio exceeded the apparent Keq in the direction of PCr production.

We also simulated the data reported by Soboll et al. in

this volume (see Fig. 1B in chapter II-3). After 20 min of mitochondrial respiration at the initial concentrations of ATP 1 mM and creatine 10 mM the system reached a value of mass action ratio 1/26, and the apparent equilibrium constant was 1/122 for the direction of PCr production, the difference between mass action ratio and equilibrium constant being 4.69 times. Computer simulation for these conditions gave the mass action ratio 1/17.63, that differs from the apparent equilibrium constant by a factor 4.3. Both qualitatively and quantitatively these results are in satisfactory agreement with the data by Soboll *et al.* (see chapter II-3).

Thus, in this work we describe the complete quantitative model of the mitochondrial creatine kinase reaction coupled to oxidative phosphorylation system in heart mitochondria. This model is based on the probability approach and, in addition to the conventional kinetic equations, includes the description of the ATP transfer from matrix by the adenine nucleotide translocase and its direct channeling to the active site of mitochondrial creatine kinase which is located on the other side of the inner membrane.

We have assumed that the probability of binding of ATP translocated across the membrane by translocase, to creatine kinase is equal to one. That means that we have strictly connected our model to the concept of direct channeling of adenine nucleotides between creatine kinase and translocase. We have developed this model for isolated motochondria *in vitro*, assuming that there is direct channelling of adenine nucleotides between creatine kinase and translocase and that there is no diffusion barrier for ADP or ATP from medium into intermembrane space of mitochondria. Our assumption of the direct channeling is based on the experimental kinetic, thermodynamic and radioisotopic data [3–12] showing that the phenomenon of the functional coupling between creatine kinase reaction and oxidative phosphorylation is perfectly preserved in mitoplasts with destroyed outer membrane [9, see also chapter III-1] and is lost in mitochondria in which the outer membrane is intact but the creatine kinase is released into the intermembrane space by KCl treatment [35]. Also, it has been shown by immunochemical methods for mitoplasts produced from heart mitochondria that mi-CK and translocase are structurally closely related to each other [14]. Further, this conclusion may also be supported by the concept developed by Wallimann *et al.* [5, 36–39] according to which mitochondrial creatine kinase forms octamers which probably are bound to the tetramers of adenine nucleotide translocator and form

one multienzyme complex, translocase-mi-CK-outer membrane pores (see chapter III-2). Thus, there are rather strong and good functional and structural evidences to believe that there is high value of the probability of direct transfer of the ATP molecules between mitochondrial creatine kinase and translocase.

There are new data showing low permeability of the outer membrane of mitochondria for ADP (and ATP?) that may create additional pool of adenine nucleotides in the intermembrane space of mitochondria (see chapters II-1, II-2 and III-1). However, this phenomenon is mostly seen in permeabilized cells or isolated mitochondria in the presence of high molecular weight substances (increased oncotic pressure). Clearly, retarded diffusion of ADP across the outer mitochondrial membrane should be accounted for when the model will be adapted for the cells *in vivo*. In the isolated mitochondria in standard *in vitro* experimental conditions the value of Km for ADP is very low (see chapter III-1) and the barrier function of the outer membrane is almost completely lost, but the functional coupling still existing, even in the mitoplasts preparation without outer membrane [9, 14, 35]. That means that aerobic phosphocreatine production from mitochondrial ATP is mostly based on functional coupling – direct channelling of adenine nucleotides between translocase and creatine kinase.

The most important results of this modeling are predicted apparent decrease of the value of the dissociation constant for ATP from ternary enzyme-substrate complex E.ATP.Cr, which is completely consistent with the experimental observations (see Table 1), and almost quantitative computer simulation of respiration-induced deviation of the creatine kinase reaction from the equilibrium. These effects may be taken to show the recycling of adenine nucleotides in the tightly coupled system mi-CK translocase-oxidative phosphorylation – an amplification effect, resulting in multiple use of small numbers of ADP or ATP and playing important role in enhancing the regulatory signal in cardiac cells *in vivo* [3, 4, 7–12, 40, 41]. One conclusion which could be made from these results is that the thermodynamic parameters of the reaction medium do not govern the coupled reactions of aerobic phosphocreatine production in mitochondria. For *in vivo* conditions this most probably means that the cytoplasmic phosphorylation potential is not an important factor in regulation of cellular respiration (see chapter III-1 in this volume).

In conclusion, the model described in this work may be useful as part of a more general mathematical model of phosphocreatine circuit [4, 5] which should also include controlled transport of ADP accross the outer mitochondrial membrane, facilitated diffusion of high energy phosphate bonds in the cytoplasmic equilibrium creatine kinase system [28] and coupled reaction in the myofibrils and at the cellular membranes. In the complete cellular system the interaction of creatine kinase with myokinase ([42], see also Savabi *et al.* and Gellerich *et al.* in this volume) will be included. This kind of mathematical modelling of coupled reactions in mitochondria may be useful for application of the control theory of metabolic channelling, developed by Kholodenko, Cascante and Westerhoff (chapter VI-1).

References

1. Kuby SA, Noltmann EA: Adenosine triphosphate-creatine transphosphorylase. In: PD Boyer, H Lardy, K Myrback (eds) The Enzymes. Academic Press, New York, 1962, vol 6, pp 515–603
2. Morrison JF, James E: The mechanism of the reaction catalyzed by adenosine triphosphate-creatine phosphotransferase. Biochem J 97: 37–52, 1965
3. Saks VA, Chernousova GB, Gukovsky DE, Smirnov VN, Chazov EI: Studies of energy transport in heart cells. Mitochondrial isoenzyme of creatine phosphokinase: kinetic properties and regulatory action of Mg^{2+} ions. Eur J Biochem 57: 273–290, 1975
4. Saks VA, Rosenstraukh LV, Smirnov VN, Chazov EI: Role of creatine phosphokinase in cellular function and metabolism. Can J Physiol Pharmacol 56: 691–706, 1978
5. Wallimann T, Wyss M, Brdiczka D, Nicolay K, Eppenberger HM: Intracellular compartmentation, structure and function of creatine kinase isoenzymes in tissues with high and fluctuating energy demands: the 'phosphocreatine circuit' for cellular energy homeostasis. Biochem J 281: 21–40, 1992
6. Saks VA, Chernousova GB, Voronkov UI, Smirnov VN, Chazov EI: Study of energy transport mechanism in myocardial cells. Circ Res 34/35, Suppl III: 138–149, 1974
7. Jacobus WE, Saks VA: Creatine kinase of heart mitochondria: changes in its kinetic properties induced by coupling to oxidative phosphorylation. Arch Biochem Biophys 219: 167–178, 1982
8. Saks VA, Kupriyanov VV, Elizarova GV, Jacobus WE: Studies of energy transport in heart cells. The importance of creatine kinase localization for the coupling of mitochondrial phosphorylcreatine production to oxidative phosphorylation. J Biol Chem 255: 755–763, 1980
9. Saks VA, Kuznetsov AV, Kupriyanov VV, Miceli MV, Jacobus WE: Creatine kinase of rat heart mitochondria. The demonstration of functional coupling to oxidative phosphorylation in an inner membrane-matrix preparation. J Biol Chem 260: 7757–7764, 1985
10. Moreadith RW, Jacobus WE: Creatine kinase of heart mitochondria. Functional coupling of ADP transfer to the adenosine nucleotide translocase. J Biol Chem 257: 899–905, 1982
11. Barbour RL, Ribaudo J, Chan SHP: Effect of creatine kinase activity on mitochondrial ADP/ATP transport. Evidence for a functional interaction. J Biol Chem 259: 8246–8251, 1984
12. Bessman SP, Geiger PJ: Transport of energy in muscle: the phosphocreatine shuttle. Science 211: 448–452, 1981
13. Muller M, Moser R, Cheneval D, Carafoli E: Cardiolipin is the membrane receptor for mitochondrial creatine phosphokinase. J Biol Chem 260: 3839–3843, 1985
14. Saks VA, Khuchua ZA, Kuznetsov AV: Specific inhibition of ATP-ADP translocase in cardiac mitoplasts by antibodies against mitochondrial creatine kinase. Biochim Biophys Acta 891: 138–144, 1987
15. Saks VA, Lipina NV, Smirnov VN, Chazov EI: Studies of energy transport in heart cells. The functional coupling between mitochondrial creatine phosphokinase and ATP-ADP translocase: kinetic evidence. Arch Biochem Biophys 173: 34–41, 1976
16. Aliev MK, Saks VA: Quantitative analysis of the 'phosphocreatine shuttle'. I. A probability approach to the description of phosphocreatine production in the coupled creatine kinase – ATP/ADP translocase – oxidative phosphorylation reactions in heart mitochondria. Biochim Biophys Acta 1143: 291–300, 1993
17. Cleland WW: Enzyme kinetics. Ann Rev Biochem 36: 77–112, 1967
18. Kenyon GL, Reed GH: Creatine kinase: structure-activity relationships. Adv Enzymol 54: 367–426, 1983
19. Vignais PV, Brandolin G, Boulay F, Dalbon P, Block MR, Gauche I: Recent developments in the study of the conformational states and the nucleotide binding sites in the ADP/ATP carrier. In: A Azzi, KA Nalecz, MJ Nalecz, L Wojtczak (eds) Anion Carriers of Mitochondrial Membranes. Springer-Verlag, Berlin, 1989, pp 133–146
20. Souverijn JHM, Huisman LA, Rosing J, Kemp A: Comparison of ADP and ATP as substrates for the adenine nucleotide translocator in rat-liver mitochondria. Biochim Biophys Acta 305: 185–198, 1973
21. Kramer R, Klingenberg M: Electrophoretic control of reconstituted adenine nucleotide translocation. Biochemistry 21: 1082–1089, 1982
22. Boyer PD, de Meis L, Carvalho MG, Hackney DD: Dynamic reversal of enzyme carboxyl group phosphorylation as the basis of the oxygen exchange catalyzed by sarcoplasmic reticulum adenosine triphosphatase. Biochemistry 16: 136–140, 1977
23. Saks VA, Chernousova GB, Vetter R, Smirnov VN, Chazov EI: Kinetic properties and functional role of particulate MM-isoenzyme of creatine phosphokinase bound to heart muscle myofibrils. FEBS Lett 62: 293–296, 1976
24. Blumenfeld LA: Physics of bioenergetic processes. (Springer ser in synergetics, vol 16). Springer, Berlin, 1983
25. Chizmadzev Yu A, Pastushenko VF, Blumenfeld LA: On dynamic theory of enzymatic catalysis. Biophysica (Rus) 21: 208–213, 1976
26. Elizarova GV: A thermodynamic description of creatine kinase reaction and quantitative characteristics of mitochondrial and myofibrillar cycles of myocardial phosphocreatine system. Dc Sc Thesis Cardiol Res Center, Moscow, 1987
27. Froehlich JP, Taylor EW: Transient state kinetics studies of sarcoplasmic reticulum adenosine triphosphatase. J Biol Chem 250: 2013–2021, 1975
28. Meyer RA, Sweeney HL, Kushmerick MJ: A simple analysis of the 'phosphocreatine shuttle'. Am J Physiol 246: C365–C377, 1984
29. Davis EJ, Lumeng L: Relationship between the phosphorylation potentials generated by liver mitochondria and respiratory state

346

under conditions of adenosine diphosphate control. J Biol Chem 250: 2275–2292, 1975

30. Jacobus WE, Moreadith RW, Vandegaer KM: Mitochondrial respiratory control. Evidence against the regulation of respiration by extramitochondrial phosphorylation potentials or by [ATP]/[ADP] ratios. J Biol Chem 257: 2397–2402, 1982

31. Bohnensack R: The role of the adenine nucleotide translocator in oxidative phosphorylation. A theoretical investigation on the basis of a comprehensive rate law of the translocator. J Bioenerg Biomembr 14: 45–61, 1982

32. Kuznetsov AV, Saks VA: Affinity modification of creatine kinase and ATP-ADP translocase in heart mitochondria: determination of their molar stoichiometry. Biochem Biophys Res Commun 134: 359–366, 1986

33. DeFuria RA, Ingwall JS, Fossel ET, Dygert MK: Microcompartmentation of the mitochondrial creatine kinase reaction. In: WE Jacobus, JS Ingwall (eds) Heart Creatine Kinase. The integration of isoenzymes for energy distribution. Williams a Wilkins, Baltimore/London, 1980, pp 135–141

34. Gellerich FN, Schlame M, Bohnensack R, Kunz W: Dynamic compartmentation of adenine nucleotides in the mitochondrial intermembrane space of rat-heart mitochondria. Biochim Biophys Acta 890: 117–126, 1987

35. Kuznetsov AV, Khuchua ZA, Vassil'eva EV, Medvedeva NV, Saks VA: Heart mitochondrial creatine kinase revisited: the outer mitochondrial membrane is not important for coupling of phosphocreatine production to oxidative phosphorylation. Arch Biochem Biophys 268: 176–190, 1989

36. Schlegel J, Zurbriggen B, Wegmann G, Wyss M, Eppenberger HM, Wallimann T: Native mitochondrial creatine kinase forms octameric structures. I. Isolation of two interconvertible mito-chondrial creatine kinase forms, dimeric and octameric mitochondrial creatine kinase: characterization, localization, and structure-function relationships. J Biol Chem 263: 16942–16953, 1988

37. Schnyder T, Engel A, Lustig A, Wallimann T: Native mitochondrial creatine kinase forms octameric structures. II. Characterization of dimers and octamers by ultracentrifugation, direct mass measurements by scanning transmission electron microscopy, and image analysis of single mitochondrial creatine kinase octamers. J Biol Chem 263: 16954–16962, 1988

38. Adams V, Bosch W, Schlegel J, Wallimann T, Brdiczka D: Further characterization of contact sites from mitochondria of different tissues: topology of peripheral kinases. Biochim Biophys Acta 981: 213–225, 1989

39. Schlegel J, Wyss M, Eppenberger HM, Wallimann T: Functional studies with the octameric and dimeric form of mitochondrial creatine kinase. Differential pH-dependent association of the two oligomeric forms with the inner mitochondrial membrane. J Biol Chem 265: 9221–9227, 1990

40. Saks VA, Belikova Yu O, Kuznetsov AV: In vivo regulation of mitochondrial respiration in cardiomyocytes: specific restrictions for intracellular diffusion of ADP. Biochim Biophys Acta 1074: 302–311, 1991

41. Saks VA, Belikova Yu O, Kuznetsov AV, Khuchua ZA, Branishte TH, Semenovsky ML, Naumov VG: Phosphocreatine pathway for energy transport: ADP diffusion and cardiomyopathy. Am J Physiol. Suppl 261: 30–38, 1991

42. Zeleznikar RJ, Heyman RA, Graeff RM, Walseth TF, Dawis SM, Butz EA, Goldberg ND: Evidence for compartmentalized adenylate kinase catalysis serving a high energy phosphoryl transfer function in rat skeletal muscle. J Biol Chem 265: 300–311, 1990